Fractal-Based
Point Processes

Fractal-Based
Point Processes

Steven Bradley Lowen

Harvard Medical School
McLean Hospital

Malvin Carl Teich

Boston University
Columbia University

WILEY-
INTERSCIENCE

A JOHN WILEY & SONS, INC., PUBLICATION

Library of Congress Cataloging-in-Publication Data:

Lowen, Steven Bradley, 1962–
    Fractal-based point processes / Steven Bradley Lowen, Malvin Carl Teich.
      p. cm.
    Includes bibliographical references and index.
    ISBN-13 978-0-471-38376-5 (acid-free paper)
    ISBN-10 0-471-38376-7 (acid-free paper)
     1. Point processes. 2. Fractals. I. Teich, Malvin Carl. II. Title.

QA274.42.L69 2005
519.2'3—dc22   2005048977

Printed in the United States of America.

10 9 8 7 6 5 4 3 2 1

Preface

Fractals and Point Processes

Fractals are objects that possess a form of self-scaling; a part of the whole can be made to recreate the whole by shifting and stretching. Many objects are self-scaling only in a statistical sense, meaning that a part of the whole can be made to recreate the whole in the likeness of their probability distributions, rather than as exact replicas. Examples of random fractals include the length of a segment of coastline, the variation of water flow in the river Nile, and the human heart rate.

Point processes are mathematical representations of random phenomena whose individual events are largely identical and occur principally at discrete times and locations. Examples include the arrival of cars at a tollbooth, the release of neurotransmitter molecules at a biological synapse, and the sequence of human heartbeats.

Fractals began to find their way into the scientific literature some 50 years ago. For point processes this took place perhaps 100 years ago, although both concepts developed far earlier. These two fields of study have grown side-by-side, reflecting their increasing importance in the natural and technological worlds. However, the domains in which point processes and fractals both play a role have received scant attention. It is the intersection of these two fields that forms the topic of this treatise.

Fractal-based point processes exhibit both the scaling properties of fractals and the discrete character of random point processes. These constructs are useful for representing a wide variety of diverse phenomena in the physical and biological sciences, from information-packet arrivals on a computer network to action-potential occurrences in a neural preparation.

Scope

The presentation begins with several concrete examples of fractals and point processes, without devoting undue attention to mathematical detail (*Chapter 1*). A brief introduction to fractals and chaos follows (*Chapter 2*). We then define point processes and consider a collection of measures useful in characterizing them (*Chapter 3*). This is followed by a number of salient examples of point processes (*Chapter 4*). With the concepts of fractals and point processes in hand, we proceed to integrate them (*Chapter 5*). Mathematical formulations for several important fractal-based point-process families are then set forth (*Chapters 6–10*). An exposition detailing how various operations modify such processes follows (*Chapter 11*). We then proceed to examine analysis and estimation techniques suitable for these processes (*Chapter 12*). Finally, we examine computer network traffic (*Chapter 13*), an important application used to illustrate the various approaches and models set forth in earlier chapters.

To facilitate the smooth flow of material, lengthy *Derivations* are relegated to *Appendix A*. *Problem Solutions* appear in *Appendix B*. For convenience, *Appendix C* contains a *List of Symbols*. A comprehensive *Bibliography* is provided.

Approach

We have been inspired by Feller's venerable and enduring *Introduction to Probability Theory and Its Applications* (1968; 1971) and Cox and Isham's concise but superb *Point Processes* (1980).

We provide an integrated exposition of fractal-based point processes, from definitions and measures to analysis and estimation. The material is set forth in a self-contained manner. We approach the topic from a practical and informal perspective — and with a distinct engineering bent. Chapters 3, 4, and 11 can serve as a comprehensive stand-alone introduction to point processes.

A number of important applications are examined in detail with the help of a canonical set of point processes drawn from biological signals and computer network traffic. This set includes action-potential sequences recorded from the retina, lateral geniculate nucleus, striate cortex, descending contralateral movement detector, and cochlea; as well as vesicular exocytosis and human-heartbeat sequences. We revisit these data sets throughout our presentation.

Other applications are drawn from a diverse collection of topics, including $1/f$ noise events in electronic devices and systems, trapping in amorphous semiconductors, semiconductor high-energy particle detectors, diffusion processes, error clustering in telephone networks, digital generation of $1/f^\alpha$ noise, photon statistics of Čerenkov radiation, power-law mass distributions, molecular evolution, and the statistics of earthquake occurrences.

Audience

Our exposition is addressed principally to students and researchers in the mathematical, physical, biological, psychological, social, and medical sciences who seek to understand, explain, and make use of the ever-growing roster of phenomena that are found to exhibit fractal and point-process characteristics. The reader is assumed to have a strong mathematical background and a solid grasp of probability theory. While not required, a rudimentary knowledge of fractals and a familiarity with point processes will prove useful.

This book will likely find use as a text for graduate-level courses in fields as diverse as statistics, electrical engineering, neuroscience, computer science, physics, and psychology. An extensive set of solved problems accompanies each chapter.

Website and Supplementary Material

Supplementary materials related to the practical aspects of data analysis and simulation are linked from the book's website. Errata are posted and readers are encouraged to contribute to the compilation. Kindly visit http://www.wiley.com/statistics/ and scroll down to the icon labeled "Download Software and Supplements for Wiley Math & Statistics Titles." Then find the entry "Lowen and Teich." Alternatively, you may directly access the authors' websites at http://cordelia.mclean.org/~lowen/ and http://people.bu.edu/teich/.

Photo Credits

We express our appreciation to the many organizations that have provided assistance in connection with our efforts to assemble the photographs used at the beginnings of each chapter: Penck (courtesy of Bildarchiv der Österreichischen Nationalbibliothek, Vienna); Richardson (courtesy of Olaf K. F. Richardson); Cantor and Poincaré (courtesy of the Aldebaran Group for Astrophysics, Prague); Poisson, Yule, Pareto, Hurst, and Erlang [from Heyde & Seneta (2001), courtesy of Chris Heyde, Eugene Seneta, and Springer-Verlag]; Lapicque (courtesy of the National Library of Medicine); Cox (courtesy of Sir David R. Cox); Fourier (courtesy of John Wiley & Sons); Haar (courtesy of Akadémiai Kiadó, Budapest); Kolmogorov (courtesy of A. N. Shiryaev); Van Ness (courtesy of John W. Van Ness); Mandelbrot (courtesy of Benoit B. Mandelbrot); Gauss (S. Bendixen portrait, 1828); Lévy and Feller [from Reid (1982), courtesy of Ingram Olkin, Constance Reid, and Springer-Verlag]; Schottky (from the Schottky family album); Rice (courtesy of the IEEE History Center, Rutgers University); Neyman [from Reid (1982), courtesy of Constance Reid and Springer-Verlag]; Bartlett (courtesy of Walter Bird and Godfrey Argent); Bernoulli [frontispiece from Fleckenstein (1969), courtesy of Birkhäuser-Verlag]; Allan (courtesy of David W. Allan); Palm (courtesy of Jan Karlqvist, from the Olle Karlqvist family album). The photographs of Lowen and Teich were provided courtesy of Jeff Thiebauth and Boston University, respectively.

We are indebted to a number of individuals who assisted us in our attempts to secure various photographs. These include Tedros Tsegaye, who helped us obtain a photograph of Conny Palm; Steven Rockey, Mathematics Librarian at the Cornell University Mathematics Library, who tracked down a photograph of Alfréd Haar in a collection of Haar's works (Szőkefalvi-nagy, 1959); and Jan van der Spiegel and Nader Engheta at the University of Pennsylvania, who valiantly attempted to secure a photograph of Gleason Willis Kenrick from the University archives.

Finally, we extend our special thanks to those individuals who kindly provided photographs of themselves: Sir David R. Cox of Nuffield College at the University of Oxford, John W. Van Ness of the University of Texas at Dallas, Benoit B. Mandelbrot of Yale University and the IBM Corporation, and David W. Allan of Fountain Green, Utah.

Acknowledgments

We greatly appreciate the efforts of the seven reviewers who carefully read an early version of the manuscript and provided invaluable feedback: Jan Beran, Patrick Flandrin, Conor Heneghan, Eric Jakeman, Bradley Jost, Michael Shlesinger, and Xueshi Yang. We acknowledge valuable discussions with Bahaa Saleh and Benoit Mandelbrot relating to the presentation of the material. Bo Ryu and David Starobinski made many helpful suggestions pertinent to the material contained in Chapter 13. Arnold Mandell brought to our attention a number of publications related to fractals in medicine. Carl Anderson, Darryl Veitch, Murad Taqqu, Larry Abbott, Luca Dal Negro, and David Bickel alerted us to many useful references in a broad variety of fields. Iver Brevik and Arne Myskja supplied information regarding Tore Engset's contributions to teletraffic theory. Wai Yan (Eliza) Wong provided extensive logistical support and crafted magnificent diagrams and figures. Anna Swan kindly assisted with translation from the Swedish. Many at Wiley, including Stephen Quigley, Susanne Steitz, and Heather Bergman have been most helpful, patient, and encouraging. We especially appreciate the attentiveness and thoroughness that Melissa Yanuzzi brought to the production process. Most of all, our families provided the love and support that nurtured the process of creating this book, from inception to publication. Our thanks go to them.

We gratefully acknowledge financial support provided by the National Science Foundation; the Center for Telecommunications Research (CTR), an Engineering Research Center at Columbia University supported by the National Science Foundation; the U.S. Joint Services Electronics Program through the Columbia Radiation Laboratory; the U.S. Office of Naval Research; the Whitaker Foundation; the Interdisciplinary Science Program at the David & Lucile Packard Foundation; the Office of National Drug Control Policy; the National Institute on Drug Abuse; the Center for Subsurface Sensing and Imaging Systems (CenSSIS), an Engineering Research Center at Boston University supported by the National Science Foundation; and the Boston University Photonics Center.

STEVEN BRADLEY LOWEN

MALVIN CARL TEICH

Boston, Massachusetts

Contents

Preface		*v*
List of Figures		*xv*
List of Tables		*xix*
Authors		*xxi*

1	*Introduction*		*1*
	1.1	*Fractals*	*2*
	1.2	*Point Processes*	*4*
	1.3	*Fractal-Based Point Processes*	*4*
		Problems	*6*

2	*Scaling, Fractals, and Chaos*		*9*
	2.1	*Dimension*	*11*
	2.2	*Scaling Functions*	*13*
	2.3	*Fractals*	*13*
	2.4	*Examples of Fractals*	*16*
	2.5	*Examples of Nonfractals*	*23*
	2.6	*Deterministic Chaos*	*25*
	2.7	*Origins of Fractal Behavior*	*32*
	2.8	*Ubiquity of Fractal Behavior*	*39*
		Problems	*46*

3   Point Processes: Definition and Measures   49
     3.1   Point Processes   50
     3.2   Representations   51
     3.3   Interval-Based Measures   54
     3.4   Count-Based Measures   63
     3.5   Other Measures   70
            Problems   79

4   Point Processes: Examples   81
     4.1   Homogeneous Poisson Point Process   82
     4.2   Renewal Point Processes   85
     4.3   Doubly Stochastic Poisson Point Processes   87
     4.4   Integrate-and-Reset Point Processes   91
     4.5   Cascaded Point Processes   93
     4.6   Branching Point Processes   95
     4.7   Lévy-Dust Counterexample   95
            Problems   96

5   Fractal and Fractal-Rate Point Processes   101
     5.1   Measures of Fractal Behavior in Point Processes   103
     5.2   Ranges of Power-Law Exponents   107
     5.3   Relationships among Measures   114
     5.4   Examples of Fractal Behavior in Point Processes   115
     5.5   Fractal-Based Point Processes   120
            Problems   126

6   Processes Based on Fractional Brownian Motion   135
     6.1   Fractional Brownian Motion   136
     6.2   Fractional Gaussian Noise   141
     6.3   Nomenclature for Fractional Processes   143
     6.4   Fractal Chi-Squared Noise   145
     6.5   Fractal Lognormal Noise   147
     6.6   Point Process from Ordinary Brownian Motion   149
            Problems   150

7   Fractal Renewal Processes   153
     7.1   Power-Law Distributed Interevent Intervals   155
     7.2   Statistics of the Fractal Renewal Process   157

7.3   *Nondegenerate Realization of a Zero-Rate Process*   *164*
       Problems   *166*

8   *Processes Based on the Alternating Fractal Renewal Process*   *171*
    8.1   *Alternating Renewal Process*   *174*
    8.2   *Alternating Fractal Renewal Process*   *177*
    8.3   *Binomial Noise*   *179*
    8.4   *Point Processes from Fractal Binomial Noise*   *182*
           Problems   *183*

9   *Fractal Shot Noise*   *185*
    9.1   *Shot Noise*   *186*
    9.2   *Amplitude Statistics*   *189*
    9.3   *Autocorrelation*   *194*
    9.4   *Spectrum*   *195*
    9.5   *Filtered General Point Processes*   *197*
           Problems   *198*

10 *Fractal-Shot-Noise-Driven Point Processes*   *201*
    10.1   *Integrated Fractal Shot Noise*   *204*
    10.2   *Counting Statistics*   *205*
    10.3   *Time Statistics*   *212*
    10.4   *Coincidence Rate*   *214*
    10.5   *Spectrum*   *215*
    10.6   *Related Point Processes*   *216*
           Problems   *219*

11 *Operations*   *225*
    11.1   *Time Dilation*   *228*
    11.2   *Event Deletion*   *229*
    11.3   *Displacement*   *241*
    11.4   *Interval Transformation*   *247*
    11.5   *Interval Shuffling*   *252*
    11.6   *Superposition*   *256*
           Problems   *261*

12 *Analysis and Estimation*   *269*
    12.1   *Identification of Fractal-Based Point Processes*   *271*

12.2 *Fractal Parameter Estimation* 273

12.3 *Performance of Various Measures* 281

12.4 *Comparison of Measures* 309

Problems 310

13 *Computer Network Traffic* 313

13.1 *Early Models of Telephone Network Traffic* 315

13.2 *Computer Communication Networks* 320

13.3 *Fractal Behavior* 324

13.4 *Modeling and Simulation* 332

13.5 *Models* 334

13.6 *Identifying the Point Process* 337

Problems 351

Appendix A Derivations 355

A.1 *Point Processes: Definition and Measures* 356

A.2 *Point Processes: Examples* 358

A.3 *Processes Based on Fractional Brownian Motion* 360

A.4 *Fractal Renewal Processes* 362

A.5 *Alternating Fractal Renewal Process* 371

A.6 *Fractal Shot Noise* 376

A.7 *Fractal-Shot-Noise-Driven Point Processes* 382

A.8 *Analysis and Estimation* 394

Appendix B Problem Solutions 397

B.1 *Introduction* 398

B.2 *Scaling, Fractals, and Chaos* 401

B.3 *Point Processes: Definition and Measures* 404

B.4 *Point Processes: Examples* 412

B.5 *Fractal and Fractal-Rate Point Processes* 427

B.6 *Processes Based on Fractional Brownian Motion* 441

B.7 *Fractal Renewal Processes* 447

B.8 *Alternating Fractal Renewal Process* 454

B.9 *Fractal Shot Noise* 459

B.10 *Fractal-Shot-Noise-Driven Point Processes* 463

B.11 *Operations* 473

B.12 *Analysis and Estimation* 486

B.13 *Computer Network Traffic* 494

Appendix C List of Symbols — 505
 C.1 Roman Symbols — 506
 C.2 Greek Symbols — 510
 C.3 Mathematical Symbols — 511

Bibliography — 513
Author Index — 567
Subject Index — 577

List of Figures

1.1	*Coastline of Iceland at different scales*	*3*
1.2	*Vehicular-traffic point process*	*5*
2.1	*Cantor-set construction*	*17*
2.2	*Realization of Brownian motion*	*20*
2.3	*Fern: a nonrandom natural fractal*	*21*
2.4	*Grand Canyon: a random natural fractal*	*22*
2.5	*Realization of a homogeneous Poisson process*	*23*
2.6	*Nonchaotic system with nonfractal attractor: time course*	*27*
2.7	*Chaotic system with nonfractal attractor: time course*	*27*
2.8	*Chaotic system with fractal attractor*	*29*
2.9	*Chaotic system with fractal attractor: time course*	*30*
2.10	*Nonchaotic system with fractal attractor*	*31*
2.11	*Nonchaotic system with fractal attractor: time course*	*32*
3.1	*Point-process representations*	*52*
3.2	*Rescaled-range analysis: pseudocode*	*60*
3.3	*Rescaled-range analysis: illustration*	*61*
3.4	*Detrended fluctuation analysis: pseudocode*	*62*
3.5	*Detrended fluctuation analysis: illustration*	*63*
3.6	*Construction of normalized variances*	*67*
4.1	*Stochastic-rate point processes*	*89*

4.2   *Cascaded point process*   *94*

5.1   *Representative rate spectra*   *118*

5.2   *Representative normalized Haar-wavelet variances*   *119*

5.3   *Normalized Daubechies-wavelet variances*   *121*

5.4   *Fractal and nonfractal point processes*   *122*

5.5   *Fractal-rate and nonfractal point processes*   *125*

5.6   *Estimated normalized-variance curves*   *127*

5.7   *Representative interval spectra*   *128*

5.8   *Representative interval wavelet variances*   *129*

5.9   *Representative interevent-interval histograms*   *130*

5.10   *Representative capacity dimensions*   *131*

5.11   *Generalized dimensions for an exocytic point process*   *132*

6.1   *Realizations of fractional Brownian motion*   *140*

6.2   *Realizations of fractional Gaussian noise*   *143*

7.1   *Power-law interevent-interval densities*   *156*

7.2   *Fractal-renewal-process spectra*   *158*

7.3   *Fractal-renewal-process coincidence rate*   *160*

7.4   *Fractal-renewal-process normalized variance*   *161*

7.5   *Fractal-renewal-process Haar-wavelet variance*   *162*

7.6   *Fractal-renewal-process counting distributions*   *163*

7.7   *Minimal covering of fractal renewal process*   *164*

7.8   *Interevent-interval density for an interneuron*   *167*

7.9   *Normalized Haar-wavelet variance for an interneuron*   *168*

8.1   *Realization of an alternating renewal process*   *173*

8.2   *Alternating fractal-renewal-process spectrum*   *178*

8.3   *Sum of alternating renewal processes: binomial noise*   *180*

8.4   *Convergence of binomial noise to Gaussian form*   *182*

9.1   *Linearly filtered Poisson process: shot noise*   *187*

9.2   *Power-law-decaying impulse response functions*   *188*

9.3   *Stable amplitude probability densities*   *193*

9.4   *Fractal-shot-noise spectra*   *196*

10.1   *Shot-noise-driven Poisson process*   *203*

10.2   *Integrated power-law impulse response function*   *205*

10.3   *Fractal-shot-noise-driven-Poisson count distributions*   *207*

10.4   *Fractal-shot-noise-driven-Poisson variances*   *210*

10.5   *Fractal-shot-noise-driven-Poisson wavelet variances*   *211*

10.6   *Fractal-shot-noise-driven-Poisson interval densities*   *213*

10.7   *Fractal-shot-noise-driven-Poisson coincidence rates*   *215*

10.8   Fractal-shot-noise-driven-Poisson spectra   216
10.9   Hawkes point process   218
10.10 Generation of Čerenkov radiation   221
11.1   Event deletion in point processes   230
11.2   Interval histograms for randomly deleted data   233
11.3   Spectra for randomly deleted data   234
11.4   Haar-wavelet variances for randomly deleted data   235
11.5   Event-time displaced point process   243
11.6   Interval histograms for event-time-displaced data   244
11.7   Spectra for event-time-displaced data   245
11.8   Haar-wavelet variances for event-time-displaced data   246
11.9   Interval-transformed point process   248
11.10 Spectra for exponentialized data   250
11.11 Haar-wavelet variances for exponentialized data   251
11.12 Randomly shuffled point process   252
11.13 Spectra for randomly shuffled data   253
11.14 Haar-wavelet variances for randomly shuffled data   254
11.15 Superposition of point processes   256
11.16 Interevent-interval density for an interneuron   265
11.17 Normalized Haar-wavelet variance for an interneuron   266
11.18 Generalized dimensions for an interneuron   267
12.1   FGP-driven-Poisson estimated Haar-wavelet variance   277
12.2   FGP-driven-Poisson estimated normalized variance   284
12.3   FGP-driven-Poisson estimated count autocovariance   286
12.4   FGP-driven-Poisson estimated rescaled range   288
12.5   FGP-driven-Poisson estimated detrended fluctuation   290
12.6   FGP-driven-Poisson estimated interval wavelet variance   292
12.7   FGP-driven-Poisson estimated interval-based spectrum   294
12.8   Fluctuations of estimated variance and wavelet variance   297
12.9   FGP-driven-Poisson estimated rate spectrum   305
13.1   M/M/1 overflow probability vs. service ratio   320
13.2   M/M/1 overflow probability vs. maximum queue length   321
13.3   Representation of major nodes of the Internet   322
13.4   Cascaded point process in computer network traffic   324
13.5   Computer-network-traffic estimated rate spectrum   326
13.6   Computer-network-traffic estimated wavelet variance   327
13.7   Computer-network-traffic statistics: BC-pOct89   340
13.8   Computer-network-traffic statistics: BC-pAug89   341

13.9   Computer-network-traffic statistics: Bartlett–Lewis   348

13.10 Computer-network-traffic statistics: Neyman–Scott   349

B.1   Length of Icelandic coastline at different scales   398

B.2   Approaching the perimeter of a circle   399

B.3   Capacity dimension for simulated point processes   430

B.4   Estimated interevent-interval density for an interneuron   451

B.5   Estimated Haar-wavelet variance for an interneuron   452

B.6   Generating a fractal-shot-noise spectrum   461

B.7   Fractal spectrum comprising two contributions   474

B.8   Block-shuffled fractal-rate-process spectrum   480

B.9   Event-time-displaced fractal-rate-process spectrum   481

B.10 Dead-time-deleted Poisson counting distributions   483

B.11 Bias in normalized-variance estimates   489

B.12 Fractal-renewal and Poisson spectral estimates   491

B.13 Fractal-renewal and Poisson wavelet-variance estimates   492

B.14 Fractal-renewal and Poisson rescaled-range estimates   493

B.15 M/M/1 queue-length histogram   495

B.16 FGPDP/M/1 queue-length histogram ($\rho_\mu = 0.9$)   496

B.17 FGPDP/M/1 queue-length histogram ($\rho_\mu = 0.5$)   497

B.18 SHUFFLED-FGPDP/M/1 queue-length histogram   498

B.19 RFSNDP/M/1 queue-length histogram ($\rho_\mu = 0.9$)   499

B.20 MODULATED-FGPDP/M/1 queue-length histogram   500

B.21 Monofractal least-squares fit to wide-range bifractal   502

B.22 Monofractal least-squares fit to narrow-range bifractal   502

List of Tables

1.1   Length of Icelandic coastline at different scales   6
1.2   Polygon approximation for perimeter of circle   7
2.1   Representative objects: measurements and dimensions   11
9.1   Classes of fractal shot noise   189
12.1   Fractal-exponent estimates from Haar-wavelet variance   278
12.2   Fractal-exponent estimates from normalized variance   285
12.3   Fractal-exponent estimates from count autocovariance   287
12.4   Fractal-exponent estimates from rescaled range   289
12.5   Fractal-exponent estimates from detrended fluctuations 291
12.6   Fractal-exponent estimates from NIWV   293
12.7   Fractal-exponent estimates from interval spectrum   295
12.8   Optimized fractal-exponent estimates from NHWV   299
12.9   Oversampled fractal-exponent estimates from NHWV   303
12.10 Optimized fractal-exponent estimates from rate spectrum 307
13.1   Computer network traffic simulation parameters   347
13.2   Ethernet-traffic interval statistics: data and simulations 350
B.1   Fractal behavior following shuffling and interval
       transformation   474

Authors

Steven Bradley Lowen received the B.S. degree in electrical engineering from Yale University in 1984, *Magna cum Laude* and with distinction in the major. He was elected to Tau Beta Pi that same year. Following two years with the Hewlett–Packard Company he entered Columbia University, from which he received the M.S. and Ph.D. degrees in 1988 and 1992, respectively, both in electrical engineering. Lowen was awarded the Columbia University Armstrong Memorial Prize in 1988 and in 1990 he was the recipient of a Joint Services Electronics Program Fellowship in the Columbia Radiation Laboratory.

He began his research career by examining fractal patterns in the sequences of action potentials traveling along auditory nerve fibers. Recognizing that efforts to understand these fractal processes were hampered by the lack of a solid theoretical framework, he set out to develop the relevant mathematical models. This effort led to the development of alternating fractal renewal processes and fractal shot noise, as well as point-process versions thereof. In connection with this effort he also investigated fractal renewal point processes and several other fractal-based processes. This body of work served as the foundation for his Ph.D. thesis, entitled *Fractal Point Processes* (Lowen, 1992), as well as the basis of a number of journal articles and the core of several chapters in this book.

After receiving the Ph.D. degree, Dr. Lowen continued his research at Columbia as an Associate Research Scientist. He then joined Boston University as a Senior Research Associate in the Department of Electrical and Computer Engineering in 1996. He was elected to Sigma Xi in 1994.

With a collection of models for fractal-based point processes in hand, Lowen focused on establishing appropriate methods for their analysis and synthesis. This work quantified the performance of fractal estimators for point processes and highlighted the practical realities of generating realizations for these processes. He also studied the interaction between dead time (refractoriness) and fractal behavior in point processes.

Concurrently, working with various collaborators, he returned to examining applications for point processes with fractal characteristics by adapting the mathematical framework he developed to a number of biomedical point processes. He demonstrated that suitably modified fractal-based point processes serve to properly characterize action-potential sequences on auditory nerve fibers. He then turned his attention to signaling in the visual system by identifying fractal models that could describe the neural firing patterns of individual cells in this system, as well as collections of such cells, and detailing how the fractal patterns affect information transmission in this network. Using a similar approach, he also examined human heartbeat patterns and investigated how different measures of these fractal data sets could serve as markers of the cardiovascular health of the subject. Finally, he explored neurotransmitter secretion at the neuromuscular junction, and developed a suitable model showing that it, too, exhibits fractal characteristics.

Dr. Lowen also applied his fractal models to physical phenomena. These included charge transport in amorphous semiconductors and noise in infrared CCD cameras; he developed multidimensional versions of his fractal-based point processes for the latter. He also devoted substantial efforts to the modeling, synthesis, and analysis of computer network traffic.

In 1999 Dr. Lowen joined McLean Hospital and the Harvard Medical School, where he is currently Assistant Professor of Psychiatry. He has brought his fractal expertise to bear on attention-deficit and hyperactivity disorder, and the analysis of data collected with functional magnetic resonance imaging (fMRI). He is currently investigating fractal and other aspects of these applications, as well as carrying out research on drug abuse.

Dr. Lowen has authored or co-authored some 30 refereed journal articles as well as a collection of book chapters and proceedings papers. He holds a number of patents, and serves as a reviewer for several technical journals and funding agencies. Over the course of his career, he has supervised three graduate students.

Malvin Carl Teich received the S.B. degree in physics from MIT in 1961, the M.S. degree in electrical engineering from Stanford University in 1962, and the Ph.D. degree from Cornell University in 1966. His bachelor's thesis comprised a determination of the total neutron cross-section of palladium metal while his doctoral dissertation reported the first observation of the two-photon photoelectric effect in metallic sodium. His first professional affiliation was with MIT's Lincoln Laboratory in Lexington, Massachusetts, where he demonstrated that heterodyne detection could be achieved in the middle-infrared region of the electromagnetic spectrum.

Teich joined the faculty at Columbia University in 1967, where he served as a member of the Electrical Engineering Department (as Chairman from 1978 to 1980), the Applied Physics Department, the Columbia Radiation Laboratory, and the Fowler Memorial Laboratory for Auditory Biophysics. Extending his work on heterodyning, he recognized that the interaction could be understood in terms of the absorption of individual polychromatic photons, and demonstrated the possibility of implementing the process in a multiphoton configuration. He developed the concept of nonlinear heterodyne detection — useful for canceling phase or frequency noise in an optical system.

During his tenure at Columbia, he also carried out extensive work in point processes, with particular application to photon statistics, the generation of squeezed light, and noise in fiber-optic amplifiers and avalanche photodiodes. Among his achievements is a description of luminescence light in terms of a photon cluster point process. This perspective led him to suggest that detector dead time could be used advantageously to reduce the variability of this process and thereby luminescence noise. This approach was incorporated in the design of the star-scanner guidance system for the Galileo spacecraft, which was subjected to high radio- and beta-luminescence background noise as a result of bombardment by copious Jovian gamma- and beta-ray emissions. In the domain of quantum optics he developed the concept of pump-fluctuation control in which the variability of a pump point process comprising a beam of electrons is reduced by making use of self-excitation in the form of Coulomb repulsion. Using a space-charge-limited version of the Franck–Hertz experiment in mercury vapor he demonstrated the validity of this concept by generating the first source of unconditionally sub-Poisson light. His work on fiber-optic amplifiers led to an understanding of the properties of the photon point process that emerges from the laser amplifier and thereby of the performance characteristics of these devices.

Teich's interest in point processes in the neurosciences was fostered by a chance encounter in 1974 with William J. McGill, then Professor of Psychology and President of Columbia University. This, in turn, led to a long-standing collaboration with Shyam M. Khanna, Director of the Fowler Memorial Laboratory for Auditory Biophysics and Professor in the Department of Otolaryngology at the Columbia College of Physicians & Surgeons. Together, Teich and Khanna carried out animal experiments over many years in which spike trains in the peripheral auditory system were recorded. Analysis of these data led to the discovery that, without exception, action-potential sequences in the auditory system exhibited fractal features. Teich and his students, including Lowen, developed suitable point-process models to accommodate these data and to offer a fresh mathematical perspective of sensory neural coding. In a collaboration with researchers at the Karolinska Institute in Stockholm, they also conducted heterodyne velocity measurements of the vibratory motion of individual sensory cells in the cochlea, discovering that these cells can vibrate spontaneously, even in the absence of a stimulus.

In 1995 Teich was appointed Professor Emeritus of Engineering Science and Applied Physics at Columbia. He joined Boston University, where he is currently teaching and pursuing his research interests as a faculty member with joint appointments in the Departments of Electrical and Computer Engineering, Physics, Cognitive and Neural Systems, and Biomedical Engineering. He is Co-Director of the Quantum Imaging Laboratory and a Member of the Photonics Center, the Hearing Research Center, the Program in Neuroscience, and the Center for Adaptive Systems. He also serves as a consultant to government and private industry.

His current efforts in the domain of quantum optics are directed toward developing imaging systems that make use of entangled photon pairs generated in the nonlinear optical process of parametric down-conversion. His work in fractals and wavelets is directed toward understanding biological phenomena such as the statistical properties of neurotransmitter exocytosis at the synapse, action-potential patterns in auditory- and visual-system neurons, and heart-rate-variability analysis of patients who suffer from cardiovascular-system dysfunction.

Teich is a Fellow of the Acoustical Society of America, the American Association for the Advancement of Science, the American Physical Society, the Institute of Electrical and Electronics Engineers, and the Optical Society of America. He is a member of Sigma Xi and Tau Beta Pi. In 1969 he received the IEEE Browder J. Thompson Memorial Prize for his paper "Infrared Heterodyne Detection." He was awarded a Guggenheim Fellowship in 1973. In 1992 he was honored with the Memorial Gold Medal of Palacký University in the Czech Republic, and in 1997 he received the IEEE Morris E. Leeds Award.

He has authored or coauthored some 300 journal articles and holds a number of patents. He is the coauthor, with Bahaa Saleh, of *Fundamentals of Photonics* (Wiley, 1991).

Among his professional activities, he served as a member of the Editorial Advisory Panel for the journal *Optics Letters* from 1977 to 1979, as a Member of the Editorial Board of the *Journal of Visual Communication and Image Representation* from 1989 to 1992, and as Deputy Editor of *Quantum Optics* from 1988 to 1994. He is currently a Member of the Editorial Board of the journal *Jemná Mechanika a Optika* and a Distinguished Lecturer of the IEEE *Engineering in Medicine and Biology Society*.

1

Introduction

Albrecht Penck (1858–1945), a German geographer and geologist known particularly for his studies of glaciation in the Alps, recognized that the length of a coastline depends on the scale at which it is measured.

Lewis Fry Richardson (1881–1953), a British mathematician and Quaker pacifist, studied turbulence, weather prediction, the statistics of wars, and the relationship between length and measurement scales.

1.1	**Fractals**	2
1.2	**Point Processes**	4
1.3	**Fractal-Based Point Processes**	4
	Problems	6

1.1 FRACTALS

What is the length of a coastline? Albrecht Penck, a Professor of Geography at the University of Vienna, observed more than a hundred years ago that the apparent length of a coastline grew larger as the size of the map increased (Penck, 1894). He concluded that this apparently simple question has a complex answer — the outcome of the measurement depends on the scale at which the coastline is measured.

Using a detailed map of a given coastline, and carefully tracing all of its bays and peninsulas, a sufficiently patient geographer could follow the features and arrive at a number for its length. A more hurried geographer, using a map of lower resolution that follows the coastline less closely, would obtain a smaller result since many of the features seen by the first geographer would be absent. In general, higher measurement precision yields a greater number of discernible details, and consequently results in a coastline of greater length. The question "What is the length of a coastline?" has no single answer.

This phenomenon is illustrated in Fig. 1.1 for a section of the Icelandic coastline between the towns of Seyðisfjörður and Höfn. The three maps, illustrated in different shades of gray, are identical; only the scale used to measure the length of the coastline differs. The measurement indicated by the white curve on the dark-gray map traces features with a scale of 0.694 km. Following all of the features of the map at this scale requires 769 segments, each of length 0.694 km, which yields an overall coastline length of 534 km. This measurement closely hugs the coastline. The medium-gray map, whose boundary (black curve) is measured with a 6.94-km scale, requires just over 45 segments, leading to a coastline length of 314 km. This coarser measurement follows the coastline more approximately; the result therefore appears more jagged and yields a shorter length. Finally, a measurement made at a scale of 69.4 km, represented on the light-gray map, yields the shortest length of the three: 133 km.

Were the scale to increase further, a minimum distance of 125 km would obtain for scales in excess of 125 km, since that is the point-to-point distance between the two towns. At the opposite extreme, if the measurement scale were to reach below 0.694 km, the length of the coastline would grow beyond 534 km, as ever smaller bays and peninsulas, rocks, and grains of sand were taken into account.

Although coastlines do not have well-defined lengths, as recognized by Penck (1894), and subsequently by Steinhaus (1954) and Perkal (1958a,b), an empirical mathematical relation connecting the measured coastline length and the measurement scale was discovered by Lewis Fry Richardson. In his *Appendix to Statistics of Deadly Quarrels*, which appeared in print some years after his death, Richardson (1961)

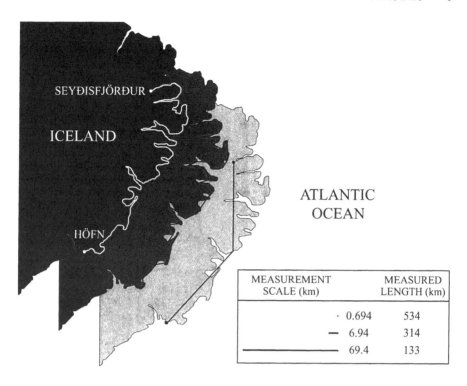

Fig. 1.1   The coastline of Iceland between Seyðisfjörður and Höfn, measured at three different scales. The finest-scale measurement is shown as the white curve on the dark-gray (top) map. The inset table indicates the measured coastline length for the three different scales. The finer the scale of the measurement, the greater the detail captured, and the larger the outcome for the length of the coastline.

showed that this relation takes the form of a power-law function,

$$d \propto s^c, \tag{1.1}$$

where d is the length of the coastline, s is the measurement scale, and c is a (negative) power-law exponent.[1] A circle, in contrast, does not fit this form, suggesting that real coastlines do not resemble simple geometrical shapes, and that Richardson's relation is not spurious.

---

[1] In *Statistics of Deadly Quarrels*, Richardson (1960) had previously demonstrated that the magnitude of a war related to its frequency of occurrence by means of a power-law function. For each tenfold increase in size, he found roughly a threefold decrease in frequency. He also determined that the occurrences of wars closely follow a Poisson process (see Sec. 4.1), albeit with a quasi-periodic modulation of the rate. He further concluded that states bordering a large number of contiguous states tended to become involved in wars more often — hence, his attention to the lengths of frontiers and coastlines. A biographical sketch of Richardson is provided by Mandelbrot (1982, Chapter 40).

Not long after Richardson's work was published, Benoit Mandelbrot (1967a, 1975) revisited the issue of coastline length, and began to lay the groundwork for what would later be called fractal analysis.[2] The dependence of a measurement outcome on the scale chosen to make that measurement is the hallmark of a **fractal object**, and coastlines are indeed fractal. The power-law relationship provided in Eq. (1.1) offers a useful description of coastlines and other fractal objects, as we will see in Chapter 2.

1.2   POINT PROCESSES

After a long day measuring coastlines, our geographer drives home for the night. Unfortunately, many other people have chosen this same hour to drive, and our geographer encounters traffic. Since this is a recurring problem, the local government has decided to charter a study of the traffic flow patterns along the road from the coast. What is the best method for describing the traffic?

Certain details of the vehicles, such as their color or the number of occupants, are irrelevant to the traffic flow. To first order, a listing of the times at which a vehicle crosses a given point on the road provides the most salient information. Such a record, in the form of a set of marks on a line, is called a **point process**.[3] The mathematical theory of point processes forms a surprisingly rich field of study despite its seeming simplicity.[4]

Figure 1.2 illustrates the process of reducing a moving set of vehicles on a roadway to a point process. The figure comprises snapshots of the same stretch of single-lane roadway, at different times as successive vehicles pass a fixed measurement location (indicated by vertical dashed line). Vehicles yet to reach the measurement location are shown as white. As each vehicle passes this location, it turns light gray, and the point-process record at the right accrues a corresponding mark at that time, indicating the passing event. Many of the vehicles (labeled by letters) appear in several of the snapshots. Dark gray indicates a vehicle that passed the dashed line elsewhere in the figure whereas black indicates a vehicle that passed this location yet earlier.

1.3   FRACTAL-BASED POINT PROCESSES

This book concerns **fractal-based point processes** — processes with fractal properties comprised of discrete events, either identical or taken to be identical.

---

[2] A photograph of Mandelbrot stands at the beginning of Chapter 7. A recent interview by Olson (2004) offers some personal reminiscences about his life and career.

[3] A point process is sometimes called a **time series**, although this latter designation usually refers to a discrete-time process.

[4] Point processes relevant to vehicular traffic flow have been studied, for example, by Chandler, Herman & Montroll (1958); Komenani & Sasaki (1958); Bartlett (1963, 1972); Newell & Sparks (1972); Bovy (1998); and Kerner (1998, 1999).

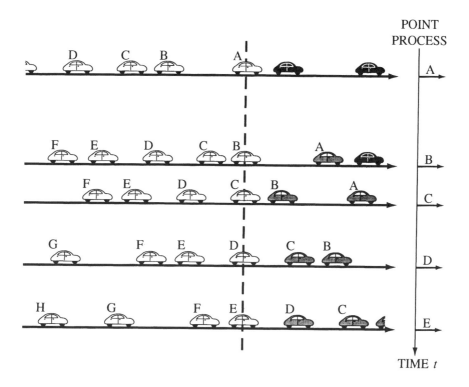

Fig. 1.2 Generation of a point process from vehicular traffic. Each horizontal line depicts vehicles traveling along the same stretch of single-lane road, but at different times. As each successive vehicle passes a measurement location (indicated by the vertical dashed line), it turns light gray in color and a corresponding mark appears in the final point-process representation (short horizontal arrow at far right), denoting the occurrence of an event at that time. In each depiction of the roadway, vehicles that have not yet passed the dashed line are shown as white, whereas those that already passed it elsewhere in the figure appear as dark gray. Vehicles shown in black passed the dashed line at a yet earlier time.

A more detailed introduction to fractals, and their connection to chaos, is provided in *Chapter 2*. We define point processes, and consider measures useful for characterizing them, in *Chapter 3*. *Chapter 4* sets forth a number of important examples of point processes. With an understanding of fractals and point processes in hand, we address their integration in *Chapter 5*. Mathematical formulations for several important fractal-based point-process families follow, in *Chapters 6–10*. An exposition of how various operations affect these processes appears in *Chapter 11*. *Chapter 12* considers techniques for the analysis and estimation of fractal-based point processes. Finally, *Chapter 13* is devoted to computer network traffic, an important application that serves as an illustration of the various approaches and models set forth in earlier chapters.

Problems

1.1 *Length of Icelandic coastline at different measurement scales*   Table 1.1 provides results for the length of a portion of the east coast of Iceland, measured between the towns of Seyðisfjörður and Höfn, at different measurement scales.

Measurement Scale s (km)	Number of Segments	Coastline Length d (km)
0.694	769	534
1.39	306	425
2.78	140	389
6.94	45.2	314
13.9	20.3	282
27.8	6.33	176
50.0	2.84	142
69.4	1.92	133

Table 1.1 Length of the Icelandic coastline between the towns of Seyðisfjörður and Höfn, determined using eight different measurement scales. The finer the scale, the greater the length. These measurements were made from a map with a resolution of 0.694 km, as determined by the edge length of the minimum pixel size. The point-to-point distance between the two towns is 125 km.

1.1.1. Plot the coastline length d vs. the measurement scale s on doubly logarithmic coordinates.

1.1.2. Use the plot generated in Prob. 1.1.1, together with Eq. (1.1), to determine the power-law exponent c that characterizes the eastern Icelandic coastline.

1.1.3. Using this same form of analysis, Richardson (1961, p. 169) showed that Eq. (1.1) indeed characterized several coastlines. He reported the following results: (1) $c \approx -0.02$ for the South African coastline between Swakopmund and Cape Santa Lucia; (2) $c \approx -0.13$ for the Australian coastline; and (3) $c \approx -0.25$ for the west coast of Great Britain. Compare the scaling exponent c you obtain for the east coast of Iceland with those determined by Richardson. Consult an atlas to estimate the relative roughness of the four coastlines and relate this to the power-law exponents c.

1.2 *Circles and fractals*   Suppose that we calculate entries for a table similar to the ones that appear in Fig. 1.1 and Table 1.1, but for a circle of unit circumference. To measure the circumference with n equal line segments, inscribe a regular polygon of n sides into the circle, and calculate the perimeter of the polygon. This procedure yields a perimeter that increases as the polygon side length decreases, as shown in Table 1.2.

Polygon Side Length	Number of Sides	Polygon Perimeter
0.033	30	0.998
0.098	10	0.984
0.187	5	0.935
0.276	3	0.827

Table 1.2   Polygon approximation for the perimeter of a circle.

1.2.1. Calculate an exact expression for the side length and total estimated perimeter as a function of the number of sides, and verify that the perimeter monotonically increases with the number of sides.

1.2.2. Is a circle a fractal? Why?

2

Scaling, Fractals, and Chaos

Georg Cantor (1845–1918), a celebrated German mathematician, founded set theory and recognized the distinction between countably infinite and uncountably infinite sets, such as the sets of rational and real numbers, respectively.

The French mathematician **Henri Poincaré (1854–1912)** established that certain deterministic nonlinear dynamical systems exhibit an acute sensitivity to initial conditions; this characteristic is now recognized as a hallmark of deterministic chaos.

2.1   Dimension   11
    2.1.1 Capacity dimension   12
2.2   Scaling Functions   13
2.3   Fractals   13
    2.3.1 Fractals, scaling, and long-range dependence   14
    2.3.2 Monofractals and multifractals   15
2.4   Examples of Fractals   16
    2.4.1 Cantor set   17
    2.4.2 Brownian motion   19
    2.4.3 Fern   21
    2.4.4 Grand Canyon river network   22
2.5   Examples of Nonfractals   23
    2.5.1 Euclidean shapes   23
    2.5.2 Homogeneous Poisson process   23
    2.5.3 Orbits in a two-body system   24
    2.5.4 Radioactive decay   24
2.6   Deterministic Chaos   25
    2.6.1 Nonchaotic system with nonfractal attractor   26
    2.6.2 Chaotic system with nonfractal attractor   28
    2.6.3 Chaotic system with fractal attractor   28
    2.6.4 Nonchaotic system with fractal attractor   30
    2.6.5 Chaos in context   31
2.7   Origins of Fractal Behavior   32
    2.7.1 Fractals and power-law behavior   32
    2.7.2 Physical laws   33
    2.7.3 Diffusion   34
    2.7.4 Convergence to stable distributions   35
    2.7.5 Lognormal distribution   36
    2.7.6 Self-organized criticality   37
    2.7.7 Highly optimized tolerance   37
    2.7.8 Scale-free networks   37
    2.7.9 Superposition of relaxation processes   38
2.8   Ubiquity of Fractal Behavior   39
    2.8.1 Fractals in mathematics and in the physical sciences   39
    2.8.2 Fractals in the neurosciences   41
    2.8.3 Fractals in medicine and human behavior   43
    2.8.4 Recognizing the presence of fractal behavior   44
    2.8.5 Salutary features of fractal behavior   45
    Problems   46

2.1 DIMENSION

The word "dimension," at least among the technically inclined, generally conjures up an image of a familiar elemental shape such as a line or rectangle. These objects have dimensions that correspond to the measurements used to quantify them: meters and square meters, respectively. Table 2.1 presents four representative "Euclidean" objects along with their dimensions (Mandelbrot, 1982).

Object	Measurement	Dimension
Point	$meters^0$	0
Line	$meters^1$	1
Square	$meters^2$	2
Cube	$meters^3$	3

Table 2.1 Representative objects: measurements and dimensions.

The union of two objects that have a particular dimension is characterized by that same dimension. Thus, any finite number of points retains a dimension of zero, and three squares connected end to end as a whole maintain a dimension of two. What happens to the dimension as we deform the objects, converting a line into a curve, say, or a square into an ellipse? Common sense suggests that the dimension of an object is robust in the face of such manipulations, and topological theory bears this out. A curve in three-dimensional space, such as a helix, maintains a dimension of unity since uncoiling the helix yields a line of that dimension.

The foregoing discussion illustrates a general property: the dimension of an object cannot exceed the dimension of the space in which it resides (the **Euclidian dimension**). An infinite collection of points immediately adjacent to each other yields a curve, and one can generate a circle from the union of all possible points equidistant from a given point. In both cases, the component objects have a dimension smaller than that of the space. However, one can form a square from a collection of smaller squares, and all squares have a dimension of two. These examples lead to another property: the dimension of each member of a group of objects (**the topological dimension**) cannot exceed the dimension of the object formed by their union.

Taken together, the two properties reveal that a collection of objects of a particular dimension, embedded in an object with another dimension, will have an overall dimension that lies between the two. For example, a collection of points (dimension zero) lying within a square (dimension two) can yield a line (dimension one). Yet, a different collection of points could yield a square of smaller size (dimension two), or a single point (dimension zero). In all cases, however, the dimension of the resulting object lies between zero and two inclusive, in accord with the properties set forth above.

2.1.1 Capacity dimension

Although the concept of dimension as discussed above has intuitive appeal, it is important to develop a more rigorous approach for quantifying dimension. We illustrate one measure, initially introduced by Pontrjagin & Schnirelmann (1932), called the **capacity dimension** or **box-counting dimension** (this technique is discussed in more detail in Sec. 3.5.4). Imagine an ellipse (including its interior) in a plane; we know this object has a dimension of two. Suppose we draw a grid over the ellipse and the surrounding region, yielding a collection of square boxes, and then count the number of squares that overlap at least one point in the ellipse. For squares of a given edge size ϵ, we obtain a number $M(\epsilon)$. (The precise value of this number depends on the alignment of the grid with respect to the ellipse, but this fact does not affect the following argument.)

Now repeat the process with squares half the size of the original ones. The number of squares required to cover the ellipse will increase roughly by a factor of four, since four new squares cover one of the original ones. Thus, $M(\epsilon/2) \approx 4M(\epsilon)$. In the limit, as the size of the squares decreases towards zero, we find

$$M(\epsilon) \rightarrow C\epsilon^{-2}, \tag{2.1}$$

where the constant C represents the area of the ellipse, and the alignment of the grid does not affect this result. To extract the exponent from Eq. (2.1), we first take the logarithm, which yields a relation linear in the exponent. Dividing by the logarithm of the inverse box size and taking the limit for small boxes yields the desired exponent:

$$\lim_{\epsilon \to 0} \frac{\ln\left[M(\epsilon)\right]}{\ln(1/\epsilon)} = 2, \tag{2.2}$$

which coincides with the dimension of an ellipse. This suggests a general method for obtaining the capacity dimension that will report the correct exponent even when it may not be readily apparent in the functional form of $M(\epsilon)$.

Now suppose that we repeat the process with a curve in a plane. In this case, the number of squares required increases linearly with $1/\epsilon$, whereupon Eq. (2.2) yields a value of unity. Similarly, a finite collection of n points requires no more than n squares, no matter how small ϵ becomes, resulting in a dimension of zero.

In general, the box-counting technique of determining the dimension proceeds by covering the set in question with "boxes," namely cubes, squares, line segments, or other forms, depending on the space within which the shape lies. The relationship between the number of boxes that contain part of the set and the size of those boxes, as the size decreases to zero, determines the capacity dimension D_0 of the set:

$$M(\epsilon) \rightarrow C\epsilon^{-D_0}. \tag{2.3}$$

Thus far, the outcome agrees with our intuition about dimension. When applied to fractals, however, we will see that this approach leads to noninteger values, although it is always bounded from above by the Euclidian dimension, and from below by the topological dimension.

2.2 SCALING FUNCTIONS

Fractals turn out to have close connections to the scaling behavior observed in some functions. A function is said to "scale" when shrinking or stretching both axes (by possibly different amounts, neither equal to unity) yields a new graph that coincides with the original. Scaling leads mathematically to power-law[1] dependencies in the scaled quantities, as we now proceed to show.

Consider a function f that depends continuously on the scale s over which we take measurements. Suppose that changing the scale by any factor a effectively changes the function by some other factor $g(a)$, which depends on the factor a but is independent of the original scale s:

$$f(as) = g(a) f(s). \tag{2.4}$$

The only nontrivial solution of this scaling equation for real functions and arguments, and for arbitrary a and s , is (see Prob. 2.5)

$$f(s) = b\,g(s), \tag{2.5}$$

with

$$g(s) = s^c \tag{2.6}$$

for some constants b and c (Lowen & Teich, 1995; Rudin, 1976). Equations (1.1) and (2.3) provide examples of this relationship.

Restricting a to a fixed value in Eq. (2.4) yields a larger set of possible solutions (Shlesinger & West, 1991):

$$g(s; a) = s^c \cos[2\pi \ln(s)/\ln(a)]. \tag{2.7}$$

2.3 FRACTALS

The concept of a fractal involves three closely related characteristics, each of which could serve as a definition in its own right. Indeed, a variety of definitions for fractals exist (Mandelbrot, 1982). Furthermore, fractals can be *deterministic* or *random*. They can also be *static*, such as the Icelandic coastline, or arise from a *dynamical process* such as Brownian motion.

First, fractals possess a form of self-scaling: parts of the whole can be made to fit to the whole in some nontrivial way by shifting and stretching. If stretching equally in all directions yields such a fit, then an object is said to be self-similar. If the fit requires anisotropic stretching, then the object is said to be self-affine (Mandelbrot,

---

[1] Power-law functions have many aliases, including "algebraic," "hyperbolic," and "allometric." When applied to distributions, the term "heavy-tailed" often (but not always) refers to the same functional form.

1982). For deterministic fractals, the fit is exact. Random fractals, in contrast, fit statistically; transformed parts resemble the whole and have similar probabilistic characteristics, although they do not precisely coincide with it. The coast of Iceland, for example, contains similar features over a range of sizes. As illustrated in Fig. 1.1, what would appear to be a simple bay on a large-scale (coarse-grained) map turns into a meandering connection of inlets and other invaginations when displayed more finely. Examining a length of coastline on a map (without intimate knowledge of the particular section under study) does not provide information about the scale of the map despite knowledge of the size of the entire object, in this case Iceland. In contrast, examining of a section of nonfractal object, such as a circle, readily yields the scale in terms of the size of the object.

Second, the statistics that are used to describe fractals scale with the measurement size employed. For example, the length of the east coast of Iceland follows the form of Eqs. (2.5) and (2.6) with an empirical fractal exponent $c \approx -0.30$, as shown in Prob. 1.1. Statistics with power-law forms are thus closely related to fractals. Indeed, we often highlight this connection by presenting various measures using logarithmic axes for both the ordinate and abscissa; power-law functions become straight lines on such doubly logarithmic graphs and their slopes provide the power-law exponents. This characteristic of fractals proves quite useful.

Third, the fractal exponent that corresponds to a particular statistic, one of the generalized dimensions (see Sec. 3.5.4), assumes a noninteger value. As their size decreases, the number of boxes required to cover the Icelandic coastline increases in such a way that the capacity dimension $D_0 \approx 1.30$ (see Secs. 2.1 and 3.5.4). This scaling exponent assumes a noninteger value lying between that of a line ($D_0 = 1$) and that of a plane ($D_0 = 2$).

2.3.1   Fractals, scaling, and long-range dependence

Fractal behavior, such as an object containing smaller copies within itself, can extend down to arbitrarily small sizes in an abstract mathematical construct. However, real-world fractals generally exhibit minimum sizes beyond which fractal behavior is not obeyed. For example, decreasing the length scale used to measure the length of a coastline will eventually lead to a breakdown in scaling behavior. The geological forces at work over kilometer scales differ from those operating over much smaller length scales, leading to different appearances over these smaller lengths. Certainly at a scale corresponding to individual atoms, the emergent features are expected to bear little resemblance to those at macroscopic length scales.

There are also limits at large scales. Fractal behavior can have a maximum scale, one that often corresponds to the size of the fractal object itself. The minimum and maximum scales that bound fractal behavior are known as the **lower cutoff** (or **inner cutoff**) and the **upper cutoff** (or **outer cutoff**), respectively.

Moreover, any data set collected from a real-life physical or biological experiment will perforce have lower and upper cutoffs, corresponding to the resolution limits of the measurement apparatus and the extent of the entire data set, respectively. These lower and upper measurement cutoffs impose limits on observable fractal behavior that sometimes prove more restrictive than those of the fractal object under study itself. In the case of the Icelandic coastline portrayed in Fig. 1.1, for example, the resolution of the map from which the measurement is constructed (0.694 km), which is determined by the edge length of the minimum pixel size (see Prob. 1.1), imposes a lower cutoff. An upper cutoff is imposed by the size of the island itself.

Generating a point process from a rate (see Chapter 4) necessarily involves some loss of information, and can also set an effective minimum scale. In both the doubly stochastic and integrate-and-reset point processes considered in Chapter 4, for example, fractal features present in the rate process over time scales shorter than the average time between events will be greatly attenuated in the resultant point process.

In a more rigorous mathematical context, fine distinctions are sometimes drawn between fractals and scaling (Flandrin, 1997; Flandrin & Abry, 1999). The term "scaling" is used when both lower and upper cutoffs exist, "fractal" denotes objects for which no small-size cutoff exists, and "long-range dependence" corresponds to the lack of a large-size cutoff.[2] Since essentially all of the applications we consider derive from limited measurements, our discussion might be more properly framed in terms of scaling rather than fractal behavior. Following common usage, however, we generally do not make this distinction. The Lévy dust, considered in Sec. 4.7, and the zero crossings of ordinary and fractional Brownian motion, considered in Sec. 6.1, are the sole exceptions. These two collections of points, which, properly speaking, are not point processes, are fractal in the strict sense of the term.

2.3.2 Monofractals and multifractals

The scaling behavior discussed thus far involves a single stretching or shifting rule, and a single exponent for each statistic. For some objects, the rule and exponents depend on the position within the object, or on the size of the component. Each such object can thus contain a range of fractal behaviors, and is therefore called a **multifractal** (Mandelbrot, 1999; Sornette, 2004). In this context, a simpler fractal object described in our earlier discussions is called a **monofractal**. Although examples of multifractals can be found, in practice relatively few point-process data sets contain sufficient information to accurately characterize their multifractal spectrum.

Perhaps the best method for attempting such a characterization leads to a multifractal spectrum by simulating a number of surrogate data sets with different parameters, and selecting the best fitting parameters as estimates of the multifractal behavior (Roberts & Cronin, 1996). This method yields good accuracy with as few as $N = 100$ points. However, its inherent parametric approach limits its usefulness in general, since the algorithm requires *a priori* knowledge of the form of the mul-

---

[2] More precisely, a process is said to have long-range dependence when its autocorrelation has an infinite integral (for continuous-time processes) or an infinite sum (for discrete-time processes) (Cox, 1984). Theoretically, a process could have long-range dependence without exhibiting power-law behavior, but this is uncommon.

tifractal spectrum. Indeed, it estimates only two parameters from the data set rather than an arbitrary form for the multifractal spectrum.

Finally, such methods typically require that the point processes themselves, rather than merely the rates of these processes, exhibit (multi)fractal behavior. Such fractal point processes (see Sec. 5.5.1) form an important subclass of the class of fractal-based point processes (see Sec. 5.5) that we explore, but they do not describe a large number of data sets.

As a consequence of these limitations, we concentrate principally on monofractals in this book. Computer network traffic is a notable exception: the availability of extensive, long data records allow a valid multifractal analysis to be carried out (see Sec. 13.3.8).

2.4 EXAMPLES OF FRACTALS

Fractals abound in many fields: **mathematics** (Mandelbrot, 1982; Stoyan & Stoyan, 1994; Peitgen, Jürgens & Saupe, 1997; Barnsley, 2000; Mandelbrot, 2001; Falconer, 2003; West, Bologna, Grigolini & MacLachlan, 2003; Doukhan, 2003); **physics** (Mandelbrot, 1982; Feder, 1988; Schroeder, 1990; Sornette, 2004); **geology** (Turcotte, 1997); **imaging science** (Peitgen & Saupe, 1988; Turner, Blackledge & Andrews, 1998; Flake, 2000); **electronic devices and systems** (Buckingham, 1983; van der Ziel, 1986, 1988; Weissman, 1988; Kogan, 1996); **complex electronic and photonic media** (Berry, 1979; Merlin, Bajema, Clarke, Juang & Bhattacharya, 1985; Kohmoto, Sutherland & Tang, 1987); **materials growth** (Kaye, 1989; Vicsek, 1992); **signal processing** (Flandrin & Abry, 1999); **engineering** (Lévy Véhel, Lutton & Tricot, 1997); **vehicular-traffic behavior** (Musha & Higuchi, 1976; Bovy, 1998); **computer networks** (Mandelbrot, 1965a; Leland, Taqqu, Willinger & Wilson, 1994; Albert, Jeong & Barabási, 1999; Park & Willinger, 2000); **biology and physiology** (Musha, 1981; Turcott & Teich, 1993; Bassingthwaighte, Liebovitch & West, 1994; West & Deering, 1994, 1995; Collins, De Luca, Burrows & Lipsitz, 1995; Turcott & Teich, 1996; Liebovitch, 1998; Vicsek, 2001; Teich, Lowen, Jost, Vibe-Rheymer & Heneghan, 2001; Shimizu, Thurner & Ehrenberger, 2002); **behavior and psychiatry** (Paulus & Geyer, 1992; West & Deering, 1995; Gottschalk, Bauer & Whybrow, 1995; Teicher, Ito, Glod & Barber, 1996; Anderson, Lowen, Renshaw, Maas & Teicher, 1999; Anderson, 2001); **neuroscience** (Verveen, 1960; Evarts, 1964; Musha, Takeuchi & Inoue, 1983; Läuger, 1988; Millhauser, Salpeter & Oswald, 1988; Teich, 1989; Lowen & Teich, 1996a; Teich, Turcott & Siegel, 1996; Teich, Heneghan, Lowen, Ozaki & Kaplan, 1997; Thurner, Lowen, Feurstein, Heneghan, Feichtinger & Teich, 1997; Lowen, Cash, Poo & Teich, 1997b); fractals also play important roles in **other fields**.

We proceed to consider four examples of fractals, one each of the possible combinations of

- *artificial* and *natural*

- *deterministic* and *random*.

2.4.1   Cantor set

The **Cantor set**, discovered by Georg Cantor (1883), provides an example of an artificial, deterministic, one-dimensional fractal structure that extends to arbitrarily small scales. One particular mathematical construction of this set has as its starting point the closed unit interval

$$C_0 \equiv [0, 1]. \tag{2.8}$$

From C_0, we form C_1, the next step in the formation of the triadic Cantor set, by removing the middle third of this interval:

$$C_1 \equiv \left[\tfrac{0}{3}, \tfrac{1}{3}\right] \cup \left[\tfrac{2}{3}, \tfrac{3}{3}\right], \tag{2.9}$$

where $\cup$ represents the set union operation. We then obtain C_2 from C_1 by removing the middle thirds of *both* segments, so that

$$C_2 \equiv \left[\tfrac{0}{9}, \tfrac{1}{9}\right] \cup \left[\tfrac{2}{9}, \tfrac{3}{9}\right] \cup \left[\tfrac{6}{9}, \tfrac{7}{9}\right] \cup \left[\tfrac{8}{9}, \tfrac{9}{9}\right]. \tag{2.10}$$

The set C_n denotes the nth stage in this process. This procedure is continued indefinitely, leading to the Cantor set itself, C, which is defined as the limit

$$C \equiv \lim_{n \to \infty} C_n. \tag{2.11}$$

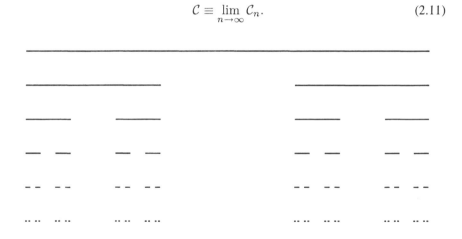

Fig. 2.1   The first six stages in the construction of a triadic Cantor set. The process begins with the unit interval; removing the middle third of each segment at a given stage yields the following stage. Continuing this process indefinitely yields the Cantor set itself as a limit.

The first six stages in the construction of a triadic Cantor set are displayed in Fig. 2.1. The Cantor set C consists of two exact copies of itself, in the intervals $\left[0, \tfrac{1}{3}\right]$ and $\left[\tfrac{2}{3}, 1\right]$, respectively, each of which is one-third the size of the whole. It also contains four copies of itself, each one-ninth the size of the whole. In fact it has 2^n copies, each 3^{-n} the size of the original set, for all nonnegative integers n. For the triadic Cantor set, increasing the length scale ϵ by a factor of 3 decreases the number of copies $N(\epsilon)$ by a factor of 2, so that $N(\epsilon) \sim \epsilon^{-D_0}$ with $D_0 \equiv \log(2)/\log(3) \doteq$

0.630930. There is, in fact, an entire family of generalized dimensions D_q (see Sec. 3.5.4) but for monofractals such as the Cantor set, these all coincide so that $D_q = D$ for all q.

A set with infinitely many copies of itself, each of vanishing size, can exhibit counterintuitive properties. Such seeming paradoxes often occur in the study of abstract fractals, and we now proceed to examine one in the light of the Cantor set: this set has a total length of zero, but just as many points as the unit interval employed as the first stage in its construction. To show this, we begin with the total length of the Cantor set. The initial stage of the Cantor set consists of the unit interval, and therefore has a Lebesgue measure or length $\mathcal{L}(\mathcal{C}_0)$ of unity. The second stage has 2 segments of length $\frac{1}{3}$, for a total length $\mathcal{L}(\mathcal{C}_1) = \frac{2}{3}$; the nth stage comprises 2^n segments each of length $1/3^n$, yielding $\mathcal{L}(\mathcal{C}_n) = (\frac{2}{3})^n$. Continuing this process indefinitely leads to the total length of the Cantor set itself as a limit:

$$\mathcal{L}(\mathcal{C}) = \lim_{n \to \infty} \mathcal{L}(\mathcal{C}_n) = \lim_{n \to \infty} \left(\tfrac{2}{3}\right)^n = 0. \tag{2.12}$$

So the Cantor set has zero total length.

Turning now to the number of points in the Cantor set, consider a ternary expansion of the points in the original interval $\mathcal{C}_0 \equiv [0, 1]$. Each point x in this interval may be represented by a corresponding sequence of digits $0.a_1 a_2 a_3 a_4 \ldots$, where

$$x = \sum_k a_k \left(\tfrac{1}{3}\right)^k, \tag{2.13}$$

with each a_k either 0, 1 or 2. Points of $\mathcal{C}_0$ contained in the open interval $\left(\frac{1}{3}, \frac{2}{3}\right)$ will not appear in $\mathcal{C}_1$, and all have a 1 in the first position after the decimal point of their ternary expansions. (The point $\frac{1}{3}$, the upper limit of the first segment, also has a 1 in the first position, but remains in $\mathcal{C}_1$ and in $\mathcal{C}$ as well. We will return to the issue of endpoints shortly.) Points with a 1 in the second position after the decimal point correspond to the middle third of both segments of $\mathcal{C}_1$, and will not appear in $\mathcal{C}_2$ or in subsequent stages. Thus, $\mathcal{C}_n$ contains only those points without a 1 in any of the first n positions of the corresponding expansions.

In the limit, then, the Cantor set $\mathcal{C}$ contains only those points that do not display a 1 in *any* position of the corresponding expansion; the ternary expansion of any point in $\mathcal{C}$ consists solely of the symbols 0 and 2. For example, the point corresponding to the expansion $0.020202\ldots$ (base 3) $= \frac{1}{4}$ belongs to $\mathcal{C}$, whereas $0.111111\ldots$ (base 3) $= \frac{1}{2}$ does not. Points in the original interval $\mathcal{C}_0$ may also be expanded in *binary* format, with each digit chosen from the set $\{0, 1\}$. Therefore, there exists a one-to-one mapping between the points in the Cantor set $\mathcal{C}$ and those in the original unit interval $\mathcal{C}_0$; simply replace the "2" symbols in the ternary expansion for the former with "1" symbols in the binary expansion of the latter. In particular, the Cantor set has uncountably many points. The endpoints, mentioned earlier, form only a countable subset of the Cantor set, and therefore do not change its cardinality.

Variants of the Cantor set described above, in which each stage in the construction removes a fraction of the points removed in constructing the ordinary Cantor set (see, for example, Rana, 1997), are known as **fat Cantor sets**. Consider removing only the middle $c/3^n$ of the segments employed in the construction of C, for example, where $0 < c < 1$. Here we remove a total of 2^{n-1} segments of width $c/3^n$ in constructing the nth stage of the fat Cantor set C^F. The total width remaining becomes

$$
\begin{aligned}
\mathcal{L}(C_n^F) &= 1 - \sum_{m=1}^{n} 2^{m-1} c/3^m \\
&= \lim_{n \to \infty} \mathcal{L}(C_n^F) \\
&= \lim_{n \to \infty} \left(1 - \sum_{m=1}^{n} 2^{m-1} c/3^m \right) \\
&= 1 - (c/3)/\left(1 - \tfrac{2}{3}\right) \\
&= 1 - c.
\end{aligned}
\tag{2.14}
$$

For $c = 1$ we recover the result that $\mathcal{L}(C) = 0$. The set C^F therefore has a Lebesgue measure of $1 - c$, but with an uncountably infinite number of points missing when compared with the unit interval C_0. In particular, in this set an infinite number of intervals of infinitesimal size exist near any point that belongs to C. For the Cantor set itself, each point in C has an infinite number of neighbors in C that are arbitrarily close to it, but C lacks intervals of any length.

2.4.2 Brownian motion

We use Brownian motion as an example of an artificial, random fractal. **Brownian motion** has a long and storied history in the annals of several scientific disciplines: biology (Brown, 1828), financial mathematics (Bachelier, 1900), physics (Einstein, 1905; Perrin, 1909), and mathematics (Wiener, 1923; Kolmogorov, 1931; Lévy, 1948). The first observation of this phenomenon appears to have been made in 1785 by Jan Ingenhousz, a Dutch physician, in the course of examining the behavior of powdered charcoal on the surface of alcohol (see Klafter, Shlesinger & Zumofen, 1996, p. 33), but the term Brownian motion arose following the Scottish botanist Robert Brown's (1828) description of the movement of pollen grains in water.

In accordance with general usage, however, we denote as Brownian motion a particular continuous-time random process known as a **Wiener–Lévy process** (Wiener, 1923; Lévy, 1948) [alternate appellations are **Wiener process** (Wiener, 1923) and **Bachelier process** (Bachelier, 1900, 1912)]. Thus, Brownian motion, like the Cantor set, is considered to be an abstract construction, although it does closely approximate much experimental data. Unlike the Cantor set, however, Brownian motion involves randomness in its definition. Different realizations of the Brownian-motion process appear different, although all are governed by the same statistical properties.

One definition of Brownian motion $B(t)$, $t \geq 0$, involves the following three properties. First, Brownian motion is a Gaussian process; this signifies that a vector $\{B(t_1), B(t_2), ..., B(t_k)\}$ for any positive integer k and any set of times $\{t_1, t_2, ..., t_k\}$ has a joint Gaussian (normal) distribution. Second, the mean is zero: $\mathrm{E}[B(t)] = 0$

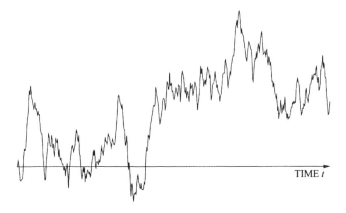

Fig. 2.2   A realization of Brownian motion. Time increases towards the right, with the origin at the left.

for all t, where $E[\cdot]$ represents expectation or mean. Third, the autocorrelation[3] of the process at two times s and t equals the smaller of the two times

$$E[B(s)\,B(t)] = \min(s,t) \tag{2.15}$$

for all s and t, where $\min(x,y)$ returns the smaller of x and y. Figure 2.2 displays a realization of Brownian motion.

We can derive a number of other characteristics from these three properties. In particular, Brownian motion contains statistical copies of itself. To see this, define a new function $B^*(t) \equiv B(at)$, a version of Brownian motion with a rescaled time axis. The random process $B^*(t)$ also belongs to the Gaussian family of random processes, with a mean of zero and an autocorrelation

$$E[B^*(s)\,B^*(t)] = E[B(as)\,B(at)] = a\min(s,t). \tag{2.16}$$

Now consider rescaling the amplitude of $B^*(t)$: define $B^\dagger(t) = a^{-1/2}\,B^*(t) = a^{-1/2}\,B(at)$. Like $B(t)$ and $B^*(t)$, $B^\dagger(t)$ is a zero-mean Gaussian process. The autocorrelation for $B^\dagger(t)$ is therefore written as

$$
\begin{aligned}
E[B^\dagger(s)\,B^\dagger(t)] &= E[a^{-1/2}\,B^*(s)\,a^{-1/2}\,B^*(t)] \\
&= a^{-1}\,E[B(as)\,B(at)] \\
&= a^{-1}\,a\min(s,t) \\
&= \min(s,t), \tag{2.17}
\end{aligned}
$$

which is identical to that of the original process $B(t)$.

---

[3] We define autocorrelation as the expectation of the product of a process at two different times or delays, while autocovariance denotes the result with the mean value removed. For zero-mean processes, the two coincide.

Since $B^\dagger(t)$ and $B(t)$ are Gaussian processes with the same mean and autocorrelation, the two processes are statistically identical (Feller, 1971). Thus, changing the time axis by a scale a and the amplitude axis by a scale a^H, with $H = \frac{1}{2}$, yields the same result, and $B(t)$ contains statistical copies of itself at any scale. In Chapter 6 we consider a generalization of Brownian motion, called fractional Brownian motion, in which the parameter H can assume any value between zero and unity.

2.4.3 Fern

We move now from abstract mathematical fractal objects, such as the Cantor set and Brownian motion, to natural fractal objects. These are ubiquitous in the real world. A simple fern provides a particularly clear example. Figure 2.3 displays the main frond of a fern (oriented vertically), which contains many sub-fronds (oriented horizontally), each a miniature copy of the whole.

Fig. 2.3 A fern, an example of a natural fractal with little randomness. The main frond comprises many sub-fronds, each a miniature copy of the whole. This fern, *Athyrium filix-femina* (Lady Fern), was collected from the backyard of the first author's residence in Massachusetts. It is well described as a deterministic fractal.

This scaling continues; each sub-frond contains sub-sub-fronds (oriented vertically again), and at the bottom of the figure there is evidence for a fourth level of detail.

Like all objects in the physical world, ferns have a minimum scale for their fractal behavior, here at the fourth level. The copies, while not perfect replicas of the whole, do not differ much from it; the fern is well described as a deterministic fractal.

2.4.4 Grand Canyon river network

Fig. 2.4 An overhead view of the Grand Canyon, a random natural fractal gouged out by the Colorado river. The main canyon contains many sub-canyons, each resembling the whole in a statistical manner. This photograph was taken from space by astronauts during U.S. Space Shuttle Flight STS61A. Image obtained from: Earth Sciences and Image Analysis, NASA–Johnson Space Center; 15 November 2004; "Astronaut Photography of Earth–Display Record." http://eol.jsc.nasa.gov/scripts/sseop/photo.pl?mission= STS61A&roll=201&frame=75

Whereas a fern provides an example of a deterministic natural fractal, most natural fractals exhibit randomness. Consider the Grand Canyon (Arizona), shown in Fig. 2.4. The main canyon, running from the top left, through the center, and exiting at the lower left, contains a number of sub-canyons along its length. While each sub-canyon appears different from the Grand Canyon itself and from the other sub-canyons, all resemble each other. The sub-canyons resemble the whole in a statistical manner. Again, the scaling continues, with the sub-canyons containing still smaller sub-sub-canyons of similar appearance within them, and so forth, down to the resolution limit of the photograph. Here the lower limit of fractal behavior derives from the measurement, rather than from the fractal object itself. The Grand Canyon thus provides an example of a random natural fractal.

2.5 EXAMPLES OF NONFRACTALS

Lest the reader gain the mistaken impression that all objects are fractals, we provide four counterexamples, again employing one each of the possible combinations of artificial and natural, and deterministic and random.

2.5.1 Euclidean shapes

Classical Euclidean shapes, such as circles, lines, and simple polyhedra have a single, well-defined scale, and therefore do not exhibit similar behavior over different scales. These artificial, deterministic objects do not reveal further detail upon magnification, nor do they possess copies of themselves. Such shapes are, therefore, not fractal.

2.5.2 Homogeneous Poisson process

Remaining in the abstract realm but turning now to random objects, we consider the one-dimensional homogeneous Poisson process (Parzen, 1962; Cox, 1962; Haight, 1967; Cox & Isham, 1980), perhaps the simplest of all point processes (see Sec. 4.1). Like Brownian motion, different realizations of this process have a different appearance, although each is governed by the same statistical properties. A single constant positive quantity, the rate, denotes the number of events (points) expected to occur in a unit interval, and this quantity completely characterizes the homogeneous Poisson process. The absence of memory completes the definition of this process; given the rate, knowledge of the entire history and future of a given realization of a homogeneous Poisson process yields no additional information about the behavior of the process at the present.

Fig. 2.5 Schematic representation of a one-dimensional homogeneous Poisson process. The time axis runs horizontally to the right, and the vertical arrows depict individual events (points) as they occur in time.

A schematic representation of a realization of a one-dimensional homogeneous Poisson process appears in Fig. 2.5. The vertical arrows depict individual events (points) as they occur, while the horizontal axis represents time. Although the intervals between the events vary they are associated with a fixed time scale via the rate parameter, in contrast to a fractal object. In particular, decreasing the time scale used to display the process yields a more sparse version of the original that appears quite different from it. Unlike a fractal process, it does not appear to be a random copy of the original. Further, the probability density function for the intervals follows an exponential form [see Eq. (4.3)], rather than the power-law form of Eqs. (2.5) and (2.6). Like the Euclidean shapes considered above, the homogeneous Poisson process is not fractal.

2.5.3   Orbits in a two-body system

The path followed by one of the bodies in a two-body orbiting system, such as the earth and the moon,[4] provides an example of a natural, deterministic, nonfractal object. Newtonian physics predicts that the two bodies will orbit about their mutual center of mass along trajectories described by perfect ellipses, and will do so indefinitely (Newton, 1687; Feynman, Leighton & Sands, 1963, vol. I, pp. 7-1–7-8). These orbits have a single scale; they do not exhibit similar behavior over different scales nor do they contain copies of themselves. The paths resemble the abstract Euclidean shapes exemplified in Sec. 2.5.1.

In contrast, the paths traced by planets in systems containing three or more bodies *do* exhibit fractal characteristics. This is particularly evident when the bodies have similar masses and are separated by similar distances. Such systems exhibit deterministic chaos (see Sec. 2.6), and chaotic systems often have fractal movement patterns.

2.5.4   Radioactive decay

Finally, for an example of a natural, random, nonfractal process, we turn to radioactive decay (Feynman et al., 1963, vol. I, pp. 5-3–5-5). A single radioactive atom will decay at some random time in the future, and, while the exact time of decay remains unknowable *a posteriori*, the probability of decay by any specified time is well known. Imagine, now, a collection of identical radioactive atoms, each undergoing decay at a random time. The registrations of these decay events form a random point process. It is associated with a single time constant or scale: the average decay time of the atoms.

The emissions resemble the homogeneous Poisson process presented in Sec. 2.5.2 provided that the observation times are sufficiently smaller than the average decay time (Rutherford & Geiger, 1910). Like the Poisson process, modifying the time scale over which the radioactive decay process is observed results in a qualitatively different process. Moreover, the probability density function for the times between decays does not follow the power-law form of Eqs. (2.5) and (2.6). Radioactive decay is not a fractal process.

---

[4] For simplicity of exposition we ignore perturbations induced by other celestial bodies and tides, as well as minor relativistic effects.

2.6 DETERMINISTIC CHAOS

Chaos and fractals are not synonymous, although the two concepts are often conflated. Chaos is, of course, an important topic in its own right (Poincaré, 1908; Devaney, 1986; Glass & Mackey, 1988; Moon, 1992; Ott, Sauer & Yorke, 1994; Strogatz, 1994; Schuster, 1995; Alligood, Sauer & Yorke, 1996; Peitgen et al., 1997; Thompson & Stewart, 2002; Ott, 2002). Even though chaos does not play a central role in the treatment of fractals, we compare and contrast these two phenomena to clarify their relationship.

Chaos describes the behavior of a deterministic nonlinear dynamical system in the presence of the following three features. First, small changes in the initial state of the system must lead to quite different results at some later time. For two identical systems beginning in slightly different states, the difference between them increases exponentially over time. Poincaré (1908) was the first to allude to this "sensitive dependence on initial conditions".[5] Second, and related to the first, the prediction of system dynamics becomes increasingly more difficult as the time of prediction moves further into the future. And third, an infinite number of unstable periodic orbits exists. With arbitrarily small continual adjustments, the dynamics of the system can be forced to follow any number of periodic paths, although in the absence of such adjustments the dynamics quickly depart from all such orbits. The diversity of behavior offered by a chaotic system has profound consequences. Conrad (1986) has categorized five functional roles that chaos might play in biological systems: search, defense, maintenance, cross-level effects, and dissipation of disturbance.

Given a dynamical system, it is customary to plot the state variables in phase space, collapsing the time information in the process. The resulting graph provides a window on the dynamics of the system. For dissipative systems, after an initial transient period system activity converges to a restricted region of phase space called the attractor of the system.

Some systems have **fractal attractors**, also known as **strange attractors**. The dynamics such systems display a rich pattern in phase space; enlarging a section of such an attractor continues to reveal new details without limit. Many (but not all) systems exhibiting chaos have strange attractors. Similarly, many (but not all) strange attractors derive from systems that are chaotic. However, neither feature necessarily implies the other. Unfortunately, the literature is rife with misconceptions pertaining to this issue.

We proceed to demonstrate the fundamental distinction between the two concepts by presenting examples of all four possibilities: chaotic and nonchaotic systems with both fractal and nonfractal attractors. To facilitate comparison between the various systems we confine ourselves to the simple class of iterated-function systems, which

---

[5] "A very small cause that escapes our notice has a considerable effect that we cannot fail to see, and we then say that the effect arises from chance . . . but it may happen that small differences in the initial conditions produce very large differences in the final phenomenon. A small error in the former then produces a very large error in the latter and prediction becomes impossible . . ."—Poincaré (1908)

follow the form

$$x_{n+1} = f(x_n) \tag{2.18}$$

or

$$\begin{array}{rlr} x_{n+1} &= f(x_n, y_n) & \text{a)} \\ y_{n+1} &= g(x_n, y_n) & \text{b)} \end{array} \tag{2.19}$$

for the one- and two-dimensional versions, respectively.

2.6.1   Nonchaotic system with nonfractal attractor

We begin with the logistic map, a particular example of Eq. (2.18) that takes the form of a quadratic recurrence relation,

$$x_{n+1} = f(x_n) = c\,x_n(1 - x_n). \tag{2.20}$$

This function maps the unit interval $[0, 1]$ to itself, and exhibits behavior that varies with the parameter c. It is a discrete version of the logistic equation of fame in ecology (Verhulst, 1845, 1847).

We begin by examining the stability of Eq. (2.20). A fixed point satisfies the equation $x_{n+1} = x_n = x_*$, which yields

$$\begin{array}{rl} x_* &= f(x_*) \\ &= cx_*(1 - x_*) \\ 0 &= x_*\big[x_* - (1 - 1/c)\big]. \end{array} \tag{2.21}$$

Eliminating the degenerate value $x_* = 0$ provides $x_* = 1 - 1/c$ for the remaining fixed point. What happens to values near the fixed point determines its stability; to assess this, we use a test value $x_n = x_* + \epsilon_n$, where ϵ_n is a value much smaller than unity. We then have

$$\begin{array}{rl} x_{n+1} &= f(x_n) \\ x_* + \epsilon_{n+1} &= f(x_* + \epsilon_n) \\ &= c\,(x_* + \epsilon_n)\big[1 - (x_* + \epsilon_n)\big] \\ &= c\,x_*(1 - x_*) + c\,\epsilon_n(1 - 2x_* - \epsilon_n) \\ &= x_* + c\,\epsilon_n(1 - 2x_* - \epsilon_n) \\ \epsilon_{n+1}/\epsilon_n &= c\,(1 - 2x_* - \epsilon_n) \\ &= 2 - c\,(1 + \epsilon_n). \end{array} \tag{2.22}$$

For the nonchaotic case, we choose $c = 2$. The fixed point then becomes $x_* = 1 - 1/c = \frac{1}{2}$, whereupon Eq. (2.22) yields $\epsilon_{n+1}/\epsilon_n = -2\epsilon_n$ indicating a rapid (quadratic, in fact) relaxation towards the fixed point. Since the fixed point is stable, the attractor of the system consists of that single point only. Thus, a plot of all possible values x_n, after transient effects have subsided, yields a single point, $x_n = \frac{1}{2}$. This zero-dimensional object has no fractal qualities whatsoever, and forms a nonfractal (non-strange) attractor. Furthermore, since all values of x_n converge rapidly to the fixed point x_*, any differences among starting values must decrease, rather than increase, over time, thereby precluding a sensitive dependence on initial conditions. The system of Eq. (2.20) with $c = 2$ therefore does not exhibit chaos. Figure 2.6 displays this convergence by showing the sequence $\{x_n\}$ that results from two different starting values, $x_0 = 0.1$ and 0.4, and illustrates the lack of chaos in this system.

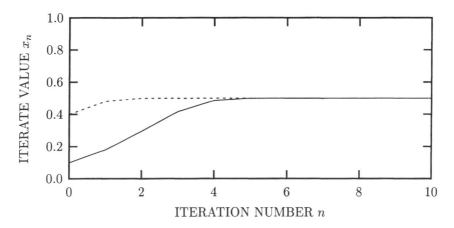

Fig. 2.6 Time course of a logistic system with parameter $c = 2$, for two different starting values: $x_0 = 0.1$ and 0.4. Although the two initial points differ widely, they both converge to the same value, the fixed point $x_* = \frac{1}{2}$. This system thus does not display sensitive dependence to initial conditions, and does not exhibit chaos.

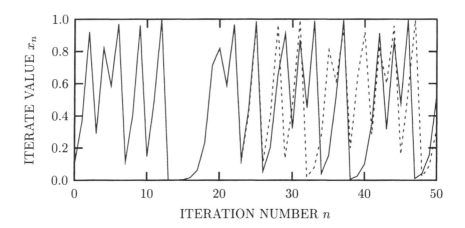

Fig. 2.7 Time course of a logistic system with parameter $c = 4$, for two different starting values: $x_0 = 0.1$ and $0.1 + 10^{-9}$. Although the two initial points differ only slightly, the iterates diverge and are completely unrelated after 30 iterations. This system does indeed display sensitive dependence to initial conditions, and exhibits chaos.

2.6.2   Chaotic system with nonfractal attractor

For the purposes of this example, we again consider the logistic map of Eq. (2.20), but now with $c = 4$. The nonzero fixed point becomes $x_* = 1 - 1/c = \frac{3}{4}$. The stability analysis of Eq. (2.22) now yields $\epsilon_{n+1}/\epsilon_n = -2 - 4\epsilon_n \approx -2$ so that deviations about the fixed point double in magnitude with each iteration. This fixed point thus does not comprise the attractor. In fact, for this value of c no limit cycles exist of any finite period and the entire interval $0 < x_n < 1$ forms the attractor (Schroeder, 1990, pp. 291–294). Except for a set of measure zero, iterates of any initial value x_0 will eventually come arbitrarily close to any specified value in the unit interval. This attractor forms a simple line segment and again has no fractal properties, establishing that for the logistic map with $c = 4$ the attractor is nonfractal.

Despite not having a fractal attractor, the system nevertheless displays chaos (Schroeder, 1990, pp. 291–294). The lack of fixed points or limit cycles suggests this, but a graphical demonstration illustrates it well. Figure 2.7 presents the sequence of iterations $\{x_n\}$ resulting from two starting values: 0.1 and a value just a bit larger, $0.1 + 10^{-9}$. Although indistinguishable at first, the difference between the two paths grows over time, and by iteration $n = 30$ the two sequences exhibit no relation to each other. This sensitivity to initial conditions illustrates the chaotic nature of the logistic system for the parameter value $c = 4$.

2.6.3   Chaotic system with fractal attractor

We next turn to the Hénon attractor (Hénon, 1976). This two-dimensional iterated-function system follows the form of Eq. (2.19) with

$$
\begin{array}{rll}
x_{n+1} &= 1.0 + ax_n^2 + by_n & \text{a)} \\
y_{n+1} &= x_n, & \text{b)}
\end{array}
\tag{2.23}
$$

with $a = -1.4$ and $b = 0.3$. We first establish the fractal nature of the attractor by simulation. Starting with $(x_0, y_0) = (1.08003, 0.305372)$, we iterate Eq. (2.23) 1 000 times, discarding these first results to eliminate any transient behavior, and then iterate a further 3 000 times and retain these values. Figure 2.8a) illustrates the attractor, which forms a boomerang shape bounded by $-1.3 < x, y < 1.3$. The initial pair $(x_0, y_0) = (1.08003, 0.305372)$ belongs to the attractor, as verified by further iterations, justifying its choice as a starting value. Enlarging a small section of Fig. 2.8a) (the box shown at the upper right) yields a banded structure [panel b)]; further enlargements yield substantially similar forms, as shown in panels c) and d). This self-similarity provides evidence of fractal characteristics, and in fact this attractor is indeed a fractal object (Peitgen et al., 1997).

We now proceed to consider the chaotic nature of the Hénon system. As before, we employ two different starting values, $(x_0, y_0) = (1.08003, 0.305372)$, as in Fig. 2.8, and $(x_0 + \epsilon_x, y_0)$ with $\epsilon_x = 10^{-7}$. Figure 2.9 shows the x values of the iterates diverging so that by $n = 43$ the two sequences have essentially no connection to each other, despite being almost identical at $n = 0$; the results resemble those for the logistic system with $c = 4$, displayed in Fig. 2.7.

This establishes the sensitivity to initial conditions of the system, and thereby provides evidence of chaos. More detailed and rigorous analysis supports this conclusion (Peitgen et al., 1997).

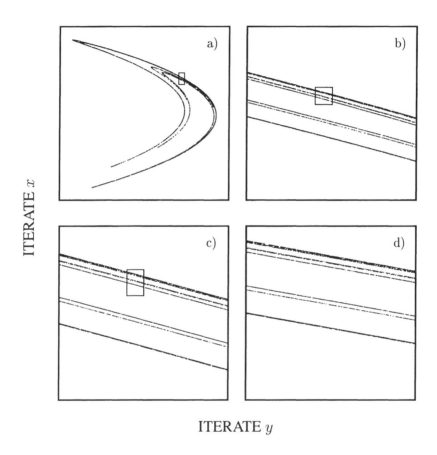

Fig. 2.8 a) Three thousand x-y pairs in the attractor of the Hénon system as shown in Eq. (2.23), with $a = -1.4$ and $b = 0.3$. An initial 1 000 iterations were discarded to eliminate transient effects. b) An enlargement of the small area within the box in panel a): 3 000 values of the attractor constrained to lie in the region $0.6 \leq x \leq 0.7$ and $0.5 \leq y \leq 0.7$. A parallel banded structure emerges. c) A further enlargement, of the area within the box in panel b): 3 000 values within $0.64 \leq x \leq 0.65$ and $0.61 \leq y \leq 0.63$. An enlargement of the upper band in panel b) yields a result similar to panel b). d) A final enlargement of the area within the box in panel c): 3 000 values within $0.644 \leq x \leq 0.645$ and $0.622 \leq y \leq 0.625$. The upper band in panel c) resolves into the same pattern as seen in the whole of panels b) and c). Hence, the attractor has similar structures over many spatial scales, suggesting that it forms a fractal object. This system is also chaotic, as illustrated by the time course displayed in Fig. 2.9.

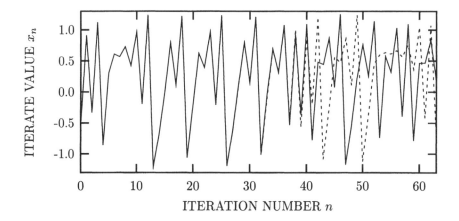

Fig. 2.9   Time course of the Hénon system with parameters $a = -1.4$ and $b = 0.3$, for two different starting values: $(x_0, y_0) = (1.08003, 0.305372)$, and $(x_0 + \epsilon_x, y_0)$ with $\epsilon_x = 10^{-7}$. Again, although the two initial points differ only slightly, the iterates diverge and appear completely unrelated by the 43 iteration. Like the logistic system with $c = 4$ (see Fig. 2.7), this system displays sensitive dependence to initial conditions, and exhibits chaos. The attractor for this system is fractal, however, as illustrated in Fig. 2.8.

2.6.4   Nonchaotic system with fractal attractor

Finally we consider a nonchaotic system which nevertheless has an attractor with fractal characteristics. We again employ the logistic map, Eq. (2.20), with the parameter $c \doteq 3.56995168804$. As previously, we begin by simulating the system to illustrate the fractal nature of the resulting attractor. Using an arbitrary starting value $x_0 = 0.31412577217182861803$, we iterate Eq. (2.20) 3 000 times after discarding the first 1 000 iterates.

The attractor is illustrated in Fig. 2.10a) — it forms a set of disconnected regions in the unit interval $0 < x < 1$. Progressive enlargements of regions of the x-axis of Fig. 2.10a) (delineated by the horizontal lines portrayed in the top three panels) yield new detail. Although different in form from that of the Hénon attractor, the evident self-similarity suggests that the attractor is fractal.

Moving now to confirm the presence or absence of chaos in this system, we again employ two different starting values: $x_0 = 0.87951016911829671$ and $x_0 = 0.89087022021791951$, chosen from the values shown in Fig. 2.10a), after 1 000 iterations to eliminate transient effects. As shown in Fig. 2.11, unlike the results for the logistic system with $c = 4$ and for the Hénon system, different starting points do not diverge. The system of Eq. (2.20) with $c \doteq 3.56995168804$ therefore does not exhibit sensitive dependence on initial conditions and is not chaotic. Other mathematical models of nonchaotic systems with fractal attractors have been set forth (Grebogi, Ott, Pelikan & Yorke, 1984), as has a physical experiment exhibiting such behavior (Ditto, Spano, Savage, Rauseo, Heagy & Ott, 1990).

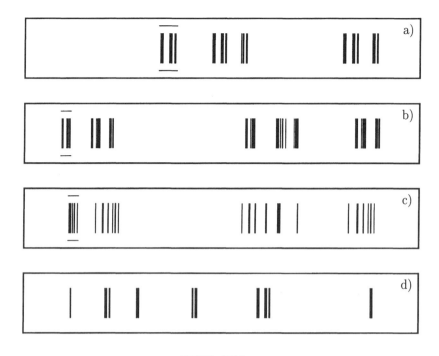

ITERATE x

Fig. 2.10 a) Three thousand values in the attractor of the logistic system [Eq. (2.20)] with parameter $c \doteq 3.56995168804$. An initial 1 000 iterations were discarded to eliminate transient effects. Each iterate is represented by a vertical line for clarity. The x axis ranges from zero to unity in this panel. Horizontal lines above and below the left-most cluster delineate the interval enlarged in the subsequent panel. b) An enlargement of the original interval delineated by the horizontal lines in panel a): 3 000 values of the attractor constrained to lie in the region $0.338 \leq x \leq 0.386$. Increased detail emerges with a structure similar to that of the whole attractor in a). c) and d) Further enlargements of the regions delineated by horizontal lines in the preceding panels: 3 000 values of the attractor in the intervals $0.3424 \leq x \leq 0.3437$ and $0.342544 \leq x \leq 0.342581$, respectively. Fresh new structures that resemble those in panels a) and b) continue to appear, suggesting that the attractor is fractal. This system is not chaotic, however, as revealed by the time course displayed in Fig. 2.11.

2.6.5 Chaos in context

Considering the results presented to this point, we see that systems can exhibit chaotic behavior or fractal (strange) attractors, or both, or neither. All four possibilities exist.

From a fundamental perspective, the term chaos describes certain nonlinear deterministic dynamical *systems* whereas the term fractal describes certain *objects*. Thus, chaos does not imply fractal nor does fractal imply chaos.

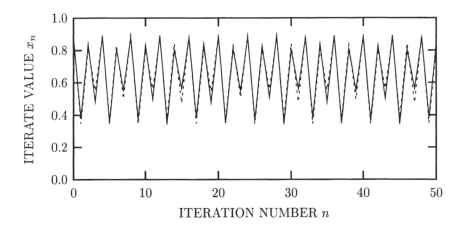

Fig. 2.11 Time course of the logistic system of Eq. (2.20) with parameter $c \doteq$ 3.56995168804, for two different starting values. The iterated values maintain a difference roughly equal to that of the starting values, ≈ 0.01, neither converging nor diverging. Hence, this system does not display sensitive dependence to initial conditions, and does not exhibit chaos. The attractor for this system is fractal, however, as illustrated in Fig. 2.10.

Moreover, the presence of significant noise or random behavior in a system generally precludes a meaningful assertion that the system is chaotic. Noise experiences the amplifying behavior of the system's sensitivity to initial conditions, so that even identical starting values experience rapidly diverging paths. Under such conditions, the concept of chaos loses its usefulness. Instead, the random nature of the system, imparted by the noise, becomes a key defining quality of its dynamics.

Since the topic of this treatise is random fractals, we do not consider chaos further.

2.7 ORIGINS OF FRACTAL BEHAVIOR

2.7.1 Fractals and power-law behavior

Why are fractal characteristics found in so many systems, both natural and synthetic? A good part of the reason turns out to be the close connection between fractals and scaling and hence between fractals and power-law behavior (see Secs. 2.2 and 2.3). Indeed, close examination reveals that the fractal behavior associated with many of the models considered throughout this book derives explicitly from the power-law relationships embodied in these models. When no such direct link exists, it turns out that other intrinsic properties of these models ultimately lead to power-law relationships. Power-law relationships can sometimes be traced to the presence of a cascade process in the underlying phenomenon.

The essential notion of a fractal has historical antecedents in theory and experiment alike (see Mandelbrot, 1982, Chapter 41). Consider Leibniz's (1646–1716) conception of fractional integro-differentiation and his definition of the straight line; Kant's (1724–1804) ruminations about the lack of homogeneity in the distribution of matter; Laplace's (1749–1827) suggestion that the scaling nature of Newton's Law of gravitation offered an axiom more natural than that of Euclid; and Weierstraß's (1815–1897) construction of a continuous, but nowhere differentiable, function.

In the empirical domain, we recall Weber's (1835) finding that the relaxation of a stretched silk thread follows a decaying power-law function of time, and Kohlrausch's (1854) observation that the decay of charge in a Leyden jar follows this very same form. We appreciate that Omori (1895) long ago recognized that the rate of aftershocks following an earthquake decays as an inverse function of time.

Indeed, power-law behavior is ubiquitous (see, for example, Malamud, 2004). It occurs in many guises, including deterministic laws, first-order statistics, second-order statistics, distributions, and nonlinear transformations. It is observed in the dynamical responses of systems and in their frequency spectra. Pareto (1896) long ago discovered that scale-invariant, power-law distributions characterize the income of individuals in many societies.[6] Behavior in accord with the **Pareto distribution**,[7] and its discrete counterpart, the **zeta distribution**, emerges in a broad array of contexts. Examples include:

- The number of species in different genera (Willis, 1922).

- The number of publications by different authors (Lotka, 1926).

- The agricultural yields of different sized plots (Fairfield-Smith, 1938).

- The energies of earthquake occurrences (Gutenberg & Richter, 1944).

- The mass densities of yarns of different lengths (Cox, 1948).

- The frequencies of word usage in natural languages (Zipf, 1949).

- The sizes of computer files (Park, Kim & Crovella, 1996).

The question posed at the beginning of this section — "Why are fractal characteristics found in so many systems?" — can thus be recast as: "Why is power-law behavior found in so many systems?"

2.7.2 Physical laws

Several key laws of classical physics take the form of deterministic power-law functions of the distance r,

$$F \propto r^c, \tag{2.24}$$

---

[6] A photograph of Pareto stands at the beginning of Chapter 7.
[7] A useful generalization of the Pareto distribution has been provided by Mandelbrot (1960, 1982), as will be elaborated subsequently.

where F is the force (or field) and r is the distance (see Feynman, 1965). Perhaps the most prominent example of this scaling relation is Newton's (1687) Law of gravitation, which provides that the gravitational field F associated with an object at a distance r follows an inverse-square law, $F \propto r^{-2}$, so that $c = -2$.

The Coulomb field associated with a charged particle also behaves in accordance with Eq. (2.24), again with $c = -2$. Other charge configurations similarly lead to power laws, but with different exponents; examples are an infinite line of charge $(c = -1)$; a charge dipole $(c = -3)$; a charge quadrupole $(c = -4)$; and the van der Waals force between a pair of dipoles $(c = -7)$. In the study of the mechanics of materials, Hooke's Law provides that the restoring force for an elastic medium also obeys Eq. (2.24), where r is the deformation and $c = +1$ (see Gere, 2001). Also, the Langmuir–Childs Law for space-charge-limited current flow in electronic devices dictates that $i \propto V^{3/2}$, where i is the current and V is the voltage (see Terman, 1947, Sec. 5–5).

Some physical processes are conveniently described in terms of power-law functions of time, as evidenced by the following examples: (1) the distance d traveled by an object falling under the force of gravity is characterized by $d \propto t^2$; (2) Kepler's Third Law of celestial mechanics specifies that the major axis b of an elliptical planetary orbit is related to the orbital period T via $b \propto T^{2/3}$; and (3) the time course of the mean photon flux density emitted by a charged particle via Čerenkov radiation varies as $h(t) \propto t^{-5}$ (see Prob. 10.6).

In quantum mechanics, the allowed energy levels E_j of many systems are proportional to some power of the quantum number j (see Saleh & Teich, 1991, Chapter 12),

$$E_j \propto j^c. \tag{2.25}$$

Examples are the hydrogen atom $(c = -2)$; the harmonic oscillator with a linear restoring force $(c = +1)$; the anharmonic oscillator with a cubic restoring force $(c = +\frac{4}{3})$; and the infinite quantum well $(c = +2)$. The rigid rotor behaves as $E_j \propto j(j + 1)$. The spatial scaling of the Lagrangian for these systems allows us to deduce these exponents directly from the form of the potential energy function (see Schroeder, 1990, pp. 66–67).

For simple physical systems, the exponents c are typically integers or rational numbers, although fractional exponents are not uncommon in semiconductor physics (see Saleh & Teich, 1991, Chapter 15). In the biological sciences, fractional exponents are more the rule than the exception, as will become apparent subsequently.

2.7.3   Diffusion

In the domain of stochastic processes, diffusion offers a straightforward route to achieving power-law dynamics (Whittle, 1962; Marinari, Parisi, Ruelle & Widney, 1983). In one-dimensional diffusion, an object moves randomly along an axis, with no preferred direction, and with motion at each instant that is independent of motion at all other times. The path of such an object coincides with Brownian motion, discussed in Sec. 2.4.2. Equation (2.15) shows that the variance of the position grows linearly with time; given the zero-mean Gaussian nature of the process, this leads immediately to a probability density for the particle position x:

$$p_x(x) = (4\pi\Delta t)^{-1/2} \exp\left(-\frac{x^2}{4\Delta t}\right),\qquad(2.26)$$

with Δ a diffusion constant. The peak height of the density decays with time as $t^{-1/2}$, an inverse power law. Given a concentration of small objects u_0 (particles, for example) clustered tightly about a starting value x_0, a simple modification of Eq. (2.26) yields a particle concentration envelope $u(x,t)$ given by

$$u(x,t) = u_0(4\pi\Delta t)^{-1/2} \exp\left(-\frac{(x-x_0)^2}{4\Delta t}\right).\qquad(2.27)$$

For **diffusion in a multidimensional Euclidean space** of (integer) dimension D_E, the motion along each of the component axes forms an independent realization of Brownian motion. The corresponding concentration profile then becomes (see, for example, Pinsky, 1984)

$$u(\mathbf{x},t) = u_0(4\pi\Delta t)^{-D_E/2} \exp\left(-\frac{|\mathbf{x}-\mathbf{x}_0|^2}{4\Delta t}\right),\qquad(2.28)$$

where $\mathbf{x}$ and $\mathbf{x}_0$ represent the general and initial position vectors, respectively. The concentration decays as $t^{-D_E/2}$, providing exponents $\frac{1}{2}$, 1, and $\frac{3}{2}$ for $D_E = 1$, 2, and 3, respectively. For diffusion on objects that are physical examples of fractals, the fractal dimension of the object replaces the Euclidean dimension D_E in Eq. (2.28), thereby offering a larger set of allowable exponents. Problems 10.8 and 10.9 address the ramifications of such diffusion processes.

Diffusion-limited aggregation (DLA) describes the aggregation and growth of structures when diffusion dominates transport (Witten & Sander, 1981). This model characterizes a broad variety of phenomena including electrodeposition, dielectric breakdown, snowflake formation, mineral-vein formation in geologic structures, and the growth of biological structures such as coral (see, for example, Vicsek, 1992; Halsey, 2000).

Subdiffusion is an important form of anomalous diffusion in which the mean-square displacement varies as t^α ($0 < \alpha < 1$) rather than as t (see Bouchaud & Georges, 1990). This process can be understood in a simple way by making use of fractional Gaussian noise (see Sec. 6.2) in a generalized Langevin equation (Kou & Xie, 2004).

2.7.4 Convergence to stable distributions

An important and far-reaching rationale for the emergence of power-law distributions has its origins in the limit theorem developed by Paul Lévy (1937, 1940) (see also Gnedenko & Kolmogorov, 1968; Feller, 1971; Mandelbrot, 1982; Christoph & Wolf, 1992; Samorodnitsky & Taqqu, 1994; Bertoin, 1998; Sato, 1999; Sornette, 2004).

Sums of identical and independent continuous random variables are characterized by **stable distributions**, which generally have power-law tails. The sole exception is the Gaussian distribution (Gauss, 1809), which emerges (via the ordinary central limit theorem) when the constituent random variables are endowed with finite second moments.[8]

Discrete analogs of the family of continuous stable distributions have recently been examined (Hopcraft, Jakeman & Tanner, 1999; Matthews, Hopcraft & Jakeman, 2003; Hopcraft, Jakeman & Matthews, 2002, 2004). These probability distributions typically follow the form

$$p(n) \sim 1/n^c, \quad 1 < c < 2, \tag{2.29}$$

for large n. They have zero mode and infinite mean in the absence of an upper cutoff. However, for counting distributions with an upper cutoff, and therefore a finite mean, sums converge to the Poisson distribution, which assumes the role played by the Gaussian for the continuous stable distributions.

2.7.5   Lognormal distribution

A closely related rationale for the presence of power-law behavior stems from the features of the **lognormal distribution** (Kolmogorov, 1941; Aitchison & Brown, 1957; Gumbel, 1958). This distribution is often used as a model for characterizing systems comprising products of random variables, via an argument that proceeds as follows. A product of random variables with finite second moments, under logarithmic transformation, becomes a sum. Application of the ordinary central limit theorem renders the sum Gaussian (normal). The original product, then, obeys the lognormal distribution since its logarithm has a normal distribution.

The lognormal distribution has a long tail and sums of independent lognormally distributed random variables retain their lognormal form (Mitchell, 1968; Barakat, 1976); although these sums ultimately do converge to Gaussian form, the convergence is exceedingly slow. Moreover, the tail of the lognormal distribution is closely mimicked by a power-law distribution over a wide range (Montroll & Shlesinger, 1982; Shlesinger, 1987; West & Shlesinger, 1989, 1990); these authors further argue that many data thought to obey an inverse power-law distribution instead obey the lognormal law over a broad range and then ultimately transition to power-law behavior at very large values of the random variable.

In the domain of discrete processes, the Poisson transform of the lognormal distribution has found widespread use in modeling the photon fluctuations of laser light transmitted through random media such as the turbulent atmosphere (Diament & Teich, 1970a; Teich & Rosenberg, 1971). The justification for using the lognormal model here is the same as that provided above: in traveling from source to receiver, the laser light encounters a large number of independent atmospheric layers with random transmittances.

---

[8] Photographs of Gauss and Lévy can be found at the beginning of Chapter 8. A biographical sketch of Lévy is provided by Mandelbrot (1982, Chapter 40).

2.7.6 Self-organized criticality

Power-law behavior arises in other ways as well. Some systems spontaneously evolve toward a critical state and thereby generate power-law distributions. A sandpile provides the canonical example of this process, called **self-organized criticality** (Bak, Tang & Wiesenfeld, 1987; Bak, 1996), and abbreviated SOC; the addition of grains of sand to the top of a sandpile results in the formation of a cone at exactly the critical angle of repose. Some added grains merely stop where they land, but many trigger avalanches with power-law varying sizes, maintaining the critical state.

Expansion-modification systems provide another example of such spontaneous evolution. In this case, two processes operate simultaneously, one creating long-range correlation, and the other destroying it; the resulting construct exhibits correlations over all scales, and therefore fractal structure (Li, 1991). As an example, each element in a binary sequence is either inverted with probability p, or duplicated with probability $1 - p$. Similar behavior may occur in a variety of artificial and natural systems, as they evolve towards complex, critical states and produce power-law behavior. In a related model, simple white noise perturbs the movement of activated neural clusters and competing dissipative and restorative forces ultimately generate $1/f$-type noise (Usher, Stemmler & Olami, 1995).

As another particular example, a collection of interconnected processes that evolves according to the logistic equation generates power-law-distributed amplitudes over a broad range of system parameters (Solomon & Richmond, 2002).

2.7.7 Highly optimized tolerance

Highly optimized tolerance (Carlson & Doyle, 1999; Doyle & Carlson, 2000; Carlson & Doyle, 2002) suggests another possible origin for power-law behavior. According to this theory, power-law behavior emerges naturally as a result of the evolution of a complex system toward optimal performance and robustness. Natural selection is said to drive the evolution for collections of living organisms, while engineering design provides the optimizing impetus for artificial systems. This evolutionary process leads to the emergence of specialized states (which would be rare in a random system without design input) concomitantly with power-law behavior.

Power-law characteristics, and hence fractal behavior, can therefore emerge naturally from system evolution via a number of different constructs (Gisiger, 2001).

2.7.8 Scale-free networks

Yet another way that power-law behavior comes into play is via **scale-free networks** (Albert & Barabási, 2002; Dorogovtsev & Mendes, 2003; Pastor-Satorras & Vespignani, 2004). For such networks, no node is typical. Some have an enormous number of connections whereas most are only weakly connected to others. Since well-connected nodes, called hubs, can have hundreds, thousands, or millions of links, there is no scale associated with the network. Connectivity in links per node is described by a probability law, known as the **degree distribution**, that typically follows a power-law form (Krapivsky, Rodgers & Redner, 2001):

$$p(n) \sim 1/n^c, \quad 2 < c < 3.5. \tag{2.30}$$

There are many ways in which scale-free networks can come into being (see, for example, Krapivsky, Redner & Leyvraz, 2000). The underlying features that lead to the formation of such networks are continual development and the preferential attachment to highly linked nodes. As new nodes are formed, the network continues to evolve; each new node tends to connect to the more highly connected existing nodes since these are most easily identified.

Examples of scale-free networks in the biological domain stretch from cellular metabolic networks, in which biochemical reactions link a collection of molecules, to the brain, in which axons and dendrites link a collection of neurons (Eguíluz, Chialvo, Cecchi, Baliki & Apkarian, 2005). Such networks are plentiful in the technological arena: important examples are air transportation systems, the Internet, and the World Wide Web (see Sec. 13.2.1). Scale-free networks are also pervasive in the social domain: examples include scientific collaborations connected by joint publications; scientific papers linked by citations; people connected by professional associations or friendships; epidemics of contagious disease linked by family members; and businesses linked by joint ventures. They have the salutary feature of being robust against accidental failures because random breakdowns selectively affect the most plentiful nodes, which are the least connected. Such networks are, however, highly vulnerable to coordinated attacks directed at the hubs, which are the most intricately connected of the nodes (Albert & Barabási, 2002).

Despite the evident diversity of these scale-free networks, their common architecture brings them under the same mathematical umbrella: the power-law distribution embodied in Eq. (2.30). The range of asymptotic power-law exponents is rather narrow and differs from that for discrete stable distributions [compare Eqs. (2.29) and (2.30)]. The convergence properties of sums of identical, independently distributed discrete zeta random variables that are suitable for characterizing scale-free networks have recently been established. The limiting form turns out to be the Poisson distribution but the convergence can be exceptionally slow (Hopcraft et al., 2004), much as with the convergence of sums of lognormal random variables to Gaussian form (see Sec. 2.7.5). Problems involving discrete scale-invariant behavior should be formulated in terms of discrete models since continuum models and mean-field approximations can lead to erroneous results (Hopcraft et al., 2004).

Not all networks are scale-free, of course. Prominent exceptions include the locations of atoms in a crystal lattice, the U.S. highway system, the power grid in the Western United States, and the neural network of the organism *Caenorhabditis elegans*.

2.7.9 Superposition of relaxation processes

Finally, we note that the observation of first- and second-order statistics with power-law behavior is often ascribed to a **superposition of relaxation processes** exhibiting a spread of time constants. Maxwell's student Hopkinson (1876) appears to have originated this explanation, suggesting that the power-law decay of the charge in a Leyden jar might be understood on the basis of various relaxation times for the different silicate components of the glass through which the discharge occurred. However, this argument was later abandoned as unworkable because of the large number of exponentials required. von Schweidler (1907) resurrected this approach by considering a large number of relaxation processes with a wide spread of time constants. He noted that the properties of the gamma function were such that a power-law function could be represented in terms of a weighted collection of exponential functions with different relaxation times.

In the context of semiconductor physics, van der Ziel (1950) used a correlation-function version of this approach to explain the inverse-frequency form of the spectrum; Halford (1968) subsequently offered a generalization of this model. This construct finds wide acceptance in the semiconductor-physics community by virtue of its connection to trapping mechanisms, which offer an exceptionally wide range of time constants (McWhorter, 1957, see also Prob. 7.10). Buckingham (1983, Chapter 6) addressed the role of the weighting functions.

Many other materials and systems, physical and biological alike, display similar power-law behavior, as shown in Chapter 5. However, the relaxation-process approach is rarely appropriate for characterizing these processes because of the enormous range of time constants required to yield $1/f$ behavior over a reasonable range of frequencies. A ratio of time constants of 10^6, for example, yields $1/f$ behavior only over four decades of frequency whereas a ratio of 10^{12} offers 10 decades (Buckingham, 1983, Chapter 6; see also Prob. 9.1 and Fig. B.6). Few systems aside from semiconductors offer the requisite range of time constants.

Another way of mitigating the presence of power-law behavior is to assume that an exponential cutoff ultimately prevails. In practice this often turns out not to be the case, however. Indeed, von Schweidler (1907) himself carried out extensive experiments seeking such a cutoff in the decay of charge in Leyden jars, but found none.

2.8 UBIQUITY OF FRACTAL BEHAVIOR

2.8.1 Fractals in mathematics and in the physical sciences

The most comprehensive treatments of fractals have principally been in mathematics and the physical sciences. Extensive treatments have appeared, for example, in the following books: Mandelbrot (1982); Feder (1988); Peitgen & Saupe (1988); Schroeder (1990); Peitgen et al. (1997); Lévy Véhel et al. (1997); Turcotte (1997); Turner et al. (1998); Flandrin & Abry (1999); Flake (2000); Barnsley (2000); Park & Willinger (2000); Mandelbrot (2001); Falconer (2003); West et al. (2003). The application of fractals in fields such as economics, finance, and hydrology is widespread (see, for example, Mandelbrot, 1982, 1997; Mandelbrot & Hudson, 2004; Henry & Zaffaroni, 2003; Montanari, 2003).

Fractal analysis in the physical sciences proves highly important, as indicated by the following examples:

- We are all keenly aware of the fractal geometry of nature, thanks to the seminal work of Benoit Mandelbrot (1982).

- The noise in many electronic components and systems exhibits fractal behavior at low frequencies (Sec. 5.4.1).

- Semiconductor layered structures comprising stacks of materials of different bandgaps have been fabricated in the form of Cantor sets (Cantor, 1883), as well as Fibonacci (1202), Thue–Morse (Thue, 1906, 1912; Morse, 1921b,a), and Rudin–Shapiro (Rudin, 1959; Shapiro, 1951) sequences. Such nonperiodic, deterministic structures can exhibit fractal electronic, thermal, and magnetic properties (see, for example, Merlin et al., 1985; Kohmoto et al., 1987; Kolář, Ali & Nori, 1991; Dulea, Johannson & Riklund, 1992).

- Photonic materials and devices consisting of layers of materials with different refractive indices have also been constructed in the form of Cantor sets, as well as Fibonacci and Thue–Morse sequences (Jaggard & Sun, 1990; Kolář et al., 1991; Liu, 1997; Jaggard, 1997; Zhukovsky, Gaponenko & Lavrinenko, 2001). For example, Hattori, Schneider & Lisboa (2000) suggested constructing a fiber Bragg grating that takes the form of a Cantor set. Such nonperiodic and deterministic photonic media can exhibit optical properties with unusual features, including: (1) optical reflection and transmission with self-similar spectra (Gellermann, Kohmoto, Sutherland & Taylor, 1994; Dal Negro, Oton, Gaburro, Pavesi, Johnson, Lagendijk, Righini, Colocci & Wiersma, 2003; Ghulinyan, Oton, Dal Negro, Pavesi, Sapienza, Colocci & Wiersma, 2005; Dal Negro, Stolfi, Yi, Michel, Duan, Kimerling, LeBlanc & Haavisto, 2004); (2) complex light dispersion (Hattori, Tsurumachi, Kawato & Nakatsuka, 1994); (3) band-edge group-velocity reduction (Dal Negro et al., 2003; Ghulinyan et al., 2005); (4) pseudo-bandgaps and omnidirectional reflection (Dal Negro et al., 2004); and (5) light emission with uncommon spectral characteristics (Dal Negro, Yi, Nguyen, Yi, Michel & Kimerling, 2005).

- Light scattered or refracted by passage through a random fractal phase screen exhibits fractal wave properties (Berry, 1979; Jakeman, 1982); Berry (1979) coined the term **diffractal** to describe the resulting wave.

- Errors in telephone networks often occur as fractal clusters (Prob. 7.7).

- The photon statistics of Čerenkov radiation exhibit fractal characteristics under certain conditions (Prob. 10.6).

- Analysis of the fractal statistics of earthquake patterns can assist in the prediction of future earthquake occurrences (Prob. 10.7).

- Computer communication networks evolve into scale-free forms and the traffic resident on these networks exhibit fractal characteristics (Chapter 13).

2.8.2   Fractals in the neurosciences

There have been fewer comprehensive treatments of fractals in the biological sciences; we explicitly note those of Bassingthwaighte et al. (1994), West & Deering (1995), and Liebovitch (1998). Fractals play an important role in biological sciences such as ecology (see, for example, Halley & Inchausti, 2004), which has often been a breeding ground for novel mathematical approaches.

In this and the following section, respectively, we examine a number of examples of fractal behavior in the neurosciences and in medicine and human behavior.

Power-law behavior is common in the neurosciences. Featured at levels from the molecular to the organism, it is manifested in many systems. Neural systems evidently benefit from the flexibility of being able to match the time scale of a current stimulus while incorporating the memories of past stimuli. We present a number of examples, emphasizing those that fall in the class of fractal-based point processes:

- Ion channels reside in biological cell membranes, permitting ions to diffuse in or out of a cell (Sakmann & Neher, 1995). Power-law behavior characterizes various features of ion-channel behavior (Liebovitch, Fischbarg & Koniarek, 1987; Liebovitch, Fischbarg, Koniarek, Todorova & Wang, 1987; Liebovitch & Tóth, 1990; Liebovitch, Scheurle, Rusek & Zochowski, 2001; Läuger, 1988; Millhauser et al., 1988). Many ion channels exhibit independent power-law-distributed closed times between open times of negligible durations, and are well described by a fractal renewal point process (Lowen & Teich, 1993c). When the open times have significant duration, the alternating fractal renewal process serves as a suitable model instead (Lowen & Teich, 1993c, 1995; Thurner et al., 1997). Moreover, the time constant attendant to the recovery of certain ion channels depends on the duration of prior activity in a power-law fashion (Toib, Lyakhov & Marom, 1998).

- Fractal behavior exists in excitable-tissue recordings for various biological systems *in vivo*, from the microscopic to the macroscopic (Bassingthwaighte et al., 1994; West & Deering, 1994). Membrane voltages vary randomly in time, often exhibiting Gaussian fluctuations with power-law spectra (Verveen, 1960; Verveen & Derksen, 1968; Stern, Kincaid & Wilson, 1997; Lowen, Cash, Poo & Teich, 1997a). Superpositions of alternating fractal renewal processes, representing collections of ion-channel openings and closings, provide a plausible model for this process (Lowen & Teich, 1993d, 1995).

- Communication in the nervous system is generally mediated by the exocytosis of multiple vesicular packets (quanta) of neurotransmitter molecules at the synapse between cells, either spontaneously (Fatt & Katz, 1952) or in response to an action potential at the presynaptic cell (Katz, 1966). Neurotransmitter packets induce miniature end-plate currents (MEPCs) at the postsynaptic membrane, and their rate of flow exhibits fractal behavior such as power-law spectra, that can be described by a fractal-based point process (Lowen et al., 1997a,b).

- Power-law behavior characterizes the second-order statistics of action-potential sequences in isolated neuronal preparations and isolated axons; the spectrum often follows a form close to $1/f$ over a broad range of frequencies (Musha, Kosugi, Matsumoto & Suzuki, 1981; Musha et al., 1983). Moreover, the spike rate in response to a step-function input in many sensory neurons follows a power-law decay during the course of adaptation (Chapman & Smith, 1963), frequently varying as $t^{-1/4}$ (Biederman-Thorson & Thorson, 1971; Thorson & Biederman-Thorson, 1974).

- Auditory nerve-fiber action potentials from essentially all *in vivo* preparations display neural-spike clusters (Teich & Turcott, 1988) and fractal-rate behavior over time scales greater than about 1 sec, under both spontaneous and driven conditions (Teich, 1989; Teich, Johnson, Kumar & Turcott, 1990; Teich, 1992; Teich & Lowen, 1994; Lowen & Teich, 1992a, 1996a; Powers & Salvi, 1992; Kelly, Johnson, Delgutte & Cariani, 1996). This behavior could arise from superpositions of fractal ion-channel transitions (Teich, Lowen & Turcott, 1991; Lowen & Teich, 1993b, 1995) or via fractal-rate vesicular exocytosis (Lowen et al., 1997a,b).

- As in the auditory system, spontaneous and driven visual-system action potentials also exhibit fractal-rate characteristics. This behavior appears in all retinal ganglion cells and lateral-geniculate-nucleus cells in the thalamus (Teich et al., 1997; Lowen, Ozaki, Kaplan, Saleh & Teich, 2001), as well as in cells of the striate cortex (Teich et al., 1996). Moreover, insect visual-system interneurons generate spike trains with fractal-rate characteristics under both spontaneous and driven conditions (Turcott, Barker & Teich, 1995). Motion-sensitive neurons in the fly visual system adapt over a wide range of time scales that are established by the stimulus rather than by the neuron (Fairhall, Lewen, Bialek & de Ruyter van Steveninck, 2001a,b).

- Fractal features appear in action-potential sequences associated with many central-nervous-system neurons operating under a broad variety of conditions, including those in the cortex, thalamus, hippocampus, amygdala, pyramidal tract, medulla, and mesencephalic reticular formation (see, for example, Evarts, 1964; Yamamoto & Nakahama, 1983; Yamamoto, Nakahama, Shima, Kodama & Mushiake, 1986; Kodama, Mushiake, Shima, Nakahama & Yamamoto, 1989; Grüneis, Nakao, Yamamoto, Musha & Nakahama, 1989; Grüneis, Nakao, Mizutani, Yamamoto, Meesmann & Musha, 1993; Lewis, Gebber, Larsen & Barman, 2001; Orer, Das, Barman & Gebber, 2003; Fadel, Orer, Barman, Vongpatanasin, Victor & Gebber, 2004; Bhattacharya, Edwards, Mamelak & Schuman, 2005).

- Networks of rat cortical neurons contained in slice cultures exhibit brief neuronal avalanches whose spatiotemporal patterns are stable and repeatable for many hours; these power-law distributed structures may serve as a substrate for memory (Beggs & Plenz, 2003, 2004).

- Power-law behavior has a strong presence in the domain of sensory perception. Although the transduction of a stimulus at the first synapse in a neural system often follows a logarithmic form, stimulus estimation and detection are usually characterized by power-law functions of the stimulus intensity with sub-unity exponents (Stevens, 1957, 1971; Barlow, 1957; McGill & Goldberg, 1968; Moskowitz, Scharf & Stevens, 1974; McGill & Teich, 1995).

- The natural course of forgetting in humans is well described by a decaying power-law function of time (Wickelgren, 1977; Wixted & Ebbesen, 1991, 1997; Wixted, 2004).

There appear to be many origins of fractal activity in the nervous system; power-law fluctuations at the level of the protein may play an underlying role.

2.8.3 Fractals in medicine and human behavior

The quantitative analysis of the fractal characteristics of biomedical signals can yield information that assists with the diagnosis of disease and with the determination of its severity. This information, in turn, can have vital implications regarding the appropriate treatment regimen, and can influence the outcome of treatment. We provide a number of examples:

- Fractal analysis of the fluctuations in human standing (Musha, 1981; Shimizu et al., 2002) reveals age-related changes not evident using conventional, non-fractal methods (Collins et al., 1995). A different constellation of changes appears in Parkinson's disease (Mitchell, Collins, De Luca, Burrows & Lipsitz, 1995). After correcting for age, the fractal dynamics of human gait (walking) reveal the severity of Huntington's disease in patients, and appear to correlate with the degree of impairment (Hausdorff, Mitchell, Firtion, Peng, Cudkowicz, Wei & Goldberger, 1997).

- Fluctuations in mood show evidence of fractal behavior in their spectra, which display quantitative differences between bipolar-disorder patients and normal controls (Gottschalk et al., 1995). That these fluctuations follow a fractal form may lead to better methods for predicting and controlling mood disorders (see Sec. 2.8.5).

- Evidence of fractal behavior in the spectrum of the human heartbeat has been known for more than two decades (Kobayashi & Musha, 1982). Fractal methods do differentiate between normal and diseased patients with some degree of success (Turcott & Teich, 1993; Peng, Mietus, Hausdorff, Havlin, Stanley & Goldberger, 1993; Peng, Havlin, Stanley & Goldberger, 1995; Turcott & Teich, 1996). However, nonfractal measures (based on a fixed time scale of about twenty seconds) are superior for indicating the presence of cardiovascular dysfunction (Thurner, Feurstein & Teich, 1998; Thurner, Feurstein, Lowen

& Teich, 1998; Ashkenazy, Lewkowicz, Levitan, Moelgaard, Bloch Thomsen
& Saermark, 1998; Teich et al., 2001).

- Fractal measures of activity have successfully quantified changes in the movement patterns of laboratory rats induced by drugs of abuse (Paulus & Geyer,
  1992), and these same fractal measures help improve the diagnosis of attention
  deficit hyperactivity disorder (ADHD) in children (Teicher et al., 1996).

- Normal prenatal development may, in fact, require that fractal activity patterns
  be established in the brain. Developmental disorders such as autism could
  possibly result from a failure in the generation of these patterns (Anderson,
  2001).

- Developmental insults, such as early abuse, quantitatively alter fractal parameters measured in experimental animals (Anderson, 2001). Evidence also exists
  that the fractal patterns of brain activity change with emotional state (Anderson
  et al., 1999), with implications for psychiatric diagnoses.

2.8.4   Recognizing the presence of fractal behavior

Fractal activity directly influences how systems operate. It is therefore important to
recognize its presence, and to understand its features, so that system performance can
be properly evaluated and controlled.

For computer network traffic (see Chapter 13) and vehicular traffic, for example,
estimates of the fractal parameters provide measures of performance and useful design guidelines. Detailed analysis of fractal activity has proven to be indispensable.
Dealing with fractal behavior in a system is not a trivial enterprise, however. Even
seemingly simple tasks, such as calculating the mean and variance of the rate for
a fractal process, offer unique challenges. The low-frequency nature of the noise
indicates that nearby values are highly correlated so that obtaining reliable estimates
often requires a prohibitive number of samples (see Chapter 12).

In some cases, fractal behavior serves as a source of unavoidable noise that diminishes system performance. An example is $1/f$-type noise in electronic components
and circuits (see Sec. 5.4.1). The presence of fractal noise places restrictions on the
information throughput of such systems, the calculation of which requires fractal
analysis.

Finally, it is important to recognize the possible presence of fractal noise to avoid
drawing erroneous conclusions. A case in point is the landmark study conducted by
Fatt & Katz in 1952, in which the authors carried out an investigation of the statistical
behavior of sequences of miniature endplate currents (MEPCs) at the neuromuscular
junction. In the course of describing the methods used to analyze their data, they
carefully noted that each segment of data selected for analysis was sufficiently short to
exclude, as they put it, the "occasional occurrences of short high-rate bursts" of events,
and to avoid "progressive changes of the mean." In fact, fractal-rate fluctuations

do exist in exocytic behavior and MEPCs (Lowen et al., 1997a,b), and the MEPCs observed by Fatt & Katz (1952) almost certainly exhibited such behavior (see Lowen et al., 1997b, for an analysis). Unaware of the presence or importance of these fluctuations, however, they removed most traces of them by selecting relatively short segments of data for analysis and, moreover, chose precisely those segments that exhibited minimal fluctuations. The observation of fractal-rate behavior requires long data sets, and the presence of both bursts and apparent trends lie at its very core.

2.8.5 Salutary features of fractal behavior

Fractal behavior is ubiquitous and its study reveals much about our surroundings. We discover, for example, that natural scenes and natural sounds exhibit fractal properties in space and time, and sensory systems have adapted to this property (Musha, 1981; Teich, 1989; Dan, Atick & Reid, 1996; Taylor, 2002; Simoncelli & Olshausen, 2001; Yu, Romero & Lee, 2005).

Given the ubiquity of fractal activity, what biological advantages might accrue from its presence?

Fractal behavior offers tolerance to noise and errors. The deleterious effects of noise diminish in importance because the concentration of power at lower frequencies assures increased predictability. New scales introduced by errors are less disruptive in fractal processes since they already exist in the initial distributions (West, 1990). Scale-free networks are robust against accidental failures, as pointed out in Sec. 2.7.8. Moreover, the presence of fractal noise can serve to optimize the throughput of a system, with examples in both neural signaling and vehicular traffic (Ruszczynski, Kish & Bezrukov, 2001).

Developmentally, internally generated fractal signals [such as neural signals arising during rapid-eye-movement (REM) sleep] provide a prenatal stimulus that mimics natural signals and assists the brain in developing normally. An animal can thus emerge at birth with its visual system attuned to the world it enters (Anderson, 2001). Search patterns executed by animals and by humans, which often have fractal properties (Cole, 1995; Viswanathan, Afanasyev, Buldyrev, Murphy, Prince & Stanley, 1996; Aks, Zelinsky & Sprott, 2002), appear optimal given the likely distributions of targets.

Finally, the salutary features of fractal behavior in medicine have been documented in a number of cases. When used for relieving pain via transcutaneous electrical nerve stimulation, $1/f$ noise outperforms white noise (Musha, 1981). This is also true for sensitizing baroreflex function in the human brain (Soma, Nozaki, Kwak & Yamamoto, 2003). The flexibility of response offered by fractal behavior may also serve as a harbinger of health (see West & Deering, 1995).

Problems

2.1 *Fractal and nonfractal objects* Comment on the fractal properties, or the lack thereof, in each of the following:

1. the aorta, all the arteries it branches into, the arterioles, and the capillaries in a rabbit;

2. a tree trunk and all its branches and twigs, as visualized in the winter when it is devoid of leaves;

3. a lock of hair;

4. a brick;

5. a sand dune in the Namibian desert, without vegetation;

6. a cumulus cloud;

7. the Himalayan mountains;

8. the path of a curve ball thrown by a major-league baseball pitcher;

9. a randomized version of C_3 — we generate the third iteration towards the Cantor set, which contains eight segments, but add an independent random value uniformly distributed over $[-0.01, +0.01]$ to the beginning and ending value of each segment.

2.2 *Logistic to tent map* Consider the logistic map, Eq. (2.20), with $c = 4$, as studied in Sec. 2.6.2.

2.2.1. Show that the substitution

$$y \equiv \pi^{-1} \arccos(1 - 2x) \qquad (2.31)$$

converts Eq. (2.20) into a tent map (Schroeder, 1990, p. 291):

$$y_{n+1} = \begin{cases} 2y_n & 0 \leq y_n \leq \frac{1}{2} \\ 2 - 2y_n & \frac{1}{2} < y_n \leq 1. \end{cases} \qquad (2.32)$$

2.2.2. Find the ratio $|\epsilon_{n+1}/\epsilon_n|$.

2.3 *Cantor variant* Imagine a variant of the Cantor set described in Sec. 2.4.1, denoted C'. At each stage in the construction of the variant set we remove the middle *half* of each remaining interval. Thus, the intervals $[\frac{0}{4}, \frac{1}{4}] \cup [\frac{3}{4}, \frac{4}{4}]$ comprise the result C'_1 after the first step in its construction.

2.3.1. What total length (Lebesgue measure) remains in the limiting set C'?

2.3.2. How many points remain in C' compared with the original unit interval?

2.3.3. What value of D_0 does C' have?

2.4   *Cantor-set membership*   Consider the point $x = 0.002002\ldots_3$, where the subscript $_3$ indicates a ternary expansion.

2.4.1. To what fraction does x correspond?

2.4.2. Does x belong to the endpoints of C?

2.4.3. Does x belong to the interior of C (in other words, in C but not an endpoint of it)?

2.4.4. Does C contain irrational numbers?

2.5   *Scaling solution*   Show that Eqs. (2.5) and (2.6) form the only solution to Eq. (2.4) for arbitrary a and x.

3

Point Processes: Definition and Measures

The famous French mathematician **Siméon Denis Poisson (1781–1840)** developed the fundamental probability distribution that bears his name; its applications are legion in a broad variety of fields.

Together with Major Greenwood, the Scottish engineer **George Udny Yule (1871–1951)** conceived an important generalization of the Poisson distribution in which the rate itself becomes a random variable.

3.1   Point Processes   50
3.2   Representations   51
3.3   Interval-Based Measures   54
      3.3.1 Marginal statistics   55
      3.3.2 Interval autocorrelation   57
      3.3.3 Interval spectrum   58
      3.3.4 Interval wavelet variance   58
      3.3.5 Rescaled range analysis   59
      3.3.6 Detrended fluctuation analysis   61
3.4   Count-Based Measures   63
      3.4.1 Marginal statistics   64
      3.4.2 Normalized variance   66
      3.4.3 Normalized Haar-wavelet variance   66
      3.4.4 Count autocorrelation   69
      3.4.5 Rate spectrum   70
3.5   Other Measures   70
      3.5.1 Coincidence rate   70
      3.5.2 Point-process spectrum   72
      3.5.3 General-wavelet variance   74
      3.5.4 Generalized dimension   74
      3.5.5 Correlation measures for pairs of point processes   77
      Problems   79

3.1 POINT PROCESSES

Some random phenomena occur at discrete times or locations, with the individual events largely identical. Examples include the events of a radioactive decay process (Sec. 2.5.4), vehicles passing a certain location on a road (Fig. 1.2), the arrival of information packets at a node of a computer communication network (Chapter 13), the occurrence of action potentials in a neural preparation (Chapter 5), and the occurrences of QRS complexes in the electrocardiogram (Sec. 12.2.2).

In all of these cases, the set of times at which the events occur comprises the salient characteristics of the process. The details of the events themselves are less important, inasmuch as one event closely resembles another. A **stochastic point process**, often abbreviated as **point process**, is a mathematical construct that represents these events as random points in a space. We use the terms "event" and "point" interchangeably.

Point-process theory grew out of studies in a number of fields: population processes, cosmic-ray showers, component durability, and queueing problems in communications engineering. The theory of point processes took shape as a discipline in the 1920s, and the literature in this area grew rapidly in the following decades (Lubberger, 1925, 1927; Lotka, 1939; Campbell, 1939; Fréchet, 1940; Palm, 1943; Feller, 1948; Wold, 1948, 1949; Bartlett, 1955; Moyal, 1962). Writing in German, the Swedish mathematician Palm (1943) coined the term *Punktprozesse*: "point process."

A comprehensive bibliography detailing some of the early landmarks of point-process theory and analysis, in the general context of stochastic processes, is available (Wold, 1965). A concise early history of the field appears in Daley & Vere-Jones (1988, Chapter 1).

Many modern books on the topic adopt a rigorous and abstract approach (Brillinger, 1981; Leadbetter, Lindgren & Rootzen, 1983; Daley & Vere-Jones, 1988; Kingman, 1993; Reiss, 1993; Baccelli & Brémaud, 2003), which has the merit of providing a great deal of generality. Other books on point processes are more didactic and applications oriented (Parzen, 1962; Cox, 1962; Cox & Lewis, 1966; Feller, 1971; Lewis, 1972; Srinivasan, 1974; Saleh, 1978; Cox & Isham, 1980; Snyder & Miller, 1991), offering specific examples useful in the physical and biological sciences. We adopt a rather informal approach to point processes, and concentrate particularly on those that exhibit fractal characteristics.

Some point processes depend on space as well as time; lightning strikes, for example, deliver more electrical activity to some areas than to others. However, we confine the treatment provided here to one-dimensional point processes; other dimensions, if present, are not incorporated into the model.

Since a variety of time variables exist, we adopt the following conventions: (1) lowercase roman italic letters (generally t) refer to absolute time, measured with respect to an origin that does not depend on the point process under study; (2) uppercase roman italic letters (generally T) refer to a duration over which events are analyzed — again, the duration does not depend on the point process under study and the analysis need not begin at the origin. Finally, the symbol τ generally represents the times between events.

3.2 REPRESENTATIONS

Figure 3.1 presents several representations useful in the analysis of point processes (Teich, Heneghan, Lowen & Turcott, 1996). Panel a) demonstrates a realization of a point process as a series of impulses occurring at specified times t_n. Since these impulses have vanishing width, they are most rigorously defined in terms of the derivative of a well-defined counting process $N(t)$ [panel b)], a monotonically increasing function of t, which starts at the origin and augments by unity when an event occurs. Accordingly, we write the point process itself as $dN(t)$, to emphasize its strict definition within the context of an integral. The point-process representation thus belongs to the family of generalized functions (Bracewell, 1986).

The set of **event times** $\{t_n\}$, or equivalently the **sequence of interevent intervals** $\{\tau_n\}$ (together with t_0), where $\tau_n = t_{n+1} - t_n$, completely describe the point process.[1] Furthermore, the **sequence of counts** depicted in Fig. 3.1c) also contains much information about the process. Here we divide the time axis into equally spaced contiguous counting durations of T sec to produce a sequence of counts $\{Z_k(T)\}$,

---

[1] To remove ambiguity, we define $N(t)$ as a right-continuous process, so that $N(t_n) = n$.

a)   POINT PROCESS $dN(t)$

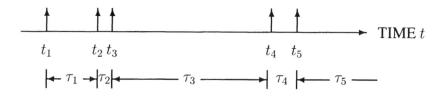

b)   COUNTING PROCESS $N(t)$

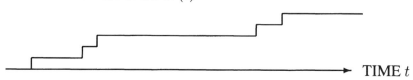

c)   GENERATION OF COUNT SEQUENCE

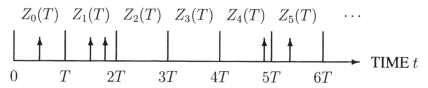

Fig. 3.1   Representations of a point process. (a) A sequence of idealized impulses, occurring at times t_n, represents the events, and form a stochastic point process $dN(t)$. We also show the interevent intervals $\tau_n = t_{n+1} - t_n$. For convenience of analysis, several alternative representations of the point process appear. (b) The counting process $N(t)$ begins at a value of zero at $t = 0$. At every event occurrence the value of $N(t)$ augments by unity. (c) The sequence of counts $\{Z_k(T)\}$, a discrete-time nonnegative integer-valued stochastic process, derives from the point process by recording the number of events in successive counting durations of length T.

where $Z_k(T) = N[(k+1)T] - N(kT)$ denotes the number of events in the kth duration. As illustrated in panel d), this sequence forms a discrete-time random process of nonnegative integers. In general, forming the sequence of counts loses information, although for an orderly point process (see below) decreasing the size of the counting duration T reduces the loss to an arbitrarily small value. An attractive feature of this representation lies in the fact that it preserves the correspondence between the discrete time axis of the counting process $\{Z_k(T)\}$ and the absolute "real" time axis of the underlying point process. Within the process of counts $\{Z_k(T)\}$, the elements $Z_k(T)$ and $Z_{k+n}(T)$ refer to the number of counts in durations separated by precisely

d)　COUNT SEQUENCE $\{Z_k(T)\}$

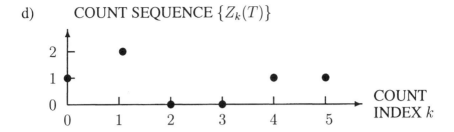

e)　INTERVAL SEQUENCE $\{\tau_k\}$

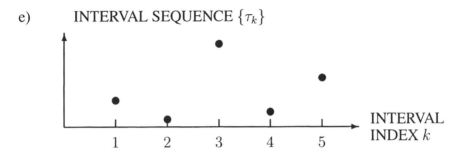

Fig. 3.1 (continued) (d) The sequence of counts $\{Z_k(T)\}$ depends on a count index k. The counting process destroys information because this representation eliminates the precise timing of events within each counting duration. Correlations in the discrete-time sequence $\{Z_k(T)\}$ can be readily interpreted in terms of real time. (e) The sequence of interevent intervals $\{\tau_k\}$ represents the time between successive events, yielding a discrete-time, positive, real-valued stochastic process. All information contained in the original point process remains in this representation, but the discrete-time axis of the sequence of interevent intervals suffers random distortion relative to the real time axis of the point process.

$(n-1)T$ sec, so that we can readily associate correlation in the process $\{Z_k(T)\}$ with correlation in the underlying point process $dN(t)$.

Much as the sequence of counts forms an auxiliary process, so does the sequence of interevent intervals. Figure 3.1e) presents the intervals $\{\tau_k\}$ drawn from the point process in panel a), indexed by interval number. In contrast to the sequence of counts, this representation preserves all of the information of the point process $dN(t)$, but eliminates the direct correspondence between absolute time and the index number. The sequence of intervals therefore affords only rough comparisons with correlations in the underlying point process, particularly for intervals with a large coefficient of variation [see Eq. (3.5)].

We restrict ourselves largely to **orderly point processes**, which essentially means that no two events occur at the same time, and that events do not localize to any single time. Formally speaking, in terms of the counting process $N(t)$ we have

$$\lim_{\epsilon \to 0} \epsilon^{-1} \Pr\{N(t + \epsilon) - N(t) > 1\} = 0 \qquad (3.1)$$

for any time t, which also implies the lack of coincident events (Daley, 1974). We also generally consider stationary processes; all statistics remain unchanged despite any shifting of the time axis,

$$\begin{aligned} \Pr\{f[N(s_1), N(s_2), \ldots, N(s_k)] < x\} = \\ \Pr\{f[N(s_1 + s), N(s_2 + s), \ldots, N(s_k + s)] < x\}, \end{aligned} \qquad (3.2)$$

for any arbitrary real-valued function $f(\cdots)$ with any number k of arguments and any offset time s. Unless explicitly stated otherwise, all point processes in this book are orderly and stationary.

For some applications, point processes as defined above do not suffice for all sets of events; customers arriving at a queue, for example, may require widely different service times (see Chapter 13). While still strongly localized, the events do not then resemble each other. In addition to its location, each event then requires a descriptive mark, such as the service time in this example. A generalization of the point-process model to a **marked point process** version accommodates problems of this type (Matthes, 1963; Cox & Isham, 1980; Sigman, 1995). All marks in a marked point process are of the same type (examples of types include integers, real numbers, vectors, and functions). Including marks adds additional information (but also complexity) to point-process models. Since most of the point processes considered in this book do not warrant this level of effort, we largely restrict ourselves to unmarked point processes.

We now proceed to describe a number of measures that prove useful in the study of point processes. In general, no one statistic, or even small group of statistics, suffices to completely characterize a point process; each provides a different view of the process and highlights different properties. A good description requires many such views. The statistics fall into three broad classes: those based on the intervals between events, as displayed in Fig. 3.1e); those based on the counting process, as shown in Fig. 3.1d); and those based on the point process as a whole, as depicted in Fig. 3.1a).

3.3   INTERVAL-BASED MEASURES

Conversion of a point process $dN(t)$ into a sequence of intervals between events $\{\tau_k\}$ reduces $dN(t)$ to a discrete-time real-valued process, for which a wide variety of analytical methods exist.

3.3.1 Marginal statistics

Perhaps the simplest statistics of a point process ignore any dependencies among event times, and focus on the marginal properties of the interevent intervals $\{\tau_k\}$. These fall into two classes.

The first includes the **probability distribution**, the **survivor function**, and the **probability density** where this derivative exists:

$$
\begin{array}{llll}
\text{distribution} & P_\tau(t) & = & \Pr\{\tau \leq t\} \\
\text{survivor} & S_\tau(t) & = & \Pr\{\tau > t\} & = & 1 - P_\tau(t) \\
\text{density} & p_\tau(t) & = & dP_\tau(t)/dt.
\end{array}
\tag{3.3}
$$

For a well-defined point process, we require that $P_\tau(t) = p_\tau(t) = 0$ for $t < 0$.

The second class comprises the **moments** $\mathrm{E}[\tau^n]$ and the statistics derived therefrom, such as:

$$
\begin{array}{ll}
\text{variance} & \mathrm{Var}[\tau] = \mathrm{E}[\tau^2] - \mathrm{E}^2[\tau] \\
\text{standard deviation} & \sigma_\tau = \sqrt{\mathrm{Var}[\tau]} \\
\text{skewness} & \mathrm{E}[(\tau - \mathrm{E}[\tau])^3]/\sigma_\tau^3 \\
\text{kurtosis} & \mathrm{E}[(\tau - \mathrm{E}[\tau])^4]/\sigma_\tau^4 - 3,
\end{array}
\tag{3.4}
$$

where these moments exist.[2] The interval **coefficient of variation** C_τ, a commonly used measure of the relative dispersion (relative width) of the intervals, is defined as

$$
C_\tau \equiv \sigma_\tau / \mathrm{E}[\tau].
\tag{3.5}
$$

As with all random variables, the **characteristic function** $\phi_\tau(\omega)$, defined as[3]

$$
\phi_\tau(\omega) \equiv \int_0^\infty p_\tau(t)\, e^{-i\omega t}\, dt,
\tag{3.6}
$$

forms a compact representation of the moments. In general we have

$$
i^n \frac{d^n}{d\omega^n}\, \phi_\tau(\omega)_{\omega=0} = \mathrm{E}[\tau^n]
\tag{3.7}
$$

---

[2] Several definitions for skewness exist (such as the difference between the mean and the median, all divided by the standard deviation), and some authors define kurtosis without the "3" subtracted. All definitions in general use have their merits, and no clear winner emerges. However, we choose to make use of the particular forms provided above for three reasons. First, the skewness and kurtosis provided in Eq. (3.4) are given by normalized versions of the third and fourth cumulants or semi-invariants, respectively. Second, these versions prove most analytically tractable. And finally, both of these definitions assume a value of zero for a Gaussian random variable.

[3] Some define the characteristic function with the argument of the exponential $i\omega t$ rather than $-i\omega t$. Since both i and $-i$ form equally valid square roots of -1, it is convention, rather than mathematics, that determines the choice. We employ the expression shown in Eq. (3.6) because it leads to simpler results.

for the moments, and

$$i^n \frac{d^n}{d\omega^n} \ln[\phi_\tau(\omega)]\Big|_{\omega=0} = C_n \qquad (3.8)$$

for the **cumulants** or **semi-invariants** C_n. The moments and cumulants determine each other, and in particular we have

$$
\begin{aligned}
\mathrm{E}[\tau] &= C_1 \\
\mathrm{E}[\tau^2] &= C_2 + C_1^2 \\
\mathrm{E}[\tau^3] &= C_3 + 3C_1 C_2 + C_1^3 \\
\mathrm{E}[\tau^4] &= C_4 + 4C_1 C_3 + 3C_2^2 + 6C_1^2 C_2 + C_1^4 \\
\mathrm{Var}[\tau] &= C_2 \\
\text{skewness} &= C_3/C_2^{3/2} \\
\text{kurtosis} &= C_4/C_2^2.
\end{aligned}
\qquad (3.9)
$$

In addition to the interevent-interval statistics, the **forward recurrence time** also proves useful. This represents the time $\vartheta(t)$ remaining to the next event of a point process, starting at an arbitrary time t independent of the process. Formally, we have

$$\vartheta(t) = t_k - t, \quad \text{where} \quad k = N(t) + 1. \qquad (3.10)$$

A simple relation exists between probability distribution functions of the interevent time and the forward recurrence time (see Prob. 3.8)

$$P_\vartheta(s) = \Pr\{\vartheta(t) \le s\} = \frac{1}{\mathrm{E}[\tau]} \int_0^s [1 - P_\tau(x)]\, dx, \qquad (3.11)$$

which yields the statistics of $\vartheta(t)$ through Eqs. (3.3) and (3.4). In particular, taking the derivative of Eq. (3.11) yields

$$p_\vartheta(s) = [1 - P_\tau(s)]/\mathrm{E}[\tau]. \qquad (3.12)$$

Thus, a normalized version of the interevent-interval survivor function provides the recurrence-time probability density.

While the interevent-interval probability distribution and survivor function exist for all point processes, some moments may not; in particular, for a probability density function that decays as $t^{-\alpha}$ for large t, moments $\mathrm{E}[\tau^n]$ for $n \ge \alpha - 1$ will not exist. For example, the density function

$$p_\tau(t) = \sqrt{t_0/\pi}\; t^{-3/2}\, \exp(-t_0/t), \qquad t > 0, \qquad (3.13)$$

with t_0 a fixed positive parameter, has infinite moments $\mathrm{E}[\tau^n]$ for all positive integers n (Feller, 1971) (see Prob. 3.6).

Densities such as these belong to the family of **heavy-tailed distributions**, for which

$$\lim_{t \to \infty} \frac{S_\tau(t + t_1)}{S_\tau(t)} = 1, \qquad t_1 \ge 0, \qquad (3.14)$$

for any fixed, finite time t, where again $\mathsf{S}_\tau(t) = 1 - P_\tau(t)$ is the interval survivor function. **Subexponential distributions**, introduced by Chistyakov (1964), form an important subclass of heavy-tailed distributions (see, for example, Embrechts, Klüppelberg & Mikosch, 1997; Sigman, 1999; Greiner, Jobmann & Klüppelberg, 1999). These distributions have survivor functions $\mathsf{S}_\tau(t)$ that obey

$$\lim_{t\to\infty} e^{\epsilon t}\,\mathsf{S}_\tau(t) = \infty, \qquad \epsilon > 0, \tag{3.15}$$

so that the tail of the survivor function tends to zero more slowly than any exponential function $e^{-\epsilon t}$. Examples of subexponential distributions include the Pareto and its variants, the lognormal, and the stretched exponential (Weibull[4]).

For a particular class of point processes, called **renewal point processes** (see Sec. 4.2), the values of each interevent interval do not depend on those before or after it. For this class of point processes only, the marginal statistics described above, and in particular the probability distribution function alone, determine the entire behavior of the processes. Generally, however, dependencies do occur among interevent intervals and this necessitates the use of several statistics for an overview of the sequence of intervals, and of the point process itself.

3.3.2 Interval autocorrelation

The **interval autocorrelation** $R_\tau(k)$, which provides further information about point processes that do not belong to the renewal point-process family, is defined as

$$R_\tau(k) \equiv \mathrm{E}[\tau_n \tau_{n+k}]. \tag{3.16}$$

For independent intervals, $R_\tau(k) = \mathrm{E}^2[\tau]$ for $k \neq 0$, confirming that the autocorrelation then provides no additional information.

A normalized version of this measure proves useful in many cases. Subtracting the value returned for independent intervals, $\mathrm{E}^2[\tau]$, and dividing by the interval variance, $\mathrm{Var}[\tau]$, yields the **interval serial correlation coefficient**

$$\varrho_\tau(k) \equiv \frac{R_\tau(k) - \mathrm{E}^2[\tau]}{\mathrm{Var}[\tau]}. \tag{3.17}$$

By construction, $\varrho_\tau(0) = 1$ for any point process. For independent intervals, $\varrho_\tau(k) = 0$ for $k \neq 0$.

Inasmuch as no direct relationship generally exists between the lag variable k and time t in seconds, these measures, as well as the other interval-based measures that follow, have limited usefulness. This restriction is relaxed when the mean interval greatly exceeds the interval standard deviation (DeBoer, Karemaker & Strackee, 1984), in which case $t \approx k\,\mathrm{E}[\tau]$ (see the beginning of Sec. 3.4).

---

[4]The Weibull distribution follows the form $P_\tau(t) = 1 - \exp[-(t/t_0)^\xi]$, typically with a shape parameter $0 < \xi < 1$ (see Gumbel, 1958, pp. 279, 302); the exponential distribution is recovered for $\xi = 1$. This distribution possesses finite moments of all orders but is nevertheless heavy-tailed because the survivor function decays more slowly than any exponential.

3.3.3 Interval spectrum

Fourier transforming the autocorrelation in Eq. (3.16) yields the **interval-based spectrum** $S_\tau(f)$:

$$S_\tau(f) = \sum_k R_\tau(k)\, e^{-i2\pi kf}, \tag{3.18}$$

where f is the (dimensionless) frequency with units of cycles per number of intervals. For independent intervals, $S_\tau(f) = \text{Var}[\tau]$ for all $f \neq 0$. Again, the independent variable f has no simple connection with its conventional counterpart (the frequency f in Hz), so this measure principally finds use in processes with small deviations from periodicity. The performance of a normalized version of this statistic is examined in Sec. 12.3.7.

3.3.4 Interval wavelet variance

A particularly appropriate method for characterizing signals with fractal behavior is via the use of wavelets (Daubechies, 1992). Fourier analysis, employed in generating the interval spectrum, decomposes a signal into a series of basis functions, all of which have different shapes: sinewaves of varying frequency, phase, and amplitude but identical duration. The basis functions differ in the number of cycles they contain. Wavelet decomposition, in contrast, employs basis functions that all have the same shape, and derive from a prototype wavelet by stretching and shifting. As with their Fourier counterpart, the inversion of wavelet transforms to return the original signal proves relatively simple.

This self-affine basis set makes wavelets well suited for analyzing signals that contains statistical copies of themselves; signals that exhibit fractal characteristics in time fall into this category. Wavelet-based methods for characterizing fractal signals yield estimates superior to those obtained by many other methods (Thurner et al., 1997; Abry, Flandrin, Taqqu & Veitch, 2000, 2003, see also Chapter 12), and wavelet analysis also enjoys the salutary property of removing nonstationarities from the signal under study (Teich et al., 1996; Abry & Flandrin, 1996; Arneodo, Grasseau & Holschneider, 1988).

The **wavelet transform** of a sequence of interevent intervals takes the form (Daubechies, 1992; Aldroubi & Unser, 1996; Akay, 1997; Abry et al., 2003)

$$W_{\psi,\tau}(k,l) = \sum_n 2^{-k/2}\, \psi(2^{-k}n - l)\, \tau_n, \tag{3.19}$$

where the continuous-time wavelet function $\psi(x)$ satisfies a number of admissibility criteria (Daubechies, 1992). We consider the **interval wavelet-transform variance** in Eq. (3.19), since the transform itself is a random variable. As a result of one of the admissibility criteria, wavelet transforms have zero mean, so that

$$\text{Var}[W_{\psi,\tau}(k,l)]$$

$$= \text{E}[W^2_{\psi,\tau}(k,l)]$$

$$= \mathrm{E}\left[\sum_n \sum_m 2^{-k}\, \psi(2^{-k}n - l)\, \psi(2^{-k}m - l)\, \tau_m\, \tau_n\right]$$

$$= 2^{-k} \sum_n \sum_m \psi(2^{-k}n - l)\, \psi(2^{-k}m - l)\, R_\tau(m - n). \qquad (3.20)$$

The performance of a normalized version of this measure, defined as

$$A_\tau(k) \equiv \frac{\mathrm{Var}[W_{\psi,\tau}(k, l)]}{\mathrm{Var}[\tau]}, \qquad (3.21)$$

is examined in Sec. 12.3.6.

For stationary point processes, the wavelet variance does not depend on the position index l, but only on the scale variable k. Under these conditions, the interval wavelet variance is directly related to the interval spectrum (see Sec. 3.3.3) via an integral transform (Heneghan, Lowen & Teich, 1999). Knowledge of one of these measures is thus equivalent to knowledge of the other. The wavelet variance exhibits the same nonlinear relationship between the lag variable k and conventional time as observed above for several other measures.

3.3.5 Rescaled range analysis

Harold Hurst studied the water flow patterns of the river Nile, and discovered the presence of long-term fluctuations in the yearly flood levels.[5] He observed that years with greater-than-average flow tended to cluster together, as did years with lower-than-average flow, but that no characteristic cluster size appeared to exist. Hurst (1951) developed **rescaled range analysis** (R/S analysis) to quantify this effect, and this statistic became the first robust method for characterizing fractal behavior in discrete-time sequences.[6] The rescaled range statistic provides information about dependencies among interevent intervals (or other sequences) in a form fundamentally different from that obtained by the interval autocorrelation $R_\tau(k)$.

One algorithm for calculating the rescaled range proceeds as follows. Begin by selecting a set of k interevent intervals that start with the first available interval. From this set, estimate the mean $\widehat{\mathrm{E}}[\tau]$ and (biased) standard deviation

$$\sqrt{\frac{k-1}{k}\, \widehat{\mathrm{Var}}[\tau]}. \qquad (3.22)$$

Next, subtract the estimated mean, divide by the biased standard deviation, and construct a running sum of this rescaled process. Now generate the rescaled range by

---

[5] Hurst (1956) and Hurst, Black & Simaika (1965) studied the water flow through other rivers as well; they also examined variations in other natural time series such as rainfall, temperature, pressure, tree-ring thickness, and sunspot numbers.

[6] The unexpected clustering results, together with the seemingly *ad hoc* character of the rescaled range statistic, led to a lack of acceptance of Hurst's work that lasted until he was in his seventies (Mandelbrot, 1982, pp. 396–398). A photograph of Hurst appears at the beginning of Chapter 12 and a biographical sketch is provided by Mandelbrot (1982, Chapter 40).

subtracting the minimum value that the sum attains from its maximum. Next, repeat this procedure for all possible contiguous blocks of k values within the entire data set, and average these values together to yield the rescaled-range estimate $\widehat{U}(k)$. Finally, repeat this procedure for a variety of lags k. Figure 3.2 provides pseudocode for this algorithm, and Fig.3.3 presents a schematic graphical calculation.

For independent intervals, we have

$$U(k) \approx \sqrt{k}, \tag{3.23}$$

where the exact relationship depends on the distribution of the $\{\tau_n\}$ (see Prob. 3.7).

While this measure now enjoys broad popularity for the study of processes that exhibit long-term correlation or large moments (Hurst, 1951; Feller, 1951; Hurst, 1956; Hurst et al., 1965; Mandelbrot, 1982; Mandelbrot & Wallis, 1969c,b; Schepers, van Beek & Bassingthwaighte, 1992), it suffers from large systematic errors for some sequences (Beran, 1994; Bassingthwaighte & Raymond, 1994; Caccia, Percival, Cannon, Raymond & Bassingthwaighte, 1997). Nevertheless, it can prove useful in some cases since it robustly handles data sets with infinite variance (Mandelbrot, 2001, Chapter 5, pp. 155–171). In Sec. 12.3.4 we examine the performance of a normalized form of this the rescaled range statistic,

$$U_2(k) \equiv U^2(k)/k. \tag{3.24}$$

```
Calculate R/S from discrete-time sequence {xₙ} of length M:
    set k = 2
    while M/k large                    /* typically require M/k ≥ 10 */
        set m = 0
        while m + k - 1 ≤ M
            estimate mean Ê[x] = Σⁿ₌ₘ^(m+k-1) xₙ/k
            estimate biased std. dev.  σ̂²ₓ = Σⁿ₌ₘ^(m+k-1) (xₙ - Ê[x])² /k
            generate normalized sequence:    yₙ = (xₙ - Ê[x]) /σ̂ₓ
            generate summed sequence:   zₙ = zₙ₋₁ + yₙ;   z₁ = y₁
            find minimum and maximum values:
                zₘᵢₙ = min(zₙ),   m ≤ n < m + k
                zₘₐₓ = max(zₙ),   m ≤ n < m + k
            find the difference between them:   U(k, m) = zₘₐₓ - zₘᵢₙ
            increment starting index:   m → m + 1
                          /* m → m + k faster, almost as accurate */
        end while
        average all values of U(k, m) to yield Û(k)
        report k and Û(k)
        increase block size k                     /* typically k → 2k */
    end while
```

Fig. 3.2   Rescaled-range analysis: Pseudocode.

SUMMED SERIES AND
LOCAL APPROXIMATIONS

RANGE

MEAN
RESCALED
RANGE

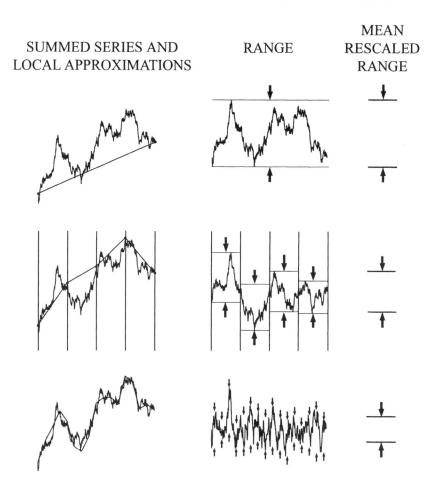

Fig. 3.3 Top row: a summed series and its approximation by a linear function through its first and last values (left column). The difference between the summed series and the linear function yields a range (middle column). After rescaling by the (biased) sample standard deviation, this single range value forms the mean rescaled range (right column). Middle row: the process repeats with the same summed series divided into four subseries, each with its own linear function, as well as range and standard deviation. Their average forms the mean rescaled range. Bottom row: the same calculation for sixteen subseries.

3.3.6   Detrended fluctuation analysis

Detrended fluctuation analysis offers yet another method for analyzing dependencies among interevent intervals (Peng et al., 1995). The algorithm for calculating the detrended fluctuation begins by constructing a running sum of the interval sequence over all indices. Next divide the summed series into blocks of length k and perform a least-squares fit on each of the data blocks, providing the trends for the individual blocks. Now detrend the sequence by subtracting the local trend in each block. Next, sum the squares of the detrended fluctuations, divide by k, and take the square root to obtain the detrended fluctuation estimate $\widehat{Y}(k)$. Finally, repeat this procedure for a variety of lags k. Figure 3.4 provides pseudocode for this algorithm, and Fig.3.5 presents a sample graphical calculation.

Like $U(k)$, for independent intervals $Y(k)$ varies as $\sqrt{k}$ for large k (see Prob. 3.7); more precisely,

$$Y(k) = \sigma_\tau \sqrt{(k^2 - 4)/15\,k} \tag{3.25}$$

for all $k > 2$ (see Sec. A.1.1). Except for the special case of Gaussian-distributed sequences, detrended fluctuation analysis exhibits significant bias and variance (Taqqu & Teverovsky, 1998). Section 12.3.5 reports the performance of a normalized form of the detrended fluctuation statistic,

$$Y_2(k_2) \equiv \frac{15\,Y^2(k+2)}{(k+2)\,\mathrm{Var}[\tau]}. \tag{3.26}$$

An extension of detrending exists, which involves the removal of higher-order polynomial trends rather than merely linear ones (Hu, Ivanov, Chen, Carpena & Stanley, 2001). Like the use of wavelets with several vanishing moments, this renders detrended fluctuation analysis insensitive to such trends in the data. Much as the interval wavelet variance is directly related to the interval spectrum, as discussed in Sec. 3.3.4, an analytical link exists between detrended fluctuation analysis and the interval spectrum (Heneghan & McDarby, 2000). Similar relations also connect the normalized count-based variance, the normalized count-based wavelet variance, and the conventional spectrum, as shown later in this chapter.

```
Calculate detrended fluctuation analysis
from discrete-time sequence {xₙ} of length M:
    generate summed sequence:   yₙ = yₙ₋₁ + xₙ;   y₁ = x₁
    set k = 3                              /* or set k = 4 */
    while M/k large               /* typically require M/k ≥ 10 */
        set m = 0
        while m + k − 1 ≤ M
            in this mth block of k values in {yₙ},
            find least-squares linear fit zₙ = an + b
                to yₙ over the range mk < n ≤ (m + 1)k
            subtract the fit from the summed sequence:   wₙ = yₙ − zₙ
            sum the squares of the remainder:   qₘ = Σₙ₌ₘₖ₊₁^(m+1)k wₙ²
            normalize the sum:   vₘ = qₘ/k
            take the square root:   Y(k, m) = √vₘ
            increment block index:   m → m + 1
        end while
        average all values of Y(k, m) to yield Ŷ(k)
        report k and Ŷ(k)
        increase block size k                /* typically k → 2k */
    end while
```

Fig. 3.4 Detrended fluctuation analysis: Pseudocode.

SUMMED SERIES AND   DETRENDED SERIES   STANDARD
LOCAL TRENDS   DEVIATION

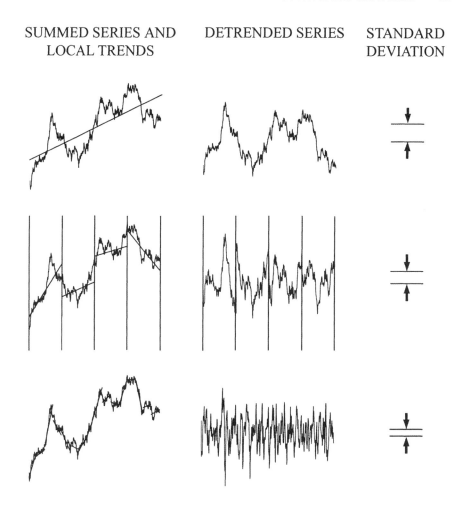

Fig. 3.5 Top row: a summed series and its least-squares fit by a linear function (left column). The difference between the summed series and the linear function (middle column). The sample standard deviation of this detrended series (right column). Middle row: the process repeats with the same summed series divided into four subseries, each with its own linear function as well as detrended series and sample standard deviation. Bottom row: the same calculation for sixteen subseries.

3.4   COUNT-BASED MEASURES

Count-based measures form the second broad class of point-process statistics, and derive from the sequence of counts $\{Z_k(T)\}$ shown in Fig. 3.1d). Perhaps the earliest application of a stochastic *counting* process was offered in the domain of legal decisions, by Poisson himself in 1837. Other early applications were Seidel's (1876) analysis of thunderstorms, blood-cell counting via microscopy carried out by Abbe (1878), Rutherford & Geiger's (1910) famous α-particle counting experiments (see also Bateman, 1910), and the well-known study of accident proneness conducted by Greenwood & Yule (1920). Researchers sometimes favor count-based measures because the information they provide corresponds to the real time of the point process. Moreover, counting measures can be used for systems that intrinsically involve integration. Count-number statistics also provide the only systematic analysis approach available for spaces of dimension greater than one.

Rate-based measures form a closely related set, where the **sample rate** $\lambda_k(T)$ is a normalized version of the sequence of counts:

$$\lambda_k(T) = Z_k(T)/T. \tag{3.27}$$

A variant of the counting procedure, called **generalized rates**, assigns fractional counts, depending on the manner in which an interevent interval spans different counting durations (Papoulis, 1991). Referring to Fig. 3.1c), for example, an interevent interval begins just before time $t = 2T$ and extends to just before $t = 5T$. Applying the generalized rate method for this case, one would set $\lambda_2(T) = \lambda_3(T) \approx 0.3/T$. Similarly, the duration from $t = T$ to $t = 2T$, with sample rate denoted $\lambda_1(T)$, would contain a small portion of the long interval discussed above (perhaps a tenth of it), plus about half of the interval that spans the divider $t = T$, plus the entire interval that spans the two events located within the duration from $t = T$ to $t = 2T$. Thus, one would set $\lambda_1(T) \approx (0.1 + 0.5 + 1)/T = 1.6/T$ using this method.

The generalized rates yield somewhat smoother estimates of the rate by reducing the quantization noise inherent in constructing the sequence of counts; they find their principal use in point processes for which the mean interval greatly exceeds the interval standard deviation, such as in heartbeat sequences (DeBoer et al., 1984). A relationship between count- and interval-based measures can be formulated in this special case, as indicated in connection with the second-order interval-based measures in Secs. 3.3.2 and 3.3.3. Since the events follow a relatively regular spacing for these point processes, results from one domain (count- or interval-based) can be readily translated into those in the other domain via the relation $t \approx k\,\mathrm{E}[\tau]$ (DeBoer et al., 1984).

While some measures, such as the spectrum, appear extensively in the literature as both interval- and count-based versions, others typically do not, despite being theoretically possible. For example, both rescaled range and detrended fluctuation analyses could serve as count-based measures, but published studies have traditionally not included such analyses in count-based form.

3.4.1 Marginal statistics

As with interval-based measures, we begin with count-based statistics that ignore any dependencies among counts, focusing on the marginal properties of the counts $\{Z_k(T)\}$ instead. These again fall into two classes. The **counting distribution** (or **probability mass function**), which is akin to the probability density function for discrete-valued random variables, forms the first:

$$p_Z(n; T) = \Pr\{Z(T) = n\}. \tag{3.28}$$

We employ the shorthand notation $p_Z(n)$ where this does not introduce confusion.

If no events occur in a time of duration T, then the time to the next event must exceed T. These two descriptions of the same outcome must have identical probabilities, so that (see Prob. 3.4)

$$\Pr\{Z(t) = 0\} = 1 - P_\vartheta(t) \tag{3.29}$$

where $P_\vartheta(t)$ again denotes the forward-recurrence-time probability distribution function. Combining Eqs. (3.11) and (3.29) yields (see Prob. 3.3)

$$p_\tau(t) = \mathrm{E}[\tau]\,\frac{d^2}{dt^2}\,\Pr\{Z(t) = 0\}, \tag{3.30}$$

thereby forming a connection between the interval-based and count-based domains.

For processes $N(t)$, and counting times T for which the number of counts greatly exceeds unity, the corresponding rate $\lambda(T)$ can assume any of a large number of possible values. In this case, the individual values of the counting distribution for $Z(T)$ [and therefore for $\lambda(T)$ as well] all become small. A continuous approximation for $\lambda(T)$, and its description by a probability density function, then becomes reasonable.

The second class of marginal count-based measures comprises the **count moments** $\mathrm{E}[Z^n(T)]$, and the statistics derived from them, such as the following:

variance	$\mathrm{Var}[Z] = \mathrm{E}[Z^2] - \mathrm{E}^2[Z]$	
factorial moments	$\mathrm{E}[Z!/(Z - k)!]$	
skewness	$\mathrm{E}[(Z - \mathrm{E}[Z])^3]/\mathrm{Var}^{3/2}[Z]$	(3.31)
kurtosis	$\mathrm{E}[(Z - \mathrm{E}[Z])^4]/\mathrm{Var}^2[Z] - 3,$	

where these moments exist, and where we suppress the explicit reference to the counting time T (see Footnote 2 on p. 55). Here $k! \equiv k \cdot (k-1) \cdots 3 \cdot 2 \cdot 1$ represents the factorial function, and we employ the notational convenience that $1/n! \equiv 0$ for n a negative integer.

For renewal point processes, as discussed in Sec. 4.2, the value of each interevent interval does not depend on those before or after it. However, for general renewal point processes the *count* sequence *does* exhibit dependence. For example, consider a point process constructed so that the smallest possible interval takes a value larger than $\tau_{\min}$. At a counting time half as large ($T = \tau_{\min}/2$), the observation that $Z_{k-1}(T) = 1$ implies that $Z_k(T) = 0$; otherwise the interval spanning $t = kT$ would not exceed $2T = \tau_{\min}$. The only point process for which the sequence of counts $\{Z_k(T)\}$ exhibits independence for all counting times T is the homogeneous Poisson process described in Sec. 4.1. For all other point processes, dependencies do occur among counts, necessitating the use of several statistics for an overview of the sequence of counts.

3.4.2 Normalized variance

As the counting duration T increases, all of the count moments provided in Eq. (3.31) increase, suggesting that some manner of normalization might prove useful. The **normalized variance** $F(T)$ is obtained by dividing the variance by the mean:

$$F(T) \equiv \frac{\mathrm{Var}[Z(T)]}{\mathrm{E}[Z(T)]}. \tag{3.32}$$

This quantity appears to have been first devised by Ugo Fano (1947) to characterize the statistical fluctuations of the number of ions generated by individual fast charged particles; it therefore garnered the appellation "Fano factor." It also goes by a number of other names including "count variance-to-mean ratio," "dispersion ratio," and "index of dispersion" (Cox & Isham, 1980). We prefer the terminology "normalized variance" for its simplicity and apt description.

Figure 3.6a) displays its construction. In general, the normalized variance is a function of the counting time. In the limit of small counting times T, the count $Z(T)$ generally takes a value of zero, and rarely unity; larger values occur with negligible frequency. The result then becomes a sequence of **Bernoulli trials**. Under these conditions,

$$
\begin{aligned}
\Pr\{Z(T) = 1\} &= & p_Z(1) & \\
\Pr\{Z(T) = 0\} &= & p_Z(0) &\approx & 1 - p_Z(1) \\
\Pr\{Z(T) > 1\} &= & \sum_{n>1} p_Z(n) &\approx & 0,
\end{aligned}
\tag{3.33}
$$

so that (see Prob. 3.2)

$$
\begin{aligned}
\lim_{T\to 0} F(T) &= \lim_{T\to 0} \frac{\mathrm{Var}[Z(T)]}{\mathrm{E}[Z(T)]} &&= \lim_{T\to 0} \frac{\mathrm{E}[Z^2(T)] - \mathrm{E}^2[Z(T)]}{\mathrm{E}[Z(T)]} \\
&\approx \lim_{T\to 0} \frac{\mathrm{E}[Z(T)] - \mathrm{E}^2[Z(T)]}{\mathrm{E}[Z(T)]} &&= \lim_{T\to 0}\{1 - \mathrm{E}[Z(T)]\} \\
&\approx \lim_{T\to 0} [1 - p_Z(1)] &&= 1
\end{aligned}
\tag{3.34}
$$

for any orderly point process. The normalized variance approaches unity, as expected for a sequence of Bernoulli trials. For the homogeneous Poisson process, $F(T) = 1$ for all counting times, as shown in Sec. 4.1.

As an estimator for finite-length data sets, the normalized variance suffers from bias for sequences of counts that exhibit dependence (Lowen & Teich, 1995; Thurner et al., 1997), as we demonstrate in Sec. 12.3.2. In contrast, the normalized Haar-wavelet variance described in the next section does not have this limitation.

3.4.3 Normalized Haar-wavelet variance

In the same way that wavelets can prove useful in characterizing the sequence of intervals derived from a fractal or fractal-rate point process (see Sec. 3.3.4), they also

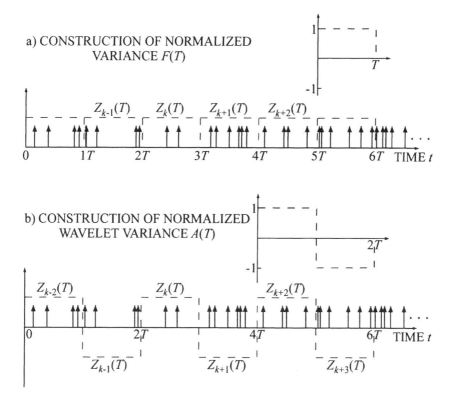

Fig. 3.6　A point process gives rise to a sequence of counts $\{Z_k(T)\}$ by counting the number of events in each contiguous time duration T. (a) Computing the variance of the number of counts, and dividing this quantity by the mean number of counts, yields the normalized variance $F(T)$. (b) Computing the variance of the *difference* in the number of counts in adjacent counting durations, and dividing this quantity by twice the mean number of counts, yields the normalized Haar-wavelet variance $A(T)$.

provide an appropriate method for the analysis of count sequences associated with such processes.

In terms of the point process, the wavelet transform becomes

$$C_{\psi,N}(a,b) = \int a^{-1/2}\, \psi[(t-b)/a]\, dN(t), \tag{3.35}$$

where the continuous wavelet transform now applies. We focus on a particular wavelet, the Haar wavelet (1910), defined by

$$\psi_{\text{Haar}}(t) = \begin{cases} 1 & \text{for } 0 \le t < \tfrac{1}{2} \\ -1 & \text{for } \tfrac{1}{2} \le t < 1 \\ 0 & \text{otherwise.} \end{cases} \tag{3.36}$$

Equation (3.35) then becomes

$$
\begin{aligned}
C_{\text{Haar},N}(a,b) &= a^{-1/2} \int_{b}^{b+a/2} dN(t) - a^{-1/2} \int_{b+a/2}^{b+a} dN(t) \\
&= a^{-1/2} \Big\{ [N(b+a/2) - N(b)] \\
&\qquad - [N(b+a) - N(b+a/2)] \Big\}.
\end{aligned} \tag{3.37}
$$

Setting $T = a/2$ and $k = 2b/a$ yields

$$
C_{\text{Haar},N}(2T,k) = (2T)^{-1/2} [Z_k(T) - Z_{k+1}(T)] \tag{3.38}
$$

$$
E\big[C_{\text{Haar},N}^2(2T,k)\big] = \frac{E\big\{[Z_k(T) - Z_{k+1}(T)]^2\big\}}{2T}. \tag{3.39}
$$

Dividing by the sample rate yields the **normalized Haar-wavelet variance** $A(T)$:

$$
\begin{aligned}
A(T) &= \frac{E\big[C_{\text{Haar},N}^2(2T,k)\big]}{E[\lambda_k(T)]} \\
&= \frac{E\big\{[Z_k(T) - Z_{k+1}(T)]^2\big\}}{2T} \frac{T}{E[Z_k(T)]} \\
&\equiv \frac{E\big\{[Z_k(T) - Z_{k+1}(T)]^2\big\}}{2E[Z_k(T)]}.
\end{aligned} \tag{3.40}
$$

We schematically illustrate the construction of this quantity from a point process in Fig. 3.6b). We initially developed this measure for the analysis of action-potential sequences recorded from the mammalian auditory nerve (Lowen & Teich, 1996a). At first we called it the "Allan factor" by virtue of its relationship to the Allan variance (Allan, 1966; Barnes & Allan, 1966), but we prefer the appellation "normalized Haar-wavelet variance," which is more descriptive.[7]

Unlike the normalized variance discussed in Sec. 3.4.2, $A(T)$ does not suffer from bias for sequences of counts that exhibit dependence (Abry & Flandrin, 1996; Thurner et al., 1997), as we demonstrate in Secs. 12.2.3, 12.3.8, and 12.4. In terms of the normalized variance, we have (Scharf, Meesmann, Boese, Chialvo & Kniffki, 1995):

$$
\begin{aligned}
2F(T) - F(2T) &= 2 \frac{\text{Var}[Z(T)]}{E[Z(T)]} - \frac{\text{Var}[Z(2T)]}{E[Z(2T)]} \\
\\
&= \frac{4E[Z^2(T)] - 4E^2[Z(T)]}{2E[Z(T)]} \\
\\
&\quad - \frac{E\big\{[Z_k(T) + Z_{k+1}(T)]^2\big\} - E^2\big\{[Z_k(T) + Z_{k+1}(T)]\big\}}{2E[Z(T)]}
\end{aligned}
$$

---

[7] Photographs of Haar and Allan appear at the beginnings of Chapters 5 and 12, respectively.

$$= \frac{4\mathrm{E}[Z^2(T)] - 4\mathrm{E}^2[Z(T)]}{2\mathrm{E}[Z(T)]}$$

$$+ \frac{-2\mathrm{E}[Z^2(T)] - 2\mathrm{E}[Z_k(T) \, Z_{k+1}(T)] + 4\mathrm{E}^2[Z(T)]}{2\mathrm{E}[Z(T)]}$$

$$= \frac{\mathrm{E}\{[Z_k(T) - Z_{k+1}(T)]^2\}}{2\mathrm{E}[Z(T)]}$$

$$= A(T). \tag{3.41}$$

In particular,

$$\lim_{T \to 0} A(T) = \lim_{T \to 0} F(T) = 1. \tag{3.42}$$

Furthermore, if the normalized variance increases monotonically, we have

$$\begin{aligned}
F(2T) &> F(T) \\
0 &> F(T) - F(2T) \\
F(T) &> F(T) + F(T) - F(2T) \\
F(T) &> A(T), \tag{3.43}
\end{aligned}$$

illustrating that the normalized variance always exceeds the normalized Haar-wavelet variance in this case.

3.4.4   Count autocorrelation

In direct parallel to the sequence of intervals, the autocorrelation of the sequence of counts provides information about their second-order properties. The **count autocorrelation** is defined as

$$R_Z(k, T) \equiv \mathrm{E}[Z_n(T)Z_{n+k}(T)]. \tag{3.44}$$

For a sequence of counts that exhibits independence at a counting time T, we have $R_Z(k, T) = \mathrm{E}^2[Z(T)]$ for $k \neq 0$. In contrast to the interval-based autocorrelation in Eq. (3.16), in this case a direct relationship exists between the lag variable k and time t in seconds: $k = t/T$.

As with the count variance, estimates of this measure suffer from bias, as we demonstrate in Sec. 12.3.3 using a normalized form of the autocovariance,

$$R_2(k) \equiv \frac{R_Z(k, T) - \mathrm{E}^2[Z(T)]}{\mathrm{Var}[Z(T)]}. \tag{3.45}$$

A generalized version of the wavelet variance defined by

$$C_{\mathrm{Haar},N}(2T, k) \equiv \mathrm{E}\left\{[Z_n(T) - Z_{n+k}(T)]^2\right\} \tag{3.46}$$

would not suffer from bias, but this quantity has not enjoyed wide use.

3.4.5 Rate spectrum

Again paralleling results for the sequence of intervals, Fourier transforming the auto-correlation in Eq. (3.44) yields the **count-based spectrum** $S_Z(f, T)$. In practice, the **rate-based spectrum**, often simply called the **rate spectrum**, proves more useful:

$$S_\lambda(f, T) = T^{-2} S_Z(f, T) = \frac{1}{T} \sum_k R_Z(k, T) e^{-i2\pi k f T}. \qquad (3.47)$$

This quantity derives from the Fourier transform of the observed sequence of rates, rather than the sequence of counts, and since $\lambda_k = Z_k(T)/T$, a factor of T^{-2} appears in Eq. (3.47). Because of this normalization, in the range $0 < f \ll 1/T$ the rate spectrum $S_\lambda(f, T)$ approaches a limiting value that does not depend on T. The count-based version, $S_Z(f, T)$, does not enjoy this property, and is therefore less useful.

We examine the performance of $S_\lambda(f, T)$ in Secs. 12.3.9 and 12.4. An estimate of the rate-based spectrum derived from a data set is often referred to as the **periodogram** (this term is also used for an estimate of the interval-based spectrum). One can reduce the variance of the rate-based periodogram by averaging it over nearby frequencies; or by partitioning the original point process into blocks of identical length, and then averaging the individual periodograms thus obtained; or by employing both methods in tandem.

3.5 OTHER MEASURES

Other measures of a point process exist that do not fall into either the interval-based or count-based categories. Some of these measures prove problematical in practice, leading to long computation times, poor statistics, or both; they are, nevertheless, of interest as theoretical constructs.

3.5.1 Coincidence rate

The coincidence rate $G(t)$ measures the correlation between events as a function of a specified time delay t, regardless of any intervening events (Kuznetsov & Stratonovich, 1956; Kuznetsov, Stratonovich & Tikhonov, 1965; Cox & Lewis, 1966):

$$
\begin{aligned}
G(t) \; &\equiv \; \lim_{\epsilon \to 0} \frac{1}{\epsilon^2} \Pr \Big\{ N(s + \epsilon) - N(s) > 0 \\
& \qquad\qquad \text{and } N(s + t + \epsilon) - N(s + t) > 0 \Big\} \qquad (3.48) \\
&= \; \mathrm{E}\left[\frac{dN(s)}{ds} \frac{dN(s + t)}{ds} \right]. \qquad\qquad\qquad (3.49)
\end{aligned}
$$

For $t = 0$, the two probabilities coincide in Eq. (3.48), which leads to an infinite value. It proves most convenient to represent this as a Dirac delta function:

$$\lim_{\substack{\epsilon>0 \\ \epsilon\to 0}} \int_{-\epsilon}^{\epsilon} G(t)\, dt = \lim_{\substack{\epsilon>0 \\ \epsilon\to 0}} \int_{-\epsilon}^{\epsilon} \delta(t)\, \mathrm{E}[\mu]\, dt = \mathrm{E}[\mu]. \tag{3.50}$$

The coincidence rate is the point-process analog of the autocorrelation used for continuous-time processes. For large delays t, the two quantities inside the expectation in Eq. (3.49) become independent, so that

$$\begin{aligned}
\lim_{t\to\infty} G(t) &= \lim_{t\to\infty} \mathrm{E}\!\left[\frac{dN(s)}{ds}\frac{dN(s+t)}{ds}\right] \\[2mm]
&= \mathrm{E}\!\left[\frac{dN(s)}{ds}\right]\mathrm{E}\!\left[\frac{dN(s+t)}{ds}\right] \\[2mm]
&= \mathrm{E}[\mu(s)]\,\mathrm{E}[\mu(s+t)] \\[2mm]
&= \mathrm{E}^2[\mu], \tag{3.51}
\end{aligned}$$

where $\mu(t) \equiv dN(t)/dt$ denotes the **instantaneous rate** of the point process $dN(t)$. In general, $\mu(t)$ varies in a random fashion, representing the local likelihood of event generation. For a stationary point process, however, the corresponding rate has statistics that do not vary with time. In this case (which describes the vast majority of point processes considered in this book), we can eliminate the explicit dependence on the time t for the marginal statistics. For the mean value, for example, we write $\mathrm{E}[\mu]$ instead of $\mathrm{E}[\mu(t)]$. A further simplification takes place in the special case when the history of the process has no effect on the instantaneous generation rate, as for the homogeneous Poisson process described in Sec. 4.1. The expectation itself then becomes superfluous, so we eliminate the explicit expectation operator as well; $\mathrm{E}[\mu]$ then becomes μ. Returning to the general point-process case, we have $\mathrm{E}[\mu] = 1/E[\tau]$, where we interpret $1/\infty$ as zero.

We can conveniently express several of the count-based measures set forth earlier in terms of the coincidence rate $G(t)$ (Cox & Isham, 1980; Thurner et al., 1997). Expressions for the normalized variance $F(T)$ (see Prob. 3.10), normalized Haar-wavelet variance $A(T)$, and autocorrelation $R_Z(k,T)$ are given by

$$F(T) = \frac{1}{\mathrm{E}[\mu]T}\int_{-T}^{T}\left\{G(t)-\mathrm{E}^2[\mu]\right\}(T-|t|)\,dt \tag{3.52}$$

$$A(T) = \frac{2}{\mathrm{E}[\mu]T}\int_{-T}^{T}[G(t)-G(2t)]\,(T-|t|)\,dt \tag{3.53}$$

$$R_Z(k,T) = \int_{-T}^{T} G(kT+t)\,(T-|t|)\,dt, \tag{3.54}$$

respectively. We can readily invert Eq. (3.52) to yield

$$G(t) = \mathrm{E}[\mu]\,\delta(t) + \mathrm{E}^2[\mu] + \frac{\mathrm{E}[\mu]}{2}\frac{d^2}{dT^2}\left[TF(T)\right]_{T=t}. \tag{3.55}$$

In practice, determining the coincidence rate from a finite set of data proves impossible. The probability that two events exist in the data set with a separation of precisely t is zero, for any *a priori* value of t. Instead, the autocorrelation $R_Z(k, T)$ provides an estimate of the coincidence rate through Eq. (3.54). For T smaller than the time scale over which $G(t)$ varies significantly, we have

$$
\begin{aligned}
R_Z(k, T) &= \int_{-T}^{T} G(kT + t) \, (T - |t|) \, dt \\
&\approx G(kT) \int_{-T}^{T} (T - |t|) \, dt = G(kT) \, T^2 \\
G(t) &\approx T^{-2} R_Z(t/T, T).
\end{aligned}
\tag{3.56}
$$

However, obtaining useful resolution in this approximation requires a small value of T, which, in turn, leads to excessive variance in this estimator of $G(t)$ for all but the largest data sets. For this reason, the coincidence rate is rarely used in practice.

3.5.2   Point-process spectrum

As with the sequence of intervals $\{\tau_k\}$ and the sequence of counts $\{Z_k(T)\}$, Fourier transformation of the coincidence rate $G(t)$ yields a spectrum $S_N(f)$, this time the spectrum of the point process $dN(t)$ itself[8] (Bartlett, 1963, 1964):

$$
S_N(f) = \int_{-\infty}^{\infty} G(t) \, e^{-i2\pi ft} \, dt.
\tag{3.57}
$$

The inverse relationship also holds

$$
G(t) = \int_{-\infty}^{\infty} S_N(f) \, e^{i2\pi ft} \, df.
\tag{3.58}
$$

Through the properties of the Fourier transform, the delta function associated with zero delay in Eq. (3.50) becomes the asymptotic value for large frequencies,

$$
\lim_{f \to \infty} S_N(f) = \mathrm{E}[\mu] = 1/\mathrm{E}[\tau].
\tag{3.59}
$$

The expression $S_N(f) / \mathrm{E}[\mu]$ therefore provides a normalized form of this spectrum. The large-delay asymptotic value for the coincidence rate in Eq. (3.51) becomes a delta function at zero frequency,

$$
\lim_{\substack{\epsilon > 0 \\ \epsilon \to 0}} \int_{-\epsilon}^{\epsilon} S_N(f) \, df = \lim_{\substack{\epsilon > 0 \\ \epsilon \to 0}} \int_{-\epsilon}^{\epsilon} \delta(f) \, \mathrm{E}^2[\mu] \, df = \mathrm{E}^2[\mu].
\tag{3.60}
$$

---

[8] We generally compute the point-process spectrum $S_N(f)$ for a theoretical construct but make use of the rate spectrum $S_\lambda(f, T)$ for actual data (see Sec. 3.4.5); the two spectra thus represent probabilistic and statistical measures, respectively.

This delta function also appears in $S_\lambda(f, T)$ and, with a different prefactor, in $S_\tau(f)$.

Combining the results obtained earlier yields (Lowen & Teich, 1993a; Lowen, 1996)

$$F(T) = \frac{2}{\pi^2 \, \mathrm{E}[\mu] \, T} \int_{0+}^{\infty} S_N(f) \sin^2(\pi f T) \, f^{-2} \, df \qquad (3.61)$$

$$A(T) = \frac{4}{\pi^2 \, \mathrm{E}[\mu] \, T} \int_{0+}^{\infty} S_N(f) \sin^4(\pi f T) \, f^{-2} \, df, \qquad (3.62)$$

as shown in Probs. 3.11 and 3.12, where the notation 0+ indicates that the integral does not include the delta function at zero frequency. We can make use of Eq. (3.61) to obtain formulas for $F(T)$ and $A(T)$ in the limit of large counting times. Using the substitution $x \equiv \pi f T$, this equation becomes

$$F(T) = \frac{2}{\pi \, \mathrm{E}[\mu]} \int_{0+}^{\infty} S_N(x/\pi T) \sin^2(x) \, x^{-2} \, dx. \qquad (3.63)$$

In the limit of large counting times, the spectrum approaches its low-frequency limit, whereupon Eq. (3.63) yields

$$\begin{aligned}
\lim_{T \to \infty} F(T) &= \frac{2}{\pi \, \mathrm{E}[\mu]} \int_{0+}^{\infty} \lim_{f \to 0} S_N(f) \sin^2(x) \, x^{-2} \, dx \\
&= \frac{2}{\pi \, \mathrm{E}[\mu]} \lim_{f \to 0} S_N(f) \int_{0+}^{\infty} \sin^2(x) \, x^{-2} \, dx \\
&= \mathrm{E}[\tau] \lim_{f \to 0} S_N(f). \qquad (3.64)
\end{aligned}$$

Substituting this limit into Eq. (3.41) gives the same result for the normalized Haar-wavelet variance, $A(T)$.

In the opposite limit of small counting times and large frequencies we have

$$\lim_{T \to 0} F(T) = \mathrm{E}[\tau] \lim_{f \to \infty} S_N(f) = 1, \qquad (3.65)$$

so that plots of the normalized Haar-wavelet variance and normalized point-process spectrum appear to be mirror images of each other.

In contrast to the coincidence rate, the estimation of $S_N(f)$ from a data set proves straightforward. A simple method (without averaging) follows from Eq. (3.49) and the Weiner–Khintchine theorem,

$$S_N(f) = \frac{1}{L} \mathrm{E}\left[\left| \sum_k e^{-i2\pi f t_k} \right|^2 \right], \qquad (3.66)$$

where $\{t_k\}$ again represents the set of times at which the events occur (rather than the times between events) and L denotes the duration of the data set. This method suffers from a major drawback: the times $\{t_k\}$ can take any of a continuous range of values, which precludes the use of the fast Fourier-transform algorithm. However, the rate spectrum set forth in Sec. 3.4.5 proves amenable to this transform, yielding an efficient and practical method for estimating the spectrum of a point process.

A direct relationship exists between the rate spectrum and the point-process spectrum,

$$S_\lambda(f, T) = \sum_{k=-\infty}^{\infty} S_N(f + k/T) \frac{\sin^2(\pi f T)}{(\pi f T + \pi k)^2}, \tag{3.67}$$

and the two quantities differ only slightly for $T \ll 1/f$ (see Prob. 3.13). For the accurate estimation of $S_N(f)$ from a given sample of a point process with duration L, we choose an integer n such that $2^{n-1}/L$ exceeds the greatest frequency of interest by a large margin, then set $T = L/2^n$ and use the fast Fourier transform to obtain frequency-domain values from the 2^n elements of $\{Z_k(T)\}$.

3.5.3 General-wavelet variance

In Sec. 3.4.3 we made use of the Haar wavelet basis (Haar, 1910) in defining the normalized wavelet variance, because the form of the Haar wavelet leads to a simple representation in terms of the counting process $\{Z_k(T)\}$. Extensions to a general wavelet basis prove possible, although at the expense of more complex computations.

We again consider the variance of the continuous wavelet transform provided in Eq. (3.35), and employ its zero-mean property to obtain (Teich et al., 1996)

$$\begin{aligned}
\mathrm{Var}[C_{\psi,N}(a,b)] &= \mathrm{E}[C_{\psi,N}^2(a,b)] \\
&= \mathrm{E}\left[\int_s \int_t a^{-1} \psi[(s - b)/a]\, \psi[(t - b)/a]\, dN(s)\, dN(t) \right] \\
&= a^{-1} \int_s \int_t \psi[(s - b)/a]\, \psi[(t - b)/a]\, G(s - t)\, ds\, dt \\
&= a \int_x G(ax) \int_y \psi(x + y)\, \psi(y)\, dy\, dx \tag{3.68} \\
&= a \int G(ax)\, C_{\psi,\psi}(1, x)\, dx \\
&= \int G(t)\, C_{\psi,\psi}(1, t/a)\, dt, \tag{3.69}
\end{aligned}$$

where $C_{\psi,\psi}(1, x)$, the continuous wavelet transform of the wavelet function itself, does not depend on the coincidence rate $G(t)$ (Teich et al., 1996). As with the spectrum of the point process $S_N(f)$, the continuous nature of the times $\{t_k\}$ precludes the use of fast transform algorithms. Again, approximating the point process $dN(t)$ with the sequence of counts $\{Z_k(T)\}$ proves useful.

3.5.4 Generalized dimension

The **generalized dimension** D_q, which is closely related to the **Rényi entropy** (Rényi, 1955, 1970; Theiler, 1990), extends the simple concept of dimension developed in

Sec. 2.1 and provides a direct measurement of the fractal properties of an object. In the context of a collection of points, we consider a point process $dN(t)$ over the range of times $0 \leq t \leq L$ and employ the sequence of counts $\{Z_k\}$ to obtain

$$D_q \equiv \frac{1}{q-1} \lim_{T \to 0} \frac{\mathrm{E}\left\{\log\left[\sum_k Z_k^q(T)\right]\right\}}{\log(T)}, \tag{3.70}$$

where the sum extends over all non-empty counts. Note that for a given value of T, k assumes a maximum value of L/T.[9]

We can calculate the generalized dimension for any real value of q. Several values of q correspond to well-known generalized dimensions (Mandelbrot, 1982, Chapter 39), such as the **capacity dimension** D_0 (Pontrjagin & Schnirelmann, 1932) first discussed in Sec. 2.1.1, which is also called the **box-counting dimension**; the **information dimension** $\lim_{q \to 1} D_q$, which is closely related to the Kolmogorov entropy; and the **correlation dimension** D_2 (Grassberger & Procaccia, 1983). We also consider D_{-1} and $D_{1/2}$ (see Prob. 5.5). As reported in Sec. 2.1, the values of D_q for any object must lie between the **topological dimension** of the object and the **Euclidean dimension** of the space in which the object resides: zero for a point, unity for a line segment, two for a square, and so on.

Points on a line, as derived from a realization of a point process for example, will exhibit values of D_q between zero (the dimension of a point) and unity (the dimension of a line). In general, D_q for a nonfractal object assumes the lower bound given by the topological dimension of the object, for all indices q; the quantity $D_q = D$ is then an integer. For a (mono)fractal object, again $D_q = D$ for all q, but in contrast to the nonfractal object, D is not integer; indeed this forms one definition of a fractal (see Sec. 2.3). Finally, for a multifractal, D_q monotonically decreases with increasing q (Theiler, 1990).

Wavelet-based methods for estimating D_q also exist (Argoul, Arneodo, Elezgaray & Grasseau, 1989; Bacry, Muzy & Arneodo, 1993; Arrault & Arneodo, 1997); these can provide localized values of the generalized dimension.

We make special mention of a fractal dimension that does not belong to the D_q family of generalized dimensions: the **Hausdorff–Besicovitch dimension** D_{HB} (Mandelbrot, 1982, pp. 362–365 and references therein). For many simple cases in which $D_q = D$ for all q (such as the Cantor set), it turns out that $D_{\mathrm{HB}} = D$. Calculating this dimension involves fewer assumptions than determining D_q; the axes of the space within which the object exists need not be specified, nor need the dimension of that embedding space. However, determining D_{HB} proves far more difficult than calculating D_q for analytic examples, and it presents significant difficulties when applied to data. We therefore deal little with the Hausdorff–Besicovitch dimension in this book.

The practical use of the generalized dimension D_q requires modification when applied to real point processes. For any orderly point process, the number of events

---

[9] If we set $q = 0$ and make the identifications $T = \epsilon$ and $Z(T) = M(\epsilon)$, we recover Eq. (2.2).

$N(L)$ occurring between the origin and a maximum time L assumes a finite value with probability one. Therefore, a minimum interevent time $\tau_{\min}$ exists; for $T < \tau_{\min}$, for all k either $Z_k(T) = 0$ or $Z_k(T) = 1$. Equation (3.70) then becomes

$$
\begin{aligned}
D_q &= \frac{1}{q-1} \lim_{T \to 0} \frac{\mathrm{E}\left\{\log\left[\sum_k 1^q\right]\right\}}{\log(T)} \\
&= \frac{1}{q-1} \lim_{T \to 0} \frac{\mathrm{E}\{\log[N(L)]\}}{\log(T)} \\
&= \frac{\mathrm{E}\{\log[N(L)]\}}{q-1} \lim_{T \to 0} \frac{1}{\log(T)} \\
&= 0,
\end{aligned}
\tag{3.71}
$$

where the sum again does not contain empty counts. Hence, for a finite collection of points, we obtain $D_q = 0$, a result not indicative of fractal characteristics.

In the context of point processes, we therefore modify the definition of the generalized dimension to accommodate scaling behavior over a range of counting times T. We recast Eq. (3.70) in terms of a **scaling equation**,

$$
\sum_k Z_k^q(T) \sim T^{(q-1)D_q},
\tag{3.72}
$$

which we can alternatively write in the form of a **generalized-dimension scaling function**:

$$
\eta_q(T) \equiv \left[\sum_k Z_k^q(T)/N(L)\right]^{\frac{1}{q-1}} \approx T^{D_q},
\tag{3.73}
$$

normalized such that $\eta_q(T \to 0) = 1$. If Eqs. (3.72) and (3.73) hold over a range of counting times T, then the resulting exponents on the right-hand sides of these equations yield the generalized dimensions D_q. This directly reveals the fractal properties of point-process sample paths. In practice, the D_q assume noninteger values only for the class of **fractal point processes**, and not for the class of **fractal-rate point processes** that largely form the focus of this book (see Sec. 5.5 and Prob. 5.5).

In the special case when $q = 0$, Eq. (3.73) reduces to

$$
\eta_0(T) \equiv \frac{N(L)}{\sum_k Z_k^0(T)} \approx T^{D_0},
\tag{3.74}
$$

which yields $D_0 = 0$ for $T < \tau_{\min}$, in accordance with Eq. (3.71), since the sum in the denominator of Eq. (3.74) is then $N(L)$.[10]

---

[10] This power-law dependence over a range of counting times recalls Eq. (1.1) rather than Eq. (2.1).

3.5.5 Correlation measures for pairs of point processes

Second-order methods prove useful for revealing correlations between sequences of events, which indicate how information is shared between pairs of point processes.[11] Although such methods may not detect all of the subtle forms of interdependence to which information-theoretic approaches are sensitive (see, for example, Kabanov, 1978; Rieke, Warland, de Ruyter van Steveninck & Bialek, 1997; Dayan & Abbott, 2001), the latter methods suffer from limitations arising from the finite sizes of many real data sets (see Lowen, Ozaki, Kaplan & Teich, 1998).

We consider two second-order measures, in turn: the **normalized Haar-wavelet covariance** and the **cross-spectrum**. They exhibit different immunity to nonstationarities, and different tradeoffs between bias and variance, just as their single-dimensional counterparts do (see Chapter 12). Both measures prove useful in the analysis of pairs of point processes.[12]

- *Normalized Haar-wavelet covariance.* We define the normalized Haar-wavelet covariance $A^{(2)}(T)$ as a generalization of the normalized Haar-wavelet variance $A(T)$ defined in Sec. 3.4.3. This measure is insensitive to constant values and can be rendered insensitive to higher-order polynomial trends by making use of other wavelets (see Sec. 3.5.3). We initially developed this measure to analyze correlations between pairs of visual-system spike trains, such as those recorded from retinal ganglion cells and cells in the lateral geniculate nucleus (Lowen et al., 2001). At first, we referred to $A^{(2)}(T)$ as the "normalized wavelet cross-correlation function," but we prefer the designation "normalized Haar-wavelet covariance" since it highlights the relationship between $A^{(2)}(T)$ and $A(T)$.

  The computation of the normalized Haar-wavelet covariance at a particular counting time T begins with the division of both point processes into contiguous counting durations T. For the first point process, we register the number of events $Z_{1,k}$ that fall within the kth duration for all indices k. Next we compute the difference between the count numbers in a given duration, $Z_{1,k}$, and the duration that immediately follows, $Z_{1,k+1}$, for all k, much as when computing the normalized Haar-wavelet variance. We then carry out the same procedure for the second point process, beginning with the number of events $Z_{2,k}$ that fall within the kth duration.

---

[11] For example, it may be of interest to study how information is shared between the point processes at the input and output of a cell. Such a pair of processes collectively forms a **bivariate point process**, a form of marked point process (see Cox & Isham, 1980, Chapter 5).

[12] They have been used to reveal unexpected correlations between pairs of visual-system spike trains (Lowen et al., 2001) and between earthquakes and geoelectrical extreme events (Telesca, Balasco, Colangelo, Lapenna & Macchiato, 2004), as examples.

In analogy with the definition of the normalized Haar-wavelet variance, we define the normalized Haar-wavelet covariance as:

$$A^{(2)}(T) \equiv \frac{\mathrm{E}\{[Z_{1,k}(T) - Z_{1,k+1}(T)]\,[Z_{2,k}(T) - Z_{2,k+1}(T)]\}}{2\{\mathrm{E}[Z_{1,k}(T)]\,\mathrm{E}[Z_{2,k}(T)]\}^{1/2}}. \qquad (3.75)$$

The normalization imposed in Eq. (3.75) gives rise to three salutary features for $A^{(2)}(T)$: (1) it is symmetric in the two point processes; (2) it reduces to the marginal normalized wavelet variance $A(T)$ if the two point processes are identical — in particular, it assumes a value of unity for all counting times T if both point processes comprise the same homogeneous Poisson process, in analogy with the normalized wavelet variance $A(T)$.

- *Cross-spectrum.* The cross-spectrum $S_N^{(2)}(f)$ is a generalization of the point-process spectrum for individual spike trains, in much the same way as the normalized Haar-wavelet covariance derives from the normalized Haar-wavelet variance (Lowen et al., 2001). Although not often used, this measure has a long history in the annals of statistics. It appears to have been first introduced by Jenkins (1961) and its use has been advanced by Brillinger (1986).

A number of definitions for the cross-spectrum have been put forward; we choose one that is real and symmetric in the two point processes, and reduces to the single-process version when the two processes coincide. An extension of Eq. (3.66) leads to the cross-spectrum

$$S_N^{(2)}(f) \equiv \frac{1}{L}\mathrm{E}\left[\mathrm{Re}\left\{\sum_k e^{-i2\pi f t_{1,k}} \sum_m e^{i2\pi f t_{2,m}}\right\}\right], \qquad (3.76)$$

where $\{t_{1,k}\}$ and $\{t_{2,k}\}$ index the events in point processes 1 and 2, respectively, occurring in a time of duration L. Equation (3.76) is indeed symmetric, and reduces to Eq. (3.66) for identical sets $\{t_{1,k}\}$ and $\{t_{2,k}\}$. Although it returns a real result, the cross-spectrum can assume negative values. For example, two Poisson processes modulated by sinewaves of identical frequency, but opposite phase, yield $S_N^{(2)}(f) < 0$ near the modulation frequency and its harmonics. For independent spike trains, we have $S_N^{(2)}(f) = 0$.

As with the single-process spectrum, for the purposes of practical estimation it proves easier to employ the rate-based version

$$S_\lambda^{(2)}(f, T) \equiv \frac{1}{L}\mathrm{E}\left[\mathrm{Re}\left\{\sum_k Z_{1,n+k}(T)e^{-i2\pi kfT} \sum_m Z_{2,n+m}(T)e^{i2\pi mfT}\right\}\right], \qquad (3.77)$$

where $\{Z_{1,n}(T)\}$ and $\{Z_{2,n}(T)\}$ describe the counts for the two point processes.

Problems

3.1 *Point-process models*   For each of the following examples, specify whether an orderly, one-dimensional point process provides a useful model. For the remainder, describe a modification of the example that would make the model apply.

1. longitude and latitude of trees on Long Island, New York;

2. the times of raindrops hitting a roof;

3. the arrival times of customers at an automatic teller machine;

4. thunderstorm occurrence times in San Diego county during February 1993;

5. the times at which cars are in the Ted Williams tunnel, which passes under Boston harbor;

6. the set of numbers $1/(n + x_n)$, where n ranges over all positive integers and $\{x_n\}$ is a set of independent exponentially distributed random variables of unit mean;

7. the times of maximum daily temperature at the summit of Mount Everest;

8. all human heartbeat times (defined as the time of maximum contraction) from anyone anywhere on the planet, as transmitted to a central recording station;

9. two random numbers selected uniformly from the unit interval;

10. the sign of the difference between the Dow Jones Industrial Average and the previous day's closing price.

3.2 *Short-time normalized variance for an orderly point process*   Justify the approximation $\mathrm{E}\big[Z^2(T)\big] \approx \mathrm{E}[Z(T)]$ in Eq. (3.34).

3.3 *Connection between interval- and count-based statistics*   Using Eqs. (3.11) and (3.29), show that Eq. (3.30) holds.

3.4 *Forward-recurrence-time distribution*   Explain why Eq. (3.29) holds.

3.5 *Skewness and kurtosis values*   As defined in Eq. (3.4), what values can the skewness and kurtosis assume? How does prohibiting negative interevent intervals change this?

3.6 *Infinite moments of the Lévy density*   Show that the random variable corresponding to the probability density function given by Eq. (3.13) has infinite moments for all positive integer orders. For which fractional orders do its moments exist?

3.7 *Rescaled range and detrended fluctuations for independent intervals*   Provide a heuristic argument showing that the rescaled range statistic $U(k)$ and detrended fluctuation analysis $Y(k)$ indeed vary as $\sqrt{k}$ for large k and independent intervals with finite variance.

3.8 *Interval- and forward-recurrence-time statistics*   Show that Eq. (3.11) is valid.

3.9 *Interval and point-process spectra*   Show that the interval-based spectrum $S_\tau(f)$ and the point-process spectrum $S_N(f)$ are proportional to each other, at low frequencies, for interevent intervals with $\text{Var}[\tau]/\text{E}^2[\tau] \ll 1$.

3.10 *Normalized variance and coincidence rate*   Prove Eq. (3.52).

3.11 *Normalized variance and point-process spectrum*   Prove Eq. (3.61).

3.12 *Normalized Haar-wavelet variance and point-process spectrum*   Show that Eq. (3.62) is valid.

3.13 *Connection between rate and point-process spectra*   Prove Eq. (3.67), and show that the two spectra approach each other for small values of the product fT. Problem 4.8 explicitly illustrates this connection for the homogeneous Poisson and gamma renewal point processes.

4

Point Processes: Examples

The French physiologist **Louis Lapicque (1866–1952)** conceived the integrate-and-reset point process; it successfully describes the generation of action potentials by a broad variety of neurons and continues to enjoy wide use today.

Sir David R. Cox (born 1924), a British statistician, studied the superposition of periodic series of events and, as part of his work in the textile industry in the 1940s, conceived the doubly stochastic Poisson process.

4.1	**Homogeneous Poisson Point Process**	82
4.2	**Renewal Point Processes**	85
4.3	**Doubly Stochastic Poisson Point Processes**	87
4.4	**Integrate-and-Reset Point Processes**	91
4.5	**Cascaded Point Processes**	93
4.6	**Branching Point Processes**	95
4.7	**Lévy-Dust Counterexample**	95
	Problems	96

Having set forth a collection of measures useful for examining point processes in Chapter 3, we now consider a number of examples. Although the examples we provide do not exhibit fractal behavior *per se*, they do play an important role in the construction of fractal and fractal-rate point processes, as we will see subsequently.

We consider the following processes, in turn[1]: homogeneous Poisson point processes (Sec. 4.1), renewal point processes (Sec. 4.2), doubly stochastic Poisson point processes (Sec. 4.3), integrate-and-reset point processes (Sec. 4.4), cascaded point processes (Sec. 4.5), branching point processes (Sec. 4.6), and Lévy dusts (Sec. 4.7). A broad range of other point processes also finds use in characterizing many diverse phenomena (see, for example, Bartlett, 1955; Parzen, 1962; Cox & Lewis, 1966; Feller, 1971; Lewis, 1972; Srinivasan, 1974; Saleh, 1978; Cox & Isham, 1980; Snyder & Miller, 1991).

4.1 HOMOGENEOUS POISSON POINT PROCESS

We begin with the one-dimensional **homogeneous Poisson process**, which arises under a broad variety of circumstances (Parzen, 1962; Cox, 1962; Haight, 1967; Cox & Isham, 1980). As indicated in Sec. 2.5.2, the definition of this process consists of two parts. First, for some fixed, constant mean rate μ, we have

$$E[N(t+s) - N(s)] = \mu t, \tag{4.1}$$

independent of the times s and t. Second, events in nonoverlapping segments do not depend on one another; formally the two differences

$$N(t_2) - N(t_1) \qquad \text{and} \qquad N(t_4) - N(t_3) \tag{4.2}$$

remain independent for any t_1, t_2, t_3, t_4, satisfying $t_1 < t_2 \leq t_3 < t_4$.

Conny Palm[2] (1943) was the first to point out that this point process is "without aftereffects." As a consequence of its "zero-memory" behavior, both the intervals

---

[1] We consider one-dimensional constructs, although most of these processes have **multidimensional point process** counterparts (see, for example, Fisher, 1972; Cox & Isham, 1980, Chapter 6).

[2] A photograph of Palm is placed at the beginning of Chapter 13.

$\{\tau_k\}$ *and* the counts $\{Z_k\}$ form sequences of independent, identically distributed random variables. Because of its simplicity, the homogeneous Poisson process serves as a benchmark against which other stochastic point processes are often compared. It plays the role that the white Gaussian process enjoys in the realm of continuous-time stochastic processes.

A number of other properties follow from the definition provided above (Cox & Isham, 1980). The times between events follow a decaying exponential probability density function[3]

$$p_\tau(t) = \begin{cases} \mu \exp(-\mu t) & \text{for } t > 0 \\ 0 & \text{otherwise,} \end{cases} \tag{4.3}$$

with associated moments

$$E[\tau^k] = k!/\mu^k. \tag{4.4}$$

In particular, $E[\tau] = \mu^{-1}$ and $\text{Var}[\tau] = \mu^{-2}$ so that $C_\tau = \sqrt{\text{Var}[\tau]}/E[\tau] = 1$; this simple result supports the use of the homogeneous Poisson process as a benchmark.

The interval-based autocorrelation and spectrum assume simple forms as a result of the independence of the intervals:

$$R_\tau(k) = \begin{cases} 2\mu^{-2}, & k = 0 \\ \mu^{-2}, & k \neq 0 \end{cases} \tag{4.5}$$

$$S_\tau(f) = \mu^{-2}\delta(f) + \mu^{-2}, \tag{4.6}$$

while rescaled range and detrended fluctuation analyses follow the forms given in Secs. 3.3.5 and 3.3.6 for independent intervals.

The number of counts over a fixed time follows the distribution set forth by Poisson[4] in 1837:

$$\Pr\{N(t+s) - N(s) = n\} = \Pr\{Z(t) = n\} = (\mu t)^n \exp(-\mu t)/n!. \tag{4.7}$$

Interestingly, this now-famous distribution aroused little interest until 1898 when von Bortkiewicz wrote a monograph providing a whole host of examples[5] for which the Poisson distribution was applicable (see Quine & Seneta, 1987, for a discussion).

The factorial moments of this distribution are

$$E\left\{\frac{[Z(t)]!}{[Z(t)-k]!}\right\} = (\mu t)^k. \tag{4.8}$$

---

[3] Despite the formal distinction between the density and distribution functions, we often refer to both simply as "distributions."

[4] The term "Poisson" conventionally denotes both the point process itself and the distribution of the number of counts in the process. Since the same term refers to two quite different mathematical constructs, we generally use the full terms to avoid confusion: "homogeneous Poisson process" (or "homogeneous Poisson" for short) and "Poisson distribution," respectively.

[5] The most celebrated among these, perhaps, is von Bortkiewicz's (1898) analysis of the number of deaths from horse kicks in the Prussian army.

In particular, $E[Z(t)] = \text{Var}[Z(t)] = \mu t$. Further, we have

$$
\begin{array}{rcll}
F(T) = A(T) & = & 1 & \text{a)} \\[4pt]
R_Z(k,T) & = & \left\{ \begin{array}{ll} \mu T + \mu^2 T^2, & k = 0 \\ \mu^2 T^2, & k \neq 0 \end{array} \right. & \text{b)} \\[10pt]
S_N(f) & = & \mu^2 \delta(f) + \mu & \text{c)} \\[4pt]
G(t) & = & \mu \delta(t) + \mu^2 & \text{d)} \\[4pt]
\text{Var}\,[C_{\psi,N}(a,b)] & = & \mu \int \psi^2(x)\,dx & \text{e)} \\[4pt]
D_q & = & 0. & \text{f)}
\end{array}
\qquad (4.9)
$$

The quantity μ that appears in Eqs. (4.1) and (4.3)–(4.9) takes the same value in each equation.

The homogeneous Poisson process successfully models a whole host of phenomena over short times, including radioactive decay (Sec. 2.5.4), the commencement of telephone conversations at large exchanges (Sec. 13.1), and the times at which falling raindrops hit the ground. These phenomena have in common the combination of events from many independent sources, so that over a short time no single source contributes significantly to the total set of events.

This broad range of applications of the homogeneous Poisson process highlights the convergence property of superpositions of point processes, which we now examine. Formally, we begin with a collection of independent counting processes $\{N_{1,k}(t)\}$, each with a mean rate $\mu_{1,k}$. Consider the sum of the first M of these processes

$$
N_{2,M}(t) \equiv \sum_{k=1}^{M} N_{1,k}(t), \qquad (4.10)
$$

which has a total rate

$$
\mu_{2,M} = \sum_{k=1}^{M} \mu_{1,k}. \qquad (4.11)
$$

Now scale the time axis by a factor of $\mu_0/\mu_{2,M}$, where μ_0 is any fixed constant rate. This yields

$$
N_{3,M}(t) \equiv N_{2,M}(t\mu_0/\mu_{2,M}) = \sum_{k=1}^{M} N_{1,k}(t\mu_0/\mu_{2,M}); \qquad (4.12)
$$

the process $N_{3,M}(t)$ has a rate μ_0 for all M.

In the limit $M \to \infty$, assuming $\mu_{2,M} \to \infty$, the superposition $N_{3,M}(t)$ approaches a homogeneous Poisson process with rate μ_0 (Palm, 1943; Cox & Smith, 1953, 1954; Khinchin, 1955; Grigelionis, 1963; Franken, 1963, 1964; Çinlar, 1972; Franken, König, Arndt & Schmidt, 1981). We can readily understand this from an intuitive point of view: for large M each of the point processes $N_{1,k}(t)$ contributes few events to $N_{3,M}(t)$ over any finite time interval $[0,t)$, and in the limit $M \to \infty$ no single process contributes more than a single event. The events are therefore completely independent of each other, whereupon the homogeneous Poisson process results.

In addition to the zero-memory and superposition approaches considered above, many other routes also lead to the homogeneous Poisson process. One example is sparse random selection from an arbitrary point process (Cox & Isham, 1980), a topic considered further in Sec. 11.2.3.

4.2 RENEWAL POINT PROCESSES

The independence property of the homogeneous Poisson process tells us that the set of intervals between adjacent events $\{\tau_n\}$ are independent and identically distributed. This provides an alternate definition of this process: an independent and identically distributed set $\{\tau_n\}$ with a probability density function specified by Eq. (4.3).

A ready generalization of the homogeneous Poisson process lies in choosing an arbitrary interevent-interval probability density function while retaining its independent and identically distributed features. The result is a **renewal point process**; the name derives from the fact that the process begins anew (undergoes a renewal) at the occurrence of each event. Renewal processes are often used to describe the behavior of parts such as light bulbs since the failure of one part results in its replacement with a replica chosen at random with an identical *a priori* lifetime distribution (Lotka, 1939; Feller, 1941; Doob, 1948; Smith, 1958; Takács, 1960; Parzen, 1962; Cox, 1962; Cox & Isham, 1980). The origins of renewal theory lie in the life tables of the citizens of London and Breslau published in the late 1600s (see Daley & Vere-Jones, 1988, Chapter 1).

As with other point processes treated in this book, we generally consider stationary versions of renewal point processes, in the sense of Eq. (3.2). For renewal point processes only, the term "equilibrium" means stationary, whereas the term "pure" denotes a renewal point processes that begins with an event.

For any renewal point process the sequence of intervals $\{\tau_k\}$ exhibits independence by construction; this leads to simple forms for second-order interval-based statistics. However, such simplicity does not extend to other measures, such as the coincidence rate, the spectrum of the point process, or statistics derived from the sequence of counts, $\{Z_k(T)\}$. Nevertheless, explicit expressions exist that quantify the characteristics of the renewal point process in terms of the interevent-interval probability density function $p_\tau(t)$.

We begin the study of stationary renewal point processes with construction of the coincidence rate. The density function $p_\tau(t)$ itself describes the probability of an event occurring at a time t given an event at the origin, with no intervening events. To obtain the probability of an event occurring at a time t given an event at the origin, with exactly one intervening event, we simply add the two (independent) random variables. The corresponding probability density is then simply the convolution of

$p_\tau(t)$ with itself

$$p_\tau^{\star 2}(t) = p_\tau(t) \star p_\tau(t) = \int_0^t p_\tau(t-s)\, p_\tau(s)\, ds, \tag{4.13}$$

where $\star$ denotes the convolution operation and $p_\tau^{\star 2}(t)$ represents $p_\tau(t)$ convolved with itself.

Continuing in this same way yields

$$p_\tau^{\star n}(t) = p_\tau^{\star(n-1)}(t) \star p_\tau(t) = \int_0^t p_\tau^{\star(n-1)}(t-s)\, p_\tau(s)\, ds \tag{4.14}$$

for precisely $n-1$ intervening events, where $p_\tau^{\star n}(t)$ represents $p_\tau(t)$ convolved with itself n times, and we employ the notational convenience $p_\tau^{\star 0}(t) = \delta(t)$, the Dirac delta function. Summing over all possible numbers of intervening events, normalizing by the conditional probability of an event at $t = 0$, and admitting negative values of t yields the coincidence rate for a renewal point process (Feller, 1971; Lowen & Teich, 1993d),

$$G(t) = \mathrm{E}[\mu] \sum_{n=0}^\infty p_\tau^{\star n}(|t|). \tag{4.15}$$

We can obtain the spectrum of a renewal point process (Lukes, 1961) via the Fourier transform of Eq. (4.15):

$$
\begin{aligned}
S_N(f) &\equiv \int_{-\infty}^\infty e^{-i2\pi ft} \, \mathrm{E}[\mu] \sum_{n=0}^\infty p^{\star n}(|t|)\, dt \\
&= \mathrm{E}[\mu] + 2\mathrm{E}[\mu]\, \mathrm{Re}\left\{ \int_0^\infty e^{-i2\pi ft} \sum_{n=1}^\infty p^{\star n}(t)\, dt \right\} \\
&= \mathrm{E}[\mu] + 2\mathrm{E}[\mu]\, \mathrm{Re}\left\{ \sum_{n=1}^\infty \int_0^\infty e^{-i2\pi ft} p^{\star n}(t)\, dt \right\} \\
&= \mathrm{E}[\mu] + \mathrm{E}^2[\mu]\, \delta(f) + 2\mathrm{E}[\mu]\, \mathrm{Re}\left\{ \sum_{n=1}^\infty \phi_\tau^n(2\pi f) \right\} \\
&= \mathrm{E}^2[\mu]\, \delta(f) + \mathrm{E}[\mu]\, \mathrm{Re}\left\{ \frac{1 - \phi_\tau(2\pi f)}{1 - \phi_\tau(2\pi f)} \right\} + 2\mathrm{E}[\mu]\, \mathrm{Re}\left\{ \frac{\phi_\tau(2\pi f)}{1 - \phi_\tau(2\pi f)} \right\} \\
&= \mathrm{E}^2[\mu]\, \delta(f) + \mathrm{E}[\mu]\, \mathrm{Re}\left\{ \frac{1 + \phi_\tau(2\pi f)}{1 - \phi_\tau(2\pi f)} \right\},
\end{aligned}
\tag{4.16}
$$

where the characteristic function of the interevent intervals $\phi_\tau(\omega)$ is defined in Eq. (3.6), $\mathrm{Re}[z]$ represents the real part of the complex expression z, and the delta function in Eq. (4.16) derives from the constant $\mathrm{E}^2[\mu]$ term in the coincidence rate. In the low-frequency limit (Lowen, 1992), this reduces to (see Prob. 4.5)

$$\lim_{f\to 0} S_N(f) = \mathrm{E}^3[\mu]\, \mathrm{Var}[\tau]. \tag{4.17}$$

Substituting Eq. (4.17) into Eq. (3.64) yields

$$\lim_{T \to \infty} F(T) = \lim_{T \to \infty} A(T) = E^2[\mu] \operatorname{Var}[\tau] = C_\tau^2. \tag{4.18}$$

Feller (1968, Sec. XIII.6, pp. 320–322) obtained this important result for renewal processes by other means.

Making use of Fourier and z transforms (Lowen, 1992) yields an expression for a type of factorial moment for renewal processes (see Sec. A.2.1)

$$\begin{aligned}
E\left\{ \frac{[Z(T) + k - 1]!}{[Z(T) - 1]!} \right\} &= E\{ Z(T) [Z(T) + 1] \cdots [Z(T) + k - 1] \} \\
&= E^{2-k}[\mu] \, k! \int_{0-}^{T} (T - t) \, G^{\star(k-1)}(t) \, dt. \tag{4.19}
\end{aligned}$$

In particular, substituting $k = 1$ into Eq. (4.19) yields

$$E[Z(T)] = E[\mu] \, T, \tag{4.20}$$

a canonical result for all point processes, while $k = 2$ and some algebra provides (see Prob. 4.6)

$$\operatorname{Var}[Z(T)] = \int_{-T}^{T} (T - |t|) \left\{ G(t) - E^2[\mu] \right\} \, dt, \tag{4.21}$$

recalling Eq. (3.52), another result general to all point processes. However, larger values of k in Eq. (4.19) apply only to renewal point processes, and not to general point processes.

Renewal point processes with interevent intervals that have power-law distributions, as considered in Chapter 7, are known as **fractal renewal point processes**. Fractal-based point processes can also be derived from collections of **alternating fractal renewal processes**, as considered in Chapter 8.

4.3 DOUBLY STOCHASTIC POISSON POINT PROCESSES

Another generalization of the homogeneous Poisson process emerges when the rate μ is modulated. The **doubly stochastic Poisson process** results from choosing $\mu(t)$ to be a positive-valued continuous-time stochastic rate process rather than a fixed constant. The resultant process is thus *doubly* random: an (unobserved) source of randomness arises from the fluctuations in the stochastic rate $\mu(t)$ while another source arises from the intrinsic Poisson event-generation fluctuations, given the rate $\mu(t)$.[6]

---

[6] The designation "mixed Poisson process," initially used by Bartlett (1955), signifies that the rate is a random variable (fixed in time) rather than a random process (varying in time). On occasion the term "compound Poisson process" appears in place of "doubly stochastic Poisson process" but this terminology is generally reserved for describing cascaded Poisson processes (see Sec. 4.5).

This stochastic point process was conceived by David Cox (1955) to describe the sequence of stops of a loom in a textile mill.[7] This sequence would ordinarily be expected to form a Poisson process with a fixed rate of stoppage. However, random variations of the quality of the material provided to the loom lead to fluctuations in the stoppage rate. Since its development, the doubly stochastic Poisson process has found wide application in a broad variety of fields (see, for example, Bartlett, 1963; Cox & Lewis, 1966; Lewis, 1972; Grandell, 1976; Saleh, 1978; Cox & Isham, 1980; Saleh & Teich, 1982; Saleh, Stoler & Teich, 1983; Teich & Saleh, 1988; Snyder & Miller, 1991; Teich & Saleh, 2000).

Formally, we have

$$\lim_{\epsilon \to 0} \epsilon^{-1} \Pr \left\{ N(t + \epsilon) - N(t) > 0 \,\middle|\, \mu(t) \right\} = \mu(t). \tag{4.22}$$

Figure 4.1 presents a realization of this point process. The fluctuations exhibited in the rate (a) appear in random form in the ensuing point-process events displayed in (b). Two statistics follow immediately from Eq. (4.22). Taking expectations of both sides yields $E[dN(t)/dt] = E[\mu]$, and integration leads to

$$E[N(t)] = E[\mu]t. \tag{4.23}$$

Similarly, employing Eq. (4.22) at two different times gives rise to the coincidence rate

$$E\left[\frac{dN(s)}{ds} \frac{dN(s+t)}{ds} \right] = E[\mu(s)\,\mu(s+t)]$$

$$G(t) = R_\mu(t) + E[\mu]\,\delta(t), \tag{4.24}$$

where $R_\mu(t)$ denotes the autocorrelation of $\mu(t)$. Taking the Fourier transform leads directly to

$$S_N(f) = S_\mu(f) + E[\mu], \tag{4.25}$$

where $S_\mu(f)$ represents the spectrum of the rate $\mu(t)$.

Other statistics of this point process derive from those of the rate process $\mu(t)$. In parallel with Eqs. (4.7) and (4.8), we have (Saleh, 1978)

$$\Pr\{Z(t) = n\} = E\{\Lambda(t) \exp[-\Lambda(t)]\}/n! \tag{4.26}$$

and

$$E\left\{ \frac{[Z(t)]!}{[Z(t) - k]!} \right\} = E\left[\Lambda^k(t)\right], \tag{4.27}$$

where we have defined the integrated rate

$$\Lambda(t) = \int_0^t \mu(s)\,ds. \tag{4.28}$$

---

[7] The process is also known as a **Cox process**. The appellation "doubly stochastic Poisson process," often abbreviated DSPP, was provided by Bartlett (1963).

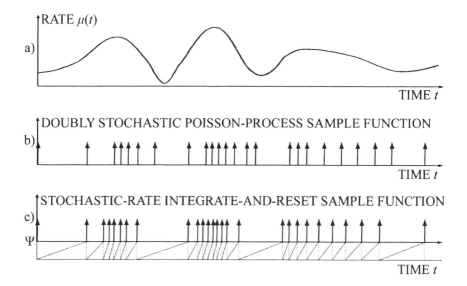

Fig. 4.1   a) Sample function of a stochastic rate $\mu(t)$. This realization serves as the rate for a Poisson point process and an integrate-and-reset point process, as considered in turn below. b) Sample function of the resulting doubly stochastic Poisson process. Events tend to occur more often when the rate $\mu(t)$ assumes larger values, although the randomness introduced by the Poisson process renders this association probabilistic. c) The integral of the rate increases until it reaches a threshold $\Psi = 1$, whereupon it resets to zero and the integration begins anew (gray). The reset times form a point process known as the stochastic-rate integrate-and-reset point process (black).

In particular,

$$\Pr\{Z(t) = 0\} = \mathrm{E}\left\{\exp\left[-\int_0^t \mu(s)\,ds\right]\right\}. \tag{4.29}$$

Equation (4.29), together with Eq. (3.30), yields the interval density

$$
\begin{aligned}
p_\tau(t) &= \mathrm{E}[\tau]\,\frac{d^2}{dt^2}\,\mathrm{E}\left\{\exp\left[-\int_0^t \mu(s)\,ds\right]\right\} \\
&= \frac{1}{\mathrm{E}[\mu]}\,\mathrm{E}\left\{\left[\mu^2(t) - \frac{d\mu(t)}{dt}\right]\exp\left[-\int_0^t \mu(s)\,ds\right]\right\}. \tag{4.30}
\end{aligned}
$$

When the rate process $\mu(t)$ exhibits fluctuations over frequency ranges that are significantly lower than the mean rate $\mathrm{E}[\mu]$, the interval density $p_\tau(t)$ takes a simpler form, and a straightforward expression for the moments of τ emerges (Saleh, 1978):

$$\mathrm{E}[\tau^n] \approx n!\,\mathrm{E}[\mu^{1-n}]/\mathrm{E}[\mu] \tag{4.31}$$

$$p_\tau(t) \approx \mathrm{E}[\mu^2 \exp(-\mu t)]/\mathrm{E}[\mu]. \tag{4.32}$$

A particularly simple result obtains when the coefficient of variation C_μ of this rate process $\mu(t)$ becomes small in comparison with unity. As $C_\mu \to 0$, the stochastic rate process $\mu(t)$ approaches a constant, deterministic value. The expectation operators in Eq. (4.32) then become superfluous, whereupon it simplifies to Eq. (4.3). Moreover, since the rate remains constant, the resulting point process $dN(t)$ becomes the homogeneous Poisson process. For rate processes $\mu(t)$ with a small, but nonzero, coefficient of variation $(0 < C_\mu \ll 1)$, the interval density approaches the exponential form

$$p_\tau(t) \approx \begin{cases} \mathrm{E}[\mu] \exp\left(-\mathrm{E}[\mu]\, t\right) & \text{for } t > 0 \\ 0 & \text{otherwise.} \end{cases} \qquad (4.33)$$

However, in this case the variation of $\mu(t)$ imparts memory to $dN(t)$, so that the *ordering* of the intervals differs from that of the homogeneous Poisson process. The resulting point process is therefore nonrenewal.

Although most doubly stochastic Poisson processes are, in fact, nonrenewal in nature, the classes of renewal point processes and doubly stochastic Poisson processes do intersect. In particular, for any distribution $P(t)$, and any positive constant time t_c, the quantity

$$\phi_\tau(\omega) = \left[1 + i\omega t_c + \int_0^\infty \left(1 - e^{-i\omega t}\right) dP(t)\right]^{-1} \qquad (4.34)$$

defines the characteristic function of an interevent-interval distribution $P_\tau(t)$. A renewal point process constructed with this interevent-interval distribution will also be a doubly stochastic Poisson point processes (Grandell, 1976). Thus, two quite different models generate identical behavior. Conversely, it is impossible to distinguish between the two models in this case, even with a full description of the process itself (the probabilities of all possible outcomes over all time).

This highlights the paucity of information generally available in point processes in comparison with continuous functions of time. The set of event times contained within a finite interval, for example, completely describes the point process during that interval. With probability one, this set has a finite dimension. In contrast, the set of continuous functions over any finite interval forms an infinite-dimensional set. As we shall see in Chapter 12, this relative sparseness makes identification of an underlying model quite difficult in practice.

Point processes whose rate functions comprise stochastic processes are examined in Chapter 10. In particular, if the rate function is **shot noise** (see Chapter 9), the associated doubly stochastic Poisson process is known as a **shot-noise-driven doubly stochastic Poisson process** (Saleh & Teich, 1982). If the rate function is **fractal shot noise** (see Chapter 9), the process is called a **fractal-shot-noise-driven doubly stochastic Poisson process** (Lowen & Teich, 1991). Point processes in this class find use in characterizing a multitude of phenomena in the physical and biological sciences.

4.4 INTEGRATE-AND-RESET POINT PROCESSES

The **integrate-and-reset**, or **integrate-and-fire**, model was introduced nearly a century ago by Lapicque (1907, 1926). This simple nonlinear construct provides a direct route for transforming a rate function into a point process.

Integrate-and-reset point processes play an important role in modeling biophysical phenomena, particularly neural spike trains, since the paradigm offers not only a suitable mathematical model, but also a plausible physiological model for the underlying behavior (Eccles, 1957; Holden, 1976; Tuckwell, 1988; Koch, 1999). These processes are closely related to **oversampled sigma-delta modulators** in the domain of signal processing, where they are used for analog-to-digital conversion (Norsworthy, Schreier & Temes, 1996).

Like the doubly stochastic Poisson process, the integrate-and-reset point process depends on a stochastic rate $\mu(t)$. However, in this case the sole source of randomness manifested in the generated point process arises from fluctuations associated with the rate $\mu(t)$; the integrate-and-reset algorithm introduces no additional randomness of its own.

The algorithm generates an event each time the integral of the rate $\mu(t)$ reaches a value of unity. It then resets the integrated value to zero whereupon the process begins anew. Formally, we have

$$t_{k+1} = \inf_{u > t_k} \left\{ u : \int_{t_k}^{u} \mu(s)\, ds = 1 \right\}, \tag{4.35}$$

where $\{t_k\}$ again represents the set of times at which the events occur (rather than the times between events). A realization of this point process is presented in Fig. 4.1, where the resulting point-process events (c) are seen to faithfully follow the fluctuations of the rate (a) to within the resolution of the point process.

The absence of additional randomness leads to trivial results for simple forms of the rate $\mu(t)$; in particular, for $\mu(t)$ a fixed constant value μ, the resulting point process comprises a perfectly periodic train of events spaced from each other by $1/\mu$. The faithfulness of the transformation also leads to a close correspondence between the second-order measures of $\mu(t)$ and those of $dN(t)$, particularly for large times (low frequencies). For example, when $f \ll \mathrm{E}[\mu]$, we obtain

$$S_N(f) \approx S_\mu(f), \tag{4.36}$$

in contrast to the corresponding result for the doubly stochastic Poisson process, Eq. (4.25), which contains an additional term, $\mathrm{E}[\mu]$, associated with the intrinsic randomness of the underlying Poisson process. Exact expressions prove difficult to obtain for the integrate-and-reset process, however, by virtue of its inherent nonlinearity.

Again, simple forms emerge for the interevent-interval statistics when the rate process $\mu(t)$ exhibits fluctuations significantly slower than the mean rate of events $\mathrm{E}[\mu]$. In the spirit of Eq. (4.32), we begin with the interevent interval probability density function for a fixed rate process μ, namely $\delta(t - 1/\mu)$, and include appropriate weighting and normalization factors,

$$
\begin{aligned}
p_\tau(t) &\approx \mathrm{E}[\mu\,\delta(t-1/\mu)]\,/\mathrm{E}[\mu] \\
&= \int_0^\infty y\,\delta(t-1/y)\,p_\mu(y)\,dy/\mathrm{E}[\mu] \\
&= \mathrm{E}[\mu]^{-1}\int_0^\infty y^2\,t^{-1}\,\delta(y-1/t)\,p_\mu(y)\,dy \\
&= \mathrm{E}[\mu]^{-1}\,t^{-3}\,p_\mu(1/t), \tag{4.37}
\end{aligned}
$$

where we have made use of a particular property of the Dirac delta function, namely $\delta(ax) = a^{-1}\delta(x)$. We note that under these conditions the output point process is generally not renewal because the reset mechanism does not erase the history of the input process, which is preserved through the fluctuations in $\mu(t)$.

We can directly obtain the moments of the interval density from the moments of the rate:

$$
\begin{aligned}
\mathrm{E}[\tau^n] &\approx \mathrm{E}[\mu]^{-1}\int_0^\infty t^{n-3}\,p_\mu(1/t)\,dt \\
&= \mathrm{E}[\mu]^{-1}\int_0^\infty y^{1-n}\,p_\mu(y)\,dy \\
&= \mathrm{E}[\mu^{1-n}]/\mathrm{E}[\mu]. \tag{4.38}
\end{aligned}
$$

In particular, Eq. (4.38) yields the interval standard deviation, and thence the coefficient of variation, for an ideal integrate-and-reset point process driven by an arbitrary stochastic rate, provided that the fluctuations of the rate process are sufficiently slow (see Prob. 4.9).

For simplicity, we chose the threshold to be unity ($\Psi = 1$) in Eq. (4.35). Any positive value would have sufficed without changing the nature of the process, serving only to divide the rate by the new threshold.

On the other hand, if the threshold varies in time, the character of the resulting integrate-and-reset point process $dN(t)$ changes substantially. The rationale for considering models of this form stems from early neurophysiological experiments in which it was demonstrated that a sequence of identical brief electric currents applied to a neuron near threshold elicited axonal action-potential responses only in a fraction of the trials, in random fashion (Blair & Erlanger, 1932, 1933). Behavior of this kind has been ascribed to fluctuations in threshold (Pecher, 1939; Verveen, 1960; Holden, 1976), also referred to as "fluctuations in excitability."

A sinusoidally varying threshold, for example, imparts its fluctuations to $dN(t)$, albeit in a nonlinear fashion. The complex interplay between the rate and the threshold, when both vary, forms a rich field of study. We consider two examples with variable thresholds in this text. The first comprises a threshold that remains fixed during each integration, but assumes an independent, unit-mean exponential random value for every event. The resulting point process is then indistinguishable from a doubly stochastic Poisson process with the same rate function, by virtue of the exponential interevent-interval density function for the homogeneous Poisson process.

The other example, which appears in Sec. 6.6, gives rise to a fractal-based point process from regular Brownian motion; it has found use in modeling the fractal-rate fluctuations of neural spike trains.

Finally, we mention a generalization of this process, known as the **leaky integrate-and-reset process** (Lapicque, 1907; Holden, 1976; Tuckwell, 1988; Park & Gray, 1992). In this case, an internal state variable x increases at a rate proportional to the instantaneous rate $\mu(t)$, while simultaneously decaying to zero with a time constant t_c. When x reaches unity, an output event is generated, x is reset to zero, and the cycle begins anew. The state equation is written as

$$dx/dt = \mu(t) - x/t_c, \tag{4.39}$$

or, equivalently,

$$x(t) = \exp(-t/t_c) \int_0^t \mu(s) \, \exp(s/t_c) \, ds. \tag{4.40}$$

In the limit $t_c \to \infty$ we recover the behavior of Eq. (4.35). The added flexibility provided by the formalism of Eqs.(4.39) and (4.40) proves useful in some applications; however, the added mathematical complexity does not warrant our considering it further here.

4.5 CASCADED POINT PROCESSES

Cascaded point processes, also known as **cluster point processes** (Neyman & Scott, 1958, 1972; Saleh & Teich, 1983) and **compound point processes** (Feller, 1968, Chapter 12), arise when each event of a point process forms the nucleus for a sequence of secondary point-process segments.

In the most general formulation, each event t_k of the primary point process $dN_1(t)$ initiates a secondary point process $dN_{2,k}(t)$, which terminates after a random number $M_k = N_{2,k}(\infty)$ of events. All secondary points, taken together as indistinguishable events, form the output point process $dN_3(t)$. Figure 4.2 illustrates this construct. The primary events can be excluded or included with the secondary processes.

The statistics for a cascaded point process necessarily depend on the details of the primary and secondary processes that comprise it. General closed-form expressions do not exist, with a single exception: the mean rate for a stationary process. Consider a primary point process $dN_1(t)$ that generates events at a mean rate $E[\mu_1]$. Each such primary event initiates a secondary process with a mean number $E[M_k]$ events. For the mean rate $E[\mu_3]$ of the cascaded point process $dN_3(t)$ itself, we arrive at

$$E[\mu_3] = E[\mu_1] \, E[M_k]. \tag{4.41}$$

Within the class of cascaded point processes, two forms have been studied extensively. For both, the homogeneous Poisson process forms the primary process and the clusters are independent of each other. In the **Neyman–Scott cluster process**

a) PRIMARY PROCESS $dN_1(t)$

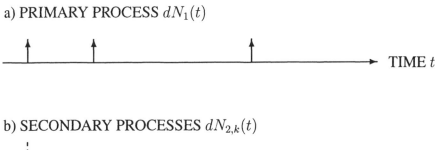

b) SECONDARY PROCESSES $dN_{2,k}(t)$

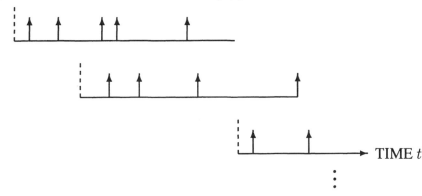

c) CASCADED POINT PROCESS $dN_3(t)$

Fig. 4.2 Generation of a cascaded point process. Each event of a primary point process $dN_1(t)$ (a) initiates a secondary point process $dN_{2,k}(t)$ (b) that terminates after a random number of events. All secondary points, taken together as indistinguishable events, form the output, which is a cascaded point process $dN_3(t)$ (c). The primary events may be excluded or included in the output. Cascaded point processes are also known as compound processes or cluster processes.

(Neyman & Scott, 1958; Bartlett, 1964; Vere-Jones, 1970; Neyman & Scott, 1972; Saleh & Teich, 1982, 1983; Daley & Vere-Jones, 1988), the times between each secondary event and its corresponding primary event are independent and identically distributed. In the **Bartlett–Lewis cascaded process**[8] (Bartlett, 1963; Lewis, 1964, 1967), each primary event initiates a segment of a renewal point process, so that the times between adjacent events from a given secondary process are independent and

---

[8] Photographs of Neyman and Bartlett can be found at the beginning of Chapter 10.

identically distributed. Both forms of cascaded processes are useful for generating fractal-rate point processes, as considered in Chapters 10 and 13.

Some cascaded point processes are isomorphic to doubly stochastic point processes, providing two different ways of viewing the same mathematical object (see, for example, Quenouille, 1949; Gurland, 1957; Bartlett, 1964; Lawrance, 1972; Neyman & Scott, 1972; Cox & Isham, 1980; Saleh & Teich, 1982; Teich & Diament, 1989; Lowen & Teich, 1991). The shot-noise-driven doubly stochastic Poisson point process, for example, is a particular Neyman–Scott cluster process, as discussed in Chapter 10.

4.6 BRANCHING POINT PROCESSES

Cascading need not be limited to two stages. Consider, for example, a shot-noise-driven doubly stochastic Poisson point process followed by a linear filter that converts the pulsatile sequence of events into a stochastic rate function suitable for driving a succeeding Poisson process, which is followed by another linear filter, and so on. The result is a cascade of shot-noise-driven doubly stochastic Poisson processes. The multifold statistics of the events at the output of an arbitrary number of such stages have been determined (Matsuo, Saleh & Teich, 1982). In accordance with expectations, the greater the number of stages, the larger the variability at the output.

The stages can also be constructed in such a way as to comprise **Thomas point processes** (Matsuo, Teich & Saleh, 1983), so that trigger events are carried forward from each stage to the next. In the limit as the number of such cascaded Thomas stages increases to infinity, while the mean number of added events per event of the previous stage becomes infinitesimal, the result converges to a **Poisson branching process** (Matsuo, Teich & Saleh, 1984). In particular, when the branching is instantaneous, the limit of continuous branching yields the **Yule–Furry branching process** with an initial Poisson population (Matsuo et al., 1984). The theory of branching processes, originally developed in connection with the survival of family names, has a long and august history in the annals of mathematics (Bienaymé, 1845; Watson & Galton, 1875; Yule, 1924; Furry, 1937; Kolmogorov & Dmitriev, 1947; Kendall, 1949, 1975; Harris, 1989).

4.7 LÉVY-DUST COUNTEREXAMPLE

We conclude this chapter with an example of a random collection of points that does *not* comprise an orderly point process. Mandelbrot (1982) coined the term **Lévy dust** to describe a particular random collection of points on a line segment. Its definition follows. Consider a finite-length segment of such a set, and count all the intervals between adjacent points that exceed a value ϵ; this number varies as ϵ^{-c} for some $c > 0$. All such intervals are independent of each other. Thus, the interval distribution exhibits scaling behavior and Lévy dusts belong to the class of one-dimensional fractal objects; indeed, they resemble randomized versions of the Cantor set.

Moreover, Lévy dusts resemble fractal renewal point processes (Chapter 7) in their interval independence and in their scaling behavior. However, these sets do not form orderly point processes. As ϵ becomes smaller, the number of intervals grows without limit; an infinite number of intervals lie in many segments. Consequently, all but an infinitesimal fraction of the intervals have a vanishing length. In particular, in any neighborhood about any point in the set, an infinite number of other points exist, violating the definition of an orderly point process.

Problems

4.1   *Normalized variance for an integrate-and-reset point process*   Consider an integrate-and-reset point process with constant rate. All interevent intervals τ_k therefore assume the same fixed value τ, for all k, for this perfectly periodic point process. To render this point process stationary, while maintaining its periodic form, we randomize the absolute times. The time between the origin and the event that follows it is taken to be uniformly distributed in the interval $(0, \tau)$. Equivalently, the forward recurrence time from any fixed time s selected independently of the point process is given by

$$p_\tau(t) = \begin{cases} 1/\tau & 0 < t < \tau \\ 0 & \text{otherwise.} \end{cases} \tag{4.42}$$

4.1.1. Find an expression for the count variance, $\mathrm{Var}[Z(T)]$, and show that it cannot exceed $\frac{1}{4}$.

4.1.2. Find an expression for the normalized variance, $F(T)$.

4.2   *Time statistics of the homogeneous Poisson process*   Let $dN(t)$ represent a homogeneous Poisson process with rate μ. Now choose a time v independently of $dN(t)$.

4.2.1. What is the probability density of the time remaining to the next event (forward recurrence time)?

4.2.2. What is the probability density of the time since the last event (the backward recurrence time)?

4.2.3. What is the probability density of the interval τ_* within which v lies?

4.2.4. Explain the difference between the expression derived immediately above and that in Eq. (4.3).

4.3   *Generalized dimensions for a homogeneous Poisson process*   As indicated in Sec. 3.5.4, the generalized dimensions D_q assume integer values for nonfractal point processes. Consider the case of a homogeneous Poisson process $dN(t)$ with rate μ. Calculate expressions for $\mathrm{E}\left[\sum_k Z_k^q(T)\right]$ for a segment of that process of duration L. For both $q = 0$ and $q = 2$, find the associated limiting forms for large and small values of T. Verify, for both values of q, that $D_q = 0$ in the sense of Eq. (3.70), and that $D_q = 1$ in the sense of Eq. (3.72), thereby confirming that the homogeneous Poisson process is a nonfractal point process.

4.4   *Renewal process with exponential interval density*   Use the steps listed below to demonstrate that a renewal point process constructed from exponentially distributed random variables, as in Eq. (4.3), satisfies Eqs. (4.1) and (4.2), and therefore must coincide with the homogeneous Poisson point process.

4.4.1. Show that the point-process spectrum follows the form of Eq. (4.9c) and use Eq. (3.59) to obtain the mean rate.

4.4.2. Show that Eq. (4.9d) provides the coincidence rate of this constructed process and that this establishes that nonoverlapping intervals are uncorrelated.

4.4.3. Extend this result to establish independence.

4.5   *Renewal-process spectrum at low frequencies*   Prove Eq. (4.17).

4.6   *Count variance for renewal point processes*   Show that substituting $k = 2$ in Eq. (4.19) indeed yields Eq. (4.21).

4.7   *Gamma renewal point process*   A gamma probability density function takes the form (Parzen, 1962; Cox & Isham, 1980)

$$p_\tau(t) = \begin{cases} [\Gamma(m)]^{-1} \tau_0^{-m} t^{m-1} \exp(-t/\tau_0) & t > 0 \\ 0 & t \le 0, \end{cases} \tag{4.43}$$

where m is the *order* of the process[9] and $\Gamma(x)$ represents the (complete) Eulerian gamma function

$$\Gamma(x) \equiv \int_0^\infty t^{x-1} e^{-t} \, dt. \tag{4.44}$$

In general, the order of the gamma density can assume any positive real value, $0 < m < \infty$.

4.7.1. Find the mean, variance, skewness, and kurtosis of the random variable associated with the probability density function in Eq. (4.43).

4.7.2. Suppose we construct a renewal point process using an interevent-interval probability density function specified by Eq. (4.43). Find the corresponding point-process spectrum.

4.7.3. For the particular case $m = 2$, find the coincidence rate $G(t)$ as well as the count-based normalized variance $F(T)$ and normalized Haar-wavelet variance $A(T)$.

4.8   *Point-process and rate spectra for gamma renewal point processes*   Equation (3.67) relates the rate spectrum $S_\lambda(f, T)$ to the point process spectrum $S_N(f)$. We examine this relation for two examples from the gamma-renewal-process family: $m = 1$ (the homogeneous Poisson process), and $m = 2$.

4.8.1. Calculate $S_\lambda(f, T)$ for the homogeneous Poisson process, and show that $S_\lambda(f, T)$ and $S_N(f)$ coincide in this particular case when $|f| < 1/T$.

---

[9] The integer-order gamma density is sometimes called the Erlang density, in honor of the Danish electrical engineer who used it to characterize waiting times associated with telephone calls (see Chapter 13 for a discussion of this issue and for a photograph of Erlang). This special form of the gamma density, along with its derivation, was known earlier (Ellis, 1844), but only in terms of a Gaussian approximation.

4.8.2. Repeat this exercise for the gamma renewal point process with m $= 2$, and show that $S_\lambda(f, T)$ and $S_N(f)$ no longer agree.

4.8.3. Verify that the two measures do coincide, however, in the limit of small counting times T.

4.9   *Integrate-and-reset process with a gamma-distributed rate*   An integrate-and-reset process is driven by a rate $\mu(t)$ that varies much more slowly than the time scale corresponding to the longest interevent interval.

4.9.1. Find an expression for the coefficient of variation C_τ for the interevent intervals, as defined in Eq. (3.5).

4.9.2. Evaluate C_τ for the special case of a gamma-distributed rate

$$
p_\mu(x) = \begin{cases} [\Gamma(m)]^{-1} \mu_0^{-m} x^{m-1} \exp(-x/\mu_0) & x > 0 \\ 0 & x \le 0, \end{cases} \tag{4.45}
$$

with m > 1 and μ_0 a fixed, deterministic rate; the gamma function is defined in Prob. 4.7.

4.9.3. Why do we require m > 1?

4.10   *Sinusoidally modulated point processes*   Suppose we have a random rate defined by

$$
\mu(t) = \mu_0 \big[1 + \cos(\omega_0 t + \theta) \big], \tag{4.46}
$$

with the random phase angle θ uniformly distributed in $(0, 2\pi]$ and μ_0 and ω_0 fixed, deterministic parameters. This renders the process ergodic and, in particular, stationary.

4.10.1. Let Eq. (4.46) be the rate of a Poisson point process. Find the mean value, coincidence rate, and count-based normalized variance for this doubly stochastic Poisson process.

4.10.2. Now let Eq. (4.46) serve as the rate for an integrate-and-reset point process, and assume that $\mu_0/\omega_0 \gg 1$. Ignoring values of μ_0 such that $2\pi\mu_0/\omega_0$ assumes a rational number, find an approximate expression for the interevent-interval probability density.

4.10.3. Attempt to calculate $E[\tau^2]$ and explain which assumption breaks down in the process. Modify Eq. (4.46) to rectify the problem.

4.11   *Cascaded point process with Poisson primaries and secondaries*   Let $dN_1(t)$ represent a homogeneous Poisson point process with rate μ_1. Suppose that every event k of $dN_1(t)$ triggers a secondary point process $dN_{2,k}(t)$, which has a duration τ_0 during which it generates events with a constant rate μ_2 as a segment of a homogeneous Poisson point process. Let all secondary points, taken together as indistinguishable events, form an output point process $dN_3(t)$. Assume that $\mu_1\tau_0 \ll 1$ so that we can safely ignore edge effects. By virtue of the memoryless property of the homogeneous Poisson process, this particular cascaded point process belongs both to the Neyman–Scott and Bartlett–Lewis families; choosing a random time or a random number of events is equivalent.

4.11.1. Find the mean, variance, and normalized variance of the number of points in an interval T such that $T/\tau_0 \gg 1$.

4.11.2. Imagine now that we modulate the primary process rate $\mu_1(t)$ to transmit information, setting $\mu_1(t) = \mu_1$ or $\mu_1(t) = 0$, and count the number of events $N_3(T)$. Here a $\mu_1(t) = \mu_1$ corresponds to a binary one, and $\mu_1(t) = 0$ corresponds to a binary zero. Find the probabilities of detecting a "one" when a "zero" was sent (an error known as a "false alarm"), and vice versa (an error known as a "miss").

5

Fractal and Fractal-Rate Point Processes

Jean-Baptiste Joseph Fourier (1768–1830) demonstrated that a time function could be constructed from a superposition of harmonic functions of different frequencies; the "Fourier transform" forms the basis of spectral analysis.

The Hungarian mathematician **Alfréd Haar (1885–1933)**, in his doctoral dissertation under David Hilbert, introduced a collection of simple orthogonal functions; these "Haar wavelets" initiated the field of time–scale analysis.

5.1	**Measures of Fractal Behavior in Point Processes**	103
	5.1.1 Spectrum	103
	5.1.2 Normalized Haar-wavelet variance	103
	5.1.3 Normalized variance	105
	5.1.4 Coincidence rate	105
	5.1.5 Count autocorrelation	106
	5.1.6 Scaling cutoffs and fractal-exponent estimates	107
5.2	**Ranges of Power-Law Exponents**	107
	5.2.1 Negative values of α	107
	5.2.2 Observed values of α	109
	5.2.3 Limited range of the normalized variance exponent	109
	5.2.4 Range of the normalized Haar-wavelet-variance exponent	111
	5.2.5 Range of the normalized general-wavelet-variance exponent	113
5.3	**Relationships among Measures**	114
5.4	**Examples of Fractal Behavior in Point Processes**	115
	5.4.1 $1/f$ noise	115
	5.4.2 Normalized rate spectrum	116
	5.4.3 Normalized Haar-wavelet variance	117
	5.4.4 Normalized Daubechies-wavelet variance	120
5.5	**Fractal-Based Point Processes**	120
	5.5.1 Fractal point processes	121
	5.5.2 Fractal-rate point processes	123
	5.5.3 Nonfractal point processes	124
	5.5.4 Identification of fractal-based point processes	125
	Problems	126

As described in Chapter 2, fractals are objects whose measures exhibit scaling. We introduced point processes, along with their appurtenant measures, in Chapter 3, and set forth various examples in Chapter 4. With the fundamental properties of fractals and point processes in hand, we are now in a position to investigate the intersection of these two concepts.

In this chapter we consider various measures that are often used to establish the presence of fractal behavior in point processes. The spectrum and normalized Haar-wavelet variance turn out to be the measures of choice, as we will show. The mathematical techniques bequeathed to us by Fourier (1822) and Haar (1910) thus play especially important roles in the analysis of fractal-based point processes. By way of example, we examine a number of point processes in the biological and physical sciences using these preferred measures.

We then compare and contrast two general classes of point processes that exhibit fractal behavior: fractal point processes and fractal-rate point processes. We conclude by touching briefly on the process of deciding which point process might best describe an observed sequence of events.

5.1 MEASURES OF FRACTAL BEHAVIOR IN POINT PROCESSES

As shown in Sec. 2.2, a measure that exhibits scaling, when considered as a function of time or frequency, indicates power-law behavior. Power-law behavior, in turn, is often a harbinger of fractal behavior.

As demonstrated in Chapter 3, various relationships exist that link different measures of a point process. In principle, explicit knowledge of one such statistic leads directly to an exact form for another. Measures so linked might thus be expected to provide the same information, although in different form.[1] Power-law behavior is generally preserved among these measures since the relationships linking them generally involve integration, differentiation, Fourier transformation, and multiplication by integer powers of the argument of the measure.

5.1.1 Spectrum

In forging interrelationships among the various measures, we begin with the point-process spectrum $S_N(f)$ introduced in Sec. 3.5.2 (the reason that we begin with this measure will become apparent subsequently). Fractal behavior suggests itself when this quantity assumes the form

$$S_N(f) \approx (f/f_S)^{-\alpha} \tag{5.1}$$

over a range of frequencies (see Sec. 2.3), where f is taken to be positive.

In general, the power-law behavior of a statistic includes an exponent ($-\alpha$ in the case at hand) that characterizes the *relative strength* of the fluctuations at different frequencies (or times), as well as a multiplicative constant (f_S^α) that indicates the *absolute strength* of the fluctuations at all times (or frequencies).

For a point process with fractal characteristics, the value of $S_N(f)$ typically becomes larger as the frequency decreases, and an increasing share of fractal fluctuations is admitted. For spectral measures, power-law exponents therefore generally take on negative values, as shown in Eq. (5.1); we discuss this issue further in Sec. 5.2.1.

For similar reasons, we also observe negative values of the power-law exponents for measures that depend on a delay parameter, such as the count-based autocorrelation (which is a function of count index k) and the coincidence rate: correlation typically decreases with increasing delay.

5.1.2 Normalized Haar-wavelet variance

Given the spectrum $S_N(f)$, for all frequencies f, we obtain the normalized Haar-wavelet variance $A(T)$, for all counting times T, via Eq. (3.62). $A(T)$, which relies

---

[1] Actually, different measures are not entirely equivalent since they are subject to different inherent mathematical limitations, such as those discussed in Sec. 5.2. Moreover, real and finite data sets are affected by bias and variance that are not the same for all measures, as considered in detail in Chapter 12. These considerations lead us to conclude that the rate spectrum and normalized Haar-wavelet variance are generally the measures of choice.

on the simplest of wavelet basis functions (Haar, 1910), is constructed in accordance with the recipe provided in Sec. 3.4.3.

We now demonstrate that power-law behavior in $S_N(f)$ leads directly to power-law behavior in $A(T)$, with a related (but different) exponent and multiplicative constant:

$$A(T) \approx (T/T_A)^\alpha. \tag{5.2}$$

We begin with a spectrum $S_N(f)$ that varies in a power-law fashion, with exponent $-\alpha$ where $0 < \alpha < 1$:

$$S_N(f) = \mathrm{E}^2[\mu]\,\delta(f) + \mathrm{E}[\mu]\left[1 + (f/f_S)^{-\alpha}\right]. \tag{5.3}$$

Inserting Eq. (5.3) into Eq. (3.62) provides

$$
\begin{aligned}
A(T) &= \frac{4}{\pi^2\,\mathrm{E}[\mu]\,T}\int_{0+}^{\infty} S_N(f)\,\sin^4(\pi f T)\,f^{-2}\,df \\
&= \frac{4}{\pi^2 T}\int_0^{\infty}\left[1 + (f/f_S)^{-\alpha}\right]\sin^4(\pi f T)\,f^{-2}\,df \\
&= \frac{4}{\pi}\int_0^{\infty}\left[1 + (\pi f_S T/x)^\alpha\right]\sin^4(x)\,x^{-2}\,dx \\
&= 1 + (4/\pi)(\pi f_S T)^\alpha\,2^\alpha(1 - 2^{\alpha-1})\,\Gamma(1-\alpha)\,\sin(\pi\alpha/2)\,/\,[\alpha(\alpha+1)] \\
&= 1 + \frac{(2\pi f_S T)^\alpha\,(2 - 2^\alpha)\,2\sin(\pi\alpha/2)\,\Gamma(1-\alpha)}{\alpha(\alpha+1)\pi} \\
&\qquad\times \frac{\sin(\pi\alpha)}{2\sin(\pi\alpha/2)\,\cos(\pi\alpha/2)}\times\frac{\pi}{\Gamma(1-\alpha)\,\Gamma(\alpha)\,\sin(\pi\alpha)} \tag{5.4} \\
&= 1 + \frac{(2 - 2^\alpha)\,(2\pi f_S T)^\alpha}{\alpha(\alpha+1)\,\Gamma(\alpha)\,\cos(\pi\alpha/2)} \\
&= 1 + \frac{(2 - 2^\alpha)\,(2\pi f_S)^\alpha}{\Gamma(\alpha+2)\,\cos(\pi\alpha/2)}\,T^\alpha \tag{5.5} \\
&= 1 + (T/T_A)^\alpha, \tag{5.6}
\end{aligned}
$$

which accords with Eq. (5.2) in the power-law regime. The notation $0+$ indicates that the integral excludes the delta function at $f = 0$, and the quantity $\Gamma(x)$ represents the (complete) Eulerian gamma function

$$\Gamma(x) \equiv \int_0^{\infty} t^{x-1}\,e^{-t}\,dt, \tag{5.7}$$

which we first met in Prob. 4.7. Both fractions following the multiplication signs in Eq. (5.4) are identically unity. For the first, this follows from the well-known double-angle trigonometric identity with an angle of $\pi\alpha/2$ whereas for the second it follows from a property of the Gamma function (Gradshteyn & Ryzhik, 1994, Eq. 8.334.3):

$$\Gamma(x)\,\Gamma(1-x)\,\sin(\pi x) = \pi. \tag{5.8}$$

Comparing Eqs. (5.5) and (5.6) provides the constant T_A in terms of α and f_S:

$$\frac{1}{T_A^\alpha} = \frac{(2\pi f_S)^\alpha (2 - 2^\alpha)}{\Gamma(\alpha + 2)\,\cos(\pi\alpha/2)}. \tag{5.9}$$

Evidently, the fractal exponent $-\alpha$ in the spectrum, Eq. (5.3), transforms to the fractal exponent $+\alpha$ in the normalized Haar-wavelet variance, Eq. (5.6). These exponents are identical in magnitude but opposite in sign. Increasing the counting time typically increases the value of $A(T)$ for a point process with fractal characteristics so that power-law exponents for this measure generally take on positive values (see Sec. 5.2.1 for a further discussion of this issue). As the counting time for the Haar-wavelet variance increases, fractal fluctuations over larger and larger time scales are accessed by this measure.

Similar results obtain when generalizing the normalized Haar-wavelet variance $A(T)$ to an arbitrary wavelet basis.

5.1.3 Normalized variance

In a similar way, we obtain the normalized variance $F(T)$ from the spectrum $S_N(f)$ using Eq. (3.61). This quantity is constructed in accordance with the approach indicated in Sec. 3.4.2.

In this case we obtain

$$
\begin{aligned}
F(T) &= \frac{2}{\pi^2\,\mathrm{E}[\mu]\,T} \int_{0+}^\infty S_N(f)\,\sin^2(\pi fT)\,f^{-2}\,df \\
&= \frac{2}{\pi^2 T} \int_0^\infty \left[1 + (f/f_S)^{-\alpha}\right] \sin^2(\pi fT)\,f^{-2}\,df \\
&= \frac{2}{\pi} \int_0^\infty \left[1 + (\pi f_S T/x)^\alpha\right] \sin^2(x)\,x^{-2}\,dx \\
&= 1 + \frac{(2\pi f_S)^\alpha}{\Gamma(\alpha + 2)\,\cos(\pi\alpha/2)}\,T^\alpha \\
&= 1 + (T/T_F)^\alpha,
\end{aligned}
$$

$$\tag{5.10}$$
$$\tag{5.11}$$

where the cutoff time T_F is implicitly defined by Eqs. (5.10) and (5.11).

5.1.4 Coincidence rate

The coincidence rate $G(t)$ is related to the point-process spectrum $S_N(f)$ through a simple Fourier transform, as provided by Eq. (3.58).

In this case, as well, a power-law form for the coincidence rate emerges, along with its associated parameter t_G:

$$G(t) = \int_{-\infty}^\infty S_N(f)\,e^{i2\pi ft}\,df$$

$$= \int_{-\infty}^{\infty} \left\{ E^2[\mu] \, \delta(f) + E[\mu] + E[\mu] \, |f/f_S|^{-\alpha} \right\} e^{i2\pi ft} \, df$$

$$= E^2[\mu] + E[\mu] \, \delta(f) + 2E[\mu] \int_0^{\infty} (f/f_S)^{-\alpha} \cos(2\pi ft) \, df$$

$$= E[\mu] \, \delta(f) + E^2[\mu] + 2E[\mu] \, f_S^{\alpha} (2\pi|t|)^{\alpha-1} \int_0^{\infty} x^{-\alpha} \cos(x) \, dx$$

$$= E[\mu] \, \delta(f) + E^2[\mu] + 2E[\mu] \, f_S^{\alpha} (2\pi|t|)^{\alpha-1} \frac{\pi}{2\Gamma(\alpha) \cos(\pi\alpha/2)}$$

$$= E[\mu] \, \delta(f) + E^2[\mu] + E^2[\mu] \frac{(2\pi f_S)^{\alpha}}{2\Gamma(\alpha) \cos(\pi\alpha/2) \, E[\mu]} |t|^{\alpha-1}$$

$$= E[\mu] \, \delta(t) + E^2[\mu] \left[1 + (|t|/t_G)^{\alpha-1} \right]. \tag{5.12}$$

5.1.5 Count autocorrelation

Finally, we determine the count autocorrelation by using Eqs. (5.12) and (3.54):

$$
\begin{aligned}
R_Z(k, T) &= \int_{-T}^{T} G(kT + t) \, (T - |t|) \, dt \\
&= \int_{-T}^{T} \left\{ E^2[\mu] \left[1 + \left(\frac{kT + t}{t_G} \right)^{\alpha-1} \right] \right\} (T - |t|) \, dt, \quad k \neq 0 \\
&= E^2[\mu] \int_{-T}^{T} (T - |t|) \, dt \\
&\quad + E^2[\mu] \int_{kT-T}^{kT+T} (s/t_G)^{\alpha-1} \left(T - |s - kT| \right) dt \\
&= E^2[\mu] \, T^2 + E^2[\mu] \, T^{1+\alpha} \, t_G^{1-\alpha} \int_{k-1}^{k+1} x^{\alpha-1} \left(1 - |x - k| \right) dx \\
&= E^2[\mu] \, T^2 + E[\mu] \, T \, \frac{E[\mu] \, T^{\alpha} \, t_G^{1-\alpha}}{\alpha(1+\alpha)} \\
&\quad \times \left[(k+1)^{\alpha+1} + (k-1)^{\alpha+1} - 2k^{\alpha+1} \right]. \tag{5.13}
\end{aligned}
$$

The case $k = 0$ reduces to the mean square $E[Z_k^2(T)]$ so it need not be considered; this permits us to ignore the delta function at $t = 0$ in the coincidence rate. Hence, we assume $k > 0$ without loss of generality.

Using the binomial theorem, for large k Eq. (5.13) yields the simplified result

$$
\begin{aligned}
R_Z(k, T) &\approx E^2[\mu] \, T^2 + E[\mu] \, T \, \frac{E[\mu] \, T^{\alpha} \, t_G^{1-\alpha}}{\alpha(1+\alpha)} \, \alpha(1+\alpha) \, k^{\alpha-1} \\
&= E^2[\mu] \, T^2 + (E[\mu] \, T) \, E[\mu] \, T^{\alpha} \, t_G^{1-\alpha} \, k^{\alpha-1} \\
&= E^2[\mu] \, T^2 + (E[\mu] \, T) \, (T/T_R)^{\alpha} \, k^{\alpha-1}. \tag{5.14}
\end{aligned}
$$

5.1.6 Scaling cutoffs and fractal-exponent estimates

Some of the relationships obtained above remain intact in the absence of either a small- or large-size cutoff. Nevertheless, dispensing with either of these necessitates additional mathematical complexity for many of these equations, and renders others meaningless. Since all data derive from limited measurements, we adhere to the argument presented in Sec. 2.3.1 and focus on the situation where both cutoffs exist. This also has the merit of ensuring stationarity (Buckingham, 1983, Chapter 6).

From a theoretical standpoint, power-law behavior in one statistic generally implies the same in various other measures. Although any measure that takes a time or frequency argument can serve to characterize fractal behavior in a point process, in practice some statistics prove more useful than others. To distinguish among the various methods, we use subscripts to denote the values of α derived from particular functions, such as α_S and α_A for the values of α obtained from theoretical plots of the spectrum $S_N(f)$ and the normalized Haar-wavelet variance $A(T)$, respectively.

Furthermore, for a given finite-length data set, each measure returns an *estimate* of α, denoted $\widehat{\alpha}$, and this stochastic value differs from the ideal value α in a random fashion. Combining notations, $\widehat{\alpha}_A$ refers to a fractal-exponent estimate obtained from an estimated normalized Haar-wavelet variance function $\widehat{A}(T)$, which, in turn, is calculated from a real, finite data set. In the context of characterizing fractal behavior in a point process, these estimates can suffer from a variety of shortcomings: excessive bias or variance in the measure itself, as mentioned in Chapter 3; a limited range of allowable power-law exponents, as discussed below; and excessive bias or variance in the resulting estimates of the power-law exponent and multiplicative constant, which we treat in detail in Chapter 12.

5.2 RANGES OF POWER-LAW EXPONENTS

5.2.1 Negative values of α

What ranges of fractal exponents are ordinarily observed in experiments? The measures set forth in Chapter 3 admit negative values of α, and the relationships considered in Sec. 5.1 essentially continue to hold, so this issue merits discussion.

Let us consider, for example, a spectrum that increases with frequency for $0 < f < f_S$:

$$S_N(f) = \mathrm{E}^2[\mu]\,\delta(f) + \mathrm{E}[\mu]\left[1 + \sqrt{f/f_S}\,\exp(-f/f_S)\right]. \qquad (5.15)$$

The spectrum $S_N(f)$, as chosen, *increases* as $\sqrt{f}$ for $0 < f \ll f_S$ so that in this frequency range, $S_N(f)$ exhibits $\alpha = -\frac{1}{2}$.

The corresponding coincidence rate, normalized variance, and normalized Haar-wavelet variance then become

$$G(t) = \mathrm{E}[\mu]\,\delta(t) + \mathrm{E}^2[\mu] + \sqrt{\pi/2}\,\mathrm{E}[\mu]\,f_S\left(1+x_n^2\right)^{-3/2}$$
$$\times \left(\sqrt{\sqrt{x_n^2+1}+1} - x_n\sqrt{\sqrt{x_n^2+1}-1}\,\right) \tag{5.16}$$

$$F(T) = 1 + \frac{\sqrt{8}}{\sqrt{\pi}\,y_n}\left(\sqrt{\sqrt{y_n^2+1}+1} - \sqrt{2}\,\right) \tag{5.17}$$

$$A(T) = 1 + \frac{\sqrt{2}}{\sqrt{\pi}\,y_n}\left(4\sqrt{\sqrt{y_n^2+1}+1} - 3\sqrt{2} - \sqrt{\sqrt{4y_n^2+1}+1}\,\right), \tag{5.18}$$

where $x_n = 2\pi f_S t$ is the normalized time for Eq. (5.16) while $y_n = 2\pi f_S T$ is the normalized time for Eqs. (5.17) and (5.18).

Over long time scales, corresponding to low frequencies $f \ll f_S$, these quantities approach

$$G(t) - \mathrm{E}^2[\mu] \;\rightarrow\; -\left(\mathrm{E}[\mu]\big/4\pi\sqrt{f_S}\right) t^{-3/2} \tag{5.19}$$

$$F(T) - 1 \;\rightarrow\; \left(2\big/\pi\sqrt{f_S}\right) T^{-1/2} \tag{5.20}$$

$$A(T) - 1 \;\rightarrow\; \left(4 - \sqrt{2}\big/\pi\sqrt{f_S}\right) T^{-1/2}, \tag{5.21}$$

respectively. (For very large times, the outcomes are effectively indistinguishable from those for the homogeneous Poisson process.) Equations (5.19)–(5.21) are indeed in accord with the results provided in Sec. 5.1, provided that Eq. (5.12) is generalized to

$$G(t) = \mathrm{E}[\mu]\,\delta(t) + \mathrm{E}^2[\mu]\left[1 + \mathrm{sgn}(\alpha)\,(t/t_G)^{\alpha-1}\right], \tag{5.22}$$

where $\mathrm{sgn}(\alpha)$ denotes the sign of α.

However, as will become apparent in Sec. 5.4, values of α generally lie above zero and negative values almost never occur in practice. Fractal behavior typically exhibits increased fluctuations as the time grows larger and the frequency grows smaller, in contradiction to $\alpha < 0$. Moreover, the high-frequency cutoff would play a far more important role in this case. Since the spectrum would increase with frequency until reaching this cutoff, most of the power would lie just below the cutoff. In effect, therefore, such a signal would not differ appreciably from narrowband noise. As a result, this characteristic would dominate the behavior of the signal and would generally obscure any fractal properties that it might have. Revealing the scaling in this putative fractal signal would require integrating it a number of times until the resulting fractal exponent became positive. However, this would radically change the nature of the signal, thereby suggesting that the narrowband noise description of the signal would prove most useful.[2]

---

[2] A similar argument could be made for the low-frequency cutoff for fractal signals with positive fractal exponents α. However, a low-frequency cutoff does not affect the behavior of a signal within a window of duration significantly less than that cutoff. For negative fractal exponents, in contrast, the dominant high-frequency oscillation appears in windows of any duration greater than the inverse of the cutoff, particularly those with windows large enough to reveal the putative fractal behavior.

We conclude that negative values of α, although not prohibited, are generally not useful for fractal-based point processes. We therefore limit ourselves to values of α that are strictly positive.

5.2.2   Observed values of α

Experience shows that values of $\alpha > 2$, although not prohibited theoretically, rarely occur in practice. Furthermore, the process of estimating large values of α is problematical. The large rate of change attendant to such values, over even just a few orders of magnitude along the abscissa, leads to very large changes along the ordinate. As an example, consider a spectrum with $\alpha = 5$ and a spectrum of 1 kW/Hz at $f = 1$ Hz; at a frequency $f = 1$ kHz, the spectrum will have fallen to 1 pW/Hz, a factor of 10^{15}.

5.2.3   Limited range of the normalized variance exponent

Mathematical constraints limit the values that power-law exponents can attain for some statistics, affecting their usefulness in characterizing fractal behavior in point processes. We begin with the normalized variance $F(T)$, which has a power-law exponent that cannot exceed unity, as we now demonstrate.

In terms of the sequence of counts $\{Z_k(T)\}$ we have, by definition,

$$F(T) \equiv \frac{\text{Var}[Z(T)]}{\text{E}[Z(T)]}, \tag{5.23}$$

which reiterates Eq. (3.32). Consider now a larger counting window of duration nT, and express the new sequence of counts $\{Z_k(nT)\}$ in terms of the original sequence as

$$Z_k(nT) = \sum_{m=kn}^{kn+n-1} Z_m(T). \tag{5.24}$$

For the mean and variance of $Z_k(nT)$ we then have

$$\text{E}[Z_0(nT)] = \text{E}\left[\sum_{m=0}^{n-1} Z_m(T)\right] = \sum_{m=0}^{n-1} \text{E}[Z_m(T)] = n\,\text{E}[Z(T)] \tag{5.25}$$

$$
\begin{aligned}
\text{Var}\,[Z_0(nT)] &= \text{E}\left[\sum_{l=0}^{n-1}\Big\{Z_l(T) - \text{E}[Z(T)]\Big\}\sum_{m=0}^{n-1}\Big\{Z_m(T) - \text{E}[Z(T)]\Big\}\right] \\
&= \sum_{l=0}^{n-1}\sum_{m=0}^{n-1} \text{E}\left[\Big\{Z_l(T) - \text{E}[Z(T)]\Big\}\Big\{Z_m(T) - \text{E}[Z(T)]\Big\}\right] \\
&\leq \sum_{l=0}^{n-1}\sum_{m=0}^{n-1} \text{E}\left[\Big\{Z_m(T) - \text{E}[Z(T)]\Big\}^2\right] \\
&= n^2\,\text{Var}[Z(T)], \tag{5.26}
\end{aligned}
$$

where we have set $k = 0$ without loss of generality for a stationary point process.

This results in an upper bound for the increase of the normalized variance:

$$F(nT) = \frac{\text{Var}[Z(nT)]}{\text{E}[Z(nT)]} \leq \frac{n^2 \, \text{Var}[Z(T)]}{n \, \text{E}[Z(T)]} = n \frac{\text{Var}[Z(T)]}{\text{E}[Z(T)]} = nF(T). \quad (5.27)$$

Thus, for an orderly, stationary point process, multiplying the counting time by an integer factor n permits the normalized variance to increase at most by that factor n. In particular, if the normalized variance follows the power-law form $F(T) = (T/T_F)^{\alpha_F}$, then the power-law exponent α_F cannot exceed unity.

Indeed, a number of point processes yield exponents that achieve the maximum value $\alpha_F = 1$. A nonfractal example is provided by an integrate-and-reset process with a rate that increases linearly between periodic resets. We illustrate this by choosing an integrate-and-reset point process with a rate given by

$$\mu(t) = \mu_0[1 + \cos(\omega_0 t + \theta)], \quad (5.28)$$

where ω_0 and μ_0 represent fixed, deterministic quantities with units of inverse time. The random variable θ, which is taken to be uniformly distributed between zero and 2π, renders the rate $\mu(t)$ and the resulting point process stationary.

For counting times T much larger that $1/\mu_0$, the number of counts will greatly exceed unity, justifying the approximations below. We then have

$$
\begin{aligned}
Z_0(T) &\approx \int_0^T \mu(t)\,dt = \int_0^T \mu_0[1 + \cos(\omega_0 t + \theta)]\,dt \\
&= \mu_0 T + \mu_0 \omega_0^{-1}[\sin(\omega_0 T + \theta) - \sin(\theta)] \\
&= \mu_0 T + 2\mu_0 \omega_0^{-1} \sin(\omega_0 T/2) \cos(\omega_0 T/2 + \theta) \quad (5.29) \\
\text{E}[Z_0(T)] &= \mu_0 T \quad (5.30) \\
\text{Var}[Z_0(T)] &\approx 4\mu_0^2 \omega_0^{-2} \sin^2(\omega_0 T/2)\,\text{E}[\cos^2(\omega_0 T/2 + \theta)] \\
&= 2\mu_0^2 \omega_0^{-2} \sin^2(\omega_0 T/2). \quad (5.31)
\end{aligned}
$$

If we further stipulate that the time scale of the sinusoid greatly exceeds that of the counting time, so that $1/\omega_0 \gg T$, then the rate approximates a linear function over the counting time T, and the results above simplify to

$$
\begin{aligned}
\text{Var}[Z_0(T)] &\approx 2\mu_0^2 \omega_0^{-2}[(\omega_0 T/2)^2 - (\omega_0 T/2)^4/3] \\
&= \mu_0^2 T^2/2 - \mu_0^2 \omega_0^2 T^4/24 \\
F(T) &\approx \mu_0 T/2 - \mu_0 \omega_0^2 T^3/24 \quad (5.32) \\
&\approx \mu_0 T/2, \quad (5.33)
\end{aligned}
$$

thereby demonstrating that this process does indeed achieve $\alpha_F = 1$. This result often emerges for nonfractal point processes with time-varying rates and when rate nonstationarities are present (see, for example Prucnal & Teich, 1979).

As with the normalized variance, values of α for the autocorrelation and coincidence rate necessarily lie below unity. Equation (3.51) establishes that for large delay times, the coincidence rate approaches a constant value $E^2[\mu]$, while Eq. (5.12) exceeds this value by $E^2[\mu]\,(t/t_G)^{\alpha-1}$. In order that this quantity vanish for large t, so that the coincidence rate can achieve the limit provided in Eq. (3.51), it is required that $\alpha < 1$. Furthermore, application of Eq. (3.57) to a coincidence rate with $\alpha > 1$ would result in a spectrum that assumes negative values at large frequencies, an impossibility. This same argument applies to the autocorrelation $R_Z(k, T)$ (taken as a function of the delay index k), through Eq. (3.56).

The generalized dimensions of point processes encountered in the treatment provided here also lie below unity, because the dimension of any object may not exceed that of the space in which it is embedded. Lines have dimensions of unity, and we focus on point processes on a line.

For the point-process spectrum $S_N(f)$ [as well as for the rate spectrum $S_\lambda(f, T)$], in contrast, no such limit exists (however, see Prob. 5.12). Indeed, we chose the spectrum as a starting point in Sec. 5.1 precisely for this reason.

5.2.4 Range of the normalized Haar-wavelet-variance exponent

A more generous maximum exponent obtains for the normalized Haar-wavelet variance. We proceed in a similar fashion, beginning with the mean-square difference in the number of counts, and rearranging the sums to obtain

$$
\begin{aligned}
&\mathrm{E}\!\left[\left\{Z_0(nT) - Z_1(nT)\right\}^2\right] \\[1mm]
&= \ \mathrm{E}\!\left[\left\{\sum_{m=0}^{n-1} Z_m(T) - \sum_{m=n}^{2n-1} Z_m(T)\right\}^2\right] \\[1mm]
&= \ \mathrm{E}\!\left[\left\{\sum_{m=1}^{n} m\left[Z_{m-1}(T) - Z_m(T)\right]\right.\right. \\
&\qquad\qquad \left.\left. + \sum_{m=n+1}^{2n-1} (2n - m)\left[Z_{m-1}(T) - Z_m(T)\right]\right\}^2\right] \\[1mm]
&\leq \ \left[\sum_{m=1}^{n} m + \sum_{m=n+1}^{2n-1} (2n - m)\right]^2 \mathrm{E}\!\left[\left\{Z_{m-1}(T) - Z_m(T)\right\}^2\right] \\[1mm]
&= \ n^4\,\mathrm{E}\!\left[\left\{Z_0(T) - Z_1(T)\right\}^2\right].
\end{aligned}
\tag{5.34}
$$

Proceeding as previously, we obtain an upper bound for the increase of the normalized Haar-wavelet variance determined by

$$
\begin{aligned}
A(nT) &= \ \mathrm{E}\!\left[\left\{Z_0(nT) - Z_1(nT)\right\}^2\right] \Big/ 2\mathrm{E}[Z_0(nT)] \\[1mm]
&\leq \ n^4\,\mathrm{E}\!\left[\left\{Z_0(T) - Z_1(T)\right\}^2\right] \Big/ 2n\,\mathrm{E}[Z_0(T)] \\[1mm]
&= \ n^3 A(T).
\end{aligned}
\tag{5.35}
$$

This indicates a maximum factor of n^3 in the growth of the normalized Haar-wavelet variance as the counting time increases by n. The corresponding power-law exponent α_A therefore cannot exceed three, which is ample to accommodate all practical fractal-based point processes (see Sec. 5.2.2).

The same nonfractal integrate-and-reset process defined by Eq. (5.28), which achieved the maximum power-law exponent of unity for the normalized variance $F(T)$ for counting times T in the range $1/\mu_0 \ll T \ll 1/\omega_0$, also yields the maximum power-law exponent for the normalized Haar-wavelet variance $A(T)$. Combining Eqs. (3.41) and (5.32) provides

$$
\begin{aligned}
A(T) &= 2F(T) - F(2T) \\
&\approx \mu_0 T - \mu_0 \omega_0^2 T^3/12 - \mu_0 T + \mu_0 \omega_0^2 T^3/3 \\
&= \mu_0 \omega_0^2 T^3/4,
\end{aligned}
\tag{5.36}
$$

indicating that this process again achieves the maximum permitted value, $\alpha_A = 3$, as do many nonstationary, nonfractal point processes.[3]

It is important to note that in deriving Eq. (5.36), we have made use of Eq. (5.32), rather than its approximation, Eq. (5.33). Using the latter yields incorrect results when linear terms dominate $F(T) - 1$. To illustrate this, suppose that over some range of counting times T, the normalized variance has a linear term that exceeds another contribution with a power-law form other than linear. To first order, we then have

$$
\begin{aligned}
F(T) &= 1 + c_1 T + c_2 T^\alpha \\
&\approx 1 + c_1 T.
\end{aligned}
\tag{5.37}
$$

Based on this approximation, Eq. (3.41) yields

$$
\begin{aligned}
A(T) &= 2F(T) - F(2T) \\
&\approx 2 + 2c_1 T - (1 + 2c_1 T) \\
&= 1.
\end{aligned}
\tag{5.38}
$$

However, the proper value of the normalized Haar-wavelet variance also contains a term that varies as T^α:

$$
\begin{aligned}
A(T) &= 2F(T) - F(2T) \\
&= 2 + 2c_1 T + 2c_2 T^\alpha - (1 + 2c_1 T + 2^\alpha c_2 T^\alpha) \\
&= 1 + (2 - 2^\alpha) c_2 T^\alpha.
\end{aligned}
\tag{5.39}
$$

The disagreement between Eqs. (5.38) and (5.39) stems from improper use of asymptotic results in intermediate calculational steps. Equation (3.41) is correct and applies exactly in all cases. We conclude that when first-order approximations yield terms in $F(T)$ that are linear in T, we must retain the higher-order terms in calculating the normalized Haar-wavelet variance by means of Eq. (3.41).

---

[3] Under some circumstances such nonstationarities can mask the presence of fractal behavior (see, for example, Turcott, Lowen, Li, Johnson, Tsuchitani & Teich, 1994).

5.2.5   Range of the normalized general-wavelet-variance exponent

Other wavelets offer even higher limits for their power-law exponents, as we now proceed to demonstrate (Teich et al., 1996; Heneghan, Lowen & Teich, 1996). We first recast Eq. (3.68) from a time domain integral into one in the frequency domain,

$$\begin{aligned}
\mathrm{Var}[C_{\psi,N}(a,b)] &= a \int_x G(ax) \int_y \psi(x+y)\,\psi(y)\,dy\,dx \\
&= a \int_x \int_f G(ax)\,|\varphi(f)|^2\,e^{i2\pi f x}\,df\,dx \\
&= \int_f S_N(f/a)\,|\varphi(f)|^2\,df,
\end{aligned} \qquad (5.40)$$

where

$$\varphi(f) = \int_x \psi(x)\,e^{-i2\pi f x}\,dx \qquad (5.41)$$

represents the Fourier transform of the wavelet $\psi(x)$.

The behavior of $\varphi(f)$ near $f = 0$ determines the convergence properties of the integral in Eq. (5.40); this is related to n_v, the number of contiguous vanishing moments of the wavelet $\psi(t)$. We define n_v as the largest integer for which

$$\int \psi(t)\,t^k\,dt = 0 \qquad (5.42)$$

for all k such that $0 \le k \le n_v$. The integral in Eq. (5.40) converges near the origin if the integrand increases more slowly than $1/f$ in that region. Given a spectrum that decays as $f^{-\alpha}$, this convergence requires that $2c > \alpha - 1$. A normalized wavelet variance constructed using a wavelet for which $|\varphi(f)| \sim f^c$ near $f \to 0$ will therefore faithfully reproduce power-law exponents α in the range $0 < \alpha < 2c + 1$.

As an example, we return to the Haar wavelet, which has

$$\begin{aligned}
\varphi(f) &= 2ie^{i\pi f}\,\sin^2(\pi f/2)\,/f \\
|\varphi(f)| &= 2\sin^2(\pi f/2)\,/f \quad \sim \quad f^1
\end{aligned} \qquad (5.43)$$

near $f = 0$, so that $c = 1$, which yields a maximum power-law exponent of $2c+1 = 3$, as previously demonstrated in Eq. (5.35).

Wavelets other than the Haar have Fourier transforms that decay as f^c with $c > 1$, and therefore appear useful for the analysis of fractal processes with $\alpha > 3$. Such wavelets typically exhibit higher regularity and therefore have a higher number of vanishing moments. However, there is an important practical caveat regarding their use: they have larger support for a given scale and hence exhibit reduced scaling ranges for finite-length data sets (Heneghan et al., 1996), as is demonstrated in Fig. 5.3. Moreover, processes with $\alpha > 3$ do not often occur in practice so that the Haar wavelet usually suffices, as demonstrated in Sec. 5.4.4. Wavelets other than the Haar also enjoy the property of being insensitive to linear or higher-order polynomial trends, but in practice nonstationarities rarely follow a polynomial form. Finally, we note that wavelets with higher regularity yield wavelet transforms with less correlation among the resulting wavelet coefficients, thereby improving the statistics of the resulting estimate (Tewfik & Kim, 1992). This also suggests the use of wavelets other than the Haar (Abry et al., 2003), but these wavelets turn out to yield *increased* variance in fractal-exponent estimates (Bardet, Lang, Oppenheim, Philippe, Stoev & Taqqu, 2003), perhaps as a result of their larger support. We conclude that the Haar wavelet is generally the wavelet of choice for the analysis of point processes.

5.3 RELATIONSHIPS AMONG MEASURES

The relationships set forth in Sec. 5.1 all follow power-law forms. They are valid for $0 < \alpha < 1$ and display simple interrelations among their exponents over some range of the independent variables. Expressions over a range of times and frequencies are also available for all measures for $\alpha = 1$. For $\alpha > 1$, the limitations exposed in Sec. 5.2 tell us that the spectrum and normalized Haar-wavelet variance are the measures of choice; they offer extended validity over the range $0 < \alpha < 3$.

The relationships among the various measures over the full range $0 < \alpha < 3$ are summarized below:

- For $0 < \alpha < 1$:

$$
\begin{aligned}
S_N(f) &= \mathrm{E}^2[\mu]\,\delta(f) + \mathrm{E}[\mu]\,[1 + (f/f_S)^{-\alpha}] & \text{a)}\\
F(T) &= 1 + (T/T_F)^\alpha & \text{b)}\\
A(T) &= 1 + (T/T_A)^\alpha & \text{c)}\\
R_Z(k,T) &= \mathrm{E}^2[\mu]\,T^2 + (\mathrm{E}[\mu]\,T)\,(T/T_R)^\alpha\,k^{\alpha-1} & \text{d)}\\
G(t) &= \mathrm{E}[\mu]\,\delta(t) + \mathrm{E}^2[\mu]\,\left[1 + (|t|/t_G)^{\alpha-1}\right] & \text{e)}
\end{aligned}
$$

$$(5.44)$$

with

$$
\begin{aligned}
(2\pi f_S\,T_F)^\alpha &= \cos(\pi\alpha/2)\,\Gamma(\alpha+2) & \text{a)}\\
(T_F/T_A)^\alpha &= 2 - 2^\alpha & \text{b)}\\
(T_F/T_R)^\alpha &= \tfrac{1}{2}\alpha(\alpha+1) & \text{c)}\\
\mathrm{E}[\mu]\,t_G^{1-\alpha}\,T_R^\alpha &= 1 & \text{d)}
\end{aligned}
$$

$$(5.45)$$

where the limits $\mathrm{E}[\mu]\,T \gg 1$ and $k \gg 1$ apply for Eq. (5.44d).

- For $\alpha = 1$:

$$
\begin{aligned}
S_N(f) &= \mathrm{E}^2[\mu]\,\delta(f) + \mathrm{E}[\mu]\,(1 + f_S/f) & \text{a)}\\
F(T) &= 1 + 2f_S T\,\ln(B/T) & \text{b)}\\
A(T) &= 1 + 4\ln(2)\,f_S T & \text{c)}\\
R_Z(k,T) &= \mathrm{E}^2[\mu]\,T^2 + 2\mathrm{E}[\mu]\,f_S\,\ln[B/(kT)]\,T^2 & \text{d)}\\
G(t) &= \mathrm{E}[\mu]\,\delta(t) + \mathrm{E}^2[\mu] + 2f_S\,\mathrm{E}[\mu]\,\ln(B/t) & \text{e)}
\end{aligned}
$$

$$(5.46)$$

where $S_N(f)$ is assumed to have a cutoff so that $F(T)$, $R_Z(k,t)$, and $G(t)$ exist.

- For $0 < \alpha < 3$:

$$
\begin{aligned}
S_N(f) &= \mathrm{E}^2[\mu]\,\delta(f) + \mathrm{E}[\mu]\,[1 + (f/f_S)^{-\alpha}] &\text{a)}\\
A(T) &= 1 + (T/T_A)^\alpha &\text{b)}
\end{aligned}
$$
$$(5.47)$$

with

$$
(2\pi f_S\, T_A)^\alpha =
\begin{cases}
\cos(\pi\alpha/2)\,\Gamma(\alpha+2)/(2-2^\alpha) & 0 < \alpha < 1\\
\pi/[2\ln(2)] & \alpha = 1\\
[-\cos(\pi\alpha/2)]\,\Gamma(\alpha+2)/(2^\alpha-2) & 1 < \alpha < 3.
\end{cases}
$$
$$(5.48)$$

Equations (5.45) and (5.48) specify the relationships among the various **fractal onset times** and **fractal onset frequencies**. Earlier recitations of such relations appeared in Lowen & Teich (1993a, 1995), Thurner et al. (1997), and Ryu & Lowen (1998).

5.4 EXAMPLES OF FRACTAL BEHAVIOR IN POINT PROCESSES

Following a brief discussion of $1/f$ noise in the context of fractal-based continuous and point processes, we examine the estimated normalized spectrum and normalized Haar-wavelet variance for six representative biological point processes and one computer network traffic trace.

5.4.1 $1/f$ noise

Many forms of data, in many fields of endeavor, behave in accordance with the power-law spectrum specified in Eq. (5.1): $S(f) \approx (f/f_S)^{-\alpha}$. Signals with spectra of this form are typically referred to as $1/f^\alpha$ **noise** or $1/f$-**type noise**. In the particular case when $\alpha = 1$, common appellations are $1/f$ **noise**, **flicker noise**, **excess noise**, and **pink noise**.[4] Sinc no strict standard for this nomenclature exists, however, all of the foregoing descriptions are also used to describe $1/f^\alpha$ noise when α is in the rough vicinity of unity.

Fluctuations of this form are ubiquitous in the natural world. The earliest observation in the physical sciences appears to have been made by Johnson (1925), who discovered excess $1/f^\alpha$ noise in the course of his studies of low-frequency circuits. Such fluctuations are also widely present in electronic materials and devices, including

---

[4] Spectra that are uniform in frequency are referred to as **white noise** in analogy with white light, which contains an equal weighting of all colors. If a spectrum obeys the form $S(f) \approx 1/f$, on the other hand, each octave is endowed with equal energy so that lower frequencies are weighed more heavily. By the same optical analogy, the red portion of the spectrum is then enhanced relative to the blue so that such spectra have come to be called **pink noise**. Spectra that follow the form $S(f) \approx 1/f^2$ should, by all rights, then be termed "red noise," but are known instead as **brown noise** by virtue of their association with ordinary *Brown*ian motion.

carbon resistors,[5] thin-film resistors, semiconductors, metal films, electrolytes, super-conductors, thermionic-emission devices, and junction devices (Bell, 1960, 1980; van der Ziel, 1988; Weissman, 1988; Buckingham, 1983; Kogan, 1996). In electronics, the range of frequencies over which such behavior is manifested can stretch over 12 orders of magnitude or more, and α typically lies between 0.8 and 1.4 (Buckingham, 1983, Chapter 6). The origins of this phenomenon remain obscure for many devices and systems. The underlying mechanism is often associated with fluctuations of the number, or the mobility, of the charge carriers, but other causes have been postulated. $1/f$-type noise is thought by some to be a surface effect whereas others attribute it to bulk behavior. In the biological sciences, $1/f^\alpha$ noise appears to have been first observed by Verveen (1960) in his studies of membrane-voltage fluctuations.

Behavior of this kind is not restricted to simple materials, components, and devices. Complex systems also exhibit $1/f^\alpha$ noise; examples stretch from fluctuations of the flood level on the river Nile (Hurst, 1951), to voltage fluctuations in the human electroencephalogram (Musha, 1981), to measurements of cerebral blood flow (West, Zhang, Sanders, Miniyar, Zuckerman & Levine, 1999), to the formation of representations in a cognitive process (Gilden, 2001), to music deemed aesthetically pleasing to the listener (Gardner, 1978; Voss & Clarke, 1978; Voss, 1989; Hsü & Hsü, 1991). The ascendancy of fractal analysis in recent years has also drawn increased attention to $1/f$ noise (Mandelbrot, 1982; Montroll & Shlesinger, 1982; Shlesinger, 1987; Schroeder, 1990; West & Deering, 1995).

Our particular interest relates to the fluctuations observed in point processes rather than in continuous processes as highlighted above. Indeed, $1/f$-type noise is ubiquitous in this domain as well.[6] Early work along these lines was strongly influenced by Toshimitsu Musha and colleagues, who examined examples as diverse as vehicular traffic flow (Musha & Higuchi, 1976), spike-discharge intervals (Musha et al., 1983), human tapping intervals (Musha, Katsurai & Teramachi, 1985), and heartbeat period in humans (Kobayashi & Musha, 1982).

5.4.2 Normalized rate spectrum

In this section and the next, we examine second-order statistics for a collection of point processes. We plot the normalized estimated rate spectrum, $\widehat{S}_\lambda(f,T)/\widehat{E}[\lambda]$ vs. normalized frequency $f/\widehat{E}[\lambda]$, in Fig. 5.1 for six biological point processes and one

---

[5] Although ubiquitous, $1/f$-type noise is not universal; it is not present, for example, in wire-wound resistors.

[6] *Interval-based* spectra are often reported for point processes since calculating the spectrum of a discrete-time sequence is straightforward and the availability of the fast Fourier transform lowers the computational cost. Strictly speaking, the descriptor "$1/f$-type noise" should be used for such results, where f has units of cycles per number of intervals (see Sec. 3.3.3), but for simplicity we use the appellation "$1/f$-type noise" for both forms of the spectrum.

computer network traffic trace. To facilitate visual comparison, we have smoothed the spectra using a suitable windowing function.[7]

Figure 5.1 displays curves for the following point processes[8]: spontaneous vesicular exocytosis at a developing *Xenopus* neuromuscular junction (SYNAPSE) (Lowen et al., 1997b, Figs. 5 and 8, pp. 5670 and 5672, cell 950315e1); action-potential sequence recorded from a cat primary afferent auditory nerve fiber driven at its characteristic frequency of 10.2 kHz (COCHLEA) (Lowen & Teich, 1992a, the companion spontaneous recording is labeled "unit I"); action-potential sequences recorded from a cat on-center X-type retinal ganglion cell (RETINA) and its associated lateral geniculate nucleus cell (GENICULATE), in response to a 4.2-Hz drifting grating with 40% contrast and a mean luminance of 50 cd/m^2 (Lowen et al., 2001, Figs. 5D and 5E, p. 388, cells y31900ret and y31900lgn); action-potential sequence recorded from a cat layer-VI standard-complex striate cortex cell (CORTEX), in response to a weak steady background luminance ≈ 0.25 cd/m^2 (Teich et al., 1996, cell 3); 20-hour sequence of heartbeats recorded from a normal human subject (HEARTBEAT) (Turcott & Teich, 1996, data set 16273 from the MIT–BIH Normal Sinus Rhythm Database; available at http://www.physionet.org/physiobank/database/nsrdb/); and one million consecutive Ethernet-packet arrivals (COMPUTER) (Leland & Wilson, 1989, 1991, data set BC-pOct89 collected at the Bellcore Morristown Research and Engineering Facility in 1989; available at http://ita.ee.lbl.gov/html/contrib/BC.html).

All of the curves follow the general form of Eq. (5.1) over a range of normalized frequencies, $\widehat{S}_\lambda(f,T)/\widehat{E}[\lambda] \sim (f/f_S)^{-\alpha}$, suggesting the presence of fractal behavior. Similar results are obtained from the interval spectrum for these particular data sets (see Fig. 5.7 and Prob. 5.2).

5.4.3 Normalized Haar-wavelet variance

To complement the estimated spectral data displayed in Fig. 5.1, we present in Fig. 5.2 the estimated normalized Haar-wavelet variance $\widehat{A}(T)$ vs. normalized counting time $T/\widehat{E}[\tau]$ for the same six biological point processes and one computer network traffic

---

[7] We smooth the measured rate and interval spectra by making use of the following procedure. We calculate the Fourier transform of the rate function (or interval sequence) and obtain its square magnitude. We then transform back into the time domain, which yields the autocorrelation. We multiply the autocorrelation by a triangular window with unity height at the origin, that decreases linearly to zero at one-eighth of the array size of the fast Fourier transform. Next we transform back to the frequency domain; this is the third Fourier transform involved in the smoothing procedure. The next step is to collect values into nonoverlapping blocks such that the largest frequency in a block, divided by the smallest, is as large as possible while lying below 1.02. Finally, all frequencies in each block are averaged and presented as a single frequency; the associated spectral values are similarly averaged. This procedure makes the graph appear progressively smoother as the frequency increases. The triangular windowing in the time domain is equivalent to subjecting the (noisy) periodogram to a moving-average $\mathrm{sinc}^2(\cdot)$ filter in the frequency domain. This procedure reduces noise, but also reduces frequency resolution. Note that smoothing is generally eschewed before rigorous parameter estimation, as pointed out in Chapter 12.

[8] These seven point processes are also examined in Figs. 5.2, 5.7–5.10, 11.2–11.4, 11.6–11.8, 11.10, 11.11, 11.13, and 11.14.

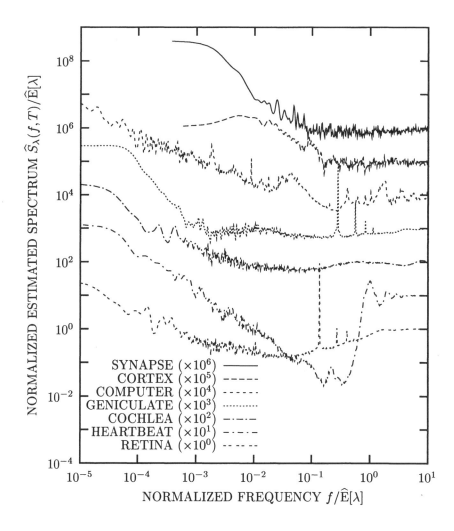

Fig. 5.1 Normalized estimated spectrum, $\widehat{S}_\lambda(f, T)/\widehat{E}[\lambda]$ vs. normalized frequency $f/\widehat{E}[\lambda]$, for six biological point processes and one computer network traffic trace. The time T is chosen such that $1/\sqrt{2} < 30\,T/E[\tau] < \sqrt{2}$, ensuring that the size of the Fourier-transform exceeds the number of intervals by a significant factor ($15\sqrt{2}$). Curves are displayed for the following point processes (see text for sources of data): vesicular exocytosis (SYNAPSE); action-potential sequence recorded from an auditory nerve fiber (COCHLEA); action-potential sequences recorded from a retinal ganglion cell (RETINA) as well as its associated lateral geniculate nucleus cell (GENICULATE); action-potential sequence recorded from a striate cortex cell (CORTEX); day-long sequence of normal human heartbeats (HEARTBEAT); and one million consecutive Ethernet-packet arrivals (COMPUTER). The curves decrease with frequency roughly as power laws, with seven negative estimated power-law exponents $-\widehat{\alpha}_S$.

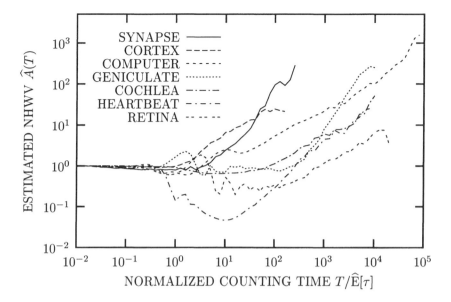

Fig. 5.2 Estimated normalized Haar-wavelet variance (NHWV), $\widehat{A}(T)$ vs. normalized counting time $T/\widehat{\mathrm{E}}[\tau]$, for the same seven point processes as displayed in Fig. 5.1. We present curves
for the following point processes (see text for sources of data): vesicular exocytosis (SYNAPSE);
action-potential sequence recorded from an auditory nerve fiber (COCHLEA); action-potential
sequences recorded from a retinal ganglion cell (RETINA) as well as its associated lateral
geniculate nucleus cell (GENICULATE); action-potential sequence recorded from a striate cortex cell (CORTEX); day-long sequence of normal human heartbeats (HEARTBEAT); and one
million consecutive Ethernet-packet arrivals (COMPUTER). The curves increase roughly as
straight lines, indicating approximate power-law dependence on the counting time, with seven
positive estimated power-law exponents $\widehat{\alpha}_A$.

trace. The calculations made use of counting times T that increased geometrically
by factors of $10^{0.1}$, providing 10 counting times per decade.

These curves are the time–scale equivalents of $1/f^\alpha$ fluctuations. They follow
the general form of Eq. (5.2) over a range of normalized counting times, $A(T) \approx
(T/T_A)^\alpha$. The normalized interval wavelet variance for these data provide further
evidence for power-law behavior (see Fig. 5.8 and Prob. 5.3). All of these results
together lend credence to the notion that the data exhibit fractal behavior, as suggested
by the results presented in Sec. 5.4.2.

In spite of its ubiquity, the fractal behavior evident in these point processes should
not be ascribed to any single physical or biological mechanism. As demonstrated in
the following chapters, behaviors in accordance with $1/f^\alpha$ and T^α are inherent in
essentially all fractal-based point processes, under suitable conditions.

5.4.4 Normalized Daubechies-wavelet variance

Of all possible wavelets, the Haar has compact support and also has the best localization in time. This temporal precision, must, of course, be traded against scale resolution by virtue of the uncertainty principle.

Ingrid Daubechies (1988) developed a family of wavelets with compact support and differing abilities to localize signals in time and scale. The Haar forms the simplest and first member of this family. Daubechies wavelets are defined in terms of discrete-time filters with n coefficients, or "taps," with n an even positive integer. Orthogonality requirements yield $n/2$ equations for these coefficients. To specify the other $n/2$ equations, and thus to determine the coefficients, Daubechies set the filter response to zero for polynomials of order less than $n/2$.

The Haar wavelet, which is equivalent to the Daubechies 2-tap wavelet, is therefore insensitive to constant values, whereas the Daubechies 20-tap wavelet, for example, is insensitive to polynomials up to and including order nine. Increasing the number of taps enhances the scale localization, but at the expense of a loss in time precision. This polynomial insensitivity is salutary inasmuch as it mitigates against some forms of nonstationarity that might be present in the point process although, as discussed in Sec. 5.2.5, nonstationarities do not always follow polynomial forms.

To illustrate the behavior of wavelets beyond the Haar, we plot the estimated normalized Daubechies-wavelet variance $\widehat{A}_W(T) \equiv \widehat{\mathrm{Var}}[C_{\psi,N}(T,\cdot)]/\widehat{\mathrm{E}}[\lambda_k(T)]$ for a sequence of geniculate action potentials, as a function of the normalized counting time $T/\widehat{\mathrm{E}}[\tau]$, in Fig. 5.3. In this case we calculate the wavelet variance using four different Daubechies wavelets: 2-tap, 8-tap, 14-tap, and 20-tap, as indicated in the figure. The four curves all begin at a normalized counting time of 0.1 by construction. For large counting times the curves all increase roughly as straight lines, yielding four exponents $\widehat{\alpha}_W$, all of which are essentially the same. For this finite-length data set, the scaling region shrinks as the wavelet support increases with the number of taps, confirming the suggestion set forth in Sec. 5.2.5 that the Haar wavelet generally suffices for the analysis of fractal-based point processes.

5.5 FRACTAL-BASED POINT PROCESSES

The statistical measures described in Sec. 5.3, and examined in Sec. 5.4 for various experimental point processes, are second-order relationships. As such, they provide important, but limited, information about the underlying point process. We now proceed to further specify these underlying point processes.

In this section we compare and contrast two mutually exclusive classes of fractal-based point processes: fractal point processes and fractal-rate point processes. Both forms are found in the physical and biological sciences. Specific models that belong to these classes are examined in Chapters 6–10, where we study their full properties. We also devote a portion of this section to re-examining and confirming the nature of the nonfractal point processes introduced in Chapter 4, and we briefly consider the identification of fractal-based point processes.

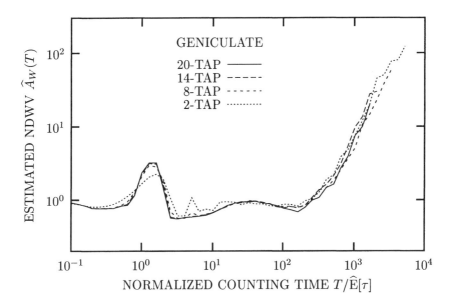

Fig. 5.3 Estimated normalized Daubechies-wavelet variances (NDWV), $\widehat{A}_W(T)$ vs. normalized counting time $T/\widehat{\mathrm{E}}[\tau]$, for an action-potential sequence recorded from a cat on-center X-type lateral geniculate nucleus cell in response to a 4.2-Hz drifting grating with 40% contrast and a mean luminance of 50 cd/m^2 (Lowen et al., 2001, cell y31900lgn). Results are shown for four Daubechies wavelet bases: 2-tap (Haar), 8-tap, 14-tap, and 20-tap. The scaling region decreases as the wavelet support increases with the number of taps.

5.5.1 Fractal point processes

We define a **fractal point process** as one that has the following properties:

- $0 < \alpha < 1$

    1. Scaling behavior in the spectrum $S_N(f)$, coincidence rate $G(t)$, autocorrelation $R_Z(k, T)$, normalized variances $F(T)$ and $A(T)$, and interval probability density function $p_\tau(t)$.

    2. Simply related exponents, as in Eq. (5.44).

    3. Generalized dimensions D_q in the sense of Eq. (3.72), with $D_q = D = \alpha$ for all q.

Because all exponents and generalized dimensions coincide, the collection of measures specified above are characterized by a single value, α, that describes the scaling behavior. Since D_q cannot exceed unity for a collection of points on a line, and the upper bound of unity leads to a degenerate point process, we have $\alpha < 1$. In addition, as discussed in Sec. 5.2.1, we require $\alpha > 0$. Taken together, these two limits yield $0 < \alpha < 1$ for a fractal point process, as indicated above.

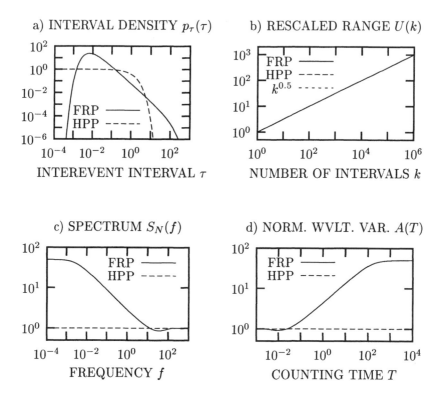

a) INTERVAL DENSITY $p_\tau(\tau)$

b) RESCALED RANGE $U(k)$

INTEREVENT INTERVAL τ

NUMBER OF INTERVALS k

c) SPECTRUM $S_N(f)$

d) NORM. WVLT. VAR. $A(T)$

FREQUENCY f

COUNTING TIME T

Fig. 5.4 Comparison of measures for a particular fractal point process, the fractal renewal process (FRP; see Chapter 7); and a point process that is devoid of fractal properties, the homogeneous Poisson process (HPP; see Sec. 4.1). We plot cartoons for four different probabilistic measures: (a) interevent-interval probability density, (b) rescaled range, (c) point-process spectrum, and (d) normalized Haar-wavelet variance. The rescaled-range measure cannot differentiate between these two processes since both belong to the family of renewal point processes. However, the other three measures exhibit nontrivial power-law variations for the fractal point process, but not for the nonfractal process.

The **intermittency**, which quantifies the unevenness of a point process, is defined as $1 - D_2$; it assumes a value of zero for a perfectly periodic point process, in which all intervals are identical, and approaches unity for a highly clustered process (Bickel, 1999). Since $D_2 = \alpha$ for fractal point processes, the intermittency is $1 - \alpha$.

In Fig. 5.4, we display the statistics of the nonfractal homogeneous Poisson process described in Sec. 4.1, together with a particular fractal point process, the **fractal renewal process** discussed in Chapter 7. This process exhibits scaling behaviors in all of the measures listed above and indeed has $D_q = \alpha$ for all q. In particular, realizations of this process are fractal.

Several of the interval-based measures described in Sec. 3.3 do *not* indicate fractal behavior for this process, which is, by definition, fractal. This is because the process is renewal (see Sec. 4.2) so that the intervals between adjacent events $\{\tau_n\}$ are independent and identically distributed. Thus, results that are indistinguishable from those of other, nonfractal renewal point processes (see Sec. 5.5.3) emerge for the interval-based autocorrelation $R_\tau(k)$, interval spectrum $S_\tau(f)$, interval wavelet variance $\mathrm{Var}[W_{\psi,\tau}(k,l)]$, rescaled range $U(k)$, and detrended fluctuation $Y(k)$. As a consequence, these interval-based measures are not suitable for determining the presence or absence of fractal behavior in general fractal-based point processes, which comprise both fractal and fractal-rate point processes (see Sec. 12.3.1). We therefore use interval-based measures judiciously in the remainder of this book.

Fractal point processes comprise hierarchies of clusters. This can arise if the interevent intervals are power-law distributed or if they exhibit long-range positive correlations. An example of a fractal point process that is distinct from the fractal renewal process arises from an infinitely divisible cascade (Castaing, 1996).

Consider a multiplicative-rate point process (Schmitt, Vannitsem & Barbosa, 1998) that begins with a constant rate over a fixed interval. Divide the interval into m subintervals, and multiply each by a random number, $W_{1,k}$, $1 \leq k \leq m$. Apply the same procedure to each subinterval, and then, in turn, to each sub-subinterval, *ad infinitum*. The weighting factors $W_{l,k}$, $1 \leq k \leq m^l$, are independent, identically distributed, unit-mean, nonnegative random variables.

A particular implementation of this process sets conditions on the weights applied to each new interval by the next stage of multiplication (Riedi, 2003). Let $Q_{l,nm} \equiv \frac{1}{m} \sum_{k=mn+1}^{m(n+1)} W_{l,k}$ denote an average, where n is any integer between 0 and m^{l-1}, inclusive. Two conditions on $Q_{l,nm}$ can be imposed for all values of l and n specified above: either $Q_{l,nm} = 1$ or $\mathrm{E}[Q_{l,nm}] = 1$. When employed as a rate for a doubly stochastic or integrate-and-reset process, as described in Secs. 4.3 and 4.4, respectively, it turns out that the resultant point process belongs to the family of fractal point processes (Bickel, 1999).

One formalism for generating a *multifractal* process emerges by changing the time axis of a monofractal process to another process with multifractal characteristics. For example, $\mathsf{B}_H(t) \equiv B_H[\mathsf{M}(t)]$, where $B_H(t)$ represents fractional Brownian motion (see Chapter 6) and $\mathsf{M}(t)$ is a nondecreasing, multifractal process (Mandelbrot, 1997, 1999). A number of variations on these processes exist (Peltier & Lévy Véhel, 1995; Benassi, Jaffard & Roux, 1997; Ayache & Lévy Véhel, 1999).

5.5.2 Fractal-rate point processes

Many point processes do not have fractional values of D_q, but nevertheless exhibit scaling behavior in other measures. Realizations of such processes are not fractals. Instead, the scaling behavior implies fractal characteristics of the *rates* associated with these point processes: the rate estimated from a realization of the process $[\lambda_k(T)]$ and the probabilistic rate that provides a mathematical description of the process $[\mu(t)]$ (Kumar & Johnson, 1993). Since the fractal behavior inheres to the rate rather than to the point process itself, we denote these as fractal-rate point processes.

We thus define a **fractal-rate point process** as one that is *not* a fractal point process and has the following properties:

- For $0 < \alpha < 1$:

  1. Scaling behavior in the spectrum $S_N(f)$, coincidence rate $G(t)$, autocorrelation $R_Z(k, T)$, and normalized variances $F(T)$ and $A(T)$.

  2. Simply related exponents, as in Eq. (5.44).

- For $1 \leq \alpha < 3$:

  1. Scaling behavior in the spectrum $S_N(f)$ and normalized Haar-wavelet variance $A(T)$.

  2. Exponents of these two measures sum to zero,

$$\alpha_A + (-\alpha_S) = 0. \qquad (5.49)$$

Fractal-rate point processes can, in principle, exhibit any positive value of α; values in excess of two rarely occur in practice, however, as discussed in Sec. 5.2.2. All of the fractal-based point processes considered in this book belong to the fractal-rate family, with the exception of the fractal renewal point process and the infinitely divisible cascade discussed in Sec. 5.5.1.

In Fig. 5.5, we display the statistics of the nonfractal **homogeneous Poisson process** described in Sec. 4.1, together with a particular fractal-rate point process, the **fractal-Gaussian-process-driven Poisson process**. This is a doubly stochastic Poisson process driven by a fractal Gaussian process; we consider it in detail in Secs. 6.3.3, 8.4, and 10.6.1, as well as in Chapter 12. For this point process, the spectrum, coincidence rate, count autocorrelation, and normalized count variances all scale with their respective arguments. However, it is a fractal-rate point process and not a fractal point process since the interevent-interval density is exponentially (rather than power-law) distributed and since D_q assumes integer values for all q (see Prob. 5.5).

Examination of Figs. 5.4 and 5.5 shows that the two interval-based measures fail to reliably reveal fractal-based behavior in a point process, whereas the other two measures do. We address this issue in greater detail in Sec. 12.3.1.

5.5.3   Nonfractal point processes

In light of the foregoing definitions, the processes described in Chapter 4 are certainly "nonfractal."

Consider first the homogeneous Poisson process, described in Sec. 4.1. Examination of the measures set forth in that section reveals that none exhibit power-law behavior or scaling, except for trivial, integer powers in the count-based autocorrelation. Thus, the homogeneous Poisson process belongs to the class of nonfractal processes.

Renewal point processes and doubly stochastic Poisson processes both have fractal-based versions, as we have seen and shall see again in subsequent chapters. In general,

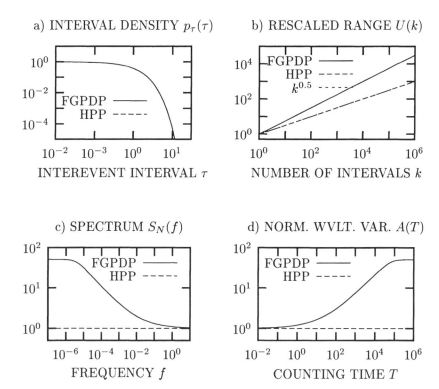

Fig. 5.5 Comparison of measures for a particular fractal-rate point process, the fractal-Gaussian-process-driven Poisson process with $\alpha = \frac{1}{2}$ (FGPDP; see Chapters 6, 8, 10, and 12); and a point process that is devoid of fractal properties, the homogeneous Poisson process (HPP; see Sec. 4.1). We plot cartoons for the same four probabilistic measures shown in Fig. 5.4: (a) interevent-interval probability density, (b) rescaled range, (c) point-process spectrum, and (d) normalized Haar-wavelet variance. For small values of the rate coefficient of variation, the interevent-interval probability density does not differentiate between these two processes [see Eq. (4.33)], although the intervals have different ordering in the two cases. However, the other three measures exhibit nontrivial power-law variations for the fractal-rate point process, but not for the nonfractal process.

however, these two classes of processes do not exhibit fractal characteristics, as is evident in Secs. 4.2 and 4.3, which leads us to term the general versions thereof as "nonfractal."

5.5.4 Identification of fractal-based point processes

A worthy goal of point-process analysis is the association of a particular point-process model with an observed point process. The ability to exclude competing models serves to narrow the range of possible mechanisms that could plausibly give rise to the data, thereby opening a window on the underlying science.

Because of the sparseness of point-process data, this is generally not an easy task. Consider the simple example of a fractal-rate point process. Rate fluctuations at frequencies significantly higher than the mean rate of the generated point events are essentially not transferred to the point process. Hence, details that could elucidate the nature of the rate process are lost. Moreover, no single statistic is sufficient to identify or characterize a fractal-based point process. Nevertheless, under certain circumstances, progress toward the identification of a fractal-based point process can be achieved by using a number of statistics in concert (see, for example, Rangarajan & Ding, 2000; Greis & Greenside, 1991).

We present a simple example that relies on the distinctions between fractal and fractal-rate point processes drawn earlier in this Section. If the estimated spectrum (Sec. 3.5.2) of the point process under study, $\widehat{S}_N(f)$, strongly indicates the presence fractal behavior, while the estimated interval spectrum (Sec. 3.3.3) of the process, $\widehat{S}_\tau(f)$, does not, then the point process in question may well belong to the fractal-renewal-process family. Further confirmation of such a hypothesis is provided by shuffling the intervals (see Sec. 11.5) and then recomputing the spectra. A renewal point process, whether fractal or not, is invariant to such shuffling since its interevent intervals are independent and identically distributed. A more direct approach to distinguishing between fractal and fractal-rate point processes relies on the generalized dimension D_q (see Prob. 5.5).

Further discussion related to the identification of point processes is deferred to Chapters 11 and 13, following the introduction of various fractal-based point-process models in Chapters 6–10.

Problems

5.1 *Rate-spectrum and wavelet-variance scaling-exponent estimates for experimental point processes* Consider the spectrum and normalized Haar-wavelet variance provided in Figs. 5.1 and 5.2.

5.1.1. Obtain estimates for the scaling exponents $\widehat{\alpha}_S$ and $\widehat{\alpha}_A$ for the COMPUTER and GENICULATE data. Show that $\widehat{\alpha}_A + (-\widehat{\alpha}_S) \approx 0$, in accordance with Eq. (5.49). Explain why the two exponents do not sum precisely to zero.

5.1.2. A plot of the normalized variance $\widehat{F}(T)$ vs. normalized counting time $T/\widehat{E}[\tau]$ for the COMPUTER and GENICULATE data appears in Fig. 5.6. The counting times T increase geometrically by factors of $10^{0.1}$, providing 10 counting times per decade. We also show theoretical curves that fit the data; these exhibit exponents $\widehat{\alpha}_F = 0.8$ and 1.0, respectively. Compare these values with $\widehat{\alpha}_S$ and $\widehat{\alpha}_A$ for the two data sets and comment on any discrepancy.

5.2 *Interval-spectrum scaling-exponent estimates for experimental point processes* Plots of the interval spectrum, $\widehat{S}_\tau(f)/\widehat{\mathrm{Var}}[\tau]$ vs. interval frequency f, are displayed in Fig. 5.7 for seven experimental point processes.

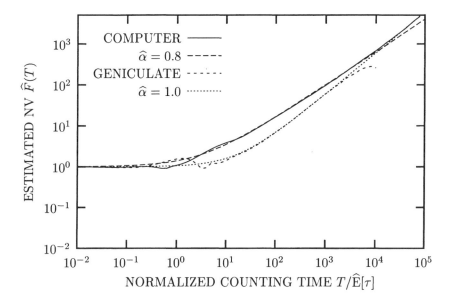

Fig. 5.6 Estimated normalized variance $\widehat{F}(T)$ vs. normalized counting time $T/\widehat{E}[\tau]$ for the COMPUTER and GENICULATE data displayed in Figs. 5.1 and 5.2. Also shown are the best fitting theoretical curves. For sufficiently large counting times, these curves increase roughly as straight lines, indicating an approximate power-law dependence on the counting time, with estimated power-law exponents $\widehat{\alpha}_F$.

5.2.1. Using the graphs provided in Figs. 5.1 and 5.7, determine whether $\widehat{\alpha}_S$ and $\widehat{\alpha}_{S_\tau}$ are less than, or greater than, unity for all data sets. In making these estimates, restrict yourself to the decreasing straight-line portions of the curves.

5.2.2. Using only the data displayed in Fig. 5.1, determine which of the point processes are likely represented by: (i) a fractal-based point process, (ii) a fractal point process, (iii) a fractal renewal process, (iv) a fractal-rate point process.

5.2.3. Now, consider the curves displayed in Fig. 5.7 in conjunction with those shown in Fig. 5.1. Using the information provided by both measures, which of the point processes are likely represented by: (i) a fractal-based point process, (ii) a fractal point process, (iii) a fractal renewal process, (iv) a fractal-rate point process?

5.2.4. Can $S_\tau(\mathsf{f})$ provide a good estimate of the fractal exponent?

5.2.5. Why is $\mathsf{f} = \frac{1}{2}$ the maximum interval frequency plotted for $S_\tau(\mathsf{f})$?

5.3 *Interval-wavelet-variance scaling-exponent estimates for experimental point processes* Figure 5.8 displays the estimated normalized interval wavelet variance, $\widehat{A}_\tau(k) = \widehat{\mathrm{Var}}[W_{\psi,\tau}(k,l)]/\widehat{\mathrm{Var}}[\tau]$ vs. number of intervals k [see Eq. (12.13)], for seven experimental point processes, calculated using the Haar wavelet.

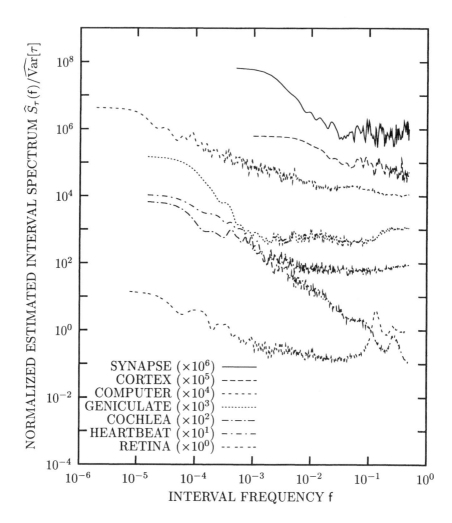

Fig. 5.7 Normalized estimated interval spectrum, $\widehat{S}_\tau(f)/\widehat{\mathrm{Var}}[\tau]$ vs. interval frequency f, for the same seven point processes as displayed in Fig. 5.1. We have smoothed the spectra to facilitate comparison (see Footnote 7 on p. 117). For sufficiently low interval frequencies, these curves decrease roughly as straight lines. This indicates an approximate power-law dependence on the interval frequency, with seven negative estimated power-law exponents $-\widehat{\alpha}_{S\tau}$.

5.3.1. Using the graphs provided in Figs. 5.2 and 5.8, determine whether $\widehat{\alpha}_A$ and $\widehat{\alpha}_{A\tau}$ are less than, or greater than, unity for all data sets. Restrict yourself to the increasing straight-line portions of the curves.

5.3.2. Using only the data displayed in Fig. 5.2, which of these point processes are likely represented by: (i) a fractal-based point process, (ii) a fractal point process, (iii) a fractal renewal process, (iv) a fractal-rate point process?

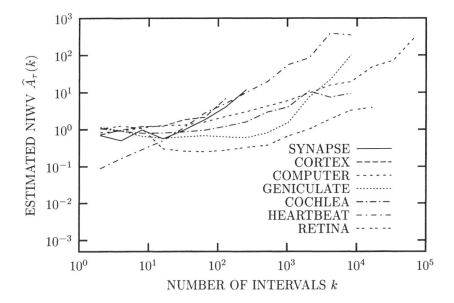

Fig. 5.8 Estimated normalized interval wavelet variance (NIWV), $\widehat{A}_\tau(k)$ vs. number of intervals k, for the same seven point processes displayed in Fig. 5.2. We employed the Haar wavelet for these calculations. The curves increase roughly as straight lines, indicating an approximate power-law dependence on the number of intervals, with seven positive estimated power-law exponents $\widehat{\alpha}_{A\tau}$.

5.3.3. Now, consider the curves displayed in Fig. 5.8 together with those shown in Fig. 5.2. Using the information provided by both measures, which of the point processes are likely represented by: (i) a fractal-based point process, (ii) a fractal point process, (iii) a fractal renewal process, (iv) a fractal-rate point process?

5.3.4. Can $\widehat{A}_\tau(k)$ provide a good estimate of the fractal exponent?

5.3.5. Why is $k = 2$ the minimum number of intervals plotted for $\widehat{A}_\tau(k)$?

5.4 *Interevent-interval histograms for experimental point processes*   In Fig. 5.9, we present plots of the normalized interevent-interval histogram, $\widehat{p}_\tau(\tau/\widehat{E}[\tau])$ vs. normalized interevent interval $\tau/\widehat{E}[\tau]$ [see Eq. (3.3)], for the seven experimental point processes considered earlier, presented on doubly logarithmic coordinates.

5.4.1. Based on the curves provided in Fig. 5.9, determine which of the underlying point processes can conceivably be represented by: (i) a fractal-based point process, (ii) a fractal point process, (iii) a fractal renewal process, (iv) a fractal-rate point process.

5.4.2. Now consider also the conclusions reached in the solutions of Probs. 5.2 and 5.3. Which of the point processes are likely represented by: (i) a fractal-based point process, (ii) a fractal point process, (iii) a fractal renewal process, (iv) a fractal-rate point process?

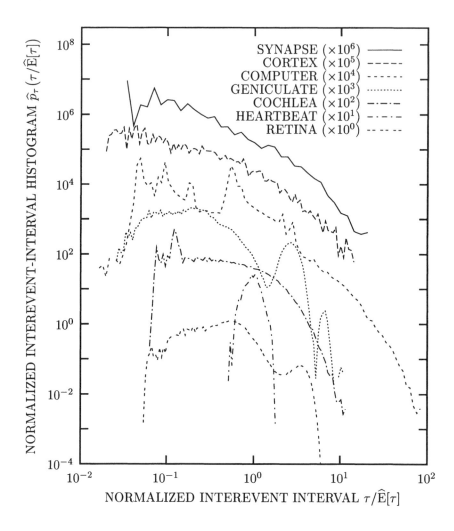

Fig. 5.9 Estimated normalized interevent-interval histogram, $\widehat{p}_\tau(\tau/\widehat{E}[\tau])$ vs. normalized interevent interval $\tau/\widehat{E}[\tau]$, for the same seven point processes displayed in Figs. 5.1 and 5.2. We constructed these histograms by employing 100 geometrically spaced bins, from the smallest to the largest interevent interval, with the exception of SYNAPSE, for which we used 45 bins to improve the presentation. When no intervals fell into a particular bin, we eliminated it; bins that contained intervals but were flanked on either side by empty bins also do not appear. Little information is lost by making use of this procedure.

5.5   *Generalized-dimension estimates for experimental point processes*   As the solutions to problems 5.2–5.4 show, the seven representative data sets discussed in this chapter clearly belong to the family of *fractal-based* point processes. Moreover, as we demonstrate in Sec. 11.5, the use of surrogate data sets (see Figs. 11.13 and

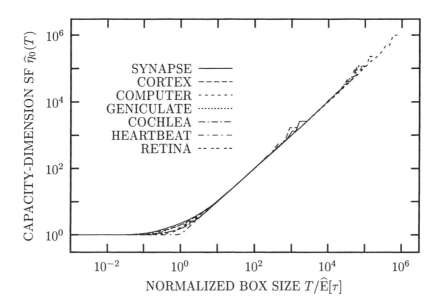

Fig. 5.10 Capacity-dimension scaling functions (SF) $\widehat{\eta}_0(T)$, based on Eq. (3.74), for the same seven point processes displayed in Figs. 5.1 and 5.2. All curves resemble those for the nonfractal homogeneous Poisson point process shown in Fig. B.3. All of the data sets displayed here are described by *fractal-rate* point processes.

11.14) will lead us to conclude that none of the seven are *fractal renewal* point processes. However, to determine whether the larger class of *fractal* point processes describes any of these data (see Sec. 5.5.1), we examine the generalized dimensions D_q considered in Sec. 3.5.4.

To facilitate comparison across different values of q, we make use of a doubly logarithmic plot of the generalized-dimension scaling function $\eta_q(T)$ provided in Eq. (3.73). These sums yield parallel curves for nonfractal and monofractal data sets and are therefore easier to visualize than the sums provided in Eq. (3.72), which have slopes that vary with q. For comparisons among data sets, it is also convenient to normalize the counting time to the mean interevent interval $\widehat{E}[\tau]$.

Figure 5.10 presents the capacity-dimension scaling function $\eta_0(T)$ for the seven canonical data sets considered earlier in this Chapter, calculated in accordance with Eq. (3.74).

5.5.1. Begin by simulating a homogeneous Poisson process (see Sec. 4.1), and a fractal renewal process with $\gamma = \frac{1}{2}$ and $B = \infty$ (see Sec. 4.2 and Ch. 7). For each process, generate 10^5 intervals, and normalize them by the estimated mean interval. Although this fractal renewal process has an infinite mean, the 10^5 intervals in the simulated realization will have a well-defined average value suitable for normalization (see Sec. 7.3). Display the capacity-dimension scaling function $\eta_0(T)$ vs. the normalized counting time $T/\widehat{E}[\tau]$ on a doubly logarithmic plot. Include the theoret-

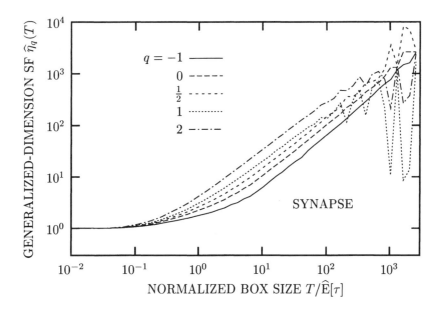

Fig. 5.11 Generalized-dimension scaling functions (SF) $\widehat{\eta}_q(T)$ for various values of q, based on Eq. (3.73), for spontaneous vesicular exocytosis at a developing *Xenopus* neuromuscular junction (SYNAPSE) (Lowen et al., 1997b, cell 950315e1). A *fractal-rate* point process describes these data. Analogous curves for the action-potential sequence at a visual-system INTERNEURON appear in Fig. 11.18.

ical form for the homogeneous Poisson process considered in Prob. 4.3, as well as a curve proportional to $T^\gamma = \sqrt{T}$ for the fractal renewal process (see Sec. 7.2.5).

5.5.2. Compare the experimental capacity-dimension scaling functions displayed in Fig. 5.10 with those shown in Fig. B.3, focusing on the slopes of the curves at small and large values of the normalized time. What conclusions can you draw? Why is the sharpness of the transition region in the vicinity of $T = \widehat{E}[\tau]$ different for the various curves?

5.5.3. Of all the data sets examined, Fig. 5.10 reveals that the capacity-dimension scaling function $\widehat{\eta}_0(T)$ for the SYNAPSE data has the most gradual transition between the slopes of zero and unity. While this curve does not indicate the presence of fractal-point-process behavior, it does not exclude the possibility that similar curves using other values of q might. Generalized-dimension scaling functions for the SYNAPSE data with $q = -1$, 0, $\frac{1}{2}$, 1, and 2 appear in Fig. 5.11. Can a fractal or multifractal point process describe these data? Why are the curves for the different values of q parallel but not coincident?

5.6 *Count autocorrelation function* Prove the last step before Eq. (5.13) in greater detail.

5.7 *Cutoff relationship for unity fractal exponent*   Observe that the expressions for $\alpha < 1$ and $\alpha > 1$ in Eq. (5.48) are identical. Show that taking the limit $\alpha \to 1$ in either of these expressions yields the result provided for $\alpha = 1$.

5.8 *Statistics for unity fractal exponent*   Starting with the spectrum provided in Eq. (5.46a), and assuming an abrupt low-frequency cutoff given by $f > 1/B$, prove the other relationships in Eq. (5.46). Consider the limit $kT/B \ll 1$. *Hint:* Prove them in the order b), c), e), and d).

5.9 *Increasing coincidence rate*   Consider a coincidence rate that *increases* in a power-law fashion with delay time ($\alpha > 1$). As set forth in Eq. (5.44e), the functional form of such a coincidence rate would increase without bound for $\alpha > 1$. In an attempt to avoid this flaw, we introduce an exponential cutoff at a large time B:

$$G(t) = \mathrm{E}[\mu]\,\delta(t) + \mathrm{E}^2[\mu] \left[1 + \mathrm{sgn}(t_G)\,(|t/t_G|)^{\alpha-1}\,e^{-|t|/B}\right]. \tag{5.50}$$

Again, sgn$(\cdot)$ denotes the sign of the argument.

Calculate the corresponding spectrum $S_N(f)$ and normalized variance $F(T)$ for $1 < \alpha < 3$, and show that one of these quantities must assume negative values for at least some times or frequencies. Since such behavior is inadmissible, what conclusions can you draw about the functional form of the coincidence rate given in Eq. (5.50)?

5.10 *Long-time-scale statistics for negative fractal exponents*   Show that Eq. (5.17) indeed approaches Eq. (5.20) in the low-frequency limit $f \ll f_S$. Use the relationships provided in Chapter 3 to derive Eqs. (5.19) and (5.21) from Eq. (5.20).

5.11 *Statistics for negative fractal exponents*   Use Eq. (5.15) (corresponding to $\alpha = -\frac{1}{2}$) to prove Eq. (5.17) for the normalized variance. Indicate how Eqs. (5.16) and (5.18) for the coincidence rate and normalized Haar-wavelet variance are obtained therefrom.

5.12 *Rate-spectrum scaling-exponent limits for data with nonstationary rates*
Calculation of the normalized variance and normalized Haar-wavelet variance for data with a nonstationary rate often produces fractal exponents that attain the maximum allowed values of 1 and 3, respectively, as shown in Secs. 5.2.3 and 5.2.4. Determine the behavior of the rate-spectrum fractal exponent for data that exhibits a nonstationary rate.

5.13 *Fractal behavior in nonstationary sets of points*   The validity of the relationships provided in Chapter 3, which are used throughout, generally requires stationarity. The Cantor set, described in Sec. 2.4.1, provides an example of a nonstationary set of points that highlights the limitations of these results for nonstationary processes (Lowen & Teich, 1995).

Consider a modification of the Cantor set construction procedure, in which we remove each closed interval from $\mathcal{C}_m$ and replace it with a single point event at its lower limit to yield a point process version thereof, $dN_m(t)$. The first three members of this series become

$$dN_0(t) = \delta(t)$$
$$dN_1(t) = \delta(t) \qquad\qquad\qquad + \delta(t - \tfrac{2}{3}) \qquad\qquad (5.51)$$
$$dN_2(t) = \delta(t) + \delta(t - \tfrac{2}{9}) + \delta(t - \tfrac{2}{3}) + \delta(t - \tfrac{8}{9}),$$

which follow from the rule

$$dN_{m+1}(t) = dN_m(t) \star [\delta(t) + \delta(t - 2/3^m)], \qquad (5.52)$$

where $\star$ denotes the convolution operation.

5.13.1. Show that the normalized variance[9] for $dN_{m+1}(t)$ does indeed exhibit scaling behavior, reflecting the fractal characteristics of this set of points.

5.13.2. Now demonstrate that the spectrum does *not* reveal scaling behavior. Comment on the applicability of the central results of fractal-based point processes to nonstationary sets of points.

---

[9] Since the collection of points under study originates via a deterministic process, it exhibits no randomness so that notions like variance and spectrum do not strictly apply. However, we treat the sets $dN_m(t)$ as if they were indeed random processes, and derive the statistical measures accordingly.

6

Processes Based on Fractional Brownian Motion

The Russian mathematician **Andrei Nikolaevich Kolmogorov (1903–1987)** substantially advanced the art of stochastic processes, particularly those involving turbulence and scaling; he formulated the concept of fractional Brownian motion in 1940.

Together with Benoit Mandelbrot, the American probabilist **John W. Van Ness (born 1936)** carried out a seminal study in 1968 that used the concept of self-similarity to advance the theory of fractional Brownian motion.

6.1	**Fractional Brownian Motion**	136
	6.1.1 Definition	137
	6.1.2 Properties	138
	6.1.3 Synthesis	139
	6.1.4 Realizations	139
	6.1.5 Rate process	140
6.2	**Fractional Gaussian Noise**	141
	6.2.1 Definition	141
	6.2.2 Properties	141
	6.2.3 Synthesis	142
	6.2.4 Realizations	142
	6.2.5 Rate process	142
6.3	**Nomenclature for Fractional Processes**	143
	6.3.1 Relationship between Hurst and scaling exponents	143
	6.3.2 Fractional integration	144
	6.3.3 Fractal Gaussian processes	144
6.4	**Fractal Chi-Squared Noise**	145
6.5	**Fractal Lognormal Noise**	147
6.6	**Point Process from Ordinary Brownian Motion**	149
	Problems	150

This chapter sets forth the properties of fractional Brownian motion, fractional Gaussian noise, and several related fractal processes. These continuous-time processes serve handily as rates for point processes. When used as the drivers for doubly stochastic or integrate-and-reset constructs, they impart their fractal characteristics to the ensuing fractal-rate point processes, thereby providing useful models for a variety of phenomena. Moreover, under suitable circumstances a number of non-Gaussian fractal-rate processes converge to these Gaussian processes. An excellent overview of the properties of fractional Brownian motion (Sec. 6.1) and fractional Gaussian noise (Sec. 6.2) has recently been provided by Taqqu (2003).

6.1 FRACTIONAL BROWNIAN MOTION

A concise mathematical description of ordinary Brownian motion (as represented by the Wiener–Lévy process) was provided in Sec. 2.4.2. The elements of **fractional Brownian motion**, an important generalization, were set forth by Kolmogorov (1940) and extensively developed by Mandelbrot & Van Ness (1968). A brief historical account of fractional Brownian motion has recently been provided by Molchan (2003).

6.1.1 Definition

Fractional Brownian motion, $B_H(t)$, is defined using the same three features that we specified for ordinary Brownian motion, $B(t)$: the amplitude distribution, the mean, and the autocorrelation. Like ordinary Brownian motion, fractional Brownian motion is Gaussian (Mandelbrot & Van Ness, 1968): a vector $\{B_H(t_1), B_H(t_2), ..., B_H(t_k)\}$, for any positive integer k and any set of times $\{t_1, t_2, ..., t_k\}$, has a joint Gaussian distribution. Fractional Brownian motion, as usually defined, also belongs to the zero-mean class of stochastic processes: $E[B_H(t)] = 0$ for all t.

The distinction between the fractional and ordinary versions thus lies in their autocorrelations. For fractional Brownian motion the autocorrelation takes the form (Kolmogorov, 1940; Mandelbrot & Van Ness, 1968)

$$E[B_H(s)\,B_H(t)] = \tfrac{1}{2}E[B_H^2(1)]\left(|t|^{2H} + |s|^{2H} - |t-s|^{2H}\right), \qquad (6.1)$$

with (Barton & Poor, 1988)

$$E[B_H^2(1)] = \Gamma(1-2H)\,\cos(\pi H)/(\pi H). \qquad (6.2)$$

The parameter H, known as the **Hurst exponent**, assumes values between zero and unity. For $H = \tfrac{1}{2}$, $B_H(t)$ is ordinary Brownian motion so that $B_{\frac{1}{2}}(t) \equiv B(t)$ (Mandelbrot & Van Ness, 1968). Hurst's (1951; 1956; 1965) pioneering work on long-range correlations in river flows, for which he developed rescaled range analysis as described in Sec. 3.3.5, also served as a precursor to fractional Brownian motion and its related processes (Mandelbrot, 1965b, 1982).

Fractional Brownian motion can be expressed in terms of ordinary Brownian motion:

$$
\begin{aligned}
\Gamma(H+\tfrac{1}{2})\,B_H(t) &= \int_{-\infty}^{0}\left[(t-s)^{H-1/2} - (-s)^{H-1/2}\right]dB(s) \\
&\quad + \int_{0}^{t}(t-s)^{H-1/2}\,dB(s) \qquad (6.3) \\
&= \int_{-\infty}^{t}(t-s)^{H-1/2}\,dB(s) \\
&\quad - \int_{-\infty}^{0}(-s)^{H-1/2}\,dB(s). \qquad (6.4)
\end{aligned}
$$

Indeed, this relationship often serves as a definition of fractional Brownian motion (Mandelbrot & Van Ness, 1968). Equation (6.3) comprises convergent integrals, whereas Eq. (6.4) exhibits more symmetry at the expense of employing divergent integrals (although with the same, convergent difference).

The standard deviation of fractional Brownian motion varies as t^H, so that this process is not stationary. However, manipulation of Eqs. (6.1), (6.3), or (6.4) reveals that fractional Brownian motion does have stationary increments:

$$\Pr\{B_H(t_2+s) - B_H(t_2) > x\} = \Pr\{B_H(t_1+s) - B_H(t_1) > x\} \qquad (6.5)$$

for all times s, t_1, t_2, and for all amplitudes x.

6.1.2 Properties

Like ordinary Brownian motion, fractional Brownian motion contains statistical copies of itself (Mandelbrot & Van Ness, 1968). Scaling the time axis of Eq. (6.1) by a factor a yields a new process $B_H(at)$ with autocorrelation

$$E[B_H(as)\,B_H(at)] = |a|^{2H}\,E[B_H(s)\,B_H(t)]; \tag{6.6}$$

scaling the amplitude directly by b yields $bB_H(t)$, with autocorrelation

$$E[bB_H(s)\,bB_H(t)] = b^2\,E[B_H(s)\,B_H(t)]. \tag{6.7}$$

By construction, both $B_H(at)$ and $bB_H(t)$ also have zero mean and belong to the Gaussian family of processes. Setting $|b| \equiv |a|^H$ makes the two autocorrelations coincide, and since the mean and autocorrelation uniquely determine a Gaussian process (Feller, 1971), $B_H(at)$ and $|a|^H B_H(t)$ are statistically identical. Thus, changing the time axis by a scale a and the amplitude axis by a scale $|a|^H$ yield the same result, and $B_H(t)$ contains statistical copies of itself at any scale. This argument generalizes the result set forth at the end of Sec. 2.4.2.

The definition provided in Sec. 6.1.1 leads to a number of other properties, including level-crossing statistics and generalizations of the spectrum. Level crossings of fractional Brownian motion, like those of ordinary Brownian motion, yield a degenerate point process; in any neighborhood about any level crossing, an infinite number of other crossings exist. The same degeneracy occurred for the Lévy dust encountered in Sec. 4.7, which this set of level crossings closely resembles. Imposing a minimum resolvable interevent time A results in a well-defined point process, for which the interevent-interval density $p_\tau(t)$ follows the Pareto form (Ding & Yang, 1995)

$$p_\tau(t) = (1 - H)\,A^{1-H}\,t^{H-2}, \quad t > A. \tag{6.8}$$

For $H \neq \frac{1}{2}$, this point process exhibits dependencies among the interevent intervals, and therefore does not belong to the renewal family of point processes.

As a result of its nonstationarity, difficulties arise in calculating the spectrum of fractional Brownian motion. The Wigner–Ville spectrum (Ville, 1948), $S_{W,X}(t, f)$, a generalization of the conventional spectrum, provides one solution to this problem. For a continuous-time real-valued process $X(t)$, this generalized spectrum takes the form

$$S_{W,X}(t, f) = \int_{-\infty}^{\infty} \left\{ E[X(t + s/2)\,X(t - s/2)] - E^2[X] \right\} e^{-i2\pi f s}\,ds. \tag{6.9}$$

For a wide-sense stationary process, the Wigner–Ville spectrum reduces to the conventional spectrum: $S_{W,X}(t, f) = S_X(f)$ for all t.

Applied to fractional Brownian motion, this transform yields (Flandrin, 1989)

$$S_{W,B_H}(t, f) = \left[1 - 2^{1-2H}\cos(4\pi f t)\right] |2\pi f|^{-(2H+1)}. \tag{6.10}$$

In a number of publications, the factor standing to the right of the brackets in Eq. (6.10) is presented out of context, as if it had arisen from a conventional spectrum. This has led to the (false) impression that fractional Brownian motion has a true spectrum that decays as $f^{-(2H+1)}$. Since this process is nonstationary, it does not possess a conventional spectrum.

6.1.3 Synthesis

A variety of synthesis techniques exist for generating discrete-time samples of fractional Brownian motion. An excellent overview of simulation methods, including their accuracies and efficiencies, has been provided by Bardet, Lang, Oppenheim, Philippe & Taqqu (2003). An early exact, but slow $[O(M^3)]$, method employed Cholesky decomposition of the complete autocorrelation matrix (Lundahl, Ohley, Kay & Siffert, 1986). Subsequent approximate methods reduced computational load $[O(M \log M)]$ at the expense of exact results; these include spectral and midpoint displacement methods (Peitgen & Saupe, 1988), improved spectral methods (Timmer & König, 1995), and wavelet approaches (Tewfik & Kim, 1992; Flandrin, 1992; Stoksik, Lane & Nguyen, 1994; Sellan, 1995; Abry & Sellan, 1996).

For $0 < H \leq \frac{1}{2}$ only, an exact and fast $[O(M \log M)]$ method is available (Lowen, 2000). It begins with a periodic discrete-time autocorrelation $R_x(n)$ proportional to $1 - |n/M|^{2H}$ for $|n| \leq M$. Succeeding steps involve transforming this function into the Fourier domain, taking the square root, randomizing the phase and amplitude, transforming back, and subtracting the first element of the resulting sequence from the rest. This final series yields M exact samples of fractional Brownian motion, with a computational load of order $M \log(M)$. Another efficient exact method relies on circulant matrices and enjoys fast computation time $[O(M \log M)]$, although minimum values for M exist (Davies & Harte, 1987).

6.1.4 Realizations

Figure 6.1 displays fractional Brownian motion for five different values of H (0.1, 0.3, 0.5, 0.7, and 0.9). The realization for $H = 0.5$ corresponds to ordinary Brownian motion (see also Fig. 2.2). To highlight the differences engendered by changing the value of H, we employ the same underlying white Gaussian random variables for generating all curves.

For convenience, we employ the improved spectral method, retaining 500 samples from a Fourier transform of length $2^{19} = 524288$, or just under 0.1%. A fraction as small as this results in negligible error. We normalize the plots so that all have a range of unity, from minimum to maximum displayed values, and we displace them from each other by unity.

Although the same statistical fluctuations appear in all five simulations, increasing the fractional-Brownian-motion parameter H leads to smoother curves. Qualitatively, larger values of H have generalized spectra that decay more quickly with frequency [see Eq. (6.10)]. Accordingly, these curves display a greater proportion of fluctuations at larger time scales than those for smaller values of H, and therefore appear smoother.

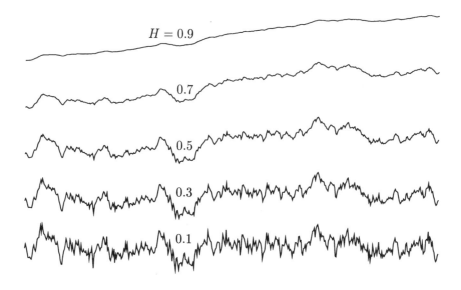

Fig. 6.1 Realizations of fractional Brownian motion for $H = 0.1, 0.3, 0.5, 0.7$, and 0.9, with H largest at the top. Corresponding values of the fractal exponent are $\alpha = 2H + 1 = 1.2$, 1.6, 2.0, 2.4, and 2.8, respectively (the relationship between H and α is provided in Sec. 6.3). The same random seed serves for all five curves, so that they all have similar features. Larger values of H and α yield smoother curves.

Generalized dimensions provide a precise description of this smoothness, and for true fractional Brownian motion (not the discrete-time approximations displayed in Fig 6.1), the various generalized dimensions yield a value of $2 - H$ for the motion itself, and $1 - H$ for its level crossings (Mandelbrot, 1982). Smaller values of H yield larger generalized dimensions, consistent with rougher curves. As H approaches unity, the generalized dimensions of fractional Brownian motion also approach unity, the value for a nonfractal, perfectly smooth curve.

6.1.5 Rate process

In serving as a rate for a point process, fractional Brownian motion exhibits two difficulties: nonstationarity and negative values. One solution to the nonstationarity issue involves imparting a cutoff at low frequencies, effectively imposing a longest time scale. An approximation of this kind yields useful results in many cases and can greatly simplify the associated theoretical and computational tasks. Other techniques are also available, including nonlinear ones such as resetting the fractional Brownian motion when it reaches a certain threshold, but these lead to more significant deviations from ideal behavior and to increased mathematical complexity.

Negative values, which do not make sense for a point-process rate, can be handled by offset, nonlinear transform, or both. The offset method, applied to a process rendered stationary by including a low-frequency cutoff, makes use of a mean value that is significantly larger than the standard deviation, so that the resulting fractional Brownian motion essentially always remains positive. The nonlinear transform method uses a nonlinear function $f(x)$ that generates a nonnegative output regardless of the input x, but this is achieved at the price of distorting the process. Candidate functions include $f(x) = (x + |x|)/2$, as well as smoother and more complex forms such as $f(x) = c \ln(1 + e^{x/c})$ for some constant $c > 0$.

6.2 FRACTIONAL GAUSSIAN NOISE

Fractional Gaussian noise $B'_H(t)$ represents the derivative of fractional Brownian motion, much as white Gaussian noise represents the derivative of ordinary Brownian motion.

6.2.1 Definition

Like white Gaussian noise, **fractional Gaussian noise** does not exist in a mathematical sense since fractional Brownian motion does not have a proper derivative. In contrast to fractional Brownian motion, where introducing a low-frequency cutoff ameliorates the effects of nonstationarity, a tractable approximation for fractional Gaussian noise is attained by introducing a high-frequency cutoff.

Perhaps the simplest method for obtaining fractional Gaussian noise in the face of the nondifferentiability of fractional Brownian motion entails convolving the latter with a rectangular filter before forming the derivative (Mandelbrot & Van Ness, 1968):

$$
\begin{aligned}
B'_{H2}(t, v) &= v^{-1} \frac{d}{dt} \int_{t-v/2}^{t+v/2} B_H(u)\, du \\
&= \frac{B_H(t + v/2) - B_H(t - v/2)}{v}.
\end{aligned}
\tag{6.11}
$$

This yields a process whose statistics resemble those of true fractional Gaussian noise, for time scales significantly greater than the separation time v. Other piecewise smooth filter functions yield similar results, but with increased mathematical complexity.

6.2.2 Properties

Combining Eqs. (6.9), (6.10), and (6.11) provides the Wigner–Ville spectrum of this filtered version of fractional Gaussian noise,

$$
S_{W, B'_{H2}}(t, v, f) = 4v^{-2} |2\pi f|^{-(2H+1)} \sin^2(\pi f v)
\tag{6.12}
$$

$$
\lim_{v \to 0} S_{W, B'_{H2}}(t, v, f) = |2\pi f|^{-(2H-1)},
\tag{6.13}
$$

where Eq. (6.13) yields the same result as that obtained for true fractional Gaussian noise $B'_H(t)$ using direct methods (Flandrin, 1989). Thus, true fractional Gaussian noise has a spectrum that follows a pure power-law decaying form, as does its approximation $B'_{H2}(t, v)$ for frequencies much lower than the effective cutoff frequency $f_S = (2\pi v)^{-1}$. Neither Eq. (6.12) nor Eq. (6.13) depend on time since both fractional Gaussian noise and its filtered version are stationary. Equation (6.12) does depend on v, but this is a filter parameter.

6.2.3   Synthesis

As a stationary process, fractional Gaussian noise has an autocorrelation with only one argument. Since the process also has a zero mean, this single-argument function completely specifies the fractional Gaussian noise process, simplifying simulation considerably.

A simpler variant of the method delineated by Lowen (2000) generates discrete-time realizations of fractional Gaussian noise for $H > \frac{1}{2}$ only (Davies & Harte, 1987). However, running sums of this process do not yield realizations of fractional Brownian motion; generating the latter over an interval $(0, t)$ from fractional Gaussian noise requires integrating over all times within that interval. Samples of fractional Gaussian noise lack information about the process at times other than the samples, and therefore cannot generate a proper integral of the underlying continuous-time fractional Gaussian noise. Similarly, differencing discrete-time samples of fractional Brownian motion does not yield samples of fractional Gaussian noise.

6.2.4   Realizations

Figure 6.2 displays fractional Gaussian noise for four different values of H (0.1, 0.3, 0.7, and 0.9), as well as for white Gaussian noise ($H = 0.5$) (see also Mandelbrot & Wallis, 1969a). To highlight the differences associated with various values of H, we again employ the same underlying white Gaussian random variables to generate all curves. Simulation methods, including the underlying random variables, follow those used for producing Fig. 6.1.

Again, increasing H leads to smoother curves, albeit with similar statistical fluctuations. In contrast to the results for fractional Brownian motion displayed in Fig. 6.1, these curves fill part of the plane near their mean values, so that the generalized dimensions assume a value of two in all cases. Similarly, the level crossing sets all have generalized dimensions of unity.

6.2.5   Rate process

Unlike fractional Brownian motion, fractional Gaussian noise is stationary. Like fractional Brownian motion, however, it can assume negative values. To enable fractional Gaussian noise to serve as a rate, the same methods used to ameliorate the negative-value shortcoming of fractional Brownian motion can be used for fractional Gaussian noise as well (see Sec. 6.1.5).

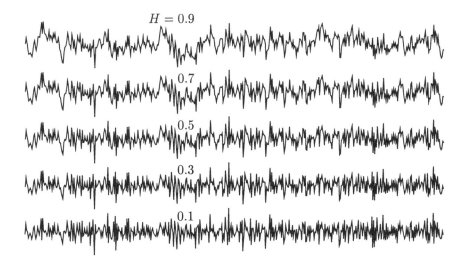

$H = 0.9$

0.7

0.5

0.3

0.1

Fig. 6.2 Realizations of fractional Gaussian noise for $H = 0.1$, 0.3, 0.5, 0.7, and 0.9, with H largest at the top. Corresponding values of the fractal exponent are $\alpha = 2H - 1 = -0.8$, -0.4, 0.0, 0.4, and 0.8, respectively (the relationship between H and α appears in Sec. 6.3). The same random seed serves for all five curves (as well as for the five curves in Fig. 6.1) so that all have similar features. The curves become somewhat smoother as H and α increase.

6.3   NOMENCLATURE FOR FRACTIONAL PROCESSES

6.3.1   Relationship between Hurst and scaling exponents

Fractional Brownian motion and fractional Gaussian noise have heretofore been defined in terms of the Hurst exponent H. Since most other processes are cast in terms of the scaling exponent α, we proceed to relate these two parameters.

We have thus far defined H in accordance with common usage: it lies between zero and unity for *both* fractional Brownian motion and fractional Gaussian noise (Mandelbrot & Van Ness, 1968; Barton & Poor, 1988). We know, however, that the spectrum of fractional Gaussian noise (FGN) varies as $f^{-\alpha} = f^{-(2H-1)}$ whereas that for a stationary version of fractional Brownian motion (FBM) varies as $f^{-\alpha} = f^{-(2H+1)}$. Thus, different relations are required to connect the exponents, depending on the process:

$$
\begin{array}{llll}
\text{FGN:} & \alpha = 2H - 1 & & -1 < \alpha < 1 \\
\text{FBM:} & \alpha = 2H + 1 & 0 < H < 1 & 1 < \alpha < 3.
\end{array} \tag{6.14}
$$

However, the relationship between H and α applicable for fractional Brownian motion is sometimes also taken to apply to fractional Gaussian noise (Mandelbrot, 1982;

Flandrin, 1992). This requires that H span different ranges for the two processes:

$$
\begin{array}{llll}
\text{FGN:} & & -1 < H < 0 & -1 < \alpha < 1 \\
& \alpha = 2H + 1 & & \\
\text{FBM:} & & 0 < H < 1 & 1 < \alpha < 3.
\end{array} \tag{6.15}
$$

Since the ranges of H differ in Eqs. (6.14) and (6.15), whereas those for α remain the same, using the latter exponent avoids confusion. Hence, we generally eschew the Hurst exponent H in favor of the scaling exponent α. We do continue to refer to fractional Brownian motion as $B_H(t)$, however.

6.3.2   Fractional integration

It is clear from Eqs. (6.14) and (6.15) that the ranges of the exponent α for fractional Gaussian noise and fractional Brownian motion differ by the integer two. This results from a simple property of integration: the power-law exponent of the spectrum increases by precisely two (differentiation results in a decrease of the exponent by precisely two). More generally, the n-fold integration of a process results in an increase in the exponent by $2n$.

Generalizing still further, x-fold Riemann–Liouville fractional integration corresponds to an increase in the exponent by $2x$ (see, for example, Pipiras & Taqqu, 2003). Such fractional integration provides a method for generating fractional Brownian motion and fractional Gaussian noise from white Gaussian noise (Barnes & Allan, 1966; Mandelbrot, 1967b; Maccone, 1981). Equations (6.3) and (6.4) represent just such an operation, with the kernels in these integrals representing fractional integration of degree $\alpha/2$ operating on the ordinary Brownian motion process $B(t)$ (Mandelbrot & Van Ness, 1968; Pipiras & Taqqu, 2003).

6.3.3   Fractal Gaussian processes

We do not intend to imply that fractional Brownian motion and fractional Gaussian noise behave identically; they certainly differ in their stationarity. However, since we focus on stationary processes throughout this book, we impose cutoffs on these two processes when necessary, thereby obviating any differences associated with the issue of stationarity.

Another distinction lies in some measures that provide useful scaling results for fractional Gaussian noise, but not for fractional Brownian motion; this is a result of the different ranges spanned by α in the two cases. For this reason, some authors have suggested employing different methods for analyzing fractional Brownian motion and fractional Gaussian noise (Raymond & Bassingthwaighte, 1999). However, if we bear in mind the limited ranges of the various measures set forth in Sec. 5.2 and, in particular, if we choose measures that prove useful for both processes, the significance of this difference is reduced.

With the differences between fractional Brownian motion and fractional Gaussian noise diminished in this context, it proves simpler and more accurate to refer to this family of processes as **fractal Gaussian processes**, indexed by a spectral power-law exponent α that may take any positive value (we specifically exclude fractional

Gaussian noise with $\alpha \leq 0$ from this class for the reasons set forth in Sec. 5.2.1). We then consider Figs. 6.2 and 6.1 as a unit; they display fractal Gaussian processes with increasing smoothness as the value of α climbs from $0+$ to 2.8.

With negative values of α eliminated, and low-frequency cutoffs for $\alpha \geq 1$ imposed, the family of fractal Gaussian processes can dutifully serve as rate processes, thereby enabling the construction of doubly stochastic Poisson point processes (Sec. 4.3) and integrate-and-reset point processes (Sec. 4.4). Indeed, the integration inherent in both of these constructions smoothes the rate sufficiently to render high-frequency cutoffs unnecessary. Gaussian processes are convenient as rates because they are ubiquitous and are fully characterized by their means and covariances; this facilitates comparison with experiment. Moreover, they emerge from fractal binomial noise and fractal shot noise in important limits, as discussed in Secs. 8.3.2 and 10.6.1, respectively.

This process has, in fact, been used as a rate for a doubly stochastic Poisson process, resulting in the **fractal-Gaussian-process-driven Poisson process** (see Fig. 5.5 as well as Secs. 8.4 and 10.6.1). Because of its widespread applicability, we have chosen it for the analysis and estimation studies carried out in Chapter 12.

In the domain of neurophysiology, for example, it is useful for modeling sequences of action potentials. In particular, it has served as a valuable point of departure for characterizing mammalian auditory-nerve action potentials for high-frequency stimuli (Teich, Turcott & Lowen, 1990; Teich, 1992; Lowen & Teich, 1993b). Low-frequency stimuli are accommodated by forming a driving function from the superposition of a fractal Gaussian process and the modulating stimulus. When modified to accommodate the effects of neural refractoriness (see Sec. 11.2.4), the dead-time-modified version of this process characterizes essentially all of the observable aspects of auditory neural spike trains elicited by a broad range of stimuli, over a broad range of time scales (Lowen & Teich, 1996b, 1997). We point out, however, that alternative fractal-based point-process constructs have also been formulated to describe the auditory neural spike train (see Secs. 6.4, 6.6, and 8.4).

In a similar manner, the **fractal-Gaussian-process-driven integrate-and-reset process** serves as an excellent model for action-potential generation in the peripheral visual system, provided that the rate process comprises a superposition of a fractal Gaussian process and the modulating stimulus, and that neural refractoriness (see Sec. 11.2.4) is accommodated (Teich & Lowen, 2003). Furthermore, imparting random displacement (see Sec. 11.3) to the fractal-Gaussian-process-driven integrate-and-reset process yields a model that serves as a good descriptor for the sequence of human heartbeats (Teich et al., 2001). Nonfractal point processes serve as suitable models only over short time scales.

6.4   FRACTAL CHI-SQUARED NOISE

The nonlinear transforms provided at the end of Sec. 6.1.5 ensure nonnegative rate functions while introducing a minimum of change in the Gaussian amplitude distribution. For some applications, however, amplitude distributions other than the Gaussian prove useful. As one example, we examine the properties of chi-squared noise.

If $\{X_k(t)\}$, $1 \leq k \leq M$, represents a collection of M independent and identically distributed Gaussian processes with zero mean, variance $\text{Var}[X]$, and autocorrelation $R_X(t)$, then $\mu(t) \equiv \sum_{k=1}^{M} X_k^2(t)$ has an autocorrelation given by

$$
\begin{aligned}
R_\mu(t) &\equiv \text{E}\left[\sum_{n=1}^{M} X_n^2(s) \sum_{m=1}^{M} X_m^2(s+t)\right] = \sum_{n=1}^{M} \sum_{m=1}^{M} \text{E}\left[X_n^2(s) X_m^2(s+t)\right] \\
&= \sum_{n=1}^{M} \left(\text{E}\left[X_n^2(s) X_n^2(s+t)\right] + \sum_{m \neq n} \text{E}\left[X_n^2(s) X_m^2(s+t)\right]\right) \\
&= \sum_{n=1}^{M} \left(2R_X^2(t) + \text{E}[X^2]\text{E}[X^2] + \sum_{m \neq n} \text{E}[X^2]\text{E}[X^2]\right) \\
&= 2MR_X^2(t) + \text{E}^2[\mu].
\end{aligned}
\tag{6.16}
$$

In deriving this result, we have made use of the independence of the component fractal Gaussian processes $X_k(t)$ that comprise $\mu(t)$, along with the well-known property

$$
\begin{aligned}
\text{E}[X_1 X_2 X_3 X_4] &= \text{E}[X_1 X_2]\text{E}[X_3 X_4] \\
&+ \text{E}[X_1 X_3]\text{E}[X_2 X_4] \\
&+ \text{E}[X_1 X_4]\text{E}[X_2 X_3]
\end{aligned}
\tag{6.17}
$$

for any four zero-mean jointly Gaussian random variables $\{X_1, X_2, X_3, X_4\}$.

The process $\mu(t)$ then has a chi-squared (χ^2) distribution with M degrees of freedom (Feller, 1971). Standard probability theory yields the statistics of this amplitude, including the probability density function

$$
p_\mu(y) = [\Gamma(M/2)]^{-1} \left(2\text{Var}[X]\right)^{-M/2} y^{M/2-1} \exp\left(-y/2\text{Var}[X]\right)
\tag{6.18}
$$

and moments

$$
\text{E}[\mu^n] = \frac{\Gamma(n+M/2)}{\Gamma(M/2)} \left(2\text{Var}[X]\right)^n.
\tag{6.19}
$$

If $X(t)$ belongs to the fractal class of continuous processes, with fractal exponent α_X in the range $\frac{1}{2} < \alpha_X < 1$, then $R_\mu(t)$ also exhibits scaling, but with the exponent $\alpha_\mu = 2\alpha_X - 1$ (Thurner et al., 1997, see also Prob. 6.6). The result is **fractal chi-squared noise**. Setting $M = 2$ yields **fractal exponential noise** since the χ^2 distribution with two degrees of freedom is the exponential distribution.

The chi-squared distribution with $2M$ degrees of freedom successfully models a whole host of phenomena, including the energy fluctuations of multimode thermal light (Mandel, 1959; Saleh, 1978) and multimode acoustic noise (McGill, 1967). Smearing the mean of a Poisson counting kernel with the chi-squared distribution yields the associated photon-counting and neural-counting distributions for these two processes. The result is the **negative binomial counting distribution** (Mandel, 1959;

McGill, 1967), the origin of which resides in Greenwood & Yule's (1920) seminal study of accident occurrences.

A related model employs the sum of the squares of positive-mean Gaussian noise processes, which generates noncentral chi-squared noise. If the components belong to the fractal class of continuous processes, the outcome is **fractal noncentral chi-squared noise**. The presence of the nonzero mean modifies the second-order characteristics of the process so that, in some cases, two power-law regions emerge with both fractal exponents, α_X and α_μ, appearing in different ranges of the associated spectrum (see Prob. 6.6). Setting $M = 2$ yields **fractal noncentral Rician-squared noise**, since the noncentral chi-squared distribution with two degrees of freedom is the noncentral Rician-squared distribution (Rice, 1944, 1945; Saleh, 1978).[1]

Fractal noncentral chi-squared noise has also been used as a rate for a doubly stochastic Poisson point process, again to model mammalian auditory-nerve action potentials (Kumar & Johnson, 1993). The Poisson transform of the noncentral chi-squared distribution, which is known as the **noncentral negative binomial distribution**, has found extensive use in photon counting and neural counting (Peřina, 1967; McGill, 1967; Teich & McGill, 1976; Li & Teich, 1993).

6.5 FRACTAL LOGNORMAL NOISE

We consider an additional example of continuous rate process with a non-Gaussian amplitude that finds use in many contexts: **fractal lognormal noise**. The term "lognormal" refers to a random quantity whose logarithm follows a Gaussian (normal) form (Aitchison & Brown, 1957; Gumbel, 1958). The exponential transform of a fractal Gaussian process follows this form precisely and, furthermore, renders the resulting process strictly positive so that it may serve as a rate without further transformation.

Specifically, let $X(t)$ represent a Gaussian process with mean $\mathrm{E}[X]$, variance $\mathrm{Var}[X]$, and autocorrelation $R_X(t)$, and define a rate $\mu(t) \equiv \exp[X(t)]$. This rate then has the lognormal probability density function

$$p_\mu(y) = \left(2\pi \mathrm{Var}[X]\right)^{-1/2} y^{-1} \exp\left(-\left\{\ln(y) - \mathrm{E}[X]\right\}^2/2\mathrm{Var}[X]\right). \quad (6.20)$$

Straightforward application of probability theory yields the moments of the rate (see Lowen et al., 1997b, and Sec. A.3.1):

$$\begin{aligned}
\mathrm{E}[\mu^n] &= \exp\left(n\,\mathrm{E}[X] + n^2\,\mathrm{Var}[X]/2\right) \\
\mathrm{E}[\mu] &= \exp\left(\mathrm{E}[X] + \mathrm{Var}[X]/2\right) \\
\mathrm{Var}[\mu] &= \exp\left(2\mathrm{E}[X]\right)\left[\exp\left(2\mathrm{Var}[X]\right) - \exp\left(\mathrm{Var}[X]\right)\right].
\end{aligned} \quad (6.21)$$

---

[1] A photograph of Rice stands at the beginning of Chapter 9.

After a fair amount of calculation, a result for the autocorrelation emerges (see Lowen et al., 1997b, and Sec. A.3.1):

$$R_\mu(t) = \mathrm{E}^2[\mu]\,\exp\{R_X(t) - \mathrm{E}^2[X]\}. \tag{6.22}$$

We point out that the exponential transform can lead to rather skewed distributions for μ. Substituting Eq. (6.21) into Eq. (3.4) yields a skewness given by

$$\mathrm{E}[(\mu - \mathrm{E}[\mu])^3]/\mathrm{Var}^{3/2}[\mu] = \left[\exp(\mathrm{Var}[X]) - 1\right]^{1/2}\left[\exp(\mathrm{Var}[X]) + 2\right]; \tag{6.23}$$

this quantity assumes large values for relatively small values of $\mathrm{Var}[X]$. As an example, $\mathrm{Var}[X] = 5$ yields a skewness of 1826. In contrast, an exponential distribution has a skewness of two, whereas a Gaussian has zero skewness.

With many of the properties of the rate $\mu(t)$ determined, we now consider point processes generated from this rate. We begin with the Poisson-process version. If we assume that the rate $\mu(t)$ exhibits fluctuations over frequency ranges significantly lower than the mean rate $\mathrm{E}[\mu]$, then closed-form expressions for the moments of the intervals τ between events exist (Lowen et al., 1997a,b):

$$\begin{aligned}
\mathrm{E}[\tau^n] &= n!\,\exp\left\{-n\,\mathrm{E}[X] + (n^2 - 2n)\,\mathrm{Var}[X]/2\right\} \\
\mathrm{E}[\tau] &= \exp(-\mathrm{E}[X] - \mathrm{Var}[X]/2) \\
\mathrm{Var}[\tau] &= \exp(-2\mathrm{E}[X])\left[2 - \exp(-\mathrm{Var}[X])\right],
\end{aligned} \tag{6.24}$$

where we have made use of Eq. (4.31). Employing Eq. (4.32) leads to the associated interevent-interval probability density (Lowen et al., 1997a),

$$\begin{aligned}
p_\tau(t) = \pi^{-1/2}\,\exp(\mathrm{E}[X] + \tfrac{3}{2}\,\mathrm{Var}[X]) \\
\times \int_{-\infty}^{\infty} \exp\left[-y^2 - t\exp\left(\mathrm{E}[X] + 2\mathrm{Var}[X] + \sqrt{2\mathrm{Var}[X]}\,y\right)\right] dy.
\end{aligned} \tag{6.25}$$

For the integrate-and-reset version, we again require that the rate process $\mu(t)$ not exhibit fluctuations over frequencies comparable to, or higher than, the mean rate $\mathrm{E}[\mu]$. Using Eq. (4.38) provides results similar to those presented in Eq. (6.24):

$$\begin{aligned}
\mathrm{E}[\tau^n] &= \exp\left\{-n\,\mathrm{E}[X] + (n^2 - 2n)\,\mathrm{Var}[X]/2\right\} \\
\mathrm{E}[\tau] &= \exp(-\mathrm{E}[X] - \mathrm{Var}[X]/2) \\
\mathrm{Var}[\tau] &= \exp(-2\mathrm{E}[X])\left[1 - \exp(-\mathrm{Var}[X])\right],
\end{aligned} \tag{6.26}$$

while combining Eqs. (4.37) and (6.20) yields

$$\begin{aligned}
p_\tau(t) = (2\pi\mathrm{Var}[X])^{-1/2}\,\exp(-\mathrm{E}[X] - \mathrm{Var}[X]/2) \\
\times \exp\left(-\{\ln(t) + \mathrm{E}[X]\}^2/2\mathrm{Var}[X]\right) t^{-2}.
\end{aligned} \tag{6.27}$$

We now turn to the specific case where $X(t)$ belongs to the family of fractal Gaussian processes. The exponential transformation in Eq. (6.22) renders nonlinear the relationship between the autocorrelation of the input process $X(t)$ and that of the rate $\mu(t)$; in particular $S_N(f)$, the spectrum of the generated point process $dN(t)$, does not follow an exact power-law decay as we assume for $S_X(f)$. This holds true both for the doubly stochastic Poisson and integrate-and-reset versions. However, when $\text{Var}[X]$ is relatively small in comparison with unity, the forms of the two spectra do not differ greatly. Conversely, one can construct a Gaussian process with an appropriate autocorrelation such that the resulting lognormal noise has precisely fractal characteristics. For example, $R_X(t) = c_1 + (\alpha - 1) \ln |t|$, for $0 < \alpha < 1$, over a large range of delay times t, yields $S_N(f) = c_2 f^{-\alpha}$ over a corresponding range of frequencies f.

The **fractal lognormal-noise-driven Poisson process** turns out to be a suitable model for describing vesicular exocytosis (see Lowen et al., 1997a,b, as well as Prob. 6.8).

6.6 POINT PROCESS FROM ORDINARY BROWNIAN MOTION

Davidsen & Schuster (2002) have recently drawn attention to a simple but plausible method for generating fractal-based point processes from ordinary Brownian motion (see also Kaulakys, 1999). Their construct resembles a conventional integrate-and-reset process but differs in that the threshold, rather than the integration rate, is taken to be a stochastic process.

This kind of behavior occurs in neurophysiology, for example, where ion-channel current fluctuations give rise to random threshold fluctuations. A variety of models have been used to introduce such "fluctuations in excitability" (Pecher, 1939; Verveen & Derksen, 1968; Holden, 1976, Chapters 1 and 4). In one well-established recipe, the threshold undergoes diffusion, with or without drift, resulting in interval statistics that obey the inverse-Gaussian density (Holden, 1976).

In the model considered by Davidsen & Schuster (2002), the rate remains fixed and the threshold process is taken to be ordinary Brownian motion. When the integrated state variable reaches the threshold, an output event is generated and the state variable is reset to some fixed value, as with a conventional integrate-and-reset process. In the case at hand, however, the threshold does not undergo a reset as a result of the generation of the output event. To ensure both a tractable process and finite interevent intervals, the threshold typically has lower and upper reflecting barriers, and these barriers are greater than the state-variable reset values. The persistence of the threshold across interevent intervals renders the process nonrenewal; the power-law exponents associated with $p_\tau(t)$ and $S_N(f)$ therefore need not coincide.

It turns out that the particular form of the integration employed does not qualitatively affect the result; a leaky integrator, for example, yields results similar to those obtained by using the linearly increasing state variable described above. This model generates fractal-based point processes with a rich variety of scaling behavior in both the interevent-interval density $p_\tau(t)$ and the spectrum $S_N(f)$ (Davidsen &

Schuster, 2002). It shows promise in characterizing a number of phenomena in the biological and physical sciences, including action-potential sequences and earthquake occurrences (see Prob. 10.7, however).

Problems

6.1 *Autocorrelation with scaled time*   Prove Eq. (6.6) using Eq. (6.1).

6.2 *Stationary increments*   Prove Eq. (6.5) using Eq. (6.1).

6.3 *Autocorrelation coefficient*   We can define an autocorrelation coefficient of sorts for fractional Brownian motion,

$$\rho(s,t) \equiv \frac{\mathrm{E}[B_H(s)\, B_H(t)]}{\left(\mathrm{E}[B_H^2(s)]\, \mathrm{E}[B_H^2(t)]\right)^{1/2}}, \tag{6.28}$$

assuming that neither $s = 0$ nor $t = 0$.

    6.3.1. Find a simplified version of Eq. (6.28), and show that it depends only on the ratio s/t.

    6.3.2. Find a further simplification for the special case $s = -t$ (Mandelbrot, 1982, p. 353). Which values of H make $B_H(t)$ and $B_H(-t)$ independent?

6.4 *Variance of ordinary Brownian motion at unity time*   Equation (6.2) provides an expression for $\mathrm{E}[B_H^2(1)]$, the variance of fractional Brownian motion at a time of unity. Show that the resulting expression for ordinary Brownian motion (where $H = \frac{1}{2}$) assumes a value of unity.

6.5 *Generation of ordinary Brownian motion*   The midpoint displacement algorithm provides a simple and fast method for generating ordinary Brownian motion, and permits the generation of additional detail (intermediate values) between points already defined. Imagine a realization of ordinary Brownian motion sampled at integer multiples of a sampling time τ_0; we thus have $B(k\tau_0)$ for all integers k. Now we wish to insert intermediate values at times $t = (k + \frac{1}{2})\tau_0$. Let $\mathcal{N}(0,1)$ denote an infinite sequence of Gaussian-distributed random variables with zero mean and unit variance, independent of each other and of the samples $B(k\tau_0)$.

    6.5.1. Given the values $B(k\tau_0)$ and $B[(k+1)\tau_0]$, and a realization of $\mathcal{N}(0,1)$, what value should we insert for $B[(k + \frac{1}{2})\tau_0]$?

    6.5.2. Show that this method does not ignore any other correlations among the inserted values.

    6.5.3. Show that this method fails for general $B_H(k\tau_0)$, remaining valid only for $H = \frac{1}{2}$.

6.6 *Doubly stochastic Poisson process driven by fractal chi-squared rate*   Let $\{X_k(t)\}$, $1 \le k \le M$, represent a collection of M independent and identically distributed Gaussian processes with variance $\mathrm{Var}[X]$ and autocorrelation $R_X(t)$. Suppose we set the mean to some large value, such that $\mathrm{E}^2[X]/\mathrm{Var}[X] \gg 1$, and we let one element of the collection [$X_1(t)$, say] serve as a rate for a doubly stochastic

Poisson point process $dN_1(t)$. Suppose further that $dN_1(t)$ has a spectrum that decays as $\sim f^{-\alpha_X}$ with $\frac{1}{2} < \alpha_X < 1$ over some large range of frequencies. Now define $\mu(t) \equiv \sum_{k=1}^{M} \{X_k(t) - \mathrm{E}[X]\}^2$, a fractal chi-squared process, and let this serve as a rate for a separate doubly stochastic Poisson point process $dN_R(t)$.

6.6.1. Show that $dN_R(t)$ has a spectrum that decays as $\sim f^{1-2\alpha_X}$ over an appreciable range of frequencies.

6.6.2. What form does the spectrum take if we do not subtract the mean when generating the rate $\mu(t)$, that is if we consider a fractal noncentral chi-squared process?

6.7 *Spectrum for Poisson process driven by exponentiated-Gaussian process*   Let $X(t)$ denote a dimensionless Gaussian process with mean $\mathrm{E}[X]$, variance $\mathrm{Var}[X]$, and autocorrelation function

$$R_X(t) \approx \mathrm{E}^2[X] + c\ln(t_0/|t|) \qquad (6.29)$$

for delay times t in the range $A \ll t \ll B$, with constants $0 < c < 1$ and $t_0 > 0$. Now let $\mu(t) \equiv \exp[X(t)]$ serve as a rate for a Poisson point process. Show that the spectrum of the resulting point process follows a $1/f$-type form for frequencies f in the range $1/B \ll f \ll 1/A$. Also, find the high-frequency asymptote, the fractal exponent α, and an expression for the cutoff frequency f_S.

6.8 *Vesicular exocytosis at the synapse*   Vesicular exocytosis is a mechanism that mediates the passage of cellular signals from one cell to another across the synapse between them (Katz, 1966). Exocytosis also occurs spontaneously, so that neurotransmitter molecules flow across the synapse even in the absence of signaling (Fatt & Katz, 1952), albeit at a substantially reduced rate. Spontaneous exocytosis appears to have its origin in random thermal fluctuations; these cause ion channels to open, which, in turn, admit sufficient calcium into the cell to trigger vesicular exocytosis (Zucker, 1993).

Transition-state theory (Berry, Rice & Ross, 1980) describes the dependence of the expected exocytic rate μ on various parameters of the cells (Hille, 2001), and predicts that it follows the Arrhenius equation,

$$\mu = \mathcal{A}\exp\{-[E_A - \mathsf{q}V]/\mathcal{R}\mathcal{T}\}, \qquad (6.30)$$

where $\mathcal{A}$ is a rate constant, E_A is a constant activation energy, q is a constant charge associated with the transition, $\mathcal{R}$ is the thermodynamic gas constant, $\mathcal{T}$ is the absolute temperature, and V is the membrane voltage of the presynaptic cell.

6.8.1. What point process would be a suitable candidate for describing the sequence of exocytic events if the membrane voltage V is fixed?

6.8.2. How would the collection of processes associated with multiple cells be described if we assume that V is constant for each cell, but varies across cells?

6.8.3. In actuality, the membrane resting voltage V does not remain fixed, but rather exhibits a Gaussian amplitude distribution with fractal ($1/f$-type) fluctuations (Verveen & Derksen, 1968; Holden, 1976; Stern et al., 1997). This suggests that the membrane voltage may be described by a fractal Gaussian process. Show how this leads to the fractal lognormal-noise-driven Poisson process as a model for the spontaneous vesicular exocytosis process in real preparations (Lowen et al., 1997a,b). We have already demonstrated that neither a fractal point process nor a multifractal point process provides a good description for these data (see Prob. 5.5.3 and Fig. 5.11).

7

Fractal Renewal Processes

Vilfredo Federigo Samaso Pareto (1848–1923), an aristocratic Italian economist associated with the University of Lausanne, discovered that scale-invariant, power-law distributions characterize the income of individuals in many societies.

Working with Jay Berger in 1963, **Benoit B. Mandelbrot (born 1924)** identified self-similar error clusters in data-transmission systems; he long ago recognized that fractals abound in many fields and set forth the principles of fractal analysis.

7.1	**Power-Law Distributed Interevent Intervals**	155
	7.1.1 **Abrupt-cutoff interevent-interval density**	155
	7.1.2 **Exponential-cutoff interevent-interval density**	156
	7.1.3 **Effect of γ on interval variability**	157
7.2	**Statistics of the Fractal Renewal Process**	157
	7.2.1 **Point-process spectrum**	157
	7.2.2 **Coincidence rate**	159
	7.2.3 **Normalized variances**	160
	7.2.4 **Counting distribution**	163
	7.2.5 **Capacity dimension**	164
7.3	**Nondegenerate Realization of a Zero-Rate Process**	164
	Problems	166

Perhaps the simplest fractal-based point process is the **fractal renewal process**, in which the intervals between successive events $\{\tau_n\}$ follow a decaying power-law (hyperbolic) probability density function. Like all renewal processes (see Sec. 4.2), these intervals are independent and identically distributed.

The power-law density function is known as **Pareto's Law**, in honor of Vilfredo Pareto who first established it in 1896. Pareto successfully used it to characterize a broad range of phenomena, the most celebrated of which is the income level of individuals. And, indeed, Pareto's Law has continued to enjoy widespread use in econometric and financial analyses, and in the evaluation of risk in trading (see, for example, Mandelbrot, 1960, 1964, 1982, 1997; Mandelbrot & Hudson, 2004).

A well-known modern application of this law lies in the statistics of errors following data transmission over a telephone line. In an approach promulgated by Berger & Mandelbrot (1963), a sequence of samples drawn from a power-law density forms a fractal renewal process that is used to model the occurrences of these errors. It had long been known that transmission-error occurrences appeared in clusters, and in clusters of clusters; these clusters were separated by relatively long periods of time during which no errors occurred. Using data provided by the German Federal Telephone Administration, Berger & Mandelbrot (1963) demonstrated that the intervals between errors could, in fact, be roughly described by a power-law distribution. Similar behavior also characterized the inter-error intervals between 255-bit *blocks* of data transmitted over telephone and high-frequency radio teletype circuits (Moriarty, 1963).

Mandelbrot (1965a) subsequently modified this model in a number of respects in an attempt to achieve improved agreement with the error data; he mandated self-similarity and, closely following Pareto, extended the duration of the upper interval to infinity. Indeed, Mandelbrot's (1965a) model characterized the inter-error intervals far better than the standard geometric-distribution model in use at the time (Gilbert, 1961). More recently, Mandelbrot (1972, 1982, pp. 282–284) further refined this model to make it more appealing from a mathematical perspective.

This chapter is devoted to the properties of fractal renewal point processes, which belong to the family of fractal point processes (see Sec. 5.5.1). Although less prevalent than their fractal-rate cousins, these processes find use in applications such as the characterization of computer cache misses (Voldman, Mandelbrot, Hoevel, Knight & Rosenfeld, 1983; Thiébaut, 1988) and the occurrences of earthquakes and their aftershocks (Lapenna, Macchiato & Telesca, 1998; Telesca, Cuomo, Lanfredi, Lapenna & Macchiato, 1999; Telesca, Cuomo, Lapenna & Macchiato, 2002a) (see Prob. 10.7, however). They are also useful in a number of other areas, some of which are considered in the form of problems at the end of this chapter.

7.1 POWER-LAW DISTRIBUTED INTEREVENT INTERVALS

With all correlations and dependencies among the intervals excluded, the fractal renewal process resets with the arrival of each event and no memory exists across events. Paradoxically, a fractal-based point process still proves possible; the scaling (fractal) behavior derives from the distribution of the intervals alone.

A probability density function that decays in a power-law form cannot conveniently persist for all values of the random variable, since the resulting probability density would have infinite area. Rather, we consider the general case in which we impose probability-density cutoffs at both small and large times, as shown in Fig. 7.1. This ensures that the resulting point process has a positive rate in the stationary (equilibrium) state.

7.1.1 Abrupt-cutoff interevent-interval density

The abrupt-cutoff power-law probability density function provides the simplest example (Lowen & Teich, 1993d):

$$p_\tau(t) = \frac{\gamma}{A^{-\gamma} - B^{-\gamma}} \times \begin{cases} t^{-(\gamma+1)} & A < t < B \\ 0 & \text{otherwise,} \end{cases} \tag{7.1}$$

where $B > A > 0$ and $\gamma > 0$. The associated moments are

$$\mathrm{E}[\tau^n] = \begin{cases} \dfrac{\gamma}{n-\gamma} (A/B)^\gamma B^n \dfrac{1 - (A/B)^{n-\gamma}}{1 - (A/B)^\gamma} & n \neq \gamma \\[2ex] \dfrac{\gamma \ln(B/A)}{A^{-\gamma} - B^{-\gamma}} & n = \gamma, \end{cases} \tag{7.2}$$

while the characteristic function is

$$\phi_\tau(\omega) = \frac{\gamma (i\omega)^\gamma}{A^{-\gamma} - B^{-\gamma}} \int_{i\omega A}^{i\omega B} e^{-x} x^{-(\gamma+1)} \, dx. \tag{7.3}$$

The **Pareto density** (1896) emerges in the special case $A = 1$ and $B \to \infty$. For $A \ll B$ and $0 < \gamma < 1$, we can express Eq. (7.3) as (Lowen, 1992)

$$1 - \phi_\tau(\omega) \approx \Gamma(1 - \gamma) (i\omega A)^\gamma \tag{7.4}$$

in the range $B^{-1} \ll \omega \ll A^{-1}$.

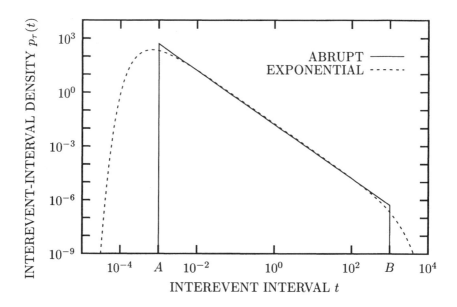

Fig. 7.1   Abrupt-cutoff (solid) and exponential-cutoff (dashed) interevent-interval probability density functions, $p_\tau(t)$ vs. t. The density functions exhibit a power-law region between $t = A$ and $t = B$. In this illustration, we set the lower and upper cutoffs at $A = 10^{-3}$ and $B = 10^3$, respectively. The power-law exponent of the density has a value $-\frac{3}{2} = -(\gamma + 1)$ so that $\gamma = \frac{1}{2}$.

7.1.2   Exponential-cutoff interevent-interval density

We can impose smooth transitions on this power-law behavior by using the interevent-interval density function (Lowen & Teich, 1993d)

$$p_\tau(t) = \frac{(AB)^{\gamma/2}}{2\mathrm{K}_\gamma\!\left(2\sqrt{A/B}\right)}\, e^{-A/t}\, e^{-t/B}\, t^{-(\gamma+1)}, \tag{7.5}$$

where $\mathrm{K}_\gamma(x)$ denotes the modified Bessel function of the second kind of order γ. It is sometimes referred to as the **generalized inverse Gaussian density** (Barndorff-Nielsen, Blaesild & Halgreen, 1978).

The associated moments then become

$$\mathrm{E}[\tau^n] = (AB)^{n/2}\, \frac{\mathrm{K}_{|\gamma-n|}\!\left(2\sqrt{A/B}\right)}{\mathrm{K}_\gamma\!\left(2\sqrt{A/B}\right)}, \tag{7.6}$$

and the corresponding characteristic function can be written as

$$\phi_\tau(\omega) = (1 + i\omega B)^{\gamma/2}\, \frac{\mathrm{K}_\gamma\!\left[2(A/B + i\omega A)^{1/2}\right]}{\mathrm{K}_\gamma\!\left(2\sqrt{A/B}\right)}. \tag{7.7}$$

For $\gamma = \frac{1}{2}$ and $B \to \infty$, Eq. (7.7) becomes the one-sided **stable distribution** of order $\frac{1}{2}$ (Feller, 1971), which was provided previously in Eq. (3.13). Combined with exponential tails, one-sided stable distributions for arbitrary values of γ between zero and unity also follow power-law forms while providing smooth transitions (Lowen, 1992).

Constructing a renewal point process using any of these random variables leads to a point process with fractal properties, as we will demonstrate shortly.

7.1.3   Effect of γ on interval variability

Whatever the nature of the cutoff, as $B \to \infty$ the character of the process changes as γ passes through unity. Values of γ smaller than unity lead to intervals with infinite mean, whereas values of γ in excess of two ensure finite variance. In the range $1 < \gamma < 2$, the interevent intervals have finite mean but exhibit wild variation about that mean as a result of the infinite variance of the intervals in this range. As a general rule of thumb, the variability decreases as γ increases.

When B is finite, so that all moments are finite, the value of γ nevertheless continues to play an important role in determining the variability of the intervals. As γ increases, the interval density becomes more concentrated near the lower cutoff, A, with proportionately fewer intervals near B. This results in a renewal process with reduced variability. Equation (7.2) highlights this effect; the mean interevent interval in the limit $A/B \ll 1$ is $\mathrm{E}[\tau] \approx \sqrt{AB}$ for $\gamma = \frac{1}{2}$ whereas it is $\mathrm{E}[\tau] \approx 3A$ for $\gamma = \frac{3}{2}$. Since $\mathrm{E}[\tau]$ is independent of B for $\gamma > 1$ and $A/B \ll 1$, relatively few intervals lie near B. Extreme events are therefore relatively less likely than for $\gamma < 1$.

7.2   STATISTICS OF THE FRACTAL RENEWAL PROCESS

7.2.1   Point-process spectrum

We begin with the point-process spectrum of the fractal renewal point process. In the mid-frequency range, $B^{-1} \ll f \ll A^{-1}$, we obtain (see Sec. A.4.1 and Lowen, 1992; Lowen & Teich, 1993d):

$$S_N(f) \to \mathrm{E}[\mu] \times \begin{cases} 2\left[\Gamma(1-\gamma)\right]^{-1} \cos(\pi\gamma/2)\,(2\pi f A)^{-\gamma} & 0 < \gamma < 1 \\[2mm] \pi\left[\ln(2\pi f A)\right]^{-2}(2\pi f A)^{-1} & \gamma = 1 \\[2mm] 2\gamma^{-2}\,(\gamma-1)\,\Gamma(2-\gamma)\left[-\cos(\pi\gamma/2)\right](2\pi f A)^{\gamma-2} & 1 < \gamma < 2 \\[2mm] -\tfrac{1}{2}\ln(2\pi f A) & \gamma = 2 \\[2mm] \gamma^{-1}\,(\gamma-2)^{-1} & \gamma > 2. \end{cases}$$

$$\tag{7.8}$$

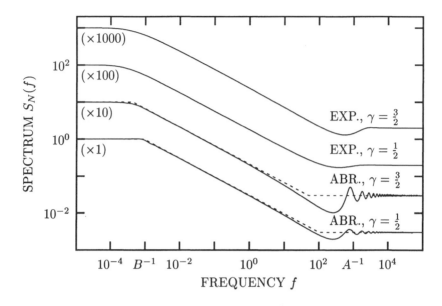

Fig. 7.2   Normalized spectra, $S_N(f)/\lim_{f\to 0} S_N(f)$, for the fractal renewal point process. In this illustration, $A = 10^{-3}$ and $B = 10^3$. The four solid curves represent spectra corresponding to different interevent-interval densities, with the following properties (top to bottom): exponential cutoffs with $\gamma = \frac{3}{2}$ ($\times 1\,000$), exponential cutoffs with $\gamma = \frac{1}{2}$ ($\times 100$), abrupt cutoffs with $\gamma = \frac{3}{2}$ ($\times 10$), and abrupt cutoffs with $\gamma = \frac{1}{2}$ ($\times 1$). The curves were obtained by using Eqs. (7.3) and (7.7) in Eq. (4.16), for the abrupt and exponential cutoffs, respectively. The dashed curves are asymptotic forms for the abrupt-cutoff interevent-interval densities, drawn from the low-, mid-, and high-frequency spectral limits represented by Eqs. (4.17), (7.8), and (3.59), respectively. All four spectra decrease with frequency as $f^{-1/2}$ in the region $B^{-1} \ll f \ll A^{-1}$, in accordance with Eq. (7.9); however, those for $\gamma = \frac{3}{2}$ depart more markedly from the asymptotic values than do those for $\gamma = \frac{1}{2}$. Spectra associated with abrupt-cutoff interval probability densities exhibit marked oscillations at higher frequencies.

Equation (7.8) reveals that the value of α associated with the spectrum depends on γ in accordance with

$$\alpha = \begin{cases} \gamma & 0 < \gamma < 1 \\ 2 - \gamma & 1 < \gamma < 2 \\ 0 & \gamma > 2. \end{cases} \tag{7.9}$$

Thus, α neither attains, nor exceeds, unity over the mid-frequency range. Indeed, this kind of behavior emerges for all power-law forms of the interevent-interval density; hence the fractal renewal point process generates $1/f^\alpha$ noise only in the range $0 < \alpha < 1$.

Figure 7.2 displays the spectra for abrupt and exponential cutoffs, normalized to the values indicated at the low-frequency limits. We generated these curves by making use of Eq. (4.16), together with Eqs. (7.3) and (7.7) for the abrupt and exponential cutoffs, respectively. The low-, mid-, and high-frequency asymptotes are set forth in

Eqs. (4.17), (7.8), and (3.59), respectively. The abrupt-cutoff probability density functions exhibit substantial oscillations in the characteristic function, which appear in the spectra. The exponential-cutoff density functions, in contrast, generate smooth transitions in the time domain and therefore nonoscillatory spectra. Interevent-interval probability density functions with $\gamma = \frac{1}{2}$ and $\gamma = \frac{3}{2}$ both translate to spectra that exhibit a fractal spectral exponent $-\alpha = -\frac{1}{2}$, but the latter depart more markedly from asymptotic values. A simulated version of the spectrum for the abrupt-cutoff case with $\gamma = \frac{3}{2}$ is shown in Fig. B.12.

We can derive a closed-form expression for the spectrum for the smooth-transition interevent-interval probability density function given in Eq. (7.5) for $\gamma = \frac{1}{2}$; we consider the normalized case $AB = 1$ to simplify the ensuing calculations (see Sec. A.4.2):

$$S_N(f) = \mathrm{E}^2[\mu]\,\delta(f) + \frac{\sinh(c)}{\cosh(c) - \cos(d)}, \tag{7.10}$$

with

$$c \equiv \sqrt{2A}\left(\sqrt{A^2 + \omega^2} + A\right)^{1/2} - 2A \tag{7.11}$$

$$d \equiv \sqrt{2A}\left(\sqrt{A^2 + \omega^2} - A\right)^{1/2}. \tag{7.12}$$

In fact, given enough patience, closed-form expressions can be derived for $\gamma = n + \frac{1}{2}$, where n is any nonnegative integer.

7.2.2 Coincidence rate

Inverse Fourier transforms of the abrupt-cutoff spectrum formulas given in Eq. (7.8) are readily calculated. This leads to approximate formulas for the coincidence rate of the power-law process in the range $A \ll |t| \ll B$, for positive values of γ (see Sec. A.4.3):

$$G(t) \rightarrow \mathrm{E}[\mu] \times \begin{cases} \pi^{-1}\sin(\pi\gamma)\,A^{-\gamma}\,t^{\gamma-1} & 0 < \gamma < 1 \\ A^{-1}\left[\ln(t/A)\right]^{-1} & \gamma = 1 \\ \gamma^{-2}\,(\gamma-1)\,A^{\gamma-2}\,t^{1-\gamma} & 1 < \gamma < 2 \\ \frac{1}{4}\,t^{-1} & \gamma = 2 \\ \mathrm{E}[\mu] & \gamma > 2. \end{cases} \tag{7.13}$$

For the exponential-cutoff interevent-interval density provided in Eq. (7.5) and $\gamma = \frac{1}{2}$, we can write the coincidence rate in the following form:

$$\begin{aligned} G(t) &= \mathrm{E}[\mu]\,\delta(t) + \mathrm{E}[\mu]\sum_{n=1}^{\infty} p^{\star n}(|t|) \\ &= (AB)^{-1/2}\,\delta(t) + (\pi B)^{-1/2}\,|t|^{-3/2}\,\exp(-|t|/B) \\ &\quad \times \sum_{n=1}^{\infty} n\exp\left[2(A/B)^{1/2}\,n - (A/|t|)\,n^2\right]. \end{aligned} \tag{7.14}$$

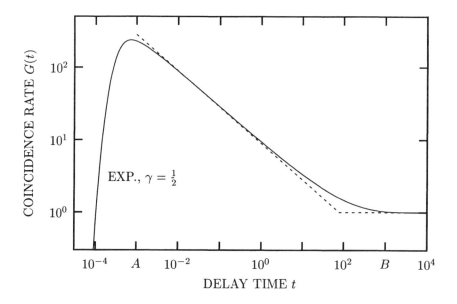

Fig. 7.3 Coincidence rate for a fractal renewal point process constructed with the exponential-cutoff probability density function specified in Eq. (7.5). The parameters are $\gamma = \frac{1}{2}$, $A = 10^{-3}$, and $B = 10^3$ (solid curve). The straight-line asymptotes derive from simplifying Eq. (7.15) in the limit $|t| \ll B$ and from Eq. (3.51).

In the limit $|t| \gg A$ and $B \gg A$, the terms comprising the sum in Eq. (7.14) vary slowly. An integral then provides a good approximation to the sum, and the coincidence rate simplifies to

$$
\begin{aligned}
G(t) &\approx (\pi B)^{-1/2} |t|^{-3/2} e^{-|t|/B} \int_0^\infty x \exp\left[2(A/B)^{1/2} x - (A/|t|) x^2\right] dx \\
&= \frac{e^{-|t|/B}}{\sqrt{4\pi A^2 B |t|}} + \frac{1}{2AB} \operatorname{erfc}\left(-\sqrt{\frac{|t|}{B}}\right),
\end{aligned}
\tag{7.15}
$$

where the complementary error function is given by

$$
\operatorname{erfc}(x) \equiv 2\pi^{-1/2} \int_x^\infty \exp(-t^2)\, dt.
\tag{7.16}
$$

The coincidence rate represented in Eqs. (7.14)–(7.16), which is applicable for exponential cutoffs and $\gamma = \frac{1}{2}$, is displayed in Fig. 7.3.

7.2.3 Normalized variances

The counting statistics of renewal point processes are often provided in terms of a special type of factorial moment, set forth in Eq. (4.19). For the fractal renewal point

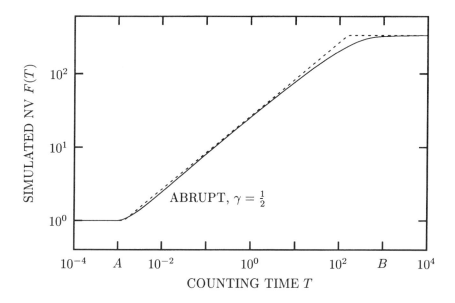

Fig. 7.4 Simulated normalized variance $F(T)$ vs. counting time T for a fractal renewal point process with abrupt cutoffs (solid curve). The parameters used to generate this curve are $\gamma = \frac{1}{2}$ ($\alpha = \frac{1}{2}$), $A = 10^{-3}$, and $B = 10^3$; 100 independent simulations were used, each of duration $L = 10^8$. Asymptotic results are shown as dashed. The mean rate is unity ($E[\mu] = E[\tau] = 1$). The associated normalized Haar-wavelet variance is shown in Fig. 7.5.

process at hand, we can cast these factorial moments in relatively simple form when $A \ll |t| \ll B$ and $0 < \gamma < 1$, even for arbitrary cutoffs.

Since $G(t) \sim t^{\gamma-1}$, we have $G^{\star k}(t) \sim t^{k\gamma-1}$, whereupon Eq. (4.19) provides

$$E\left\{\frac{[Z(T) + k]!}{[Z(T) - 1]!}\right\} \sim T^{k\gamma+1}. \qquad (7.17)$$

The constants of proportionality depend on the details of the interevent-interval probability density function.

In particular, for the abrupt-cutoff fractal renewal point process with arbitrary γ, we can readily obtain expressions for the normalized variance and normalized Haar-wavelet variance in the range $A \ll T \ll B$. Substituting Eq. (7.13) into Eqs. (3.52) and (3.53) yields $F(T)$ and $A(T)$, respectively (see Sec. A.4.4):

$$F(T) \to \begin{cases} 2\big[\pi\gamma(\gamma+1)\big]^{-1}\sin(\pi\gamma)\,A^{-\gamma}\,T^{\gamma} & 0 < \gamma < 1 \\ A^{-1}\big[\ln(T/A)\big]^{-1}T & \gamma = 1 \\ 2\big[\gamma^2(2-\gamma)(3-\gamma)\big]^{-1}(\gamma-1)\,A^{\gamma-2}\,T^{2-\gamma} & 1 < \gamma < 2 \\ \frac{1}{2}\ln(T/A) & \gamma = 2 \\ 1 & \gamma > 2 \end{cases}$$

$$(7.18)$$

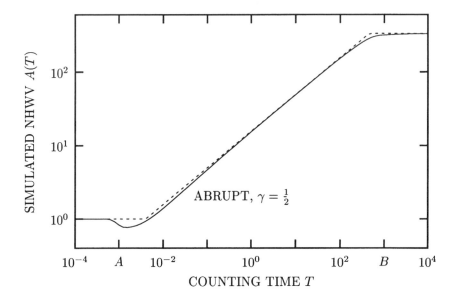

Fig. 7.5 Simulated normalized Haar-wavelet variance $A(T)$ vs. counting time T for a fractal renewal point process with abrupt cutoffs (solid curve). The parameters are the same as those specified in Fig. 7.4, which displays the associated normalized variance. We show asymptotic results as dashed lines. A simulated version of $A(T)$ for $\gamma = \frac{3}{2}$, which also corresponds to $\alpha = \frac{1}{2}$, appears in Fig. B.13. The dip in the curve derives from the abrupt cutoff in the interevent-interval density for small intervals.

while

$$
A(T) \rightarrow
\begin{cases}
4(1 - 2^{\gamma-1}) \left[\pi\gamma(\gamma + 1)\right]^{-1} \sin(\pi\gamma)\, A^{-\gamma} T^{\gamma} & 0 < \gamma < 1 \\
2\ln(2)\, A^{-1} \left[\ln(T/A)\right]^{-2} T & \gamma = 1 \\
4(1 - 2^{1-\gamma}) \left[\gamma^2(2 - \gamma)(3 - \gamma)\right]^{-1} (\gamma - 1)\, A^{\gamma-2}\, T^{2-\gamma} & 1 < \gamma < 2 \\
\frac{1}{2}\ln(T/2A) & \gamma = 2 \\
1 & \gamma > 2.
\end{cases}
$$

$$(7.19)$$

   Since it is difficult to obtain useful analytic forms for the normalized variance $F(T)$ and the normalized Haar-wavelet variance $A(T)$ over the full range of counting times T, we present simulations for these quantities as functions of T in Figs. 7.4 and 7.5, respectively. The solid curves represent simulated results for $\gamma = \frac{1}{2}$ ($\alpha = \frac{1}{2}$); the central asymptotes (dashed lines) represent Eqs. (7.18) and (7.19), respectively. A cartoon version of the normalized Haar-wavelet variance was presented earlier, as the solid curve in Fig. 5.4d).

7.2.4 Counting distribution

We present simulated counting distributions in Fig. 7.6 for $\gamma = \frac{1}{2}$ (solid curve) and $\gamma = \frac{3}{2}$ (dashed curve), when the mean count $E[Z] = 10$. Again, we employ simulations because tractable analytic results cannot be obtained. We display the curves on doubly logarithmic coordinates to highlight the different count ranges spanned by the two curves. Although the *interevent-interval* standard deviations are identical for the two values of γ, the variances of the associated *counting* distributions are very different. Estimating these values from the simulated counting distributions yields $\mathrm{Var}[Z(10\,E[\tau])] \doteq 777.703$ for $\gamma = \frac{1}{2}$ and $\mathrm{Var}[Z(10\,E[\tau])] \doteq 28.9254$ for $\gamma = \frac{3}{2}$. As expected, the variance is larger for the smaller value of γ. We also plot a Poisson distribution of the same mean for comparison (dotted curve). These fractal-renewal-process counting distributions are distinctly non-Gaussian. However, the renewal nature of the process and the finite cutoffs assure us that they converge to Gaussian form for $T/B \gg 1$; Eq. (4.18) applies in that domain. Identical counting distributions obtain when A, B, and T are all multiplied by a common factor, since the determining parameters are *ratios* between the times rather than the times themselves.

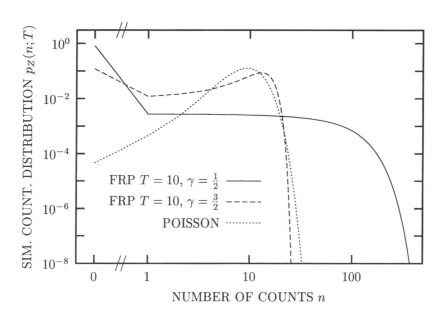

Fig. 7.6 Simulated counting distributions, $p_Z(n; T)$ vs. number of counts n, for two fractal renewal point processes with abrupt cutoffs. The parameters used to generate these curves were $E[\tau] = 1$, $T = 10$, and $B/A = 10^6$. These values, in turn, give rise to the following exact (not rounded) values for the remaining parameters. For $\gamma = \frac{1}{2}$ we have: $A = 10^{-3}$, $B = 10^3$, and $\mathrm{Var}[\tau] = 332.667$; for $\gamma = \frac{3}{2}$ we have: $A = 0.333667$, $B = 333667.0$, and again $\mathrm{Var}[\tau] = 332.667$.

7.2.5 Capacity dimension

Calculation of the capacity dimension (see Secs. 2.1.1 and 3.5.4) yields the expected result: for the parameter ranges $0 < \gamma < 1$ and $0 < A \ll B < \infty$, the points generated by the fractal renewal process do indeed form a fractal set with dimension γ, in the sense of Eq. (3.72) (Lowen & Teich, 1993d).

Consider a realization of the process and a covering of it using segments of length B, as shown in Fig. 7.7. For minimal covering, place the beginning of each segment on the first uncovered event. The empty space between coverings is thus the residual waiting time for a pure renewal point process at time B. This construction closely resembles that for fixed dead time, which is illustrated in Fig. 11.1d) and discussed in Sec. 11.2.4.

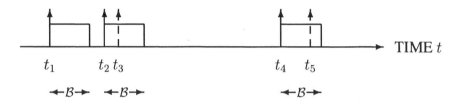

FRACTAL RENEWAL PROCESS

Fig. 7.7 Realization of a fractal renewal point process and its minimal covering. For this particular illustration, three segments suffice to cover the set. Events occurring at t_3 and t_5 lie within a duration B of the prior event at which a segment initiates, and therefore do not require additional segments.

Let $W(B)$ represent the expected value of the time between coverings, including the coverings themselves. Wald's Lemma (Feller, 1971) then provides

$$W(B) = \mathrm{E}^2[\tau] \int_{0-}^{B} G(t)\,dt, \tag{7.20}$$

where the notation $0-$ indicates that the range of the integral spans the delta function contribution to $G(t)$ at $t = 0$. For the range $A \ll B \ll B$, the approximation $G(t) \sim t^{\gamma-1}$ yields $W(B) \sim B^{\gamma}$. The number of intervals required to cover the fractal renewal point process thus scales as $B^{-\gamma}$, and the capacity dimension is therefore γ.

7.3 NONDEGENERATE REALIZATION OF A ZERO-RATE PROCESS

Both versions of the fractal renewal point process set forth in Sec. 7.1, namely the abrupt- and exponential-cutoff forms, can be extended to $B \to \infty$. The associated interevent-interval probability density functions then become

$$p_\tau(t) = \begin{cases} \gamma A^\gamma t^{-(\gamma+1)} & t > A \\ 0 & t \leq A, \end{cases} \tag{7.21}$$

and

$$p_\tau(t) = \frac{1}{\Gamma(\gamma)} A^\gamma e^{-A/t} t^{-(\gamma+1)}, \tag{7.22}$$

respectively. Equation (7.21) is often called the **generalized Pareto density**.

Focusing on the case $0 < \gamma < 1$, both interevent-interval probability density functions are well defined; however, the moments of τ, $\mathrm{E}[\tau^n]$, are infinite for all positive integers n. In particular, $\mathrm{E}[\tau] = \infty$, so that $\mathrm{E}[\mu] = 1/\mathrm{E}[\tau] = 0$, indicating that the resulting renewal point process has zero rate in the stationary (equilibrium) case. Since the rate cannot assume negative values, this implies, with probability one, that no events can occur in any finite interval.

However, we can extend the framework considered earlier for the positive-rate stationary fractal renewal point process and, in fact, obtain nontrivial results for the zero-rate nonstationary case. For renewal point processes that are not in equilibrium, but rather begin with an event, the zero-rate argument does not apply and the resulting statistics can indeed assume nondegenerate values. A segment of such a nonstationary renewal point processes can therefore contain a positive number of events, even though the mean number of events in a comparable segment of the stationary process assumes a value of zero. For the interevent-interval probability density functions specified in Eqs. (7.21) and (7.22), the probability of observing zero events in a segment of length T can still become vanishingly small as the ratio T/A increases.

In general, a fractal renewal point process that begins with the occurrence of an event, with an associated interevent-interval probability density function $p(t) \sim t^{-(\gamma+1)}$, has a residual waiting time that approaches a limiting density (Feller, 1971). Specifically, suppose that we can cast the interevent-interval survivor function in the form

$$S_\tau(t) = 1 - P_\tau(t) = \int_t^\infty p(v)\, dv = t^{-\gamma} L(t), \tag{7.23}$$

where $L(t)$ is a "slowly varying" function such that, for any $x > 0$,

$$\lim_{t\to\infty} L(xt)/L(t) = 1. \tag{7.24}$$

Equations (7.21) and (7.22) both fall in this category. Recall, now, from Eq. (3.10) that $\vartheta(t)$ denotes the random interval between the deterministic time t and the next event in the fractal renewal point process. This random interval $\vartheta(t)$ then has a probability density function given by (Feller, 1971)

$$p_\vartheta(s) = \frac{t^\gamma \sin(\pi\gamma)}{\pi} \frac{s^{-\gamma}}{s+t}. \tag{7.25}$$

Thus, when a fractal renewal point process with zero mean rate begins at the occurrence of an event, the resulting process has a nonzero effective rate for all finite times. We conclude that any experiment will, of necessity, record a process with a positive expected rate, and the results derived above will also apply to this process.

Any realization of an infinite-mean fractal renewal point process that begins with an event, and which we observe for a finite time, will exhibit largest and smallest intervals, which we label B^* and A^*, respectively. Given the power-law exponent of the distribution, and only the values A^* and B^*, the other intervals will follow a power-law distribution between them, and will exhibit the same power-law exponent. The observed process will therefore have the same statistics as a finite-mean fractal renewal point process, with cutoff times $A < A^*$ and $B > B^*$, and the results derived above will apply to this process with the *a posteriori* values of A^* and B^*. Although we cannot know the values A^* and B^* *a priori*, the sample spectrum will decay in a power-law fashion, whatever these values may be.

Problems

7.1   *Distinct processes with a common fractal exponent*   Use the abrupt-cutoff power-law probability density function provided in Eq. (7.1) to construct two different fractal renewal point processes that have the following parameters: fractal exponent $\alpha = \frac{1}{2}$, fractal onset frequency $f_S = 10$ Hz, and $B/A = 10^6$. Find the mean rate $\mathrm{E}[\mu]$ for both.

7.2   *Characteristic function for the exponential-cutoff interval density*   Find an approximate expression for the characteristic function along the lines of Eq. (7.4), but for the smoother interevent-interval density function provided in Eq. (7.5). Note that the modified Bessel function of the second kind varies as $\mathrm{K}_\gamma(z) \approx 2^{\gamma-1}\Gamma(\gamma)z^{-\gamma}$ for small arguments z (Gradshteyn & Ryzhik, 1994, Secs. 8.445 and 8.485).

7.3   *Deriving the point-process spectrum from the characteristic function*   Use Eq. (7.4) to reproduce the $0 < \gamma < 1$ condition in Eq. (7.8).

7.4   *Relation of mean rate and fractal onset frequency*   Consider the abrupt-cutoff power-law density in Eq. (7.1) for $A \ll B$.
    7.4.1. Find an equation that relates the mean rate $\mathrm{E}[\mu]$ to the fractal onset frequency f_S in the range $1 < \gamma < 2$.
    7.4.2. Find an inequality that relates these quantities in the range $0 < \gamma < 1$.

7.5   *Limiting form for the characteristic function*   Show that Eq. (7.4) obtains for the stated limits.

7.6   *Simulation time*   To obtain good estimates of $F(T)$ and $A(T)$ for presentation in Figs. 7.4 and 7.5, we made use of 100 independent simulations of fractal renewal point processes, each of duration 10^8. The simulation of the point processes and the calculation of these statistics took about 24 hours of computation time on a personal computer with a clock speed of 1.6 GHz. The simulated point processes were represented by floating-point numbers at four bytes per interevent interval; for fastest execution times these numbers were stored in memory for the calculation of $F(T)$ and $A(T)$. Why were curves for $\gamma = \frac{3}{2}$ with the same values for A and B not included in these figures?

7.7   *Error clustering in telephone networks*   In the 1960s, researchers began to recognize that data errors following information transmission over telephone lines could not be properly described by a memoryless binary symmetric channel, with its attendant geometric distribution of inter-error intervals and binomial distribution of error counts. Gilbert (1961) attempted to improve the state of affairs by considering a channel that switched between two states that suffered different error probabilities. This added variability allowed some qualitative features of the data to be modeled, but this approach fell far short of providing detailed agreement.

Shortly thereafter, Berger & Mandelbrot (1963) and Mandelbrot (1965a) made a substantial advance. These authors recognized that a fractal renewal point process, with power-law rather than geometric inter-error intervals, provided a far superior model for characterizing the data errors. It had long been known that such errors appeared to occur in clusters, and in clusters of clusters, a feature that is the hallmark of a fractal point process.

Consider a system that exhibits errors that obey this fractal-renewal-process model. Suppose now that we add another source of noise, independent of the first, that takes the form of a homogeneous Poisson process with a large mean time between events τ_{HPP}. The overall noise process then comprises the superposition of error events, which are clustered, and events associated with the homogeneous Poisson process, which are not clustered. If the presence (or absence) of an error cluster is verified every τ_{clk} seconds, determine the process that characterizes the error events.

7.8 *Action-potential statistics in an insect visual-system interneuron: Counter-example* The curve in Fig. 7.8 displays the estimated interevent-interval density

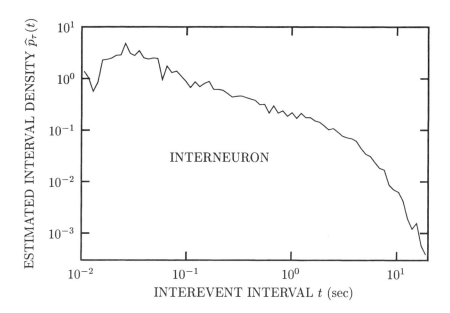

Fig. 7.8 Estimated interevent-interval density, $\widehat{p}_\tau(t)$ vs. interevent interval t, for an action-potential sequence recorded from the descending contralateral movement detector, a visual-system INTERNEURON in the locust (Turcott et al., 1995, Fig. 2, pp. 261–262, cell ADA062). The normalized Haar-wavelet variance for these same data appears in Fig. 7.9.

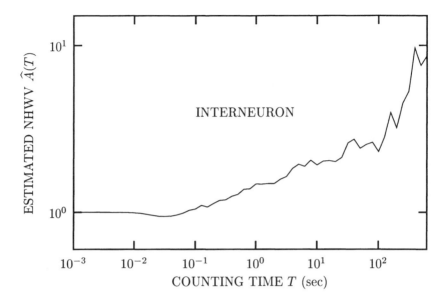

Fig. 7.9 Estimated normalized Haar-wavelet variance $\widehat{A}(T)$ vs. counting time T (sec), for an action-potential sequence recorded from the descending contralateral movement detector, a visual-system INTERNEURON in the locust (Turcott et al., 1995, Fig. 2, pp. 261–262, cell ADA062). Unlike the display provided in Fig. 5.2, the abscissa reports the counting time in unnormalized form. The interevent-interval density for these same data appears in Fig. 7.8.

function for a spontaneous sequence of action potentials recorded from a visual-system INTERNEURON in the locust, the descending contralateral movement detector (Turcott et al., 1995), plotted on doubly logarithmic coordinates. The curve in Fig. 7.9 shows the estimated normalized Haar-wavelet variance $\widehat{A}(T)$, plotted as a function of the counting time T, for these same data.

7.8.1. Determine the values of $\widehat{\gamma}$, $\widehat{A}$, and $\widehat{B}$ that characterize the interval data in Fig. 7.8. What is the corresponding value of $\widehat{\alpha}$ for a fractal renewal point process?

7.8.2. What is the origin of the slight dip below unity observed in the data in Fig. 7.9? Determine the values of $\widehat{\alpha}_A$ and $\widehat{T}_A$ that characterize this plot.

7.8.3. Why does the value of $\widehat{\alpha}_A$ observed from the normalized Haar-wavelet variance differ so drastically from the value predicted for a fractal renewal point process?

7.9 *Molecular evolution*   The numbers of differences in amino-acid sequences in related organisms appear to be roughly proportional to the time since the organisms diverged in their joint evolutionary history (Zuckerkandl & Pauling, 1962). This leads to the notion of a molecular clock (Zuckerkandl & Pauling, 1965). While this process lacks extensive data, the existing data are adequate to exclude the homogeneous Poisson process as a viable model (Gillespie, 1994). Review the evidence presented by Gillespie (1994); West & Bickel (1998); Bickel & West (1998a,b); and Bickel

(2000), and provide a rationale for the use of a fractal renewal point process to model the available data.

7.10   *Trapping in amorphous semiconductors*   A number of approaches have been used to investigate the relationship between trapping processes and $1/f$ noise in semiconductors (McWhorter, 1957; Stepanescu, 1974; Scher & Montroll, 1975; Tiedje & Rose, 1980; Orenstein, Kastner & Vaninov, 1982; Kastner, 1985; Hooge, 1995, 1997). In particular, the multiple trapping model forges a connection between traps that are exponentially distributed over a large range of energies and a transient current that decays as a power-law function of time. Once emitted by a trap, a carrier is available to conduct current for a very brief interval of time before it falls into another trap. For any particular carrier, the times spent in successive traps are independent. Consider an amorphous semiconductor with localized states (traps) whose energies are exponentially distributed, with parameter E_0, between the limits E_L and E_H. If the random variable E represents the trap energy relative to the conduction band edge, the probability density function for the trap energy $p_E(E)$ is

$$p_E(E) = \begin{cases} c \exp(-E/E_0) & E_L < E < E_H \\ 0 & \text{otherwise,} \end{cases} \tag{7.26}$$

where c is a normalization constant.

7.10.1. Determine the value of c in terms of the remaining parameters of the model.

7.10.2. For a trap at energy E, the corresponding mean waiting time $q(E) = \mathrm{E}[\tau]$ is given by $\tau_0 \exp(E/\kappa T)$, where τ_0 is the average vibrational period of the atoms in the semiconductor, κ is Boltzmann's constant, and T is the absolute temperature of the material. Show how to recast the multiple trapping model for a single carrier in terms of the fractal renewal point process by using the definition of $q(E)$ given above and Eq. (7.26) (Lowen & Teich, 1992b; Lowen, 1992).

7.10.3. Given the conditional mean $q(E)$, each trap is assumed to hold carriers for times that follow an exponential density function:

$$p_\tau\big[t|q(E)\big] = \frac{1}{q(E)} e^{-t/q(E)}. \tag{7.27}$$

Average the density provided in Eq. (7.27) over all possible values of $q(E)$ to determine the unconditional trapping-time density. Obtain an asymptotic form for the case $A \ll t \ll B$.

8

Processes Based on the Alternating Fractal Renewal Process

Carl Friedrich Gauss (1777–1855), a celebrated German mathematician, arrived at the limit theorem for random variables that bears his name.

The French mathematician **Paul Lévy (1886–1971)** developed a family of "stable" probability distributions and processes that have widespread applicability.

8.1	**Alternating Renewal Process**	174
	8.1.1 Amplitude statistics	174
	8.1.2 Autocorrelation	175
	8.1.3 Spectrum	175
8.2	**Alternating Fractal Renewal Process**	177
	8.2.1 Spectrum	177
	8.2.2 Autocovariance	178
	8.2.3 $1/f$-type noise from Markov processes	178
8.3	**Binomial Noise**	179
	8.3.1 Fractal binomial noise	181
	8.3.2 Convergence to a Gaussian process	181
8.4	**Point Processes from Fractal Binomial Noise**	182
	Problems	183

As implied by its name, the **alternating renewal process** $X(t)$ alternates between two values, a and b, as portrayed in Fig. 8.1. We set these values to unity and zero, respectively, in accordance with the usual convention, dictated by algebraic convenience (we exclude the degenerate case $a = b$). The results for these particular values of a and b are linked to those for general values by straightforward linear transforms. The dwell times in the upper and lower states derive from two separate distributions; they are also taken to be independent, thereby endowing the process with its renewal character.

Because of its apparent similarity to a Morse-code sequence, many authors call the bistable step waveform associated with the alternating renewal process a **random telegraph signal** (RTS) (others use the appellation **on–off process**). Gleason Willis Kenrick provided its correlation function and spectrum in 1929.[1] Rice (1944, 1945) subsequently studied this process extensively, with a particular emphasis on transition times that follow a Poisson process (exponentially distributed dwell times).

The alternating renewal process often serves as a useful mathematical model for describing **burst noise** in semiconductor devices (Machlup, 1954). Noise with these characteristics, first observed in junction transistors (Montgomery, 1952), also afflicts other kinds of electronic devices and integrated circuits under certain conditions (see, for example, Buckingham, 1983, Chapter 7). It is often ascribed to the presence of microplasmas, defects such as crystallographic dislocations, or to the on–off switching associated with various conduction paths. The random telegraph signal enjoys a broad variety of other applications.

A special case of the alternating renewal process is the **alternating fractal renewal process**, in which one or both of the dwell-time distributions follows a fractal (power-law) form, as indicated in the caption to Fig. 8.1. This process differs from the

---

[1] Kenrick's work was perhaps the first to use the correlation function of a random process to determine its spectrum. The Fourier-transform relation between these two quantities later came to be called the Wiener–Khinchin theorem (Wiener, 1930; Khinchin, 1934).

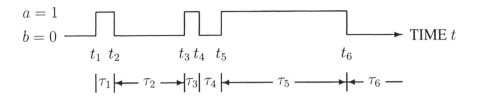

Fig. 8.1 The *alternating renewal process* switches between two values, chosen for convenience to be $a = 1$ and $b = 0$. In general, the odd and even interevent intervals derive from two different distributions. All interevent intervals are independent. In the *alternating fractal renewal process*, one or both of the dwell-time distributions follows a fractal (power-law) form.

fractal renewal process considered in Chapter 7 by the presence of two time-interval distributions and by the fact that the process is finite-valued rather than composed of Dirac δ functions. The power-law-distributed dwell times lead to a $1/f$-type spectrum (Lowen & Teich, 1993d).

Various forms of this process serve as plausible models for semiconductor $1/f$-type noise arising from fluctuations associated with the capture and release of carriers at traps (McWhorter, 1957; Stepanescu, 1974; Buckingham, 1983; Lowen & Teich, 1992b; Sikula, 1995). It also proves useful for characterizing the behavior of individual ion channels in biological membranes (Schick, 1974; Lowen & Teich, 1993d,c; Liebovitch et al., 2001). This process finds its place in other arenas such as dynamics in systems with fractal boundaries (Aizawa & Kohyama, 1984; Aizawa, 1984; Arecchi & Lisi, 1982; Arecchi & Califano, 1987; Sapoval, Baldassarri & Gabrielli, 2004), fluorescence fluctuations of nanoparticles diffusing through a region of focused laser light (Zumofen, Hohlbein & Hübner, 2004), the analysis of rainfall data (Schmitt et al., 1998), computer network traffic (as discussed in Chapter 13), and the generation of fractal test signals (as considered in Prob. 8.6). After considering the statistical properties of the alternating renewal process, we turn to the properties of sums thereof. The result is binomial noise, which, as the number of constituent processes increases without limit, converges to a Gaussian form. Similarly, the sum of alternating *fractal* renewal processes gives rise to fractal binomial noise, which, in turn, converges to the fractal Gaussian process discussed in Sec. 6.3.3 (Lowen & Teich, 1993d, 1995). This property provides a simple and plausible rationale for the fractal Gaussian behavior of nerve-membrane voltage fluctuations (Verveen, 1960; Verveen & Derksen, 1968): currents flowing through individual ion channels embedded in the nerve membrane behave as on–off processes with intervals between channel openings and closings that are power-law distributed. The sum of large numbers of these converges to a fractal Gaussian process.

Sums of independent and identically distributed random variables with *finite* second moments converge to normal form via Gauss' (1809) **central limit theorem**. There is a widespread perception that the central limit theorem always holds although this is, of course, not the case.[2] Independent and identically distributed constituent variables with *infinite* second moments converge instead to the family of stable distributions set forth by Paul Lévy (1940, 1954; see also Samorodnitsky & Taqqu, 1994; Bertoin, 1998; Sato, 1999). This is sometimes called the **noncentral limit theorem**. The associated distributions have power-law tails and differ dramatically in character from the normal distribution.

Finally, we briefly consider the properties of doubly stochastic and integrate-and-reset point processes driven by fractal binomial noise (Lowen & Teich, 1993b, 1995; Thurner et al., 1997); these constructs find particular use in neuroscience because of the prevalence of fractal binomial noise and fractal Gaussian processes in neurobiology.

8.1   ALTERNATING RENEWAL PROCESS

The dwell times in the $X(t) = a = 1$ state, corresponding to interevent intervals τ_n with odd indices n in Fig. 8.1, all derive from the same distribution. Similarly, times for which $X(t) = b = 0$, corresponding to even indices n in this figure, also share a distribution, possibly different from that describing the $X(t) = a = 1$ dwell times. All interevent intervals are independent of each other.

We use the notations τ_a and τ_b to refer to the dwell times for which $X(t) = a = 1$ and $X(t) = b = 0$, respectively. For a well-defined process in the stationary case, we require that $p_{\tau a}(t) = p_{\tau b}(t) = 0$ for $t < 0$, effectively prohibiting negative dwell times; and that the expected mean dwell times $\mathrm{E}[\tau_a]$ and $\mathrm{E}[\tau_b]$ both assume finite values. We further require that the sequence of transitions forms an orderly point process, so that Eq. (3.1) holds for the transition times of $X(t)$.

8.1.1   Amplitude statistics

The marginal moments of $X(t)$ are simple to evaluate. The expected value $\mathrm{E}[X(t)]$ becomes simply the ratio of the mean time spent in the $X(t) = 1$ state ($\mathrm{E}[\tau_a]$) to the mean time spent in both states ($\mathrm{E}[\tau_a] + \mathrm{E}[\tau_b]$). Since $X(t)$ can only take on values of zero or unity, it belongs to the family of **Bernoulli random variables** (Feller, 1968); in particular, $X^c(t) = X(t)$ for any positive real number c.

We therefore have

$$r \equiv \mathrm{E}[X^n(t)] = \frac{\mathrm{E}[\tau_a]}{\mathrm{E}[\tau_a] + \mathrm{E}[\tau_b]} \tag{8.1}$$

---

[2] As Poincaré (1908) observed with respect to the central limit theorem: "All the world believes it firmly because the mathematicians imagine that it is a fact of observation and the observers imagine that it is a theorem of mathematics."

for all positive moments n. Further results, akin to those in Eqs. (3.4) and (3.31) include

$$\text{variance} \qquad \text{Var}[X] = r(1-r)$$

$$\text{skewness} \qquad \frac{\text{E}\left[(X - \text{E}[X])^3\right]}{\text{Var}^{3/2}[X]} = \frac{1 - 2r}{\sqrt{r(1-r)}} \qquad (8.2)$$

$$\text{kurtosis} \qquad \frac{\text{E}\left[(X - \text{E}[X])^4\right]}{\text{Var}^2[X]} - 3 = \frac{1}{r(1-r)} - 6.$$

8.1.2 Autocorrelation

Closed forms for the autocorrelation $R_X(t)$ do not exist in general, but rather involve an infinite sum of convolutions of arbitrary complexity (Lowen, 1992). However, simple results do obtain in the limits of very small and very large delay times t.

For small delay times, the probability of a transition between the states $X(t) = 1$ and $X(t) = 0$ becomes vanishingly small, so that

$$\lim_{t \to 0} R_X(t) = \lim_{t \to 0} \text{E}[X(s)\, X(s+t)] = \text{E}[X^2(s)] = \text{E}[X(s)]$$

$$= r. \qquad (8.3)$$

For large delay times, on the other hand, the two values of $X(t)$ become independent, whereupon

$$\lim_{t \to \infty} R_X(t) = \lim_{t \to \infty} \text{E}[X(s)\, X(s+t)]$$

$$= \text{E}[X(s)]\, \text{E}[X(s+t)] = \text{E}^2[X(s)]$$

$$= r^2. \qquad (8.4)$$

Indeed, a similar argument holds for any well-behaved real-valued process.

8.1.3 Spectrum

As shown in Sec. A.5.1, the spectrum for an arbitrary alternating renewal process takes the form (Rice, 1983; Lowen, 1992)

$$S_X(f) = \text{E}[X]\,\delta(f) + \frac{2(2\pi f)^{-2}}{\text{E}[\tau_a] + \text{E}[\tau_b]} \text{Re}\left\{ \frac{\left[1 - \phi_{\tau a}(2\pi f)\right]\left[1 - \phi_{\tau b}(2\pi f)\right]}{1 - \phi_{\tau a}(2\pi f)\, \phi_{\tau b}(2\pi f)} \right\}, \qquad (8.5)$$

where $\phi_{\tau a}(\omega)$ and $\phi_{\tau b}(\omega)$ are the characteristic functions associated with the dwell-time distributions for τ_a and τ_b, respectively.

When the means and variances exist, this spectrum approaches an asymptotic form in the low-frequency limit given by (see Sec. A.5.2),

$$\lim_{f \to 0} S_X(f) = \frac{\text{E}^2[\tau_a]\,\text{Var}[\tau_b] + \text{E}^2[\tau_b]\,\text{Var}[\tau_a]}{\left(\text{E}[\tau_a] + \text{E}[\tau_b]\right)^3}, \qquad (8.6)$$

whereas in the high-frequency limit we have

$$S_X(f) \to \frac{2\big(\mathrm{E}[\tau_a] + \mathrm{E}[\tau_b]\big)^{-1}}{(2\pi f)^2} \quad \text{as} \quad f \to \infty. \tag{8.7}$$

In the special case when $p_{\tau a}(t) = p_{\tau b}(t) \equiv p_\tau(t)$, with $p_\tau(t)$ arbitrary, the spectrum set forth in Eq. (8.5) simplifies to (Aizawa, 1984)

$$S_X(f) = \frac{\delta(f)}{2} + \frac{(2\pi f)^{-2}}{\mathrm{E}[\tau]} \, \mathrm{Re}\left\{\frac{1 - \phi_\tau(2\pi f)}{1 + \phi_\tau(2\pi f)}\right\}. \tag{8.8}$$

We turn now to a special case in which there is extreme asymmetry in the dwell times, such that the times τ_b spent in state $X(t) = b$ greatly exceed the times τ_a spent in state $X(t) = a$. As shown in Sec. A.5.3, the spectrum then simplifies to a form closely related to that provided in Eq. (4.16) for the renewal point process (Lowen, 1992). More formally, given a randomly selected pair of dwell times τ_a and τ_b, and for frequencies $f \ll 1/\mathrm{E}[\tau_a]$, the relation $\Pr\{\tau_a \ll \tau_b\} \approx 1$ implies that

$$S_X(f) \approx \frac{\mathrm{E}[\tau_a]}{\mathrm{E}[\tau_b]} \, \delta(f) + \frac{\mathrm{E}[\tau_a^2]}{\mathrm{E}[\tau_b]} \, \mathrm{Re}\left\{\frac{1 + \phi_{\tau b}(2\pi f)}{1 - \phi_{\tau b}(2\pi f)}\right\}. \tag{8.9}$$

From a geometrical perspective, if $X(t) = a = 1$ only infrequently, then $X(t)$ remains in state $b = 0$ except for relatively brief visits to state $a = 1$. As examined in more detail in Sec. A.5.3, such an alternating renewal process $X(t)$ then resembles a renewal point process, albeit with thin, unit-height rectangles in place of the infinite-height delta functions that comprise the renewal point process. Under these circumstances, a marked version of the renewal point process provides a useful approximation to the alternating renewal process. The marks take values given by the times τ_a spent in the state $X(t) = a$, and are therefore independent and identically distributed in accordance with $p_{\tau a}(t)$.

In fact, the alternating renewal process resembles the renewal point process by construction, so it is not surprising that they share many characteristics in common. The symmetric alternating renewal process and the renewal point process, by definition, have identical transition number statistics. They differ only in the type of transition; the former has two types that alternate. For the extreme asymmetric alternating renewal process, where τ_a almost always lies well below τ_b, as discussed above, a different correspondence exists. Here the state transitions occur in closely spaced pairs separated by negligible widths, so that the alternating renewal process transition-*pair* number statistics closely follow the renewal point-process *single*-transition number statistics. Moreover, the spectrum (and thus the coincidence rate) for the extreme asymmetric alternating renewal process is proportional to that for the renewal point process, except in the high-frequency (short-time) limit where the alternating-renewal-process spectrum varies as $S_X(f) \approx 2(2\pi f)^{-2}/\mathrm{E}[\tau_b]$.

Finally, for a Markovian system where both dwell times follow exponential distributions, Rice (1944, 1945) demonstrated that the general result for the spectrum provided in Eq. (8.5) assumes the familiar Lorentzian form,

$$S_X(f) = E[X]\,\delta(f) + \frac{2\left(E[\tau_a] + E[\tau_b]\right)^{-1}}{(2\pi f)^2 + (2\pi f_S)^2}\,, \tag{8.10}$$

where $2\pi f_S \equiv 1/E[\tau_a] + 1/E[\tau_b]$.

8.2 ALTERNATING FRACTAL RENEWAL PROCESS

We turn now to the properties of the alternating fractal renewal process, a special alternating renewal processes in which one or both of the dwell-time distributions follows a fractal (power-law) form (see Fig. 8.1).

8.2.1 Spectrum

Consider the case when the dwell times in both states, τ_a and τ_b, have identical, abrupt-cutoff power-law distributions of the form displayed in Eq. (7.1). As shown in Sec. A.4.1, in the mid-frequency range ($A^{-1} \ll f \ll B^{-1}$) the alternating-fractal-renewal-process spectrum becomes (Lowen, 1992)

$$S_X(f) \to \frac{E[\mu]}{4} \times \begin{cases} 2\Gamma(1-\gamma)\,\cos(\pi\gamma/2)\,A^\gamma\,(2\pi f)^{\gamma-2} & 0 < \gamma < 1 \\ \frac{1}{2}A\,f^{-1} & \gamma = 1 \\ 2(\gamma-1)^{-1}\,\Gamma(2-\gamma)\left[-\cos(\pi\gamma/2)\right]A^\gamma\,(2\pi f)^{\gamma-2} & 1 < \gamma < 2 \\ 2A^2\left[-\ln(2\pi fA)\right] & \gamma = 2 \\ \gamma(\gamma-2)^{-1}A^2 & \gamma > 2. \end{cases} \tag{8.11}$$

The spectrum $S_X(f)$ follows a power-law form, but the exponent in the range $0 < \gamma < 1$ in Eq. (8.11) differs from that in Eq. (7.8) for the renewal-point-process spectrum $S_N(f)$.

Equation (8.11) establishes that the fractal exponent α depends on γ in accordance with

$$\alpha = \begin{cases} 2 - \gamma & 0 < \gamma < 2 \\ 0 & \gamma > 2. \end{cases} \tag{8.12}$$

This relationship differs from that provided in Eq. (7.9) for the fractal renewal point process; the fractal exponent α now extends over a larger range $0 < \alpha < 2$. The alternating fractal renewal process thus can serve as a source of $1/f^\alpha$ noise for $0 < \alpha < 2$, over the frequency range $B^{-1} \ll f \ll A^{-1}$ (see Prob. 8.6).

We present the normalized spectrum, $S_X(f)/S_0$, in Fig. 8.2. To facilitate comparisons among the plots, we divide by $S_0 \equiv \lim_{f \to 0} S_X(f)$ so that the normalized spectrum becomes unity at the low-frequency limit. The asymptotes are specified by Eqs. (8.6), (8.7), and (8.11), and the spectrum follows a $1/f^\alpha$ form over a range of frequencies. Taqqu & Levy (1986) set forth a closely related process that has no upper cutoffs.

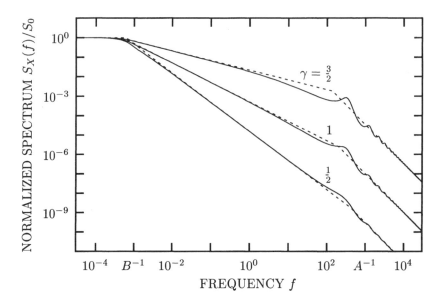

Fig. 8.2 Normalized spectrum $S_X(f)/S_0$ for the alternating fractal renewal process (solid curves). Both dwell times, τ_a and τ_b, derive from identical, abrupt-cutoff power-law distributions that take the form of Eq. (7.1), with $A = 10^{-3}$ and $B = 10^3$. The remaining parameter, the power-law exponent γ, takes the values $\frac{3}{2}$ (top), 1 (middle), and $\frac{1}{2}$ (bottom). Asymptotic forms from Eqs. (8.6), (8.11), and (8.7) highlight the low-, mid-, and high-frequency limits, respectively (dashed lines). All spectra decrease with frequency as $f^{-\alpha} = f^{\gamma-2}$ in the region $B^{-1} \ll f \ll A^{-1}$, in accordance with Eq. (8.11), and exhibit oscillations associated with the abrupt cutoffs in the probability densities.

8.2.2 Autocovariance

The autocovariance (defined as the autocorrelation minus the square of the mean), like the coincidence rate, can display power-law behavior as a result of the power-law-varying spectrum and the properties of the Fourier transform. For $1 < \gamma < 2$ in the region $A \ll |t| \ll B$, Eq. (8.11) leads to (Lowen & Teich, 1993d)

$$R_X(t) - \mathrm{E}^2[X] \approx (4\gamma)^{-1} A^{\gamma-1} |t|^{1-\gamma}. \tag{8.13}$$

For $0 < \gamma \leq 1$ in the region $A \ll |t| \ll B$, on the other hand, the autocovariance does not exhibit power-law behavior.

8.2.3 $1/f$-type noise from Markov processes

A chain of Markov alternating renewal processes, with rates that scale geometrically, provides another method for generating a process with a $1/f$-type spectrum (Lowen, Liebovitch & White, 1999). A variant of this construction imparts fractal behavior by embedding on–off processes endowed with different time scales, one inside another

(Krishnam, Venkatachalam & Capone, 2000). This latter construction begins with a largest-time-scale Markov alternating renewal process, with its "on" state corresponding to another Markov alternating renewal process with a shorter time scale, and so on, for a number of stages. In both constructions, the overall fractal behavior emerges from the combined scaling set of Markov processes.

8.3   BINOMIAL NOISE

We turn now to an analysis of the sum of M statistically identical and independent copies of the alternating renewal process. We begin by examining the amplitude distribution of the sum, which we denote X_Σ:

$$X_\Sigma(t) \equiv \sum_{n=1}^{M} X_n(t). \tag{8.14}$$

Figure 8.3 displays an example of the sum for $M = 4$ such processes.

Since the marginal distribution of each of the component processes follows a Bernoulli form with parameter

$$r \equiv \Pr\{X = 1\} = \mathrm{E}[X] = \frac{\mathrm{E}[\tau_a]}{\mathrm{E}[\tau_a] + \mathrm{E}[\tau_b]}, \tag{8.15}$$

as provided in Eq. (8.1), their sum comprises a binomial random variable with parameters r and M (Feller, 1968). The resulting process, $X_\Sigma(t)$, thus has a marginal amplitude that follows the **binomial distribution**:

$$\Pr\{X_\Sigma = n\} = \frac{M!}{n!\,(M - n)!}\, r^n (1 - r)^{M-n}, \tag{8.16}$$

with moments and related quantities given by

mean	$\mathrm{E}[X_\Sigma]$	$=$	Mr
variance	$\mathrm{Var}[X_\Sigma]$	$=$	$Mr(1 - r)$
skewness	$\dfrac{\mathrm{E}\big[(X_\Sigma - \mathrm{E}[X_\Sigma])^3\big]}{\mathrm{Var}^{3/2}[X_\Sigma]}$	$=$	$\dfrac{1 - 2r}{\sqrt{Mr(1 - r)}}$
kurtosis	$\dfrac{\mathrm{E}\big[(X_\Sigma - \mathrm{E}[X_\Sigma])^4\big]}{\mathrm{Var}^2[X_\Sigma]} - 3$	$=$	$\dfrac{1}{M}\left[\dfrac{1}{r(1 - r)} - 6\right].$

$$\tag{8.17}$$

We now examine the second-order properties of X_Σ, beginning with the autocorrelation. Using the independence of the component alternating renewal processes $X_n(t)$ that comprise $X_\Sigma(t)$, we obtain

$$R_{X_\Sigma}(t) \equiv \mathrm{E}\left[\sum_{n=1}^{M} X_n(s) \sum_{m=1}^{M} X_m(s + t)\right]$$

$$
\begin{aligned}
&= \sum_{n=1}^{M} \sum_{m=1}^{M} \mathrm{E}[X_n(s)\, X_m(s+t)] \\
&= \sum_{n=1}^{M} \left\{ \mathrm{E}[X_n(s)\, X_n(s+t)] + \sum_{m \neq n} \mathrm{E}[X_n(s)\, X_m(s+t)] \right\} \\
&= \sum_{n=1}^{M} \left\{ R_X(t) + \sum_{m \neq n} \mathrm{E}[X]\, \mathrm{E}[X] \right\} \\
&= M R_X(t) + M(M-1)\, \mathrm{E}^2[X], \quad\quad\quad (8.18)
\end{aligned}
$$

M ALTERNATING RENEWAL PROCESSES $X_n(t)$

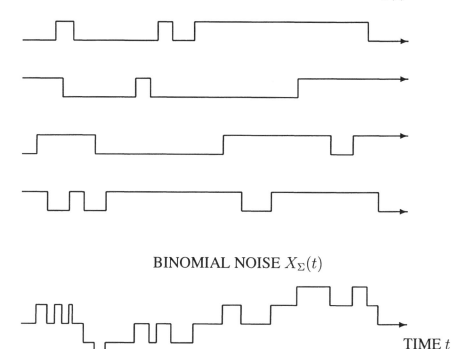

BINOMIAL NOISE $X_\Sigma(t)$

TIME t

Fig. 8.3 The addition of M statistically independent and identical alternating renewal processes $X_n(t)$ leads to a process $X_\Sigma(t)$ with similar second-order characteristics but with a binomial amplitude distribution: *binomial noise*. We illustrate sample functions for four alternating renewal processes, along with their arithmetic sum, which is binomial noise. Since the binomial-noise process comprises all transitions that occur within each of the individual alternating renewal processes, it exhibits greater fluctuations than the constituent processes. The sum of alternating fractal renewal processes results in *fractal binomial noise*.

so that

$$R_{X\Sigma}(t) - E^2[X_\Sigma] = M\{R_X(t) - E^2[X]\}. \tag{8.19}$$

Thus, the statistics of the binomial noise process $X_\Sigma(t)$ closely resemble those of the component alternating renewal processes $X_n(t)$. Because binomial noise exhibits all of the transitions occurring within each of the individual alternating renewal processes, as mentioned above, $X_\Sigma(t)$ exhibits M times more fluctuations per unit time than each of the constituent processes $X_n(t)$.

8.3.1 Fractal binomial noise

If the $X_n(t)$ exhibit fractal characteristics, and thus belong to the class of alternating fractal renewal processes, then $X_\Sigma(t)$ becomes fractal binomial noise. This process therefore has the same second-order statistics as its component alternating fractal renewal processes, and thus belongs to the fractal class of processes, as well as having a binomial amplitude distribution.

Finally, we mention an extension to the alternating fractal renewal process, in which $X(t)$ assumes random amplitudes during the on state, all independent and drawn from the same distribution (Yang & Petropulu, 2001) (see Prob. 8.5). The aggregation of a number of such processes provides a generalized form of fractal binomial noise, with similar temporal characteristics but arbitrary amplitude statistics.

8.3.2 Convergence to a Gaussian process

Consider again a sum of alternating renewal processes that gives rise to binomial noise, as set forth at the beginning of Sec. 8.3. As the number of constituent processes that comprise the sum increases, the skewness and kurtosis diminish, as shown in Eq. (8.17). Since the second moment of $X_n(t)$ exists, the central limit theorem applies and $X_\Sigma(t)$ approaches a Gaussian process as M increases. However, since the mean and variance of $X_\Sigma(t)$ increase without limit as M increases, we construct a process with bounded statistics for arbitrary values of M by using the usual linear conversion

$$X_M^*(t) \equiv [Mr(1-r)]^{-1/2} \sum_{n=1}^{M} [X_n(t) - Mr], \tag{8.20}$$

where r is the Bernoulli parameter set forth in Eq. (8.15). The quantity $X_M^*(t)$, as defined in Eq. (8.20), has zero mean by construction, and unit variance for all values of M, and thus remains well defined as $M \to \infty$. For delay times t in the range $A \ll t \ll B$, $X_M^*(t)$ has a covariance $R_{XM*}(t)$ proportional to that of the fractal Gaussian process discussed in Sec. 6.3.3. Since the mean and covariance define a Gaussian process, $X_M^*(t)$ indeed converges to a fractal Gaussian process.

We illustrate the convergence graphically in Fig. 8.4 for three values of the number of constituent alternating renewal processes M: 10, 100, and 1 000. The small number of processes results in a blocky appearance for the $M = 10$ curve; larger values of M lead to smoother plots. For simplicity of presentation, we employed Markov

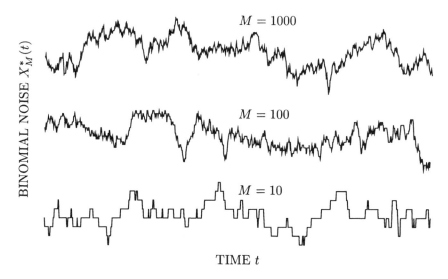

Fig. 8.4 Zero-mean unit-variance versions of binomial noise $X_M^*(t)$ constructed from sums of different numbers of constituent alternating renewal processes: $M = 10$, 100, and 1 000. The plots are displaced by different amounts. For simplicity, the alternating renewal processes $X_n(t)$ all belong to the Markov family; each has a symmetric, exponentially decaying dwell-time density with a mean equal to $\frac{1}{10}$ of the horizontal extent of the graph. The discrete nature of the process is readily apparent for small numbers of constituent processes (bottom plot); it diminishes as the number increases (middle), and becomes undetectable at the resolution used for still larger numbers (top).

alternating renewal processes for this figure. Similar results obtain for fractal binomial noise over times longer than the upper cutoff B of the constituent alternating fractal renewal processes. Over significantly shorter times, however, most of the constituent processes do not change state, which leads to a proportional decrease in the apparent number of processes, independent of M. Within the scaling range $A \ll t \ll B$, this apparent number increases with the time t in a power-law fashion.

8.4 POINT PROCESSES FROM FRACTAL BINOMIAL NOISE

Since binomial noise (and, in particular, fractal binomial noise) has a well-defined, finite integral and never assumes negative values, it can directly serve as the rate for a point process. We briefly consider several point processes constructed from fractal binomial noise.

If the binomial noise modulates a Poisson process, we refer to the resulting doubly stochastic Poisson point process as a **fractal-binomial-noise-driven Poisson process**. The relations provided in Sec. 4.3, together with those set forth earlier in this chapter, yield the statistical properties of this point process. In particular, the spectrum differs from that of fractal binomial noise only by an additive constant so that the fractal features of the underlying alternating fractal renewal processes are preserved, at least up to the frequency at which the constant becomes important. In the limit where the binomial noise converges to a Gaussian process, as described in Sec. 8.3.2, the process converges to the **fractal-Gaussian-process-driven Poisson process**[3] (see Secs. 6.3.3 and 10.6.1, as well as Fig. 5.5 and Chapter 12). The linearity of the Poisson transform yields relatively tractable results; however, it does introduce additional variability into the point process.

Transforms other than the Poisson can, of course, be implemented. If fractal binomial noise serves as the input to an integrate-and-reset process, instead of to a Poisson process, the **fractal-binomial-noise-driven integrate-and-reset process** results (see Sec. 4.4). The integrate-and-reset operation does not introduce additional variability so that the spectrum of the point process closely follows that of the fractal binomial noise, at least for sufficiently slow rate fluctuations. However, the highly nonlinear integrate-and-reset construct makes it difficult to obtain useful approximations in many regimes. As a final example, we mention the **fractal-binomial-noise-driven gamma process**, which has been considered as a model for characterizing sequences of action potentials in the mammalian visual system (Teich et al., 1997).

Problems

8.1 *Spectrum for symmetric alternating renewal process*   Starting with Eq. (8.5), show that if τ_a and τ_b have identical density functions, the corresponding alternating renewal process has a spectrum given by Eq. (8.8).

8.2 *Asymptotic limits of Lorentzian spectrum*   Show that the Lorentzian spectrum of Eq. (8.10) approaches the limits shown in Eqs. (8.6) and (8.7) for low and high frequencies, respectively.

8.3 *Spectrum from dwell-time characteristic functions*   Demonstrate that the substitution of characteristic functions corresponding to exponentially distributed dwell times in Eq. (8.5) yields a spectrum that accords with Eq. (8.10).

8.4 *Power-law autocovariance from power-law spectrum*   Show that Eq. (8.11) leads to Eq. (8.13).

8.5 *Alternation between arbitrary values*   We specified for simplicity that the alternating fractal renewal process $X_1(t)$ switches between two particular values, zero and unity; however, other values can be used as well. Suppose we define a new process $X_2(t)$ that has the same transition times as $X_1(t)$, but assumes values of b and a instead of zero and unity, respectively.

8.5.1. Express $E[X_2^n]$ in terms of $E[X_1^m]$, for $0 < m \leq n$.

8.5.2. Express the autocorrelation $R_{X2}(t)$ in terms of $R_{X1}(t)$ and $E[X]$.

---

[3] Although the fractal-binomial-noise-driven Poisson process is highly effective as a model in many applications, it is often replaced by this limiting form.

8.5.3. Now consider an alternating fractal renewal process $X_2(t)$ where a takes on random values, remaining the same during each "on" period, but independent of the values taken during other "on" periods (Yang & Petropulu, 2001). Set $b = 0$ for simplicity. What condition must be imposed on a in order that the autocorrelation $R_{X2}(t)$ have a meaningful value for all times t?

8.5.4. Assuming that this condition is satisfied, find an expression for $R_{X2}(t)$.

8.6   *Digital generation of $1/f^\alpha$ noise*   Simulation of the alternating fractal renewal process generates a synthetic signal with a spectrum that varies as $1/f^\alpha$. Given a random number X that is uniformly distributed in the interval $0 < X < 1$, the transformed random variable

$$Y = P_Y^{-1}(X) \tag{8.21}$$

has the probability distribution $P_Y(y)$, where P_Y^{-1} denotes the inverse of the function $P_Y(y)$.

8.6.1. Given a source of computer-generated random numbers $\{X_U\}$ uniformly distributed in the unit interval $(0 < X_U < 1)$, show how to generate a test signal with a $1/f^\alpha$ spectrum over the range $f_L \ll f \ll f_H$.

8.6.2. A Gaussian-distributed test signal generally proves more useful than one with only two states, such as that considered above. Suppose now that there is a further constraint requiring that both the skewness and kurtosis assume values within ϵ of those for a Gaussian random variable. To this end, we construct fractal binomial noise by summing the values of M of these processes. We may also employ different values of A or B for the dwell times in the two states to generate an asymmetric process. What values for the asymmetry and for M satisfy these constraints with the smallest number of processes M?

9

Fractal Shot Noise

While engaged in studies under Max Planck in Berlin, the German physicist **Walter Schottky (1886–1976)** characterized the stochastic properties of the random current arising from irregular electron arrivals at an anode; he bestowed on this process the name "shot noise."

Steven O. Rice (1907–1986), an American electrical engineer, studied the mathematical properties of shot noise in fine detail, demonstrating that its amplitude probability density often approaches a Gaussian form as the driving rate of the process increases without limit.

9.1	**Shot Noise**	186
9.2	**Amplitude Statistics**	189
9.3	**Autocorrelation**	194
9.4	**Spectrum**	195
9.5	**Filtered General Point Processes**	197
	Problems	198

Fractal shot noise, like a fractal Gaussian process, is a continuous-time process. Since it is everywhere nonnegative, it can serve as the rate of a doubly stochastic Poisson process, or an integrate-and-reset process, thereby generating associated point processes. Because many characteristics of these point processes derive from the properties of the underlying continuous rate process, we devote this chapter to the properties of fractal shot noise. Extensive discussion of the ensuing fractal-shot-noise-driven point processes is provided in Chapter 10. The material presented in this chapter and the next derives principally from Lowen & Teich (1989a,b, 1990, 1991).

The mean and variance of the classic shot-noise process were established by Campbell (1909b,a) at the beginning of the last century. Not long thereafter, Walter Schottky (1918) defined and extensively studied this process, and bestowed on it the name "shot effect." Twenty five years later, Rice (1944, 1945) carried out a classic detailed study of shot noise in which he demonstrated an important general feature of the process: the probability density function of its amplitude usually approaches a Gaussian form when the impulse response function has a finite duration and the emissions are dense. The classic shot-noise process serves as a remarkably useful construct in many fields of endeavor and it has been extensively studied (see, for example, Rice, 1944, 1945; Gilbert & Pollak, 1960; Picinbono, 1960; Saleh & Teich, 1982; Davenport & Root, 1987; Papoulis, 1991; Lax, 1997).

As schematically illustrated in Fig. 9.1, **shot noise** results from driving a memoryless, linear filter by a train of impulses derived from a homogeneous Poisson point process. The constant rate of event production μ characterizes the homogeneous Poisson process, and the **impulse response function** $h(t)$ characterizes the linear filter.

9.1 SHOT NOISE

As indicated in the schematic provided in Fig. 9.1, we define the shot-noise amplitude $X(t)$ in terms of an infinite sum of impulse response functions. The impulse response functions themselves are assumed to be deterministic. They can have stochastic components, however, so that we write

$$X(t) \equiv \sum_{k=-\infty}^{\infty} h(K_k, t - t_k). \qquad (9.1)$$

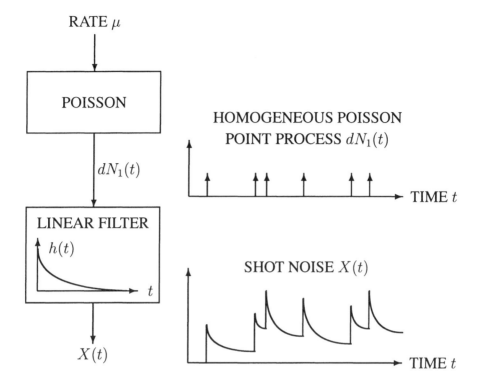

Fig. 9.1   A linearly filtered Poisson point process gives rise to shot noise. The quantity μ represents the rate of the Poisson process, $h(t)$ is the impulse response function of the linear filter, and $X(t)$ is the shot-noise amplitude. Fractal shot noise results when $h(t)$ takes the form of a decaying power-law function.

The random event times t_k belong to a homogeneous Poisson point process of rate μ, and $\{K_k\}$ is a random sequence that serves as an index for the impulse response functions $h(K, t)$. We take the elements of the random sequence $\{K_k\}$ as identically distributed, and independent of each other and of the Poisson process. Shot noise endowed with such an additional degree of randomness in its impulse response function (see, for example, Gilbert & Pollak, 1960; Picinbono, 1960) is known as **generalized shot noise**.

Fractal shot noise forms an important special case (Lowen & Teich, 1989b,a, 1990) in which the impulse response function assumes a general decaying power-law form,

$$h(K, t) \equiv \begin{cases} Kt^{-\beta} & A < t < B \\ 0 & \text{otherwise,} \end{cases} \tag{9.2}$$

as portrayed in Fig. 9.2. We refer to this process as *fractal* shot noise because power-law functions often characterize one or another of its properties in addition to the

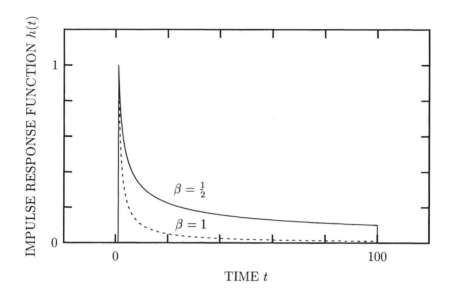

Fig. 9.2   Power-law impulse response functions $h(t)$ vs. time t, with lower cutoff time $A = 1$, upper cutoff time $B = 100$, deterministic amplitude $K = 1$, and two decay exponents: $\beta = \frac{1}{2}$ (solid curve) and $\beta = 1$ (dashed curve).

impulse response function; these properties include the amplitude probability density, autocorrelation, and spectrum.

The parameters A, B, and β are deterministic, fixed, and nonnegative. In general, the range of the impulse response function may extend down to $A = 0$ and up to $B = \infty$, and β may take any finite positive value. The formalism presented here assumes that all component impulse response functions have the same duration and power-law shape, but need not have the same amplitudes K. We consider a more general multifractal version of this impulse response function in Sec. 13.3.8.

Markedly different behavior obtains for different ranges of the parameters, as delineated in Table 9.1. For some parameters, the process exhibits a $1/f$-type spectrum. For a square-integrable impulse response function, the process converges to a Gaussian form by virtue of the central limit theorem. Conversely, when the impulse response function has infinite tail area, the resulting shot-noise process assumes a value of infinity with probability one; in fact, a degenerate process results even when the impulse response function is normalized to constant area (see Sec. A.6.1). On the other hand, impulse response functions with infinite area near the origin result in a process that is not degenerate but rather follows ($B = \infty$) or approaches ($B < \infty$) a stable distribution with infinite mean.

In the fractal shot-noise processes considered above, the impulse response functions themselves vary in a power-law fashion, as provided in Eq. (9.2). Another version of fractal shot noise may be constructed by endowing relatively simple impulse response functions, such as those that are rectangular, with variable duration

	$A > 0$ $B = \infty$	$A = 0$ $B = \infty$	$A = 0$ $B < \infty$	$A > 0$ $B < \infty$
$0 < \beta < \frac{1}{2}$ $(2 > \alpha > 1)$				$X \to$ Gaussian $S \sim 1/f^\alpha$
$\frac{1}{2} \leq \beta < 1$ $(1 \geq \alpha > 0)$	$\Pr\{X = \infty\} = 1$ no $S(f)$		$\mathrm{E}[X^2] = \infty$ $S \sim 1/f^\alpha$	
$\beta = 1$			$\mathrm{E}[X] = \infty$ no $S(f)$	$X \to$ Gaussian $S(f)$ not $1/f^\alpha$
$\beta > 1$ $(0 < \zeta < 1)$	$X \to$ Gaussian $S(f)$ not $1/f^\alpha$	$X =$ stable no $S(f)$	$X \to$ stable no $S(f)$	

Table 9.1 Characteristics of the amplitude probability density functions and spectra for fractal shot noise with various values of the parameters A, B, and β. A variety of different features emerge. For $B = \infty$ and $\beta \leq 1$, the impulse response function has a tail of infinite area; this yields a degenerate process for which finite amplitude values do not occur. Other regions in the parameter space yield well-defined shot-noise processes $X(t)$, although some have infinite variance or infinite mean. In particular, for $A = 0$ and $\beta > 1$, the distribution of X either is $(B = \infty)$ or approaches $(B < \infty)$ a stable random variable with parameter ζ, assuming that $\mathrm{E}[K^\zeta]$ exists. In other regions, both the mean and variance exist, and X converges to a Gaussian random variable as $\mu \to \infty$, assuming that $\mathrm{E}[K^2]$ exists. Finally, for $B < \infty$ and $\beta < 1$, the shot-noise process exhibits a spectrum that decays as $1/f^\alpha$.

and amplitude, and ascribing decaying power-law distributions to these parameters (see Prob. 9.2). Although distinct in their construction, the two formalisms yield similar results (Ryu & Lowen, 2000; Masoliver, Montero & McKane, 2001). Indeed, with a proper choice of distributions their amplitude statistics can be made to coincide (Gilbert & Pollak, 1960; Picinbono, 1960; Lowen & Teich, 1990). This variant of fractal shot noise is useful in modeling computer network traffic (see Chapter 13).

9.2 AMPLITUDE STATISTICS

Standard shot-noise theory provides the characteristic function $\phi_X(\omega)$ of the shot-noise process X (Doob, 1953; Saleh & Teich, 1982; Davenport & Root, 1987):

$$\phi_X(\omega) \equiv \mathrm{E}\left[e^{-i\omega X}\right] = \int_K \exp\left(-\mu \int_{t=-\infty}^{\infty} \left\{1 - \exp\left[-i\omega h(K, t)\right]\right\} dt\right) p_K(y) \, dy,$$

$$(9.3)$$

where $p_K(y)$ represents the probability density function of the impulse-response-function amplitudes K. For the specific case of a power-law decaying impulse response function, as provided in Eq. (9.2), with deterministic amplitudes K, we obtain

$$
\begin{aligned}
\ln\left[\phi_X(\omega)\right] &= -\mu \int_A^B \left[1 - \exp\left(-i\omega K\, t^{-\beta}\right)\right] dt \\
&= -\frac{\mu\,(i\omega K)^{1/\beta}}{\beta} \int_{i\omega K B^{-\beta}}^{i\omega K A^{-\beta}} \frac{1 - e^{-u}}{u^{1+1/\beta}}\, du \qquad (9.4) \\
&= \mu A\left[1 - \exp\left(-i\omega K A^{-\beta}\right)\right] \\
&\quad - \mu B\left[1 - \exp\left(-i\omega K B^{-\beta}\right)\right] \\
&\quad + \mu(i\omega K)^{1/\beta}\,\Gamma\left(1 - 1/\beta,\, i\omega K A^{-\beta}\right) \\
&\quad - \mu(i\omega K)^{1/\beta}\,\Gamma\left(1 - 1/\beta,\, i\omega K B^{-\beta}\right). \qquad (9.5)
\end{aligned}
$$

Here $\Gamma(x,\, a)$ represents the incomplete Eulerian gamma function

$$
\Gamma(x,\, a) \equiv \int_a^\infty t^{x-1}\, e^{-t}\, dt, \qquad (9.6)
$$

and integration by parts yields Eq. (9.5) from Eq. (9.4).

Derivatives of the logarithm of this function lead to the cumulants C_n of X (Rice, 1944, 1945), as defined in Eq. (3.8),

$$
\begin{aligned}
C_n &\equiv i^n \frac{d^n}{d\omega^n} \ln\left[\phi_X(\omega)\right]_{\omega=0} \\
&= \mu\, \mathrm{E}\left[\int_{-\infty}^\infty h^n(t)\, dt\right] \\
&= \mu\, \mathrm{E}[K^n] \int_A^B t^{-n\beta}\, dt, \qquad (9.7)
\end{aligned}
$$

which become (Lowen & Teich, 1990)

$$
C_n = \mu\, \mathrm{E}[K^n] \times
\begin{cases}
\dfrac{A^{1-n\beta} - B^{1-n\beta}}{n\beta - 1} & \beta \neq 1/n \\[2ex]
\ln(B/A) & \beta = 1/n.
\end{cases}
\qquad (9.8)
$$

The nth cumulant assumes an infinite value if $\mathrm{E}[K^n]$ does, or if $A = 0$ and $\beta \geq 1/n$, or if $B = \infty$ and $\beta \leq 1/n$. The moments and cumulants determine each other, as shown in Eq. (3.9).

Two general approaches are available for obtaining the probability density $p_X(x)$ for the shot-noise-process amplitude X. The first method involves carrying out the inverse Fourier transform of Eq. (9.3) [in a form much like Eq. (9.15)], which rarely proves feasible. The second method involves constructing an integral equation (Gilbert & Pollak, 1960). Note that if $B < \infty$, then $\Pr\{X = 0\} = \exp[-\mu(B - A)] > 0$, and thus the density has a delta function at $x = 0$.

For deterministic K, the amplitude probability density function then follows the integral equation (Lowen & Teich, 1989a, 1990):

$$p_X(x) = \begin{cases} 0 & x < 0 \\ \exp[-\mu(B - A)]\,\delta(x) & x = 0 \\ 0 & 0 < x \le KB^{-\beta} \\ \dfrac{\mu K^{1/\beta}}{\beta x} \displaystyle\int_{KB^{-\beta}}^{x} p_X(x - u)\,u^{-1/\beta}\,du & KB^{-\beta} < x < KA^{-\beta} \\ \dfrac{\mu K^{1/\beta}}{\beta x} \displaystyle\int_{KB^{-\beta}}^{KA^{-\beta}} p_X(x - u)\,u^{-1/\beta}\,du & x \ge KA^{-\beta}. \end{cases}$$

$$(9.9)$$

If $B = \infty$, then Eq. (9.9) simplifies to

$$p_X(x) = \frac{\mu K^{1/\beta}}{\beta x} \int_0^{\min(x,\,KA^{-\beta})} p_X(x - u)\,u^{-1/\beta}\,du, \qquad (9.10)$$

where $\min(x, y)$ returns the smaller of x and y. The integral equation Eq. (9.10) admits a family of solutions, all proportional to each other; imposing the requirement that $\int_0^\infty p_X(x)\,dx = 1$ provides the correct one.

For finite C_1 and C_2, the amplitude probability density function $p_X(x)$ satisfies the conditions of the central limit theorem, and therefore approaches a Gaussian density as $\mu \to \infty$. This always obtains for $A > 0$ and $B < \infty$ [see Eq. (9.7)], as displayed in the right-most column of Table 9.1. The first and second cumulants provide the mean and variance, respectively, of the resulting amplitude density, so that the limiting form becomes

$$p_X(x) \to (2\pi C_2)^{-1/2} \exp\left[-\frac{(x - C_1)^2}{2C_2}\right]. \qquad (9.11)$$

In fact, the vector $\{X(t_1), X(t_2), ..., X(t_k)\}$, for any positive integer k and any set of times $\{t_1, t_2, ..., t_k\}$, possesses a joint Gaussian distribution, so that $X(t)$ becomes a Gaussian process as $\mu \to \infty$. For finite μ and for values of x close to the mean of the process (C_1), we can expand the amplitude probability density about the asymptotic Gaussian result (Rice, 1944, 1945), which yields (Lowen & Teich, 1990)

$$\begin{aligned} p_X(x) \approx\ & (2\pi C_2)^{-1/2} \exp\left[-\frac{(x - C_1)^2}{2C_2}\right] \\ & \times \left[1 - \frac{C_3}{2C_2^2}(X - C_1) + \frac{C_3}{6C_2^3}(X - C_1)^3\right]. \end{aligned} \qquad (9.12)$$

For $\beta > 1$, $A \to 0$, and $B \to \infty$, a much simpler form obtains directly from the characteristic function (Lowen & Teich, 1990) (see Sec. A.6.2). The resulting expression, written as

$$\phi_X(\omega) = \exp\left\{-\mu\,\mathrm{E}[K^\zeta]\,\Gamma(1 - \zeta)\,(i\omega)^\zeta\right\}, \qquad (9.13)$$

with $\zeta \equiv 1/\beta$, follows the general form

$$\phi(\omega) = \exp\left[-(ic\omega)^\zeta\right], \tag{9.14}$$

for a constant c. The shot noise X then has a one-sided stable distribution (Lévy, 1937, 1940; Pollard, 1946; Feller, 1971) for all μ, with an associated parameter ζ that lies between zero and unity. Stochastic values of K that assume positive and negative values make stable distributions with other than one-sided forms possible (Petropulu, Pesquet, Yang & Yin, 2000). The Gaussian and other stable distributions share the property that two random variables taken from the same distribution, and added together, result in a new random variable whose distribution differs from the original one only by a scaling constant; the Gaussian differs from the stable distributions encountered here in that the latter have infinite means.

Further, if $A = 0$ and $\beta > 1$, but $B < \infty$, the amplitude X converges to a stable random variable as $\mu \to \infty$ (see Sec. A.6.2). The vector $\{X(t_1), X(t_2), ..., X(t_k)\}$ for any integer $k > 1$ and any set of times $\{t_1, t_2, ..., t_k\}$ does not have a joint stable distribution, so $X(t)$ is not a stable process. However, $X(t_1)$ and $X(t_2)$ do become jointly stable as the separation $t_2 - t_1$ approaches infinity (Petropulu et al., 2000).

Two methods exist for obtaining the associated amplitude probability density function $p_X(x)$. The first involves the use of a Fourier integral (Rice, 1944, 1945; Feller, 1971):

$$p_X(x) = \frac{1}{\pi} \operatorname{Re} \int_0^\infty \exp\left\{i\omega x - \mu \operatorname{E}[K^\zeta] \Gamma(1 - \zeta) (i\omega)^\zeta\right\} d\omega, \tag{9.15}$$

whereas the second makes use of an infinite sum (Humbert, 1945; Pollard, 1946; Feller, 1971):

$$p_X(x) = \frac{1}{\pi x} \sum_{n=1}^{\infty} \frac{(-1)^{n+1} \Gamma(1 + n\zeta) \sin(\pi n\zeta)}{n!} \left[\frac{\mu \Gamma(1 - \zeta) \operatorname{E}[K^\zeta]}{x^\zeta}\right]^n. \tag{9.16}$$

Grouping adjacent terms (of opposite sign) and simplifying improves the convergence properties of the sum (Weiss, Dishon, Long, Bendler, Jones, Inglefield & Bandis, 1994). For large values of X the sum converges quickly, whereas for small values of X the integral proves more useful.

For the particular case $\zeta = \frac{1}{2}$, the amplitude probability density function is described by the well-known closed-form expression (Lévy, 1940; Feller, 1971)

$$p_X(x) = \frac{\mu \operatorname{E}[K^{1/2}]}{2} x^{-3/2} \exp\left(-\frac{\mu^2 \pi \operatorname{E}^2[K^{1/2}]}{4x}\right), \tag{9.17}$$

which is identical to Eq. (3.13) with $t_0 = (\mu^2 \pi/4) \operatorname{E}^2[K^{1/2}]$. Lévy distributions arise in many contexts, such as the gravitational field produced by a random distribution of masses in one dimension and the electric field at the growing edge of a quantum wire, among others (see, for example, Good, 1961; Parzen, 1962; Lowen & Teich, 1989a). The force acting on a star as a result of the gravitational attraction of neighboring stars obeys a three-dimensional, spherically symmetric $\frac{3}{2}$-stable distribution,

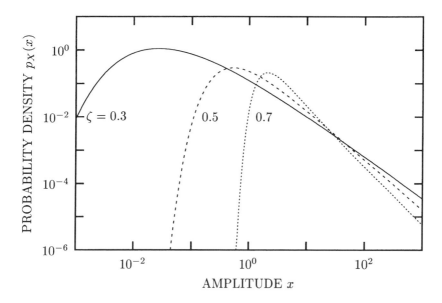

Fig. 9.3 One-sided stable amplitude probability density $p_X(x)$ vs. x provided in Eqs. (9.15)–(9.17), for three values of the associated parameter: $\zeta = 0.3$, 0.5, and 0.7 ($A = 0$, $B = \infty$, $K = 1$, $\mu = 1$). Long power-law tails are present for all values of ζ.

which is traditionally called the **Holtsmark distribution** (Holtsmark, 1919, 1924; Chandrasekhar, 1943).

Figure 9.3 displays stable amplitude probability density functions for three values of the parameter ζ. All have power-law tails, as provided by Eq. (9.16). For large x the first term dominates; in the limit $x \to \infty$, after simplification we obtain (Lowen & Teich, 1990)

$$p_X(x) \approx \mu\,\zeta\,\mathrm{E}[K^\zeta]\,x^{-(1+\zeta)}. \tag{9.18}$$

For other infinite-area impulse response functions the resulting shot noise has trivial amplitude properties. For $0 < \beta \leq 1$ and $B = \infty$, the shot-noise process X assumes an infinite value with probability one (Lowen & Teich, 1990) (see Sec. A.6.1). This degenerate case arises because the infinite area of each impulse response function lies in its tail, which persists throughout time. These tails accumulate to produce an unbounded sum. Even normalizing the impulse response functions to constant area as B increases toward infinity yields a degenerate process, in this case one with zero variance (Lowen & Teich, 1991) (see Sec. A.6.1). In contrast, for $\beta > 1$ the infinite area occurs only during an infinitesimal interval immediately following the onset of the impulse response function; this leads to a well-defined shot noise process $X(t)$ in spite of the fact that the mean is infinite.

9.3 AUTOCORRELATION

The autocorrelation of the process $X(t)$ assumes the simple form (Rice, 1944, 1945)

$$R_X(t) \equiv E[X(s)\,X(s+t)] = E^2[X] + \mu R_h(t), \tag{9.19}$$

where the autocorrelation of $h(K,t)$ itself obeys the equation

$$
\begin{aligned}
R_h(t) &\equiv E\left[\int_{-\infty}^{\infty} h(K,s)\,h(K,s+|t|)\,ds\right] \\
&= E[K^2]\int_{A}^{B-|t|}(s^2+|t|s)^{-\beta}\,ds. \tag{9.20}
\end{aligned}
$$

When $|t| \geq B - A$, $R_h(t) = 0$ so that $R_X(t) = E[X]^2$.

The integral in Eq. (9.20) becomes infinite for parameters in the following ranges:

$$
\begin{array}{lll}
\beta \leq \tfrac{1}{2} & \text{and} & B = \infty, \\
\beta \geq 1 & \text{and} & A = 0, \\
\beta \geq \tfrac{1}{2} & \text{and} & A = 0 \quad \text{and} \quad t = 0;
\end{array} \tag{9.21}
$$

in those cases $R_X(t)$ does not exist (Lowen & Teich, 1990).

For parameter values outside the ranges set forth in Eq. (9.21), we can develop useful approximations to $R_h(t)$, and therefore to $R_X(t)$. For $\beta < \tfrac{1}{2}$ and $A \ll |t| \ll B$, we have

$$
\begin{aligned}
R_h(t) &= E[K^2]\int_{A}^{B-|t|} s^{-\beta}\,(s+|t|)^{-\beta}\,ds \\
&\approx E[K^2]\int_{0}^{B} s^{-2\beta}\,ds \\
&= (1-2\beta)^{-1}\,E[K^2]\,B^{-2\beta}, \tag{9.22}
\end{aligned}
$$

where the approximations derive from the specified range of t, and from the domination of the integrand by the tail. Thus, for these values of β and this range of t, the autocorrelation essentially remains fixed with respect to t.

For $\tfrac{1}{2} < \beta < 1$, and the same range of t, neither A nor B is important so that

$$
\begin{aligned}
R_h(t) &= E[K^2]\int_{A}^{B-|t|} s^{-\beta}(s+|t|)^{-\beta}\,ds \\
&\approx E[K^2]\int_{0}^{\infty} s^{-\beta}(s+|t|)^{-\beta}\,ds \\
&= \Gamma(1-\beta)\,\Gamma(2\beta-1)\,[\Gamma(\beta)]^{-1}\,E[K^2]\,|t|^{1-2\beta}; \tag{9.23}
\end{aligned}
$$

the exponent $1 - 2\beta$ lies between -1 and 0. For $\beta > 1$ and $A \ll |t| \ll B$ we obtain

$$
R_h(t) = E[K^2]\int_{A}^{B-|t|} s^{-\beta}\,(s+|t|)^{-\beta}\,ds
$$

$$\approx \; \mathrm{E}[K^2] \int_A^\infty s^{-\beta} |t|^{-\beta} \, ds$$

$$= \; (\beta - 1)^{-1} \, \mathrm{E}[K^2] \, A^{1-\beta} \, |t|^{-\beta}, \tag{9.24}$$

where the approximations derive from the specified range of t, and from the domination of the integrand by the area near the origin. Closed-form expressions exist for particular values of β, and for large values of $|t|$ when $\beta > 1$ (see Sec. A.6.3).

9.4 SPECTRUM

Carson's theorem (Carson, 1931) provides the spectrum $S_X(f)$ of the fractal shot-noise process X in terms of μ and the Fourier transform $\mathcal{F}$ of the impulse response function defined in Eq. (9.2) (Rice, 1944, 1945). We focus on the domain $\beta < 1$ and $B < \infty$ where $1/f^\alpha$ spectral behavior prevails. Denoting the Fourier transform $\mathcal{F}$ by $H(f)$, we obtain (Lowen & Teich, 1990)

$$H(f) \; \equiv \; \mathcal{F}\{h(t)\} = K \int_A^B t^{-\beta} \, e^{-i2\pi f t} \, dt \tag{9.25}$$

$$= \; K\left[\Gamma(1 - \beta, \, i2\pi f A) - \Gamma(1 - \beta, \, i2\pi f B)\right] (i2\pi f)^{\beta - 1}, \tag{9.26}$$

where $\Gamma(x, a)$ again represents the (incomplete) Eulerian gamma function defined in Eq. (9.6).

For $B < \infty$ and $\mathrm{E}[K^2] < \infty$ the autocovariance $R_X(t) - \mathrm{E}^2[X]$ has a finite integral. Carson's theorem then applies, which yields (Lowen & Teich, 1990)

$$S_X(f) \; = \; \mathrm{E}^2[X] \, \delta(f) + \mu \mathrm{E}\left[|H(f)|^2\right] \tag{9.27}$$

$$= \; \mathrm{E}^2[X] \, \delta(f)$$

$$+ \mu \mathrm{E}[K^2] \left|\Gamma(1 - \beta, \, i2\pi f A) - \Gamma(1 - \beta, \, i2\pi f B)\right|^2$$

$$\times \, (2\pi f)^{-2(1-\beta)}. \tag{9.28}$$

This spectrum appears as the solid curve in Fig. 9.4.

For $0 < \beta < 1$, fitting Eq. (9.28) with a spectrum of the form $S(f) \sim f^{-\alpha}$ provides

$$\alpha = 2(1 - \beta). \tag{9.29}$$

In particular, for $1/B \ll f \ll 1/A$, and this range of β, the first and second incomplete gamma functions approach the complete gamma function and zero, respectively, whereupon

$$S_X(f) \approx \mu \mathrm{E}[K^2] \, \Gamma^2(\alpha/2) \, (2\pi f)^{-\alpha}. \tag{9.30}$$

The abrupt cutoff in the time domain leads to oscillations in the frequency domain (observe the solid curve in Fig. 9.4). The special case $\alpha = 1$ was initially examined by Schönfeld (1955) and developed by van der Ziel (1979). Inspection of Eq. (9.25)

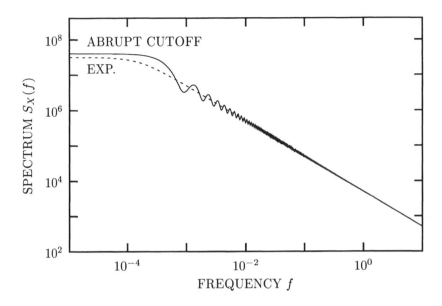

Fig. 9.4 Spectrum $S_X(f)$ vs. frequency f for fractal shot noise with different cutoffs: abrupt (solid curve) and exponential (dashed curve). The two processes do not otherwise differ, sharing the parameters $\beta = \frac{1}{2}$, $A = 0$, $B = 1\,000$, $K = 100$, and $\mu = 1$. For sufficiently high frequencies, both curves exhibit $1/f^\alpha$ behavior, with exponent $\alpha = 1$. The impulse response function with an abrupt cutoff gives rise to oscillations in the frequency domain whereas the exponential transition does not. Their low-frequency values approach different asymptotes since their mean values differ; Eq. (9.31) yields values of 4×10^7 and $10^7 \pi \approx 3.14 \times 10^7$ for the abrupt-cutoff and exponential-cutoff impulse response functions, respectively.

reveals that the spectrum approaches a constant value in the limit $f \to 0$, for any impulse response function (Lowen & Teich, 1990):

$$\lim_{f \to 0} S_X(f) = \lim_{f \to 0} \mu E\left[|H(f)|^2\right] = \mu E[K^2]\left(E[X]/E[K]\right)^2. \tag{9.31}$$

A power-law impulse response function with exponential transitions,

$$h_2(K,t) = K \exp(-A/t) \, \exp(-t/B) \, t^{-\beta}, \tag{9.32}$$

yields a smoother transition near the cutoff frequency $f \approx (2\pi B)^{-1}$, at the expense of more complex expressions for other quantities. In particular, for $A \to 0$ the impulse response function is the same as that considered by Buckingham (1983, Chapter 6), which leads to (Lowen & Teich, 1989b, 1990)

$$
\begin{aligned}
H_2(f) &= K \int_0^\infty t^{-\beta} \exp(-t/B - i2\pi ft) \, dt \\
&= \Gamma(\alpha/2) \, B^{\alpha/2} \, K (1 + i2\pi fB)^{-\alpha/2} \\
S_{X2}(f) &= E[X]^2 \, \delta(f) + \mu E[K^2] \, \Gamma^2(\alpha/2) \, B^\alpha [1 + (2\pi fB)^2]^{-\alpha/2}. \quad (9.33)
\end{aligned}
$$

In the high-frequency limit $f \gg 1/B$, Eq. (9.30) applies for this impulse response function as well. The spectrum for a shot-noise process with an exponential-transition impulse response function is displayed as the dashed curve in Fig. 9.4.

While Eq. (9.28) promises $1/f$-type behavior for $\beta < 1$ and $B = \infty$, the process is actually degenerate for this set of parameters, as shown in Sec. A.6.1. Thus, $1/f^\alpha$ behavior emerges only for $\beta < 1$ and $B < \infty$. We note that long-duration memory exists in a generalized sense for other parameter ranges, notably for $\beta > 1$ and $A = 0$ (Petropulu et al., 2000).

We emphasize that the parameter α is fundamentally different in character from the parameter ζ for the stable probability distribution encountered in Eq. (9.13) and thereafter. The quantity α describes the properties of a signal over time, with no reference to its amplitude. The quantity ζ, on the other hand, characterizes only the amplitude distribution and is therefore unrelated to the time course of a signal. Considering a signal such as a fractal rate, the spectrum $S(f)$ evaluated along the abscissa (horizontal axis) decays in a power-law fashion as $f^{-\alpha}$. Considering, now, the graph of a different signal with a stable amplitude distribution, the probability $\Pr\{X > x\}$ of observing a large value x occurring on the ordinate (vertical axis) decays in a power-law fashion as $x^{-\zeta}$. Figuratively speaking, therefore, α and ζ describe orthogonal properties. Moreover, the stable distributions described here obey $\zeta < 1$, and therefore have infinite means and no spectra. No single process examined in the present work exhibits nontrivial values of both ζ and α simultaneously.

9.5   FILTERED GENERAL POINT PROCESSES

Although this chapter has been directed toward setting forth the properties of a linearly filtered homogeneous Poisson process, this particular focus does not represent a fundamental limitation of the approach. Indeed, the filtering of any orderly point process yields a well-behaved continuous-valued process, although the properties of such filtered general point processes are usually more difficult to derive when the underlying process is non-Poisson (Lukes, 1961). Filtered versions of many types of point processes have been examined (see, for example, Parzen, 1962; Weiss, 1973; Grandell, 1976; Snyder & Miller, 1991).

Our final considerations in this chapter are devoted to two measures for which tractable results for filtered general point processes are readily established: the mean and spectrum. We further restrict ourselves to linear filtering in which the random filter functions $h(K,t)$ are independent of the point process $dN(t)$.

If $\mu(t)$ represents the rate of the point process, the mean value of the resulting generalized process $X(t)$ is simply

$$\mathrm{E}[X] = \mathrm{E}[\mu]\,\mathrm{E}\left[\int_0^\infty h(K,t)\,dt\right]. \tag{9.34}$$

The square of this quantity appears as a delta function in the spectrum $S_X(\omega)$ of the generalized process $X(t)$. For other frequencies, linear systems theory (Papoulis,

1991) provides a spectrum given by

$$S_X(f) = \mathrm{E}\left[|H(f)|^2\right] S_N(f). \tag{9.35}$$

These formulas are readily applied to the results provided in Chapter 4 for several classes of point processes. In the special case when $dN(t)$ is a homogeneous Poisson process, $S_N(f) = \mu$ and we recover Eq. (9.27).

Problems

9.1 *Sums of fractal-shot-noise processes*   Let $X_m(t)$, for $1 \le m \le M$, each denote an independent shot-noise process with rate μ_m and impulse response function $h_m(K,t)$. Define $X_R(t)$ as the sum of all the $X_m(t)$.

9.1.1. Suppose that the impulse response functions coincide for all m, so that $h_m(K,t) = h_1(K,t)$ for all indices m. Show that $X_R(t)$ also belongs to the shot-noise family of processes and find its rate μ_R.

9.1.2. Now remove the identity among the impulse response functions. Assuming that k is fixed and deterministic within each component process $X_m(t)$, show that it is still possible to describe $X_R(t)$ as a shot noise process.

9.1.3. If the $X_m(t)$ are each fractal shot noise processes for all m with $0 < \alpha < 2$, under what conditions does $X_R(t)$ belong to the fractal shot-noise family of processes?

9.1.4. Now suppose that $h_m(K,t) = \exp(-c_m t)$, where K is deterministic and set to unity for simplicity. Given this set of impulse response functions, is it possible to generate an approximation to fractal shot noise for $0 < \alpha < 2$? If so, plot a representative spectrum.

9.2 *Rectangular impulse response function with power-law-distributed duration* As indicated at the end of Sec. 9.1, we need not limit the functional form of $h(K,t)$ to a multiplicative decomposition such as $Kh(1,t)$. Consider the rectangular impulse response function

$$h(K,t) = \begin{cases} c & 0 < s < K \\ 0 & \text{otherwise.} \end{cases} \tag{9.36}$$

9.2.1. Setting $c = 1$ for simplicity, find an expression for the autocorrelation of the resulting shot-noise process $X(t)$.

9.2.2. Now suppose that K has a probability density function that assumes the generalized Pareto form

$$p_K(t) = \begin{cases} (\beta - 1)A^{\beta-1}t^{-\beta} & t > A \\ 0 & t \le A, \end{cases} \tag{9.37}$$

with $2 < \beta < 3$. Find an expression for the autocorrelation $R_X(t)$ for $t > A$, and in particular, identify the associated range of α.

9.3 *Decaying-power-law mass distributions*   In many systems of aggregated particles the mass distribution $P_\mathcal{M}(m)$ obeys a power-law form over some range of masses m, such that (Witten & Sander, 1981; Grassberger, 1985; Takayasu, Nishikawa & Tasaki, 1988)

$$\Pr\{\mathcal{M} \geq m\} = cm^{-D}, \tag{9.38}$$

where c is a normalizing constant. The power-law exponent D typically falls in the range $0 < D < 1$. In some systems the number of particles in any given region is Poisson distributed and their masses are independent of each other and of the number of particles. Suppose we express the mass $\mathcal{M}$ as the amplitude of a power-law impulse response function and then invert Eq. (9.38). Show how this procedure yields a fractal shot-noise process that provides appropriate values for the total mass in a specified region, and find the corresponding range of β.

10

Fractal-Shot-Noise-Driven Point Processes

In 1939, the Polish-born mathematician **Jerzy Neyman (1894–1981)** conceived the "Neyman Type-A" probability distribution and, with Elizabeth Scott in 1958, developed its generalization: the cluster point process.

Maurice Stevenson Bartlett (1910–2002), a British statistician, constructed the shot-noise-driven doubly stochastic Poisson point process and showed that it is a particular Neyman–Scott cluster point process.

10.1	Integrated Fractal Shot Noise	204
10.2	Counting Statistics	205
	10.2.1 Counting distribution	205
	10.2.2 Count moments	207
	10.2.3 Normalized variance	208
	10.2.4 Normalized Haar-wavelet variance	210
10.3	Time Statistics	212
10.4	Coincidence Rate	214
10.5	Spectrum	215
10.6	Related Point Processes	216
	10.6.1 Point process in the Gaussian limit of fractal shot noise	216
	10.6.2 Fractal-shot-noise-driven integrate-and-reset point process	217
	10.6.3 Hawkes point process	217
	10.6.4 Bartlett–Lewis fractal point process	218
	Problems	219

In this chapter we consider two classes of point processes for which fractal shot noise serves as the rate. Fractal shot noise is a continuous stochastic process described in detail in the previous chapter. We focus on a point-process generation mechanism that is Poisson; however, we also briefly consider an integrate-and-reset generation mechanism at the end of the chapter. The properties of these two classes of point processes are closely related, as will become apparent subsequently.

Conceived by Bartlett (1964) in the context of ecology, the **shot-noise-driven doubly stochastic Poisson point process** (abbreviated **shot-noise-driven Poisson process**) results when any form of shot noise serves as the rate for a Poisson-event generator. Bartlett developed a two-dimensional version of this process and recognized it as a particular **Neyman–Scott cluster process** (see Sec. 4.5) comprising Poisson primary and Poisson secondary event sequences [see Neyman & Scott (1958); Vere-Jones (1970); Lawrance (1972)]. An extensive list of applications has come to the fore for this family of point processes, in fields as diverse as entomology, astrophysics, visual science, geophysics, neurophysiology, and photon statistics, among others.[1] The probability distribution that universally emerges in the long counting-time limit, the Neyman Type-A distribution (Neyman, 1939), has also found extensive application.[2]

The generation of this process is schematically illustrated in Fig. 10.1. A primary homogeneous Poisson point process $dN_1(t)$ with mean rate μ comprises the first stage. These events then pass through a linear filter with impulse response function $h(t)$.

---

[1] Examples can be found in Neyman & Scott (1958); Vere-Jones (1970); Neyman & Scott (1972); Teich & Saleh (1981b, 1987, 1988, 1998); Saleh & Teich (1982, 1983); Saleh, Tavolacci & Teich (1981); Teich, Saleh & Peřina (1984).

[2] See, for example, McGill (1967); Teich (1981); Teich & Saleh (1981a, 1987, 2000); Saleh & Teich (1985a); Saleh et al. (1983); Teich, Prucnal, Vannucci, Breton & McGill (1982a,b); Prucnal & Teich (1982); Teich, Tanabe, Marshall & Galayda (1990).

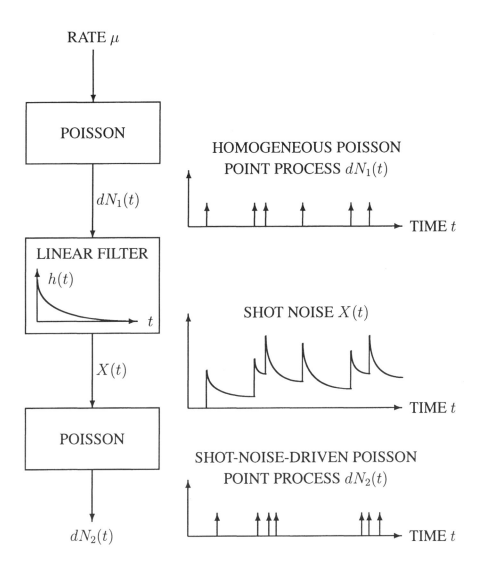

Fig. 10.1 The shot-noise-driven doubly stochastic Poisson point process arises from two Poisson processes mediated by a linear filter. The quantity μ represents the rate of the first Poisson point process, $h(t)$ is the impulse response function of the linear filter, and $X(t)$ is the continuous-time shot-noise amplitude at the output of the linear filter. The stochastic process $X(t)$ serves as the rate function for a second Poisson point process. The overall output $dN_2(t)$ is a shot-noise-driven doubly stochastic Poisson point process. If $h(t)$ decays in a power-law fashion, $X(t)$ is fractal shot noise and $dN_2(t)$ is a fractal-shot-noise-driven Poisson process.

This filter produces a shot noise $X(t)$ at its output. This shot noise, in turn, serves as the input to a second Poisson process that generates events $dN_2(t)$ at the time-varying rate determined by $X(t)$. Since the rate of the point process $dN_2(t)$ coincides with the shot-noise driving process $X(t)$, $dN_2(t)$ incorporates the variance imparted by this rate and therefore does not belong to the family of homogeneous Poisson point processes. The primary-process rate μ and the impulse response function $h(t)$ completely characterize $dN_2(t)$.

In this chapter we develop the properties of the fractal-based form of $dN_2(t)$, which is officially known as the **fractal-shot-noise-driven doubly stochastic Poisson point process**, which we abbreviate as the **fractal-shot-noise-driven Poisson process**. In accordance with the results established in Chapter 9, the linear-filter impulse response function decays in a power-law fashion, which gives rise to the fractal-shot-noise stochastic rate $X(t)$. The impulse response function $h(K, t)$ is taken to contain a stochastic component. We can also describe the fractal version of the shot-noise-driven Poisson process in terms of a two-stage **fractal Neyman–Scott cluster process**, in which each event of a primary Poisson point process directly generates a random number of events in a secondary Poisson point process (see Sec. 4.5). The two formulations are isomorphic (see Lowen & Teich, 1991, Appendix A). In the last sections of the chapter, several related processes are briefly described; these include the **fractal-shot-noise-driven integrate-and-reset process**, the **Hawkes point process**, and the **fractal Bartlett–Lewis cascaded process**.

The fractal-shot-noise-driven Poisson process enjoys a broad variety of applications, including the modeling of earthquake occurrence times, Čerenkov photon statistics, diffusion processes, action-potential statistics, and computer network traffic (considered in Secs. 13.5.5 and 13.6). Various applications are discussed by Vere-Jones (1970); Lowen & Teich (1991); Teich et al. (1997, 1990); Ryu & Lowen (1995, 1997, 1998), as well as in the problems at the end of this chapter.

10.1 INTEGRATED FRACTAL SHOT NOISE

The time integral of the shot-noise process $X(t)$ forms an auxiliary random process $X_T(t)$,

$$X_T(t) \equiv \int_t^{t+T} X(u)\, du. \tag{10.1}$$

Conveniently, the time integral of a shot-noise process forms another shot-noise process, with corresponding impulse response function

$$h_T(K, t) \equiv \int_t^{t+T} h(K, u)\, du. \tag{10.2}$$

A representative pair of impulse response functions, $h(t)$ and $h_T(t)$, is shown in Fig. 10.2.

The integrated process finds use in establishing the properties of the fractal-shot-noise-driven Poisson process $dN_2(t)$. The first-order moment generating function

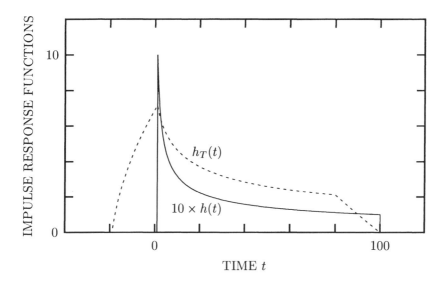

Fig. 10.2 Representation of a particular power-law-decaying impulse response function, $h(t)$ ($\times 10$) vs. time t (solid curve), and its associated integrated impulse response function, $h_T(t)$ (dotted curve). The parameters are $\beta = \frac{1}{2}$, $A = 1$, $B = 100$, $K = 1$, and $T = 20$. The integrated version $h_T(t)$ is derived from $h(t)$ via Eq. (10.2); for the parameters shown it is proportional to the average of $h(t)$ over the following 20 time units.

of the integrated shot-noise process $X_T(t)$ turns out to yield all of the first-order statistics of $dN_2(t)$, including its counting and time-interval distributions (Saleh & Teich, 1982). Closed-form results are available for some special cases (Lowen & Teich, 1991, Sec. IIIC and Appendix B).

10.2 COUNTING STATISTICS

10.2.1 Counting distribution

A recurrence relation provides the counting distribution, $p_Z(n; T) = \Pr\{Z(T) = n\}$, for any shot-noise-driven Poisson process (Saleh & Teich, 1982). The first step is to determine the probability of zero events occurring in a specified time duration T:

$$p_Z(0; T) = \exp\left(\mu E\left[\int_{-\infty}^{\infty} \{\exp[-h_T(K, t)] - 1\} \, dt\right]\right), \qquad (10.3)$$

where the expectation is over K. For $n > 0$ we write

$$p_Z(n + 1; T) = \frac{1}{n + 1} \sum_{k=0}^{n} c_k \, p_Z(n - k; T), \qquad (10.4)$$

where the coefficients c_k are given by

$$c_k \equiv \frac{\mu}{k!}\, \mathrm{E}\left[\int_{-\infty}^{\infty} \left[h_T(K,t)\right]^{k+1} \exp[-h_T(K,t)]\, dt\right]. \tag{10.5}$$

Detailed expressions for the coefficients c_k, which permit the counting distribution to be calculated, are provided in Sec. A.7.1 for a fractal-shot-noise-driven Poisson process with a deterministic impulse response function.

The **Neyman Type-A counting distribution** emerges from Eqs. (10.3)–(10.5) in the limit of a deterministic, delta-function impulse response function of area a, $h(t) = \mathsf{a}\,\delta(t)$:

$$p_Z(0; T) \;=\; \exp\{\mu T\left(e^{-\mathsf{a}} - 1\right)\} \tag{10.6}$$

$$p_Z(n+1; T) \;=\; \frac{1}{n+1}\, \mu T \sum_{k=0}^{n} \frac{\mathsf{a}^{k+1} e^{-\mathsf{a}}}{k!}\, p_Z(n-k; T). \tag{10.7}$$

Although this distribution corresponds to the instantaneous generation of multiple secondary events, and thereby violates the assumption of an orderly point process (see Sec. 3.2), we nevertheless consider this simplification to illustrate how the Neyman Type-A counting distribution arises in appropriate limits. This distribution also applies for arbitrary $h(t)$ in the domain $T \gg (B - A)$ since all secondary events born of a single primary event, although splayed out over a time $(B - A)$, are fully captured within the counting time T [see Lowen & Teich, 1991, Eq. (21)].

With the help of these results, we plot $p_Z(n; T)$ in Fig. 10.3 for $\beta = \frac{1}{2}$ ($\alpha = 1$). Representative results for other parameters appear in Lowen & Teich (1991, Figs. 3–5). The counting distributions displayed in Fig. 10.3 interpolate between the Poisson distribution [Eq. (4.7)] and the Neyman Type-A distribution [Eq. (10.7)], in the short- and long-counting-time limits, respectively.

In fact, the two-parameter Neyman Type-A distribution serves as an excellent approximation for a broad variety of counting distributions associated with shot-noise-driven Poisson processes for arbitrary values of $T/(B - A)$. Good agreement over a substantial range of parameters obtains by matching the means and variances of the Neyman Type-A and the exact distributions, as has been explicitly demonstrated for rectangular and exponential impulse response functions (Teich & Saleh, 1987).

Finally, we note that the Neyman Type-A distribution also provides a good approximation for counting distributions associated with the Thomas point process, for these same impulse response functions (Teich & Saleh, 1987). This latter process incorporates the events of the primary homogeneous Poisson process (see Fig. 10.1), along with the secondary events, into the final process. The statistical properties of the Thomas process, which does not belong to the family of doubly stochastic Poisson processes, have been investigated by Matsuo et al. (1983) (see also Secs. 4.5 and 4.6); Thomas (1949) initially developed the exact two-parameter counting distribution that emerges in the limit of large counting times. Various properties and applications of these distributions, as well as doubly stochastic versions thereof, have been studied by Teich (1981).

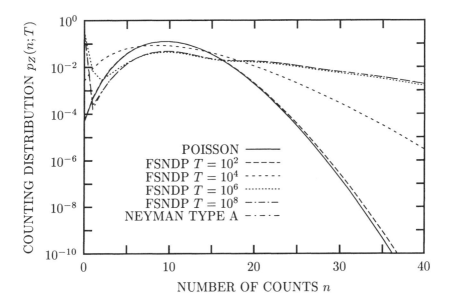

Fig. 10.3 Counting distribution $p_Z(n; T)$ vs. number of counts n for the fractal-shot-noise-driven Poisson process (FSNDP) with $\beta = \frac{1}{2}$, $A = 1$, $B = 10^5$, $a = 10$, and $E[Z] = 10$. Curves are shown for four values of the counting time: $T = 10^2$, 10^4, 10^6, and 10^8. An entire family of counting distributions can be presented while preserving $E[Z] = 10$ since the fractal-shot-noise-driven Poisson process has five parameters. For small values of the counting time ($T/A \rightarrow 0$), the distribution approaches the Poisson whereas for large values of the counting time ($T \gg B$), it approaches the Neyman Type-A for any A and β. Since the fractal renewal process depends on only three parameters, it is not possible to provide a similar display in Fig. 7.6.

10.2.2 Count moments

Although general expressions for the count moments prove complex, a relatively simple result emerges for the factorial moments. This again takes the form of a recurrence relation (Saleh & Teich, 1982):

$$E\left\{\frac{[Z(t)]!}{[Z(t) - (n+1)]!}\right\} = \sum_{k=0}^{n} b_k \binom{n}{k} E\left\{\frac{[Z(t)]!}{[Z(t) - (n-k)]!}\right\}, \qquad (10.8)$$

with

$$E\left\{\frac{[Z(t)]!}{[Z(t)]!}\right\} \equiv 1 \qquad \text{and} \qquad b_k \equiv \mu E\left[\int_{-\infty}^{\infty} \left[h_T(K, t)\right]^{k+1} dt\right]. \qquad (10.9)$$

Explicit formulas for the particular case of the fractal shot-noise-driven Poisson process exist in this case as well (Lowen & Teich, 1991) (see Sec. A.7.1).

The first factorial moment is the mean number of counts, and rearranging the first two factorial moments yields the variance:

$$
\begin{aligned}
\mathrm{E}[Z(T)] &= \mathrm{E}\left\{\frac{[Z(t)]!}{[Z(t)-1]!}\right\} = b_0 \\
&= \mu\mathrm{E}\left[\int_{-\infty}^{\infty} h_T(K,t)\,dt\right] = \mu T\,\mathrm{E}\left[\int_0^{\infty} h(K,t)\,dt\right] \\
&= \mu a T;
\end{aligned}
\qquad (10.10)
$$

$$
\begin{aligned}
\mathrm{Var}[Z(T)] &= \mathrm{E}\left\{\frac{[Z(t)]!}{[Z(t)-2]!}\right\} + \mathrm{E}[Z(T)] - \mathrm{E}^2[Z(T)] \\
&= b_0 + b_1 = \mu a T + \mu\mathrm{E}\left\{\int_{-\infty}^{\infty} [h_T(K,t)]^2\,dt\right\} \\
&= \mu a T + 2\mu\int_0^T (T-u) \\
&\qquad \times \mathrm{E}\left[\int_{-\infty}^{\infty} h(K,t)\,h(K,t+u)\,dt\right]\,du.
\end{aligned}
\qquad (10.11)
$$

The quantity a is the expected value of the area of the impulse response function,

$$
a \equiv \mathrm{E}\left[\int_0^{\infty} h(K,t)\,dt\right]. \qquad (10.12)
$$

Equation (9.7) provides that $\mathrm{E}[X] = \mu a$.

10.2.3 Normalized variance

As provided in Sec. 3.4.2, the ratio of the count variance to the count mean yields the normalized variance (Saleh & Teich, 1982):

$$
F(T) = 1 + \frac{2}{aT}\int_0^T (T-u)\,\mathrm{E}\left[\int_{-\infty}^{\infty} h(K,t)\,h(K,t+u)\,dt\right]\,du. \qquad (10.13)
$$

As with any shot-noise-driven Poisson process, this quantity does not depend on the rate μ of the driving Poisson process.

Using the impulse response function provided in Eq. (9.2) for fractal shot noise, the normalized variance for the fractal-shot-noise-driven Poisson process becomes (Lowen & Teich, 1991):

$$
F(T) = 1 + \frac{2\mathrm{E}[K^2]}{aT}\int_0^{\min(T,B-A)} (T-u)\int_A^{B-u} (t^2 + ut)^{-\beta}\,dt\,du. \qquad (10.14)
$$

Equation (10.14) does not in general reduce to a closed-form expression, although closed-form results exist for $\beta = \frac{1}{2}$ and $\beta = 2$ (see Sec. A.7.2). In other cases, approximations must suffice.

For $T \ll A$, and for any B and β, the normalized variance varies linearly with T (Lowen & Teich, 1991),

$$F(T) \approx 1 + \frac{E[K^2]}{a} \left[\int_A^B t^{-2\beta} \, dt \right] T = 1 + \frac{Var[X]}{E[X]} T, \qquad (10.15)$$

as shown in Sec. A.7.2.

In the range $A \ll T \ll B$, $F(T)$ approaches a number of simple forms (Lowen & Teich, 1991) that depend on β (see Sec. A.7.2):

$$F(T) \approx 1 + \frac{E[K^2]}{E[K]} \times \begin{cases} \dfrac{1-\beta}{1-2\beta} B^{-\beta} T & 0 \le \beta < \frac{1}{2} \\[2ex] \frac{1}{2} B^{-1/2} \ln(B/T) \, T & \beta = \frac{1}{2} \\[2ex] \dfrac{\Gamma(\alpha/2)\,\Gamma(1-\alpha)}{(1+\alpha)\,\Gamma(1-\alpha/2)} B^{-\alpha/2} T^\alpha & \begin{array}{l} \frac{1}{2} < \beta < 1 \\ (1 > \alpha > 0) \end{array} \\[2ex] \dfrac{\ln^2(T/A)}{\ln(B/A)} & \beta = 1 \\[2ex] (\beta-1)^{-1} A^{1-\beta} & \beta > 1. \end{cases}$$

$$\qquad (10.16)$$

In the domain $\frac{1}{2} < \beta < 1$, we cast the expression in terms of $\alpha \equiv 2(1-\beta)$, rather than in terms of β, to highlight the scaling behavior of this measure over this range of exponents ($0 < \alpha < 1$).

Finally, for $T \gg B$, and for any A and β, the normalized variance approaches a constant value (see Sec. A.7.2; Lowen & Teich, 1991) given by

$$F(T) \approx 1 + \frac{E[K^2]}{E^2[K]} a. \qquad (10.17)$$

For a deterministic impulse response function, this reduces to $F(T) \approx 1 + a$. This result for $T \gg B$ is consistent with the Neyman Type-A counting distribution set forth in Eq. (10.7), as required for any shot-noise-driven Poisson process (Saleh & Teich, 1982).

We plot the normalized variance $F(T)$ in Fig. 10.4 for a range of power-law exponents β, as a function of the counting time T.

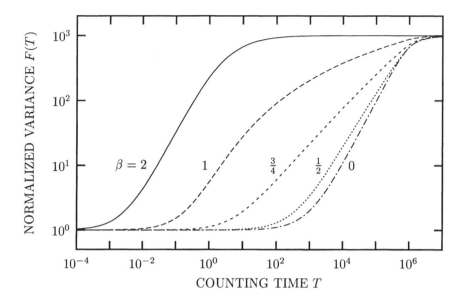

Fig. 10.4 Normalized variance $F(T)$ vs. counting time T provided in Eq. (10.14) for five values of the power-law exponent β: 0, $\frac{1}{2}$, $\frac{3}{4}$, 1, and 2. The remaining parameters are fixed at $A = 1$, $B = 10^6$, and a $= 10^3$ (K is chosen to be deterministic). For $\frac{1}{2} < \beta < 1$, the normalized variance grows as $T^{2-2\beta} = T^{\alpha}$, in accordance with Eq. (10.16).

10.2.4 Normalized Haar-wavelet variance

We calculate the normalized Haar-wavelet variance $A(T)$ set forth in Sec. 3.4.3 by using Eq. (10.14) in conjunction with Eq. (3.41). The calculations depend on the value of T, and three domains emerge: $T \le (B - A)/2$, $(B - A)/2 < T \le (B - A)$, and $T > (B - A)$. Rather than using Eq. (10.14), we could alternatively insert Eqs. (10.15)–(10.17) for $F(T)$ directly into Eq. (3.41) to provide a parallel set of equations for $A(T)$.

Effecting such a direct substitution requires caution, however, as discussed at the end of Sec. 5.2.4. As shown by Eqs. (5.37)–(5.39), the use of asymptotic formulas for $F(T)$ in Eq. (3.41) yields invalid results when linear terms dominate $F(T) - 1$. This caution specifically applies to Eq. (10.15) for all values of β and to Eq. (10.16) for $\beta < \frac{1}{2}$. For substitution into Eq. (3.41) we must therefore either use Eq. (10.14) directly, or employ versions of Eqs. (10.15)–(10.17) that retain higher-order terms.

Using these methods, we carry out calculations in Sec. A.7.3 that lead to the following results for the normalized Haar-wavelet variance.

For $T \ll A$, the dependence of $A(T)$ on T is quadratic:

$$A(T) \approx 1 + (3a)^{-1}\, E[K^2]\, (A^{-2\beta} + B^{-2\beta})\, T^2. \tag{10.18}$$

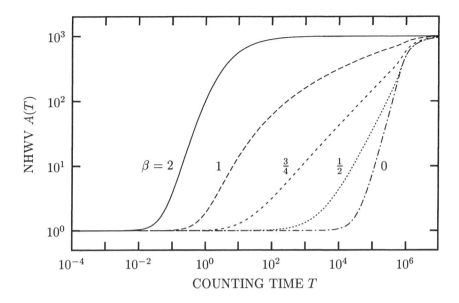

Fig. 10.5 Normalized Haar-wavelet variance $A(T)$ vs. counting time T derived via Eqs. (10.14) and (3.41) for five values of the power-law exponent β: 0, $\frac{1}{2}$, $\frac{3}{4}$, 1, and 2. The remaining parameters are fixed at $A = 1$, $B = 10^6$, and $a = 10^3$ (K is chosen to be deterministic). For $0 < \beta < 1$, the normalized Haar-wavelet variance grows as $T^{2-2\beta} = T^\alpha$, in accordance with Eq. (10.19).

In the range $A \ll T \ll B$, $A(T)$ approaches a simple form that depends on β (see Sec. A.7.3):

$$A(T) \approx 1 + \frac{\mathrm{E}[K^2]}{\mathrm{E}[K]} \times \begin{cases} \dfrac{(2^\alpha - 2)\,\Gamma(\alpha/2)\,\Gamma(2-\alpha)}{(\alpha^2 - 1)\,\Gamma(1-\alpha/2)}\, B^{-\alpha/2}\, T^\alpha & \begin{matrix} 0 < \beta < \frac{1}{2} \\ (2 > \alpha > 1) \end{matrix} \\[2.5ex] \ln(2)\, B^{-1/2}\, T & \begin{matrix} \beta = \frac{1}{2} \\ (\alpha = 1) \end{matrix} \\[2.5ex] \dfrac{(2 - 2^\alpha)\,\Gamma(\alpha/2)\,\Gamma(2-\alpha)}{(1 - \alpha^2)\,\Gamma(1-\alpha/2)}\, B^{-\alpha/2}\, T^\alpha & \begin{matrix} \frac{1}{2} < \beta < 1 \\ (1 > \alpha > 0) \end{matrix} \\[2.5ex] \dfrac{\ln^2(T/A)}{\ln(B/A)} & \beta = 1 \\[2.5ex] (\beta - 1)^{-1}\, A^{1-\beta} & \beta > 1. \end{cases}$$

$$(10.19)$$

We have cast some of these expressions in terms of $\alpha \equiv 2(1 - \beta)$, rather than in terms of β, to highlight the scaling behavior of $A(T)$. Scaling extends over the range $0 < \beta < 1$ ($0 < \alpha < 2$). This range exceeds that over which scaling extends for the normalized variance $F(T)$ inasmuch as the latter cannot increase faster than T^1, as discussed in Sec. 5.2.3: $\frac{1}{2} < \beta < 1$ $(1 > \alpha > 0)$.

Finally, for $T \gg B$, and for any A and β, Eqs. (3.41) and (10.17) provide that $A(T) = 2F(T) - F(2T) = F(T)$ so that $A(T)$ is constant. This is, of course, the domain in which the Neyman Type-A counting distribution prevails.

The normalized Haar-wavelet variance $A(T)$ is displayed in Fig. 10.5 for a range of power-law exponents β, as a function of the counting time T.

10.3 TIME STATISTICS

The probability densities for the forward recurrence time, $p_\vartheta(t)$, and the interevent time, $p_\tau(t)$, are determined from the probability that zero events occur in an interval of duration T, $p_Z(0; T)$, as its first two derivatives (see Secs. 3.3.1 and 3.4.1). Combining Eqs. (3.30) and (10.3) provides

$$p_\tau(t) = \frac{1}{\mathrm{E}[X]} \frac{d^2}{dt^2} \exp\left(\mu\mathrm{E}\left[\int_{-\infty}^{\infty} \{\exp[-h_t(K, u)] - 1\} \, du\right]\right), \qquad (10.20)$$

where $X(t)$ represents the shot-noise process that emerges at the output of the linear filter (see Fig. 10.1).

Section A.7.4 provides detailed expressions for $p_\vartheta(t)$ and $p_\tau(t)$ for a deterministic impulse response function. With the help of these results, we plot the interevent-interval probability density function $p_\tau(t)$ in Fig. 10.6 for several values of the power-law exponent β. Representative results for other parameters, and for the forward-recurrence-time density, are presented in Lowen & Teich (1991, Figs. 9–12).

The significant differences among the curves in Fig. 10.6 reflect the varying degree of clustering that fractal-shot-noise-driven Poisson processes can exhibit. A large degree of clustering is accompanied by an increase in the probability of very short and very long interevent times, at the expense of times near the mean, relative to an exponential density of the same mean. Figure 10.6 shows that the clustering increases as β increases. To explain this, we observe that the clustering has its origin in the variations of the fractal shot-noise rate $X(t)$. For larger values of β, particularly $\beta > 1$, the majority of the area a of the impulse response function lies in a small region near the onset time A, with proportionately less area in the tail. For smaller values of β, in contrast, the value of the impulse response function changes far less over its duration $B - A$. Thus, the fractal shot-noise rate $X(t)$ exhibits greater variations for large values of β, and the fractal-shot-noise-driven Poisson process therefore concomitantly exhibits more clustering, assuming that all other parameters remain constant.

For $\tau = 0$ and all values of β we have (Lowen & Teich, 1991)

$$p_\tau(0) = 1/\mathrm{E}[\tau] + \mathrm{E}[\tau] \, \mathrm{Var}[X], \qquad (10.21)$$

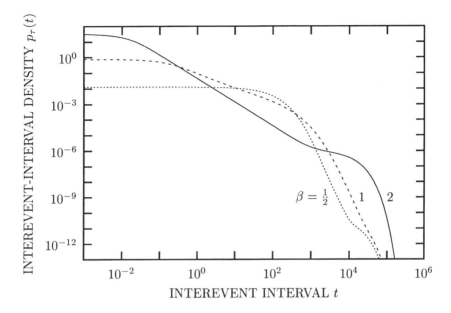

Fig. 10.6 Interevent-interval probability density function, $p_\tau(t)$ vs. t, for the fractal-shot-noise-driven Poisson process. We display results for three values of the power-law exponent: $\beta = \frac{1}{2}, 1$, and 2. The remaining parameters are fixed at $A = 1$, $B = 10^5$, $a = 100$, $\mu = 10^{-4}$, and $\mathrm{E}[\tau] = 100$ (K is chosen to be deterministic). The probability density exhibits a range of power-law behaviors as the time t and the power-law exponent β vary.

whereas in the limit $\tau \to \infty$ the probability density function approaches exponential form (Lowen & Teich, 1991),

$$p_\tau(t) \to \frac{\mu(1 - e^{-a})^2}{a} \exp\left[-\mu t(1 - e^{-a})\right],\qquad (10.22)$$

by virtue of Eqs. (3.30) and (10.6). The exponential nature of the primary Poisson process $dN_1(t)$ carries over to the interevent-interval statistics for large interevent times τ. For $\beta > 1$, the concentration of the area a of the impulse response function near the onset time A results in tight clustering of the events of $dN_2(t)$ following primary events. The long intervals thus essentially derive from the primary process $dN_1(t)$, and $p_\tau(t)$ exhibits an exponential tail, albeit with reduced amplitude in comparison with a homogeneous Poisson process. Even for $\beta < 1$, interevent intervals longer than $B - A$ usually derive from the primary process, particularly for small values of $\mu(B - A)$, where $X(t) = 0$ for significant periods of time. This, too, results in an exponential tail.

10.4   COINCIDENCE RATE

The coincidence rate $G(t)$ for a doubly stochastic Poisson process closely follows the autocorrelation of its driving rate, as provided by Eq. (4.24). Since the fractal-shot-noise-driven Poisson process belongs to the family of doubly stochastic Poisson processes, this equation applies here.

Inserting Eq. (9.20) for the autocorrelation of the $X(t)$ into Eq. (4.24) provides an expression for the coincidence rate:

$$
\begin{aligned}
G(t) \quad = \quad & \mathrm{E}[X]\,\delta(t) + \mathrm{E}^2[X] \\
& + \begin{cases} \mu\mathrm{E}[K^2]\displaystyle\int_A^{B-|t|} (u^2 + |t|u)^{-\beta}\,du & |t| < B - A \\[2mm] 0 & |t| \geq B - A. \end{cases}
\end{aligned}
\tag{10.23}
$$

Closed-form expressions for $G(t)$ exist for $\beta = \frac{1}{2}$, 1 and 2 (Lowen & Teich, 1991).

When the delay time t is small, $G(t)$ approaches a constant value for any power-law exponent β:

$$
G(t) = \mathrm{E}^2[X] + \mu\mathrm{E}[K^2]\int_A^B t^{-2\beta}\,dt.
\tag{10.24}
$$

In the region $A \ll |\tau| \ll B$, the coincidence rate takes a variety of forms for different values of β (Lowen & Teich, 1991, Appendix G):

$$
G(t) = \mathrm{E}^2[X] + \mu\mathrm{E}[K^2] \times
\begin{cases}
\dfrac{(1-\beta)^2}{(1-2\beta)}\,B^{-1} & 0 \leq \beta < \frac{1}{2} \\[3mm]
\frac{1}{4}B^{-1}\ln(B/|t|) & \beta = \frac{1}{2} \\[3mm]
\dfrac{\alpha\,\Gamma(1+\alpha/2)\,\Gamma(1-\alpha)}{2\Gamma(1-\alpha/2)}\,B^{-\alpha}\,|t|^{\alpha-1} & \begin{array}{c}\frac{1}{2} < \beta < 1 \\ (1 > \alpha > 0)\end{array} \\[3mm]
\dfrac{\ln(|t|/A)}{\ln^2(B/A)}\,|t|^{-1} & \beta = 1 \\[3mm]
(\beta - 1)\,A^{\beta-1}\,|t|^{-\beta} & \beta > 1.
\end{cases}
\tag{10.25}
$$

In the domain $\frac{1}{2} < \beta < 1$, we express the results in terms of $\alpha \equiv 2(1 - \beta)$ rather than in terms of β, again to emphasize the scaling behavior of this measure.

The coincidence rate $G(t)$, as a function of the delay time t, appears in Fig. 10.7. To avoid the constant term $\mathrm{E}^2[X]$ from obscuring the variation in $G(t)$, we actually graph the expression $G(t) - \mathrm{E}^2[X]$ rather than $G(t)$ itself.

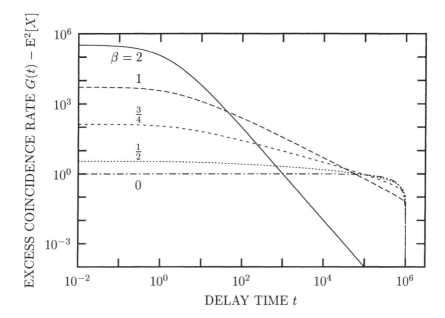

Fig. 10.7 Excess coincidence rate, $G(t) - E^2[X]$ vs. delay time t, obtained from Eq. (10.23) for five values of the power-law exponent: $\beta = 0$, $\frac{1}{2}$, $\frac{3}{4}$, 1, and 2. The remaining parameters are set at fixed values for all curves: $A = 1$, $B = 10^6$, $\mu = 1$, and a $= 10^3$ (K is chosen to be deterministic). In accordance with Eq. (10.25), the functions corresponding to $\beta > \frac{1}{2}$ exhibit approximate power-law behavior, with various exponents, over a good portion of their range. Note the abrupt drop near $t = B - A \approx 10^6$, where $G(t) - E^2[X] \to 0$ in accordance with Eq. (10.23).

10.5 SPECTRUM

We turn now to the spectrum of the fractal-shot-noise-driven Poisson process. In accordance with Eq. (4.25), the spectrum of $dN_2(t)$ differs from that of $X(t)$ only by an additive constant term, $E[X]$. In the domain $\beta < 1$ and $A \ll B < \infty$, where $S_X(f)$ behaves as $1/f$-type noise (see Table 9.1), we insert Eq. (9.30) in Eq. (4.25) to obtain

$$
\begin{aligned}
S_{N2}(f) &= S_X(f) + E[X] \\
&\approx \mu E[K^2]\,\Gamma^2(\alpha/2)\,(2\pi f)^{-\alpha} + (2/\alpha)\mu E[K]\,B^{\alpha/2}, \quad (10.26)
\end{aligned}
$$

where we have employed the approximation $A \to 0$, valid for $A \ll B$ and $\beta < 1$.

Figure 10.8 illustrates the spectrum for the fractal-shot-noise-driven Poisson process using the same impulse response functions (abrupt cutoff and exponential), and the same parameters, as those used to generate the fractal-shot-noise spectrum displayed in Fig. 9.4. We can define a crossover frequency by equating the two terms

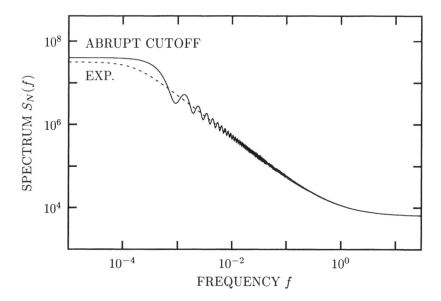

Fig. 10.8 Spectrum $S_N(f)$ vs. frequency f for the fractal-shot-noise-driven Poisson process. The parameter values are the same as those used to generate the curves for the fractal-shot-noise spectrum, $S_X(f)$, displayed in Fig. 9.4: $\beta = \frac{1}{2}$ ($\alpha = 1$), $A = 0$, $B = 1\,000$, $K = 100$, and $\mu = 1$. For sufficiently high frequencies, the spectrum exhibits $1/f^\alpha$ behavior, with $\alpha = 1$. The impulse response function with an abrupt cutoff in the time domain results in oscillations in the frequency domain, whereas an exponential transition yields a smooth curve. In the high-frequency limit, both curves approach the asymptotic value, $\lim_{f \to \infty} S_N(f) = E[X] = 2\,000\sqrt{10} \doteq 6325$.

on the right-hand side of Eq. (10.26):

$$\mu E[K^2]\,\Gamma^2(\alpha/2)\,(2\pi f_S)^{-\alpha} = (2/\alpha)\mu E[K]\,B^{\alpha/2}$$
$$\alpha\,\Gamma^2(\alpha/2)/2 = \left(E[K]/E[K^2]\right)(2\pi f_S)^\alpha\,B^{\alpha/2}. \quad (10.27)$$

10.6 RELATED POINT PROCESSES

10.6.1 Point process in the Gaussian limit of fractal shot noise

Under suitable conditions, the probability density of the driving fractal-shot-noise rate $X(t)$ converges to a Gaussian form [see Eqs. (9.11) and (9.12), as well as Table 9.1], and indeed $X(t)$ becomes a Gaussian process. As provided by the central limit theorem, this takes place in the limit $\mu \to \infty$ if $E[K^k] < \infty$ for all k, $A > 0$, and $B < \infty$ for $\beta \leq 1$ (Lowen & Teich, 1990). Over this range of β, the spectrum varies as $1/f^{2-2\beta} = 1/f^\alpha$, and $X(t)$ converges to a **fractal Gaussian process** (see Sec. 6.3.3). The resulting point process $dN_2(t)$ then becomes a **fractal-Gaussian-process-driven Poisson process** (see Fig. 5.5, Secs. 6.3.3 and 8.4, and Chapter 12).

10.6.2   Fractal-shot-noise-driven integrate-and-reset point process

The discussion thus far has centered on point processes produced by fractal shot noise driving a Poisson generator. However, fractal shot noise can also serve as the rate function for other generation mechanisms, such as the integrate-and-reset point process set forth in Sec. 4.4 (Thurner et al., 1997). The **fractal-shot-noise-driven integrate-and-reset process**, as an example, suitably characterizes spontaneous action-potential generation in the visual system (Teich & Lowen, 2003).

Many of the properties of the fractal-shot-noise-driven integrate-and-reset process readily derive from the results set forth in Sec. 4.4, together with the results obtained earlier in this chapter and in Chapter 9. In particular, the statistics over time scales that are substantially longer than $E[\tau]$ (corresponding to frequencies much lower than $1/E[\tau]$) virtually coincide with those of the fractal-shot-noise-driven Poisson process. This conclusion follows because Poisson processes introduce few fluctuations over these time scales, while integrate-and-reset processes introduce none.

The statistics that are manifested over times comparable to, or less than, $E[\tau]$ depend largely on the amplitude distribution of the fractal shot noise; Sec. 9.2 and Eqs. (4.37) and (4.38) prove useful for these calculations. Over all time scales, fractal shot noise itself provides useful results for the second-order statistics of the fractal-shot-noise-driven integrate-and-reset process, through Eq. (4.36).

10.6.3   Hawkes point process

We now consider the nontrivial, critical, self-exciting point process displayed in Fig. 10.9. A classical **Hawkes point process** (Hawkes, 1971) comprises a Poisson process whose output drives a linear filter $h(t)$ to produce a continuous stochastic process $X(t)$. The sum of this process and an external input (a constant μ_0 in the example at hand) forms a function $\mu(t)$ that serves as the rate for the original Poisson process. The stochastic process $X(t)$ therefore shares some similarity with shot noise. Because the resulting point process $dN(t)$ modulates itself in a feedback loop, it is a special **self-exciting point process**.

In order that the process be stationary, the area of the impulse response function $h(t)$ must lie below unity; were that not the case, the rate would grow exponentially over time. The external input μ_0 can, instead, be a time-varying function that integrates to a finite value. It is then possible to have an impulse response function with unity area, but it turns out that the ensuing process has trivial characteristics for general forms of $h(t)$. However, the selection of fractal forms for $h(t)$ (more precisely, those with heavy tails) results in nontrivial critical Hawkes point processes, which themselves have fractal properties (Brémaud & Massoulié, 2001).

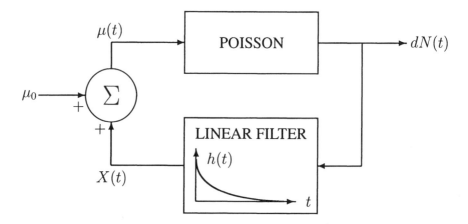

Fig. 10.9 A stochastic rate process $\mu(t)$ modulates the rate of a Poisson process, yielding a classical Hawkes point process $dN(t)$. This point process serves as the input to a linear filter with impulse response function $h(t)$. A continuous-time stochastic process $X(t)$, which is a form of generalized shot noise, emerges from the output of this filter. The sum of $X(t)$ and an external constant rate μ_0 provides the stochastic rate process $\mu(t)$, completing the loop.

10.6.4   Bartlett–Lewis fractal point process

We conclude this chapter by examining a process constructed from nonfractal renewal processes to which fractal behavior is imparted in a different way. Consider a Bartlett–Lewis-type cascaded point process (see Sec. 4.5) constructed from a primary Poisson point process $dN_1(t)$, each event of which initiates a segment of a secondary Poisson point process $dN_{2,k}(t)$ which, in turn, terminates after a certain number of secondary events M_k. The superposition of all secondary events forms the final point process $dN_3(t)$. Primary events (illustrated as dashed vertical lines in the secondary processes in Fig. 4.2) can be excluded or included in the final process. A number of variations on this theme have appeared in the literature (Grüneis, 1984; Grüneis & Baiter, 1986; Grüneis & Musha, 1986; Grüneis, 1987, 2001). This model finds application in characterizing computer network traffic, as discussed in Secs. 13.5.4 and 13.6.

Fractal behavior in this process arises from the imposition of a power-law distribution on the number of events M_k associated with each secondary process $dN_{2,k}(t)$ before its termination. Specifically, let this number follow the distribution

$$\Pr\{M_k = m\} = m^z \Big/ \sum\nolimits_{l=1}^{M_{\max}} l^z, \tag{10.28}$$

where M_k ranges from a minimum of one event to a maximum of $M_{\max}$ events. A single segment of a Poisson process of mean interevent interval $\mathrm{E}[\tau_2]$, and of fixed, deterministic duration with M_k events, has a spectrum given by (Grüneis & Musha, 1986)

$$2(2\pi f\,\mathrm{E}[\tau_2])^{-2} \left\{\mathrm{Re}\big[(1 + i2\pi f\,\mathrm{E}[\tau_2])^{1-M_k}\big] - 1\right\}. \tag{10.29}$$

If M_k now becomes a random variable taking values with probabilities given by Eq. (10.28), and if each event in the primary process $dN_1(t)$, with mean interevent interval $E[\tau_1]$, initiates such a secondary process, the resulting cascaded process $dN_3(t)$ exhibits a spectrum of the form

$$S_{N3}(f) = \left(E[M_k]/E[\tau_1]\right)^2 \delta(f) + E[M_k]/E[\tau_1]$$

$$+ \frac{2\sum_{l=1}^{M_{max}} l^z \left\{\text{Re}\left[(1 + i2\pi f\, E[\tau_2])^{1-l}\right] - 1\right\}}{E[\tau_1]\,(2\pi f\, E[\tau_2])^2 \sum_{l=1}^{M_{max}} l^z}. \quad (10.30)$$

The mean interevent interval of the secondary processes, $E[\tau_2]$, and the mean duration of the secondary processes, $E[\tau_2]\,E[M_k]$, then play the roles of the cutoffs A and B associated with the impulse response function $h(t)$ of the fractal-shot-noise-driven Poisson process. The spectrum remains relatively constant for frequencies outside the reciprocals of these two times. However, for frequencies well within these limits, scaling behavior emerges for certain values of the event-number distribution exponent z (Grüneis & Musha, 1986). The spectrum then turns out to follow the form

$$\alpha = \begin{cases} 0 & z \leq -3 \\ z + 3 & -3 < z < -1 \\ 2 & -1 \leq z. \end{cases} \quad (10.31)$$

This relationship indicates that the process at hand generates $1/f^\alpha$ noise over the extended range $0 < \alpha < 2$.

Both the fractal Bartlett–Lewis cascaded process and the fractal Neyman–Scott cluster process are plausible models for describing computer network traffic, as discussed extensively in Chapter 13.

Problems

10.1 *Interevent-interval density function for large intervals*   Equation (10.22) shows that the interevent-interval probability density decays as an exponential for large interevent intervals τ. Show heuristically that this holds for the simple case $\tau > B - A$.

10.2 *Normalized variance for rectangular impulse response functions*   Evaluate Eq. (10.14) explicitly for the case $\beta = 0$, where the impulse response functions reduce to rectangles. Find the limit of the resulting expression when $A \ll T \ll B$, and show that this limit agrees with Eq. (10.16).

10.3 *Interval density function associated with a single impulse response*   Figure 10.6 displays a set of interevent-interval probability densities $p_\tau(t)$ for several values of the parameter β. For the particular curve associated with $\beta = 2$, the density $p_\tau(t)$ decays as a power-law function of the interval. Consider a single impulse response function, and show that the density $p_\tau(t)$ of the resulting process indeed follows this power-law form. Find the slope and extent of this scaling region. Show that this behavior holds for general $\beta > 1$.

10.4   *Designing a fractal-shot-noise-driven Poisson process*   By appropriate selection of the parameters β, A, B, μ, and K, we can design a fractal-shot-noise-driven Poisson process that exhibits a $1/f^\alpha$ spectrum, a cutoff frequency f_S, and a rate of events with a particular mean value.

    10.4.1. Identify the appropriate equations in the text that relate these three design values to the five parameters of the fractal-shot-noise-driven Poisson process.

    10.4.2. Suppose we also choose values for the coefficient of variation and skewness of the rate, for example, in an attempt to fully specify the process. Using the specific case $\alpha = 1$; $f_S = 1$; an average rate of unity; and fixed, deterministic K as an example, discuss the constraints on the five design values. One might wish to employ a rate with a large coefficient of variation to yield an appreciable amount of fluctuation in the numbers of events generated in $dN_2(t)$, and also a small skewness to better approximate a fractal Gaussian process. Show that attempting to simultaneously specify both large values of the rate coefficient of variation and small values of the rate skewness leads to conflicting requirements and unspecified parameters.

    10.4.3. Rather than specifying higher moments of the rate, we instead seek to use the following two constraints: $B/A = 10^3$ for an appreciable range of frequencies following the $1/f$ spectrum, and $A = 1/f_S$ to ensure that the spectrum indeed follows this form up to $f = f_S$. Design a fractal-shot-noise-driven Poisson process with the specific values given here and in Prob. 10.4.2, and find the coefficient of variation and the skewness of the rate.

10.5   *Impulse response functions without cutoffs*   Impulse response functions without cutoffs lead to fractal-shot-noise processes with infinite moments (see Chapter 9). Suppose that we nevertheless employ such a shot-noise-process as a rate for a Poisson process, generating a point-process output $dN_2(t)$. Describe this point process, and comment on its orderliness.

10.6   *Photon statistics of Čerenkov radiation*   Charged particles traveling faster than the group velocity of light in a transparent medium emit photons, often in the visible range. Čerenkov was the first to systematically examine this phenomenon in a series of experiments conducted during the years 1934–1938 (see, for example, Čerenkov, 1934, 1937, 1938).

    One can use electromagnetic theory to show that the fractal-shot-noise-driven Poisson process provides a useful model for describing the light produced by Čerenkov radiation arising from a sparse random stream of charged particles. Consider a charged particle traveling along the positive x-axis through a transparent, non-ferromagnetic medium of refractive index n, at a speed $v > c/n$ where c is the speed of light in free space, as shown in Fig. 10.10. We define the quantity $J \equiv [(nv/c)^2 - 1]^{1/2}$; it is a function of the degree to which the particle velocity exceeds the Čerenkov limit c/n. We calculate the electric and magnetic fields at a distance d from the x-axis, where we choose the arbitrary point in the x-z plane $\{-Jd,\, 0,\, d\}$ for algebraic

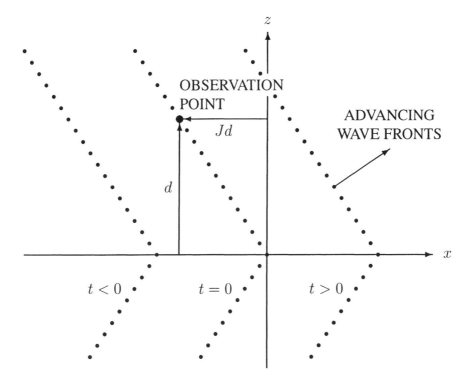

Fig. 10.10 A charged particle moving faster than the speed of light in a medium emits Čerenkov radiation. At the point $\{-Jd, 0, d\}$, the photon flux density decays as an inverse power-law function of time. We illustrate three wavefronts for a particle traveling along the x-axis, corresponding to times $t < 0$, $t = 0$, and $t > 0$. The particle passes the origin at $t = 0$.

simplicity. We assume that the particle does not experience substantial deceleration while significantly close to this observation point. In accordance with the Frank–Tamm theory (see Jelley, 1958; Zrelov, 1968), we obtain scalar and vector potentials that satisfy the Lorentz gauge condition

$$\phi_L \;=\; 2qn^{-2}\left[(x - vt)^2 - J^2(y^2 + z^2)\right]^{-1/2} \tag{10.32}$$
$$\mathbf{A}_L \;=\; n^2 c^{-1}\,\mathbf{v}\,\phi_L, \tag{10.33}$$

respectively, where q represents the charge of the particle.

The corresponding electric and magnetic fields associated with this single charged particle traveling through the medium are written in terms of the scalar and vector potentials as

$$\mathbf{E} = -\nabla\phi_L - c^{-1}\partial\mathbf{A}_L/\partial t \tag{10.34}$$

and

$$\mathbf{H} = \mathbf{B} = \nabla \otimes \mathbf{A}_L, \tag{10.35}$$

where $\otimes$ denotes the vector cross product. The Poynting vector,

$$\mathbf{S} \equiv (4\pi)^{-1} c\, \mathbf{E} \otimes \mathbf{H}, \tag{10.36}$$

characterizes the energy flux density (intensity) and its direction; its magnitude specifies only the energy flux density.

The spectrum of the light may be calculated by Fourier-transform methods (see Jelley, 1958; Zrelov, 1968), or via Eqs. (9.28) and (4.25). We denote $E[\nu]$ as the mean frequency of the light, so that the mean photon energy is $hE[\nu]$, where h is Planck's constant. This permits us to convert the time-varying intensity generated by a single traveling charged particle into an approximate expression for the time-varying photon flux density generated by the particle (see Saleh & Teich, 1991, Chaps. 5 and 11):

$$h(t) \approx |\mathbf{S}| / hE[\nu]. \tag{10.37}$$

10.6.1. Over what range of times does the foregoing description apply?

10.6.2. Derive $h(t)$ as defined in Eq. (10.37), and cast it in the form of a simple power-law impulse-response function as in Eq. (9.2), assuming that the refractive index of the medium departs significantly from unity. Obtain power-law exponents and cutoff values in terms of quantities previously given, and give an approximate form with a single power-law exponent.

10.6.3. What changes if the index of refraction differs only slightly from unity?

10.6.4. Under what conditions does a stream of charged particles yield a sequence of photons well described by the fractal-shot-noise-driven Poisson process?

10.7 *Earthquake occurrences*   A full description of earthquake activity requires a marked spatiotemporal point process, so that the time and location of the epicenters, as well as the total energy dissipated, is specified for each seismic event. The analysis of earthquake patterns reveals that earthquakes obey power-law statistics in their magnitude distributions, in their spatial clustering, and in their second-order time statistics (see, for example, Kagan & Knopoff, 1987; Lapenna et al., 1998; Telesca et al., 1999; Telesca, Cuomo, Lapenna & Macchiato, 2002b, which provides a recent overview). However, in keeping with the approach used throughout this book, we pay no heed to the spatial and energy information; rather, we treat all seismic activity within a specified area, and above a limiting energy, as the events of an unmarked point process.

The statistician David Vere-Jones (1970) used a version of the fractal-shot-noise-driven Poisson process to model shallow (< 100 km) earthquakes of magnitude > 4.5 that occurred between January 1942 and September 1961 in New Zealand.[3] He obtained good fits to both the mean rate of earthquakes (about 22 per year), and the variance–time curve, over a range of 0.1 to 1.5 years. From the latter statistic at 0.1 year, and from the count-based autocorrelation $R_Z(1, 0.1\text{ year})$, he concluded

---

[3] An often-used alternative model for earthquake occurrences is self-organized criticality (see Sec. 2.7.6); however, recent results indicate that earthquake data are not in good accord with this theory (Yang, Du & Ma, 2004).

that the average cluster contained six earthquakes and that the expected number of earthquakes remaining in a cluster decayed with time as $t^{-1/4}$. He further obtained a cluster start time of 2.3 days after a primary event, and assumed that no mechanism terminated the clusters at any specific time after that.

10.7.1. From the information provided above, identify the fractal-shot-noise-driven Poisson process employed by Vere-Jones in terms of the parameters used in this chapter.

10.7.2. The value of β used in the model (which is close to but greater than unity), together with $B = \infty$, leads to a long tail in the seismic activity. How much time must pass after a cluster starts so that, with 0.8 probability, no events still remain in the cluster?

10.8 *Diffusion* Consider a collection u_0 of infinitesimal particles, all initially at some point $\mathbf{x}_0$ of a Euclidean space of (integer) dimension $D_E < 4$ at a starting time $t = 0$. The concentration u at some other point $\mathbf{x}$ and some later time t will then vary in proportion to a Gaussian density with a variance that increases with time in a power-law fashion (see, for example, Pinsky, 1984),

$$u(\mathbf{x}, t) = u_0 \, (4\pi\Delta\,t)^{-D_E/2} \, \exp\left(-\frac{|\mathbf{x} - \mathbf{x}_0|^2}{4\Delta\,t}\right), \tag{10.38}$$

where Δ is the diffusion constant. We assume that the particles have some lifetime t_1, resulting in $u(\mathbf{x}, t) \approx 0$ for $t > t_1$. Now suppose that some external process deposits packets of concentration at random times t that form a homogeneous Poisson process, and let $u_\Sigma(\mathbf{x}, t)$ represent the linear sum of the decaying concentrations arising from all of the deposited events. Finally, suppose that secondary events occur in a random fashion, with the generation probability of an event at a particular time t and location $\mathbf{x}$ proportional to the accumulated concentration $u_\Sigma(\mathbf{x}, t)$ at that time and location, independent of other generated events.

10.8.1. Show how the fractal-shot-noise-driven Poisson process provides a useful model for the resulting secondary event process.

10.8.2. What values of β are likely operative? What dimensionality leads to exact $1/f$ noise?

10.8.3. Demonstrate how the approach can be generalized to the case where the packets arrive at different points $\mathbf{x}$, and need not all have the same initial concentration u_0.

10.8.4. Provide an example of how physical constraints can make diffusion unrealistic as a model even though it yields a mathematically plausible fractal exponent.

10.9 *Semiconductor high-energy particle detectors* A typical high-energy particle detector consists of a lightly doped p–n junction with a large reverse bias voltage applied across it (Knoll, 1989). Energetic charged particles enter the detector, usually along the p–n axis, and create electron-hole pairs within a large part of the semiconductor depletion region. The higher the energy of the particle, the greater the number of electron-hole pairs produced. The high reverse-bias field then sweeps these carriers out of the depletion region of the diode, electrons toward the n region and holes toward the p region. This occurs before many of the electrons and holes recombine although some of the carriers do recombine, thereby reducing the detected charge created by the original energetic charged particle. A description of the recombination process therefore proves useful.

Consider a single energetic particle entering the detector at time $t = 0$. Assume that the electron-hole pairs are created instantaneously throughout the semiconductor depletion region, distributed as a three-dimensional Poisson point process, and that they begin diffusing immediately after their creation. Whenever an electron and a hole approach within some critical radius, the two carriers either recombine, thereby annihilating each other immediately, or first form an exciton and later recombine. In either case the carriers no longer carry current and effectively vanish.

10.9.1. If we ignore the drift current, with what exponents do the concentrations of electrons and holes decay?

10.9.2. How does drift affect the exponents?

10.9.3. Cast the recombination process in terms of an impulse response function.

10.9.4. Finally, show how the fractal-shot-noise-driven Poisson process may help in understanding the total recombination process in a working particle detector.

10.10 *Trapping in amorphous semiconductors: Revisited*   An alternative approach to the problem of trapping in amorphous semiconductors, initially considered in Prob. 7.10, treats all conductance changes as events in an auxiliary point process (Azhar & Gopala, 1992). Analysis of current flow in an AC128 germanium transistor reveals relatively fast current fluctuations during conduction events, separated by somewhat longer intervals between conduction events. For a fixed counting time, the mean and variance of the numbers of conduction-event onsets assume similar values; the same holds for the numbers of conductance changes within a conduction event. Furthermore, all the secondary events taken together, which form the auxiliary point process, have a spectrum that follows a $1/f$-type form. Considering the evidence presented, suggest a plausible model for the sequence of conductance-change events.

11

Operations

The famous Swiss mathematician **Jakob Bernoulli (1654–1705)** conceived of a sequence of statistical trials, each of which yields one of two mutually exclusive outcomes ("heads" or "tails"), with probabilities that are fixed across trials.

The celebrated probabilist **William Feller (1906–1970)**, who studied at Zagreb and Göttingen, made significant advances in many areas of probability, including Brownian motion, diffusion theory, and dead-time-modified counting processes.

11.1	Time Dilation	228
11.2	Event Deletion	229
	11.2.1 General results	229
	11.2.2 Decimation	231
	11.2.3 Random deletion	231
	11.2.4 Dead-time deletion	236
11.3	Displacement	241
	11.3.1 Interval displacement	242
	11.3.2 Event-time displacement	242
11.4	Interval Transformation	247
	11.4.1 Interval normalization	249
	11.4.2 Interval exponentialization	249
11.5	Interval Shuffling	252
	11.5.1 Block shuffling	255
	11.5.2 Bootstrap method	255
	11.5.3 Identification of fractal-based point processes	255
11.6	Superposition	256
	11.6.1 Superposition of doubly stochastic Poisson point processes	258
	11.6.2 Superposition of renewal point processes	259
	Problems	261

A number of transformations exist by means of which one or more point processes $\{dN_1(t), \ldots, dN_n(t)\}$ are converted into a new point process $dN_R(t)$. We focus on six such operations in the context of fractal and fractal-rate point processes:

- **Dilation**, which involves contracting or expanding the time axis of a counting process.

- **Deletion**, which eliminates selected events of a point process according to a specified rule, examples of which include:

  1. Retaining every ℓ th event with all others eliminated (**decimation**);

  2. Subjecting each event to an independent Bernoulli trial in which it is deleted with a fixed probability (**random deletion** or **thinning**);

  3. Eliminating events if they follow other events more closely than a specified time interval (**dead-time deletion** or **refractoriness deletion**).

- **Displacement**, where we modify the occurrence time of each event of a point process in a specified manner, for example by jitter.

- **Interval transformation**, where the ordering of the intervals remains unchanged but we transform the interval density to a different form, for example an exponential density.

- **Shuffling**, where we randomize the interevent-interval ordering in a particular way, while the interval probability density is constrained:

1. We randomly reorder the intervals and preserve the original interval density (**full shuffling**);

2. We randomly reorder the intervals within blocks of a realization and retain the original interval density (**block shuffling**);

3. We select intervals with replacement, thereby generating a renewal process (**bootstrap method**).

- **Superposition**, which forms a new point process from the sum of a collection of point processes.

Transformations such as these play important roles in the identification of point processes. In some cases, they are inherent in the measured events. This can occur because: (1) they are intrinsic to the underlying process; (2) they are unavoidably imposed by the detection/measurement system recording the events. Commonly encountered operations in the physical and biological sciences include Bernoulli random deletion, dead-time deletion, displacement, and superposition.[1]

In other cases, experimenters deliberately use such transformations to create surrogate data sets that are useful for elucidating the underlying nature of an observed point process[2] (Schiff & Chang, 1992; Theiler, Eubank, Longtin, Galdrikian & Farmer, 1992; Ott et al., 1994). The shuffling and exponentialization operations, for example, prove valuable for determining whether the fractal behavior observed in a sequence of action potentials stems from the form of the interevent-interval density, the ordering of the intervals, or both.[3] Examples that illustrate these operations appear throughout this chapter.

Operations such as decimation and dead-time deletion are also sometimes used to deliberately reduce the variability of a point process [see, for example, Saleh & Teich (1985b) and Teich & Cantor (1978), respectively].

We devote this chapter to determining the effect of each of these transformations on the relevant point-process statistics, and illustrating how these operations affect the nature of fractal and fractal-rate point processes.

---

[1] Examples in which these four transformations are intrinsic to the underlying process appear in Teich & Khanna (1985); Teich, Matin & Cantor (1978); Teich et al. (2001); and Palm (1943), respectively. Examples in which they are imposed in the course of measurement appear in Teich & Saleh (1982); Teich & Vannucci (1978); Teich, Khanna & Guiney (1993); and Abeles, de Ribaupierre & de Ribaupierre (1983), respectively.

[2] The creation of surrogate data resembles the creation of "knockout mice," a biological procedure developed in the mid-1980s to study gene function (see Evans, Smithies & Capecchi, 2001). Knockout animals are created by replacing a specific natural gene with an inactive or mutated allele. The behavior or performance of the "surrogate mouse" provides information about the role played by the gene.

[3] Other operations also create useful surrogates. Phase randomization in the Fourier domain, for example, preserves the spectral magnitude and therefore the second-order properties of a point process. Since this procedure removes other temporal structure, it yields information about the presence or absence of deterministic chaos in a system [see, for example, Turcott & Teich (1996) and Teich et al. (2001)].

11.1 TIME DILATION

Time dilation is, perhaps, the simplest of operations that can be carried out on a point process (Papangelou, 1972). The time axis t of a counting process is expanded or contracted by a factor c that is, respectively, larger or smaller than unity:

$$N_R(t) = N_1(t/c). \tag{11.1}$$

Forming the point process as the derivative of the counting process introduces an additional factor[4] of $1/c$,

$$dN_R(t) = c^{-1} dN_1(t/c). \tag{11.2}$$

This factor carries through to the statistics of the point process in a straightforward manner. Four cases exist, depending on whether the measure or its arguments (if any) has dimensions of real time (sec) or frequency (Hz). The simplest case obtains when the statistic neither has units, nor takes arguments with units, as is true for the interval-based skewness, kurtosis, rescaled range, and generalized dimensions. For these statistics, dilation causes no change.

We next consider the case where the measure itself has no units but takes arguments that do. These follow relations similar to those for the point process as a whole:

$$
\begin{array}{llll}
\text{interval distribution} & P_{\tau R}(t) & = & P_{\tau 1}(t/c) \\
\text{normalized count variance} & F_R(T) & = & F_1(T/c) \\
\text{normalized wavelet count variance} & A_R(T) & = & A_1(T/c) \\
\text{count autocorrelation} & R_{ZR}(k, T) & = & R_{Z1}(k, T/c).
\end{array} \tag{11.3}
$$

Some interval statistics do have units, but do not take arguments with units, which leads to multiplicative factors for the measures themselves:

$$
\begin{array}{llll}
\text{interval moments} & E[\tau_R^n] & = & c^n\, E[\tau_1^n] \\
\text{interval variance} & \text{Var}[\tau_R] & = & c^2\, \text{Var}[\tau_1] \\
\text{interval autocorrelation} & R_{\tau R}(k) & = & c^2\, R_{\tau 1}(k) \\
\text{interval spectrum} & S_{\tau R}(\mathsf{f}) & = & c^2\, S_{\tau 1}(\mathsf{f}),
\end{array} \tag{11.4}
$$

and similarly for the interval wavelet variance.

Finally, for statistics that have units and also take arguments with units, multiplicative factors appear in the measures themselves as well as in their arguments:

$$
\begin{array}{llll}
\text{interval probability density} & p_{\tau R}(t) & = & c^{-1} p_{\tau 1}(t/c) \\
\text{coincidence rate} & G_R(t) & = & c^{-2} G_1(t/c) \\
\text{point-process spectrum} & S_{NR}(f) & = & c^{-1} S_{N1}(cf) \\
\text{general-wavelet variance} & \text{Var}[C_{\psi, NR}(a, b)] & = & c^{-1} \text{Var}[C_{\psi, N1}(a/c, b/c)].
\end{array} \tag{11.5}
$$

---

[4] However, the delta functions that comprise $dN_R(t)$ do not change since $a\delta(at) = \delta(t)$ for all positive, finite a, as discussed in Sec. 4.4.

The foregoing establishes that dilation simply alters the relevant time scales. It does not convert a fractal point process into a fractal-rate point process, nor the opposite. All of the fractal-based point processes we consider therefore retain their form under time dilation and maintain their scaling exponents, although intrinsic times (and frequencies) such as the mean interval and fractal onset time change by the factor c.

11.2 EVENT DELETION

Event deletion forms another class of operations on point processes. We consider three subclasses of deletion that differ in how events are selected for elimination: (1) **Decimation**, in which every ℓ th event survives while all others do not; (2) **Random deletion**, where each event survives with a probability r that is independent of the survivals of other events and of the point process itself; and (3) **Dead-time deletion**, in which the probability of survival of an event $r(t)$ resets to zero following each surviving event, and thence increases monotonically to unity. Figure 11.1 schematically illustrates how these three types of event deletion operate on a representative point process.

11.2.1 General results

How does the deletion of events affect a fractal or fractal-rate point process? Assuming that the deletion probability does not itself exhibit significant long-term fluctuations,[5] the deleted process, although altered, remains a member of the fractal family of point processes with the same fractal exponent.

To understand how this comes about, consider a fractal or fractal-rate point process with mean interevent time $\mathrm{E}[\tau_1]$, fractal exponent α_1, and fractal onset frequency f_{S1}. Over a range of frequencies, the spectrum follows the form of the mid-frequency term in Eq. (5.44a):

$$S_{N1}(f) \approx \mathrm{E}[\mu_1] \, (f/f_{S1})^{-\alpha_1}. \tag{11.6}$$

Suppose now that we retain some proportion r of the events in $dN_1(t)$ to form $dN_R(t)$, where the selection probability of these points is devoid of low-frequency components. Decimation and random deletion, discussed in Secs. 11.2.2 and 11.2.3, respectively, do not selectively alter the low-frequency components of the process so that the following considerations apply to these deletion processes. On the other hand, dead-time deletion, discussed in Sec. 11.2.4, violates this condition so that

---

[5] We mention this caveat since such behavior can affect the fractal characteristics of a point process. Suppose, for example, that $dN_1(t)$ belongs to the homogeneous Poisson-process family and that the deletion process selects events from $dN_1(t)$ with a probability $r(t)$ described by a fractal Gaussian process. The resultant deleted process $dN_R(t)$ then belongs to the fractal-Gaussian-process-driven Poisson-process family and hence exhibits the fractal characteristics imparted to it by $r(t)$.

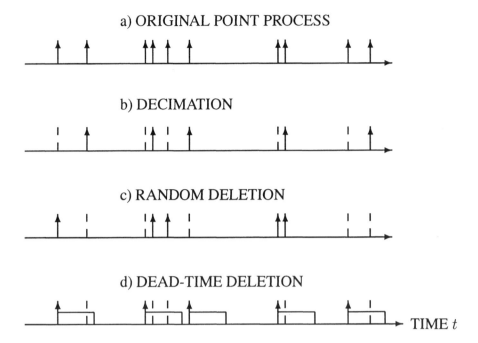

Fig. 11.1 The effect of different types of event deletion on a representative point process. Arrows indicate surviving events; dashed lines indicate deleted events. (a) *Original point process.* (b) *Decimated point process:* Every ℓth event survives, here with $\ell = 2$. (c) *Randomly deleted point process:* Events are randomly deleted, here with survival probability $r = \frac{1}{2}$ so that roughly half of the events survive the deletion process; no particular pattern emerges since each event deletion is independent of the others and of the form of the point process. (d) *Dead-time-modified point process:* We choose a fixed dead time in this illustration so that $r(t)$ switches abruptly from zero to unity at the expiration of the dead-time interval τ_f; the rectangles beginning at each surviving event serve to illustrate the duration of this interval and do not form part of the resulting point process.

the following results do not necessarily apply. The point-process spectrum for a fractal-rate process subjected to dead time is instead provided in Eq. (11.30).

Since the mean rate changes by the factor r, we have $E[\tau_R] = E[\tau_1]/r$. Consider the spectrum at a particular frequency f that lies within the scaling range of frequencies. This quantity derives from the squared magnitude of a Fourier transform, so that reducing the number of events to a fraction r of the initial number results in a decrease in the Fourier transform by this same fraction. The spectrum thus changes by the factor r^2 at intermediate frequencies, which provides

$$
\begin{aligned}
S_{NR}(f) &= r^2 S_{N1}(f) \\
&\approx r^2 E[\mu_1] \, (f/f_{S1})^{-\alpha_1} \\
&= E[\mu_R] \, (f/f_{SR})^{-\alpha_1},
\end{aligned} \tag{11.7}
$$

with fractal onset frequency $f_{SR} = f_{S1} r^{1/\alpha}$.

We conclude that the fractal exponent of the process remains unchanged while the fractal onset *frequency* changes by the factor $r^{1/\alpha}$, which serves to reduce the magnitude of the fractal component. The onset *times* of other second-order measures thus increase by the factor $r^{-1/\alpha}$. At higher frequencies no correlation exists among events in the point process so that the spectrum varies linearly (rather than quadratically) with r, whereupon

$$\lim_{f \to \infty} S_{NR}(f) = \mathrm{E}[\mu_R] = r\mathrm{E}[\mu_1] = r \lim_{f \to \infty} S_{N1}(f). \qquad (11.8)$$

11.2.2 Decimation

The deletion of the initial $\ell - 1$ of every ℓ events (retaining every ℓ th event) generally changes the nature of a point process and tends to reduce fluctuations in all measures. The quantity ℓ is known as the **decimation parameter**. We illustrate this operation in Fig. 11.1b) for $\ell = 2$. Consider, for example, a homogeneous Poisson process $dN_1(t)$: in accordance with the results provided in Sec. 4.1, the coefficient of variation of the intervals, $C_\tau = \sqrt{\mathrm{Var}[\tau]}/\mathrm{E}[\tau]$, assumes a value of unity, as does the normalized variance of the counts, $F(T)$, for all counting times T. Retaining the ℓ th, 2ℓ th, 3ℓ th,... events of such a process results in a gamma renewal process of order $\ell = m$ (see Prob. 4.7 and, for example, Parzen, 1962; Cox, 1962; Cox & Isham, 1980). The coefficient of variation of the intervals becomes $1/\sqrt{\ell}$ whereas the normalized variance $F(T)$ becomes $1/\ell$ for large counting times T. The diminution of both of these quantities indicates that the point process has reduced variability.

Similar results obtain for an arbitrary point process; orderly deletion generally modifies the form of the point process and reduces the fluctuations. Two notable exceptions exist, however. A renewal point process, under decimation, remains within the fold of the renewal-process family despite the fact that it becomes a different process. Integrate-and-reset processes form the second exception. The deletion of the first $\ell - 1$ of every ℓ events in such a process yields results identical to those obtained by decreasing the rate of the process by the same factor ℓ.

For an arbitrary fractal-based point process $dN_1(t)$, decimation leads to a process $dN_R(t)$ that continues to belong to the family of fractal-based point processes, with parameters given by Eq. (11.7) and its associated results, despite possible changes to the form of the point process.

11.2.3 Random deletion

We now consider the consequences of subjecting every event of a point process $dN_1(t)$ to a Bernoulli trial (van der Waerden, 1975), which it survives with probability r or fails to survive with probability $(1 - r)$ (Palm, 1943; Parzen, 1962; Cox & Isham, 1980). We depict this operation in Fig. 11.1c) for $r = \frac{1}{2}$.

The **Burgess variance theorem** (Burgess, 1959) leads to a simple relationship between the normalized count variance for an arbitrary point process and that of its randomly deleted cousin (Teich & Saleh, 1982):

$$F_R(T) - 1 = r[F_1(T) - 1].$$ (11.9)

We can readily extend Eq. (11.9) to the collection of measures set forth in Chapter 3. Beginning with Eq. (3.41), we immediately obtain an analogous relation between the normalized Haar-wavelet variances:

$$A_R(T) - 1 = r[A_1(T) - 1].$$ (11.10)

Continuing along these same lines, Eq. (3.55) reveals that the coincidence rate requires multiplication by a factor of r^2, since both $E[\mu]$ and $F(T) - 1$ decrease by a factor of r, but the delta function at $t = 0$ is subjected only to a factor of r, so that

$$G_R(t) - E[\mu_R]\,\delta(t) = r^2\{G_1(t) - E[\mu_1]\,\delta(t)\}.$$ (11.11)

This result, in turn, propagates to the count autocorrelation $R_Z(k, T)$ via Eq. (3.54), which provides

$$R_{ZR}(k, T) - E[\mu_R]\,T = r^2\{R_{Z1}(k, T) - E[\mu_1]\,T\},$$ (11.12)

and to the point-process spectrum $S_N(f)$ through Eq. (3.57), which gives rise to

$$S_{NR}(f) - E[\mu_R] = r^2\{S_{N1}(f) - E[\mu_1]\}.$$ (11.13)

Fractal onset times increase by $r^{-1/\alpha}$, and fractal onset frequencies decrease by $r^{1/\alpha}$, maintaining the validity of Eq. (5.45).

   As with decimation, random deletion generally alters the nature of a point process. This operation causes the resulting process $dN_R(t)$ to ultimately move toward the homogeneous Poisson process, whatever its original form (Rényi, 1956; Kallenberg, 1975; Wescott, 1976). Random deletion thus serves to reduce the variability of a point process when its fluctuations are greater than those of the benchmark homogeneous Poisson, and to increase the variability for a process whose fluctuations lie below those of the Poisson (Teich & Saleh, 1982). This outcome stands in contrast to that for decimation, which always results in a reduction of variability, as discussed in Sec. 11.2.2.

   This tendency toward the homogeneous Poisson process can be observed to some extent in Fig. 11.2, which displays the estimated normalized interevent-interval histogram following random deletion for the seven canonical point processes shown in Fig. 5.9. The retention probability $r = \frac{1}{4}$. In particular, the HEARTBEAT interval histogram, which is quite narrow in Fig. 5.9 by virtue of the clocklike sequence of events comprising the point process, is substantially broadened and displays a nearly exponential form in Fig. 11.2. Nevertheless, examination of the COMPUTER and GENICULATE data reveals that their features are more-or-less conserved under random deletion. In particular, deletion highlights the phase-locked character of the GENICULATE point process by increasing the relative number of interevent intervals at large multiples of the stimulus period (see, for example, Teich et al., 1993, for a discussion of phase locking).

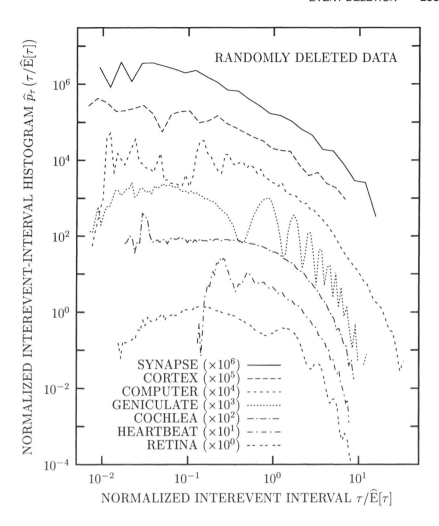

Fig. 11.2 Estimated normalized interevent-interval histogram following random deletion, $\widehat{p}_\tau(\tau/\widehat{E}[\tau])$ vs. normalized interevent interval $\tau/\widehat{E}[\tau]$, for the same seven point processes displayed in Fig. 5.9. The retention probability $r = \frac{1}{4}$. We reduced the histogram data to a smaller number of bins for the CORTEX and SYNAPSE data sets because they had only 441 and 688 surviving intervals, respectively. Random deletion ultimately drives point processes toward Poisson form. This can be observed to some extent in the interevent-interval histograms, most noticeably for the HEARTBEAT data. Nevertheless, the principal features of the COMPUTER and GENICULATE data largely remain. In particular, the deletion operation highlights the phase-locked character of the GENICULATE point process.

The perfectly periodic point process $dN_1(t)$, which has interevent intervals $\tau_k = \tau_0$ for all indices k, provides a useful illustration for how random deletion can engender an increase in variability (Teich & Saleh, 1982). The coefficient of variation of the

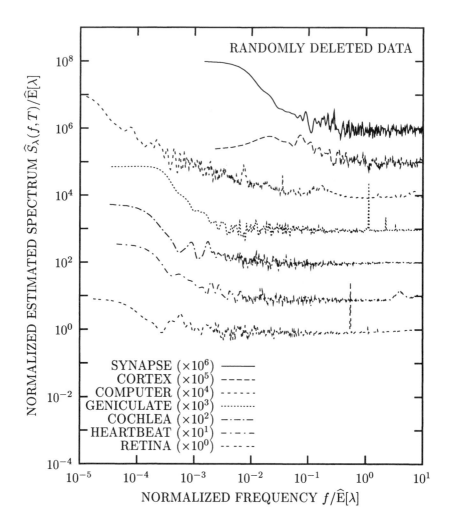

Fig. 11.3 Normalized estimated rate spectra following random deletion, $\widehat{S}_\lambda(f, T)/\widehat{E}[\lambda]$ vs. normalized frequency $f/\widehat{E}[\lambda]$, for the six biological point processes and one computer network traffic trace illustrated in Fig. 5.1. The retention probability $r = \frac{1}{4}$. The fractal variability diminishes in magnitude for each of the point processes. The counting time T is chosen such that $1/\sqrt{2} < 30\,T/E[\tau] < \sqrt{2}$, as it is for all spectra displayed in this chapter, thereby ensuring that the size of the Fourier-transform exceeds the number of intervals by a significant factor $(15\sqrt{2})$.

intervals, $C_\tau = \sqrt{\mathrm{Var}[\tau]}/E[\tau]$, is of course zero, as is the normalized variance of the counts, $F(T)$, for counting times $T = n\tau_0$ where n is a positive integer. The fact that both of these quantities vanish indicates the absence of randomness in this initial point process. We now subject this process to random deletion by retaining

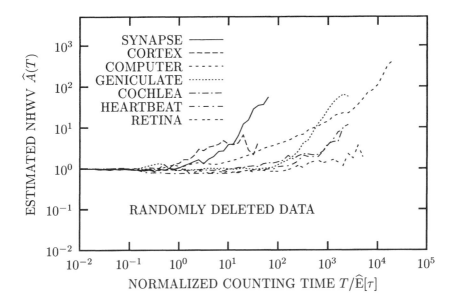

Fig. 11.4 Estimated normalized Haar-wavelet variance (NHWV) following random deletion, $\widehat{A}(T)$ vs. normalized counting time $T/\widehat{\mathrm{E}}[\tau]$, for the same seven point processes as displayed in Fig. 5.2. The retention probability $r = \frac{1}{4}$. As in Fig. 11.3, the fractal variability diminishes in magnitude for each of the point processes.

each event with probability r, independent of the other events and of the time at which the event occurs. The resulting point process $dN_R(t)$, which belongs to the renewal point-process family, obeys

$$
\begin{aligned}
\Pr\{\tau_R = n\tau_0\} &= r\,(1-r)^{n-1} \\
\Pr\{\tau_R \leq n\tau_0\} &= 1 - (1-r)^n \\
\Pr\{\tau_R \leq t\} &= 1 - (1-r)^{\mathrm{int}(t/\tau_0)} \\
&= 1 - \exp[\mathrm{int}(t/\tau_0)\log(1-r)], \quad (11.14)
\end{aligned}
$$

where $\mathrm{int}(c)$ represents the integer part of the real number c.

In the limit of large values of t/τ_0 and small values of r (signifying that only a small fraction of the events survives), we have $\mathrm{int}(t/\tau_0) \approx t/\tau_0$ and $\log(1-r) \approx -r$, whereupon Eq. (11.14) assumes an exponential form,

$$
\begin{aligned}
\Pr\{\tau_R \leq t\} &\approx 1 - \exp(-tr/\tau_0) \\
p_{\tau R}(t) &\approx \left(1/\mathrm{E}[\tau_R]\right)\exp\left(-t/\mathrm{E}[\tau_R]\right), \quad (11.15)
\end{aligned}
$$

where $\mathrm{E}[\tau_R] = \tau_0/r$, as a result of the properties of the geometric distribution. Since $dN_R(t)$ has an exponentially decaying interevent-interval density, and belongs to the family of renewal point processes, it is, by definition, a homogeneous Poisson process. Thus, under the influence of heavy random deletion the perfectly periodic integrate-and-reset process becomes a homogeneous Poisson process. The coefficient of variation of the intervals, C_τ, and the normalized variance of the counts, $F(T)$, grows from zero to unity, which unequivocally demonstrates that a substantial increase in variability can be introduced by the operation of random deletion.

As indicated in Sec. 11.2.1, and implicit in the results derived above, random deletion reduces the magnitude of the fractal variability inherent in a point process; this operation converts an arbitrary original fractal-based point process $dN_1(t)$ into its deleted cousin $dN_R(t)$ via Eq. (11.7) and associated formulas. Figures 11.3 and 11.4 demonstrate this reduction for the rate spectra and normalized Haar-wavelet variances, respectively, associated with the canonical data sets displayed in Figs. 5.1 and 5.2. The reduction agrees with the analytical results provided in Eqs. (11.13) and (11.10), respectively.

Two exceptions exist to this rule, whereby random deletion does not change the character of a point process. A renewal point process retains its renewal form under random deletion, although it becomes a different renewal process (see Prob. 11.9). And a doubly stochastic Poisson process remains a doubly stochastic Poisson process (Teich & Saleh, 1982).

11.2.4 Dead-time deletion

In the presence of **dead time**, called **refractoriness** in the biological literature, the probability of event deletion depends on the history of the point process in a more complex manner than it does for decimation. In its most general form, dead-time deletion invokes the history all event occurrences prior to the event at hand. Treating this general problem turns out to be quite unwieldy, however, so that it is common to consider dead-time-deletion models that reach back only to the previous n events (nth-order dead time). Typically, $n = 1$ or 2.

In the following, we consider only first-order dead time ($n = 1$), also known as "one-memory dead time." This case further divides into two types: one in which the surviving process $dN_R(t)$ depends on the last event of the original process $dN_1(t)$ (**paralyzable dead time** or **extended dead time**), and one in which $dN_R(t)$ depends on the last event of the output process $dN_R(t)$ (**nonparalyzable dead time** or **nonextended dead time**).[6] Both types of dead time belong to the more general family of **type-p dead time** (Albert & Nelson, 1953; Parzen, 1962; Bharucha-Reid, 1997). An enormous body of literature exists on dead-time effects, and Müller compiled a rather remarkable bibliography on the topic in 1981.

We focus on nonparalyzable dead time, by far the most prevalent form encountered in physical and biological systems. The probability of survival of an event in $dN_1(t)$,

---

[6] There is a further division, depending on whether the counter is in a "blocked" ("ordinary"), "unblocked" ("shifted" or "free"), or "equilibrium" ("stationary") state at the beginning of the counting duration (see, for example, Müller, 1973, 1974; Libert, 1976). The various results differ minimally when the mean number of events recorded during a sampling time is much greater than unity (Libert, 1976).

which depends only on the time to the last event in $dN_R(t)$, is specified by the so-called **recovery function** $r(t)$. In the special case of **fixed dead time** (called **absolute refractoriness** in the biological literature), $r(t)$ follows the form of a step function:

$$r(t) = \begin{cases} 0 & t < \tau_f \\ 1 & t \geq \tau_f, \end{cases} \tag{11.16}$$

where τ_f is the dead-time period. The deletion of events subjected to fixed dead time is illustrated in Fig 11.1d). More generally, **relative dead time** (known as **relative refractoriness** in the biological literature) occurs, whereupon $r(t)$ assumes a more general form. Relative dead time is also descriptively referred to as **sick time** (Teich & Diament, 1980). Yet another variation on the theme involves **stochastic dead time**, in which the dead time is absolute but random (see, for example, Cantor, Matin & Teich, 1975; Teich et al., 1978).

Even with these simplifications, general results are difficult to obtain. We therefore further limit ourselves to a few particular cases that serve as useful models for the kinds of dead-time effects often encountered in point processes and that prove amenable to analysis. In particular, we consider (1) general point processes modified by fixed nonparalyzable dead time; (2) homogeneous Poisson processes modified by relative nonparalyzable dead time; (3) Poisson-based processes modified by relative nonparalyzable dead time; and (4) integrate-and-reset processes modified by relative nonparalyzable dead time.

- The imposition of **fixed dead time** on a **general point process** closely resembles the minimal-covering procedure used to calculate the capacity dimension in Sec. 7.2.5 (see Fig 7.7). Indeed, the same formalism yields a general result for the mean interevent time. For a general point process, we again make use of Wald's Lemma (Feller, 1971) to obtain the expected value of the time between events in the resulting point process,

$$E[\tau_R] = E^2[\tau_1] \int_{0-}^{\tau_f} G_1(t)\, dt. \tag{11.17}$$

The quantity $G_1(t)$ denotes the coincidence rate of the point process $dN_1(t)$, and the notation $0-$ indicates that the range of integration spans the delta function in $G_1(t)$ at $t = 0$. Examples of point processes in this class have been considered by Cantor & Teich (1975); Teich & McGill (1976); Teich & Vannucci (1978); Vannucci & Teich (1978, 1981); Saleh et al. (1981); and Prucnal & Teich (1983).

- We next consider the **dead-time-modified homogeneous Poisson process**, which has a broad range of applicability in many fields of endeavor. The fixed-dead-time version of this process has a long history in the annals of probability (Morant, 1921; Palm, 1943; Jost, 1947; Parzen, 1962; DeLotto, Manfredi & Principi, 1964; Müller, 1973; Cantor & Teich, 1975; Müller, 1981). Feller (1948) provided an early comprehensive analysis. As a result of the absence of memory in the original process $dN_1(t)$, the dead-time deleted version, $dN_R(t)$, is renewal in nature.

Following the occurrence of an event in $dN_R(t)$, the effective rate $\mu_R(t)$ of $dN_R(t)$ simply becomes $\mu_1 r(t)$. The probability of zero events occurring in $dN_R(t)$ between an event and a later time t, which is equivalent to the survivor function $1 - P_{\tau R}(t)$ of $dN_R(t)$, decays as an exponential transform of the integrated rate, as shown in Eqs. (4.26) and (4.28):

$$1 - P_{\tau R}(t) = \exp\left[-\mu_1 \int_0^t r(u)\, du\right]. \tag{11.18}$$

The mean interevent time for the resulting point process is obtained from Eq. (11.18):

$$
\begin{aligned}
\mathrm{E}[\tau_R] &= \int_0^\infty t p_{\tau R}(t)\, dt \\
&= \int_0^\infty [1 - P_{\tau R}(t)]\, dt \tag{11.19} \\
&= \int_0^\infty \exp\left[-\mu_1 \int_0^t r(u)\, du\right] dt, \tag{11.20}
\end{aligned}
$$

where Eq. (11.19) follows from integration by parts and resembles Eq. (3.11) for $s \to \infty$.

We define an effective dead time τ_e as the difference between the mean interevent intervals for the two processes,

$$\tau_e \equiv \mathrm{E}[\tau_R] - \mathrm{E}[\tau_1] = \mathrm{E}[\tau_R] - 1/\mu_1. \tag{11.21}$$

Rearranging Eq. (11.21) yields the expectation of the effective rate,

$$
\begin{aligned}
\mathrm{E}[\tau_R] &= 1/\mu_1 + \tau_e = 1/\mathrm{E}[\mu_R] \\
\mathrm{E}[\mu_R] &= \frac{\mu_1}{1 + \mu_1 \tau_e}. \tag{11.22}
\end{aligned}
$$

- More generally, we examine the **dead-time-modified driven Poisson process**. Suppose that the initial point process $dN(t)$ maintains its Poisson character but has a varying rate $\mu_1(t)$, as considered by Vannucci & Teich (1978, 1981). We assume that $\mu_1(t)$ varies slowly in comparison with the mean interevent time of the resulting process, $\mathrm{E}[\tau_R]$, and that the standard deviation of $\mu_1(t)$ is much smaller than its mean. The instantaneous rate then follows the form of Eq. (11.22) without the expectation,

$$\mu_R(t) = \frac{\mu_1(t)}{1 + \mu_1(t)\tau_e}. \tag{11.23}$$

To simplify the ensuing algebra, we define two functions,

$$
\begin{aligned}
x(t) &\equiv 1 + \mu_1(t)\,\tau_e \\
y(t) &\equiv \frac{x(t) - \mathrm{E}[x]}{\mathrm{E}[x]},
\end{aligned}
\tag{11.24}
$$

whereupon

$$
\begin{aligned}
\mu_R(t)\,\tau_e &= \frac{\mu_1(t)\,\tau_e}{1 + \mu_1(t)\,\tau_e} \\[2mm]
&= 1 - \frac{1}{1 + \mu_1(t)\,\tau_e} \\[2mm]
&= 1 - \frac{1}{x(t)} \\[2mm]
&= 1 - \frac{1}{x(t) - \mathrm{E}[x] + \mathrm{E}[x]} \\[2mm]
&= 1 - \frac{1}{\mathrm{E}[x]}\frac{1}{1 + y(t)} \\[2mm]
&= 1 - \frac{1}{\mathrm{E}[x]}\sum_{n=0}^{\infty}[-y(t)]^n.
\end{aligned}
\tag{11.25}
$$

The convergence of the power series in Eq. (11.25) follows from the assumed small value of $y(t)$, which, in turn, derives from the small degree of relative fluctuations in $\mu_1(t)$. Retaining terms to second order in this perturbation analysis leads to

$$
\begin{aligned}
\mu_R(t)\,\tau_e &\approx 1 - \frac{1}{\mathrm{E}[x]}\left[1 - y(t) + y^2(t)\right] \\[2mm]
&= 1 - \frac{1}{\mathrm{E}[x]} + \frac{x(t) - \mathrm{E}[x]}{\mathrm{E}^2[x]} - \frac{\{x(t) - \mathrm{E}[x]\}^2}{\mathrm{E}^3[x]} \\[2mm]
\mu_R(t) &\approx \frac{\mathrm{E}[\mu_1]}{1 + \mathrm{E}[\mu_1]\,\tau_e} + \frac{\mu_1(t) - \mathrm{E}[\mu_1]}{(1 + \mathrm{E}[\mu_1]\,\tau_e)^2} - \tau_e\frac{\{\mu_1(t) - \mathrm{E}[\mu_1]\}^2}{(1 + \mathrm{E}[\mu_1]\,\tau_e)^3}.
\end{aligned}
\tag{11.26}
$$

Finally, forming the expectation of Eq. (11.26) leads to a second-order approximation for the mean rate,

$$
\mathrm{E}[\mu_R] \approx \frac{\mathrm{E}[\mu_1]}{1 + \mathrm{E}[\mu_1]\,\tau_e} - \frac{\mathrm{Var}[\mu_1]\,\tau_e}{(1 + \mathrm{E}[\mu_1]\,\tau_e)^3}.
\tag{11.27}
$$

More exact results are available for the specific case when $r(t)$ takes the form of a delayed exponential function (Lowen, 1996).

In preparation for computing the autocovariance, we rearrange Eq. (11.27) to provide

$$\mu_R - \mathrm{E}[\mu_R] \approx \frac{\mu_1(t) - \mathrm{E}[\mu_1]}{\left(1 + \mathrm{E}[\mu_1]\,\tau_e\right)^2} - \tau_e \frac{\left\{\mu_1(t) - \mathrm{E}[\mu_1]\right\}^2 - \mathrm{Var}[\mu_1]}{\left(1 + \mathrm{E}[\mu_1]\,\tau_e\right)^3}. \quad (11.28)$$

The second fraction in Eq. (11.28) becomes insignificant when calculating the autocovariance, however, since it only appears in third- and higher-order terms; we thus obtain (Lowen, 1996)

$$\mathrm{E}\!\left[\left\{\mu_R(s) - \mathrm{E}[\mu_R]\right\}\left\{\mu_R(s+t) - \mathrm{E}[\mu_R]\right\}\right]$$

$$\approx\ \mathrm{E}\!\left\{\frac{\mu_1(s) - \mathrm{E}[\mu_1]}{\left(1 + \mathrm{E}[\mu_1]\,\tau_e\right)^2} \cdot \frac{\mu_1(s+t) - \mathrm{E}[\mu_1]}{\left(1 + \mathrm{E}[\mu_1]\,\tau_e\right)^2}\right\}$$

$$R_{\mu R}(t) - \mathrm{E}^2[\mu_R]\ \approx\ \left(\mathrm{E}[\mu_R]/\mathrm{E}[\mu_1]\right)^4 \left\{R_{\mu 1}(t) - \mathrm{E}^2[\mu_1]\right\}. \quad (11.29)$$

Pursuing the formalism provided in Eq. (11.7) then leads to

$$\begin{aligned}
S_{NR}(f)\ &\approx\ \left(\mathrm{E}[\mu_R]/\mathrm{E}[\mu_1]\right)^4 S_{N1}(f)\\
&=\ \left(\mathrm{E}[\mu_R]/\mathrm{E}[\mu_1]\right)^4 \mathrm{E}[\mu_1]\,(f/f_{S1})^{-\alpha}\\
&=\ \mathrm{E}[\mu_R]\,(f/f_{SR})^{-\alpha}, \quad (11.30)
\end{aligned}$$

with fractal onset frequency $f_{SR} = f_{S1}(\mathrm{E}[\mu_R]/\mathrm{E}[\mu_1])^{3/\alpha}$. This result contrasts with that provided just after Eq. (11.7) for random deletion, which is $f_{SR} = f_{S1}(\mathrm{E}[\mu_R]/\mathrm{E}[\mu_1])^{1/\alpha}$. The distinction arises because dead time preferentially deletes closely spaced intervals relative to sparsely populated ones. This serves to reduce high rates more than low rates and thereby reduces the fluctuations in the rate. This, in turn, leads to a more regular process than would emerge by deletion of the same number of events without regard to the event timings. Hence, given the same initial point process, dead-time-modified processes tend to have lower fractal onset frequencies (and higher fractal onset times) than randomly deleted processes of the same rate. The relationships in Eq. (5.45) remain valid since the cutoff time T_A (as well as T_F, t_G, and T_R for $\alpha < 1$) increases by the inverse factor, namely $(\mathrm{E}[\mu_R]/(\mathrm{E}[\mu_1]))^{-3/\alpha}$.

Dead time effects are principally short term while fractal effects are most important in the long term, so they conveniently decouple, at least for the doubly stochastic Poisson process. We can readily incorporate the two effects into a single formula by using the dead-time factor $(1 + \mathrm{E}[\mu]\tau_e)^3$ in conjunction with the fractal term, and adding in the dead-time result. As an example, let $A_1(T)$ denote the normalized Haar-wavelet variance for a dead-time-modified homogeneous Poisson process, and $A_2(T)$ denote the normalized Haar-wavelet variance for a fractal doubly stochastic Poisson process without dead time. The overall normalized Haar-wavelet variance $A_3(T)$ then becomes (Lowen, 1996)

$$A_3(T) = A_1(T) + (1 + \mathrm{E}[\mu]\tau_e)^3 \left[A_2(T) - 1\right]. \quad (11.31)$$

- Finally we turn to the **dead-time-modified integrate-and-reset process**. For a constant rate μ_1, we again have $\mu_R(t) = \mu_1 r(t)$, but in this case Eq. (4.35) provides

$$
\int_0^{\tau_R} \mu_R(t)\, dt \;=\; \mu_1 \int_0^{\tau_R} r(t)\, dt \;=\; 1
$$
$$
\int_0^{\tau_R} r(t)\, dt \;=\; \tau_1
$$
$$
r(\tau_R) \;=\; d\tau_1/d\tau_R. \tag{11.32}
$$

Following the procedure used earlier, but now for the integrate-and-reset process, we obtain

$$
\mu_R(t) \;\approx\; \mathrm{E}[\mu_R] + \frac{d\mu_R}{d\mu_1}\left\{\mu_1(t) - \mathrm{E}[\mu_1]\right\}^2
$$
$$
R_{\mu R}(t) - \mathrm{E}^2[\mu_R] \;\approx\; \left(\frac{d\mu_R}{d\mu_1}\right)^2 \left\{R_{\mu 1}(t) - \mathrm{E}^2[\mu_1]\right\}. \tag{11.33}
$$

We calculate the derivative in Eq. (11.33) with the help of Eq. (11.32), which gives rise to

$$
\frac{d\mu_R}{d\mu_1} \;=\; \frac{d(1/\tau_R)}{d\mu_1} = -\tau_R^{-2}\frac{d\tau_R}{d\mu_1} = -\tau_R^{-2}\left[\frac{d\mu_1}{d\tau_R}\right]^{-1}
$$
$$
=\; -\tau_R^{-2}\left[\frac{d(1/\tau_1)}{d\tau_R}\right]^{-1} = -\frac{1}{\tau_R^2}\left[-\frac{r(\tau_R)}{\tau_1^2}\right]^{-1}
$$
$$
=\; \left(\mu_R/\mu_1\right)^2 r^{-1}(\tau_R). \tag{11.34}
$$

Finally, combining Eqs. (11.33) and (11.34) yields

$$
R_{\mu R}(t) - \mathrm{E}^2[\mu_R] \approx \left(\mathrm{E}[\mu_R]/\mathrm{E}[\mu_1]\right)^4 r^{-2}(\mathrm{E}[\tau_R])\left\{R_{\mu 1}(t) - \mathrm{E}^2[\mu_1]\right\}. \tag{11.35}
$$

In particular, for fixed dead time, as specified in Eq. (11.16), we have $r(\tau_R) = 1$ for all possible τ_R, which yields results identical to those provided in Eq. (11.29), and thus also to those given in Eq. (11.30). In general, for integrate-and-reset deletion processes the changes in fractal onset times and frequencies depend on the form of $r(t)$.

11.3 DISPLACEMENT

The **displacement** operation, sometimes called **translation**, signifies a shifting of the individual events of a point process, typically by a random amount (Cox, 1963).

Random jitter results in a loss of phase coherence, so it can mask features of interest in the unmodified point process. Although the magnitude of the displacement can in principle be made to depend on a variety of properties of the point process, such as its local mean rate or some complex function of its history (Harris, 1971), constructs of this kind have limited usefulness and are difficult to analyze. Rather, we focus on two simple approaches to displacement, based on the intervals between events and on the events themselves, respectively.

11.3.1 Interval displacement

In the displaced-interval approach, each interevent interval is multiplied by a random quantity that is close to unity, such as

$$\tau_{kR} = \tau_{k1}\Big[1 + \sigma\mathcal{N}_k(0,1)\Big] \tag{11.36}$$

where σ is the dimensionless relative scale of the displacement (akin to a standard deviation) and $\{\mathcal{N}_k(0,1)\}$ is a sequence of independent, zero-mean, unity-variance Gaussian random variables (Thurner et al., 1997). Typically we choose $\sigma \ll 1$ so that, with rare exception, $\tau_{kR} > 0$.

Rather than relying on multiplication, we could instead add a random value to each interevent interval, for example via

$$\tau_{kR} = \tau_{k1} + \sigma\mathcal{N}_k(0,1)\,\mathrm{E}[\tau_{k1}]. \tag{11.37}$$

We have not found the additive approach to be useful, however, because it either requires very small values of σ (in which case large values of τ_{kR} experience little change) or it leads to many negative values of τ_{k1}.

For displacement, as specified in either Eq. (11.36) or (11.37), we can rectify the problem of negative τ_{k1} in a number of ways; we specify two. The first approach involves specifying a minimum interevent interval, and replacing all values inferior to it with that minimum value. Alternatively, we can reorder the events t_{kR} in the resulting point process so that the intervals all become positive, but this approach introduces additional complexity. For any method, displacing the intervals yields a new point process that precisely preserves the number of interevent intervals in the original process, but not its duration.

11.3.2 Event-time displacement

In the displaced event-time approach, we add a random quantity to each event time so that, for example,

$$t_{kR} = t_{k1} + \sigma\mathcal{N}_k(0,1)\,\mathrm{E}[\tau_{k1}], \tag{11.38}$$

where the definitions of σ and $\{\mathcal{N}_k(0,1)\}$ are given in conjunction with Eqs. (11.36) and (11.37). Note that this differs from Eq. (11.37). Multiplication by a random value, as formulated in Eq. (11.36), is not viable in this case since it leads to increasingly larger displacements as t increases. Again, we can set $\Pr\{\tau_{k1} < \sigma\mathrm{E}[\tau_{k1}]\} \ll 1$ so

a) ORIGINAL POINT-PROCESS REALIZATION $dN_1(t)$

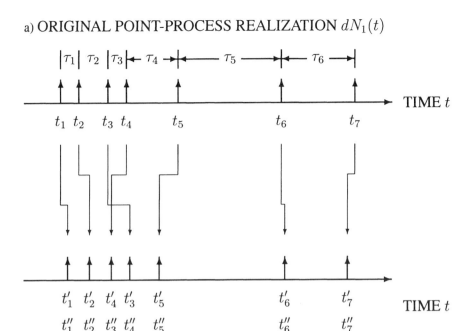

b) DISPLACED POINT-PROCESS REALIZATION $dN_R(t)$

Fig. 11.5 Original and displaced point-process realizations. (a) An original point process realization $dN_1(t)$ comprises seven event times $\{t_k\}$ and six concomitant interevent intervals $\{\tau_k\}$. A random displacement of the event times $\{t_k\}$ in the original realization shown in a) yields a new set of event times $\{t'_k\}$; after reordering in increasing temporal order, these become the event times $\{t''_k\}$ of the displaced realization shown in b). The six interevent intervals $\{\tau'_k\}$ of this displaced realization derive from the relation $\tau'_k = t''_{k+1} - t''_k$. In general the start times differ ($t'_1 \neq t_1$), in contrast to the outcome for interval shuffling (Sec. 11.5).

that $\tau_{kR} > 0$ with only rare exception, and then set a minimum interevent interval; sort t_{kR} into increasing order; or use a combination of these methods to ensure that τ_{kR} does not assume negative values.

A schematic illustration of displacing the event times of a point process appears in Fig. 11.5. Event displacement yields a new point process that exactly preserves the number of interevent intervals in the original process, and also closely preserves its duration. In general, both methods (with reordering, if required) destroy whatever structure might have been present in the original point process $dN_1(t)$ at time scales smaller than $\sigma \mathrm{E}[\tau_{k1}]$, replacing it with (locally) homogeneous-Poisson behavior at those scales. To understand how this comes about, consider a fractal-Gaussian-process-driven integrate-and-reset process $dN_1(t)$. Now apply the transformation

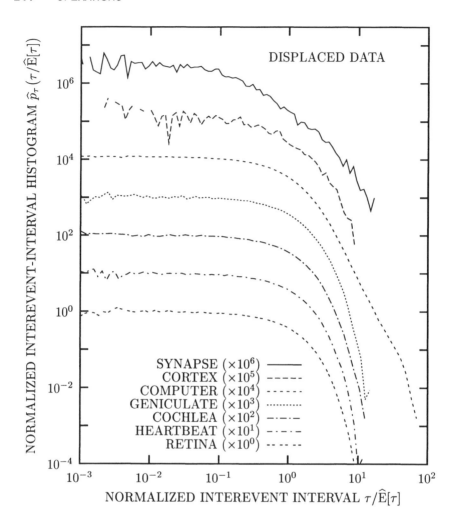

Fig. 11.6 Estimated normalized interevent-interval histogram following random displacement, $\widehat{p}_\tau(\tau/\widehat{E}[\tau])$ vs. normalized interevent interval $\tau/\widehat{E}[\tau]$, for the same seven point processes displayed in Fig. 5.9. We add a random value to each event time in the data set. These values are drawn from a zero-mean Gaussian-distributed distribution with a standard deviation equal to ten times the mean interevent interval of the data set under study. All random variables are independent of each other. We reduced the histogram data to the same number of bins (namely 100) for all data sets. Most of the curves assume a form that is essentially exponential, confirming that the structure in the original histograms reflects the existence of particular interval orderings over short time scales.

of Eq. (11.38) with a large value of σ, such as 10; each event then experiences a random shift of the order of $10E[\tau_{k1}]$. In particular, the integrate-and-reset character of the point process does not survive in the resulting process $dN_R(t)$. However,

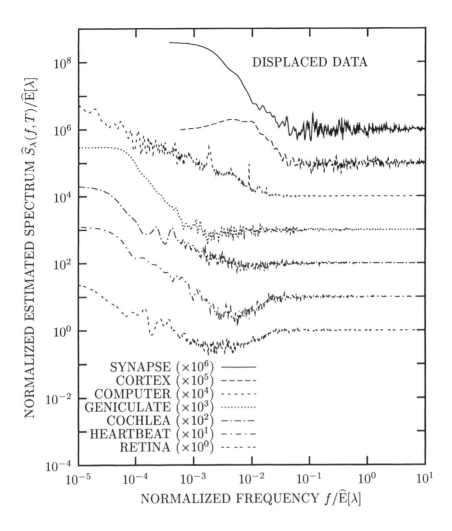

Fig. 11.7 Normalized estimated spectra following random displacement, $\widehat{S}_\lambda(f, T)/\widehat{E}[\lambda]$ vs. normalized frequency $f/\widehat{E}[\lambda]$, for the six biological point processes and one computer network traffic trace illustrated in Fig. 5.1. We add a random value to each event time in the data set. These values are drawn from a zero-mean Gaussian-distributed distribution with a standard deviation equal to ten times the mean interevent interval of the data set under study. All random variables are independent of each other. This operation eliminates all structure at high frequencies for all of the point processes, whereas the fractal variability at low frequencies remains intact.

fractal behavior, and indeed all long-term fluctuations, suffer little change so that the statistics of $dN_R(t)$ closely resemble those of $dN_1(t)$ over time scales significantly greater than $10E[\tau_{k1}]$.

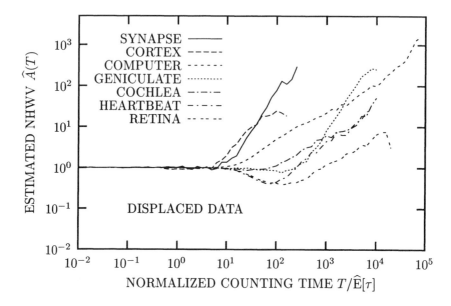

Fig. 11.8 Estimated normalized Haar-wavelet variance (NHWV) following random displacement, $\widehat{A}(T)$ vs. normalized counting time $T/\widehat{E}[\tau]$, for the same seven point processes as displayed in Fig. 5.2. We add a random value to each event time in the data set. The values are drawn from a zero-mean Gaussian-distributed distribution with a standard deviation equal to ten times the mean interevent interval of the data set under study. All random variables are independent of each other. This operation eradicates all structure at small values of the normalized counting time for all of the point processes, whereas the fractal variability at large values of the normalized counting time remains intact.

We now proceed to establish the Poisson character of $dN_R(t)$ over small times scales. Consider a large number n of adjacent counting intervals of short duration T, such that $nT/10E[\tau_{k1}] \ll 1$. The displacements, which greatly exceed the duration of the counting intervals nT, ensure that the resulting counts $Z_k(T)$ are independent and identically distributed, given the same overall local rate. In the limit as $T \to 0$ with nT fixed, the counts $Z_k(T)$ approach independent Bernoulli random variables with identical properties. This satisfies Eqs. (4.1) and (4.2), indicating that the point process $dN_R(t)$ over this interval nT becomes identical to a homogeneous Poisson process.

Under these conditions, the interevent-interval histogram approaches an exponential form. The results presented in Fig. 11.6 are, indeed, close to exponential. This is not unexpected since the events are shifted by Gaussian random numbers with zero mean and standard deviations equal to ten times the mean interevent interval. Locally, then, the process is nearly Poisson. The aggregate displaced interevent-interval histogram should then become the original histogram smeared by an exponential. If the rate remains relatively constant, as it is for the lower four data sets displayed in

Fig. 11.6, the smearing process obliterates all traces of the original interevent-interval histogram, leaving essentially a pure exponential in its place. On the other hand, if the rate exhibits large variations, some regions of the point processes will be dominated by short intervals while other regions will be dominated by long ones. The interevent-interval histogram will then deviate from an exponential form, especially for the larger intervals. This occurs in the upper three data sets displayed in Fig. 11.6 because of the wide variation in their rates. The deviation from exponential form is particularly apparent for the COMPUTER data because it has nearly a thousand times as many intervals as the SYNAPSE and CORTEX data, thereby making the effect more apparent.

In Figs. 11.7 and 11.8 we illustrate the effects of event-time displacement, implemented according to the prescription provided in Fig. 11.5, on the spectra and normalized Haar-wavelet variances of the canonical set of point processes shown in Figs. 5.1 and 5.2. The fractal behavior at low frequencies (large values of the normalized counting time) remains intact; however, all structure at high frequencies (small values of the normalized counting time) is eliminated and the process becomes Poisson-like, as evidenced by the fact that $\widehat{A}(T) = 1$ in this range.

11.4 INTERVAL TRANSFORMATION

We turn now to interval transformation, in which the relative ordering of the interevent intervals remains unchanged but their overall probability distribution is modified. This operation creates a surrogate data set testing the hypothesis that behavior in a particular measure derives from the interevent-interval distribution.

Figure 11.9 illustrates the procedure of interval transformation. This procedure maintains the ordering of the set of interevent intervals $\{\tau_n\}$, but each interval is resized by drawing it from an arbitrary probability distribution. Like displacement, interval transformation yields a new point process that exactly preserves the number of interevent intervals in the original process, and also closely preserves its duration. However, the particular original point process realization illustrated in Fig. 11.9 has interevent intervals that exhibit a relatively low coefficient of variation; the interval-transformed realization has a greater proportion of smaller and larger intervals at the expense of values near the mean. The nature of the point process has changed. A simple method for carrying out this transformation on a set of M interevent intervals includes the following five steps[7] (steps 3 and 4 can precede steps 1 and 2 to save memory):

---

[7] Alternatively, one could construct a monotonically increasing function and apply it to each interevent interval τ_n in the original realization to generate the corresponding interval in the new realization, choosing a form for the function that yields the correct empirical fit for the distribution of $\{\tau_n'\}$. However, aside from requiring intervention tailored to each data set, this procedure does not admit more than one possible realization. Randomness is not present in the generated realization in this case.

1. Generate M independent random numbers with the desired distribution, and call this array $\{\tau'_n\}$.

2. Sort $\{\tau'_n\}$ into ascending order.

3. Generate an auxiliary array of integers in ascending order, ranging from 1 to M, and call this array $\{k_n\}$. Thus, $k_n = n$ at this point.

4. Sort the original array $\{\tau_n\}$ into ascending order, and perform the same operations on the auxiliary array $\{k_n\}$.

5. Sort the auxiliary array $\{k_n\}$ back into ascending order, and perform the same operations on the generated array $\{\tau'_n\}$.

a) ORIGINAL POINT PROCESS REALIZATION $dN_1(t)$

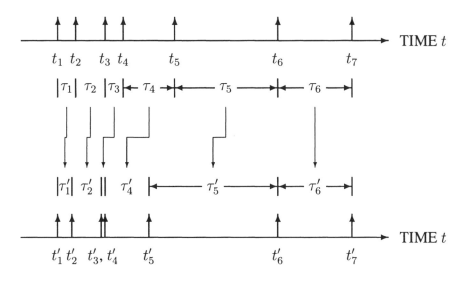

b) INTERVAL-TRANSFORMED POINT PROCESS REALIZATION $dN_R(t)$

Fig. 11.9 Interval-transformed point process. (a) The original point-process realization has seven event times $\{t_n\}$ and therefore six interevent intervals $\{\tau_n\}$. We replace each interevent interval τ_n with a new random variable drawn from a predetermined distribution (for example, exponential), and preserve the relative ordering of the intervals. This yields the interevent intervals $\{\tau'_n\}$ of the interval-transformed realization (b). As with the interval-transformed realization, the two realizations share the same start time ($t'_1 = t_1$). We then employ the relation $\tau'_n = t'_{n+1} - t'_n$ to yield the seven event times $\{t'_n\}$ of the interval-transformed realization. The label for the interevent interval τ'_3 has been omitted because of lack of space.

11.4.1 Interval normalization

Distribution transformation has long been applied to interval and count data because it confers a number of benefits[8] (Tukey, 1957; Kendall & Stuart, 1966). The principal goals of such transformations are generally to stabilize the variance so that conventional analysis measures can be used (Prucnal & Teich, 1980); convert nonadditive noise into additive form so that well-established detection/estimation techniques apply (Prucnal & Saleh, 1981); and to cast the distribution into a symmetrical form for ease of calculation. A transformation suitable for effecting one of these features often turns out to effect the others as well (Tukey, 1957). Most often, the transformation is to a Gaussian distribution by virtue of the extensive body of available results; it is then called a **normalizing transformation**.

11.4.2 Interval exponentialization

We now consider the effects of transforming the intervals to an exponential distribution, which we term **exponentialization**. In Figs. 11.10 and 11.11, we illustrate the consequences of carrying out this operation on the spectra and normalized Haar-wavelet variances of the canonical set of point processes shown in Figs. 5.1 and 5.2. (We do not display exponentialized interevent-interval histograms since they are exponential by construction.)

To understand the role played by exponentialization, we consider the fractal-Gaussian-process-driven Poisson process described in Sec. 6.3.3. When the coefficient of variation of the rate is small in comparison with unity ($C_\mu \ll 1$), the rate does not depart significantly from its mean value; in particular, it remains nearly constant over times corresponding to a few interevent intervals. Under these conditions, as shown in Eq. (4.33), the interevent-interval density for this process closely follows an exponential form,[9] modulated only minimally by the range of rate values (compare with the argument in Sec. 11.3.2). The sequence of events then closely resembles a homogeneous Poisson process over those time scales. The interval distribution therefore does not contribute to the fractal characteristics of this process; rather, this behavior arises from the ordering of the intervals. Hence, one method for testing whether the interval distribution induces fractal behavior in a point process involves

---

[8] Approximate normalizing transformations do, in fact, exist for nonparalyzable and paralyzable dead-time-modified Poisson counting systems (Teich, 1985).

[9] The sequence of action potentials generated in primary afferent auditory nerve fibers, either firing spontaneously or stimulated by high-frequency tones, is well described by a dead-time-modified form of this point process (Teich et al., 1990; Lowen & Teich, 1997). The observation of an interevent-interval density described by a delayed exponential, while ignoring other statistics of the point process, has often led researchers to the (improper) conclusion that the action potentials follow a dead-time-modified homogeneous Poisson process (see, for example, Kiang, Watanabe, Thomas & Clark, 1965, Chapter 8); for a discussion of this issue, see Teich & Khanna (1985); Teich (1992); Lowen & Teich (1992a); Teich & Lowen (1994). Indeed, any number of distinctly non-Poisson point processes can be constructed with exponential interevent-interval densities (see, for example, Moran, 1967; Lawrance, 1972; Glass & Mackey, 1988).

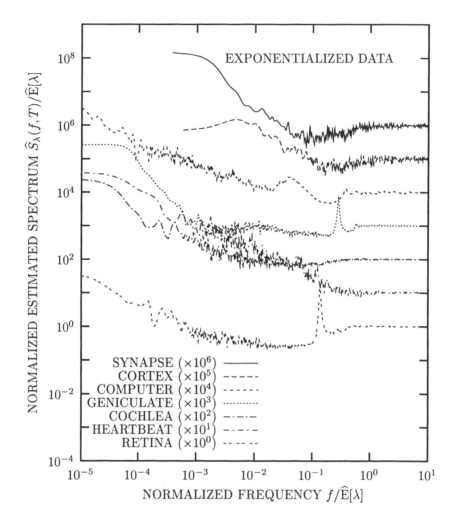

Fig. 11.10 Normalized estimated spectrum after exponentialization, $\widehat{S}_\lambda(f, T)/\widehat{E}[\lambda]$ vs. normalized frequency $f/\widehat{E}[\lambda]$, for the six biological point processes and one computer network traffic trace illustrated in Fig. 5.1. There is a substantial reduction of structure at high frequencies for all of the point processes while the fractal variability at low frequencies remains essentially intact or, in the case of the HEARTBEAT point process, is enhanced.

replacing the intervals of the original realization with ones having an exponential form and observing whether the fractal characteristics change.

By way of example, consider a point-process realization exhibiting fractal behavior that depends on both the relative ordering *and* the distribution of the interevent intervals; the fractal-lognormal-noise-driven Poisson process described in Sec. 6.5 behaves in this way. Exponentialized realizations for this process yield results in

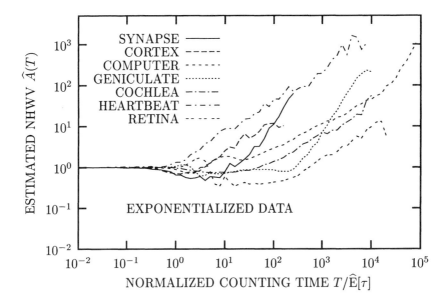

Fig. 11.11 Estimated normalized Haar-wavelet variance (NHWV) following exponential-ization, $\widehat{A}(T)$ vs. normalized counting time $T/\widehat{E}[\tau]$, for the same seven point processes as displayed in Fig. 5.2. Structure at small values of the normalized counting time is reduced for all of the point processes, whereas the fractal variability at large values of the normal-ized counting time remains essentially intact or, in the case of the HEARTBEAT point process, increases.

substantial agreement with each other, but with characteristics that differ from those of the original realization, thereby demonstrating that the distribution of the interevent intervals plays a role in the manifestation of fractal behavior in this process. That is not to say, however, that this is a fractal point process; the fact that changing the in-terval distribution changes the fractal characteristics of a point process does not mean that it is a fractal object. We have, in fact, definitively shown in Fig. 5.11 that the synapse data, which are well-described by a fractal-lognormal-noise-driven Poisson process, are characterized by a fractal-rate point process.

The exponentialized data displayed in Figs. 11.10 and 11.11 are, for the most part, devoid of the sharp spectral components and Haar-wavelet-variance oscillations evident in the original data.[10] However, the power-law behavior of the curves remains essentially intact in all cases, demonstrating that the fractal behavior of all seven point processes arises principally from the relative ordering of the intervals and not from the distribution of the interevent intervals themselves. Behavior of this kind is the hallmark of a fractal-rate point process.

---

[10] The consequences of event-time displacement are similar (see Figs. 11.7 and 11.8). Both are associated with a jittering of the event times, which results in a concomitant loss of phase coherence.

11.5 INTERVAL SHUFFLING

Interval shuffling provides a complementary method for introducing additional randomness into a point process. The shuffling operation randomly reorders the set of interevent intervals $\{\tau_n\}$, thereby destroying any correlations or dependencies that might have existed among them.

Figure 11.12 illustrates how to implement the shuffling operation. The original point process depicted in (a) happens to have events clustered towards the left, corresponding to a rate that effectively decreases with time. The shuffled version in (b) exhibits no such global behavior. Although any possible rearrangement of the original sequence $\{\tau_n\}$ can occur in principle, those with long-term structure, such

a) ORIGINAL POINT PROCESS REALIZATION $dN_1(t)$

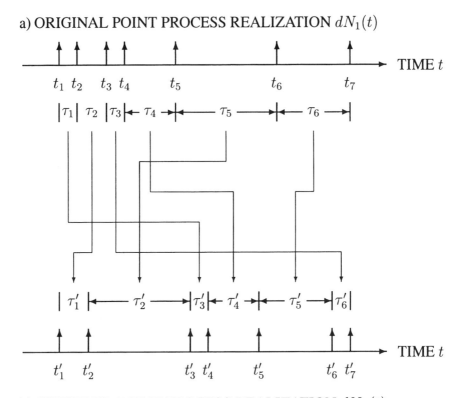

b) SHUFFLED POINT PROCESS REALIZATION $dN_R(t)$

Fig. 11.12 Shuffled point process. (a) The original point process realization has seven event times $\{t_n\}$ and therefore six interevent intervals $\{\tau_n\}$. (b) Randomly reordering the interevent intervals $\{\tau_n\}$ of the original realization yields the interevent intervals $\{\tau_n'\}$ of the shuffled realization. For the particular shuffling illustrated, we have $\tau_1' = \tau_2$, $\tau_2' = \tau_5$, $\tau_3' = \tau_1$, $\tau_4' = \tau_4$, $\tau_5' = \tau_6$, and $\tau_6' = \tau_3$. Assuming the same start time ($t_1' = t_1$) and employing the relation $\tau_n' = t_{n+1}' - t_n'$ yields the seven event times $\{t_n'\}$ of the shuffled realization.

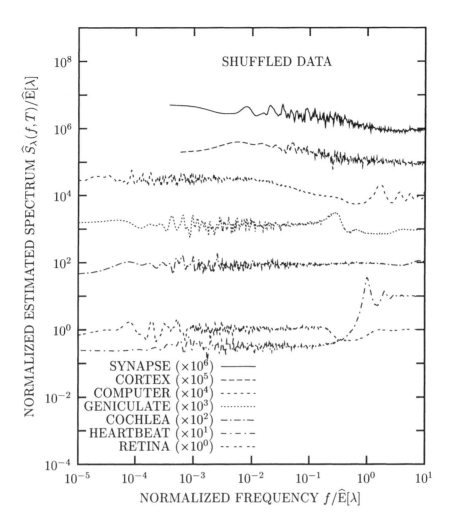

Fig. 11.13 Normalized estimated spectrum after random shuffling, $\widehat{S}_\lambda(f, T)/\widehat{\mathrm{E}}[\lambda]$ vs. normalized frequency $f/\widehat{\mathrm{E}}[\lambda]$, for the six biological point processes and one computer network traffic trace illustrated in Fig. 5.1. The curves are flat for low frequencies, indicating that no vestiges of fractal behavior remain in the shuffled processes.

as a decreasing rate or fractal fluctuations in the sizes of the intervals, prove much less likely. The end result resembles a renewal point process.

Rather than generating new point processes, shuffling typically finds use as a non-parametric method for testing a hypothesis in a point-process realization (Schiff & Chang, 1992; Theiler et al., 1992; Theiler & Prichard, 1996; Schreiber & Schmitz, 1996). A series of independent shufflings of such a realization yields a set of statistically independent reorderings of the same set of interevent intervals $\{\tau_n\}$, and

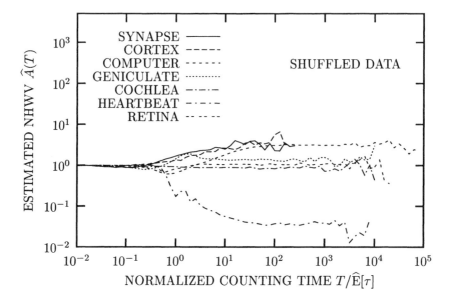

Fig. 11.14 Estimated normalized Haar-wavelet variance (NHWV) following random shuffling, $\widehat{A}(T)$ vs. normalized counting time $T/\widehat{E}[\tau]$, for the same seven point processes as displayed in Fig. 5.2. The flattening of all of the curves indicates the absence of fractal behavior in all of the shuffled processes.

reveal the likelihood that a particular feature of the realization occurred by chance. For example, the rate spectrum $S_\lambda(f, T)$ can be used to test for nonrenewal behavior. If a number of independent shufflings yield estimates of the spectrum that are in substantial agreement with each other, but differ from that of the original unshuffled process, then the original point process likely has dependent interevent intervals and therefore does not derive from a renewal point process.

Like displacement, interval shuffling destroys any structure present in the original point process $dN_1(t)$, but in this case over time scales smaller than the shuffling, replacing it with (locally) renewal behavior at those scales. In contrast to interval transformation, the relative ordering of the interevent intervals is destroyed but the overall interval distribution is maintained. Indeed, the two operations are complementary, each preserving what the other destroys. The shuffling operation tests the hypothesis that behavior in a particular measure derives from the ordering of the interevent intervals.

In Figs. 11.13 and 11.14 we illustrate the effects of random shuffling on the spectra and normalized Haar-wavelet variances of the canonical set of point processes shown in Figs. 5.1 and 5.2. (We do not display shuffled interevent-interval histograms since they are identical to the original histograms.) The results are dramatic. The shuffling eliminates all vestiges of fractal behavior in all of the processes; it results in a flattening of all curves for both measures. The surrogates considered previously, random deletion (Figs. 11.3 and 11.4), event-time displacement (Figs. 11.7 and 11.8), and exponentialization (Figs. 11.10 and 11.11), behave quite differently.

However, shuffling an interval-exponentialized realization, or exponentializing a shuffled realization, essentially yields a homogeneous Poisson process (Prob. 11.1).

11.5.1 Block shuffling

Finally, we mention two variations on the theme of shuffling. The first variation, known as **block shuffling**, consists of shuffling intervals within sections of a realization, while keeping intervals from different sections from commingling. For example, one can divide a realization into blocks of k interevent intervals and shuffle the intervals within each block separately. Or, one can divide the realization into contiguous, nonoverlapping periods of duration T and separately shuffle the intervals that begin in each period. These block-shuffling methods preserve fractal and other long-term characteristics over time scales larger than the average block time, while destroying all dependencies over shorter time scales.

11.5.2 Bootstrap method

The second variation selects intervals with replacement, rather than rearranging the intervals; in this case, some of the interevent intervals of the original realization may not appear in the shuffled version, while others can have multiple copies. This approach, called the **bootstrap method**, enjoys some mathematical advantages since it generates renewal processes based on an interevent-interval distribution estimated from the intervals at hand (Efron, 1982; Efron & Tibshirani, 1993). Nevertheless, shuffling without replacement proves superior in some cases, since the set of intervals remains identical. In particular, the mean rate, the total duration of the realization, and all first-order statistics of the interevent intervals do not change.

11.5.3 Identification of fractal-based point processes

With the collection of surrogate-data techniques in hand, we revisit the issue of fractal-based point-process identification. Associating a model with an observed point process is often useful for elucidating mechanisms that underlie the data. We took a first step in this direction in Sec. 5.5.4 (see also Probs. 5.2, 5.3, 5.4, and 5.5) where we showed the possibility of discriminating between fractal and fractal-rate point processes.

What information about the nature of the seven canonical point processes portrayed in Figs. 5.1 and 5.2 might be provided to us by the surrogate-data curves exhibited in Secs. 11.3–11.5?

Fractal behavior manifests itself at low frequencies and large normalized counting times. The displaced surrogates shown in Figs. 11.7 and 11.8 reveal that the fractal behavior does not arise from the details of the local occurrence times for any of the point processes. The exponentialized surrogates presented in Figs. 11.10 and 11.11

inform us further that the fractal behavior is not a result of the particular form of the interevent-interval distribution in any of the point processes.

And, finally, the shuffled surrogates displayed in Figs. 11.13 and 11.14 assure us that, for all of the point processes, the fractal behavior is associated with the ordering of the intervals. The large counting-time asymptotes of $\widehat{A}(T)$ in Fig. 11.14 reveal the clustering characteristics inherent in the interevent-interval distribution, as understood from Eq. (4.18).

We conclude that all seven data sets represent fractal-rate point processes, and not fractal point processes. This conclusion accords with that reached by examining the generalized dimensions D_q of these point processes (see Prob. 5.5). We revisit the issue of fractal-based point-process identification in Secs. 12.1 and 13.6.

11.6 SUPERPOSITION

The superposition $N_R(t)$ of a set of M counting processes, $N_1(t) \ldots N_M(t)$, is formed from their addition. In terms of the point-process formalism, $dN_R(t)$ includes every event in all of the component point processes $dN_k(t)$, as illustrated in Fig. 11.15.

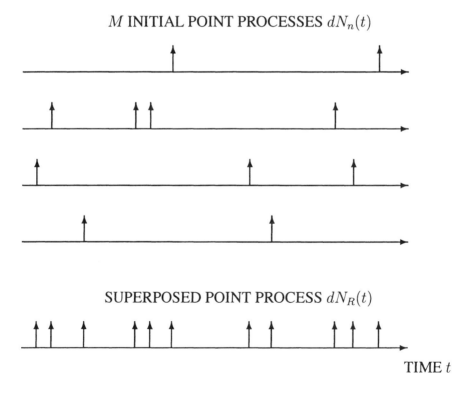

M INITIAL POINT PROCESSES $dN_n(t)$

SUPERPOSED POINT PROCESS $dN_R(t)$

TIME t

Fig. 11.15 The superposition of M point processes $dN_n(t)$ leads to a process comprising all of the events of the constituent processes. The rate of the superposed process equals the sum of the rates of the constituent point processes, and thus exceeds each of the individual rates.

We therefore have

$$
N_R(t) = \sum_{k=1}^{M} N_k(t)
$$

$$
dN_R(t) = \sum_{k=1}^{M} dN_k(t).
$$

(11.39)

Palm[11] (1943) appears to have been the first to examine the superposition of point processes; Çinlar (1972) studied this operation extensively.

In broad terms, the superposition of identical fractal and fractal-rate point processes yields results that resemble those for event deletion, in that the characteristic times of the process change but the fractal exponent remains the same.

The most fruitful mathematical approach for examining this problem appears to use the coincidence rate of the resulting point process $dN_R(t)$. We begin with the definition of this quantity for $dN_R(t)$, as presented in Eqs. (3.48) and (3.49):

$$
\begin{aligned}
G_R(t) &\equiv \lim_{\epsilon \to 0} \epsilon^{-2} \Pr\{N_R(s+\epsilon) - N_R(s) > 0 \\
&\qquad\qquad \text{and } N_R(s+t+\epsilon) - N_R(s+t) > 0\} \\[4pt]
&= \mathrm{E}\left[\frac{dN_R(s)}{ds} \frac{dN_R(s+t)}{ds}\right] \\[4pt]
&= \mathrm{E}\left[\sum_{m=1}^{M} \frac{dN_m(s)}{ds} \sum_{n=1}^{M} \frac{dN_n(s+t)}{ds}\right] \\[4pt]
&= \sum_{m=1}^{M} \sum_{n=1}^{M} \mathrm{E}\left[\frac{dN_m(s)}{ds} \frac{dN_n(s+t)}{ds}\right] \\[4pt]
&= \sum_{m=1}^{M} \mathrm{E}\left[\frac{dN_m(s)}{ds} \frac{dN_m(s+t)}{ds}\right] \\[4pt]
&\qquad + \sum_{m=1}^{M} \sum_{n \neq m} \mathrm{E}\left[\frac{dN_m(s)}{ds} \frac{dN_n(s+t)}{ds}\right] \\[4pt]
&= \sum_{m=1}^{M} \mathrm{E}\left[\frac{dN_m(s)}{ds} \frac{dN_m(s+t)}{ds}\right] \\[4pt]
&\qquad + \sum_{m=1}^{M} \sum_{n \neq m} \mathrm{E}\left[\frac{dN_m(s)}{ds}\right] \mathrm{E}\left[\frac{dN_n(s+t)}{ds}\right] \\[4pt]
&= \sum_{m=1}^{M} G_m(t) + \sum_{m=1}^{M} \sum_{n \neq m} \mathrm{E}[\mu_m] \mathrm{E}[\mu_n]
\end{aligned}
$$

(11.40)

---

[11] A photograph of Palm appears at the beginning of Chapter 13.

$$= \sum_{m=1}^{M} \left\{ G_m(t) - \mathrm{E}^2[\mu_m] \right\} + \sum_{m=1}^{M} \sum_{n=1}^{M} \mathrm{E}[\mu_m] \, \mathrm{E}[\mu_n]$$

$$= \sum_{m=1}^{M} \left\{ G_m(t) - \mathrm{E}^2[\mu_m] \right\} + \mathrm{E}^2[\mu_R], \tag{11.41}$$

where the independence of the M point processes yields Eq. (11.40) and the last step leading to Eq. (11.41) results from the fact that the sum of the rates is equal to the rate of the sum. The second-order properties of the superposed process $dN_R(t)$ are simply given by the sum of those for the constituent processes. Equation (11.41) simplifies for constituent processes with identical statistics, whereupon

$$G_R(t) = MG_1(t) + \mathrm{E}^2[\mu_R] \, (1 - 1/M). \tag{11.42}$$

Turning now to the frequency domain, and considering point processes that have identical fractal characteristics, we have

$$\begin{aligned}
S_R(f) &= MS_1(f) + \mathrm{E}^2[\mu_R] \, (1 - 1/M) \, \delta(f) \\
&= M \{ \mathrm{E}^2[\mu_1] \, \delta(f) + \mathrm{E}[\mu_1] \left[1 + (f/f_{S1})^{-\alpha} \right] \} \\
&\quad + \mathrm{E}^2[\mu_R] \, (1 - 1/M) \, \delta(f) \\
&= \mathrm{E}^2[\mu_R] \, \delta(f) + \mathrm{E}[\mu_R] \left[1 + (f/f_{S1})^{-\alpha} \right], \tag{11.43}
\end{aligned}$$

which differs from $S_1(f)$ only in the mean rate. In particular, fractal onset times and frequencies remain unchanged under point-process superposition, as do fractal exponents.

11.6.1 Superposition of doubly stochastic Poisson point processes

In general, superposition changes the nature of a point process (Palm, 1943). Doubly stochastic Poisson processes provide the sole exception; superposing M identical processes in this class yields a new point process that differs from the originals only in that its rate is a factor of M larger. Indeed, for a set of arbitrary point processes $dN_1(t), dN_2(t), \ldots$, as M increases $dN_R(t)$ tends towards a doubly stochastic point process with a rate process identical to that of the constituent processes (Çinlar, 1972; Cox & Isham, 1980).

The homogeneous Poisson point process forms a special case. The superposition of this process $dN_1(t)$, and a fractal-rate doubly stochastic Poisson processes $dN_2(t)$, becomes a different doubly stochastic Poisson processes $dN_R(t)$, which is closely related to $dN_2(t)$. The resulting process $dN_R(t)$ shares fractal exponents with $dN_2(t)$, and differs only in its fractal onset times and frequencies.

Setting terms in the spectra of $dN_2(t)$ and $dN_R(t)$ equal, Eq. (5.44a) yields

$$(\mathrm{E}[\mu_1] + \mathrm{E}[\mu_2]) \, (f/f_{SR})^{-\alpha} = \mathrm{E}[\mu_2] \, (f/f_{S2})^{-\alpha}$$

$$(f_{SR}/f_{S2})^{\alpha} = \frac{\mathrm{E}[\mu_2]}{\mathrm{E}[\mu_1] + \mathrm{E}[\mu_2]}. \tag{11.44}$$

Straightforward application of Eq. (5.45) then yields

$$(T_{F2}/T_{FR})^\alpha = (T_{A2}/T_{AR})^\alpha = (T_{R2}/T_{RR})^\alpha = \frac{E[\mu_2]}{E[\mu_1] + E[\mu_2]}, \qquad (11.45)$$

and $t_{GR} = t_{G2}$.

More generally, consider the superposition of a number of independent, identical fractal-based point processes with aggregate rate $E[\mu_2]$, and a number of independent point processes lacking long-term correlations with aggregate rate $E[\mu_1]$. The former converge to a fractal-based doubly stochastic Poisson point process whereas the latter converge to a homogeneous Poisson point process. Together, the superposition yields results similar to those of Eqs. (11.44) and (11.45).

11.6.2 Superposition of renewal point processes

For the particular case of superposed renewal point processes, the resulting process $dN_R(t)$ does not belong to the class of renewal point processes, but we may still solve for the interevent interval density function (Cox & Isham, 1980; Ryu & Lowen, 1996).

Consider M independent and identical renewal point processes, with a mean interevent time $E[\tau_1]$, interevent-interval probability density function $p_{\tau 1}(t)$, and associated survivor function $S_{\tau 1}(t) \equiv \Pr\{\tau_1 > t\} = \int_t^\infty p_{\tau 1}(u)\,du$. The forward recurrence time $\vartheta_1(t)$, defined in Eq. (3.10), is the time that remains to the next event from an arbitrary starting time. This quantity has a survivor function $S_{\vartheta 1}(t)$ given by Eq. (3.11). In terms of $p_{\tau 1}(t)$, we therefore have

$$
\begin{aligned}
S_{\vartheta 1}(t) &= E[\mu_1] \int_t^\infty S_{\tau 1}(u)\,du \\
&= E[\mu_1] \int_{u=t}^\infty \int_{v=u}^\infty p_{\tau 1}(v)\,dv\,du \\
&= E[\mu_1] \int_{v=t}^\infty \int_{u=t}^v p_{\tau 1}(v)\,dv\,du \\
&= E[\mu_1] \int_t^\infty (v - t) p_{\tau 1}(v)\,dv, \qquad (11.46)
\end{aligned}
$$

which denotes the probability of zero arrivals in the renewal process during a period of duration t that starts at a random time independent of the process.

For M independent identical renewal point processes, the probability of zero events occurring in any of the M processes in a time t is simply the product of the probabilities from the individual processes. Therefore, the forward recurrence-time survivor function for the superposed process, $S_{\vartheta R}(t)$, becomes

$$
\begin{aligned}
& E[\mu_R] \int_t^\infty (v - t)\, p_{\tau R}(v)\,dv \\
&= S_{\vartheta R}(t)
\end{aligned}
$$

$$= \left[S_{\vartheta 1}(t) \right]^M$$

$$= E^M[\mu_1] \left[\int_t^\infty (v - t)\, p_{\tau 1}(v)\, dv \right]^M. \tag{11.47}$$

After two differentiations and some algebra (Ryu & Lowen, 1996), we obtain the interevent-interval probability density function for the superposed process:

$$p_{\tau R}(t) = E[\tau_1]^{-(M-1)} \left[\int_t^\infty (v - t)\, p_{\tau 1}(v)\, dv \right]^{M-2}$$

$$\times \left\{ (M - 1) \left[\int_t^\infty p_{\tau 1}(v)\, dv \right]^2 \right.$$

$$\left. + p_{\tau 1}(t) \int_t^\infty (v - t)\, p_{\tau 1}(v)\, dv \right\}. \tag{11.48}$$

Two limiting conditions emerge. First, for small interevent times, such that $P_{\tau 1}(t) \ll 1/M$, the probability of two events occurring in the same constituent process approaches zero. In this case we have

$$S_{\vartheta R}(t) = \left[S_{\vartheta 1}(t) \right]^M$$

$$\approx \left(1 - E[\mu_1]\, t \right)^M$$

$$\approx 1 - M\, E[\mu_1]\, t = 1 - E[\mu_R]\, t$$

$$\approx \exp\!\left(-E[\mu_R]\, t \right). \tag{11.49}$$

This result is the same as that obtained for a homogeneous Poisson process with the same rate; in fact, $dN_R(t)$ resembles the homogeneous Poisson process over these short time scales. Second, for long times, such that $S_{\tau 1}(t) \ll 1$, we similarly find that $S_{\tau R}(t) \ll 1$. In particular, $P_{\tau 1}(t) = 1$ implies that $P_{\tau R}(t) = 1$.

We now specialize to the particular case of a fractal renewal point process with abrupt cutoffs, in which case the interevent-interval probability density function assumes the form of Eq. (7.1); we focus on the limit $A \ll t \ll B$:

$$p_{\tau 1}(t) = \frac{\gamma}{A^{-\gamma} - B^{-\gamma}} \times \begin{cases} t^{-(\gamma+1)} & \text{for } A < t < B \\ 0 & \text{otherwise.} \end{cases} \tag{11.50}$$

For $0 < \gamma < 1$, to first order the forward recurrence-time survivor function follows the form

$$S_{\vartheta 1}(t) \approx 1 - \gamma^{-1}(t/B)^{1-\gamma}, \tag{11.51}$$

so that

$$S_{\vartheta R}(t) \approx \left[1 - \gamma^{-1}(t/B)^{1-\gamma} \right]^M \approx 1 - M\gamma^{-1}(t/B)^{1-\gamma}. \tag{11.52}$$

Substituting Eq. (11.52) into Eq. (3.12), and taking another derivative, yields the corresponding interevent interval probability density function

$$p_{\tau R}(t) \approx \gamma A^\gamma t^{-(\gamma+1)}, \tag{11.53}$$

identical to $p_{\tau 1}(t)$ over that time scale. Taken together, Eqs. (11.52) and (11.53) indicate that for $0 < \gamma < 1$ and $A \ll t \ll B$, the time statistics for the superposed process resemble those of a single fractal renewal process, with the upper cutoff B replaced by the smaller value $M^{1/(\gamma-1)}B$. The superposition process modifies the interval probability density only slightly for moderate values of M. Achieving an approximate Poisson-process form requires on the order of $M = (B/A)^{1-\gamma}$ fractal renewal processes.

The situation changes for larger values of γ. For $\gamma > 1$, the same procedure yields

$$S_{\vartheta 1}(t) \approx \gamma^{-1}(t/A)^{1-\gamma} \tag{11.54}$$

$$S_{\vartheta R}(t) \approx \gamma^{-M}(t/A)^{M(1-\gamma)} \tag{11.55}$$

$$p_{\tau R}(t) \approx c\gamma^{1-M} A^c t^{-(c+1)}, \tag{11.56}$$

where we define $c \equiv 1 + M(\gamma - 1)$ and note that $c > 1$. Thus, the survivor function of the superposition decays progressively more quickly as we add more fractal renewal processes, approaching the exponential limit of the Poisson process far more quickly than in the domain $0 < \gamma < 1$. For $\gamma = 1$, intermediate results obtain; the superposition approaches Poisson form slowly as M increases, but not as slowly as when $\gamma < 1$.

Problems

11.1   *Shuffled and exponentialized point processes*   Show that shuffling a fractal-based point process and transforming its intervals to an exponentially distributed form, in either order, must result in a homogeneous Poisson process.

11.2   *Superposed fractal-based point processes*   Consider two fractal-based point processes, $dN_1(t)$ and $dN_2(t)$, whose superposition yields another process $dN_R(t)$.
   11.2.1. Under what conditions does $dN_R(t)$ precisely follow the power-law form of Eq. (5.44)?
   11.2.2. Under those conditions, find α_R, $E[\mu_R]$, and f_{SR} in terms of the related parameters for $dN_1(t)$ and $dN_2(t)$.
   11.2.3. Under what set of more-relaxed conditions does $dN_R(t)$ closely approximate a fractal form?

11.3   *Interval transformation and shuffling as surrogates*   Both interevent-interval transformation (Sec. 11.4) and shuffling (Sec. 11.5) modify fractal-based point processes in ways that yield insights into the processes under study. Suppose that we evaluate the spectrum of a point process before and after shuffling, and compare the results. For simplicity, we consider three classes of results: (1) shuffling leaves the spectrum essentially unchanged; (2) shuffling effectively eliminates the fractal character of the process; and (3) shuffling decreases the fractal onset frequency, thereby leading to a process with decreased, but still present, fractal content. We also examine the same three classes for the interevent-interval transformation. Consider each of the nine possible pairs of classes. Determine which pairs cannot occur and, for the others, describe processes that could yield those results.

11.4   *Fractal content of superposed point processes*   We can quantify the fractal content in a fractal-based point process, discussed in Prob. 11.3, with a dimensionless quantity such as $c_f \equiv f_S/\mathrm{E}[\mu]$, which remains constant under time dilation. Consider a fractal-based point process $dN_R(t)$, formed by the superposition of two component fractal-based point processes $dN_1(t)$ and $dN_2(t)$ with the same fractal exponent α. Under what conditions does c_{fR} equal or exceed both c_{f1} and c_{f2}?

11.5   *Statistics of a dilated and deleted point process*   We can scale the time axis in a fractal-rate point process $dN_1(t)$ to effectively double its rate, yielding $dN_2(t)$, and we can then delete half of the events in $dN_2(t)$ to yield a new process $dN_R(t)$ closely related to the first. Assume that the deletion process does not selectively affect the low-frequency components of the point process.

11.5.1.   How do the mean rate $\mathrm{E}[\mu]$, fractal exponent α, and fractal onset frequency f_S compare for $dN_R(t)$ and $dN_1(t)$?

11.5.2.   Suppose that $dN_1(t)$ is (a) a homogeneous Poisson process or (b) an integrate-and-reset process with constant rate; and that we delete events either (1) randomly and independently and retaining half of them or (2) by decimation in which every other event is retained. What type of point-process results in each of the four cases?

11.6   *Homogeneous Poisson process modified by fixed and stochastic dead time*
Suppose a homogeneous Poisson process $dN_1(t)$ with rate μ_1 experiences a fixed dead time of duration τ_f, followed by an exponentially distributed stochastic dead time of expected duration τ_r. The resulting process $dN_R(t)$ becomes a dead-time-modified homogeneous Poisson process. Determine the nature of this point process, as well as the mean and variance of the interevent interval, the rate, the interevent-interval probability density, and the associated characteristic function. Finally, for the special case $\tau_r = 0$, find the spectrum.

11.7   *Forward-recurrence-time survivor function for a fractal renewal process*
Prove Eqs. (11.51)–(11.56) in greater detail.

11.8   *Modification of a point process by block shuffling and event-time displacement*   Consider a doubly stochastic Poisson process $dN_1(t)$ driven by a fractal rate. The spectrum of this process, $S_{N1}(f)$, follows Eq. (5.44a) over a wide range of frequencies.

11.8.1.   Now suppose we divide the time axis into contiguous blocks of duration T. Effect a block shuffle of the process by taking all interevent intervals that begin in each block and shuffle them with each other, but not with intervals beginning in other blocks. For $f_S T = 100$, sketch the spectrum $S_{NR}(f)$ of the resulting block-shuffled process $dN_R(t)$.

11.8.2. Instead of block shuffling, we now add to each event time t_k a Gaussian-distributed random number with zero mean and standard deviation equal to $100/f_S$, as illustrated in Eq. (11.38). We then sort the displaced event times back into ascending order and define $dN_R(t)$ in terms of the resulting sorted set of event times. Sketch the spectrum $S_{NR}(f)$ of this displaced process $dN_R(t)$.

11.9 *Properties of a randomly deleted renewal process*   Let $dN_1(t)$ denote a renewal point process with interevent-interval probability density $p_\tau(t)$.

11.9.1. Show that independently deleting events in $dN_1(t)$ with a probability r yields another renewal point process $dN_R(t)$.

11.9.2. Find the interevent-interval probability density of $dN_R(t)$.

11.9.3. Using characteristic functions, examine the behavior of $dN_R(t)$ for the specific case where $dN_1(t)$ has the abrupt power-law probability density exhibited in Eq. (7.1). Focus on the medium-time limit $A \ll t \ll B$. How do results for $0 < \gamma < 1$ and $\gamma \geq 1$ differ?

11.10 *Dead-time-modified and decimated Poisson counting distributions*   The counting distribution for the dead-time-modified homogeneous Poisson process proves useful in many contexts, including nuclear, neural, and photon counting (DeLotto et al., 1964; Ricciardi & Esposito, 1966; Müller, 1973; Cantor & Teich, 1975; Libert, 1976; Teich & Cantor, 1978; Müller, 1981; Prucnal & Teich, 1983). The formula is considerably more complex than Eq. (4.7) for the Poisson counting distribution as a result of correlation between the numbers of events in adjacent counting windows, engendered by the occurrence of the dead-time periods that straddle them. The counts $\{Z_k(T)\}$ are therefore not independent.

For the nonparalyzable dead-time counter unblocked at the beginning of the counting interval (see Footnote 6 on p. 236), the exact result is (Ricciardi & Esposito, 1966; Müller, 1974; Prucnal & Teich, 1983):

$$p_Z(n;T) = \Pr\{N(t+s) - N(s) = n\}$$

$$= \begin{cases} \sum_{m=0}^{n} \dfrac{[\mu_1 T(1 - n\tau_e/T)]^m}{m!} \exp[-\mu_1 T(1 - n\tau_e/T)] \\ \quad - \sum_{m=0}^{n-1} \dfrac{[\mu_1 T(1 - (n-1)\tau_e/T)]^m}{m!} \exp[-\mu_1 T(1 - (n-1)\tau_e/T)] \\ \hfill \text{for } 0 \leq n < T/\tau_e, \\[2ex] 1 - \sum_{m=0}^{n-1} \dfrac{[\mu_1 T(1 - (n-1)\tau_e/T)]^m}{m!} \exp[-\mu_1 T(1 - (n-1)\tau_e/T)] \\ \hfill \text{for } T/\tau_e \leq n < T/\tau_e + 1, \\[2ex] 0 \hfill \text{for } n \geq T/\tau_e + 1, \end{cases}$$

$$\hfill (11.57)$$

where μ_1 is the driving rate of the process before the imposition of the fixed dead-time period τ_e. The rate after dead-time modification is $\mathrm{E}[\mu_R] = \mu_1/(1 + \mu_1\tau_e)$ [see Eq. (11.22)].

Results for various kinds of dead-time counters are available (Müller, 1973, 1974; Libert, 1976); exact formulas for *equilibrium* (stationary) and *blocked* (ordinary) counting distributions resemble Eq. (11.57) for the *unblocked* (shifted or free) result provided above, but they are more complex. In the usual situation when the mean count greatly exceeds unity, the differences among the blocked, unblocked, and equilibrium counting results are, in fact, insubstantial and all three of these processes may essentially be viewed as stationary equivalents of each other (Libert, 1976).

Decimation of a homogeneous Poisson process with decimation parameter m yields the integer-order gamma renewal process of index m (see Sec. 11.2.2 and Prob. 4.7). In this case, the counting distribution takes the form of a finite sum of Poisson distributions, provided that the process is turned on at $t = 0$ (Parzen, 1962; Cox, 1962; Cox & Isham, 1980; Teich et al., 1984, Eq. (A26)):

$$p_Z(n; T) = \sum_{l=mn}^{mn+m-1} \frac{(\mu_1 T)^l \exp(-\mu_1 T)}{l!}, \qquad n \geq 0; \qquad (11.58)$$

the rate of the initial Poisson process is μ_1 and the decimated rate $E[\mu_R] \approx \mu_1/m$.[12] This formula comes about because the integer-order gamma process arises from the homogeneous Poisson process by permitting every mth event to survive, while deleting intermediate events, as illustrated in Fig. 11.1b).

11.10.1. Simulate (or plot) the counting distribution for the nonparalyzable dead-time-modified Poisson process for $\mu_1 T = 15$ and $\mu_1 \tau_e = 0, 0.2, 0.5$, and 1.0. Confirm that the count mean and count variance agree with $E[\mu_R] T = \mu_1 T/(1 + \mu_1 \tau_e)$ and $\mathrm{Var}[n] \approx \mu_1 T/(1 + \mu_1 \tau_e)^3$, respectively.

11.10.2. Now simulate (or plot) the counting distribution using values of μ_1 that are adjusted such that the count mean *after* dead-time modification is $E[\mu_R] T = 15$ for all values of $\mu_1 \tau_e$. Compare the distributions.

11.10.3. Explain qualitatively how the count mean and variance would differ if the dead-time operation were imposed on a Poisson process with a slowly varying fractal rate instead of on a homogeneous Poisson process.

11.10.4. Simulate the counting distribution for a decimated Poisson process with $\mu_1 T = 15$ and m = 1, 0.5, 1.5, and 2.0. Confirm that the same curve emerges from Eq. (11.58) for m = 1 and 2 (this equation is suitable only for integer values of m). Confirm that the count mean and count variance follow $E[\mu_R] T \approx \mu_1 T/m$ and $\mathrm{Var}[n] \approx \mu_1 T/m^2$, respectively.

11.10.5. Finally, simulate the decimated-Poisson process using values of μ_1 that are adjusted such that the count mean *after* decimation is $E[\mu_R] T = 15$ for all values of m. Compare the distributions.

11.11 *Amplification of fractal behavior via exponentialization*   The fractal behavior inherent in a 20-hour sequence of heartbeats recorded from a normal human

---

[12] This approximation asymptotically achieves equality as time increases for this renewal process, which begins with an event. For a starting time selected randomly with respect to the process, the equality holds for all times $t > 0$.

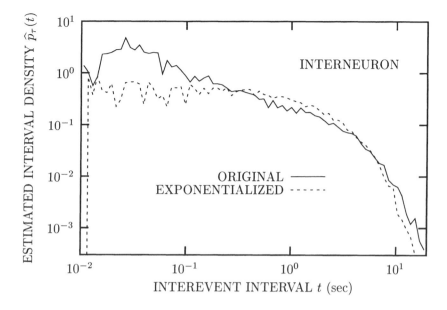

Fig. 11.16   Estimated interevent-interval density, $\widehat{p}_\tau(t)$ vs. interevent interval t, for an action-potential sequence recorded from the descending contralateral movement detector, a visual-system INTERNEURON in the locust (Turcott et al., 1995, Fig. 2, pp. 261–262, cell ADA062) (solid curve). The dotted curve represents results for an exponentialized version of these data. We show the normalized Haar-wavelet variance for these same data in Fig. 11.17.

subject (HEARTBEAT) (Turcott & Teich, 1996, data set 16273) increases considerably when we exponentialize the original point process. Comparing Figs. 11.10 and 11.11 with Figs. 5.1 and 5.2 illustrates this effect. Explain why this comes about.

11.12   *Action-potential statistics in an insect visual-system interneuron: Revisited*
The solid curve in Fig. 11.16 displays the estimated interevent-interval density for a spontaneous sequence of action potentials observed from a locust visual-system interneuron (the descending contralateral movement detector, Turcott et al., 1995). The solid curve in Fig. 11.17 shows the estimated normalized Haar-wavelet variance $\widehat{A}(T)$, plotted as a function of the counting time T, for the same set of data. These curves were first presented in Figs. 7.8 and 7.9, and Figs. B.4 and B.5, in connection with Prob. 7.8, where we demonstrated that a fractal renewal point process fails to describe this spike train in spite of the fact that the estimated interevent-interval density follows a decaying power-law form.

   11.12.1.   The dotted curve shown in Fig. 11.16 represents the exponentialized interevent-interval surrogate data discussed in Sec. 11.4.2. The shuffled surrogate discussed in Sec. 11.5 is identical to the original data since the interevent-interval density, by construction, ignores interval ordering. What conclusions can we draw from these curves with respect to the nature of the underlying point process?

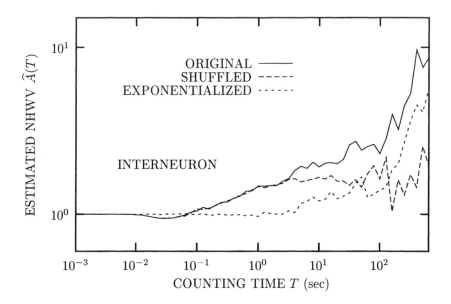

Fig. 11.17 Estimated normalized Haar-wavelet variance $\widehat{A}(T)$ vs. counting time T (sec), for an action-potential sequence recorded from the descending contralateral movement detector, a visual-system INTERNEURON in the locust (Turcott et al., 1995, Fig. 2, pp. 261–262, cell ADA062) (solid curve). Dashed and dotted curves represent results for the shuffled and exponentialized versions of these data, respectively. Unlike the display in Fig. 5.2, the abscissa reports the counting time in unnormalized form. The interevent-interval density appears in Fig. 11.16.

11.12.2. The dashed and dotted curves shown in Fig. 11.17 represent, respectively, the shuffled and exponentialized normalized Haar-wavelet variance surrogate data. What conclusions about the underlying point process can we draw from the behavior of these curves?

11.12.3. Verify that the sum of the shuffled and exponentialized *excess* normalized Haar-wavelet variances, $\widehat{A}(T) - 1$, closely approximates the *excess* normalized Haar-wavelet variance of the original data. What does this signify about the ability of these two surrogates to achieve their intended goals?

11.12.4. Based on all of this information, speculate on the form of a suitable point-process model for characterizing the spike train from the descending contralateral movement detector in the locust.

11.12.5. Figure 11.18 displays generalized-dimension scaling functions for the interneuron data for $q = -1, 0, \frac{1}{2}, 1$, and 2. Do these curves support your speculation with respect to a suitable model for the point process, as considered in Prob. 11.12.4?

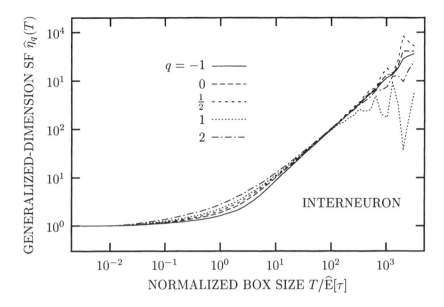

Fig. 11.18 Generalized-dimension scaling functions (SF) $\widehat{\eta}_q(T)$ for various values of q, based on Eq. (3.73), for an action-potential sequence recorded from the descending contralateral movement detector, a visual-system INTERNEURON in the locust (Turcott et al., 1995, cell ADA062). The curves resemble each other, and also those for the nonfractal homogeneous Poisson process (see Fig. B.3). The interevent-interval density and the normalized Haar-wavelet variance for these same data, along with their surrogates, are displayed in Figs. 11.16 and 11.17, respectively. A fractal-rate point process describes these data. Analogous curves for neurotransmitter exocytosis at a SYNAPSE appear in Fig. 5.11.

12

Analysis and Estimation

In the 1950s, **Harold Edwin Hurst (1880–1978)**, a British physicist who spent a great deal of his life in Egypt, developed rescaled range analysis, a statistical measure that revealed long-term dependence and the absence of a characteristic scale in the outflows of the River Nile.

Working with James A. Barnes in the 1960s, the American physicist **David W. Allan (born 1936)** introduced a measure of variability based on differences between successive numbers of events; computation of the Allan variance makes use of Haar wavelets.

12.1	Identification of Fractal-Based Point Processes	271
12.2	Fractal Parameter Estimation	273
	12.2.1 Nonparametric estimation	273
	12.2.2 Example: Fractal exponent of the human heartbeat	274
	12.2.3 Simulation and fractal-exponent estimation	275
12.3	Performance of Various Measures	281
	12.3.1 Limitations of measures not based on counts	281
	12.3.2 Normalized variance	282
	12.3.3 Count autocorrelation	285
	12.3.4 Rescaled range	287
	12.3.5 Detrended fluctuation analysis	289
	12.3.6 Interval wavelet variance	291
	12.3.7 Interval spectrum	294
	12.3.8 Normalized Haar-wavelet variance	296
	12.3.9 Rate spectrum	304
12.4	Comparison of Measures	309
	Problems	310

Given a segment of a fractal-based point process, it is often desirable to identify the point-process model that predicts the data as closely as possible; this can help to elucidate the mechanisms that underlie the data. In particular, we often wish to estimate the fractal exponent that characterizes the data.

Because of the sparseness of point-process data, however, the identification of a point process is a difficult enterprise. Under many circumstances it cannot be achieved even in principle. Nevertheless, a few situations exist in which we can identify the underlying point process by using a number of statistics in concert. More often, as we will demonstrate in Sec. 12.1 by example, only partial identification can be achieved by analyzing the point process.

Since the estimation problem is far more amenable to solution than is the identification problem, we devote the lion's share of this chapter to investigating procedures for estimating fractal-related parameters associated with an arbitrary fractal-based point process. As discussed in Sec. 12.2.1, we take a nonparametric approach to estimation, making no assumptions about the nature of the point process. We devote particular attention to obtaining estimates for the fractal exponent $\widehat{\alpha}$; the bias, standard deviation, and root-mean-square error of the estimators are of principal interest. We discuss heart rate variability analysis as an example in which careful estimation of the fractal exponent is important.

We obtain our results by making use of simulations of a fractal-Gaussian-process-driven Poisson process, which we introduced in Sec. 6.3.3, using a typical set of parameters. We make use of this process because of its widespread applicability, as discussed in Sec. 12.2.3. The simulations enable us to compare the performance of a large collection of estimators, as detailed in Sec. 12.3. Within the bounds of our study, as discussed in Sec. 12.4, the normalized Haar-wavelet variance $\widehat{A}(T)$ and the rate spectrum $\widehat{S}_\lambda(f, T)$ emerge as measures of choice for estimating fractal exponents associated with unidentified fractal-based point processes.

For the particular set of parameters examined, we achieve optimal fractal-exponent estimation using the normalized Haar-wavelet variance $\widehat{A}(T) - 1$ with five geometrically spaced counting times per decade, weighting of these counting times by $1/\sqrt{T}$, and oversampling of the point process by a factor of two. Optimal estimation using the rate-based spectrum is achieved by with $\widehat{S}_\lambda(f, T) - \widehat{E}[\mu]$ and calls for the use of uniformly spaced frequency intervals. The performance of the normalized Haar-wavelet variance slightly exceeds that of the rate-based spectrum, albeit at increased computational cost.

Although we expect the estimation results established in this chapter to prove useful in most circumstances, we caution the reader that the *a priori* information available within the realm of all possible experimental scenarios spans far too wide a range for one approach to yield the best results in all cases.

12.1 IDENTIFICATION OF FRACTAL-BASED POINT PROCESSES

We have discussed the identification problem from a simple perspective in Secs. 5.5.4 and 11.5.3, and in a number of Problems sprinkled throughout the text.[1] We will again revisit the point-process identification issue in Sec. 13.6, in connection with computer network traffic.

The identification of the underlying point-process associated with an observed point process is possible only in special cases. More often, the analysis of a point process leads to the identification of only some of its features. Such partial identification is highlighted by the following examples:

- If the estimated rate spectrum (Sec. 3.4.5) of the process, $\widehat{S}_\lambda(f, T)$, strongly indicates the presence fractal behavior, whereas the estimated interval spectrum (Sec. 3.3.3) of the process, $\widehat{S}_\tau(f)$, does not, then the point process under study may well belong to the family of fractal renewal processes, as discussed in Sec. 5.5.4.

  We can confirm this presumption by shuffling the intervals and then recomputing the spectra, which provides a nonparametric test of this hypothesis (see Sec. 11.5). A renewal process, whether fractal or not, is invariant to such shuffling since its interevent intervals are independent and identically distributed. Surrogate data analysis permits us to achieve a good measure of discrimination between fractal and fractal-rate point processes, as discussed in Sec. 11.5.3. It also generally permits the separation of the fractal and nonfractal features of a point process.

---

[1] See, in particular, Probs. 5.2, 5.3, 5.4, 5.5, 6.8, 7.7, 7.8, 7.9, 7.10, 10.6, 10.7, 10.8, 10.9, 11.3, and 11.12.

The estimated generalized dimension $\widehat{D}_q$ (Sec. 3.5.4) frequently offers a means for definitively discriminating between fractal-rate and fractal point processes (Prob. 5.5).

- In the domain of doubly stochastic Poisson processes (Sec. 4.3), a fractal-binomial-noise-driven Poisson process $dN(t)$ (Sec. 8.4) comprising a small number M of component alternating fractal renewal processes $X(t)$ exhibits a particular interevent interval probability density $p_\tau(t)$. When the underlying fractal binomial noise $X_\Sigma(t)$ (Sec. 8.3.1) assumes a value of zero, no events contribute to $dN(t)$. Otherwise, $dN(t)$ comprises a locally homogeneous Poisson process, with relatively few intervals spanning changes in the rate; this yields an interval density that consists of sums of exponentials.

  The overall interval density $p_\tau(t)$ therefore approximates exponential decay for short times and power-law decay for longer times. This behavior is most evident when the mean rate of $dN(t)$ greatly exceeds the mean switching rates of the constituent alternating renewal processes $X(t)$, and it requires that at least the off times for $X(t)$ have a power-law distribution.

- Similarly, the interevent-interval density of a fractal-shot-noise-driven Poisson process (Sec. 10.3) can sometimes reveal the statistics of the times between the primary events. When the rate of the primary Poisson process is sufficiently small in comparison with the durations of the impulse response functions, significant periods of time elapse between the termination of one impulse response function and the arrival of the next primary event. During this period, no intervals can occur. Therefore, $p_\tau(t)$ essentially follows the form of the primary Poisson process, as provided in Eq. (10.22).

We also note some of the manifold difficulties associated with point-process identification:

- In general, no single statistic is sufficient to identify or characterize a fractal-based point process.

- Even the partial identification of a fractal-based point process generally requires a large quantity of data. Consider the fractal-shot-noise-driven Poisson process discussed above as an example. The interval density $p_\tau(t)$ for $\beta = \frac{1}{2}$ (dotted curve in Fig. 10.6) differs from the homogeneous-Poisson-process density [Eq. (4.3)] only for the largest 10^{-10} of the intervals.

- A general difficulty associated with identifying fractal-rate point processes stems from the fact that the process of generating point events from a rate process destroys information about fluctuations in the rate that occur over time scales shorter than the local interevent interval. Hence, the sampling eradicates detail that is an intrinsic part of the rate process (see Sec. 5.5.4).

  As an example, one can choose parameters for which it is quite easy to distinguish segments of: (1) a fractal Gaussian process (Sec. 6.3.3), (2) fractal binomial noise (Sec. 8.3.1), and (3) fractal shot noise (Chapter 9). However, when used as rates for a Poisson process, these three diverse rate functions can yield essentially indistinguishable fractal-rate point processes (but see Prob. 12.1).

As another example, given the superposition of a fractal doubly stochastic Poisson process and a homogeneous Poisson process, it is nearly impossible to distinguish the individual component processes.

- The identification of fractal-based point processes is confounded by the large variety of forms that they take, many of which have quite similar statistics. One can easily construct a collection of distinct processes that do not belong to any of the families that we have considered. For example, we can apply the block-shuffling operation (Sec. 11.5.1) to a fractal-shot-noise-driven Poisson process.

The identification problem proves a bit less difficult for the family of integrate-and-reset point processes (Sec. 4.4), by virtue of their deterministic kernel. Nevertheless, just as with the Poisson kernel, all information resident in the rate over time scales significantly shorter than the local interevent time is not carried forward to the ensuing point process.

12.2 FRACTAL PARAMETER ESTIMATION

We turn now to parameter estimation for fractal-based point processes. This is a far easier task than identification.

Well-established techniques are available for estimating various conventional measures of a point process, such as the mean and variance of the interevent intervals, the mean rate of the process, and more complex measures such as the spectrum (see, for example, Cox & Lewis, 1966). We expressly consider the *estimation of fractal parameters* in point processes, a topic that has received scant attention.

12.2.1 Nonparametric estimation

We cast the estimation problem as follows: Given a segment of a fractal-rate point process, we seek an estimate $\widehat{\alpha}$ of the true fractal exponent α of the entire process from which the segment was extracted. We often seek an estimate of the onset time or frequency as well. Several effects contribute to estimation error for finite-length segments, regardless of the methods used (Lowen & Teich, 1995; Thurner et al., 1997).

The fractal exponent provides a measure of the relative strengths of fluctuations over various time or frequency scales; a particular fractal exponent leads to a particular distribution of power over these scales. Variance in the estimated values stems from the inherent randomness of the strengths of fluctuations in finite data sets. A collection of finite realizations of a fractal-based point process with the same parameters will exhibit fractal fluctuations of varying strengths, thereby leading to a distribution of estimated fractal exponents and onset times or frequencies.

Whereas fluctuations lead to variance in estimators, cutoffs lead to bias. Cutoffs arise from limitations in the measurement process: noise and finite precision lead to a minimum practical time scale, whereas the limited duration of any data set defines a maximum time scale (see Sec. 2.3.1). Cutoffs can also derive from the data itself, such as when behavior with different characteristics (nonfractal or even fractal with a different exponent) occurs over adjacent time-scale ranges.

Although algorithms exist that can accurately compensate for the effects of variance and bias, they presuppose detailed *a priori* knowledge of the process, which violates the principle of estimating an unknown signal. We therefore do not attempt to compensate for these effects in this manner, except for very simple cases justified by previous knowledge of the system that generates the data. Indeed, we do not attempt to specifically estimate parameters for a fractal point process; rather we employ the formalism of a fractal-rate point process, which is encountered far more frequently.

In particular, we do not consider maximum-likelihood techniques because they require an exact model, despite the fact that they yield the best possible performance in a certain sense. By definition, applying one model to data generated by another model no longer yields the maximum-likelihood estimate of a parameter; instead it returns an estimate of unknown utility. Furthermore, even when knowledge of a model does permit the use of maximum-likelihood methods, the loss of accuracy entailed in employing a nonparametric estimation technique is small (Veitch & Abry, 1999). We therefore eschew maximum-likelihood and related techniques in favor of a nonparametric approach.

A closely related issue pertains to selecting the range of times or frequencies over which to calculate a fractal measure. Given a monofractal model, a χ^2-approach provides a powerful tool for automatically selecting the appropriate power-law region (Abry et al., 2000). Indeed, selection of the appropriate scaling range proves both important and nontrivial in the general case (see Abry et al., 2000, 2003, for a discussion of this issue). However, since this selection also depends on the decision criterion (χ^2 limit vs. range of dependent variable), we provide estimates of fractal exponents over several different ranges of time or frequency.

As an aside, such *a priori* knowledge also plays an important role in estimating fractal exponents for discrete-time processes. Some estimators have superb characteristics when applied to data generated from a restricted class of models but perform poorly in general, whereas robust estimators yield good (but not superb) results in many cases (Taqqu, Teverovsky & Willinger, 1995). Often a speed/accuracy tradeoff exists as well. Again, detailed knowledge of the underlying model permits the use of an estimation technique tailored to the data at hand.

12.2.2 Example: Fractal exponent of the human heartbeat

The human heartbeat provides an example of the usefulness of estimating the fractal exponent of a point process. Fluctuations in the sequence of heartbeat interevent intervals, over time scales ranging from minutes to hours, can help assess the presence of, and likelihood of acquiring, cardiovascular disease (Malik et al., 1996). This noninvasive approach has come to be called **heart rate variability** (HRV) analysis

(Hon & Lee, 1965), whether attention is directed to fluctuations of the actual heart rate or of the interevent intervals.

A whole host of heart-rate-variability measures have been developed and examined over the years. Although the vast majority of these are nonfractal (scale dependent) in nature, the fractal exponents $\widehat{\alpha}$ associated with a number of measures have become a part of the armamentarium used for the analysis of the heartbeat point process. These measures include the interval-based spectrum $\widehat{S}_\tau(\mathsf{f})$ (Kobayashi & Musha, 1982; Turcott & Teich, 1996), the normalized interval-based Haar-wavelet variance $\widehat{A}_\tau(k)$ (Thurner et al., 1998; Ashkenazy et al., 1998), the rate spectrum $\widehat{S}_\lambda(f, T)$ (Turcott & Teich, 1996), and the normalized Haar-wavelet count variance $\widehat{A}(T)$ (Turcott & Teich, 1993, 1996).

Heart-rate-variability measures that can discriminate patients with congestive heart failure from normal subjects have received particular attention. Teich et al. (2001) recently carried out such a study using 16 measures, of which five were scaling exponents: $\widehat{\alpha}_{A\tau}$, $\widehat{\alpha}_{S\tau}$, $\widehat{\alpha}_A$, $\widehat{\alpha}_Y$, and $\widehat{\alpha}_U$. The results of this investigation revealed that the interval-based Haar-wavelet variance at a scale near 32 heartbeat intervals, $\widehat{A}_\tau(32)$, along with its interval-based spectral counterpart near 1/32 cycles/interval, $\widehat{S}_\tau(1/32)$, are the most reliable of the measures, even for electrocardiogram records just minutes long. However, some evidence suggests that the scaling exponents $\widehat{\alpha}_{A\tau}$ and $\widehat{\alpha}_Y$ outperform scale-dependent measures for predicting *mortality* following myocardial infarction (Ashkenazy et al., 2001). Hence, heart-rate-variability analysis provides an example in which careful estimation of the fractal exponent is an important task.

12.2.3   Simulation and fractal-exponent estimation

To illustrate how the fractal exponent of a point process may be estimated using nonparametric techniques, we turn to a specific, but important, example: the **fractal-Gaussian-process-driven Poisson process** introduced in Sec. 6.3.3. This point process has widespread applicability and therefore provides a good testbed for our analysis.

We generate the fractal Gaussian driving process via a simple spectral method (Peitgen & Saupe, 1988) that enables us to use the fast Fourier transform with an array of size $M = 2^{17} = 131\,072$. After the inverse Fourier transform, we discard half of the array to reduce periodicity effects, leaving $M/2 = 2^{16} = 65\,536$ elements. We use a simulation duration $L_0 = 1.1 \times 10^4$. Each element of the resulting fractal Gaussian process therefore corresponds to a duration of $2L_0/M \doteq 0.167847$, and serves as the rate for a (locally) homogeneous Poisson process for that same duration. Thus, every $2L_0/M$ time units, the rate changes to the value specified by the next element in the array. While this method yields a good approximation to a fractal Gaussian process over long time scales, the piecewise-constant construction essentially eliminates fluctuations over time scales significantly shorter than $2L_0/M$.

As discussed in Secs. 6.1 and 6.2, Gaussian processes in general, and fractal Gaussian processes in particular, can assume negative values. These are not permitted for point-process rate functions and must be eliminated. Eschewing nonlinear methods for dealing with this issue, we instead choose a mean rate that yields positive elements in all simulations. For the example at hand, we chose the parameters as follows: mean rate $E[\mu] = 100$, duration $L_0 = 1.1 \times 10^4$, fractal exponent $\alpha = 0.8$, and onset frequency $f_S = 0.2$. The resulting rate has a coefficient of variation $C_\mu \doteq 0.176181$. For $M/2 = 65\,536$ and 100 simulations, the complementary error function provides that all rates lie above zero with a probability greater than 0.95, and indeed this was the case for our simulations.[2] For some of the interval-based measures we examine, it proves simpler to employ data sets for which the number of intervals is an integral power of two. The simulations have an expected number of events equal to $1\,100\,000$; of the 100 simulations generated, the one with the least number of events has $1\,067\,365$. We therefore retain the first $2^{20} = 1\,048\,576$ of the intervals in each of the 100 simulations. A further rationale for the choice of these particular values will emerge as the chapter unfolds. These truncated simulations have total durations L that vary; by construction, none can exceed the value $L_0 = 11\,000$ employed in the original simulations. We measure $\widehat{E}[L] = 10\,511.2$ (yielding $\widehat{E}[\tau] = \widehat{E}[L]/N = 10\,511.2/1\,048\,576 \doteq 0.010024$) and $\widehat{\sigma}_L = 187.0$. Finally, within simulations the variability in the rate slightly favors small and large intervals at the expense of the mean, yielding an average interval coefficient of variation $\widehat{E}[C_\tau] = 1.0327$. For the homogeneous Poisson process, in contrast, such variability does not occur, and we have $C_\tau = 1$ exactly (see Sec. 4.1).

We begin by examining the estimation of the fractal exponent via the normalized Haar-wavelet variance $\widehat{A}(T)$. As discussed in Sec. 3.4.3, Allan and Barnes introduced the unnormalized version of this measure for discrete-time processes in 1966. The presentation in this section is intended to serve as an example that provides a general format for carrying out the analysis. The performance of other measures set forth in Chapter 3 will follow in subsequent sections of this chapter. We will carry out a more thorough study of fractal-exponent estimation via $\widehat{A}(T)$ in Sec. 12.3.8.

To first order, the estimated normalized Haar-wavelet variance should increase as a power-law function of the counting time, which, according to Eq. (5.2), obeys

$$\widehat{A}(T) \approx (T/T_A)^\alpha. \tag{12.1}$$

A representation that is more suitable in many cases follows the relation

$$\widehat{A}(T) \approx 1 + (T/T_A)^\alpha, \tag{12.2}$$

as provided in Eq. (5.6). We performed the simulations detailed above, and estimated the normalized Haar-wavelet variance $\widehat{A}(T)$ at a sequence of counting times T for each of the 100 runs separately. As with prior displays of the normalized Haar-wavelet variance, we chose these to increase geometrically by factors of $10^{0.1}$, thereby providing 10 counting times per decade of the overall counting-time range considered.

---

[2] The strong correlations among the rates makes this probability still closer to unity.

NORMALIZED HAAR-WAVELET VARIANCE

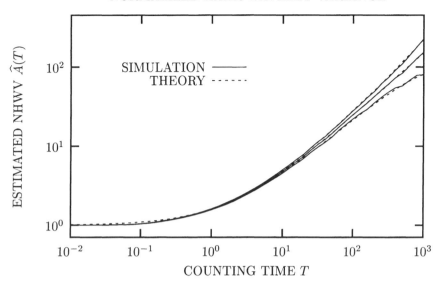

Fig. 12.1 Estimated normalized Haar-wavelet variance $\widehat{A}(T)$ vs. counting time T for a fractal-Gaussian-process-driven Poisson process. Using different random seeds, we simulated 100 runs of this process and estimated the normalized Haar-wavelet variance for each. We present the mean value (center solid curve) and mean $\pm$ one standard deviation (upper and lower solid curves) for these simulations. The corresponding three theoretical (dashed) curves derive from Eqs. (5.44), (5.45), (12.21), and (12.25). Note the increase in the uncertainty as the counting time T increases, and the slight dip in the simulation outcome near $T = 0.1$.

Figure 12.1 presents the corresponding simulation results. The solid curves correspond to values calculated from these individual $\widehat{A}(T)$ estimates: mean (center solid curve) and mean $\pm$ one standard deviation (upper and lower solid curves).

The dashed curves in Fig. 12.1 display theoretical results. Equations (5.44) and (5.45) yield $T_A \doteq 1.8939$; the dashed curve at the center corresponds to the representation provided in Eq. (12.2). Results established in Sec. 12.3.8, our more thorough study of estimation via the normalized Haar-wavelet variance, provide the upper and lower dashed curves, which represent theoretical values for the mean $\pm$ one standard deviation [see Eq. (12.25)].

As with all finite data sets, increasing the counting time T yields fewer members of the sequence of counts, $\{Z_k(T)\}$, which, in turn, leads to increased variability in the estimated normalized Haar-wavelet variance, $\widehat{A}(T)$. This results in a wider separation between the mean $\pm$ one standard deviation curves in Fig. 12.1 for larger values of the counting time T. For this reason, employing values of $\widehat{A}(T)$ near $T = L$ has dubious value; we typically set an upper limit $T \leq T_{\max} \equiv L/10$ for all count-based statistics.

Time Range		$\widehat{\alpha}_A$ from $\widehat{A}(T)$			$\widehat{\alpha}_{A-1}$ from $\widehat{A}(T) - 1$		
$T_{\min}$	$T_{\max}$	Bias	SD	RMSE	Bias	SD	RMSE
10^0	10^1	-0.321	0.019	0.321	0.007	0.028	0.029
10^0	10^2	-0.193	0.020	0.194	0.001	0.023	0.023
10^0	10^3	-0.128	0.039	0.134	-0.009	0.039	0.040
10^1	10^2	-0.082	0.052	0.097	0.000	0.056	0.056
10^1	10^3	-0.061	0.077	0.098	-0.018	0.078	0.080
10^2	10^3	-0.060	0.197	0.206	-0.046	0.199	0.204

Table 12.1 Performance of fractal-exponent estimates $\widehat{\alpha}$ for 100 different simulations of a fractal-Gaussian-process-driven Poisson process, obtained via estimates of the normalized Haar-wavelet variance $\widehat{A}(T)$. We chose the counting times to increase geometrically by factors of $10^{0.1}$, thereby providing 10 counting times per decade of the overall counting-time range used to obtain the estimate; the first two columns lists this range. The next three columns represent the bias, standard deviation (SD), and resulting root-mean-square error (RMSE) for calculations carried out on the logarithm of the estimated normalized Haar-wavelet variance $\widehat{A}(T)$; the final three columns display the same calculations carried out on $\widehat{A}(T) - 1$. Best results (least root-mean-square error) obtain by using the counting-time range $10^1 \leq T \leq 10^2$ for $\widehat{A}(T)$, and by using $10^0 \leq T \leq 10^2$ for $\widehat{A}(T) - 1$.

Apart from statistical deviations about the mean, the mean value itself also deviates from ideal fractal behavior, as a result of the method for constructing the fractal Gaussian process used in the simulation. As indicated above, we employ a discrete-time approximation; this leads to a generated process that remains fixed for periods of $2L/M$, or approximately 0.160 time units, bearing in mind that L differs among the simulations. For counting times T significantly less than $2L/M$, the rate rarely changes between counts $Z_k(T)$ in adjacent counting windows. This leads to results that differ little from those of a homogeneous Poisson process, and provides an explanation for why the simulation (solid curve) lies closer to unity than the simple theoretical result (dashed curve) for small counting times T (this slight difference appears in many other measures presented in this chapter as well). Hence, for this process with small counting times, $A(T) \approx 1$ provides a somewhat better approximation than does Eq. (12.2). We can obtain further improvement by making use of an empirical fit that more precisely accommodates the piecewise-constant construction of the fractal Gaussian process, namely

$$A(T) = 1 + \frac{(T/T_A)^\alpha}{1 + (T/T_1)^{-2}}, \qquad (12.3)$$

with $T_1 \doteq 0.124$. Equation (12.3) follows the simulation results almost perfectly (not shown).

The deviations from ideal scaling behavior displayed in graphical form in Fig. 12.1 directly carry over to estimates $\widehat{\alpha}$ of the fractal exponent. We examined the statistics of the fractal exponents estimated from individual simulations, rather than from the aggregate results shown in Fig. 12.1. Specifically, for each simulation we calculated the logarithms of the estimated normalized Haar-wavelet variance $\widehat{A}(T)$ and of the counting time T, obtained the least-squares best fit of a straight line to this curve over a specified range of counting times T, and called the resulting slope $\widehat{\alpha}$. (Fitting the logarithms of the values to a linear function turns out to be superior to fitting the original values to a power-law function; see Sec. 12.3.8.)

We then determined the statistics of $\widehat{\alpha}$ over the 100 simulations, namely its sample bias

$$\widehat{\mathrm{E}}[\widehat{\alpha} - \alpha],$$

standard deviation (SD)

$$\widehat{\mathrm{Var}}^{1/2}[\widehat{\alpha}],$$

and root-mean-square error (RMSE)

$$\widehat{\mathrm{E}}^{1/2}[(\widehat{\alpha} - \alpha)^2],$$

which is equal to the square root of the sum of the squares of the standard deviation and the bias. We followed this procedure for various ranges of the counting time, spanning 10^0 through 10^3 in decade steps. Nonfractal properties of the data set, such as dead time, often heavily influence behavior for counting times substantially smaller than T_A; for this reason, we do not consider $T_{\min} < 1$. Finally, we repeated the entire procedure for the logarithm of $[\widehat{A}(T) - 1]$, rather than of $\widehat{A}(T)$. Table 12.1 presents the results of this procedure.

Examination of the third column in Table 12.1 reveals the strong negative bias that emerges by using $\widehat{A}(T)$ directly. The form of Eq. (12.2) implies that the slope on a doubly logarithmic plot will lie significantly below α, because of the presence of the unity term; this occurs for all ranges of the counting time, but is most apparent for the shorter ones. This observation accords with the curved appearance of Fig. 12.1. Were $\widehat{A}(T)$ to follow a pure power-law form, as in Eq. (12.1), Fig. 12.1 would behave as a straight line. The standard deviation of $\widehat{\alpha}$ (fourth column) increases with counting time, in concert with the increase in the standard deviation of $\widehat{A}(T)$. Taken together, the bias and standard deviation of $\widehat{\alpha}$ result in a root-mean-square error that never lies below 0.09, largely as a result of the bias, which, in turn, arises from the constant unity term.

Subtracting a constant term of unity yields better estimates for the fractal exponent. The magnitude of the bias decreases markedly, although the bias still proves significant for the largest and smallest counting times. For the smallest of these, the departure of $\widehat{A}(T)$ from the simple form of Eq. (12.2), resulting from the discrete-time approximation employed for the fractal Gaussian process, gives rise to the slight dip in the simulation outcome (relative to theory) in Fig. 12.1 near $T = 0.1$. This, in turn, leads to an estimated slope *larger* than α, and therefore an estimated fractal exponent $\widehat{\alpha} > \alpha$.

Turning to the largest counting times, all bias calculations including $T = 10^3$ exhibit negative values. In all three cases the magnitude of the bias exceeds the standard error (standard deviation divided by the square root of the number of trials) by a factor of two or three, and the bias values appear approximately to follow a Gaussian distribution, making this difference highly significant. This likely arises from periodicity effects in the simulated fractal Gaussian process array; the circular nature of the original array of size M reduces the variance below the value that would obtain for a true (nonperiodic) fractal Gaussian process. While eliminating half of the array reduces this effect, it nevertheless remains statistically significant, particularly for longer times.[3] This deviation also appears at the right-most edge of Fig. 12.1, where the simulated mean $\widehat{A}(T)$ curve lies slightly below its theoretical value.

Intermediate counting times yield excellent bias values, which are quite small for $10^0 \leq T \leq 10^2$; this range of counting times yields the best overall performance as well, with a root-mean-square error of only 0.023. Finally, we note that the bias values, while significantly different from zero, do not significantly affect the root-mean-square error values; they cause an increase of less than four percent of the value that would obtain from the standard deviation alone.

Using the functional form provided in Eq. (12.3) to model the results in Fig. 12.1 would provide still better estimates $\widehat{\alpha}$. However, as indicated in Sec. 12.2.1, this violates the spirit of analyzing an unknown process. Indeed, subtracting unity from the estimated normalized Haar-wavelet variance is truly acceptable only if we know that the process approximately follows the form of Eq. (12.2). In fact, replacing the Poisson kernel with an integrate-and-reset process yields an estimated normalized Haar-wavelet variance that more closely follows Eq. (12.1) for intermediate counting times. Subtracting unity from such values of $\widehat{A}(T)$ would then lead to a *positive* bias in the estimate $\widehat{\alpha}$, and would likely even lead to negative values for $\widehat{A}(T) - 1$ over some range of counting times. Certainly, progressively more complex models for $\widehat{A}(T)$ yield progressively better estimates $\widehat{\alpha}$ when extensive *a priori* knowledge of the process exists [see Bardet, Lang, Moulines & Soulier (2000) and Bardet et al. (2003) for a sophisticated treatment]. However, these models become progressively less tenable in the absence of such knowledge. In this chapter we make use of both Eqs. (12.1) and (12.2) in estimating the fractal exponent.

We conclude this section by noting that other simulation methods can provide results that follow Eq. (12.2) even more closely. In later sections of this chapter, where we evaluate the performance of various estimation approaches, it turns out that the slight difference between simulated and ideal behavior does not affect the performance measures significantly more (or less) than it does for the normalized Haar-wavelet variance examined here. These differences give rise to estimated fractal exponents with expected values that differ from the nominal value of 0.8. This leads to a small apparent bias, but as shown in the right-most three columns of Table 12.1,

---

[3] A further systematic source of error derives from the fact that the expectation of the logarithm differs from the logarithm of the expectation. This difference, which arises in all exponent estimates based on doubly logarithmic plots, can prove important when statistics become sparse, such as at the longest counting times for $\widehat{A}(T)$. In this particular case, explicit forms are available for bias correction (Veitch & Abry, 1999). However, these forms are valid only for Gaussian-distributed counts, and the calculated values overestimate the bias by a factor of three for our simulations.

we may generally neglect the effects of the bias, and therefore the effects of non-ideal fractal behavior, for the purposes of evaluating various point-process measures.

Furthermore, the results developed in this section illustrate some of the issues involved in analyzing non-ideal point processes. The simulation approach presented here is therefore instructive since most real data sets depart from ideal fractal behavior to some extent.

We now proceed to use the same simulated data to evaluate the performance of various other measures.

12.3 PERFORMANCE OF VARIOUS MEASURES

12.3.1 Limitations of measures not based on counts

Of the collection of measures presented in Chapter 3, some prove more useful than others for estimating the parameters of fractal behavior such as fractal exponents and onset times or frequencies. As suggested in Sec. 5.5.1, we can cast aside interval-based measures from consideration when analyzing a general fractal-based point process. These measures generally prove less useful than count-based and point-process-based measures because they reliably reveal fractal behavior only in special cases.

To establish this, we examine the results of applying interval-based measures to two specific fractal-based point processes, and demonstrate that all such measures fail with one process or the other.

First we consider a fractal renewal process (Chapter 7). As a member of the renewal point-process family, the interevent intervals are independent so that all of the measures set forth in Secs. 3.3.2–3.3.6 return simple, nonfractal forms. These measures thus *cannot* distinguish between a fractal renewal process and a nonfractal one such as the homogeneous Poisson process. Figure 5.4b) illustrates the indistinguishability of the rescaled range statistic for these two processes. The interevent interval density (or distribution) set forth in Sec. 3.3.1 forms the sole exception. Since it readily reveals fractal behavior in fractal renewal processes it *can* distinguish the two forms.

As the second example, we consider a doubly stochastic Poisson process driven by a fractal Gaussian process (Sec. 6.3.3) whose rate has a low coefficient of variation, $C_\mu \ll 1$. In this case, all of the measures set forth in Secs. 3.3.2–3.3.6 *can* serve to discriminate between this process and a homogeneous Poisson process. However, the interevent-interval density and distribution *cannot* distinguish between them; although the interval ordering differs, quite similar interval densities characterize the two processes [see Eqs. (4.33) and (4.3)]. Figure 5.5a) illustrates the indistinguishability of the interval densities for these two processes.

We conclude that the statistics in Secs. 3.3.2–3.3.6 fail for the former process whereas those in Sec. 3.3.1 fail for the latter, so that no interevent-interval statistic proves useful in all cases. Even in cases where dependencies among intervals do exist, the measures in Secs. 3.3.2–3.3.6 can suffer from problems of interpretation since these measures describe the point process in terms of a time axis that is warped with respect to real time, as shown in Fig. 3.1e). In particular, the interval-based auto-correlation and spectrum reliably describe fractal behavior only when the coefficient of variation of the intervals is small (DeBoer et al., 1984; Turcott & Teich, 1996).

Nevertheless, some interval-based measures can be useful for analyzing point processes whose fractal characteristics derive only from the relative ordering of the interevent intervals, and not from their distribution. This holds for the general doubly stochastic Poisson process, results for which we report throughout this chapter. We therefore include four often-used interval-based measures within our purview: rescaled range analysis (Sec. 3.3.5), detrended fluctuation analysis (Sec. 3.3.6), the interval-based wavelet variance (Sec. 3.3.4), and the interval-based spectrum (periodogram) (Sec. 3.3.3). We continue to bear in mind, however, that these measures do not always yield reliable results for general fractal-based point processes.

The point-process-based measures presented in Sec. 3.5 also prove impractical in the general case. Generalized dimensions for fractal-rate point processes assume integer (nonfractal) values, as discussed in Sec. 3.5.4, whereas the remaining measures set forth in Sec. 3.5 all have count-based analogs (some with superior statistics) that prove far easier to calculate. For example, estimates of the coincidence rate $G(t)$ based on a finite-length data set containing N points comprise $N(N-1)/2$ Dirac delta functions representing the delay times that happened to occur in that particular data set, and another N delta functions at zero delay time. Increasing the number of points in the data set yields a greater number of delta functions, but does not cause the coincidence-rate estimate to converge to a smooth form. One solution is to bin the events. Rather than using the point process $dN(t)$ to generate the coincidence rate $G(t)$, this procedure effectively makes use of the counting process $\{Z_k(T)\}$ to construct an autocorrelation $R_Z(k,T)$, which yields smooth results for finite data sets. Indeed, Eq. (3.54) illustrates how the autocorrelation derives from a smoothed version of the coincidence rate (see Prob. 12.4).

We are left with four count-based measures whose merit we wish to assess for the estimation problem at hand: the normalized variance, normalized Haar-wavelet variance, autocorrelation, and rate spectrum. In addition we study the performance of the four interval-based measures indicated immediately above. We proceed to examine these in turn.

12.3.2 Normalized variance

As mentioned in Sec. 3.4.2, the normalized variance suffers from bias and thus proves problematical as a measure. To explicitly demonstrate this, we begin with the estimate of the count-sequence variance,

$$\widehat{\mathrm{Var}}[Z(T)] = (M-1)^{-1} \sum_{n=0}^{M-1} \left\{ Z_n(T) - \widehat{\mathrm{E}}[Z(T)] \right\}^2$$

$$= (M-1)^{-1} \sum_{n=0}^{M-1} \left\{ Z_n(T) - M^{-1} \sum_{m=0}^{M-1} Z_m(T) \right\}^2$$

$$= (M-1)^{-1} \sum_{n=0}^{M-1} Z_n^2(T)$$

$$- M^{-1}(M-1)^{-1} \sum_{m=0}^{M-1} \sum_{n=0}^{M-1} Z_m(T) Z_n(T)$$

$$= M^{-1} \sum_{n=0}^{M-1} Z_n^2(T)$$

$$- M^{-1}(M-1)^{-1} \sum_{m=0}^{M-1} \sum_{n \neq m}^{M-1} Z_m(T) Z_n(T), \quad (12.4)$$

where $M = \text{int}(L/T)$ represents the number of counts and $\text{int}(x)$ is the largest integer not exceeding x.

The count-sequence variance has an expected value given by

$$E\{\widehat{\text{Var}}[Z(T)]\}$$

$$= M^{-1} \sum_{n=0}^{M-1} E[Z_n^2(T)]$$

$$- M^{-1}(M-1)^{-1} \sum_{m=0}^{M-1} \sum_{n \neq m}^{M-1} E[Z_n(T) Z_m(T)] \quad (12.5)$$

$$= E[Z^2(T)] - 2 \sum_{k=1}^{M} \frac{(M-k) R_Z(k,T)}{M(M-1)}.$$

To evaluate this expression, we make use of Eqs. (5.13) (for $k = 0$), (5.14), and (5.45), which yield

$$E\{\widehat{\text{Var}}[Z(T)]\}$$

$$= E^2[Z(T)] + E[Z(T)] + E[Z(T)] (T/T_F)^\alpha$$

$$- \sum_{k=1}^{M} \frac{2(M-k)}{M(M-1)} \Big\{ E^2[Z(T)] + E[Z(T)]$$

$$+ \tfrac{1}{2} E[Z(T)] (T/T_F)^\alpha \left[(k+1)^{\alpha+1} + (k-1)^{\alpha+1} - 2k^{\alpha+1} \right] \Big\}$$

$$= E[Z(T)] (T/T_F)^\alpha$$

$$\times \Big\{ 1 - \sum_{k=1}^{M} \frac{(M-k)}{M(M-1)} \left[(k+1)^{\alpha+1} + (k-1)^{\alpha+1} - 2k^{\alpha+1} \right] \Big\}$$

$$\approx E[Z(T)] (T/T_F)^\alpha$$

$$\times \Big\{ 1 - \int_{x=1}^{M} \frac{(M-x)}{M(M-1)} \left[(x+1)^{\alpha+1} + (x-1)^{\alpha+1} - 2x^{\alpha+1} \right] dx \Big\}$$

$$\approx\ \mathrm{E}[Z(T)]\,(T/T_F)^{\alpha}$$

$$\times \left\{ 1 - \int_{x=0}^{M} \frac{(M-x)}{M^2} \left[\alpha(\alpha+1)\, x^{\alpha-1} \right]\, dx \right\}$$

$$=\ \mathrm{E}[Z(T)]\,(T/T_F)^{\alpha} \left[1 - M^{\alpha-1} \right]$$

$$=\ \mathrm{E}[Z(T)]\,(T/T_F)^{\alpha} \left[1 - (T/L)^{1-\alpha} \right]. \tag{12.6}$$

The bias in Eq. (12.6) carries over to the estimate of the normalized variance $\widehat{F}(T)$ as a similar bias. The result is a spuriously low fractal-exponent estimate $\widehat{\alpha}$. This appears in the simulation results presented in Fig. 12.2 and Table 12.2, which illustrate the substantial departure of this statistic from precise T^{α} behavior.

Moreover, as indicated in Sec. 5.2.3, the normalized variance cannot rise faster than T^1. As a result of the inherent bias in its estimate, and its inability to reveal fractal exponents greater than unity, we do not recommend use of the normalized variance $\widehat{F}(T)$. The normalized Haar-wavelet variance $\widehat{A}(T)$ is a far superior statistic, as comparison of Fig. 12.2 with 12.1, and Table 12.2 with 12.1, affirms.

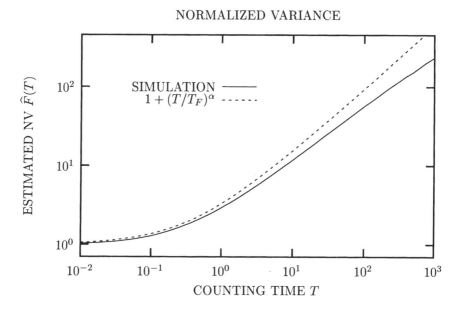

NORMALIZED VARIANCE

Fig. 12.2 Estimated normalized variance $\widehat{F}(T)$ vs. counting time T, based on the same simulations as those used to generate Fig. 12.1 (a fractal-Gaussian-process-driven Poisson process). We present the mean value (solid curve) along with the simple theoretical form provided in Eq. (5.44b) (dashed). The simulation results deviate from this ideal fractal behavior, especially for long counting times T, illustrating the substantial bias inherent in the normalized variance statistic.

Time Range		$\widehat{\alpha}_F$ from $\widehat{F}(T)$			$\widehat{\alpha}_{F-1}$ from $\widehat{F}(T) - 1$		
$T_{\min}$	$T_{\max}$	Bias	SD	RMSE	Bias	SD	RMSE
10^0	10^1	-0.191	0.022	0.192	-0.051	0.018	0.054
10^0	10^2	-0.151	0.028	0.153	-0.073	0.026	0.077
10^0	10^3	-0.159	0.050	0.167	-0.112	0.048	0.122
10^1	10^2	-0.126	0.039	0.132	-0.096	0.037	0.103
10^1	10^3	-0.167	0.074	0.183	-0.151	0.074	0.168
10^2	10^3	-0.221	0.141	0.262	-0.216	0.141	0.258

Table 12.2 Performance of fractal-exponent estimates $\widehat{\alpha}$ obtained via estimates of the normalized variance $\widehat{F}(T)$. See Table 12.1 for details. Best results (least root-mean-square error) obtain by using $10^1 \leq T \leq 10^2$ for $\widehat{F}(T)$, and by using $10^0 \leq T \leq 10^1$ for $\widehat{F}(T) - 1$. We find that $\widehat{\alpha} < \alpha$ in all cases; the deviation increases with increasing values of T. Bias is problematic for this estimator. For all ranges of counting time, the normalized Haar-wavelet variance $\widehat{A}(T)$ returns significantly better results.

12.3.3 Count autocorrelation

The autocorrelation $R_Z(k, T)$ of the counting sequence $\{Z_k(T)\}$, as a function of the delay k, forms a windowed version of the coincidence rate $G(t)$, as demonstrated by Eq. (3.54). Like the coincidence rate, the limit for large delays dominates the fractal portion for all but the smallest delays. Rearranging Eq. (5.44d) yields

$$R_Z(k, T)/\mathrm{E}[Z(T)] = \mathrm{E}[Z(T)] + (T/T_R)^\alpha \, k^{\alpha-1}. \tag{12.7}$$

Since $\alpha < 1$ (a condition that is required for this measure to demonstrate any fractal behavior), the last term decreases with increasing k, making estimates of its form difficult for all but small values of k.

We therefore use instead the autocovariance (autocorrelation minus the square of the mean), which eliminates the dominant constant term. Unlike other measures considered in this chapter, the count-based autocorrelation (or autocovariance) takes two arguments; analyses as a function of delay number k require a counting time T as a parameter. Choosing a long counting time leads to loss of information at shorter time scales. In contrast, small counting times lead to excessive noisiness in the resulting computed autocovariance and, as mentioned earlier, nonfractal behavior often dominates for these small times. We choose $T = 1$ as a compromise for this simulation, which proves close to optimal.

Figure 12.3 and Table 12.3 present the simulation results for the normalized autocovariance, that is, the autocorrelation minus the square of the mean, normalized by the variance, as a function of the delay number k:

$$R_2(k) \equiv \frac{R_Z(k, T) - \mathrm{E}^2[Z(T)]}{\mathrm{Var}[Z(T)]}. \tag{12.8}$$

Unlike the presentations in Tables 12.1 and 12.2, we present only one set of columns here; we must subtract the square-mean number of counts or essentially constant behavior results. This estimator performs poorly in comparison with the normalized Haar-wavelet variance $\widehat{A}(T)$; it is inferior even to the normalized variance $\widehat{F}(T)$.

Since the autocovariance has two arguments, one can consider this quantity as a function of the counting time T with a fixed delay number k. Examination of Eqs. (5.44d) and (12.7) indeed reveals a T^α dependency. However, rearrangement of these equations yields a diminished (and therefore relatively more noisy) version of the normalized variance:

$$
\begin{aligned}
R_Z(k,T)/\mathrm{E}[Z(T)] - \mathrm{E}[Z(T)] &= (T/T_R)^\alpha\, k^{\alpha-1} \\
&= [F(T) - 1]\, (T_F/T_R)^\alpha\, k^{\alpha-1} \\
&= \tfrac{1}{2}\alpha(\alpha+1)\, k^{\alpha-1}\, [F(T) - 1]. \quad (12.9)
\end{aligned}
$$

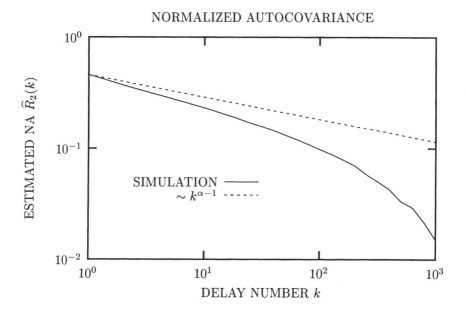

NORMALIZED AUTOCOVARIANCE

Fig. 12.3 Plot of $\widehat{R}_2(k) = \{\widehat{R}_Z(k,T) - \widehat{\mathrm{E}}^2[Z(T)]\}/\widehat{\mathrm{Var}}[Z(T)]$ vs. number of intervals k, the estimated normalized autocovariance, based on the same simulations as those used to generate Figs. 12.1 and 12.2 (a fractal-Gaussian-process-driven Poisson process). We obtained values of the autocovariance and variance for each simulation run, and then averaged the ratio of these two quantities (solid curve). This normalization provides results in the spirit of Figs. 12.1 and 12.2. We chose $T = 1$ for this simulation. A simple theoretical form proportional to $k^{\alpha-1}$, following Eq. (5.44d), also appears (dashed curve). The large disagreement between the simulation results and the theoretical behavior, especially at large delays k, illustrates the bias inherent in the count-based autocovariance.

Delay Range		$\widehat{\alpha}_{R2}$ from $\{\widehat{R}_Z(k,T) - \widehat{E}^2[Z(T)]\}/\widehat{Var}[Z(T)]$		
$k_{\min}$	$k_{\max}$	Bias	SD	RMSE
10^0	10^1	-0.097	0.034	0.103
10^0	10^2	-0.139	0.058	0.150
10^0	10^3	-0.259	0.106	0.280
10^1	10^2	-0.185	0.100	0.211
10^1	10^3	-0.360	0.163	0.395
10^2	10^3	-0.611	0.417	0.739

Table 12.3 Performance of fractal-exponent estimates $\widehat{\alpha}$ obtained via estimates of the count-based autocovariance, $\widehat{R}_Z(k,T) - \widehat{E}^2[Z(T)]$, for different ranges of the delay k. See Table 12.1 for details. Best results (least root-mean-square error) obtain for the shortest delay ranges; however, the bias is excessive for all delays and the standard deviation is large for all but the shortest delays. Results returned by the normalized variance $\widehat{F}(T)$ are better, and those returned by the normalized Haar-wavelet variance $\widehat{A}(T)$ are uniformly superior. This estimator does not appear to be useful.

Equation (12.9) evidently suffers from all of the deficiencies of the normalized variance $F(T)$; moreover, it is afflicted by a multiplicative factor $k^{\alpha-1}$, always less than unity, that decreases the useful term while leaving the constant, nonvarying terms unchanged.

We conclude that the autocorrelation $\widehat{R}_Z(k,T)$ is deficient as a statistic when considered as a function either of the delay k or the counting time T. We do not recommend its use.

12.3.4 Rescaled range

The rescaled range $U(k)$ of the interevent-interval sequence $\{\tau_k\}$, as a function of the delay k, yields information about dependencies among the intervals. Although it fails to reveal fractal behavior in the fractal renewal process (see Sec. 12.3.1), and presents information on a warped time axis, it enjoys substantial use as a measure of fractal activity and is therefore worth examining. Hurst introduced this measure in 1951, as discussed in Sec. 3.3.5. The rescaled range varies as $\sqrt{k}$ for independent intervals and, more generally as k^H, where $H = (1+\alpha)/2$, for a fractal process with fractal exponent α (see Sec. 6.3.1).

To facilitate the comparison of this measure with others, we examine the statistics of

$$U_2(k) \equiv U^2(k)/k. \tag{12.10}$$

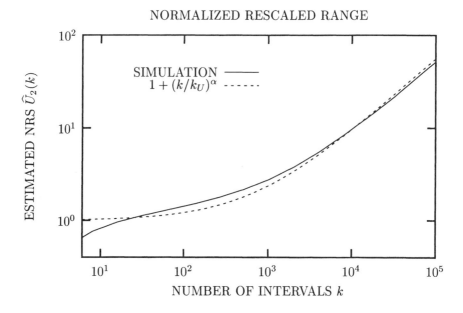

NORMALIZED RESCALED RANGE

Fig. 12.4   Normalized estimated rescaled range $\widehat{U}_2(k) \equiv \widehat{U}^2(k)/k$, as a function of the number of intervals k, based on the same simulations as those used to generate Figs. 12.1–12.3 (a fractal-Gaussian-process-driven Poisson process). Results begin at $k = 4$, and increase by a factor of 2 per displayed point thereafter. We obtained values of the R/S statistic for each simulation run, normalized them so that they came close to unity for independent intervals [converted to $\widehat{U}_2(k)$], and then averaged all 100 runs together (solid curve). A theoretical form akin to those shown in Figs. 12.1–12.3 also appears (dashed curve), employing $k_U = 671.3$, which minimizes the sum of squares of the differences in the logarithms. A simpler form, proportional to k^α, provides an even worse fit to the data (not shown). The deviation of the simulation results from these curves, even for large interval numbers k, illustrates the large bias inherent in the rescaled range statistics.

For a fractal-based point process, $U_2(k)$ varies as k^α, much like the normalized Haar-wavelet variance $A(T)$ which varies as T^α. In the context of the simulations employed in this chapter, for small values of k the underlying fractal rate changes little so that the intervals are relatively independent in these local neighborhoods. We therefore expect that $U(k) \sim \sqrt{k}$ so that $U_2(k) \sim k^0$ for small k. On the other hand, large values of k should encompass significant fractal fluctuations, so that $U(k) \sim k^H$ and $U_2(k) \sim k^\alpha$.

   We expect the value of k where these two behaviors meet (the onset value) to be of the order of the fractal onset times for the normalized variance and normalized Haar-wavelet variance (about 1 time unit), multiplied by the average rate (100 per time unit); the onset should therefore occur at about $k = 100$. Figure 12.4 and Table 12.4 present simulation results for the normalized estimated rescaled range $\widehat{U}_2(k)$. The averaged results presented in Fig. 12.4 follow the theoretical power-law form k^α

Time Range		$\widehat{\alpha}_{U2}$ from $\widehat{U}^2(k)/k$			$\widehat{\alpha}_{U2-1}$ from $\widehat{U}^2(k)/k - 1$		
$k_{\min}$	$k_{\max}$	Bias	SD	RMSE	Bias	SD	RMSE
10^2	10^3	-0.537	0.008	0.537	-0.194	0.016	0.195
10^2	10^4	-0.386	0.010	0.386	-0.154	0.012	0.155
10^2	10^5	-0.282	0.015	0.282	-0.116	0.015	0.117
10^3	10^4	-0.265	0.022	0.266	-0.112	0.025	0.114
10^3	10^5	-0.175	0.026	0.177	-0.077	0.027	0.082
10^4	10^5	-0.069	0.101	0.122	-0.033	0.105	0.110

Table 12.4 Performance of fractal-exponent estimates $\widehat{\alpha}$ obtained via normalized estimates of the rescaled range (R/S), $\widehat{U}_2(k) \equiv \widehat{U}^2(k)/k$, as a function of the number of intervals k. See Table 12.1 for details. Best results (least root-mean-square error) obtain by using $10^4 \leq k \leq 10^5$ for $\widehat{U}_2(k)$, and by using $10^3 \leq k \leq 10^5$ for $\widehat{U}_2(k) - 1$. Results obtained by using $\widehat{A}(T)$, the normalized Haar-wavelet variance, are uniformly superior to those based on the normalized rescaled range. This estimator does not appear to be useful.

only approximately, seriously diverging from it for smaller values of k; the function $1 + (k/k_U)^\alpha$ with $k_U \approx 671.3$ provides a better fit to $\widehat{U}_2(k)$ but still differs considerably from it. Employing $\widehat{U}_2(k)$ instead of $\widehat{U}(k)$ permits us to subtract the short-delay asymptote (which is approximately unity), but doing so improves the performance only marginally inasmuch as the asymptote bears only a slight likeness to the simulation. Overall, the normalized Haar-wavelet variance $\widehat{A}(T)$ proves substantially superior to the rescaled range for estimating the fractal exponent $\widehat{\alpha}$.

12.3.5　Detrended fluctuation analysis

The detrended fluctuation $Y(k)$ of the interevent-interval sequence $\{\tau_k\}$, as a function of the delay k, yields information about dependencies among the intervals related to those provided by the rescaled range $U(k)$. Like the rescaled range, it fails to reveal fractal behavior in the fractal renewal process (see Sec. 12.3.1) but we nevertheless investigate its performance. Taqqu & Teverovsky (1998) have pointed out that this measure exhibits significant bias and variance, except for the special case of Gaussian-distributed sequences.

The detrended fluctuation resembles the rescaled range in that $Y(k)$ typically varies as k^H, where $H = (1 + \alpha)/2$ for a process with fractal exponent α. Using Eq. (3.25) to proceed along lines similar to those followed in Sec. 12.3.4, we define a normalized version of this measure:

$$Y_2(k_2) \equiv \frac{15\, Y^2(k+2)}{(k+2)\, \mathrm{Var}[\tau]}.$$

(12.11)

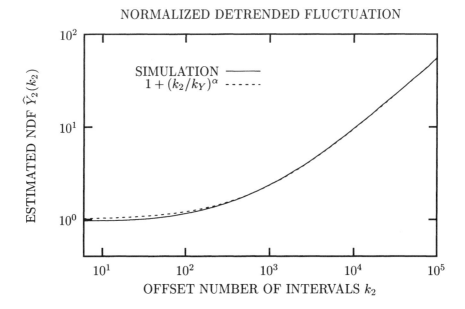

Fig. 12.5 Normalized estimated detrended fluctuation shown in Eq. (12.11), as a function of the offset number of intervals k_2, based on the same simulations as those used to generate Figs. 12.1–12.4 (a fractal-Gaussian-process-driven Poisson process). We define $k_2 \equiv k + 2$, the offset delay number. Results begin with $k_2 = 5$ ($k = 3$), and increase by factors of $10^{0.1}$ thereafter (excluding multiple copies of identical values of k_2). We obtained values of the detrended fluctuation statistic for each simulation run, normalized them to yield values equal to unity for independent intervals as specified in Eq. (12.11), and then averaged all 100 runs together (solid curve). The simple form provided in Eq. (12.12) fits the simulated data very well (dashed curve).

We present the simulation results for the normalized estimated detrended fluctuation $\widehat{Y}_2(k)$ in Fig. 12.5 and Table 12.5. Overall, we expect this measure to vary as

$$Y_2(k_2) \approx 1 + (k_2/k_Y)^\alpha. \qquad (12.12)$$

Equation (12.12) does indeed fit the simulations presented in Fig. 12.5 very well, with only one free parameter, k_Y, chosen to minimize the sum of squares of the differences in the logarithms. This takes a value of $k_Y = 684.6$, which is within an order of magnitude of our original estimate in Sec. 12.3.4.[4]

---

[4] Although Eq. (12.11) reduces to $Y_2(k_2) = 1 - 4/k_2^2$, given Eq. (3.25), for our simulations it nevertheless yields results that closely approach those of Eq. (12.12). We therefore employ it rather than a simpler form without the $+2$ offset. The origin of the offset may lie in the warping involved in translating the process from time in seconds to dimensionless interval number, which, in turn, may affect the representations of the relevant correlations. A number of *ad hoc* methods have been devised to address this issue (see, for example, Buldyrev et al., 1995), but they do not provide as good a fit to our simulations.

Time Range		$\widehat{\alpha}_{Y2}$ from $\widehat{Y}_2(k_2)$			$\widehat{\alpha}_{Y2-1}$ from $\widehat{Y}_2(k_2) - 1$		
$k_{2\min}$	$k_{2\max}$	Bias	SD	RMSE	Bias	SD	RMSE
10^2	10^3	-0.490	0.011	0.490	0.123	0.025	0.125
10^2	10^4	-0.336	0.013	0.337	0.046	0.017	0.049
10^2	10^5	-0.225	0.019	0.226	0.020	0.020	0.028
10^3	10^4	-0.189	0.027	0.191	-0.001	0.031	0.031
10^3	10^5	-0.111	0.034	0.116	-0.007	0.035	0.036
10^4	10^5	-0.052	0.087	0.101	-0.014	0.089	0.091

Table 12.5 Performance of fractal-exponent estimates $\widehat{\alpha}$ obtained using normalized estimates of the detrended fluctuation as defined in Eq. (12.11), as a function of offset delay number $k_2 \equiv k + 2$. See Table 12.1 for details. Best results (least root-mean-square error) obtain by using $10^4 \le k_2 \le 10^5$ for $\widehat{Y}_2(k_2)$, and by using $10^2 \le k_2 \le 10^5$ for $\widehat{Y}_2(k_2) - 1$. This measure yields errors nearly as small as those returned by the normalized Haar-wavelet variance $\widehat{A}(T)$; it is substantially superior to the rescaled range statistic examined in Fig. 12.4 and Table 12.4.

Employing $\widehat{Y}_2(k_2)$ instead of $\widehat{Y}(k)$ again permits the short-delay asymptote of about unity to be subtracted, and in this case the excellent fit of Eq. (12.12) to the simulated data imparts substantial improvement with this procedure. Detrended fluctuation analysis thus yields results only slightly inferior to those provided by the normalized Haar-wavelet variance $\widehat{A}(T)$ for the point process under study.

12.3.6   Interval wavelet variance

The wavelet variance $\text{Var}[W_{\psi,\tau}(k,l)]$ of the interevent-interval sequence $\{\tau_k\}$ as a function of the delay k, as set forth in Sec. 3.3.4, provides another window on the second-order properties of this sequence. Although different in construction from the rescaled range $U(k)$ and detrended fluctuation $Y(k)$, it nevertheless has much in common with these two measures. As before, we consider the merits of this measure, bearing in mind the same cautions regarding its inability to discriminate fractal renewal processes from nonfractal point processes (see Sec. 12.3.1). We choose the Haar wavelet to simplify processing and to minimize the variance of the resulting estimate (see Sec. 5.2).

Unlike rescaled range and detrended fluctuation analysis, the wavelet variance $\text{Var}[W_{\psi,\tau}(k,l)]$ typically varies as k^α rather than as k^H for a process with fractal exponent α. Hence, we need only divide by the variance of interevent intervals to render this quantity dimensionless, as provided in Eq. (3.21):

$$A_\tau(k) \equiv \frac{\text{Var}[W_{\psi,\tau}(k,l)]}{\text{Var}[\tau]}. \tag{12.13}$$

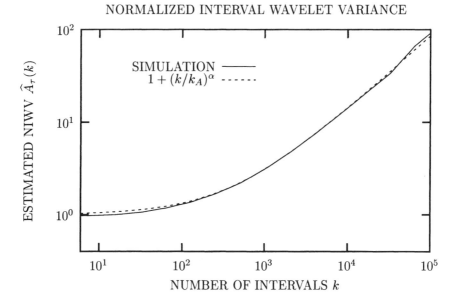

Fig. 12.6 Normalized estimated interval-based wavelet variance $\widehat{A}_\tau(k)$ shown in Eq. (12.13), as a function of wavelet scale (number of intervals) k, based on the same simulations as those used to generate Figs. 12.1–12.5 (a fractal-Gaussian-process-driven Poisson process). Results begin with $k = 2$, and increase by factors of 2 thereafter. We obtained values of the wavelet variance statistic for each simulation run, normalized them to yield values equal to unity for independent intervals as specified in Eq. (12.13), and then averaged all 100 runs together (solid curve). The simple theoretical form shown in Eq. (12.14) fits the simulated data very well (dashed curve). The increased difference between the two curves for the largest numbers of intervals derives from the stochastic fluctuations in the simulations themselves. Changing only the random seeds used in the simulations, but leaving all other parameters intact, yields an averaged $\widehat{A}_\tau(k)$ that exhibits different fluctuations, albeit of similar magnitude (not shown).

We again assume that the warping required to transform time in seconds to dimensionless interval number only marginally affects the results. As a parallel to Eq. (12.12), we expect the normalized interval-based wavelet variance to follow the form

$$A_\tau(k) \approx 1 + (k/k_A)^\alpha. \tag{12.14}$$

Equation (12.14) indeed provides an excellent fit to the simulated data, as shown in Fig. 12.6, by choosing the one free parameter to be $k_A \doteq 389.2$. This value minimizes the sum of squares of the differences in the logarithms. Again, the parameter k_A lies within an order of magnitude of our original estimate in Sec. 12.3.4. As with the rescaled range and detrended fluctuation, using $\widehat{A}_\tau(k)$, rather than $\widehat{\mathrm{Var}}[W_{\psi,\tau}(k,l)]$, permits us to subtract the short-delay asymptote near unity. As with $Y_2(k_2)$, Eq. (12.14) fits the simulated data extremely well and offers a considerable decrease in the estimation error; the results are only slightly inferior to

Time Range		$\widehat{\alpha}_{A\tau}$ from $\widehat{A}_\tau(k)$			$\widehat{\alpha}_{A\tau-1}$ from $\widehat{A}_\tau(k) - 1$		
$k_{\min}$	$k_{\max}$	Bias	SD	RMSE	Bias	SD	RMSE
10^2	10^3	-0.454	0.024	0.454	0.057	0.055	0.079
10^2	10^4	-0.269	0.020	0.270	0.016	0.025	0.030
10^2	10^5	-0.179	0.029	0.182	0.008	0.029	0.030
10^3	10^4	-0.147	0.059	0.158	-0.005	0.067	0.067
10^3	10^5	-0.083	0.054	0.099	-0.001	0.056	0.056
10^4	10^5	0.029	0.298	0.299	0.054	0.306	0.311

Table 12.6 Performance of fractal-exponent estimates $\widehat{\alpha}$ obtained via normalized estimates of the interval-based wavelet variance as defined in Eq. (12.13), as a function of wavelet scale k. See Table 12.1 for details. Best results (least root-mean-square error) obtain by using $10^3 \leq k \leq 10^5$ for $\widehat{A}_\tau(k)$, and by using $10^2 \leq k \leq 10^4$ or 10^5 for $\widehat{A}_\tau(k) - 1$. This measure yields errors comparable to those generated by normalized detrended fluctuation analysis, and nearly as small as those returned by the normalized Haar-wavelet variance.

those provided by the normalized Haar-wavelet variance $\widehat{A}(T)$ (compare Tables 12.6 and 12.1). For the Poisson-based process at hand, multiplying the abscissa by half the average interevent time[5] yields a plot that is nearly coincident with that of the normalized Haar-wavelet variance $\widehat{A}(T)$. In particular, we obtain a more precise value for the fractal onset number:

$$k_A = \frac{T_A}{\mathrm{E}[\tau]/2} \doteq 1.8939/0.005 = 378.78, \tag{12.15}$$

which is in excellent accord with the rough estimate of $k_A \doteq 389.2$ obtained empirically.

Agreement of this kind between $\widehat{A}(T)$ and $\widehat{A}_\tau(k)$ does not emerge in general, however. Consider, for example, a point process whose interval standard deviation lies well below the interval mean. For small numbers of intervals k, the interval-based wavelet variance $A_\tau(k)$ can assume arbitrarily small values, as a result of the small relative standard deviation of the intervals. However, $A(T)$ must approach an asymptote of unity for small values of T, so it can diverge significantly from $A_\tau(2T/\mathrm{E}[\tau])$. The human-heartbeat point process provides a case in point; examination of Figs. 5.2 and 5.8 reveals precisely this difference. The underlying reason for this disparity lies in the integrate-and-reset kernel associated with useful models for the heartbeat process. Poisson kernels, in contrast, yield plots of $\widehat{A}(T)$ and $\widehat{A}_\tau(k)$ that nearly coincide, as shown in the simulations presented here.

---

[5] The factor of a half arises from the definition of the count-based normalized Haar-wavelet variance $A(T)$, for which T encompasses half the duration of the Haar wavelet [see Eq. (3.40)].

12.3.7 Interval spectrum

The interval-based spectral estimate (periodogram) $\widehat{S}_\tau(f)$ forms the last interval-based measure we consider. As with other such measures, we note its most serious limitation, its inability to discriminate fractal renewal processes from nonfractal point processes (see Sec. 12.3.1).

Like the wavelet variance $\mathrm{Var}[W_{\psi,\tau}(k,l)]$, but unlike the rescaled range and detrended fluctuation, the interval-based periodogram typically varies as $1/f^\alpha$ rather than as $1/f^H$ for a process with fractal exponent α. Normalizing by $\mathrm{Var}[\tau]$, the estimated high-frequency asymptote as discussed in Sec. 3.3.3, yields a dimensionless form that achieves a value of unity for large interval frequencies f. We again assume that the warping required to transform time in seconds to dimensionless interval number only marginally affects the results. In another parallel to Eq. (12.12), we expect

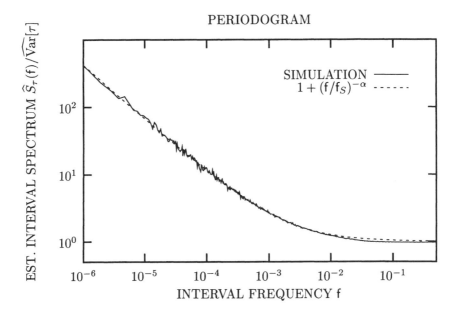

Fig. 12.7 Normalized estimated interval-based spectrum (periodogram) $S_\tau(f)/\widehat{\mathrm{Var}}[\tau]$ vs. interval frequency f, based on the same simulations as those used to generate Figs. 12.1–12.6 (a fractal-Gaussian-process-driven Poisson process). For each simulation, we calculated the (fast) Fourier transform of the data, using all 2^{20} intervals in a single transform without windowing. Next, we took the square of the magnitude, and divided by the sample variance of the intervals of that simulation. We then collected values into blocks whose frequencies differed by less than a factor of 1.02, and plotted a single point comprising the mean of the frequencies and the mean of the spectral estimate. Finally, we averaged all 100 curves together. This appears as the solid curve. Equation (12.16) generates a simple theoretical curve (dashed) which fits the simulated data very well.

Freq. Range		$\widehat{\alpha}_{ST}$ from $\widehat{S}_\tau(f)$			$\widehat{\alpha}_{ST-\text{Var}[\tau]}$ from $\widehat{S}_\tau(f) - \widehat{\text{Var}}[\tau]$		
f_{min}	f_{max}	Bias	SD	RMSE	Bias	SD	RMSE
10^0	10^1	0.008	0.606	0.606	0.095	0.843	0.848
10^0	10^2	−0.032	0.137	0.141	−0.174	0.179	0.250
10^0	10^3	−0.168	0.037	0.172	−0.070	0.052	0.087
10^0	10^4	−0.436	0.015	0.436	0.309	0.377	0.487
10^1	10^2	−0.054	0.227	0.234	−0.230	0.320	0.394
10^1	10^3	−0.189	0.040	0.193	−0.059	0.057	0.082
10^1	10^4	−0.446	0.015	0.447	0.324	0.389	0.506
10^2	10^3	−0.256	0.067	0.264	−0.010	0.089	0.090
10^2	10^4	−0.483	0.016	0.483	0.384	0.448	0.589
10^3	10^4	−0.562	0.023	0.563	0.644	0.746	0.985

Table 12.7 Performance of fractal-exponent estimates $\widehat{\alpha}$ for 100 different simulations of a fractal-Gaussian-process-driven Poisson process obtained from estimates of the interval spectrum (periodogram) $\widehat{S}_\tau(f)$. The simulations were the same as those employed to produce Table 12.1. We calculated the Fourier transform, and took the square magnitude. The first two columns specify the range of frequencies employed for calculating $\widehat{S}_\tau(f)$. We express frequency in terms of its product with the number of retained intervals N; thus, the first row corresponds to the frequency range $1/N \le f \le 10/N$. The next three columns represent the bias, standard deviation (SD), and resulting root-mean-square error (RMSE) for calculations employing the interval spectrum $\widehat{S}_\tau(f)$. We next sought to remove the effects of the high-frequency asymptote $\widehat{\text{Var}}[\tau]$, and re-calculate the fractal exponent. Since the interval spectrum fluctuates about this asymptote, many of the resulting differences would lie below zero, making calculation of the logarithm meaningless. We therefore employed averaging and thresholding. Specifically, we averaged the value of $\widehat{S}_\tau(f)$ at a frequency $f = n/N$ with the $n/16$ values following it in frequency, and then subtracted the estimated asymptote $\widehat{\text{Var}}[\tau]$. Finally, to avoid the occasional nonpositive number that would still result, we replaced all numbers less than $\varepsilon\widehat{\text{Var}}[\tau]$ with $\varepsilon\widehat{\text{Var}}[\tau]$, where $\varepsilon = 2^{-12} = 0.000244140625$. We estimated the fractal exponent employing these modified values; these results appear in columns 6–8. Subtracting the high-frequency asymptote does indeed effectively remove the bias, returning reduced root-mean-square errors. Best results (least root-mean-square error) obtain by using $10^0 \le Nf \le 10^2$ for $\widehat{S}_\tau(f)$, and by using $10^1 \le Nf \le 10^3$ for $\widehat{S}_\tau(f) - \widehat{\text{Var}}[\tau]$.

the normalized interval-based periodogram to follow the form

$$S_\tau(f) \approx 1 + (f/f_S)^{-\alpha}. \qquad (12.16)$$

Figure 12.7 shows that Eq. (12.16) indeed fits the simulated data well, with $f_S = 0.001893$ yielding the least mean-square difference in the logarithms. As with the other interval-based measures, using $\widehat{S}_\tau(f)/\widehat{\text{Var}}[\tau]$ instead of $\widehat{S}_\tau(f)$ makes it possible to subtract unity from the resulting statistic. Since Eq. (12.16) fits the data well, this operation reduces the bias in the subsequent estimates of α. The root-mean-square estimation error indeed improves significantly, although it does not achieve nearly the results of the normalized Haar-wavelet variance $\widehat{A}(T)$ (compare Tables 12.7 and 12.1).

The link that we demonstrated between $A(T)$ and $A_\tau(k)$ also holds between the interval-based and rate-based periodograms, and for the same reason: the simulations employ a Poisson kernel. Again, this connection suggests a more accurate estimate of a fractal onset value, the interval frequency in this case

$$f_S = \text{E}[\tau]f_S \doteq 0.010024 \times 0.2 = 0.0020048, \tag{12.17}$$

which agrees well with the value 0.001893 estimated above. We reiterate that the close link between interval-based and rate-based measures fails to hold in the general case.

12.3.8   Normalized Haar-wavelet variance

We now consider again, in greater detail, the use of the normalized Haar-wavelet variance $\widehat{A}(T)$ for obtaining the fractal exponent $\widehat{\alpha}$. Our preliminary study of the properties of this statistic appeared in Sec. 12.2.3.

In contrast to the normalized variance $\widehat{F}(T)$ and the autocorrelation $\widehat{R}_Z(k,T)$, considered in Secs. 12.3.2 and 12.3.3, respectively, the normalized Haar-wavelet variance $\widehat{A}(T)$ has the distinct merit that it is free of bias, as shown in Sec. 12.2.3.[6] This statistic offers another important advantage in addition, and that is its wavelet origin. Values computed at scales that differ by factors of two are nearly independent, as are values computed for a given counting time but in nonoverlapping counting windows (Tewfik & Kim, 1992).

One consequence of this wavelet property is that a plot of $\widehat{A}(T)$ appears noisier than that of $\widehat{F}(T)$ for a given set of data. Figure 12.8 illustrates this, where we graphically present $\widehat{A}(T)$ and $\widehat{F}(T)$ from an individual simulation run. We used averages of multiple runs such as these to generate Figs. 12.1 and 12.2. Figure 12.8 demonstrates that, despite its noisier appearance, $\widehat{A}(T)$ proves superior to $\widehat{F}(T)$ for quantifying fractal behavior. In particular, it has the distinct merits that its deviations from ideal fractal behavior have zero mean and are nearly independent. Tables 12.1 and 12.2, and Figs. 12.1 and 12.2, demonstrate this superiority for our canonical point-process model.

Although normalized variances employing other wavelet bases have been developed (Teich et al., 1996; Heneghan et al., 1996), fractal-based point processes rarely exhibit fractal exponents exceeding two (see Sec. 5.2.2). This, together with the reduced effective scaling range of more complex wavelets, usually renders them less

---

[6] As discussed in Sec. 12.2.3, the putative residual bias evident in the sixth column of Table 12.1 is apparently the result of small deviations from ideal fractal behavior that arise in the simulation itself, and that $\widehat{A}(T) - 1$ faithfully reports.

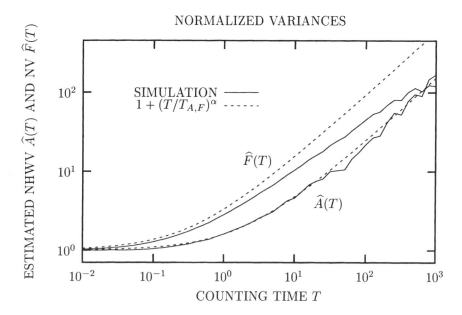

Fig. 12.8 Estimated normalized Haar-wavelet variance $\widehat{A}(T)$ (lower solid curve) and estimated normalized variance $\widehat{F}(T)$ (upper solid curve) vs. counting time T. Averaging these curves with 99 others that are similar yields the results displayed in Figs. 12.1 and 12.2. We provide mean theoretical results from Eqs. (5.44c) and (5.44b), respectively, for comparison (dashed curves). The estimate of the normalized Haar-wavelet variance $\widehat{A}(T)$ exhibits greater fluctuations than that of the normalized variance $\widehat{F}(T)$, yet it exhibits significantly superior performance.

useful than the Haar wavelet, as discussed in Sec. 5.2.5. Such wavelets do prove insensitive to linear or higher-order polynomial trends, unlike the Haar. However, the trends observed in experimental point processes rarely follow exact polynomial forms, so that this insensitivity does not provide a significant advantage. Finally, although there is near independence between wavelet transforms at scales that differ by a factor of two or more, and at different times within the same scale, a small residual correlation does remain. Wavelets other than the Haar do serve to reduce this correlation (Tewfik & Kim, 1992), but the concomitant reduced effective scaling range again typically outweighs any advantage gained by the slight decrease in the within-scale correlation. In the following, we therefore restrict our attention to the Haar version of the count-based wavelet variance, $A(T)$.

We now examine three sampling issues that prove important in obtaining optimal estimates of the fractal exponent $\widehat{\alpha}$ from $\widehat{A}(T)$: (1) **counting-time increments**; (2) **counting-time weighting**; and (3) **oversampling**. We examine these in turn.

Counting-time increments. We first examine the values of T that offer the most accurate estimates of the fractal exponent, when analyzing a fixed-length data set using the normalized Haar-wavelet variance $\widehat{A}(T)$. Wavelet theory provides a solution. For a discrete-time sequence, a wavelet transform sampled at scales that increase by powers of two retains all of the information in the data; an inverse transform of such a dyadic wavelet transform returns exactly the original sequence. Two caveats apply in the current circumstances, however. First, computing the variance at a given scale collapses all of the transform results at that scale into a single number, thereby eliminating much of the original information. Second, our interest lies in point-process data rather than in discrete-time sequences; however, the difference between the two proves unimportant at longer counting times where fractal behavior becomes significant.

Despite these caveats, optimal sampling turns out not to differ greatly from dyadic sampling. We investigated this issue by recalculating the results shown in Table 12.1 using various counting-time increments. In particular, we reexamine the root-mean-square error of $\widehat{\alpha}$ obtained from $\widehat{A}(T) - 1$, which made use of factor of $10^{0.1}$ increments of the counting time, as reported in the right-most column of Table 12.1. Results appear in the top half of Table 12.8 for four counting-time increments: factors of $10^{0.3}$, $10^{0.2}$, $10^{0.1}$, and $10^{0.05}$. Comparing the entries in these four columns reveals that close to optimal results obtain with a spacing of $10^{0.2} \doteq 1.584893$ while nearly dyadic spacing ($10^{0.3} \doteq 1.995262 \approx 2$) yields noticeably inferior performance. A spacing of slightly less than a factor of two appears to yield the best tradeoff between estimation accuracy and computational load.

Counting-time weighting. The optimal weighting for different counting times follows from the accuracy of the normalized Haar-wavelet variance $\widehat{A}(T)$ at a given value of T, which we now derive. From the ideal value provided in Eq. (3.40), we obtain its estimate

$$\widehat{A}(T) = \frac{\widehat{E}\{[Z_k(T) - Z_{k+1}(T)]^2\}}{2\widehat{E}[Z_k(T)]} = \frac{\widehat{E}[C_{\text{Haar},N}^2(2T, k)]}{2\widehat{E}[\lambda_k(T)]}. \tag{12.18}$$

The statistics of $\widehat{A}(T)$ prove difficult to analyze since the calculation involves the division of two random quantities. However, the numerator exhibits far more relative variation than the denominator, since estimates of second-order quantities such as autocorrelations and autocovariances generally fluctuate more than estimates of the mean, a first-order statistic. Moreover, for a given data set, the proportional error in estimating the denominator remains constant for all counting times, since this error derives directly from the same estimate of the mean rate regardless of counting time. After taking the logarithm, this error becomes a constant offset, irrelevant for the estimation of α.

We therefore focus on the numerator alone, since replacing the denominator with its expected value yields the same result:

$$\begin{aligned}
\widehat{A}(T) &= \frac{\widehat{E}[C_{\text{Haar},N}^2(2T, k)]}{2\widehat{E}[\lambda_k(T)]} \\
&\approx \frac{\widehat{E}[C_{\text{Haar},N}^2(2T, k)]}{2E[\lambda_k(T)]}
\end{aligned}$$

Time Range			RMSE of $\widehat{\alpha}_{A-1}$ from $\widehat{A}(T) - 1$			
$T_{\min}$	$T_{\max}$	Weighting	$10^{0.3}$	$10^{0.2}$	$10^{0.1}$	$10^{0.05}$
10^0	10^1	Equal	0.033	0.031	0.029	0.028
10^0	10^2		0.025	0.022	0.022	0.023
10^0	10^3		0.052	0.045	0.040	0.038
10^1	10^2		0.080	0.056	0.056	0.056
10^1	10^3		0.116	0.088	0.080	0.077
10^2	10^3		0.258	0.223	0.203	0.196
10^0	10^1	$T^{-1/2}$	0.033	0.030	0.029	0.028
10^0	10^2		0.022	0.019	0.019	0.019
10^0	10^3		0.023	0.021	0.019	0.019
10^1	10^2		0.080	0.053	0.053	0.054
10^1	10^3		0.082	0.059	0.055	0.053
10^2	10^3		0.232	0.201	0.186	0.182

Table 12.8 Root-mean-square error of $\widehat{\alpha}$ based on $\widehat{A}(T) - 1$, using the same simulated data and counting-time ranges used in Table 12.1. This table provides results for different counting-time increments within those time ranges (left-most two columns). The counting times increase geometrically by factors of $10^{0.3}$, $10^{0.2}$, $10^{0.1}$ (used in Table 12.1), and $10^{0.05}$, thereby providing, respectively, $3\frac{1}{3}$, 5, 10, and 20 counting times per decade of the time range. The top half of the table represents equal weighting, whereas the bottom half represents weighting inversely proportional to the square root of the counting time. Comparison of the entries in the four RMSE columns shows that estimator accuracy generally improves with decreasing counting-time increments (that is, as the number of counting times per decade increases). Exceptions to this rule occur in a few cases, such as in the second row, where the fluctuations in the individual simulations result in estimates of the normalized Haar-wavelet variance that happen, on the whole, to lie further from ideal behavior for increments of $10^{0.05}$ than $10^{0.2}$. A different simulation, with a different set of 100 random seeds, would yield slightly different results. Counting times incremented by factors of $10^{0.2}$ provide nearly optimal results at half the computational burden of increments by factors of $10^{0.1}$, and at a quarter the burden of increments by factors of $10^{0.05}$. Comparing the two weighting schemes reveals that accurately weighted results equal or exceed equally weighted results, as expected. Finally, comparing rows shows that a counting-time range of $10^0 \leq T \leq 10^2$ proves best for equal weighting, and that increasing the range improves the accurately weighted results with few exceptions.

$$= \ \widehat{E}\left[C_{\mathrm{Haar},N}^2(2T,k)\right]/2E[\mu]T. \tag{12.19}$$

Beginning with the bias, Eq. (3.39) yields

$$\widehat{E}\left[C_{\mathrm{Haar},N}^2(2T,k)\right] \ = \ [2T(M-1)]^{-1}\sum_{k=0}^{M-2}[Z_k(T) - Z_{k+1}(T)]^2$$

$$\mathrm{E}\left\{\widehat{\mathrm{E}}\left[C_{\mathrm{Haar},N}^2(2T,k)\right]\right\} = [2T(M-1)]^{-1}\sum_{k=0}^{M-2}\mathrm{E}\left\{[Z_k(T)-Z_{k+1}(T)]^2\right\}$$

$$= (2T)^{-1}\mathrm{E}\left\{[Z_0(T)-Z_1(T)]^2\right\} \tag{12.20}$$

$$= \mathrm{E}\left[C_{\mathrm{Haar},N}^2(T,k)\right], \tag{12.21}$$

where $M \equiv \mathrm{int}(L/T)$, L represents the duration of the data segment, and $\mathrm{int}(\cdot)$ returns the largest integer not greater than its argument. Equation (12.20) follows from the stationarity of the point process.

Estimates of the Haar-wavelet variance therefore have zero bias, and all error in its estimate, $\widehat{\mathrm{E}}\left[C_{\mathrm{Haar},N}^2(2T,k)\right]$, must derive from its variance. To obtain the variance, we make use of two further simplifications. First, we assume that wavelet-transform results are exactly independent although, as we mentioned at the beginning of this section, theoretical results show that some correlation does remain in the transformed values, albeit a small amount (Tewfik & Kim, 1992). Second, we consider large counting times T, in particular $T \gg \mathrm{E}[\tau]$, so that the counting statistics become nearly Gaussian. With these simplifications, we obtain

$$[4T^2(M-1)^2]\,\mathrm{Var}\left\{\widehat{\mathrm{E}}\left[C_{\mathrm{Haar},N}^2(T,k)\right]\right\}$$

$$= \sum_{k=0}^{M-2}\sum_{l=0}^{M-2}\left[\mathrm{E}\left\{[Z_k(T)-Z_{k+1}(T)]^2\,[Z_l(T)-Z_{l+1}(T)]^2\right\}\right.$$

$$\left. - \mathrm{E}\left\{[Z_k(T)-Z_{k+1}(T)]^2\right\}\mathrm{E}\left\{[Z_l(T)-Z_{l+1}(T)]^2\right\}\right]$$

$$[4T^2(M-1)]\,\mathrm{Var}\left\{\widehat{\mathrm{E}}\left[C_{\mathrm{Haar},N}^2(T,k)\right]\right\}$$

$$\approx \left[\mathrm{E}\left\{[Z_0(T)-Z_1(T)]^4\right\} - \mathrm{E}^2\left\{[Z_0(T)-Z_1(T)]^2\right\}\right] \tag{12.22}$$

$$\approx \left[3\mathrm{E}^2\left\{[Z_0(T)-Z_1(T)]^2\right\} - \mathrm{E}^2\left\{[Z_0(T)-Z_1(T)]^2\right\}\right] \tag{12.23}$$

$$= 2\mathrm{E}^2\left\{[Z_0(T)-Z_1(T)]^2\right\}$$

$$= 2T^2\,\mathrm{E}^2\left[C_{\mathrm{Haar},N}^2(2T,k)\right], \tag{12.24}$$

where Eqs. (12.22) and (12.23) derive from the whiteness and Gaussian assumptions, respectively.

Combining Eqs. (12.21) and (12.24) then gives rise to the following simple result for the square of the coefficient of variation of the Haar-wavelet variance estimate: $1/[2(M-1)]$. Incorporating the assumption that the rate estimate exhibits much smaller fluctuations than does the Haar-wavelet variance [Eq. (12.19)] leads to a squared coefficient of variation for the estimated normalized Haar-wavelet variance that takes the form

$$\frac{\text{Var}[\widehat{A}(T)]}{\text{E}^2[\widehat{A}(T)]} \approx \frac{T}{2(L-T)}, \tag{12.25}$$

where again $M \equiv \text{int}(L/T)$. We use this result to generate the dashed curves in Fig. 12.1. Taking the logarithm of $\widehat{A}(T)$ yields

$$
\begin{aligned}
\ln[\widehat{A}(T)] &= \ln\left\{ A(T)\left[1 + \frac{\widehat{A}(T) - A(T)}{A(T)}\right] \right\} \\
&= \ln[A(T)] + \ln\left[1 + \frac{\widehat{A}(T) - A(T)}{A(T)}\right] \\
&\approx \ln[A(T)] + \frac{\widehat{A}(T) - A(T)}{A(T)} \tag{12.26}
\end{aligned}
$$

$$
\begin{aligned}
\text{Var}\left\{\ln[\widehat{A}(T)]\right\} &\approx \text{Var}\left[\frac{\widehat{A}(T) - A(T)}{A(T)}\right] \\
&= \text{Var}\left[\frac{\widehat{A}(T)}{A(T)}\right] \\
&\approx \frac{T}{2(L-T)}. \tag{12.27}
\end{aligned}
$$

Equation (12.26) makes use of the fact that $\ln(1 + x) \approx x$ for $x \ll 1$, and remains valid for $T/L \ll 1$; Eq. (12.27) results from Eq. (12.25).

Equation (12.27) provides a simple but fundamental result: the variance of the logarithm of the normalized Haar-wavelet variance estimate increases linearly with counting time T. This makes good sense; since each independent wavelet-transform value has a duration proportional to T, the number of independent samples in a fixed length of data varies inversely with T. Thus, the variance of the estimate, which is inversely proportional to the number of samples, varies in direct proportion to T.

For best accuracy in estimating α, then, weighting functions should vary inversely with the square root of T. We incorporated this weighting function in the entries reported in the lower half of Table 12.8. The results are indeed superior to those obtained using equal weighting, in all cases. Furthermore, for this weighting increasing the range of times used in estimating α almost always yields an improved estimate, as expected with a proper weighting function. Employing this $T^{-1/2}$ weighting function, and counting times incremented by factors of $10^{0.2}$, we obtain the best performance by using the largest possible range of counting times ($10^0 \leq T \leq 10^2$); the root-mean-square error is 0.019.

We note that the foregoing analysis fails without the use of logarithms. Rearranging Eq. (12.25) yields the variance of the estimate itself, as a function of counting time T:

$$\text{Var}[\widehat{A}(T)] \approx \text{E}^2[\widehat{A}(T)] \frac{T}{2(L-T)}$$

$$\approx \ [(T/T_A)^\alpha]^2 \ T/2L$$

$$= \ \tfrac{1}{2}T_A^{-2\alpha}L^{-1}T^{1+2\alpha}$$

$$\sim \ T^{1+2\alpha}. \tag{12.28}$$

Thus, fitting a power-law function directly to values of $\widehat{A}(T)$ results in a variance that increases as $T^{1+2\alpha}$. This implies that the optimal weighting function for estimating α itself depends on α, which violates the spirit of estimating an unknown signal. We thus employ logarithms and linear fits, rather than direct values of $\widehat{A}(T)$ and power-law fits.

Oversampling. We conclude this section by considering oversampled versions of the normalized Haar-wavelet variance estimate. Figure 3.6b) illustrates the method we have used to estimate the normalized Haar-wavelet variance to this point; the first counting duration begins at $t = 0$, with subsequent counting durations immediately following each other in turn.

Reiterating the construction of the normalized Haar-wavelet variance we obtain

$$\widehat{A}(T) \quad = \quad (M-1)^{-1} \sum_{k=0}^{M-2} \left[Z_k(T) - Z_{k+1}(T) \right]^2 \Big/ \left[2TN(L)/L \right]$$

$$\left[2T(M-1)N(L)/L \right] \widehat{A}(T)$$

$$= \quad \sum_{k=0}^{L/T-2} \left[2N(kT+T) - N(kT) - N(kT+2T) \right]^2, \tag{12.29}$$

again with $M = \text{int}(L/T)$. However, the time $t = 0$ bears no particular significance to the point process $dN(t)$; beginning the counting durations at $t = \tfrac{1}{2}T$ rather than at $t = 0$ is equally valid. In fact, an average of the two yields improved statistics:

$$2\left[2T(M-1)N(L)/L \right] \widehat{A}_{2\times}(T)$$

$$= \quad \sum_{k=0}^{L/T-2} \left[2N(kT+\tfrac{3}{2}T) - N(kT+\tfrac{1}{2}T) - N(kT+\tfrac{5}{2}T) \right]^2$$

$$+ \quad \sum_{k=0}^{L/T-2} \left[2N(kT+T) - N(kT) - N(kT+2T) \right]^2, \tag{12.30}$$

where the notation $\widehat{A}_{2\times}(T)$ indicates the average of the variances estimated over two different sets of counting durations. Continuing this process, one can average results over four sets of durations that differ in their starting times by $\tfrac{1}{4}T$, yielding $\widehat{A}_{4\times}(T)$:

$$4\left[2T(M-1)N(L)/L \right] \widehat{A}_{4\times}(T)$$

$$= \quad \sum_{k=0}^{L/T-2} \left[2N(kT+\tfrac{7}{4}T) - N(kT+\tfrac{3}{4}T) - N(kT+\tfrac{11}{4}T) \right]^2$$

Time Range		RMSE of $\widehat{\alpha}_{A-1}$ from $\widehat{A}(T) - 1$			
T_{min}	T_{max}	$10^{0.2}, 1\times$	$10^{0.2}, 2\times$	$10^{0.2}, 4\times$	$10^{0.05}, 1\times$
10^0	10^1	0.030	0.027	0.025	0.028
10^0	10^2	0.019	0.018	0.017	0.019
10^0	10^3	0.021	0.018	0.017	0.019
10^1	10^2	0.053	0.046	0.044	0.054
10^1	10^3	0.059	0.048	0.046	0.053
10^2	10^3	0.201	0.164	0.154	0.182

Table 12.9 Root-mean-square error of $\widehat{\alpha}$ based on $\widehat{A}(T) - 1$, using the same simulated data and time ranges as employed for Table 12.1, but with different amounts of oversampling. Weighting varies as $T^{-1/2}$ in all cases. The results for counting-time increments of $10^{0.2}$ (labeled 1×) and $10^{0.05}$ (labeled 1×) coincide with those from Table 12.8. Other columns also derive from the same data, but make use of staggered counting durations. Computing the column labeled "$10^{0.2}, 2\times$" uses a set of counting durations that begins at $t = 0$, in conjunction with a second set that begins at $t = \frac{1}{2}T$; the average over all counting durations, from both sets, forms the estimate of the normalized Haar-wavelet variance. Similarly, the column labeled "$10^{0.2}, 4\times$" makes use of four sets of counting times, beginning at $t = 0$, $t = \frac{1}{4}T$, $t = \frac{1}{2}T$, and $t = \frac{3}{4}T$. For each row, double sampling decreases the error, while quadruple sampling brings little further improvement. Doubly sampled versions at a counting-time increment of $10^{0.2}$ consistently yield results superior to those obtained using singly sampled versions at a counting-time increment of $10^{0.05}$, although the latter imposes twice the computational load.

$$+ \sum_{k=0}^{L/T-2} \left[2N(kT + \tfrac{3}{2}T) - N(kT + \tfrac{1}{2}T) - N(kT + \tfrac{5}{2}T) \right]^2$$

$$+ \sum_{k=0}^{L/T-2} \left[2N(kT + \tfrac{5}{4}T) - N(kT + \tfrac{1}{4}T) - N(kT + \tfrac{9}{4}T) \right]^2$$

$$+ \sum_{k=0}^{L/T-2} \left[2N(kT + T) - N(kT) - N(kT + 2T) \right]^2. \qquad (12.31)$$

The results presented in Table 12.9 illustrate the improvement in fractal-exponent estimation accuracy obtained by using such oversampling. Having established in Table 12.8 that a weight of $T^{-1/2}$, a counting time increment of $10^{0.2}$, and subtracting unity from $\widehat{A}(T)$ optimizes the estimation process, we display results in Table 12.9 for this choice of parameters at 2× and 4× oversampling. We also reproduce from Table 12.8 results for increments of $10^{0.2}$ and $10^{0.05}$ without oversampling (1×). For all six choices of counting-time ranges, 2× oversampling turns out to provide a reduced root-mean-square error, whereas 4× oversampling further improves performance only slightly. Both oversampling schemes yield performance superior to that obtained by using a counting-time increment of $10^{0.05}$ (right-most column of

Table 12.8), despite the fact that this latter increment has a computational load equal to that of an increment of $10^{0.2}$ at $4\times$ oversampling. Holding the ratio of oversampling constant, the performance largely resembles that portrayed in Table 12.8, thereby confirming the choice of $10^{0.2}$ for the counting-time increment, both for oversampling and for the original method. Similar results also obtain for estimates that are based on $\widehat{A}(T)$ rather than $\widehat{A}(T) - 1$ (not shown).

Overall, we conclude that optimal fractal-exponent estimation mandates the following, with the choices largely independent of each other: (1) a counting-time increment of $10^{0.2}$, (2) estimates based on $\widehat{A}(T) - 1$, (3) weighting in accordance with $T^{-1/2}$, and (4) counting-time ranges of 10^0–10^2 or 10^3.

12.3.9 Rate spectrum

The rate-based spectrum $S_\lambda(f, T)$ for a point process has an estimate known as the rate-based periodogram (see Sec. 3.4.5). Like the normalized Haar-wavelet variance examined in Secs. 12.2.3 and 12.3.8, this measure has no bias. Furthermore, errors about the true value are multiplicative, independent, and have the same statistics for all frequencies f (Oppenheim & Schafer, 1975, p. 547).[7] This multiplicative character suggests that the logarithm of the periodogram renders the errors additive with similar statistics. The independence and uniformity of the statistics with frequency suggest equal weighting.

What set of frequencies provides the best results? Although any frequency spacing will yield information about the spectrum of the underlying point process, linear spacing is far superior from a practical perspective: it permits the use of fast Fourier transform techniques, thereby vastly reducing the computational load. We make use of the rate spectrum rather than the point-process spectrum since fast Fourier transforms call for discrete-time data. As shown in Eq. (3.67), the two measures bear a close resemblance to each other for frequencies much lower than the inverse of the counting time. For the resemblance to hold, the Fourier transform must have at least $2fL$ elements, where f is the largest frequency of interest, and L is the duration of the data under study.

Since Fourier transforms operate on periodic sequences, the discontinuity between the first and last counting windows can introduce spurious components into the periodogram, thereby blurring fractal features. We examine this effect in detail in Sec. A.8.1. The effect of this discontinuity decays with frequency as f^{-2}, making good estimation of processes with $\alpha > 2$ difficult. Although values of α in excess of two rarely occur, as indicated in Sec. 5.2.2, the use of a simple Hanning window (Oppenheim & Schafer, 1975, p. 242) serves to increase the theoretical limit from $\alpha = 2$ to $\alpha = 6$. Figure 12.9 displays the rate spectrum estimate obtained with a Hanning window (dashed curve) and with a rectangular window (no windowing, solid curve). The simulated data were identical to those used to produce Figs. 12.1–12.7.

---

[7] While this reference presents a proof only for white Gaussian noise, these results hold to an excellent approximation for the situation at hand.

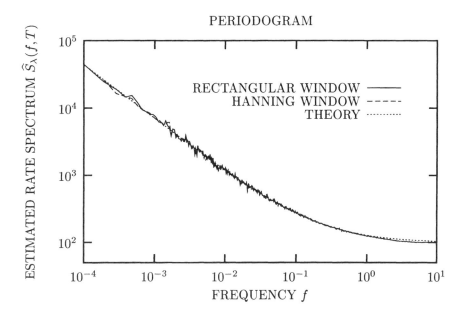

Fig. 12.9 Estimated rate spectrum (periodogram) $\widehat{S}_\lambda(f,T)$ vs. frequency f, based on the same simulations as those used to generate Figs. 12.1–12.7. For each simulation, we calculated its rate estimate using a counting time of $T = L/2^{18} \approx 0.0401$ time units (bearing in mind that L differs among the simulations). We then calculated the square magnitude of the fast Fourier transform of this rate estimate. Next, we collected values into blocks whose frequencies differed by less than a factor of 1.02, and plotted a single point comprising the mean of the frequencies and the mean of the spectral estimate. Finally, we averaged all 100 curves together, both ordinate and abscissa. We used two different windows: a Hanning (raised cosine) window, and a rectangular window equal to the duration L of the simulation (no windowing). The theoretical results directly follow Eq. (5.44a). The Hanning window yields a single large spurious value at $f = 1/L$ which we removed since it arises from the window itself. Otherwise, the results for both windows closely follow the theoretical results. The dip in the simulated curve near $f = 3$ derives from the discrete nature of the rate process used to generate the simulations (see Sec. 12.2.3).

We calculated the square magnitude of the Fourier transform of the rate estimate for each simulation, then averaged over frequencies within a factor of 1.02 of each other and, finally, averaged all 100 curves together. Except for a large peak arising from the cosine term in the Hanning window that we removed, the estimated spectra for both windows closely follow the theoretical form associated with Eq. (5.44a) (dotted curve). We conclude that windowing does not provide an advantage for these particular simulations, most likely because the fractal exponent $\alpha = 0.8$ lies below 2.

We next examined the fractal-exponent statistics based on the collection of individual simulations. In analogy with the method used for estimating the normalized

Haar-wavelet variance, we carried out a least-squares fit on the logarithm of the periodogram vs. the logarithm of frequency, over several ranges of frequencies, using equal weighting. For fits that include high frequencies ($10^3/L$ and $10^4/L$), the high-frequency asymptote $\lim_{f \to \infty} S_\lambda(f, T) = \mathrm{E}[\mu]$ causes the rate of decay to decrease with frequency, thereby leading to negative bias. Fits limited to low frequencies, on the other hand, suffer from excessive variance since fewer frequencies exist in these ranges. Overall, the best performance obtains for $1/L \leq f \leq 10^3/L$, but with a root-mean-square error of 0.103, it remains poor.

In an attempt to improve matters, we also performed fits to the periodogram after subtracting the estimate of the high-frequency asymptote, $\widehat{\mathrm{E}}[\mu]$. However, as a result of the variance inherent in the periodogram, this occasionally yielded nonpositive values. To rectify this, we employed averaging and thresholding. Specifically, we averaged the value of $\widehat{S}_\lambda(f, T)$ at a frequency $f = n/L$ with the $n/16$ values following it in frequency, and then subtracted the estimated asymptote $\widehat{\mathrm{E}}[\mu]$. Finally, we replaced all numbers less than $\varepsilon\widehat{\mathrm{E}}[\mu]$ with $\varepsilon\widehat{\mathrm{E}}[\mu]$, where $\varepsilon = 2^{-12} = 0.000244140625$, and performed a least-squares fit on the modified values. Indeed, removal of the high-frequency asymptote leads to a substantial improvement in the results. Larger frequency ranges generally yielded better accuracy, as was the case for the normalized Haar-wavelet variance with proper weighting. However, the smallest root-mean-square error, 0.059, occurred for the frequency range $10/L \leq f \leq 10^3/L$. The results of these simulations appear in Table 12.10. We note that our results are somewhat idiosyncratic, given the parameters used for averaging (16 adjacent frequency bins and $\varepsilon = 2^{-12}$); as with the interval spectrum, changing these values would likely lead to a different result.

While analytical results for the bias prove difficult to obtain in general, relatively simple results are available for the standard deviation, given a number of assumptions (Lowen & Teich, 1995). We proceed to derive this result. Starting with the sequence of counts $\{Z_k(T)\}$ obtained from a fractal-based point process, we take its Fourier transform. Each element of this transform comprises a sum of a relatively large number of terms (the counts), multiplied by trigonometric functions of magnitude not exceeding unity. This suggests the use of the central limit theorem, in which case the transform approaches a circularly symmetric complex Gaussian process. We assume that the Gaussian limit holds exactly. Converting to a spectrum estimate yields

$$\widehat{S}(n/L) = \mathrm{E}[\mu]\left[1 + (n/f_S L)^{-\alpha}\right]\exp(\epsilon_n), \tag{12.32}$$

where n denotes the discrete index in the fast Fourier transform and ϵ_n represents some small error term near zero. For estimates of white noise, the errors for different indices n are independent and have a standard deviation equal to the mean (Oppenheim & Schafer, 1975); we assume that this holds for the fractal-rate process under study as well. We then have

$$
\begin{aligned}
\mathrm{E}[\exp(\epsilon_n)] &= 1 \\
\mathrm{E}\big\{[\exp(\epsilon_n) - 1]^2\big\} &= 1 \\
\mathrm{E}\big\{[\exp(\epsilon_n) - 1][\exp(\epsilon_m) - 1]\big\} &= 0, \quad n \neq m.
\end{aligned}
\tag{12.33}
$$

Freq. Range		$\widehat{\alpha}_S$ from $\widehat{S}_\lambda(f,T)$			$\widehat{\alpha}_{S-\mathrm{E}[\mu]}$ from $\widehat{S}_\lambda(f,T) - \widehat{\mathrm{E}}[\mu]$			Calc.
$f_{\min}$	$f_{\max}$	Bias	SD	RMSE	Bias	SD	RMSE	SD
10^0	10^1	-0.026	0.530	0.531	0.023	0.607	0.608	0.578
10^0	10^2	-0.003	0.130	0.130	-0.151	0.142	0.207	0.147
10^0	10^3	-0.096	0.036	0.103	-0.055	0.041	0.069	0.045
10^0	10^4	-0.322	0.014	0.322	0.076	0.028	0.081	0.014
10^1	10^2	0.007	0.196	0.196	-0.172	0.224	0.283	0.222
10^1	10^3	-0.108	0.040	0.115	-0.044	0.041	0.059	0.050
10^1	10^4	-0.330	0.015	0.331	0.081	0.029	0.086	0.014
10^2	10^3	-0.149	0.071	0.165	-0.016	0.064	0.066	0.077
10^2	10^4	-0.361	0.016	0.362	0.101	0.033	0.106	0.016
10^3	10^4	-0.439	0.022	0.439	0.172	0.060	0.182	0.025

Table 12.10 Performance of fractal-exponent estimates $\widehat{\alpha}$ for 100 different simulations of a fractal-Gaussian-process-driven Poisson process obtained using estimates of the rate spectrum (periodogram) $\widehat{S}_\lambda(f,T)$. The simulations were the same as those employed to produce Table 12.1. We counted the number of events falling into 2^{15} windows of equal length $T = L/2^{15} \approx 0.320$ (bearing in mind that L differs among the simulations), calculated the Fourier transform, and took the square magnitude. The first two columns specify the range of frequencies employed for calculating $\widehat{S}_\lambda(f,T)$. We express frequency in terms of its product with the duration of the simulation, L; thus, the first row corresponds to the frequency range $1/L \leq f \leq 10/L$. The next three columns represent the bias, standard deviation (SD), and resulting root-mean-square error (RMSE) for calculations employing the rate spectrum $\widehat{S}_\lambda(f,T)$. We next sought to remove the effects of the high-frequency asymptote $\widehat{\mathrm{E}}[\mu]$. Since the rate spectrum fluctuates about this asymptote, many of the resulting differences would lie below zero, making calculation of the logarithm meaningless. We therefore employed averaging and thresholding. Specifically, we averaged the value of $\widehat{S}_\lambda(f,T)$ at a frequency $f = n/L$ with the $n/16$ values following it in frequency, and then subtracted the estimated asymptote $\widehat{\mathrm{E}}[\mu]$. Finally, to ensure positive values, we replaced all numbers less than $\varepsilon\widehat{\mathrm{E}}[\mu]$ with $\varepsilon\widehat{\mathrm{E}}[\mu]$, where $\varepsilon = 2^{-12}$. We estimated the fractal exponent employing these modified values; the corresponding results appear in columns 6–8. Subtracting the high-frequency asymptote indeed reduces the bias significantly, returning reduced root-mean-square errors. Moreover, increasing the range generally improves the results, as we observed in the properly weighted outcomes for the normalized Haar-wavelet variance presented in the lower half of Table 12.8. The rightmost column presents the predicted theoretical values for the standard deviation; results for the corresponding simulations stand two columns to its left. The best results (minimum root-mean-square error) for $\widehat{S}_\lambda(f)$ obtain by using $1/L \leq f \leq 10^3/L$, whereas for $\widehat{S}_\lambda(f) - \widehat{\mathrm{E}}[\mu]$ the optimal range is $10/L \leq f \leq 10^3/L$. While the normalized Haar-wavelet variance $\widehat{A}(T)$ yields superior results, it does incur a larger computational burden.

Proceeding with the analysis, we estimate α from $\widehat{S}(n/L)$. Subtracting the constant term $\text{E}[\mu]$ from Eq. (12.32), and taking logarithms of both independent and dependent variables, we define

$$
\begin{aligned}
x_n &\equiv \ln(n) \\
y_n &\equiv \ln\{\widehat{S}(n/L) - \text{E}[\mu]\} \\
&= \ln\left(\text{E}[\mu]\right) + \alpha \ln(f_S L) - \alpha \ln(n) + \epsilon_n.
\end{aligned}
\tag{12.34}
$$

The estimate of α simply becomes the covariance between $\{x\}$ and $\{y\}$ divided by the variance of $\{x\}$ (Lowen & Teich, 1993a):

$$
\begin{aligned}
\widehat{\alpha}_S &= \frac{(M-1)^{-1} \sum\limits_{n=1}^{M} x_n y_n - M^{-1}(M-1)^{-1} \left(\sum\limits_{n=1}^{M} x_n\right)\left(\sum\limits_{m=1}^{M} y_m\right)}{(M-1)^{-1} \sum\limits_{n=1}^{M} x_n^2 - M^{-1}(M-1)^{-1} \left(\sum\limits_{n=1}^{M} x_n\right)^2} \\[2mm]
&= \alpha + \frac{\left(\sum\limits_{n=1}^{M} \epsilon_n\right)\left(\sum\limits_{m=1}^{M} \ln(m)\right) - M \sum\limits_{n=1}^{M} \epsilon_n \ln(n)}{M \sum\limits_{n=1}^{M} \ln^2(n) - \left(\sum\limits_{n=1}^{M} \ln(k)\right)^2},
\end{aligned}
\tag{12.35}
$$

with $M = \text{int}(L/T)$, the number of counts as defined in Eq. (12.4). Equation (12.35) has a mean of α by construction, as it should. After some algebra, an expression for the variance emerges:

$$
\text{Var}[\widehat{\alpha}_S] = \text{Var}[\epsilon] \cdot \left[\sum\limits_{n=1}^{M} \ln^2(n) - M^{-1}\left(\sum\limits_{n=1}^{M} \ln(n)\right)^2\right]^{-1}.
\tag{12.36}
$$

The variance of ϵ depends on the distribution of the errors $\{\epsilon_n\}$ themselves, while the rest of Eq. (12.36), which is a deterministic function of M, approaches $1/M$ as $M \to \infty$. For the circularly symmetric Gaussian form assumed for the Fourier transform of $\{Z_k(T)\}$, the spectrum estimate $\widehat{S}(n/L)$ has an exponential distribution, which provides

$$
\begin{aligned}
\text{Var}[\epsilon] &= \int_0^\infty \ln^2(t)\,\exp(-t)\,dt \\
&= \tfrac{1}{6}\pi^2 + C_{\text{Euler}}^2 \\
&\doteq 1.978,
\end{aligned}
\tag{12.37}
$$

where $C_{\text{Euler}} \doteq 0.5772156649$ denotes Euler's constant.

The right-most column in Table 12.10 presents the expected theoretical values for the standard deviation, calculated from Eqs. (12.36) and (12.37). Here we substitute

$f_{\max}L$ and $f_{\min}L$ for the upper limit (M) and lower limit (unity), respectively, in the sums in Eq. (12.36), and replace M^{-1} with $f_{\max}L + 1 - f_{\min}L$. Results for the corresponding simulations for each frequency range stand two columns to the left. While none of the assumptions employed in the derivation provided above holds exactly, the calculated and simulated values nevertheless agree quite well, especially for more constrained frequency ranges ($f_{\max}L < 10^4$).

12.4   COMPARISON OF MEASURES

Based on the results provided in Sec. 12.3, we conclude that the normalized Haar-wavelet variance and the rate spectrum prove to be the most useful measures for estimating the exponent of an unidentified fractal-based point process. Furthermore, as shown in Sec. 5.2, these two measures are valid over the full range of fractal exponents normally encountered, $0 < \alpha < 2$.

Detrended fluctuation and the interval-based wavelet variance also perform well as estimators for the point process at hand, and the optimization techniques we used for the normalized Haar-wavelet variance would presumably provide comparable improvements for these two measures as well. Nevertheless, as with all interval-based approaches, they cannot reliably detect fractal behavior, let alone quantify it, so we do not recommend their general use.

The normalized Haar-wavelet variance returns the best performance, with a root-mean-square error of 0.018, in comparison with 0.059 for the rate spectrum, under optimal conditions for the estimation problem at hand (appropriate weighting, choice of counting time increment, oversampling, and subtraction of short time and high-frequency asymptotes). However, the former measure has significantly more computational burden than the latter. The ultimate choice of measure thus appears to depend on the relative costs of estimation error and processing time.

A number of other approaches exist for estimating fractal exponents (see, for example, Beran, 1992, 1994; Taqqu et al., 1995), but these expect real-valued, discrete-time sequences, generally with Gaussian statistics. Although counting turns a point process into a positive-integer-valued discrete-time sequence, the question then becomes the choice of counting time. Short counting times yield statistics far from a Gaussian form, whereas long counting times fail to capture the short-time-scale information inherent in a point process; this issue bears some similarity to the choice of range for a dependent variable, mentioned at the beginning of this chapter. In general, discrete-time methods do not appear to offer the performance and robustness desired in a fractal-exponent estimator for point processes. However, two notable exceptions exist. First, for extremely long point-process realizations, the very large numbers of events renders the choice of counting time far less critical, so that conversion of a data set to a discrete-time form becomes relatively straightforward. Second, for calculation of the rate-based spectral estimate, we employ an effective counting time L/M, with L again the duration of the realization and M the size of the (fast) Fourier transform used. In this case, the choice of counting time (or, equivalently, M) proves relatively unimportant as long as the counting time does not exceed the Nyquist limit for frequencies of interest.

Other measures considered in this Chapter have made use of counting sequences as intermediate steps, but have avoided the limitations associated with choosing a fixed value of counting time. Little advantage appears to accrue from using traditional fractal methods based on discrete-time sequences, which confirms our conclusion that the normalized Haar-wavelet variance and the rate spectrum are the measures of choice.

We offer a final caveat in closing this chapter. We chose the simulation set analyzed because of its generic nature. The fractal-Gaussian-process-driven Poisson process characterizes a number of observed point processes; furthermore, superpositions of many fractal-based point processes converge to it. The value $\alpha = 0.8$ offers a significant range of variation of the various measures with time and frequency, yet it lies sufficiently below unity to preclude problems with measures that fail for $\alpha \geq 1$.[8] However, just as *a priori* information heavily influences the identification of a point process, as described in Sec. 12.1, so too does prior information affect the choice of an optimal measurement statistic.

While we expect that the results established here will prove useful in many circumstances, the *a priori* information available in all possible experiments spans far too great a range for one approach to yield optimal results in all cases. Different applications will surely involve different ranges of α; the estimation of other fractal-based parameters, such as fractal onset times and frequencies, are certainly of interest, and more subtle features of various measures will come to the fore. Rather than attempting to catalog such a vast parameter space, we have chosen instead to direct our presentation to investigating how well various statistical measures function for estimating the fractal exponent of an unidentified fractal-based point processes, and, most importantly, to setting forth a collection of techniques and mathematical relations that should find use in a broad variety of applications.

Problems

12.1   *Discriminating fractal-rate point processes via their interevent-interval densities*   Suppose we have a realization of a fractal-rate point process, and we know that it derives from a Poisson process driven by either: (1) a fractal Gaussian process, (2) fractal binomial noise, or (3) fractal shot noise. Suppose further that processes (2) and (3) approach process (1) fairly closely, but not exactly, of course. The measures used in this chapter to quantify fractal behavior cannot distinguish among these three possibilities since all three processes generate power-law forms akin to those provided in Eq. (5.44) (see Sec. 12.1). Discuss how the interevent-interval probability density (or distribution) might assist us in determining which of the three rate processes is

---

[8] We previously examined the normalized Haar-wavelet variance and spectrum for this process, as well as for the fractal-Gaussian-process-driven integrate-and-reset process (with jitter), for $\alpha = 0.2$, 0.8, and 1.5 (Thurner et al., 1997).

responsible for generating the realization at hand. How does the situation differ if the three rate processes drive an integrate-and-reset kernel instead?

12.2 *Robustness/error tradeoff in estimation* In Sec. 12.2 we pointed out that a tradeoff exists between robustness and error. For nearly all estimation tasks, some estimators provide excellent results for a restricted class of data, whereas others yield useful results for a wider range of data at the cost of somewhat greater overall error. Discuss this issue in the context of estimating the fractal exponents for a fractal renewal process and for a fractal-shot-noise-driven Poisson process.

12.3 *Bias/variance tradeoff in estimation* In addition to the tradeoff between robustness and error discussed in Prob. 12.2, a tradeoff also exists between bias and variance in many estimators. Demonstrate this tradeoff explicitly for the estimation of $\widehat{\alpha}$ from the normalized Haar-wavelet variance $\widehat{A}(T)$ by computing the correlation coefficient between the absolute values in the third and fourth columns of Table 12.1, which represent the bias and standard deviation, respectively. Repeat this calculation for the estimation of $\widehat{\alpha}$ from $\widehat{A}(T) - 1$ by making use of the sixth and seventh columns of Table 12.1. What might account for the difference in the correlation coefficients obtained in the two cases?

12.4 *Coincidence-rate and spectrum estimation* Discuss the problems involved in attempting to construct an estimate of the coincidence rate $G(t)$, and show that addressing these problems leads to the count-based autocorrelation $R_Z(k, T)$ or a similar measure. Also show that the point-process spectrum $S_N(f)$ does not suffer from the same shortcoming. Why then do we employ the rate spectrum $S_\lambda(f, T)$ instead of $S_N(f)$?

12.5 *Bias in normalized-variance estimates* For a fractal-based point process with $0 < \alpha < 1$, Eq. (12.6) indicates that the normalized variance will decrease for counting times T near the duration of the recording L.

 12.5.1. Use this equation to find the counting time at which the normalized variance achieves a maximum, and express this counting time in terms of the fractal onset time T_F, the duration of the recording L, and the fractal exponent α.

 12.5.2. Simulate a fractal-based point process, say the fractal-Gaussian-process-driven Poisson process, and calculate the normalized variance for the largest value of T possible. Plot the results of the simulations (include both mean and mean ± 1-standard deviation values) as well as the predictions of Eqs. (5.44b) and (12.6), choosing a plotting format that highlights the differences between the two predictions.

12.6 *Effect of averaging on spectral estimates* To obtain useful nonparametric estimates of the spectrum (rather than estimates of the fractal exponent α that we have been heretofore pursuing), it is common to average n adjacent values of the estimated spectrum, with n a large number that does not vary with frequency. One can also divide the data into blocks, compute spectrum estimates of each block separately, and then average these estimates across blocks. Explain why we do not use this approach in estimating the fractal exponent.

12.7 *Asymptote subtraction for improved fractal-exponent estimation*   All of the count-based measures employed in this chapter attain a constant value in the high-frequency/short-time limit. For example, Eq. (3.59) shows that the point-process spectrum approaches the mean event rate at high frequencies, whereas Eq. (3.42) indicates that both the normalized variance and the normalized Haar-wavelet variance attain a value of unity for short counting times. It is tempting to subtract these limits from the associated measures to extend the useful range of fractal scaling and to thereby improve the performance of the fractal-exponent estimators. We have, in fact, done just that for the periodogram in Sec. 12.3.9, and for the normalized Haar-wavelet variance in Secs. 12.2.3 and 12.3.8. Cite two reasons for caution in applying this approach to real data sets.

12.8 *Fractal behavior in a simulated fractal renewal process*   Simulate a number of runs of a fractal renewal process (see Chapter 7) and an equal number of runs of a homogeneous Poisson process (see Sec. 4.1), with the same mean interevent interval (choose $E[\tau] = E[\mu] = 1$ for simplicity). Select $\gamma = \frac{3}{2}$, which leads to $\alpha = \frac{1}{2}$ by virtue of Eq. (7.9). This value of γ lies within the range $1 < \gamma < 2$, which limits variation, as described in Sec. 7.1.3. Choose abrupt cutoffs with $B/A = 10^6$ to ensure the presence of fractal behavior over a wide range of times and frequencies; this yields $A = 1.001001/3 = 0.333667$, $B = 0.333667 \times 10^6$, and $L = 10^6$, which gives an expected number of events $E[N(L)] = 10^6$ [see Eq. (7.2)]. Carry out 100 simulations of this process and average the results to obtain accurate statistics.[9]

Show that the rate periodogram (spectrum estimate) and the normalized Haar-wavelet variance accurately characterize fractal behavior in the fractal renewal process and properly reveal its absence in the homogeneous Poisson process. Also show that the rescaled range statistic, an interval-based measure, cannot distinguish between these two renewal processes and therefore does not reliably detect fractal behavior in general.

---

[9] Producing a smooth version of this curve that accurately follows its expected value requires inordinate simulation resources (see Prob. 7.6).

13

Computer Network Traffic

In the course of his studies of telephone traffic, **Agner Krarup Erlang (1878–1929)**, a Danish mathematician, conceived of a number of important point processes and established the fundamental framework for queueing theory.

The Swedish mathematician **Conny Palm (1907–1951)** advanced the approach set forth by Erlang by incorporating realistic features of telephone traffic, such as the clustering of calls and the superposition of traffic on multiple channels.

13.1	Early Models of Telephone Network Traffic	315
	13.1.1 Queueing theory	316
13.2	Computer Communication Networks	320
	13.2.1 Scale-free networks	321
	13.2.2 Static representation	322
	13.2.3 Vertical layers	323
13.3	Fractal Behavior	324
	13.3.1 Early evidence	325
	13.3.2 Second-order statistics	325
	13.3.3 Queueing-theory analysis	327
	13.3.4 Predictability	329
	13.3.5 Origins	329
	13.3.6 Cutoffs	330
	13.3.7 Rate-process and point-process descriptions	330
	13.3.8 Multifractal features	331
13.4	Modeling and Simulation	332
	13.4.1 Analysis and synthesis	332
	13.4.2 Simulation approaches and model complexity	332
	13.4.3 Equivalent models in different guises	333
13.5	Models	334
	13.5.1 Fractal renewal point process	334
	13.5.2 Alternating fractal renewal process	334
	13.5.3 Fractal-Gaussian-process-driven Poisson process	335
	13.5.4 Fractal Bartlett–Lewis point process	335
	13.5.5 Fractal Neyman–Scott point process	336
13.6	Identifying the Point Process	337
	13.6.1 Compute multiple statistical measures	337
	13.6.2 Compute statistical measures for multiple data sets	342
	13.6.3 Identify characteristic features	342
	13.6.4 Compare with other point processes	343
	13.6.5 Formulate and simulate candidate models	345
	13.6.6 Compare model simulations with data	351
	Problems	351

In this, the final chapter of the book, we show how the various approaches and models developed in previous chapters can be used to analyze **computer network traffic**, a process that is at the same time complex and rich in fractal behavior.[1] The mathematical study of computer network traffic is called **teletraffic theory**. This

---

[1] This chapter is not designed to provide a comprehensive introduction to computer network traffic in general, nor to its fractal characteristics in particular. For the latter, we refer the reader to the comprehensive tome compiled by Park & Willinger (2000), the excellent article by Abry, Baraniuk, Flandrin, Riedi & Veitch (2002), and the didactic book chapter authored by Willinger, Paxton, Riedi & Taqqu (2003).

theory encompasses various features of queueing theory, stochastic processes, control theory, optimization theory, and graph theory. In practice it proves useful for ensuring network stability and for the optimization of resource allocation. Teletraffic theory enables us to evaluate routing protocols and switch designs and offers a point of departure when planning network expansion. Agner Krarup Erlang is widely known as the "father of teletraffic theory."

We begin in Sec. 13.1 with a brief review of early Poisson-based approaches to modeling **telephone network traffic**, as initially set forth by Erlang (1909), Engset (1915), and Palm (1937). In the course of this review, we provide an elementary introduction to queueing theory. In Sec. 13.2, we examine modern **computer communication networks**, which carry information in the form of packets,[2] and contrast these systems with telephone networks. We devote Sec. 13.3 to an examination of the fractal nature of computer network traffic. Various salient issues pertaining to modeling and simulation are set forth in Sec. 13.4. In Sec. 13.5 we consider a number of fractal-based point-processes that have served as models for computer network traffic.

Finally, in Sec. 13.6 we offer the reader a didactic step-by-step approach designed to assist in the identification of an unknown fractal-based point process. Using computer network traffic as an example, we demonstrate that the data sets we examine follow the form of a fractal-rate point process, and closely resemble the biological point process recorded at the striate cortex. Bearing in mind the tradeoff between model accuracy and parsimony, we conclude that two point-process models are good candidates for describing computer network traffic: a Neyman–Scott cluster process and a Bartlett–Lewis cascaded process. We examine the performance of these two models in some detail, and compare and contrast their predictions with two classic Ethernet-traffic data sets.

13.1 EARLY MODELS OF TELEPHONE NETWORK TRAFFIC

In the early years of telephone service, the subscribers in a town typically connected to a common exchange, staffed by an operator who routed all calls to their intended destinations. Routing calls within the town required only a simple connection at the exchange and rarely led to delay. However, call requests to numbers at other exchanges required the use of shared lines to those exchanges, and to additional lines and exchanges for very long-distance calls. When all lines to another exchange were busy, someone wishing to place a call through it would have to wait until one of the lines became free.

Service could, of course, be improved by installing an individual line for each customer, but the cost of doing so would be exorbitant. The intelligent design of

---

[2]Packets comprise small blocks of bits that travel together over a computer network, independently of other packets. Typically, the information in a file or data stream comprises a large number of packets.

any telephone network offers the engineer the following challenge: how to route calls among exchanges with a specified degree of reliability — and within a certain budget.

The design of an efficient system requires detailed knowledge of the offered load of call traffic. As part of a comprehensive examination of the applications of probability theory to telephone traffic in his native Denmark, Erlang carried out the first analyses of inter-exchange telephone traffic in 1909, 1917, and 1920.[3] He argued that the calls initiated by any one person form a negligible part of the aggregate call traffic at a large exchange. He also reasoned that different people initiate calls largely independently. Taken together, these heuristic arguments suggested that the homogeneous Poisson process (see Sec. 4.1) provides a suitable model for the aggregate traffic. And, indeed, this does turn out to be the case under many circumstances.

An extension of some of Erlang's results was provided by the Norwegian mathematician Engset, both in an unpublished manuscript completed in 1915 [Myskja (1998a) provides commentary on this manuscript] and in a paper published in 1918 [Jensen (1992) provides commentary on this paper].[4] In 1943, the Swedish mathematician Palm offered a number of significant generalizations of Erlang's results. He introduced such key features as slow rate modulations associated with daily, weekly, and yearly cycles; sudden increases in traffic following popular sporting events or major disasters; and the complexities of traffic that span multiple exchanges. The incorporation of these considerations played a crucial role in the design of efficient telephone networks.

13.1.1 Queueing theory

Queueing theory provides a suitable point of departure for studying simple telephone networks (Cohen, 1969; Cooper, 1972; Kleinrock, 1975; Asmussen, 2003). This mathematical formalism describes the utilization of a resource on which demands are made in a random fashion. The arrival times, and the magnitude of the resource requested per demand, may be random, and the resource itself may also vary in time. Demands that cannot be immediately met are queued (stored in a buffer) or declined.

For didactic purposes, we begin by considering the simple homogeneous Poisson-process model of call arrivals at a telephone exchange. Upon arrival, each call is queued. Resources can only be provided for storing a finite number of unprocessed call requests. As telephone lines come available, operators connect calls to their intended destinations in the order in which they arrived. The call durations follow an exponential distribution, corresponding to the interevent intervals of a homogeneous Poisson process. We consider the case of a single outgoing telephone line.

To model this call-activity sequence, we make use of the following construct:

---

[3] For a brief discussion of these papers, see Brockmeyer, Halstrøm & Jensen (1948, pp. 101–104). The 1917 paper is widely considered to be Erlang's most important.
[4] Engset (1915) highlighted the importance of the *truncated binomial distribution*, an extension of Erlang's (1917) *B formula*.

1. The **queue length** or **buffer occupancy** $Q(t)$ assumes integer values between a minimum of zero and a maximum of Q_m. The quantity Q_m is known as the **maximum queue length** or **buffer size**.

2. Calls arrive at times $t_{a,k}$ corresponding to a homogeneous Poisson process $N_a(t)$ with fixed, deterministic rate μ_a, where the label a denotes that it represents the **arrival process**, and k indexes the arrival times.

3. The service times are independent and identically distributed exponential random variables with mean duration $1/\mu_s$; the corresponding auxiliary homogeneous Poisson process $N_s(t)$ has a fixed, deterministic rate μ_s and corresponding event times $t_{s,k}$. The label s denotes that it represents the **service process**, and k again serves as an index.

4. When $Q(t) < Q_m$, $Q(t)$ increments by unity at each $t_{a,k}$.

5. When $Q(t) = Q_m$, the events of $N_a(t)$ correspond to dropped calls.

6. When $Q(t) > 0$, $Q(t)$ decrements by unity at each $t_{s,k}$.

In a handy notation developed by Kendall (1953), this model is called an M/M/1/Q_m queue (Kleinrock, 1975; Gross & Harris, 1998). The first symbol describes the **arrival process**, "M" for "Markov" in this case, indicating independent arrivals and therefore a homogeneous Poisson process. The exponentially distributed duration of each call corresponds to a homogeneous Poisson **service process**, so that "M" stands as the second symbol as well. The "1" that stands as the third symbol signifies the **number of servers** (outgoing lines). Finally, the last symbol "Q_m" characterizes the **maximum queue length**; by convention, the omission of this symbol signifies that $Q_m = \infty$. Other queueing models comprise different arrival or service processes, including those that are deterministic ("D") or general ("G"), and allow for an arbitrary number of servers.[5]

We now proceed to write a state equation for this model. Let $p_Q(n,t) \equiv p(n,t)$ represent the **queue-length distribution**, the probability that $Q(t) = n$. Except for the boundary cases $n = 0$ and $n = Q_m$, a constant rate of change μ_a associated with the arrival process $N_a(t)$ carries the queue-length distribution from $p(n,t)$ to $p(n+1,t)$, which concomitantly decreases $p(n,t)$. For this component we therefore have $dp(n,t)/dt = -\mu_a\, p(n,t)$. Similarly, an arrival when $Q(t) = n-1$ increases $p(n,t)$ via the term $+\mu_a\, p(n-1,t)$. The service process provides analogous contributions: $-\mu_s\, p(n,t) + \mu_s\, p(n+1,t)$. Recognizing that $p(n-1,t) = 0$ for $n = 0$ and $p(Q_m + 1,t) = 0$ for $n = Q_m$ accommodates the boundary cases. Combining

---

[5] Poisson-arrival and exponential-service processes have traditionally provided a good description for the public switched telephone network. These assumptions can no longer be fully justified, however, because of the vast changes that have taken place in the voice telephone network in recent years, such as its increased use for internet connections and facsimile transmission (see, for example, Duffy, McIntosh, Rosenstein & Willinger, 1994).

all terms, including those for the boundary cases, then leads to a rate equation known as a **forward Kolmogorov equation**:

$$
\frac{dp\,(n,t)}{dt} = \begin{cases} -\mu_a\,p\,(n,t) & +\mu_s\,p\,(n+1,t) & n=0 \\ -\mu_s\,p\,(n,t) & +\mu_a\,p\,(n-1,t) & n=Q_m \\ -(\mu_a+\mu_s)\,p\,(n,t)+\mu_a\,p\,(n-1,t)+\mu_s\,p\,(n+1,t) & & 0<n<Q_m. \end{cases}
$$
(13.1)

Under steady-state conditions, the left-hand side of Eq. (13.1) is zero for all n. A bit of algebra then leads directly to the **geometric queue-length distribution** (Erlang, 1917; Palm, 1943),

$$
p_Q(n,t) \to \frac{(1-\rho_\mu)\,\rho_\mu^n}{1-\rho_\mu^{Q_m+1}},
$$
(13.2)

where the **service ratio** (also called **server utilization**) is defined as

$$
\rho_\mu \equiv \frac{\mu_a}{\mu_s}.
$$
(13.3)

For the special case of infinite buffer size, we recover the $M/M/1/\infty \equiv M/M/1$ queue, in which case Eq. (13.2) reduces to

$$
p_\infty(n,t) = (1-\rho_\mu)\,\rho_\mu^n.
$$
(13.4)

For a service ratio $\rho_\mu = 0.9$, we display this geometric queue-length distribution as the dashed straight line in Fig. B.15 (semilogarithmic coordinates), and as the dotted curve in Fig. B.16 (doubly logarithmic coordinates).

Three measures turn out to be useful for assessing queueing-system performance: the **mean queue length** (or **mean number of waiting calls**), the **mean waiting time** spent in the buffer, and the **overflow probability**. We consider these measures in turn.

The mean number of waiting calls follows directly from the distribution provided in Eq. (13.2) (Palm, 1943):

$$
\begin{aligned}
E[Q] &= \sum_{n=0}^{\infty} n p_Q(n) \\
&= \sum_{n=0}^{Q_m} n \frac{(1-\rho_\mu)\,\rho_\mu^n}{1-\rho_\mu^{Q_m+1}} \\
&= \frac{\rho_\mu - (Q_m+1-\rho_\mu Q_m)\,\rho_\mu^{Q_m+1}}{(1-\rho_\mu)\left(1-\rho_\mu^{Q_m+1}\right)}.
\end{aligned}
$$
(13.5)

Straightforward algebra yields the higher-order moments of this distribution as well.

An intuitive but nontrivial result, known as **Little's law** (Little, 1961), provides that the mean waiting time for a single server is simply the mean number of waiting calls multiplied by the mean service time:

$$
E[\tau_w] = \frac{E[Q]}{\mu_s}.
$$
(13.6)

Results for multiple servers are somewhat more complex, although still quite tractable, since call-traffic sharing occurs across lines; before any call encounters a delay, all lines must be occupied. We readily modify Eq. (13.6) to yield an approximate result for M servers:

$$\mathrm{E}[\tau_w] \approx \frac{\mathrm{E}[Q]}{M\mu_s}. \tag{13.7}$$

Nevertheless, we emphasize that Eqs. (13.1)–(13.5) change form for M servers. Transition rates among different queue occupancy probabilities $p(n,t)$ vary with n for $n < M$; not all M lines carry calls if fewer than M calls reside in the buffer. In particular, $\mathrm{E}[Q]$ no longer follows the form set forth in Eq. (13.5).

The third performance measure is the probability P_B that an arriving call fails to enter the buffer because it is full. This quantity is known as the **buffer overflow probability** (or **call-drop probability** or **blocking probability**). Setting $n = Q_m$ in Eq. (13.2) for the single server yields

$$
\begin{aligned}
P_B &= \lim_{t \to \infty} p_Q(Q_m, t) \\
&= \frac{(1 - \rho_\mu)\rho_\mu^{Qm}}{1 - \rho_\mu^{Qm+1}} \tag{13.8} \\
&= \frac{1 - \rho_\mu}{\rho_\mu^{-Qm} - \rho_\mu}. \tag{13.9}
\end{aligned}
$$

The proportion of arrivals that finds the queue full equals the proportion of times that the queue is full. Said differently: *Poisson arrivals see time averages*, often captured by the acronym *PASTA* (Wolff, 1982). For large buffer sizes Q_m, the term ρ_μ^{-Qm} in the denominator of Eq. (13.9) dominates ρ_μ for $\rho_\mu < 1$, so that $\rho_\mu^{-Qm} - \rho_\mu \to \rho_\mu^{-Qm}$ [this approximation understates P_B by the factor $1/(1 - \rho_\mu^{Qm-1}) \approx \rho_\mu^{Qm-1}$]. The overflow probability then reduces to

$$P_B \approx (1 - \rho_\mu)\rho_\mu^{Qm} \sim \rho_\mu^{Qm}. \tag{13.10}$$

Equation (13.10) reveals that Poisson arrival and service processes give rise to an overflow probability that decreases with decreasing service ratio ρ_μ as a power-law function, and decreases with increasing maximum queue length Q_m as an exponential function.

Figures 13.1 and 13.2 display the behavior of the overflow probability P_B set forth in Eq. (13.10), as a function of the service ratio ρ_μ and of the maximum queue length Q_m, respectively. Relatively modest values of the maximum queue length yield quite small overflow probabilities. Erlang first presented these results, as well as exact results for M independent servers (telephone lines), in 1917.

For the M/M/1/Q_m queue, Eq. (13.2) shows that $p_Q(n) \sim \rho_\mu^n$ while Eq. (13.10) tells us that $P_B(Q_m) \sim \rho_\mu^{Qm}$. We conclude that for fixed ρ_μ, both the queue-length distribution and the overflow probability follow a geometric distribution. Indeed, for

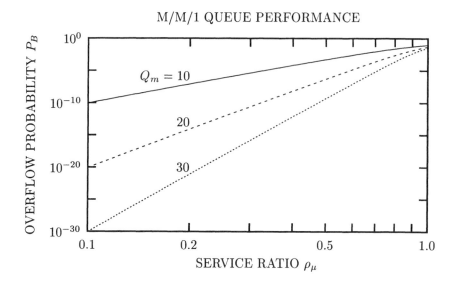

Fig. 13.1   Buffer overflow probability P_B as a function of the service ratio $\rho_\mu \equiv \mu_a/\mu_s$, for three values of the maximum queue length: $Q_m = 10$ (solid curve), 20 (dashed curve), and 30 (dotted curve). The roughly straight-line behavior on this doubly logarithmic plot represents the power-law relation between P_B and ρ_μ inherent in Eq. (13.10).

$Q_m \to \infty$, Eqs. (13.4) and (13.10) provide

$$
\begin{aligned}
p_\infty(Q_m) &= (1 - \rho_\mu)\,\rho_\mu^{Q_m} \\
P_B &\approx (1 - \rho_\mu)\,\rho_\mu^{Q_m},
\end{aligned}
\tag{13.11}
$$

respectively, where $p_\infty(n)$ represents the queue-length distribution for an M/M/1/$\infty \equiv$ M/M/1 queue. As discussed in Probs. 13.3 and 13.5, these equations demonstrate that the infinite-buffer queue-length distribution $p_\infty(n)$, evaluated at $n = Q_m$ where Q_m is the buffer size, provides an approximation for the overflow probability of the M/M/1/Q_m queue:

$$
P_B \approx p_\infty(Q_m).
\tag{13.12}
$$

13.2   COMPUTER COMMUNICATION NETWORKS

Modern computer communication networks differ greatly from their voice-based precursors. Indeed, they are possibly the most complex of all systems contrived by humans. Data travel as small blocks of digital bits, in the form of packets, rather than as entities such as entire telephone conversations or files. No master scheduler directs the functioning of routers in the network; rather, each router passes packets

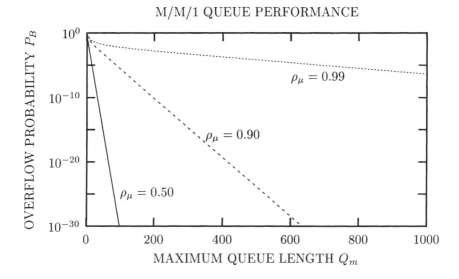

Fig. 13.2 Buffer overflow probability P_B as a function of the maximum queue length Q_m, for three values of the service ratio: $\rho_\mu \equiv \mu_a/\mu_s = 0.50$ (solid curve), 0.90 (dashed curve), and 0.99 (dotted curve). The roughly straight-line behavior on this semilogarithmic plot represents the exponential relation between P_B and Q_m inherent in Eq. (13.10).

on to other routers based largely on local activity and availability. The network itself dynamically allocates the routes over which the packets travel. As a consequence, packets flow smoothly around a blocked router, whereas a corresponding failure in a voice network might easily disable a large section of the network.

13.2.1 Scale-free networks

Both the Internet and the World Wide Web[6] behave as a scale-free networks [see Sec. 2.7.8 and Albert & Barabási (2002); Dorogovtsev & Mendes (2003); Pastor-Satorras & Vespignani (2004); Song, Havlin & Makse (2005)]. Such networks abound in the domain of computer communications — power-law distributions describe: (1) the number of edges emanating from a vertex in the Internet graph (Faloutsos, Faloutsos & Faloutsos, 1999; Aiello, Chung & Lu, 2001); (2) the number of exchanged emails per email address (Ebel, Mielsch & Bornholdt, 2002); (3) the number of web pages per website (Huberman & Adamic, 1999); and (4) the number of hyperlinks per web page in the virtual World Wide Web (Albert et al., 1999).

---

[6]The nodes of the Internet are the physical routers and computers while the edges are the connecting cables and wires. The nodes of the World Wide Web are web documents while the edges are the directed hyperlinks (URLs) that connect them.

Proper design of network topologies, and avoiding the deleterious effects of coordinated attacks against network hubs, require that we understand and accommodate the scaling nature of the network.

13.2.2 Static representation

Even a static representation of the Internet proves difficult to analyze. Figure 13.3 shows one representation of the major ISP (Internet Service Provider) nodes of the Internet, indicated as small squares. The angular position around the circle indicates the geographical longitude of the node while the distance from the center to each node varies inversely with the traffic carried by that node. The Internet comprises more than 100 000 separate networks with more than 100 million hosts. There are millions of routers, billions of web locations, and tens of billions of catalogued documents resident on the World Wide Web.

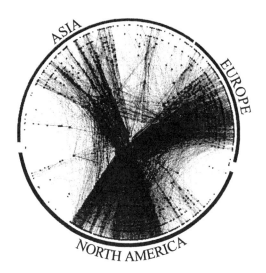

Fig. 13.3 Snapshot of ISPs (Internet Service Providers) constructed from data collected during the period 21 April 2003 through 8 May 2003. The angular position around the circle represents the geographical longitude of the ISP node (represented by a small square) while the distance from the center to each node varies inversely with the traffic carried by that node. The graph reflects more than 1 million IP (Internet Protocol) addresses and more than 2 million IP links that are, roughly speaking, aggregated into a topology of 11 000 ISPs. Adapted from http://www.caida.org/analysis/topology/as_core_network/, which provides details of this representation.

13.2.3 Vertical layers

In conjunction with the horizontal complexity of the Internet described above, information in computer communication networks is transmitted in a vertically rich manner, usually using a five-layer TCP/IP (Transmission Control Protocol/Internet Protocol) suite. Each layer relies on the layer below it for executing more primitive functions, while providing services to the layer above it. The highest layer corresponds to applications such as HTTP, whereas the lowest layer handles the physical transfer of bits over the medium.

We thus consider teletraffic in terms of five layers, each with its own set of tasks and protocols (conventions and rules):

- *Application Layer*: Execution of individual applications such as HTTP (hypertext transfer protocol), FTP (file-transfer protocol), TELNET (telephone network remote connection), or SSH (secure shell).

- *Transport Layer*: Delivery of events within those applications, such as individual file transfers within an HTTP session (TCP is the transport protocol for TCP/IP).

- *Internetwork Layer*: Transmission through the Internet of blocks of packets within those file transfers (IP is the internetwork protocol for TCP/IP).

- *Link Layer*: Transmission of individual packets within those blocks of packets on individual links.

- *Physical Layer*: Transmission of individual bits within those individual packets on a particular link.

In general, different vertical layers exhibit different statistics. Users initiating HTTP sessions, for example, might well follow a homogeneous Poisson process, at least over time scales of an hour or less (Feldmann, Gilbert & Willinger, 1998). On the other hand, the initiation times for individual HTTP commands, such as requests for documents or images, would likely follow a different statistical pattern. For example, individual packet arrivals might be characterized by a fractal-based point process as a result of power-law-distributed file sizes (see Sec. 13.3.5).

Figure 13.4 displays a highly schematized picture of information transmission on such a multi-layered structure, in the form of a cascaded point process. The primary point process $dN_1(t)$ in a) might represent the arrivals of HTTP file-transfer requests to a server. Each secondary point process $dN_{2,k}(t)$ in b) would then describe the resulting packet transfers for the corresponding files measured at a nearby downstream node, with the number of packets or temporal duration of each secondary process corresponding to the extent of the associated file-transfer flow. The total packet traffic process $dN_3(t)$ displayed in c) might then comprise the superposition of all packet arrival times at that nearby node.

a) PRIMARY PROCESS $dN_1(t)$

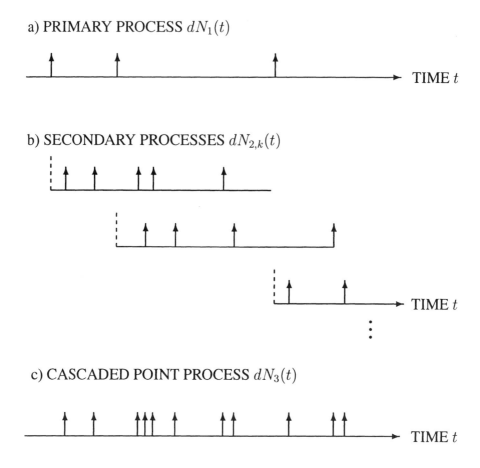

b) SECONDARY PROCESSES $dN_{2,k}(t)$

c) CASCADED POINT PROCESS $dN_3(t)$

Fig. 13.4 Partial schematic for computer network traffic based on a cascaded point process (see Fig. 4.2 and Sec. 4.5). Each event of a primary point process $dN_1(t)$ (displayed in a) initiates a secondary point process $dN_{2,k}(t)$ that terminates after a random number of events or a random duration (displayed in b). All secondary points, taken together as indistinguishable events, form the cascaded-point-process output $dN_3(t)$ (displayed in c). Special cases of cascaded point processes include the fractal Bartlett–Lewis process (Sec. 10.6.4) and the fractal Neyman–Scott cluster process (Chapter 10).

13.3 FRACTAL BEHAVIOR

Designers of the first computer communication networks attempted to emulate the approach used for voice networks, borrowing equations and even terminology from telephony. Telephone lines became links in computer networks, exchanges became servers, and calls became, variously, data streams, files, or packets. However, early results proved disappointing. Small increases in buffer size did not dramatically reduce the overflow probability for computer communication networks, as would be expected on the basis of Fig. 13.2.

Examining the packet streams revealed that computer network traffic arrived in unpredictable bursts of activity over many time scales. To accommodate this behavior, researchers proceeded to formulate increasingly complex Markov models, but with limited success. These models relied on the implicit assumption that fluctuations in the offered load resemble those of a homogeneous Poisson process for time scales beyond a manageable cutoff time. But no such cutoff appeared to exist. Moreover, as described in Sec. 13.2, the topology of the Internet and the dynamics of the World Wide Web are constantly in flux. Unusual features such as these have far-reaching implications for network engineering (Taubes, 1998).

13.3.1 Early evidence

In 1993, Leland and colleagues presented a seminal paper, followed a year later by an extended version (Leland et al., 1994), in which they demonstrated that the rate of Ethernet traffic varied as a fractal process with long-range dependence; these authors further suggested that Poisson behavior does not obtain at any useful time scale. Many subsequent measurements of computer communication traffic have vetted this early finding (see, for example, Willinger et al., 2003, and references therein), demonstrating that fractal behavior over a large range of time scales is present in many different kinds of traffic: Ethernet local-area-network (LAN) traffic (Leland et al., 1994); wide-area-network (WAN) traffic (Paxson & Floyd, 1995), variable-bit-rate (VBR) video traffic (Beran, Sherman, Taqqu & Willinger, 1995); and World Wide Web (WWW) traffic (Crovella & Bestavros, 1997).

Soon after the first of these results appeared, traffic models based on fractional Brownian motion (Norros, 1995) revealed that classical Poisson-based techniques provided seriously flawed predictions for such systems. The queue-length distributions and overflow probabilities turned out to decrease far more slowly with buffer size than expected on the basis of the exponential functions displayed in Fig. 13.2. Markov models can generate highly variable traffic loads ("burstiness") over short time scales but the variability always diminishes as the time scale increases. Traffic with fractal characteristics, on the other hand, exhibits significant fluctuations at all time scales, with concomitant high-rate periods of all durations.

13.3.2 Second-order statistics

To illustrate the fractal nature of computer network traffic, we analyze the classic Ethernet local-area-network (LAN) data set BC-pOct89. The data comprise the arrival times and durations of the first 1 million packets recorded on the main Ethernet cable at the Bellcore (BC) Morristown Research and Engineering Facility over a period of about 29 minutes beginning at 11:00 AM on 5 October 1989 (Leland & Wilson, 1989, 1991). We initially examine the rate spectrum and the normalized Haar-wavelet variance, measures that prove to be highly useful for parameter estimation, as discussed in Sec. 12.4. We subsequently examine a whole raft of statistical measures for these data (see Sec. 13.6.1).

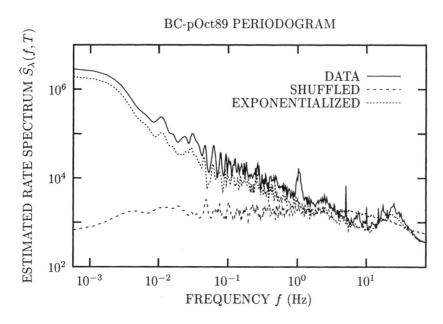

Fig. 13.5 Estimated rate spectrum (periodogram) $\widehat{S}_\lambda(f, T)$ vs. frequency f for the BC-pOct89 data set (solid curve), as well as for its exponentialized (dotted curve) and shuffled (dashed curve) surrogates. We smoothed the spectral estimate using the procedure reported in Footnote 7 on p. 117. The more-or-less straight-line decrease of the solid curve suggests that BC-pOct89 has a fractal rate. Since exponentialization leaves the fractal behavior only slightly changed, while shuffling destroys it, we conclude that the fractal behavior derives from the ordering of the intervals rather than from their distribution. Periodic components are in evidence at a number of frequencies.

As shown in Fig. 13.5, the estimated rate spectrum (solid curve) decreases in a power-law fashion over a broad range of frequencies f, confirming the presence of $1/f$-type noise and fractal behavior.[7] The periodogram of the exponentialized intervals (dotted curve; see Sec. 11.4.2) resembles the periodogram of the original data. In contrast, a shuffled version of the data (dashed curve; see Sec. 11.5) yields a periodogram that is devoid of power-law behavior. These results collectively indicate that the relative ordering of the intervals, rather than their distribution, is responsible for the fractal character of the data. Using the same reasoning we conclude that the broad spectral feature near $f = 30$ Hz derives largely from the interval ordering.

---

[7] The ordinate and abscissa are unnormalized in the BC-pOct89 periodogram displayed in Fig. 13.5. These same data appear in Fig. 5.1 with both the ordinate and abscissa normalized, and in Fig. 13.7i) with only the ordinate normalized.

BC-pOct89 NORMALIZED HAAR-WAVELET VARIANCE

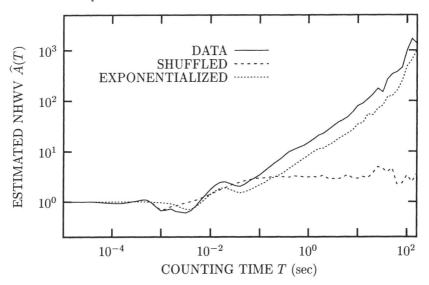

Fig. 13.6 Estimated normalized Haar-wavelet variance $\widehat{A}(T)$ vs. counting time T for the BC-pOct89 data set (solid curve). As with $\widehat{S}_\lambda(f, T)$, shown in Fig. 13.5, the more-or-less straight-line behavior suggests that BC-pOct89 has a fractal rate. Again, the surrogate data indicate that the interval ordering, rather than the interval distribution, is responsible for the fractal character of the data.

The estimated normalized Haar-wavelet variance (see Sec. 3.4.3) displayed in Fig. 13.6 (solid curve) also follows a power-law form over a broad range of counting times T, thereby confirming the conclusions drawn from the periodogram in Fig. 13.5. Computing this statistic for the two surrogate data sets also confirms that the relative ordering of the intervals, rather than their distribution, generates the fractal behavior. The broad bump in $\widehat{A}(T)$ near $T = 0.02$ sec corresponds to the spectral feature near $f = 30$ Hz in Fig. 13.5. The refractory behavior evident near $T = 0.005$ sec corresponds to frequencies that lie above the upper limit of the periodogram in Fig. 13.5.

13.3.3 Queueing-theory analysis

As a consequence of its fractal character, the second-order statistics of teletraffic do not follow Markov predictions, as shown in Sec. 13.3.2. Nor does the queueing behavior, as we now proceed to demonstrate.

For negligibly small buffers, fractal behavior has little impact on queueing performance since short-term (nonfractal) fluctuations overwhelm the buffer resources (Grossglauser & Bolot, 1996; Ryu & Elwalid, 1996). At the opposite extreme, exceptionally large buffers rarely overflow. For intermediate buffer sizes, however, the fractal nature of the traffic adversely affects queueing performance.

The queue-length distribution provides a useful window on network performance for this commonly encountered situation. Queue-length distributions resulting from fractal arrivals, or heavy-tailed service times, decay slowly with queue length in comparison with Markov predictions, often as power-law or Weibull functions (see, for example, Cohen, 1969, 1973; Norros, 1994; Brichet, Roberts, Simonian & Veitch, 1996; Roughan, Veitch & Rumsewicz, 1998; Asmussen, 2003). This has important implications for computer network traffic and for the design of computer communication networks (Erramilli, Narayan & Willinger, 1996).

A queue-length histogram that follows a decaying power-law form appears as the solid curve in Fig. B.16 (the solution to Prob. 13.6). This simulated result derives from the FGPDP/M/1 queue ($\rho_\mu = 0.9$), for which a fractal-Gaussian-process-driven doubly stochastic Poisson process (FGPDP) describes the arrivals, and the service times are exponential. We focus on this particular queue because of the ubiquity and importance of this arrival process (see Secs. 6.3.3, 8.4, 10.6.1, 13.5.3, and Chapter 12). Moreover, these results closely approximate those for the rectangular fractal-shot-noise-driven Poisson process (RFSNDP) (see Fig. B.19, the solution to Prob. 13.8), a plausible model for computer network traffic as discussed in Secs. 13.5.5 and 13.6. This latter queue-length histogram is also equivalent to a queue comprising Poisson flow arrivals and heavy-tailed service times, as discussed in Sec. 13.4.3.

These power-law queue-length histograms differ sharply from their M/M/1 geometric cousins (dotted curves in Figs. B.16 and B.19), which emerge when a homogeneous Poisson process describes arrivals at the queue, and the service times are exponential. The arrival process evidently imparts its fractal character to the resulting FGPDP/M/1 and RFSNDP/M/1 queue-length histograms, yielding power-law forms for these relations (straight lines on doubly logarithmic plots, as shown in Figs. B.16 and B.19).

The net result is a far larger range of possible queue lengths for the fractal queues. No characteristic size exists beyond which overflow probabilities decrease dramatically with increasing buffer size. Ensuring that fractal traffic reaches its destination, rather than encountering buffer overflows, thus demands far larger buffers than those needed for traffic based on Markov processes. Furthermore, the buffer-size requirements depend critically on the value of the fractal exponent that characterizes the offered traffic. Finally, we cannot fully describe the network by the service ratio ρ_μ, since this quantity effectively fluctuates.

In addition to the first-order "quality-of-service" measures of network performance considered above, second-order queueing statistics also prove important for characterizing fractal computer network traffic (Park, 2000). These include the standard deviations of the queue waiting time ("jitter") and of the message-loss probability, which, in many cases, can greatly exceed their mean values by virtue of the large fluctuations imparted by the fractal rate.

Finally, we note that the multilayered structure of commonly used protocols adds complexity to quality-of-service specifications (Park, 2000). Each layer has its own communication structure, and therefore a different set of statistics to specify. For example, specifications for the application layer might include file transfer rates for FTP or latency for an SSH session, but for the link layer they may involve packet-drop probabilities.

13.3.4   Predictability

The discussion provided in Sec. 13.3.3 shows that buffer overflows in computer communication networks stem from the persistence of fractal-rate fluctuations; a rate above the mean will likely remain so for some time. However, this same persistence leads to predictability in the traffic flow and can therefore be used to facilitate the allocation of resources to meet future needs. This approach can be useful in dynamically configuring network topologies; reallocation times can easily exceed buffer overflow times yet still lie well below the duration of long-term fluctuations.

Indeed, researchers have reported the feasibility of using predictive congestion control for fractal computer network traffic (Tuan & Park, 2000). However, different models yield vastly different values for the predictability,[8] and different estimates of fractal exponents also lead to varying results. Careful model choice and parameter estimation prove crucial in taking advantage of the predictability of such traffic.

13.3.5   Origins

As mentioned in Sec. 2.7.1, fractal behavior in computer network traffic is often ascribed to the power-law-distributed nature of file sizes (Park et al., 1996; Crovella & Bestavros, 1997). Imagine transferring a collection of such files over a computer communication network. Transfer via TCP (transmission control protocol) or UDP (user datagram protocol), in conjunction with flow and reliability control mediation, yields traffic with fractal properties (Park et al., 1996). In many cases, however, the fractal nature of computer network traffic appears to depend on all of the features involved. Eliminating flow and reliability controls by using UDP alone, for example, gives rise to output traffic that lacks much of the fractal structure of the input traffic (Park, Kim & Crovella, 2000). Moreover, the exponents of the file-size distributions do not always linearly predict the fractal exponents of the ensuing computer network traffic (Park et al., 2000). Evidently, the flow-control process can impart considerable complexity to the network traffic, beyond that of a simple fractal model.

Furthermore, large numbers of data-transfer processes occur concurrently within the same network, dividing network resources among them. Higher throughput for one application means less throughput for others. Features of one data stream, including the fractal exponents of its flow, may therefore ultimately derive from features of other streams. This interconnection becomes most important over times smaller than the round-trip time of the network under consideration. The melding of various fractal components may well lead to the multifractal characteristics seen over these shorter time scales (Willinger et al., 2003).

---

[8] We consider the predictability of fractional Brownian motion in Prob. 6.3.

Requests for power-law distributed traffic, mediated via lower-level transfer protocols, can lead to fractal behavior in communication networks. However, higher-level elements of the traffic stream can generate fractal fluctuations more directly. For example, VBR video traffic appears to exhibit fractal characteristics in a unified manner, and to lack meaningful discrete elements that themselves have power-law statistics (Garrett & Willinger, 1994). This fractal behavior may in turn derive from fractal characteristics in the input video stream.[9] Simple explanations do not always apply for such a richly complex system as the Internet.

13.3.6 Cutoffs

As discussed in Secs. 2.3.1 and 12.2.1, cutoffs play an important role in modeling fractal point processes; computer network traffic is no exception. The lion's share of the research in this connection concerns the long-time limit, and much of the mathematical framework that has been developed depends on the fractal behavior extending to infinite times. However, there are two reasons why fractal characteristics cannot extend to arbitrarily large times for real network traffic: (1) daily, weekly, and yearly rhythms exist, interfering with pure fractal behavior; and (2) all data sets truly have finite duration. Moreover, the absence of cutoffs leads to unwieldy mathematics. None of the results that we have set forth in this or earlier chapters depend on fractal activity extending to infinite times. Following the approaches specified in Secs. 2.3.1 and 12.2.1, we continue this tradition and employ finite cutoffs. Nevertheless, compelling mathematical reasons exist in some cases suggesting that outer cutoffs should be eliminated (Mandelbrot, 1997), especially in computer network traffic (Willinger, Alderson & Li, 2004).

13.3.7 Rate-process and point-process descriptions

As illustrated by the canonical data set labeled BC-pOct89, fractal fluctuations often form the salient characteristic of computer network traffic while effects at shorter time scales are less significant. Consider, for example, the normalized Haar-wavelet variance presented in Fig. 13.6 for these data. The plot manifests little evidence of fractal activity for counting times below about 10 msec, yet fully 98% of the interevent intervals lie below this value. Furthermore, some (although certainly not all) computer communication data sets have mean interevent intervals that lie far below the timestamp resolution. For example, the World Cup 1998 access log (Arlitt & Jin, 1998) comprises some 1 352 804 107 requests collected over 88 days from 30 April through 26 July 1998, at a resolution of one second. This translates to an average rate of 17.8 requests per second over the entire log; daily averages exceeding 81 requests per second (30 June); and correspondingly higher local rates over shorter time scales. Indeed, reconstructing the underlying point process for the World Cup log proves impossible.

---

[9]Natural scenes exhibit *spatial* fractal behavior in their own right, as discussed in Sec. 2.8.5.

We conclude that rate-based models, known as **fluid-flow models** in the context of computer network traffic, are highly useful. Many key results in fractal network traffic make use of this formulation, including early results relating to fractional Brownian motion (Norros, 1995).

We do not suggest, however, that rate-based models are superior to point-process models. As will become apparent in Sec. 13.6, data sets BC-pOct89 and BC-pAug89, which comprise experimental point processes, are readily analyzed as such.

13.3.8 Multifractal features

Computer network traffic comprises a multitude of events over a large range of time scales. In the BC-pOct89 data set analyzed in Figs. 13.5 and 13.6, for example, fully 5% of the intervals lie below 104 μsec; since the total duration of this data set exceeds 29 minutes, it effectively spans more than seven orders of magnitude in time.

With such a wide range of scales, detecting two or more scaling exponents becomes feasible.[10] Indeed, researchers have detected multifractal properties in wide-area network traffic over short time scales (Riedi & Lévy Véhel, 1997; Lévy Véhel & Riedi, 1997; Mannersalo & Norros, 1997). In particular, Feldmann et al. (1998) demonstrated that a conservative cascade model (Mandelbrot, 1974) could be used to characterize such traffic; however, they argued that multifractal properties exist only for (short) time scales that lie below the typical packet round-trip time (Riedi & Willinger, 2000). Taken together, this suggests that wide-area network traffic flow behaves as a multiplicative process over these short times, but becomes additive over longer time scales (Riedi & Willinger, 2000). Using this flow as a rate leads to the multiplicative-rate point process discussed in Sec. 5.5.1 (Schmitt et al., 1998). Interestingly, and in contrast, local-area network traffic appears to exhibit only a single fractal exponent and is therefore monofractal (Taqqu, Teverovsky & Willinger, 1997).

Other multifractal formalisms include processes with fractal exponents that vary with time, or across different realizations of the random process (Abry et al., 2000). We can readily construct a multifractal version of the fractal-shot-noise-driven Poisson process set forth in Chapter 10. Generalizing the fractal impulse response function defined in Eq. (9.2) provides

$$h(K, t) \equiv \begin{cases} f(K) \, t^{-\beta(K)} & A < t < B \\ 0 & \text{otherwise,} \end{cases} \tag{13.13}$$

where $\beta(K)$ and $f(K)$ are functions of the random variable K that describe the fractal exponent and its relative strength, respectively, for that particular impulse response function. In spirit, this approach resembles that used to generate fractal behavior from

---

[10]This stands in contrast to essentially all of the other fractal point-process data examined in earlier chapters, which had far shorter durations. In general, we found that characterizing one fractal exponent involved sufficient difficulty that other exponents, were they present, remained essentially undetectable.

a continuous superposition of Lorentzian spectra representing relaxation processes with time constants distributed over a range of values (see Sec. 2.7.9).

To analyze all putative multifractal processes, it proves helpful to access an extended range of statistical measures, such as the higher-order moments of wavelet transforms (Abry et al., 2000), $E\left[|C_{\psi,N}(T,\cdot)|^q\right]$, and generalized dimensions D_q for many values of q (see Sec. 3.5.4).

13.4   MODELING AND SIMULATION

We now examine a number of salient issues pertaining to the mechanics of modeling and simulation. In Sec. 13.4.1 we contrast the use of models for analysis and synthesis; in Sec. 13.4.2 we provide a discussion of simulation methods and model complexity; and in Sec. 13.4.3 we highlight the fact that equivalent models can appear in different guises.

13.4.1   Analysis and synthesis

Models prove useful for both the analysis and synthesis of computer network traffic. The use of models for analysis helps us visualize the effects of various parameters on traffic flow and, in particular, assists us in identifying sources of fractal characteristics in the traffic stream.

The synthesis of computer network traffic, on the other hand, provides synthetic data that is invaluable for studying and testing yet-to-be-developed computer-communication-network protocols and topologies. It is hard to overstate the value of synthesis because of the prodigious volume of traffic data required to meaningfully evaluate a new network design. For example, establishing a message-loss probability of 10^{-9} with reasonable precision requires far more than 10^9 packets for a memory-less data stream, using a brute-force approach. Although good approximate methods requiring less data do exist, simulation sizes remain considerable. The use of realistic (fractal) traffic data, with its concomitant clusters of high- and low-activity periods, further increases data requirements. Evaluating such a network over a range of parameters easily involves terabytes of data. In the face of such vast requirements, the synthesis of data from fractal models is often the only viable alternative for evaluating performance, particularly for novel networks that have not yet been implemented.

13.4.2   Simulation approaches and model complexity

What methods prove best for simulating computer network traffic? One of the first issues that arises is whether to impose a feedback loop from the network under evaluation to the simulated incoming traffic. In other words, should the simulated traffic source change what it offers on the basis of network parameters such as queue length or number of dropped messages? In many real-world applications, such as video and audio streaming (see Sec. 13.3.5), the offered load does not depend on the state of the network. In other applications, however, network performance does affect the input traffic; users encountering excessive delays often terminate their connections and wait until the network becomes less busy.

We do not address this issue explicitly for our network traffic models since none of the processes we have considered to this point has provision for such feedback. But we note that implementing feedback of this kind is not difficult since every model has parameters that directly control the output rate. Indeed, as discussed in Sec. 13.3.5, evidence exists that some of the fractal behavior inherent in computer network traffic may derive from the flow-control process (Park et al., 2000).

After determining the broad class of simulation, the issue becomes the level of detail to incorporate in the model. A tradeoff always exists between reality and simplicity; the ideal model captures all salient features of a data set on the one hand, and yet derives simply from a few underlying principles on the other hand. Useful models must strike a balance on the continuum between these two ideals. In the context of computer network traffic, the vast quantities of data shift the optimal model strongly toward parsimony. Complex models do indeed capture more features of the data, but they also tend to assume an *ad hoc* nature and require significant efforts to program. Indeed, simulation execution times can grow out of bounds so rapidly that models including even modest complexity can become useless.

13.4.3   Equivalent models in different guises

Identical network traffic can sometimes appear in different guises, as we briefly mentioned in Sec. 13.3.3. We saw such a duality in the context of general point processes in Sec. 4.5: under certain conditions, cascaded and doubly stochastic representations offer two different formalisms for the same underlying point process.

Consider, for example, a collection of data flows that follow a Poisson arrival process and exhibit heavy-tailed durations. The overall flow is then fractal shot noise with a rectangular impulse response function. We expect similar service times for all packets since they are restricted to a maximum size and most are at or near that maximum. The constituent packets consequently behave as a fractal-shot-noise-driven Poisson process (or a closely related integrate-and-reset version thereof). This leads us to recognize that the two processes are therefore different descriptions of precisely the same traffic. We conclude that the rectangular fractal-shot-noise-driven Poisson process (fractal Neyman–Scott process) is equivalent to a queue comprising Poisson flow arrivals and heavy-tailed service times. The connection is most valuable since this latter queue has been studied extensively in the literature. Moreover, in some cases one formulation may prove conceptually simpler than another, or it may offer faster simulations.

13.5 MODELS

A brief overview of the structure of computer communication networks appeared in Sec. 13.2, and we considered the fractal character of the resident traffic flow in Sec. 13.3. We discussed a number of salient issues pertaining to modeling and simulation in Sec. 13.4.

With this background, we are now in a position to consider several of the models presented in earlier chapters in the context of computer network traffic. These models, which offer different balances between reality and simplicity, as discussed in Sec. 13.4.2, prove useful in elucidating the behavior of computer network traffic. The mere fact that we consider *several* models, however, highlights the difficulties associated with identifying a unique fractal-based point process for a given collection of data. New models continue to be set forth (see, for example, Field, Harder & Harrison, 2004a,b).

13.5.1 Fractal renewal point process

A fractal renewal point process (Chapter 7) serves as a suitable model for the activity associated with a single network-traffic application. The superposition of a number of these processes (see Sec. 11.6.2) then represents the aggregate traffic from a collection of such applications, and thus provides a useful model for computer network traffic (Ryu & Lowen, 1996, 1998). The power-law decaying interevent-interval distribution imparts fractal fluctuations to the simulated teletraffic. A useful generalization of this approach, which takes the form of a marked-point-process model, accommodates messages of different (often power-law distributed) sizes (Levy & Taqqu, 2000).

13.5.2 Alternating fractal renewal process

The alternating fractal renewal process (Chapter 8) also leads to useful models for computer network traffic (Ryu & Lowen, 1996, 1998). Rather than each message forming a point event, as considered in Sec. 13.5.1, periods during which $X(t) = 1$ correspond to a message (such as a TCP connection) whereas periods during which $X(t) = 0$ correspond to inter-message quiet. The sum of a number of such alternating fractal renewal processes then represents messages independently generated by several applications. This sum, which is fractal binomial noise (see Sec. 8.3.1), serves as the rate process for packet generation.

Both Poisson and integrate-and-reset packet-generation mechanisms prove useful. Specific results are available for the queue-length distribution (Boxma, 1996) and the buffer overflow probability (Ryu & Lowen, 1997, 1998) for these point processes, as well as for their rate-based approximations (Heath, Resnick & Samorodnitsky, 1998; Jelenković & Lazar, 1999). Both approaches illustrate the sometimes paradoxical effects of fractal behavior: for $1 < \gamma < 2$, the dwell times for $X(t)$ have mean values that are finite, yet the average quantity of data residing in a buffer, waiting to be transmitted, becomes infinite (Boxma, 1996). Also, in some cases buffer overflow probabilities turn out not to depend significantly on the rate at which messages leave the buffer (Heath et al., 1998). The alternating fractal renewal process may be particularly suitable for modeling HTTP activity; file sizes for this application are generally power-law distributed (Feldmann, Gilbert, Willinger & Kurtz, 1998) and users often alternate between web-page downloading $[X(t) = 1]$ and viewing $[X(t) = 0]$.

The extended alternating fractal renewal process (Yang & Petropulu, 2001; Yu, Petropulu & Sethu, 2005), in which the packet-generation rate alternates between zero and a random value, with all such values independent of each other, adds flexibility to the rate process for modeling the burstiness of computer network traffic. A related model that lacks the explicit final Poisson process has also been extensively investigated (Mandelbrot, 1969; Taqqu & Levy, 1986; Levy & Taqqu, 2000).

13.5.3 Fractal-Gaussian-process-driven Poisson process

The fractal-Gaussian-process-driven Poisson process is ubiquitous because many fractal-based point processes, as well as superpositions thereof, converge to it (see Secs. 6.3.3, 8.4, 10.6.1, 11.6.1, 13.3.3, and Chapter 12). Kurtz (1996) showed that similar behavior is observed in computer network traffic for the fractal-shot-noise-driven Poisson processes considered in Sec. 13.5.5.

If a number of traffic sources aggregate to produce an overall traffic stream, flow control will not significantly affect any one source by itself. Over long time scales, then, the fractal Gaussian process (Sec. 6.3.3) should provide a useful model for the rate of the resulting system. A number of queueing results based on such processes exist in the literature (Norros, 1995; Lévy Véhel & Riedi, 1997).

13.5.4 Fractal Bartlett–Lewis point process

The vertical-layer structure discussed in Sec. 13.2.3 suggests that cascaded-point-process models (see Fig. 13.4) may be useful for characterizing computer network traffic. We consider two such models, in turn: the fractal Bartlett–Lewis point process and the fractal Neyman–Scott point process.

The Bartlett–Lewis point process introduced in Sec. 4.5 makes use of a primary homogeneous Poisson process; the secondary processes comprise segments of renewal processes with independent intervals drawn from identical distributions. The homogeneous Poisson process does indeed provide a good description for session arrivals in some forms of computer network traffic, at least over time scales of an hour or less; examples include FTP and TELNET (Paxson & Floyd, 1995), as well as HTTP (Feldmann et al., 1998). In the fractal version of the Bartlett–Lewis model introduced by Grüneis and colleagues (see, for example, Grüneis, 1984; Grüneis & Baiter, 1986; Grüneis, 2001), which was discussed in Sec. 10.6.4, the number of events M_k in each secondary process follows a power-law form. The power-law-distributed nature of file sizes (Park et al., 1996; Crovella & Bestavros, 1997) accords with this model.

Hohn, Veitch & Abry (2003) considered the fractal Bartlett–Lewis model in the context of computer network traffic. To match the measured interevent-interval histogram, they drew renewal-process segments from identical gamma distributions[11] of order m = 0.60 (see Prob. 4.7). They selected the number of events M_k in each secondary process to follow a power-law form with no upper scaling cutoff, and chose the exponent such that M_k had finite mean but infinite variance. The model of Hohn et al. (2003) thus has five parameters: the primary-process mean interevent interval $E[\tau_1]$, the secondary-segment mean interevent interval $E[\tau_2]$, the order of the gamma distribution m for this interevent interval, the mean number of events $E[M_k]$ in a secondary segment, and the power-law exponent z that characterizes the distribution of these secondary events. Simulations based on the model accord well with many features of network traffic; in particular, wavelet analysis reveals good agreement with measured first- and second-order statistics for a variety of packet traces.

This model, as well as the fractal Neyman–Scott model considered in the next section (Sec. 13.5.5), are promising candidates for characterizing computer network traffic; we consider both in greater detail in Sec. 13.6.

13.5.5   Fractal Neyman–Scott point process

The final teletraffic model we consider makes use of a fractal Neyman–Scott cluster point process. As with the Bartlett–Lewis process considered above, the primary events derive from a homogeneous Poisson process corresponding to the start times of traffic flows (see Fig. 13.4), but in this model the primary events initiate impulse response functions $h(t)$ with power-law-varying durations (see Prob. 9.2 and Ryu & Lowen, 1995, 1997, 1998). Adding a second Poisson process gives rise to a form of the fractal-shot-noise-driven Poisson process set forth by Lowen & Teich (1991) and studied in Chapter 10. It is a special Neyman–Scott cluster process, as discussed in Sec. 4.5; related models have also been considered by Cox (1984), Mikosch, Resnick, Rootzén & Stegeman (2002), and Latouche & Remiche (2002). The power-law feature of the impulse response function captures the burstiness of the traffic.

In the general case, no direct correspondence is patently obvious between the form of the power-law impulse response function and any particular feature of the network or traffic. For FTP and several other specific forms of traffic, however, file sizes follow a power-law distribution (Paxson & Floyd, 1995; Park et al., 1996; Crovella & Bestavros, 1997). As suggested by Ryu & Lowen (2002), we can therefore make the fractal-shot-noise-driven Poisson process quite realistic by positing a rectangular impulse response function $h(t)$ with a random cutoff time B characterized by a decaying power-law distribution (see Prob. 13.8). This rectangular fractal-shot-noise-driven Poisson model closely mimics observed FTP traffic.

Moreover, this approach permits the use of a queueing representation, as considered in Sec. 13.3.3 (see Fig. B.19, the solution to Prob. 13.8). As discussed in

---

[11] Inasmuch as the primary and secondary processes *both* give rise to the form of the interevent-interval distribution, its deviation from exponential form should, properly speaking, not be attributed solely to the secondary process.

Sec. 13.4.3, the queue-length histogram is identical to that for a queue comprising Poisson flow arrivals, heavy-tailed service times, and locally Poisson packet arrivals. In particular, the M/G/1/∞ queue represents shot noise with a fixed-height impulse response function $h(t)$. Specifying a power-law form for the service-time distribution G yields fractal shot noise with this fixed-height impulse response function (Likhanov, Tsybakov & Georganas, 1995).

The fractal Neyman–Scott model is similar to, but distinct from, the fractal Bartlett–Lewis model considered in the previous section (Sec. 13.5.4). Since both are promising candidates for characterizing computer network traffic, we examine them in greater detail in the following section (Sec. 13.6).

13.6 IDENTIFYING THE POINT PROCESS

As discussed in Secs. 5.5.4, 11.5.3, and 12.1, identifying a fractal-based point process is not an easy endeavor. In this, the final section of the book, we set forth a step-by-step approach toward identifying an arbitrary fractal-based point process, using computer network traffic as a didactic testbed.

We offer the following steps as a possible blueprint for the analysis of unknown fractal-based point processes.

13.6.1 Compute multiple statistical measures

We begin by gathering a whole range of measures from the experimental point process and presenting them in a single graphic. For the case at hand, Ethernet-traffic data set BC-pOct89, we present nine statistical measures as the solid curves in Fig. 13.7.

While the statistics of the data set itself are vital, calculating the same statistics for surrogate data sets yields further information that is highly valuable for elucidating the nature of the point process. We employ two surrogates. The first, a shuffled surrogate, comprises the same interevent intervals as the original data set, but rearranged into a random order. As described in Sec. 11.5, a shuffled surrogate has marginal interevent-interval statistics that coincide with those of the original data. Since shuffling generally destroys any dependencies among the intervals, thereby rendering them independent, the other statistics mimic those of a renewal point process. In short, shuffling destroys the long-term properties of a data set while preserving its short-term qualities.

The second surrogate achieves essentially the reverse. As described in Sec. 11.4, we construct this surrogate by transforming the interevent intervals from their original form into a specified distribution, while preserving their relative ordering and the mean of the interevent intervals. In particular, we transform the intervals through exponentialization, which yields an exponential interevent-interval density. An exponentialized data set roughly resembles a Poisson process, but with a variable rate.

Figure 13.7a) directly displays the sequence of interevent intervals for the first five seconds of the data set. The average of the two event times flanking an interval forms the abscissa for that interval, while the interval duration determines the ordinate. We express the duration in terms of its mean, $\tau(t)/\widehat{E}[\tau]$, so that a value of unity corresponds to an interval that equals this average. While no strong pattern emerges from this panel, we see the large variability in the rate, as well as evidence for preferred intervals at and below unity. We do not present shuffled or exponentialized versions of these data since distinguishing among the three types of points would prove difficult.

While the individual intervals betray mainly short-term effects, the longer-term properties of the point process are more readily revealed by variations in the rate, presented in normalized form in panel b). We use a counting time of 0.3 sec to compute the normalized rate, $\lambda_k/\widehat{E}[\lambda]$, so that the 41 windows shown along the abscissa span 12.3 sec. A value of unity indicates a local rate equal to that of the data set as a whole. The relatively low rate over the first ten or so windows corresponds to the relative preponderance of long intervals in the first three seconds of panel a). As expected, the shuffled surrogate shows far less variability than the original data, since shuffling destroys inter-interval dependencies. The exponentialized surrogate yields results resembling those of the original data.

The interevent-interval histogram displayed in panel c), $\widehat{p}(\tau/\widehat{E}[\tau])$, provides an estimate of the underlying interval probability density (see Sec. 3.3.1). We normalize both the abscissa and ordinate by the mean interevent interval, which yields dimensionless quantities on both axes. The data roughly follow a decaying exponential form, punctuated by a number of peaks. The shuffled surrogate has precisely the same histogram, by construction. The exponentialized form follows a straight line, as it must on this semilogarithmic plot.

The next four panels, d)–g), present interval-based measures that examine dependencies across interevent intervals. All of these measures employ normalization such that independent intervals yield values of unity on the ordinate, and all follow a power-law form (straight line on these plots) for interevent intervals exhibiting fractal correlations.

Panel d) presents the interval-based normalized rescaled range (NR/S), $\widehat{U}_2(k) \equiv \widehat{U}^2(k)/k$, as defined in Secs. 3.3.5 and 12.3.4. This measure follows a power-law form for the original data, indicating fractal behavior. The exponentialized surrogate yields similar results, but lies slightly above the original data. The shuffled surrogate approaches a value of unity, but not closely; the difference stems from the well-known bias inherent in this statistic.

In panel e) we display the normalized detrended fluctuation (NDF), $\widehat{Y}_2(k_2) \equiv 15\,\widehat{Y}^2(k_2)/k_2\mathrm{Var}[\tau]$, where $k_2 = k + 2$, as defined in Secs. 3.3.6 and 12.3.5. Rather than plotting the number of intervals on the abscissa, we offset this by two, as explained in Sec. 12.3.5. This measure also exhibits power-law behavior, and perhaps more closely follows the canonical fractal form of Eq. (12.2). Results for the exponentialized data again lie slightly above those of the original data, while the shuffled version yields values quite close to unity, as expected.

We show the normalized interval-based Haar-wavelet variance (NIWV), $\widehat{A}_\tau(k) \equiv \widehat{\mathrm{Var}}[W_{\psi,\tau}(k,l)]/\widehat{\mathrm{Var}}[\tau]$, in panel f). This measure, considered in Secs. 3.3.4 and

12.3.6, yields results quite similar to those of the normalized detrended fluctuation shown in panel e).

Panel g) displays the last of the interval measures, the normalized interval spectrum (NIS), $\widehat{S}_\tau(\mathsf{f})/\widehat{\mathrm{Var}}[\tau]$, considered in Secs. 3.3.3 and 12.3.7. We smoothed this measure as described in Footnote 7 on p. 117. Like the three preceding interval-based measures, the normalized interval spectrum follows a power-law form for both the original and exponentialized data. However, a bump appears in the spectrum at about $\mathsf{f} = 0.05$, corresponding to a conventional frequency $f = \mathsf{f}/\mathrm{E}[\tau] \approx 28$ Hz. The shuffled data fluctuates closely about unity. All three curves approach a value of unity for large frequencies, as imposed by the normalization.

The normalized Haar-wavelet variance (NHWV) shown in panel h), $\widehat{A}(T)$, derives from the sequence of counts rather than from the intervals (see Secs. 3.4.3, 12.2.3, and 12.3.8). The abscissa therefore corresponds precisely to conventional time. All curves achieve a value of unity at small counting times, by construction. For the original data, this measure follows a power-law form for the most part. However, a few bumps appear at shorter times; these are consistent with a periodic component in the neighborhood of 28 Hz, as we also inferred from the interval spectrum in panel g) [see also Prob. 4.10.1 in conjunction with Eq. (3.41)]. In contrast to panels d) through f), the curve for the exponentialized data lies slightly below that of the original data, although it otherwise resembles it in most respects. The shuffled surrogate yields a normalized Haar-wavelet variance that dips slightly below unity at small counting times and then increases to a value of about three at much larger counting times. A renewal point process with an interval coefficient of variation $C_\tau = 1.8$ (see Table 13.2) would yield a similar curve, in accordance with Eq. (4.18). Shuffling retains the interevent-interval statistics, in particular preserving the empirical value $C_\tau = 1.8$ from the original data.

Finally, panel i) presents the normalized rate spectrum (NRS), $\widehat{S}_\lambda(f, T)/\widehat{\mathrm{E}}[\lambda]$, considered in Secs. 3.4.5 and 12.3.9. This measure also has an abscissa that corresponds exactly to a conventional quantity, in this case frequency. To permit the frequency to extend as high as 100 Hz, we employed a Fourier transform of size 2^{19}, the minimum for a data set of this duration. As with the interval-based spectrum in panel g), we smoothed this measure as described in Footnote 7 on p. 117. Normalization by the mean rate forces all curves to attain an asymptote of unity for sufficiently large frequencies; however, the spectrum reaches this asymptote only for frequencies greater than about 10^4 Hz (not shown). (Unnormalized versions of the rate spectrum appear elsewhere — see Footnote 7 on p. 326.) This measure yields results that roughly resemble those for the interval spectrum displayed in panel g), with power-law behavior for both the original and exponentialized data. A peak appears at about 28 Hz, in concert with those observed in panels g) and h). The normalized rate spectrum for the shuffled data achieves a value of about three at low frequencies, in mirror image to the normalized Haar-wavelet variance, as expected on the basis of Eq. (4.17).

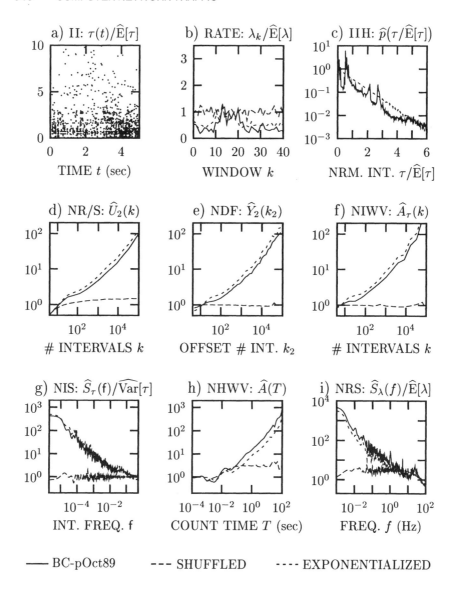

Fig. 13.7 Nine statistical measures for the classic computer network traffic data set BC-pOct89 (solid curves). The data comprise the arrival times of the first 1 million packets recorded on the main Ethernet cable at the Bellcore (BC) Morristown Research and Engineering Facility over a period of some 29 minutes beginning at 11:00 AM on 5 October 1989 (Leland & Wilson, 1989, 1991). Results for the shuffled and exponentialized surrogates appear as the dashed and dotted curves, respectively. We describe the measures and surrogates in the text.

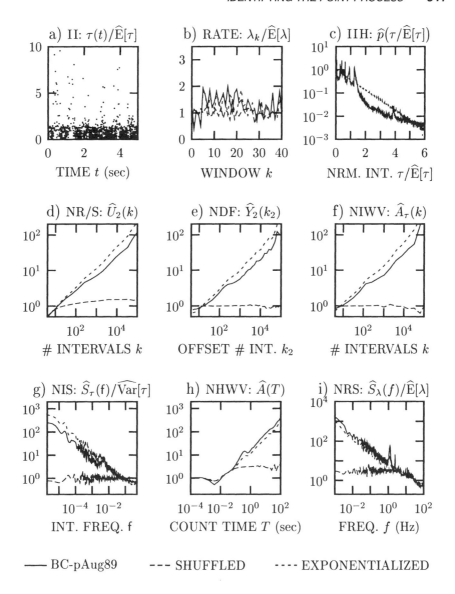

Fig. 13.8 Nine statistical measures for the classic computer network traffic data set BC-pAug89 (solid curves). The data comprise the arrival times of the first 1 million packets recorded on the main Ethernet cable at the Bellcore (BC) Morristown Research and Engineering Facility over a period of some 52 minutes beginning at 11:25 AM on 29 August 1989 (Leland & Wilson, 1989, 1991). Results for the shuffled and exponentialized surrogates appear as the dashed and dotted curves, respectively. The statistics resemble those presented in Fig. 13.7 for data set BC-pOct89.

13.6.2 Compute statistical measures for multiple data sets

Since a similar, but independent, Ethernet-traffic data set is available, we present the same nine statistical measures displayed in Fig. 13.7 as the solid curves in Fig. 13.8. The object of investigating multiple data sets is to establish which features of the data are general characteristics and which appear to vary from one data set to another.

The statistics for BC-pAug89 resemble those for BC-pOct89 in broad outline (compare Figs. 13.7 and 13.8), although they differ in some small details, as we proceed to highlight.

Periodicities are more dominant in the first two panels of Fig. 13.8 than in Fig. 13.7. Panel a) in Fig. 13.8 displays the occurrence of rather long interevent intervals roughly every second or so, the presence of which, in turn, leads to roughly periodic fluctuations of the rate estimate in panel b), based on 0.3-sec counting windows. This periodic component also appears as a peak in the normalized rate spectrum in panel i) at about 1 Hz; a second harmonic also appears.[12] The peaks and valleys in the normalized Haar-wavelet variance displayed in panel h) derive from this as well. However, this strong periodicity is specific to the time at which the data record begins; other times within BC-pAug89 also display this feature, but less strongly. The beginning of this data set also differs from the remainder in that the local rate exceeds the mean. Normalized rates in excess of unity in panel b), and a preponderance of small intervals in panel a), both accord with this observation. Conversely, some portions of data set BC-pOct89 appear more periodic than its beginning. In fact, the periodic component manifested at 28 Hz in BC-pOct89 is totally absent in BC-pAug89.

The interevent-interval histogram displayed in panel c) of Fig. 13.8 reveals different peaks at various normalized interevent times; however, it exhibits an exponential tail in the long-time limit (not shown). Similar behavior appears in panel c) of Fig. 13.7, although the peaks are localized at different interevent times; it, too, exhibits an exponential tail in the long-time limit.

We conclude that data sets BC-pOct89 and BC-pAug89 have similar principal features, although they differ in many details.

13.6.3 Identify characteristic features

Taking the collection of these observations into account, two characteristic features emerge with respect to the data presented in Figs. 13.7 and 13.8: (1) the presence of a fractal rate, and (2) an estimated interevent-interval distribution that does not impart fractal properties to the data. A signature of the first feature is the nearly constant, nonzero slopes of the solid curves in panels d)–i) of both figures. Because shuffling destroys this property (dashed curves), we conclude that the point process belongs to the class of fractal-rate point processes, and not to the class of fractal point processes. This conclusion is confirmed by the behavior of the generalized dimensions D_q for

---

[12] Since data set BC-pAug89 has a duration that is approximately twice that of BC-pOct89, we doubled the Fourier transform size to 2^{20} to allow the abscissa in panel i) to reach 100 Hz.

these data (see Sec. 3.5.4 and Prob. 5.5), which exhibit integer values (not shown). The capacity-dimension scaling function presented in Fig. 5.10 explicitly illustrates this for D_0.

While estimates of the fractal exponents vary considerably, depending on the measure and the method of estimation (tables not shown), $\widehat{\alpha} = 0.8$ and 0.7 are good compromises for Figs. 13.7 and 13.8, respectively. Obtaining accurate estimates is particularly challenging because of the significant deviations from canonical forms, such as those set forth in Eq. (5.44). As an example, the peaks and valleys near $T = 10^{-2}$ in the estimated normalized Haar-wavelet variance displayed in panel h) of both figures are confounding short-time effects.

As both panels c) show, the exponentialized interevent-interval densities behave as decaying exponential functions (dotted straight lines on these semilogarithmic plots), as they must. The original histograms of both figures (solid curves) exhibit peaks and valleys imparted by preferred intervals in the local traffic flow, and depart significantly from exponential behavior. Furthermore, some slight evidence of a positive curvature exists, particularly near $\tau/\widehat{E}[\tau] = 4$ and 2, in Figs. 13.7c) and 13.8c), respectively. Power-law curves displayed on a semilogarithmic plot would, in fact, exhibit just such a positive second derivative. However, for the largest intervals shown, the interevent-interval histograms for these data coincide with the exponential form engendered by exponentialization. Indeed, exponential behavior persists at far larger intervals, as demonstrated in Fig. 5.9 for data set BC-pOct89.[13] Thus, while an exponential distribution provides only a fair model of the BC-pOct89 and BC-pAug89 interevent-interval histograms, a power-law distribution would be significantly worse. We conclude that the estimated interevent-interval histograms are nonfractal.

Furthermore, for both Figs. 13.7 and 13.8, the fractal behavior in panels d)–i) (solid curves) is destroyed by shuffling (dashed curves), but modified only slightly by exponentialization (dotted curves), thereby confirming the absence of power-law tails in the interevent-interval histograms. The behavior of these surrogates demonstrates conclusively that the interval distribution does not contribute significantly to the fractal nature of the computer network traffic at hand. Indeed, these observations validate the use of the interval-based measures displayed in panels d)–g) for the analysis of these data, as discussed in Sec. 12.3.1.

13.6.4 Compare with other point processes

Comparing the BC-pOct89 and BC-pAug89 COMPUTER data with the collection of other experimental point-process data examined in Chapter 5 yields a number of interesting parallels and contrasts. Our goal of identifying the point process at hand is also furthered by comparing the COMPUTER surrogates with the surrogates of other

---

[13] Furthermore, the largest interval exceeds the mean by a factor of ≈ 87 for Fig. 13.7c), and a factor of ≈ 109 for Fig. 13.8c) (see Table 13.2). This is reasonable for 999999 intervals with an exponential tail $[\ln(999999) \approx 14]$, but not for a putative power-law distribution that imparts significant fractal behavior to a point process.

experimental point-process data examined in Chapter 11. All of the data sets that we investigated turned out to be fractal-rate point processes. The following comparisons prove useful in identifying the COMPUTER point process:

- The *normalized rate spectra* for the COMPUTER data displayed in Fig. 5.1 [and in Figs. 13.7i) and 13.8i)] reveal spectral features of various widths, sporadically located over a broad range of frequencies. A number of other point processes exhibit similar behavior.

- The *normalized Haar-wavelet variance* curves presented in Fig. 5.2 [and in Figs. 13.7h) and 13.8h)] demonstrate that the fractal exponent of the COM-PUTER data has a value $\widehat{\alpha} \approx 0.8$, which is below unity; fractal exponents for the CORTEX, COCHLEA, RETINA, and INTERNEURON (the latter appears in Fig. 11.17) also lie below unity. All of the other data sets have fractal exponents in excess of unity.

- Results gleaned from the corresponding *interval-based spectra*, displayed in Fig. 5.7 [also Figs. 13.7g) and 13.8g)], and *interval-based wavelet variances*, shown in Fig. 5.8 [also Figs. 13.7f) and 13.8f)], offer a broad confirmation of the count-based results reported above. As discussed in Secs. 12.3.1 and 13.6.3, these measures are suitable for use in the analysis of fractal-rate point processes.

  Interval-based measures typically appear smoother than their count-based counterparts. This arises because interval frequency and interval number do not precisely track real frequency and time, respectively. This results in a loss of phase coherence and a concomitant attenuation of narrow local features. The increased smoothness does not signify that interval-based measures are in any way superior to count-based measures, however. In fact, we have already seen in the counting domain that while the normalized variance $\widehat{F}(T)$ appears substantially smoother than the normalized Haar-wavelet variance $\widehat{A}(T)$, the former is significantly inferior to the latter for purposes of estimation (see Fig. 12.8).

- The COMPUTER *interevent-interval histograms* displayed in Fig. 5.9 [and in Figs. 13.7c) and 13.8c)] reveal a number of idiopathic features, of various widths and at sporadic intervals. Several other point processes behave similarly. The COMPUTER interevent-interval histograms most closely resemble those associated with the SYNAPSE and CORTEX.

- For all data sets, the *normalized rate spectra* and *normalized Haar-wavelet variances* for the *randomly deleted surrogates*, shown in Figs. 11.3 and 11.4, respectively, resemble the original curves, displayed in Figs. 5.1 and 5.2, respectively, but the random deletion dilutes the local features.

- For all data sets, the *normalized rate spectra* and *normalized Haar-wavelet variances* for the *shuffled surrogates*, shown in Figs. 11.13 and 11.14, respectively, are devoid of the fractal behavior displayed in Figs. 5.1 and 5.2, indicating that all of the point processes we examined are fractal-rate in nature. Using Eq. (4.18) in conjunction with Figs. 11.14 and 11.17, these surrogates reveal clustered underlying interevent-interval histograms ($C_\tau > 1$) for the COMPUTER, SYNAPSE, CORTEX, and INTERNEURON data, and anticlustered histograms ($C_\tau < 1$) for the HEARTBEAT and COCHLEA data.

- For all data sets, comparison of the *normalized rate spectra* and *normalized Haar-wavelet variances* for the *exponentialized surrogates*, portrayed in Figs. 11.10 and 11.11, respectively, with the corresponding original curves, shown in Figs. 5.1 and 5.2, respectively, reveals a reduction of sharp spectral and temporal features. This follows from the jittering of occurrence times imparted by exponentialization, which results in a loss of phase coherence.

- Taken together, these observations lead us to conclude that the *Ethernet-traffic* COMPUTER *point process* shares an essential similarity with all of the other point processes we have investigated, although it *most closely resembles the striate* CORTEX *point process*.

13.6.5 Formulate and simulate candidate models

In Sec. 13.5 we considered a number of fractal-based point processes as candidate models for computer network traffic. We highlighted the family of cascaded point processes, schematized in Fig. 13.4, since this class of models accommodates the vertical-layer structure of the Internet in a parsimonious way (see Sec. 13.2.3).[14] In particular, we devoted considerable attention to the fractal Bartlett–Lewis point process (Sec. 13.5.4) and the fractal Neyman–Scott point process (Sec. 13.5.5). Although these models are distinct, they nevertheless share many features in common.

In this section we simulate these two point processes using parameters appropriate for the classic Ethernet-traffic data set BC-pOct89 (Leland & Wilson, 1989, 1991). We thereby obtain simulated collections of statistical measures, including surrogates, analogous to those shown in Figs. 13.7 and 13.8. These, in turn, enable us to compare the model results with the original data.

Both simulated processes have primary events $dN_1(t)$ that comprise homogeneous Poisson processes (see Fig. 13.4). The difference in the two cascade models lies in the manner in which the secondary events $dN_{2,k}(t)$ are generated. While an exponential distribution clearly does not do justice to the structure of the interevent-interval histogram displayed in Fig. 13.7c), the accurate modeling of all of its periodicities and favored intervals would require a significant investment in terms of both model adjustment and simulation time. In keeping with the spirit of parsimony discussed in Sec. 13.4.2, we therefore employ a homogeneous Poisson process for the genera-

---

[14] As emphasized in the solution to Prob. 11.12.4, details regarding the underlying physical or biological phenomena play an important role in framing a proper point-process model.

tion of secondary events as well as for the primary process.[15] This should still yield reasonable results for longer-term effects.

For the Bartlett–Lewis process, fractal behavior in $dN_3(t)$ arises from the power-law distribution of the numbers of events M in each secondary process $dN_{2,k}(t)$. This distribution, together with the mean rates of primaries and secondaries, define the process. Constraints include the mean rate of the process, the fractal onset time, and the fractal exponent. For ease of simulation, we chose the number distribution for the secondary processes to be $\Pr\{M > n\} = (n+1)^{\alpha-1}$ for all $n \geq 0$. For $\alpha = 0.8$, this distribution has a finite mean of approximately $E[M] \doteq 5.27908$, but infinite variance. Dividing the measured rate of BC-pOct89 (568 sec^{-1}) by $E[M]$ yields the primary rate, $\mu_1 \doteq 107.595$/sec, which we round to 110/sec for simplicity. Finally, a secondary rate of $\mu_2 = 160$/sec yields approximately correct values for the fractal onset times and frequencies. Note that the secondary rate does not affect the overall rate of the point process, since the M_k events eventually appear in $dN_3(t)$ regardless of this rate. With this combination of values, an average of 3.62937 secondary processes exist at any given time. Table 13.1 provides a summary of the parameters used in carrying out the simulations along with values derived from the models.

For the Neyman–Scott process, we chose a fractal-shot-noise-driven Poisson process with a rectangular impulse-response function $h(K, t)$ of constant height c and varying duration K (see Prob. 9.2.1). With this construct, each secondary process has a constant rate while it exists; both primary and secondary processes are therefore again homogeneous Poisson processes. We chose the same simple power-law distribution for K as we used in Prob. 9.2.2, a generalized Pareto form; this imparts fractal characteristics to the overall point process $dN_3(t)$. This process closely resembles the Bartlett–Lewis point process discussed above. The principal distinction between the two is that the Bartlett–Lewis process specifies the random *number* of events in each secondary process $dN_{2,k}(t)$, whereas the Neyman–Scott process instead specifies its *duration*.

With the form of the secondary-process duration established, there remain four parameters: the primary rate; and c, A, and β from the impulse response function $h(K, t)$. Since the point process has but three constraints (mean rate, fractal exponent, and fractal onset time), a free parameter remains. However, fractal behavior cannot exist below A, since by definition no impulse-response functions exist below this cutoff. If A lies below the fractal onset times, then fractal behavior will be suppressed between the onset time and A, reaching its asymptote at times somewhat larger than A.[16] Since this does not mimic the data at hand, we obviate this problem by choosing

---

[15] Our model differs from that considered by Hohn et al. (2003) in a number of other respects as well. Ideally, we would obtain results for various candidate secondary point processes $dN_{2,k}(t)$, and then fit the resulting interevent-interval histogram of $dN_3(t)$ (the process as a whole) to the data, adjusting the secondary processes $dN_{2,k}(t)$ as necessary. Invoking the argument of parsimony, we do not carry out this procedure.

[16] Figure B.9 shows a similar effect, in the frequency domain, although it arises from a different origin (random displacement of the events).

	Units	Bartlett-Lewis	Neyman-Scott
Primary Parameters			
Power-Law Exponent α		0.8	0.8
Primary-Process Rate μ_1	(sec^{-1})	110	70
Simulation Duration	(sec)	1760	1760
Secondary-Process Rate μ_2	(sec^{-1})	160	—
Secondary-Process Amplitude c	(sec^{-1})	—	140
Secondary-Process Cutoff A	(msec)	—	10
Derived Expected Values			
Concurrent Secondary Processes		3.63	4.2
Secondary-Process Duration	(msec)	33.0	60.0
Events per Secondary Process		5.28	8.4
Total Aggregate Rate	(sec^{-1})	581	588
Total Number of Events	$(\times 10^6)$	1.02	1.03

Table 13.1 Parameters used for simulating realizations of the Bartlett–Lewis and Neyman–Scott point processes. The entries that apply to both simulations, in the upper portion of the table, derive from the target data set, BC-pOct89; we adjusted the other entries to fit the data (see text for details). The five entries in the lower portion of the table are expected results based on the theoretical properties of the models.

$A = 0.01$ sec, well below estimated fractal onset times. We also fix $\beta = 3 - \alpha = 2.2$, in accord with Eq. (B.196). This determines the average area a of the impulse response function:

$$a = c\,\text{E}[K] = c\,(\beta - 1)(\beta - 2)^{-1}A = 0.06\,c. \tag{13.14}$$

Together with Eq. (10.10), this fixes the product $\mu_1 c$. Finally, we adjust these two quantities so that they produce a fractal onset time that resembles that of data set BC-pOct89. We used a primary rate $\mu_1 = 70/\text{sec}$, and a height $c = 140/\text{sec}$. Table 13.1 summarizes the values used. With these parameters, an average of 4.2 secondary processes exist at any given time.

With the parameters for both processes established, we simulated the two cascaded point processes. We present collections of simulated statistical measures for the fractal Bartlett–Lewis and fractal Neyman–Scott point processes in Figs. 13.9 and 13.10, respectively. The results exhibited strong sensitivity to the random seed chosen (not shown), as expected for a fractal-based point process.[17] Changing the parameters slightly had the same effect since it effectively shifted the random numbers used

---

[17] Figure 12.1 displays the effects of using various random seeds in a different context.

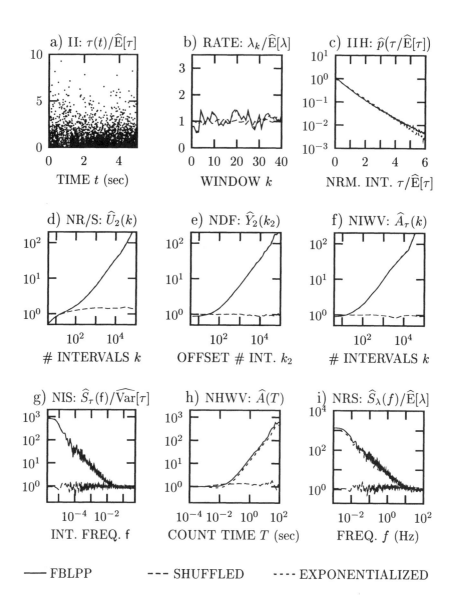

Fig. 13.9 Nine statistical measures for a simulated fractal Bartlett–Lewis point process (FBLPP) with parameters chosen to model the classic computer network traffic data set BC-pOct89 (solid curves). Results for the shuffled and exponentialized surrogates appear as the dashed and dotted curves, respectively. The statistics nicely mimic those shown for BC-pOct89 in Fig. 13.7, particularly those that portray fractal features. The results are quite similar to those generated by the fractal Neyman–Scott point process (see Fig. 13.10).

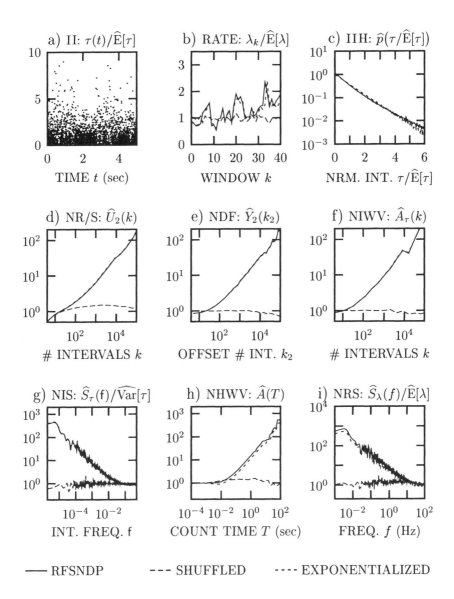

Fig. 13.10 Nine statistical measures for a simulated fractal Neyman–Scott point process (a rectangular fractal-shot-noise-driven Poisson process, RFSNDP) with parameters chosen to model the classic computer network traffic data set BC-pOct89 (solid curves). Results for the shuffled and exponentialized surrogates appear as the dashed and dotted curves, respectively. The statistics mimic those shown for BC-pOct89 in Fig. 13.7 quite well, particularly those that portray fractal features. The results are very similar to those generated by the fractal Bartlett–Lewis point process (see Fig. 13.9).

Interval Statistics	pOct89	pAug89	Bartlett-Lewis	Neyman-Scott
Total Duration L (sec)	1760	3143	1760	1760
Number of Intervals $[N(L) - 1]$	999 999	999 999	1 279 212	1 114 851
Minimum Interval (μsec)	16	20	0.000765	0.00108
Maximum Interval (msec)	154	342	36.0	51.3
Mean Interval $\widehat{E}[\tau]$ (msec)	1.76	3.14	1.38	1.58
Mean Rate $\widehat{E}[\mu]$ (sec^{-1})	568	318	727	634
Interval Standard Deviation $\widehat{\sigma}_\tau$ (msec)	3.20	5.64	1.59	1.89
Interval Coefficient of Variation $\widehat{C}_\tau$	1.82	1.80	1.16	1.20
Interval Skewness $\widehat{C}_3/\widehat{C}_2^{3/2}$	9.77	9.35	3.07	3.40
Interval Kurtosis $\widehat{C}_4/\widehat{C}_2^2$	170	153	17.5	21.9
Interval Serial Correlation Coefficient $\left\{\widehat{R}_\tau(1) - \widehat{E}^2[\tau]\right\}\Big/ \widehat{\text{Var}}[\tau]$	0.180	0.200	0.122	0.138

Table 13.2 Representative estimated interval statistics (see Sec. 3.3 for definitions) for two classic Ethernet-traffic data sets: BC-pOct89 and BC-pAug89. The data comprise the arrival times of the first 1 million packets recorded on the main Ethernet cable at the Bellcore (BC) Morristown Research and Engineering Facility on the mornings of 5 October 1989 and 29 August 1989, respectively (see Leland & Wilson, 1989, 1991). Although the mean rate of BC-pOct89 is nearly a factor of two greater than that of BC-pAug89, as shown in row 6, the normalized interval statistics, based on ratios of moments, agree quite closely (see rows 8–11). The skewness, and especially the kurtosis, greatly exceed the values corresponding to a Gaussian distribution, which are zero for the definitions we employ (see Footnote 2 on p. 55). We also include simulated point-process results for two models of computer network traffic: the fractal Bartlett–Lewis cascade point process and the fractal Neyman–Scott cluster point process. The statistics for these two model processes agree with each other quite closely; they also agree reasonably well with the statistics of the two computer network traffic data sets.

among the different stochastic quantities, yielding completely different secondary-point-process durations, for example. On the whole, obtaining precise fits to the target data set, BC-pOct89, proved quite difficult, and certainly beyond any sort of automated minimization procedure such as Marquardt-Levenberg. Rather than search extensively for a particularly lucky set of parameters and random seed, we used round numbers for the parameters, and a default seed. Thus, while we could precisely match the durations of the simulations to that of the BC-pOct89 data set (1759.6 sec), the total number of events varied considerably from the target of 999 999; we generated 1 279 212 and 1 114 851 events for the Bartlett–Lewis and Neyman–Scott processes, respectively. Table 13.2 presents a collection of interval statistics derived from these simulations, along with those for the canonical data sets BC-pOct89 and BC-pAug89.

13.6.6 Compare model simulations with data

The similarity in construction between the Bartlett–Lewis and Neyman–Scott cascaded point processes carries forth to their results, which closely resemble each other, as demonstrated in Figs. 13.9 and 13.10, as well as in Table 13.2. By construction, both lack the local bumps, peaks, and valleys evident in the statistics of the computer network traffic data shown in Figs. 13.7 and 13.8 (features such as these also appear in biological point processes, as mentioned in Sec. 13.6.4). As a result, the two model simulations resemble each other a bit more closely than either does the target data set. Lacking explicit mechanisms for generating these effects, both simulations change little with exponentialization.

Aside from this, both simulations share all of the characteristic features of the original data discussed in Sec. 13.6.3. Furthermore, as shown in Table 13.2, the interval statistics for the two model processes agree reasonably well with those of the two computer network traffic data sets, BC-pOct89 and BC-pAug89. Even the interval coefficients of variation for the Bartlett–Lewis and Neyman–Scott simulations, $C_\tau \doteq 1.16$ and 1.20, respectively, are not inordinately different from those of data sets BC-pOct89 and BC-pAug89, $C_\tau \doteq 1.82$ and 1.80, respectively. Interval shuffling has essentially the same effect on the simulations as it does on the original data (compare dashed curves in Figs. 13.7–13.10).

It appears that the Bartlett–Lewis and Neyman–Scott constructs yield statistics that accord quite well with those obtained from the canonical Ethernet-traffic COMPUTER data sets that we studied. Interestingly, of all the point processes we investigated, the COMPUTER point process most closely resembles that observed at the striate CORTEX (see Sec. 13.6.4). In fact, the Neyman–Scott cascade process has also been effectively used to model the sequence of action potentials recorded from striate CORTEX neurons (Teich et al., 1996).

Problems

13.1 M/M/1/∞ *queue-length distribution*   How do Eqs. (13.1) and (13.2) change when the buffer size is infinite?

13.2 M/M/M/Q_m *queue-length distribution*   Describe the changes required for Eq. (13.1) when M servers handle requests from the same buffer.

13.3 *Buffer overflow probability approximations*   For some queueing models with finite-size buffers, the calculation of the buffer overflow probability P_B is mathematically intractable. In such cases one typically solves the relevant equations assuming infinite buffer size (see Prob. 13.1), and then chooses some other representation for buffer overflow. Consider Eq. (13.1) under the simplification of infinite buffer size, and denote the resulting queue-length distribution by $p_\infty(n)$. A number of approximations for the overflow probability are commonly employed, including $p_\infty(Q_m)$, $p_\infty(Q_m + 1)$, $\sum_{n=Q_m}^{\infty} p_\infty(n)$, and $\sum_{n=Q_m+1}^{\infty} p_\infty(n)$. Derive forms for each of these quantities for the M/M/1 queue and show that for large values of Q_m, the approximate result $p_\infty(Q_m)$ lies closest to the true result $p_Q(Q_m)$.

13.4 M/M/1 *queue buffer design* Consider a homogeneous Poisson process with a mean interevent interval of 10 msec providing an arrival stream to a queue. Suppose that the service times follow an exponential distribution with a mean value of 9 msec and that there is a single server. For this traffic, determine the minimum buffer sizes that give rise to overflow probabilities of no more than 10^{-3}, 10^{-6}, and 10^{-9}.

13.5 M/M/1/∞ *queue simulation* Simulate the queue specified in Prob. 13.4 for 10^6 seconds ($\approx 10^8$ arrivals), but assume now that $Q_m \to \infty$. Plot the estimated queue-length histogram $\widehat{p}_\infty(n)$. To ensure stationarity, include an additional 10^4 seconds at the beginning and discard it. Show that the histogram follows a geometric form. Use this plot in conjunction with the results obtained in Prob. 13.4 to confirm that Eq. (13.12) provides a good estimate for the overflow probability in the M/M/1/Q_m queue.

13.6 *Fractal-Gaussian-process-driven Poisson process queue simulation*
Figures 12.1–12.7 and 12.9 provide results derived from simulations of a Poisson process driven by a fractal Gaussian process (FGPDP) with a mean rate $E[\mu] = 100$, duration $L = 10^4$, fractal exponent $\alpha = 0.8$, onset frequency $f_S = 0.2$, and fractal-Gaussian-process array size $M = 2^{17}$ (of which we used half). We considered this point process in Secs. 6.3.3, 8.4, and 10.6.1, as well as in Chapter 12.

   13.6.1. Using this same process, but with L (and M) increased by a factor of 100, simulate the associated G/M/1 queue assuming exponentially distributed service times with a mean value $1/\mu_s = 0.009$. Plot the estimated queue-length histogram for this process on doubly logarithmic coordinates. Compare it with the theoretical M/M/1 result and with a decaying power-law distribution.

   13.6.2. Now repeat the simulation, changing the mean service time to $1/\mu_s = 0.005$ while leaving everything else unchanged. Plot the result on semilogarithmic coordinates this time, and explain why it does not follow a fractal form.

13.7 *Shuffled-fractal-process queue simulation* Consider the traffic-process simulation described in Prob. 13.6. Randomly shuffle this simulation and repeat the queueing analysis with an average service time of 0.009. Show that the result agrees well with that obtained for the M/M/1 queue (see Prob. 13.5).

13.8 *Fractal-shot-noise-driven Poisson process queue simulation* Simulate a fractal-shot-noise-driven Poisson process $N(t)$ (see Chapter 10), where the impulse response functions have a rectangular shape of constant height c, and a duration that obeys a decaying power-law distribution (see Prob. 9.2). Specifically, let the probability that the impulse-response-function duration K exceeds a value x take the form

$$\Pr\{K > x\} = \begin{cases} (A/x)^{\beta-1} & x > A \\ 1 & x \leq A, \end{cases} \tag{13.15}$$

with $\beta = 2.2$, $A = 1$, and $c = \frac{10}{3}$. Let the primary Poisson process rate $\mu = 5$. This yields an expected interevent interval at the secondary Poisson process of $E[\tau] = 0.01$. Assume that all impulse response functions are independent and identically distributed. Again, set the duration of the simulation L to 10^6 for an expected number of events $E[N(L)] = 10^8$; include an additional duration of 10^4 at the beginning of the simulation, and discard these results. Simulate the associated G/M/1 queue using $N(t)$ as the arrival process and assume exponentially distributed service times with a mean value of $1/\mu_s = 0.009$. Plot the estimated queue-length histogram and compare it with the results displayed in Fig. B.16.

13.9 *Modulated-fractal-process queue simulation*   We have considered queue-length histograms for several fractal-rate arrival processes (see Figs. B.16, B.17, and B.19). We wish to investigate how a periodically modulated arrival process changes the character of these histograms.

Consider, as a simple example, a Poisson point process $dN_1(t)$ driven by a periodic, deterministic rate of the form

$$\mu(t) = \mu_0[1 + a\cos(\omega_0 t)], \tag{13.16}$$

where a is the modulation depth and we posit $\mu_0/\omega_0 \gg 1$ to ensure that a large number of events occur within each period of the modulated waveform. We can impose such modulation in the following manner: Generate a new point process $dN_3(t)$ from a homogeneous Poisson process $dN_2(t)$ by multiplying the event times of $dN_2(t)$ by a suitable nonlinear function of $\cos(\omega_0 t)$, chosen so that $dN_3(t)$ has the same statistics as $dN_1(t)$. Said differently, we can warp the time axis of the (unmodulated) point process in a periodic manner to generate a result that mimics a sinusoidally modulated inhomogeneous process.

We can impose such periodic time warping on any arbitrary point process. Begin with the fractal-Gaussian-process-driven Poisson process (FGPDP) considered in Prob. 13.6. Carry out the time warping discussed above and generate a point process that mimics a sinusoidally modulated (inhomogeneous) version of the fractal-Gaussian-process-driven Poisson process. Now let the modified point process serve as the arrival process for a G/M/1 queue, which we denote MODULATED-FGPDP/M/1. For the service process, assume exponentially distributed service times with a mean value $1/\mu_s = 0.009$. Use a modulation period $2\pi/\omega_0 = 1$ min and a modulation depth $a = 1$, as defined by Eq. (13.16). Simulate and plot the estimated queue-length histogram for this queueing problem, and compare your result with those obtained in Prob. 13.6.1 (Fig. B.16).

13.10 *Estimating two fractal exponents*   Consider a data set that gives rise to a normalized Haar-wavelet variance with two separate power-law regions. The first exhibits a fractal exponent α_1, and extends from T_{A1} to T_{A2}; the second exhibits a fractal exponent $\alpha_2 > \alpha_1$, and extends from T_{A3} to T_{A4}.

13.10.1. The accurate estimation of α_1 and α_2 requires that T_{A2}/T_{A1} and T_{A4}/T_{A3} both exceed 10^3. We also set $T_{A1} = 10\,E[\tau]$ to ensure a practical process, and require a total duration $L \geq 10\,T_{A4}$ to achieve a reasonably small variance near T_{A4}. How many events must a simulated data set with these properties contain on average?

13.10.2. One can always fit a monofractal form to a bifractal data set. Using the minimum suitable values found in Prob. 13.10.1, and the exponents $\alpha_1 = 0.4$ and $\alpha_2 = 0.8$, calculate the corresponding ideal normalized Haar variance. Plot this bifractal curve, and find the monofractal curve that minimizes the mean-square error on a doubly logarithmic plot. Compare the two curves and comment on the difference, bearing in mind the implications of Eq. (12.25). Repeat this exercise for $T_{A2}/T_{A1} = T_{A4}/T_{A3} = 10$.

Appendix A
Derivations

A.1	**Point Processes: Definition and Measures**	356
	A.1.1 Detrended fluctuation analysis for renewal process	356
A.2	**Point Processes: Examples**	358
	A.2.1 Moments for renewal process	358
A.3	**Processes Based on Fractional Brownian Motion**	360
	A.3.1 Fractal lognormal noise	360
A.4	**Fractal Renewal Processes**	362
	A.4.1 Spectrum in the mid-frequency range	363
	A.4.2 Spectrum for $\gamma = \frac{1}{2}$	368
	A.4.3 Coincidence rate in the medium-time limit	369
	A.4.4 Normalized variances in the medium-time limit	370
A.5	**Alternating Fractal Renewal Process**	371
	A.5.1 Alternating-renewal-process spectrum	371
	A.5.2 Low-frequency limit of the spectrum	374
	A.5.3 Spectrum under extreme dwell-time asymmetry	375
A.6	**Fractal Shot Noise**	376
	A.6.1 Infinite-area tail	376
	A.6.2 Approach to stable form	377
	A.6.3 Autocorrelation	379
A.7	**Fractal-Shot-Noise-Driven Point Processes**	382
	A.7.1 Integrals for counting statistics	382
	A.7.2 Expressions for normalized variance	384
	A.7.3 Expressions for normalized Haar-wavelet variance	390
	A.7.4 Integrals for time statistics	392
A.8	**Analysis and Estimation**	394
	A.8.1 Fourier-transform effects	394

A.1   POINT PROCESSES: DEFINITION AND MEASURES

A.1.1   Detrended fluctuation analysis for renewal process

To prove Eq. (3.25), we begin by normalizing the sequence of interevent intervals $\{\tau_n\}$. The detrending process removes linear trends from the summed series $\{y_n\}$, which is tantamount to removing the mean from the interevent intervals themselves. With the mean value of the interevent intervals $E[\tau]$ rendered irrelevant, we set it to zero for algebraic convenience. Furthermore, we divide by the standard deviation to generate a zero-mean unit-variance sequence $\{x_n\}$ by defining

$$x_n \equiv \frac{\tau_n - E[\tau]}{\sigma_\tau}. \tag{A.1}$$

Without loss of generality we consider $\{x_n\}$ instead of $\{\tau_n\}$ in the following.

For an arbitrary set of ordered pairs $\{(t_n, y_n)\}$, $1 \le n \le k$, classical statistical theory (Press, Teukolsky, Vetterling & Flannery, 1992) yields the residual errors after subtracting the trends. If we define

$$
\begin{aligned}
z_n &= a t_n + b \\
w_n &= y_n - z_n
\end{aligned}
\tag{A.2}
$$

and

$$\chi^2 = \sum_{n=1}^{k} w_n^2 \tag{A.3}$$

(see Fig. 3.4), and find a and b that minimize χ^2, we obtain

$$\chi^2 = S_{yy} + \frac{S_{tt} S_y^2 + S S_{ty}^2 - 2 S_t S_y S_{ty}}{S S_{tt} - S_t^2}, \tag{A.4}$$

where we have used the notation of Press et al. (1992) and defined

$$
\begin{aligned}
S &\equiv \sum_{n=1}^{k} 1 & S_t &\equiv \sum_{n=1}^{k} t_n & S_{tt} &\equiv \sum_{n=1}^{k} t_n^2 \\
S_y &\equiv \sum_{n=1}^{k} y_n & S_{ty} &\equiv \sum_{n=1}^{k} t_n y_n & S_{yy} &\equiv \sum_{n=1}^{k} y_n^2.
\end{aligned}
\tag{A.5}
$$

Substituting $t_n = n$ into Eq. (A.5) we obtain

$$
\begin{aligned}
S &= \sum_{n=1}^{k} 1 = k \\
S_t &= \sum_{n=1}^{k} t_n = \sum_{n=1}^{k} n = k(k+1)/2 \\
S_{tt} &= \sum_{n=1}^{k} t_n^2 = \sum_{n=1}^{k} n^2 = k(k+1)(2k+1)/6 \\
S_{ty} &= \sum_{n=1}^{k} t_n y_n = \sum_{n=1}^{k} n y_n.
\end{aligned}
\tag{A.6}
$$

To establish the link with detrended fluctuation analysis, we need to substitute n for t_n and take the expectation of Eq. (A.4). (We cannot take the expectation of y_n

and then perform a least-squares fit since $E[y_n] = 0$ by construction; rather, the fit must precede the expectation.) Taking this expectation yields

$$E[\chi^2] = E[S_{yy}] + \frac{S_{tt}\,E[S_y^2] + S\,E[S_{ty}^2] - 2S_t\,E[S_y S_{ty}]}{SS_{tt} - S_t^2}, \tag{A.7}$$

where the deterministic nature of t_n permits us to move sums not involving y_n outside the expectations.

We now proceed to simplify the expectations in Eq. (A.7). The key step involves the independence of the sequence $\{x_n\}$. In particular, consider the expectation of the product of two terms y_m and y_n, with $m < n$; we have

$$
\begin{aligned}
E[y_m y_n] &= E\left[\sum_{p=1}^{m}\sum_{q=1}^{n} x_p x_q\right] \\
&= \sum_{p=1}^{m}\sum_{q=1}^{n} E[x_p x_q] \\
&= \sum_{p=1}^{m}\{E[x_p x_q] - E[x_p]\,E[x_q]\} + \sum_{\substack{p=1 \\ }}^{m}\sum_{\substack{q=1 \\ q\neq p}}^{n} E[x_p]\,E[x_q] \\
&= \sum_{p=1}^{m}\{1 - 0 \times 0\} + \sum_{\substack{p=1}}^{m}\sum_{\substack{q=1 \\ q\neq p}}^{n} 0 \times 0 \\
&= m, \tag{A.8}
\end{aligned}
$$

and the reason for constructing a zero-mean unit-variance sequence $\{x_n\}$ now becomes apparent. In general, we have $E[y_m y_n] = \min(m, n)$, the smaller of m and n. Employing Eq. (A.8) and the last line of Eq. (A.6), we obtain

$$
\begin{aligned}
E[S_{yy}] &= E\left[\sum_{n=1}^{k} y_n^2\right] = \sum_{n=1}^{k} E[y_n^2] = \sum_{n=1}^{k} n \\
&= k(k+1)/2 \tag{A.9} \\
E[S_y^2] &= E\left[\sum_{m=1}^{k}\sum_{n=1}^{k} y_m y_n\right] = \sum_{m=1}^{k}\sum_{n=1}^{k} E[y_m y_n] = \sum_{m=1}^{k}\sum_{n=1}^{k} \min(m, n) \\
&= k(k+1)(2k+1)/6 \tag{A.10} \\
E[S_y S_{ty}] &= \sum_{m=1}^{k}\sum_{n=1}^{k} m\,\min(m, n) \\
&= k(k+1)(5k^2 + 5k + 2)/24 \tag{A.11} \\
E[S_{ty}^2] &= \sum_{m=1}^{k}\sum_{n=1}^{k} mn\,\min(m, n) \\
&= k(k+1)(2k+1)(2k^2 + 2k + 1)/30. \tag{A.12}
\end{aligned}
$$

Finally, substituting Eqs. (A.12) and (A.6) into Eq. (A.7), after a fair amount of algebra we obtain

$$E[\chi^2] = (k^2 - 4)/15. \tag{A.13}$$

Normalizing Eq. (A.13) by the number of values k, generalizing to arbitrary variance, and taking the square root yields the final result presented in Eq. (3.25).

A.2 POINT PROCESSES: EXAMPLES

A.2.1 Moments for renewal process

In this section we derive expressions for the count probabilities and moments of a stationary renewal point process $dN(t)$.

Let s be any time selected independently of $dN(t)$, and recall that the random variable $\vartheta(s)$ denotes the time remaining between s and the next event in $dN(t)$. We reiterate Eq. (3.12), which provides

$$
\begin{aligned}
p_\vartheta(s) &= [1 - P_\tau(s)]/E[\tau] \\
&= E[\mu] \int_s^\infty p_\tau(u)\, du.
\end{aligned}
\tag{A.14}
$$

The associated characteristic function becomes

$$
\begin{aligned}
\phi_\vartheta(\omega) &\equiv \int_0^\infty p_\vartheta(t)\, e^{-i\omega t}\, dt \\
&= E[\mu] \int_{t=0}^\infty \int_{v=t}^\infty p_\tau(v)\, e^{-i\omega t}\, dv\, dt \\
&= E[\mu] \int_{v=0}^\infty p_\tau(v) \int_{t=0}^v e^{-i\omega t}\, dt\, dv \\
&= (i\omega)^{-1} E[\mu] \int_{v=0}^\infty p_\tau(v) \left[1 - e^{-i\omega v}\right] dv \\
&= (i\omega)^{-1} E[\mu] \left[1 - \phi_\tau(\omega)\right].
\end{aligned}
\tag{A.15}
$$

To simplify the notation, we can set $s = 0$ without loss of generality for a stationary renewal point process, thereby permitting the use of $Z(T)$ instead of $N(t)$. With this Ansatz, consider the probability density for the nth event following the origin occurring at a time T. As a result of the renewal nature of $dN(t)$, this becomes $p_\tau(v)$ convolved with itself n times, all convolved with $p_\vartheta(v)$ as the time to the first event. The integral of this probability density yields the probability that the nth event occurs by the time T, which is equivalent to the probability that at least n events have occurred in $Z(T)$.

We thus arrive at

$$\Pr\{Z(T) > n\} = \int_0^T p_\vartheta \star p_\tau^{\star n}(t)\, dt, \tag{A.16}$$

so that

$$
\begin{aligned}
\Pr\{Z(T) = n\} \;=\;& \Pr\{Z(T) > n-1\} - \Pr\{Z(T) > n\} \\[4pt]
=\;& \begin{cases}
0 & n < 0 \\[4pt]
1 - \int_0^T p_\vartheta(t)\,dt & n = 0 \\[4pt]
\int_{0-}^T p_\vartheta \star \big[p_\tau^{\star(n-1)} - p_\tau^{\star n}\big](t)\,dt & n > 0,
\end{cases}
\end{aligned}
\tag{A.17}
$$

which is the probability distribution for the number of counts $Z(T)$. We have used the notation $0-$ for the lower limit in Eq. (A.17) and subsequently to explicitly include delta functions that may occur at $t = 0$.

Suppose we now carry out a Fourier transform on Eq. (A.17):

$$
\begin{aligned}
f_1(\omega, n) \;\equiv\;& \int_0^\infty e^{-i\omega T}\,\Pr\{Z(T) = n\}\,dT \\[4pt]
=\;& \begin{cases}
0 & n < 0 \\[4pt]
(i\omega)^{-1} - \mathrm{E}[\mu]\,(i\omega)^{-2}\big[1 - \phi_\tau(\omega)\big] & n = 0 \\[4pt]
\mathrm{E}[\mu]\,(i\omega)^{-2}\big[1 - \phi_\tau(\omega)\big]^2 \phi_\tau^{n-1}(\omega) & n > 0,
\end{cases}
\end{aligned}
\tag{A.18}
$$

and following this take the z transform, which yields a time and event-number generating function $f_2(\omega, z)$:

$$
\begin{aligned}
f_2(\omega, z) \;\equiv\;& \sum_{n=0}^\infty f_1(\omega, n)\, z^{-n} \\[4pt]
=\;& (i\omega)^{-1} - \mathrm{E}[\mu]\,(i\omega)^{-2}\big[1 - \phi_\tau(\omega)\big] \\
& + \mathrm{E}[\mu]\,(i\omega)^{-2}\big[1 - \phi_\tau(\omega)\big]^2 \sum_{n=1}^\infty z^{-n}\,\phi_\tau^{n-1}(\omega) \\[4pt]
=\;& (i\omega)^{-1} - \mathrm{E}[\mu]\,(i\omega)^{-2}\big[1 - \phi_\tau(\omega)\big] \\
& + \mathrm{E}[\mu]\,(i\omega)^{-2}\big[1 - \phi_\tau(\omega)\big]^2 z^{-1} \sum_{m=0}^\infty \big[\phi_\tau(\omega)/z\big]^m \\[4pt]
=\;& (i\omega)^{-1} - \mathrm{E}[\mu]\,(i\omega)^{-2}\big[1 - \phi_\tau(\omega)\big] \\
& + \mathrm{E}[\mu]\,(i\omega)^{-2}\big[1 - \phi_\tau(\omega)\big]^2 \big[z - \phi_\tau(\omega)\big]^{-1}.
\end{aligned}
\tag{A.19}
$$

Next, take k derivatives with respect to z

$$
\frac{\partial^k}{\partial z^k} f_2(\omega, z) = (-1)^k\, \mathrm{E}[\mu]\, k!\, (i\omega)^{-2}\big[1 - \phi_\tau(\omega)\big]^2 \big[z - \phi_\tau(\omega)\big]^{-(k+1)}.
\tag{A.20}
$$

Set $z = 1$ to obtain

$$
(-1)^k\, \frac{\partial^k}{\partial z^k} f_2(\omega, z)\bigg|_{z=1} = \mathrm{E}[\mu]\, k!\, (i\omega)^{-2}\big[1 - \phi_\tau(\omega)\big]^{1-k},
\tag{A.21}
$$

and carry out an inverse Fourier transform. This yields

$$
\int_{\omega=-\infty}^{\infty} e^{+i\omega T} \left[(-1)^k \frac{\partial^k}{\partial z^k} f_2(\omega, z) \Big|_{z=1} \right] \frac{d\omega}{2\pi}
$$

$$
= \int_{\omega=-\infty}^{\infty} e^{+i\omega T} \, \mathrm{E}[\mu] \, k! \, (i\omega)^{-2} \left[1 - \phi_\tau(\omega) \right]^{1-k} \frac{d\omega}{2\pi}
$$

$$
= \mathrm{E}[\mu] \, k! \int_{t=0-}^{T} \int_{v=0-}^{t} G^{\star(k-1)}(v) \, \mathrm{E}^{1-k}[\mu] \, dv \, dt
$$

$$
= \mathrm{E}^{2-k}[\mu] \, k! \int_{0-}^{T} (T - t) \, G^{\star(k-1)}(t) \, dt, \tag{A.22}
$$

where we have made use of Eq. (4.15), and again use the notation $0-$ to include delta functions.

However, the two Fourier transforms cancel, so we also obtain

$$
\int_{\omega=-\infty}^{\infty} e^{+i\omega T} \left[(-1)^k \frac{\partial^k}{\partial z^k} f_2(\omega, z) \Big|_{z=1} \right] \frac{d\omega}{2\pi}
$$

$$
= (-1)^k \frac{\partial^k}{\partial z^k} \sum_{n=0}^{\infty} \Pr\{Z(T) = n\} z^{-n} \Big|_{z=1}
$$

$$
= \sum_{n=1}^{\infty} \Pr\{Z(T) = n\} \frac{(n + k - 1)!}{(n - 1)!} z^{-(n+k)} \Big|_{z=1}
$$

$$
= \mathrm{E}\left\{ \frac{[Z(T) + k - 1]!}{[Z(T) - 1]!} \right\}. \tag{A.23}
$$

Equating Eqs. (A.22) and (A.23) yields the result provided in Eq. (4.19):

$$
\mathrm{E}\left\{ \frac{[Z(T) + k - 1]!}{[Z(T) - 1]!} \right\} = \mathrm{E}^{2-k}[\mu] \, k! \int_{0-}^{T} (T - t) \, G^{\star(k-1)}(t) \, dt. \tag{A.24}
$$

A.3 PROCESSES BASED ON FRACTIONAL BROWNIAN MOTION

A.3.1 Fractal lognormal noise

Following the method employed by Lowen et al. (1997b), we derive expressions for the moments and autocorrelation of fractal lognormal noise. To simplify notation, we reiterate the first and second cumulants:

$$
\begin{aligned}
C_1 &\equiv \mathrm{E}[X] \\
C_2 &\equiv \mathrm{Var}[X].
\end{aligned} \tag{A.25}
$$

For the moments of μ we have

$$
\mathrm{E}[\mu^n] = \int_0^{\infty} p_\mu(y) \, y^n \, dy
$$

$$
\begin{aligned}
&= \int_{-\infty}^{\infty} p_X(x) \exp(nX)\, dx \\
&= \int_{-\infty}^{\infty} (2\pi C_2)^{-1/2} \exp\left[nx - \frac{(x - C_1)^2}{2C_2}\right] dx \\
&= (2\pi C_2)^{-1/2} \int_{-\infty}^{\infty} \exp\left[nC_1 + n^2\frac{C_2}{2} - \frac{[x - (C_1 + nC_2)]^2}{2C_2}\right] dx \\
&= \exp(nC_1 + n^2 C_2/2)\,(2\pi C_2)^{-1/2} \int_{-\infty}^{\infty} \exp(-z^2/2C_2)\, dz \\
&= \exp(nC_1 + n^2 C_2/2),
\end{aligned}
\tag{A.26}
$$

in agreement with Eq. (6.21).

For the autocorrelation of the rate, we write

$$
\begin{aligned}
R_\mu(t) &\equiv \mathrm{E}[\mu(s)\,\mu(s+t)] \\
&= \mathrm{E}\{\exp[X(s)]\,\exp[X(s+t)]\} \\
&= \mathrm{E}\{\exp[X(s) + X(s+t)]\}.
\end{aligned}
\tag{A.27}
$$

To proceed, we divide $X(s+t)$ into two parts, one proportional to $X(s)$ and one uncorrelated with it

$$
X(s+t) - C_1 = f(t)\,[X(s) - C_1] + X_\perp(s,t).
\tag{A.28}
$$

Since $X(s)$ is a Gaussian process, so too is $X_\perp(s,t)$, and since the two are uncorrelated, they are also independent. Substituting Eq. (A.28) into Eq. (A.27) then yields

$$
\begin{aligned}
R_\mu(t) &= \mathrm{E}\{\exp[X(s) + X(s+t)]\} \\
&= \mathrm{E}\Big\{\exp\big[X(s) + C_1 + f(t)\,[X(s) - C_1] + X_\perp(s,t)\big]\Big\} \\
&= \mathrm{E}\Big[\exp\big\{[1 + f(t)]\,X(s)\big\}\Big]\,\mathrm{E}\Big[\exp\big\{[1 - f(t)]\,C_1\big\}\Big] \\
&\quad \times \mathrm{E}\Big[\exp\big\{X_\perp(s,t)\big\}\Big],
\end{aligned}
\tag{A.29}
$$

where the last step leading to Eq. (A.29) derives from the independence of $X(s)$ and $X_\perp(s,t)$.

Next we find expressions for $f(t)$ as well as for the mean and variance of $X_\perp(s,t)$. Taking the expectation of Eq. (A.28) yields

$$
\begin{aligned}
C_1 - C_1 &= f(t)\,[C_1 - C_1] + \mathrm{E}[X_\perp(s,t)] \\
\mathrm{E}[X_\perp(s,t)] &= 0.
\end{aligned}
\tag{A.30}
$$

We then multiply Eq. (A.28) by $X(s) - C_1$ and take expectations,

$$
\begin{aligned}
\mathrm{E}\{[X(s+t) - C_1]\,[X(s) - C_1]\} \\
= f(t)\,\mathrm{E}\{[X(s) - C_1]\,[X(s) - C_1]\} + \mathrm{E}\{X_\perp(s,t)\,[X(s) - C_1]\} \\
R_X(t) - C_1^2 = f(t)\,C_2 + C_1\,\mathrm{E}[X_\perp(s,t)] \\
f(t) = [R_X(t) - C_1^2]\,/C_2,
\end{aligned}
\tag{A.31}
$$

where we have made use of the independence of $X(s)$ and $X_\perp(s,t)$, as well as Eq. (A.30). Rearranging Eq. (A.28) yields the variance of $X_\perp(s,t)$:

$$
\begin{aligned}
X_\perp(s,t) &= [X(s+t) - C_1] - f(t)[X(s) - C_1] \\
X_\perp^2(s,t) &= [X(s+t) - C_1]^2 + f^2(t)[X(s) - C_1]^2 \\
&\quad - 2f(t)[X(s) - C_1][X(s+t) - C_1] \\
E[X_\perp^2(s,t)] &= C_2 + f^2(t)C_2 - 2f(t)[R_X(t) - C_1^2] \\
\mathrm{Var}[X_\perp(s,t)] &= [1 - f^2(t)]C_2,
\end{aligned}
\tag{A.32}
$$

where we have made use of Eq. (A.31).

Finally, we substitute Eqs. (A.30)–(A.32) into Eq. (A.29). To evaluate the ensuing expressions, consider Eq. (A.26). Any fixed expression can substitute for n in the second line, so that

$$
\begin{aligned}
R_\mu(t) &= E\left[\exp\{[1 + f(t)]X(s)\}\right] E\left[\exp\{[1 - f(t)]C_1\}\right] \\
&\quad \times E\left[\exp\{X_\perp(s,t)\}\right] \\
&= \exp\{[1 + f(t)]C_1 + [1 + f(t)]^2 C_2/2\} \exp\{[1 - f(t)]C_1\} \\
&\quad \times \exp\{[1 - f^2(t)]C_2/2\} \\
R_\mu(t) &= \exp\{2C_1 + [1 + f(t)]C_2\} \\
&= \exp\{2[C_1 + C_2/2]\} \exp\{f(t)C_2\} \\
&= E^2[\mu] \exp\{R_X(t) - E^2[X]\},
\end{aligned}
\tag{A.33}
$$

in agreement with Eq. (6.22).

A.4 FRACTAL RENEWAL PROCESSES

In preparation for the results that follow, we first obtain expressions for three quantities involving the square root of negative unity. Using the De Moivre relation for $\theta = \pi/2$ provides

$$
\begin{aligned}
\exp(i\pi/2) &= \cos(\pi/2) + i\sin(\pi/2) \\
&= i.
\end{aligned}
\tag{A.34}
$$

Raising both sides of Eq. (A.34) to the same power yields

$$
\begin{aligned}
i^x &= \exp(ix\pi/2) \\
\mathrm{Re}\{i^x\} &= \cos(x\pi/2).
\end{aligned}
\tag{A.35}
$$

The second expression is obtained by taking logarithm of both sides of Eq. (A.34):

$$
\ln(i) = i\pi/2.
\tag{A.36}
$$

We can generally ignore multiplicative factors inside the logarithm when the argument otherwise assumes a large or small value. This is demonstrated via

$$
\begin{aligned}
\lim_{x \to 0} \frac{\ln(cx)}{\ln(x)} &= \lim_{x \to 0} \frac{\ln(c) + \ln(x)}{\ln(x)} \\
&= 1 + \lim_{x \to 0} \frac{\ln(c)}{\ln(x)} \\
&= 1, \qquad\qquad\qquad\qquad\text{(A.37)}
\end{aligned}
$$

so that $\ln(cx) \approx \ln(x)$ as $x \to 0$ for any finite value c. A similar result obtains in the limit $x \to \infty$. Exceptions occur in cases where we must distinguish two forms that would otherwise appear identical, such as those in Eqs. (7.18) and (7.19) for $\gamma = 2$.

Third, we derive an expression for the square root of a complex number:

$$
\begin{aligned}
\sqrt{a + ib} &= c + id \\
a + ib &= c^2 - d^2 + i2cd \\
a &= c^2 - d^2 \\
b &= 2cd \\
a^2 + b^2 &= c^4 + d^4 - 2c^2 d^2 + 4c^2 d^2 = (c^2 + d^2)^2 \\
\sqrt{a^2 + b^2} &= c^2 + d^2 \\
\sqrt{a^2 + b^2} + a &= 2c^2 \\
\sqrt{a^2 + b^2} - a &= 2d^2
\end{aligned}
$$

$$
\begin{aligned}
c &= \sqrt{\frac{\sqrt{a^2 + b^2} + a}{2}} \\
&\qquad\qquad\qquad\qquad\qquad\text{(A.38)} \\
d &= \sqrt{\frac{\sqrt{a^2 + b^2} - a}{2}}.
\end{aligned}
$$

Thus, the quantity $c + id$ forms a solution to $\sqrt{a + ib}$.

A.4.1 Spectrum in the mid-frequency range

In this section we obtain approximate expressions for the spectrum of the fractal renewal point process, as well as for the symmetric alternating fractal renewal process, in the mid-frequency range, $B^{-1} \ll f \ll A^{-1}$.

Since the power-law tail of the interevent-interval probability density function determines the fractal behavior of the process, the results are insensitive to the precise form of the density. We make use of the abrupt-cutoff power-law form to simplify the calculations; similar results would result when using any power-law-varying density. To simplify the notation, we express the results in terms of the radian frequency $\omega \equiv 2\pi f$; this does not, of course, alter the validity of the arguments.

We begin with Eq. (7.3), the characteristic function for the abrupt-cutoff power-law probability density provided in Eq. (7.1). Substituting $y \equiv x/i\omega A$, we have

$$
\begin{aligned}
\phi_\tau(\omega) &= \frac{\gamma}{1 - (A/B)^\gamma} \int_1^{B/A} e^{-i\omega A y}\, y^{-(\gamma+1)}\, dy \\
&\to \frac{\gamma}{1 - (A/B)^\gamma} \int_1^{B/A} e^0\, y^{-(\gamma+1)}\, dy \qquad\qquad \text{(A.39)} \\
&= \frac{\gamma}{1 - (A/B)^\gamma} \int_1^{B/A} y^{-(\gamma+1)}\, dy \\
&= \frac{\gamma}{1 - (A/B)^\gamma} \frac{1 - (B/A)^{-\gamma}}{\gamma} \\
&= 1, \qquad\qquad\qquad\qquad\qquad\qquad\qquad\quad \text{(A.40)}
\end{aligned}
$$

where Eq. (A.39) derives from the mid-frequency assumption $\omega/2\pi = f \ll A^{-1}$. Thus, to zeroth order we have $\phi_\tau(\omega) \approx 1$.

Use of this approximation leads to

$$
\text{Re}\left\{ \frac{1 + \phi_\tau(\omega)}{1 - \phi_\tau(\omega)} \right\} \approx \text{Re}\left\{ \frac{1 + 1}{1 - \phi_\tau(\omega)} \right\} = 2\,\frac{\text{Re}\{1 - \phi_\tau(\omega)\}}{|1 - \phi_\tau(\omega)|^2} \qquad \text{(A.41)}
$$

for the nontrivial part of the point-process spectrum, and to

$$
\text{Re}\left\{ \frac{1 - \phi_\tau(\omega)}{1 + \phi_\tau(\omega)} \right\} \approx \text{Re}\left\{ \frac{1 - \phi_\tau(\omega)}{1 + 1} \right\} = \tfrac{1}{2}\text{Re}\{1 - \phi_\tau(\omega)\} \qquad \text{(A.42)}
$$

for the symmetric alternating fractal renewal process. To obtain asymptotic expressions, we expand $1 - \phi_\tau(\omega)$ into a series of powers of ω until we obtain the first term with a nonzero real part. Substituting these results into Eqs. (A.41) and (A.42) yields Eqs. (7.8) and (8.11), respectively, as we will show forthwith.

To carry the calculation forward, we employ Eq. (7.3) evaluated at $\omega = 0$ and subtract Eq. (7.3) again:

$$
1 - \phi_\tau(\omega) = \gamma(i\omega A)^\gamma \left[1 - (A/B)^\gamma\right]^{-1} \int_{i\omega A}^{i\omega B} (1 - e^{-x})\, x^{-(\gamma+1)}\, dx. \qquad \text{(A.43)}
$$

Since $B^{-1} \ll f = \omega/2\pi \ll A^{-1}$, we have $A/B \to 0$ and $\omega B \to \infty$. Defining $z \equiv i\omega A$, we therefore obtain

$$
\begin{aligned}
1 - \phi_\tau(\omega) &\to \gamma z^\gamma \int_z^\infty (1 - e^{-x})\, x^{-(\gamma+1)}\, dx \\
&= \gamma \int_1^\infty (1 - e^{-zy})\, y^{-(\gamma+1)}\, dy. \qquad\qquad \text{(A.44)}
\end{aligned}
$$

We proceed to calculate express results for various ranges of the exponent γ.

- For $0 < \gamma < 1$ we make use of l'Hôpital's rule to evaluate the limit

$$
\begin{aligned}
\lim_{z \to 0} \frac{1 - \phi_\tau(\omega)}{z^\gamma} &= \lim_{z \to 0} \frac{\gamma \int_1^\infty (1 - e^{-zy}) y^{-(\gamma+1)} \, dy}{z^\gamma} \\
&= \lim_{z \to 0} \frac{\gamma \int_1^\infty e^{-zy} y^{-\gamma} \, dy}{\gamma z^{\gamma-1}} \\
&= \lim_{z \to 0} \frac{\gamma z^{\gamma-1} \int_z^\infty e^{-x} x^{-\gamma} \, dx}{\gamma z^{\gamma-1}} \\
&= \int_0^\infty e^{-x} x^{-\gamma} \, dx \\
&= \Gamma(1 - \gamma),
\end{aligned}
\tag{A.45}
$$

so that

$$
\begin{aligned}
1 - \phi_\tau(\omega) &\to \Gamma(1 - \gamma) \, (i\omega A)^\gamma \\
|1 - \phi_\tau(\omega)| &\to \Gamma(1 - \gamma) \, (\omega A)^\gamma \\
\mathrm{Re}\{1 - \phi_\tau(\omega)\} &\to \Gamma(1 - \gamma) \, \cos(\pi\gamma/2) \, (\omega A)^\gamma.
\end{aligned}
\tag{A.46}
$$

- For $\gamma = 1$, we again use l'Hôpital's rule and evaluate

$$
\begin{aligned}
\lim_{z \to 0} \frac{1 - \phi_\tau(\omega)}{-z \ln(z)} &= \lim_{z \to 0} \frac{\int_z^\infty (1 - e^{-x}) x^{-2} \, dx}{-\ln(z)} \\
&= \lim_{z \to 0} \frac{-(1 - e^{-z}) z^{-2}}{-1/z} \\
&= \lim_{z \to 0} \frac{1 - e^{-z}}{z} \\
&= 1,
\end{aligned}
\tag{A.47}
$$

which, with the help of the results set forth at the beginning of Sec. A.4, leads to

$$
\begin{aligned}
1 - \phi_\tau(\omega) &\to -(i\omega A) \ln(i\omega A) \\
&= -(i\omega A) \ln(i) - (i\omega A) \ln(\omega A) \\
&= (\pi/2)(\omega A) - i(\omega A) \ln(\omega A) \\
|1 - \phi_\tau(\omega)| &\to -(\omega A) \ln(\omega A) \\
\mathrm{Re}\{1 - \phi_\tau(\omega)\} &\to (\pi/2)(\omega A).
\end{aligned}
\tag{A.48}
$$

- For $\gamma > 1$ in general, the dominant term becomes linear in ω. The limit

$$
\begin{aligned}
\lim_{z \to 0} \frac{1 - \phi_\tau(\omega)}{z} &= \lim_{z \to 0} \frac{\gamma \int_1^\infty (1 - e^{-zy}) \, y^{-(\gamma+1)} \, dy}{z} \\
&= \lim_{z \to 0} \frac{\gamma \int_1^\infty e^{-zy} y^{-\gamma} \, dy}{1} \\
&= \gamma \int_1^\infty y^{-\gamma} \, dy \\
&= \frac{\gamma}{\gamma - 1}
\end{aligned}
\tag{A.49}
$$

implies that

$$
\begin{aligned}
1 - \phi_\tau(\omega) &\to \frac{\gamma}{\gamma - 1} i\omega A \\
|1 - \phi_\tau(\omega)| &\to \frac{\gamma}{\gamma - 1} \omega A,
\end{aligned}
\tag{A.50}
$$

for all $\gamma > 1$. To evaluate the spectrum we still require a first term with a nonzero real part. We continue by expanding the quantity

$$
1 - \phi_\tau(\omega) - \frac{\gamma}{\gamma - 1} i\omega A.
\tag{A.51}
$$

For $1 < \gamma < 2$, the next term arises from the limit

$$
\begin{aligned}
\lim_{z \to 0} &\frac{1 - \phi_\tau(\omega) - \gamma z/(\gamma - 1)}{z^\gamma} \\
&= \lim_{z \to 0} \frac{\gamma \int_1^\infty (1 - e^{-zy}) \, y^{-(\gamma+1)} \, dy - \gamma z/(\gamma - 1)}{z^\gamma} \\
&= \lim_{z \to 0} \frac{\gamma \int_1^\infty e^{-zy} y^{-\gamma} \, dy - \gamma/(\gamma - 1)}{\gamma z^{\gamma - 1}} \\
&= \lim_{z \to 0} \frac{-\gamma \int_1^\infty e^{-zy} y^{1-\gamma} \, dy}{\gamma(\gamma - 1) z^{\gamma - 2}} \\
&= \lim_{z \to 0} \frac{-\gamma z^{\gamma - 2} \int_z^\infty e^{-x} x^{1-\gamma} \, dx}{\gamma(\gamma - 1) z^{\gamma - 2}} \\
&= -(\gamma - 1)^{-1} \int_0^\infty e^{-x} x^{1-\gamma} \, dx \\
&= -(\gamma - 1)^{-1} \Gamma(2 - \gamma),
\end{aligned}
\tag{A.52}
$$

so that

$$\mathrm{Re}\{1 - \phi_\tau(\omega)\} \to (\gamma - 1)^{-1}\Gamma(2 - \gamma)\left[-\cos(\pi\gamma/2)\right](\omega A)^\gamma. \quad (A.53)$$

- For $\gamma = 2$ we again obtain logarithmic correction terms,

$$
\begin{aligned}
\lim_{z\to 0}\frac{1 - \phi_\tau(\omega) - \gamma z/(\gamma - 1)}{z^\gamma \ln(z)} &= \lim_{z\to 0}\frac{2z^2\displaystyle\int_z^\infty(1 - e^{-x})\,x^{-3}\,dx - 2z}{z^2 \ln(z)} \\
&= \lim_{z\to 0}\frac{\displaystyle\int_z^\infty(1 - e^{-x})\,x^{-3}\,dx - z^{-1}}{\ln(z)/2} \\
&= \lim_{z\to 0}\frac{-(1 - e^{-z})\,z^{-3} + z^{-2}}{z^{-1}/2} \\
&= \lim_{z\to 0}\frac{(z - 1 + e^{-z})}{z^2/2} \\
&= 1, \quad\quad\quad (A.54)
\end{aligned}
$$

which results in

$$\mathrm{Re}\{1 - \phi_\tau(\omega)\} \to -(\omega A)^2 \ln(\omega A). \quad (A.55)$$

- Finally, for $\gamma > 2$ the power-law exponents do not depend on γ, but are constant at the square of ω:

$$
\begin{aligned}
\lim_{z\to 0}\frac{1 - \phi_\tau(\omega) - \gamma z/(\gamma - 1)}{z^2} &\\
&= \lim_{z\to 0}\frac{\gamma\displaystyle\int_1^\infty(1 - e^{-zy})\,y^{-(\gamma+1)}\,dy - \gamma z/(\gamma - 1)}{z^2} \\
&= \lim_{z\to 0}\frac{\gamma\displaystyle\int_1^\infty e^{-zy}\,y^{-\gamma}\,dy - \gamma/(\gamma - 1)}{2z} \\
&= \lim_{z\to 0}\frac{\gamma\displaystyle\int_1^\infty e^{-zy}\,y^{1-\gamma}\,dy}{2} \\
&= -\tfrac{1}{2}\gamma\int_1^\infty y^{1-\gamma}\,dy \\
&= -\tfrac{1}{2}\gamma(2 - \gamma)^{-1}, \quad\quad (A.56)
\end{aligned}
$$

which gives

$$\mathrm{Re}\{1 - \phi_\tau(\omega)\} \to \tfrac{1}{2}\gamma(2 - \gamma)^{-1}(\omega A)^2. \quad (A.57)$$

A.4.2 Spectrum for $\gamma = \frac{1}{2}$

For the particular case $\gamma = \frac{1}{2}$, the smooth-transition interevent-interval probability density function provided in Eq. (7.5) simplifies to

$$p_\tau(t) = \sqrt{A/\pi}\, \exp\left(2\sqrt{A/B}\right)\, \exp(-A/t)\, \exp(-t/B)\, t^{-3/2}, \tag{A.58}$$

and the corresponding characteristic function becomes

$$\phi_\tau(\omega) = \exp\left[2\sqrt{A/B} - 2(A/B + i\omega A)^{1/2}\right]. \tag{A.59}$$

The derivative of Eq. (A.59) at $\omega = 0$ yields the mean interevent interval,

$$\frac{d}{d\omega}\phi_\tau(\omega) = \phi_\tau(\omega)\left[-(A/B + i\omega A)^{-1/2}\right]iA$$

$$\mathrm{E}[\tau] = i\,\frac{d}{d\omega}\phi_\tau(\omega)\bigg|_{\omega=0} = \sqrt{AB}. \tag{A.60}$$

For simplicity, we set the mean interval to unity, so that $AB = 1$. Equation (A.59) then becomes

$$\begin{aligned}
\phi_\tau(\omega) &= \exp\left[2A - 2\left(A^2 + i\omega A\right)^{1/2}\right] \\
&= \exp\left[2A - \sqrt{2A}\left(\sqrt{A^2 + \omega^2} + A\right)^{1/2}\right. \\
&\quad \left. - i\sqrt{2A}\left(\sqrt{A^2 + \omega^2} - A\right)^{1/2}\right] \\
&= \exp(-c)\left[\cos(d) - i\sin(d)\right], \tag{A.61}
\end{aligned}$$

where we have defined

$$c \equiv \sqrt{2A}\left(\sqrt{A^2 + \omega^2} + A\right)^{1/2} - 2A \tag{A.62}$$

$$d \equiv \sqrt{2A}\left(\sqrt{A^2 + \omega^2} - A\right)^{1/2}, \tag{A.63}$$

making use of Eq. (A.38) set out at the beginning of Sec. A.4.

Reiterating Eq. (4.16) and substituting Eq. (A.61) into it yields

$$\begin{aligned}
&S_N(f) - \mathrm{E}^2[\mu]\,\delta(f) \\
&= \mathrm{E}[\mu]\,\mathrm{Re}\left[\frac{1 + \phi_\tau(2\pi f)}{1 - \phi_\tau(2\pi f)}\right] \\
&= \mathrm{Re}\left\{\frac{1 + e^{-c}[\cos(d) - i\sin(d)]}{1 - e^{-c}[\cos(d) - i\sin(d)]} \times \frac{e^c - [\cos(d) + i\sin(d)]}{e^c - [\cos(d) + i\sin(d)]}\right\} \\
&= \mathrm{Re}\left\{\frac{e^c - 2i\sin(d) - e^{-c}[\cos^2(d) + \sin^2(d)]}{e^c - 2\cos(d) + e^{-c}[\cos^2(d) + \sin^2(d)]}\right\}
\end{aligned}$$

$$= \frac{e^c - e^{-c}}{e^c - 2\cos(d) + e^{-c}}$$

$$= \frac{\sinh(c)}{\cosh(c) - \cos(d)}, \tag{A.64}$$

which reproduces Eqs. (7.10)–(7.12).

A.4.3 Coincidence rate in the medium-time limit

Equation (7.13) derives from Eq. (7.8) via the Fourier-transform-pair relations that comprise Eqs. (3.57) and (3.58). Because our focus is on the mid-scale limit, we typically ignore delta functions at zero time or frequency, as well as limits for large times and frequencies.

Results for the regions $0 < \gamma < 1$ and $1 < \gamma < 2$ follow directly from Eqs. (5.44) and (5.45). It thus remains to consider the ranges $\gamma > 2$, $\gamma = 2$, and $\gamma = 1$.

- For $\gamma > 2$, the spectrum remains relatively constant in the limit $1/B \ll f \ll 1/A$, differing little from its low-frequency limit; it therefore resembles the spectrum for a homogeneous Poisson process provided in Eq. (4.9c). The coincidence rate thus follows the form in Eq. (4.9d), which appears in Eq. (7.13) for $\gamma > 2$.

- For $\gamma = 2$, we begin with the coincidence rate given in Eq. (7.13) to obtain

$$
\begin{aligned}
S_N(f) &= 2\mathrm{E}[\mu] \int_A^\infty \cos(2\pi f t) \tfrac{1}{4} t^{-1} dt \\
&= \frac{\mathrm{E}[\mu]}{2} \int_{2\pi f A}^\infty \cos(x)\, x^{-1} dx. \tag{A.65}
\end{aligned}
$$

The cosine factor in the integrand ensures that the integral converges for large x; in fact, since the integral diverges near $x = 0$ and $2\pi f A \ll 1$, the contribution of this factor for large values of x becomes negligible. For small values of x, the cosine term does not vary significantly, and also becomes unimportant. Bearing these arguments in mind, we then have to first order

$$
\begin{aligned}
S_N(f) &= \frac{\mathrm{E}[\mu]}{2} \int_{2\pi f A}^\infty \cos(x)\, x^{-1} dx \\
&\approx \frac{\mathrm{E}[\mu]}{2} \int_{2\pi f A}^1 x^{-1} dx \\
&= \frac{\mathrm{E}[\mu]}{2} \big[-\ln(2\pi f A)\big]. \tag{A.66}
\end{aligned}
$$

This agrees with Eq. (7.8), and thus establishes the validity of Eq. (7.13) for $\gamma = 2$.

- For $\gamma = 1$, we employ a similar argument but center it on the forward Fourier transform to obtain

$$
\begin{aligned}
G(t) &= 2\pi \mathrm{E}[\mu] \int_0^\infty \cos(2\pi f t) \left[\ln(2\pi f A)\right]^{-2} (2\pi f A)^{-1} df \\
&= \frac{\mathrm{E}[\mu]}{A} \int_0^\infty \cos(x) \left[\ln(Ax/t)\right]^{-2} x^{-1} dx \\
&\approx \frac{\mathrm{E}[\mu]}{A} \int_0^1 \cos(x) \left[\ln(Ax/t)\right]^{-2} x^{-1} dx \\
&\approx \frac{\mathrm{E}[\mu]}{A} \int_0^1 \left[\ln(Ax/t)\right]^{-2} x^{-1} dx \\
&= \frac{\mathrm{E}[\mu]}{A} \int_{-\infty}^{\ln(A/t)} y^{-2} dy \qquad\qquad\qquad (A.67) \\
&= \frac{\mathrm{E}[\mu]}{A} \left[\frac{-1}{\ln(A/t)}\right] \\
&= \mathrm{E}[\mu] A^{-1} \left[\ln(t/A)\right]^{-1}, \qquad\qquad\qquad (A.68)
\end{aligned}
$$

which accords with Eq. (7.13) for $\gamma = 1$. Equation (A.67) makes use of the substitution $y = \ln(Ax/t)$.

A.4.4   Normalized variances in the medium-time limit

Proceeding to the normalized variance $F(T)$ in Eq. (7.18), Eqs. (5.44) and (5.45) again provide results for $0 < \gamma < 1$ and $1 < \gamma < 2$. It therefore remains to consider the ranges $\gamma > 2$, $\gamma = 2$, and $\gamma = 1$:

- For $\gamma > 2$, results for the homogeneous Poisson process continue to apply, and we obtain the result provided in Eq. (4.9a).

- For $\gamma = 2$, we employ Eq. (3.55) to obtain

$$
G(t) - \mathrm{E}^2[\mu] = \frac{\mathrm{E}[\mu]}{2} \frac{d^2}{dT^2} \left[T \ln(T/A)/2\right]_{T=t} = \tfrac{1}{4}\mathrm{E}[\mu] t^{-1}, \qquad (A.69)
$$

in accordance with Eq. (7.13).

- For $\gamma = 1$, a similar approach leads to

$$
\begin{aligned}
G(t) - \mathrm{E}^2[\mu] &= \frac{\mathrm{E}[\mu]}{2} \frac{d^2}{dT^2} \left\{T A^{-1} \left[\ln(T/A)\right]^{-1} T\right\}_{T=t} \\
&= \frac{\mathrm{E}[\mu]}{2A} \left[\frac{2}{\ln(t/A)} - \frac{3}{\ln^2(t/A)} + \frac{2}{\ln^3(t/A)}\right] \\
&\approx \frac{\mathrm{E}[\mu]}{2A} \frac{2}{\ln(t/A)} \\
&= \mathrm{E}[\mu] A^{-1} \left[\ln(t/A)\right]^{-1}, \qquad\qquad\qquad (A.70)
\end{aligned}
$$

which is identical to the result given in Eq. (7.13) for $\gamma = 1$.

Results for the normalized Haar-wavelet variance $A(T)$ in Eq. (7.19) follow directly from Eqs. (5.44), (5.45), and (4.9a), except for $\gamma = 1$ and $\gamma = 2$. We employ Eq. (3.41) for these two cases:

- For $\gamma = 2$, we have

$$
\begin{aligned}
A(T) &= 2F(T) - F(T) \\
&= \ln(T/A) - \tfrac{1}{2}\ln(2T/A) \\
&= \ln(T/A) - \tfrac{1}{2}\ln(T/A) - \tfrac{1}{2}\ln(2) \\
&= \tfrac{1}{2}\ln(T/A) - \tfrac{1}{2}\ln(2) \\
&= \tfrac{1}{2}\ln(T/2A),
\end{aligned}
\tag{A.71}
$$

in agreement with Eq. (7.19).

- For $\gamma = 1$, we obtain

$$
\begin{aligned}
A(T) &= 2F(T) - F(T) \\
&= 2A^{-1}\big[\ln(T/A)\big]^{-1} T - A^{-1}\big[\ln(2T/A)\big]^{-1}(2T) \\
A(T) &= 2A^{-1}T\left[\frac{1}{\ln(T/A)} - \frac{1}{\ln(2T/A)}\right] \\
&= 2A^{-1}T\,\frac{\ln(2T/A) - \ln(T/A)}{\ln(T/A)\,\ln(2T/A)} \\
&= 2A^{-1}T\,\frac{\ln(2)}{\ln(T/A)\,\ln(2T/A)} \\
&\approx 2\ln(2)\,A^{-1}\big[\ln(T/A)\big]^{-2} T,
\end{aligned}
\tag{A.72}
$$

which also accords with the result provided in Eq. (7.19).

A.5 ALTERNATING FRACTAL RENEWAL PROCESS

A.5.1 Alternating-renewal-process spectrum

We obtain the spectrum of the alternating renewal process by calculating a sequence of quantities, each from the preceding: the probability distribution function of the counting process $N(t)$, the characteristic function of the forward recurrence time, the autocorrelation, and finally the spectrum.

Assume that at $t = 0$ the alternating fractal renewal process lies in state 1, so that $X(0) = 1$. As a first step, we seek the probability that the count exceeds a certain even, nonnegative value $2n$. For this to occur, the interevent interval that encompasses $t = 0$ must end, as well as the next n intervals of both types. The probability that

$N(t) > 2n$ is then the probability that the aggregate of the $2n + 1$ intervals does not exceed t. Thus,

$$\Pr\{N(t) > 2n \,|\, X(0) = 1\}$$

$$= \Pr\left\{ \vartheta_a(0) + \sum_{k=0}^{n-1} [\tau_{ak} + \tau_{bk}] < t \,\Big|\, X(0) = 1 \right\}$$

$$= \int_0^t p_{\vartheta a}(s) \star p_{\tau a}^{\star n}(s) \star p_{\tau b}^{\star n}(s) \, ds, \qquad (A.73)$$

where ϑ again denotes the forward recurrence time.

We next require an expression for the characteristic function of the forward recurrence time. Employing Eq. (3.12), we have

$$\phi_\vartheta(\omega) = \int_{t=0}^\infty \exp(-i\omega t) \, p_\vartheta(t) \, dt$$

$$= \int_{t=0}^\infty \exp(-i\omega t) \, [1 - P_\tau(t)] \, \mathrm{E}[\mu] \, dt$$

$$= \mathrm{E}[\mu] \int_{t=0}^\infty \exp(-i\omega t) \int_{u=t}^\infty p_\tau(u) \, du \, dt$$

$$= \mathrm{E}[\mu] \int_{u=0}^\infty p_\tau(u) \int_{t=0}^u \exp(-i\omega t) \, dt \, du$$

$$= (i\omega)^{-1} \mathrm{E}[\mu] \int_{u=0}^\infty p_\tau(u) \, [1 - \exp(-i\omega t)] \, du$$

$$= (i\omega)^{-1} \mathrm{E}[\mu] \, [1 - \phi_\tau(\omega)]. \qquad (A.74)$$

Taking a Fourier transform of the convolution in Eq. (A.73), and substituting $\omega = 2\pi f$, yields a simple expression involving characteristic functions

$$\int_0^\infty \exp(-i2\pi ft) \Pr\{N(t) > 2n \,|\, X(0) = 1\} \, dt$$

$$= \int_0^\infty \exp(-i2\pi ft) \int_0^t p_{\vartheta a}(s) \star p_{\tau a}^{\star n}(s) \star p_{\tau b}^{\star n}(s) \, ds \, dt$$

$$= \mathrm{E}[\mu_a] \, (i2\pi f)^{-2} \, [1 - \phi_{\tau a}(2\pi f)] \, \phi_{\tau a}^n(2\pi f) \, \phi_{\tau b}^n(2\pi f), \quad (A.75)$$

where one factor of $(i2\pi f)^{-1}$ arises from Eq. (A.74), and the other from the integration in Eq. (A.73). Similarly, for an odd positive number of intervals $2n + 1$, we obtain

$$\int_0^\infty \exp(-i2\pi ft) \Pr\{N(t) > 2n + 1 \,|\, X(0) = 1\} \, dt$$

$$= \mathrm{E}[\mu_a] \, (i2\pi f)^{-2} \, [1 - \phi_{\tau a}(2\pi f)] \, \phi_{\tau a}^n(2\pi f) \, \phi_{\tau b}^{n+1}(2\pi f). \quad (A.76)$$

Proceeding to the autocorrelation leads to

$$R_X(t) \equiv \mathrm{E}[X(0) \, X(t)]$$

$$
\begin{aligned}
&= \Pr\{X(0) = 1 \text{ and } X(t) = 1\} \\
&= \Pr\{X(0) = 1\}\, \Pr\{X(t) = 1 \mid X(0) = 1\} \\
&= \mathrm{E}[X]\, \Pr\{N(t) \text{ is even}\} \\
&= \mathrm{E}[X]\, \Pr\left\{ \sum_{n=0}^{\infty} N(t) = 2n \right\} \\
&= \mathrm{E}[X] \sum_{n=0}^{\infty} \Pr\{N(t) = 2n\} \\
&= \mathrm{E}[X] \sum_{n=0}^{\infty} \Big(\Pr\{N(t) > 2n - 1\} - \Pr\{N(t) > 2n\} \Big) \\
&= \mathrm{E}[X] \left[1 + \sum_{n=0}^{\infty} \Big(\Pr\{N(t) > 2n + 1\} - \Pr\{N(t) > 2n\} \Big) \right]. \quad \text{(A.77)}
\end{aligned}
$$

Finally, taking the Fourier transform yields the spectrum

$$
\begin{aligned}
&S_X(f)/\mathrm{E}[X] \\
&= \frac{1}{\mathrm{E}[X]} \int_{-\infty}^{\infty} \exp(-i2\pi ft)\, R_X(t)\, dt \\
&= \frac{2}{\mathrm{E}[X]} \mathrm{Re}\left\{ \int_{0}^{\infty} \exp(-i2\pi ft)\, R_X(t)\, dt \right\} \\
&= 2\mathrm{Re}\left\{ \int_{0}^{\infty} \exp(-i2\pi ft) \left[1 \right.\right. \\
&\qquad\qquad \left.\left. + \sum_{n=0}^{\infty} \Big(\Pr\{N(t) > 2n + 1\} - \Pr\{N(t) > 2n\} \Big) \right] dt \right\} \\
&= \delta(f) + 2\mathrm{Re}\left\{ \int_{0}^{\infty} \exp(-i2\pi ft) \right. \\
&\qquad\qquad \left. \times \sum_{n=0}^{\infty} \Big(\Pr\{N(t) > 2n + 1\} - \Pr\{N(t) > 2n\} \Big) dt \right\} \\
&= \delta(f) + 2\mathrm{Re}\left\{ \sum_{n=0}^{\infty} \mathrm{E}[\mu_a]\, (i2\pi f)^{-2} \left[1 - \phi_{\tau a}(\omega) \right] \right. \\
&\qquad\qquad \left. \times \left[\phi_{\tau a}^{n}(2\pi f)\, \phi_{\tau b}^{n+1}(2\pi f) - \phi_{\tau a}^{n}(2\pi f)\, \phi_{\tau b}^{n}(2\pi f) \right] \right\} \\
&= \delta(f) + 2\mathrm{E}[\mu_a]\, (2\pi f)^{-2}\, \mathrm{Re}\left\{ \left[1 - \phi_{\tau a}(2\pi f) \right] \right. \\
&\qquad\qquad \left. \times \left[1 - \phi_{\tau b}(2\pi f) \right] \sum_{n=0}^{\infty} \phi_{\tau a}^{n}(2\pi f)\, \phi_{\tau b}^{n}(2\pi f) \right\}
\end{aligned}
$$

$$= \delta(f) + \frac{2E[\mu_a]}{(2\pi f)^2} \text{Re}\left\{ \frac{[1 - \phi_{\tau a}(2\pi f)][1 - \phi_{\tau b}(2\pi f)]}{1 - \phi_{\tau a}(2\pi f)\, \phi_{\tau b}(2\pi f)} \right\}, \quad \text{(A.78)}$$

which leads directly to Eq. (8.5).

A.5.2 Low-frequency limit of the spectrum

To determine the spectrum as $f \to 0$, we use Eq. (8.5) [or, equivalently, Eq. (A.78)], retaining terms to second order in frequency at each stage. To simplify the notation, we use radian frequency $\omega \equiv 2\pi f$, and the quantities s, t, u, and v to represent fixed constants. We then have

$$\phi_{\tau a}(\omega) \approx 1 - i\omega E[\tau_a] - \frac{\omega^2}{2} E[\tau_a^2], \quad \text{(A.79)}$$

along with the analogous expression for $\phi_{\tau b}(\omega)$. Substituting Eq. (A.79) into Eq. (8.5), and ignoring the delta function which does not appear in the limit, yields

$$\left(E[\tau_a] + E[\tau_b] \right) \lim_{\omega \to 0} S_X(\omega)/2$$

$$= \lim_{\omega \to 0} \omega^{-2} \text{Re}\left\{ \frac{[1 - \phi_{\tau a}(\omega)][1 - \phi_{\tau b}(\omega)]}{1 - \phi_{\tau a}(\omega)\, \phi_{\tau b}(\omega)} \right\}$$

$$= \lim_{\omega \to 0} \text{Re}\left\{ \frac{\omega^{-2}\left(i\omega E[\tau_a] + \frac{\omega^2}{2} E[\tau_a^2] \right)\left(i\omega E[\tau_b] + \frac{\omega^2}{2} E[\tau_b^2] \right)}{i\omega E[\tau_a] + i\omega E[\tau_b] + \frac{\omega^2}{2} E[\tau_a^2] + \frac{\omega^2}{2} E[\tau_b^2] + \omega^2 E[\tau_a] E[\tau_b]} \right\}$$

$$= \lim_{\omega \to 0} \text{Re}\left\{ \frac{-2E[\tau_a] E[\tau_b] + i\omega\left(E[\tau_a] E[\tau_b^2] + E[\tau_b] E[\tau_a^2] \right)}{2i\omega\left(E[\tau_a] + E[\tau_b] \right) + \omega^2\left(E[\tau_a^2] + E[\tau_b^2] + 2E[\tau_a] E[\tau_b] \right)} \right\}$$

$$= \lim_{\omega \to 0} \text{Re}\left\{ \frac{s + i\omega t}{i\omega u + \omega^2 v} \right\}$$

$$= \lim_{\omega \to 0} \text{Re}\left\{ \frac{s + i\omega t}{i\omega u + \omega^2 v} \times \frac{v + u/(i\omega)}{v + u/(i\omega)} \right\}$$

$$= \lim_{\omega \to 0} \text{Re}\left\{ \frac{tu + sv + i\omega tv + su/(i\omega)}{u^2 + \omega^2 v^2} \right\}$$

$$= \lim_{\omega \to 0} \frac{tu + sv}{u^2 + \omega^2 v^2}$$

$$= \frac{tu + sv}{u^2}$$

$$\left(E[\tau_a] + E[\tau_b] \right)^3 \lim_{\omega \to 0} S_X(\omega)$$

$$= (tu + sv)/2$$

$$= \left(E[\tau_a] E[\tau_b^2] + E[\tau_b] E[\tau_a^2] \right) \left(E[\tau_a] + E[\tau_b] \right)$$

$$\quad - E[\tau_a] E[\tau_b] \left(E[\tau_a^2] + E[\tau_b^2] + 2E[\tau_a] E[\tau_b] \right)$$

$$
\begin{aligned}
&= \mathrm{E}[\tau_a]^2 \mathrm{E}[\tau_b^2] + \mathrm{E}[\tau_a] \mathrm{E}[\tau_b] \mathrm{E}[\tau_a^2] + \mathrm{E}[\tau_a] \mathrm{E}[\tau_b] \mathrm{E}[\tau_b^2] + \mathrm{E}[\tau_b]^2 \mathrm{E}[\tau_a^2] \\
&\quad - \mathrm{E}[\tau_a] \mathrm{E}[\tau_b] \mathrm{E}[\tau_a^2] - \mathrm{E}[\tau_a] \mathrm{E}[\tau_b] \mathrm{E}[\tau_b^2] - 2\mathrm{E}[\tau_a]^2 \mathrm{E}[\tau_b]^2 \\
&= \mathrm{E}[\tau_a]^2 \mathrm{E}[\tau_b^2] + \mathrm{E}[\tau_b]^2 \mathrm{E}[\tau_a^2] - 2\mathrm{E}[\tau_a]^2 \mathrm{E}[\tau_b]^2 \\
&= \mathrm{E}[\tau_a]^2 \operatorname{Var}[\tau_b] + \mathrm{E}[\tau_b]^2 \operatorname{Var}[\tau_a], \quad\quad\quad\quad\quad\quad\text{(A.80)}
\end{aligned}
$$

which accords with Eq. (8.6).

A.5.3 Spectrum under extreme dwell-time asymmetry

Consider an alternating renewal process $X(t)$ for which the times τ_b spent in the state $X(t) = b$ greatly exceed the times τ_a spent in state $X(t) = a$. More formally, given a randomly selected pair of inter-transition intervals τ_a and τ_b, we assume that the relation $\Pr\{\tau_a \ll \tau_b\} \approx 1$ holds. In this case, the sum of the dwell times, $\tau_a + \tau_b$, will have marginal statistics that are nearly the same as those of τ_b, and the process will closely resemble a filtered version of a renewal point process $dN(t)$ constructed solely from the longer intervals τ_b.

Linear systems theory (Papoulis, 1991) leads directly to the spectrum. Citing Eq. (9.35) in an approximate sense, we have

$$
S_X(f) \approx \mathrm{E}\!\left[|H(f)|^2\right] S_N(f), \quad\quad\quad\quad\quad\quad\text{(A.81)}
$$

where $S_N(f)$ denotes the spectrum of the renewal point process $dN(t)$ constructed from τ_b, $H(f)$ represents the Fourier transform of the filter (whose form we will establish shortly), and $S_X(f)$ is the spectrum of the resulting alternating renewal process. This approximation remains valid for all nonzero frequencies, but it does not hold for $f = 0$; at this frequency the difference between point processes and real-valued processes requires us to invoke other methods.

Specifying the impulse response function $h(t)$ to be a rectangular filter of (random) duration τ_a, we have

$$
\begin{aligned}
H(f) &= \int_{-\infty}^{\infty} \exp(-i2\pi ft)\, h(t)\, dt \\
&= \int_{0}^{\tau_a} \exp(-i2\pi ft)\, dt \\
&= \left[1 - \exp(-i2\pi f\tau_a)\right] / (i2\pi f) \\
&= \exp(-i\pi f\tau_a) \sin(\pi f\tau_a) / (\pi f) \\
|H(f)|^2 &= \sin^2(\pi f\tau_a) / (\pi f)^2 \\
&\approx \tau_a^2 \\
\mathrm{E}\!\left[|H(f)|^2\right] &\approx \mathrm{E}[\tau_a^2], \quad\quad\quad\quad\quad\quad\text{(A.82)}
\end{aligned}
$$

where we have made use of the approximation $\sin(x) \approx x$ for small arguments x, which is valid in the domain $f\tau_a \ll 1$. Finally, substituting Eqs. (4.16) and (A.82) into Eq. (A.81) leads to

$$S_X(f) \approx \frac{E[\tau_a^2]}{E[\tau_b]} \operatorname{Re}\left\{\frac{1 + \phi_{\tau b}(2\pi f)}{1 - \phi_{\tau b}(2\pi f)}\right\}. \tag{A.83}$$

For the contribution at $f = 0$, we note that Eq. (8.5) contains a term $E[X]\delta(f)$; in the limit considered here, $E[X]$ approaches $E[\tau_a]/E[\tau_b]$, which agrees with Eq. (8.9).

A.6 FRACTAL SHOT NOISE

A.6.1 Infinite-area tail

If the impulse response function $h(t)$ has infinite area in its tail, the resulting shot noise process is degenerate (Lowen & Teich, 1990, Appendix A). Such an impulse response function has the property

$$\int_c^\infty h(t)\, dt = \infty \tag{A.84}$$

for any finite real number c. We rewrite Eq. (9.3), considering deterministic K for simplicity, to obtain

$$
\begin{aligned}
\ln[\phi_X(\omega)] &= -\mu \int_{-\infty}^\infty \{1 - \exp[-i\omega h(t)]\}\, dt \\
&= -\mu \int_{-\infty}^c \{1 - \exp[-i\omega h(t)]\}\, dt \\
&\quad -\mu \int_c^\infty \{1 - \exp[-i\omega h(t)]\}\, dt \\
&= -\mu f(c) - \mu \int_c^\infty \{1 - \exp[-i\omega h(t)]\}\, dt \\
&\approx -\mu f(c) - \mu \int_c^\infty \{1 - [1 - i\omega h(t)]\}\, dt \tag{A.85} \\
&= -\mu f(c) - i\omega\mu \int_c^\infty h(t)\, dt \\
&= -\mu f(c) - i\omega\mu\infty, \tag{A.86}
\end{aligned}
$$

where $f(c)$ denotes the value of the integral below c. We choose the value of c to be sufficiently large so that the argument of the exponential lies close to zero, permitting the approximation $\exp(-x) \approx 1 - x$ in Eq. (A.85).

Of the two terms in Eq. (A.86), the second has infinite absolute value, indicating an infinite shot-noise process amplitude X, unless the first term cancels the second. However, the integrands in Eqs. (A.85) and (A.86) include a unity term, so $f(c)$ must contain a real component of comparable magnitude to its imaginary component. If the imaginary components cancel, then a real component of infinite magnitude must also exist, again leading to an infinite shot-noise process. In any case, therefore, an impulse-response function with an infinite-area tail leads to a shot-noise process X with infinite amplitude. A stochastic amplitude K does not affect this conclusion.

Even a normalized version of this impulse response function leads to a degenerate shot noise process (Lowen & Teich, 1991, Appendix D). Consider the fixed-area family defined by

$$h_B(t) \equiv \begin{cases} \dfrac{a\,h(t)}{\int_{-\infty}^{B} h(u)\,du} & t < B \\ 0 & \text{otherwise,} \end{cases} \tag{A.87}$$

where the quantity a denotes a specified area as in Eq. (10.12). By construction, $h_B(t)$ has total area a. We again rewrite Eq. (9.3) for deterministic K, yielding

$$\begin{aligned} \ln\left[\phi_X(\omega)\right] &= -\mu \int_{-\infty}^{\infty} \left\{1 - \exp[-i\omega h_B(t)]\right\} dt \\ &= -\mu \int_{-\infty}^{B} \left\{1 - \exp\left[-\frac{i\omega\,a\,h(t)}{\int_{-\infty}^{B} h(s)\,ds}\right]\right\} dt \\ &\approx -\mu \int_{-\infty}^{B} \left\{1 - \left[1 - \frac{i\omega\,a\,h(t)}{\int_{-\infty}^{B} h(s)\,ds}\right]\right\} dt \tag{A.88} \\ &= -\mu \frac{i\omega\,a\int_{-\infty}^{B} h(t)\,dt}{\int_{-\infty}^{B} h(s)\,ds} \\ &= -i\omega\mu a. \tag{A.89} \end{aligned}$$

As B approaches ∞, the argument of the exponential decreases without limit, enabling the approximation $\exp(-x) \approx 1 - x$ to be used in Eq. (A.88). Equation (A.89) comprises a power series in ω, with terms of the power series identified with the cumulants of the shot-noise amplitude X (see Sec. 9.2). The lack of a second-order term (or indeed of any higher-order terms) indicates that the variance of the process assumes a value of zero; the amplitude remains fixed at the constant value μa. We conclude that a normalized impulse response function with an infinite-area tail converges to a constant value, and thus leads to a degenerate process.

A.6.2   Approach to stable form

For fractal shot noise with $\beta > 1$, $A = 0$, and $B < \infty$, the amplitude of the shot-noise process X is not a stable random variable, but it does approach one as the rate μ of the driving Poisson process approaches infinity (Lowen & Teich, 1990, Appendix B).

To demonstrate this, we employ l'Hôpital's rule to find the limit $A \to 0$ in Eq. (9.5), which yields

$$\begin{aligned} \ln\left[\phi_X(\omega)\right] &= -\mu B\left[1 - \exp\left(-i\omega K B^{-\beta}\right)\right] \\ &\quad - \mu(i\omega K)^{1/\beta}\,\Gamma\left(1 - 1/\beta,\, i\omega K B^{-\beta}\right). \tag{A.90} \end{aligned}$$

As B approaches ∞, we obtain

$$
\begin{aligned}
\ln\left[\phi_X(\omega)\right] &\approx -\mu B\left[1 - \left(1 - i\omega K B^{-\beta}\right)\right] \\
&\quad - \mu(i\omega K)^{1/\beta}\,\Gamma(1 - 1/\beta) \\
&\approx -i\mu\omega K B^{1-\beta} - \mu(i\omega K)^{1/\beta}\,\Gamma(1 - 1/\beta) \\
&\approx -\mu(i\omega K)^{1/\beta}\,\Gamma(1 - 1/\beta),
\end{aligned}
\tag{A.91}
$$

thereby validating the equivalent result shown in Eq. (9.13), but for deterministic K only.

For random K, we turn to Lebesgue measure theory (Lowen & Teich, 1990). Suppose that a stochastic impulse response function $h_1(K_1, t)$ and a deterministic impulse response function $h_2(t)$ obey the relation

$$
\mathrm{E}\left[\mathcal{L}\{t : h_1(K_1, t) > x\}\right] = \mathcal{L}\{t : h_2(t) > x\}
\tag{A.92}
$$

for all amplitudes x, where $\mathcal{L}$ denotes the Lebesgue set measure. Shot-noise processes constructed from these two impulse response functions will then have identical first-order statistics (Gilbert & Pollak, 1960). For fractal shot noise, in particular, any stochastic impulse response function $h_1(K_1, t)$ satisfying

$$
\mathrm{E}\left[\mathcal{L}\{t : h_1(K_1, t) > x\}\right] =
\begin{cases}
\infty & x < 0 \\
B - A & 0 \le x \le K_2 B^{-\beta} \\
(x/K_2)^{-1/\beta} - A & K_2 B^{-\beta} < x < K_2 A^{-\beta} \\
0 & x \ge K_2 A^{-\beta}
\end{cases}
\tag{A.93}
$$

for some β, A, B, and K_2, has identical first-order statistics as the deterministic impulse response function (Lowen & Teich, 1990)

$$
h_2(K_2, t) =
\begin{cases}
K_2 t^{-\beta} & A \le t < B \\
0 & \text{otherwise.}
\end{cases}
\tag{A.94}
$$

In general, finding a nontrivial ensemble of impulse response functions for which the equivalent impulse response function follows the form of Eq. (A.94) proves difficult. However, for the particular case $A = 0$ and $B = \infty$ we find

$$
\begin{aligned}
\mathrm{E}\left[\mathcal{L}\{t : h_1(K_1, t) > x\}\right] &= \mathrm{E}\left[\mathcal{L}\{t : K_1 t^{-\beta} > x\}\right] \\
&= \mathrm{E}\left[\mathcal{L}\{t : t < K_1^{1/\beta} x^{-1/\beta}\}\right] \\
&= \mathrm{E}\left[K_1^{1/\beta}\right] x^{-1/\beta}
\end{aligned}
\tag{A.95}
$$

for all amplitudes x. For the deterministic power-law impulse response function, we have

$$
\begin{aligned}
\mathcal{L}\{t : h_2(K_2, t) > x\} &= \mathcal{L}\{t : K_2 t^{-\beta} > x\} \\
&= \mathcal{L}\{t : t < K_2^{1/\beta} x^{-1/\beta}\} \\
&= K_2^{1/\beta} x^{-1/\beta},
\end{aligned}
\tag{A.96}
$$

again for all amplitudes x. Thus, the stochastic ensemble of impulse response functions in Eq. (A.95) and the deterministic impulse response function in Eq. (A.96) exhibit identical first-order amplitude statistics, provided

$$\mathrm{E}\left[K_1^{1/\beta}\right] = K_2^{1/\beta}, \tag{A.97}$$

so that

$$K_2 \equiv \mathrm{E}^\beta\left[K_1^{1/\beta}\right]. \tag{A.98}$$

For $A > 0$ or $B < \infty$, Eqs. (A.95) and (A.96) are no longer in accord for all x, so that the equivalent impulse response function does not have the form of Eq. (A.96). But for the case of interest, we indeed have $A = 0$ and $B = \infty$, permitting $\mathrm{E}\left[K^{1/\beta}\right]^\beta$ to be used in place of deterministic K in all first-order statistics, including Eq. (9.13).

For finite B, the process still approaches a stable form for large values of μ. However, merely increasing μ leads to a degenerate characteristic function; normalization becomes necessary. To demonstrate convergence to a particular form, we therefore consider the limit $\mu \to \infty$, $K \to 0$, with the dimensionless product $\mu^\beta K$ fixed at a value of ω_0^{-1}. Considering the above limit, Eq. (A.90) becomes

$$
\begin{aligned}
\ln\left[\phi_X(\omega)\right] \quad &\to \quad -\mu B\left[1 - \exp\left(-i\omega\, \omega_0^{-1}\, \mu^{-\beta}\, B^{-\beta}\right)\right] \\
&\quad - \mu(i\omega\, \omega_0^{-1}\, \mu^{-\beta})^{1/\beta}\, \Gamma\left(1 - 1/\beta,\; i\omega\, \omega_0^{-1}\, \mu^{-\beta}\, B^{-\beta}\right) \\
&= \quad -x\left[1 - \exp\left(-ix^{-\beta}\, \omega/\omega_0\right)\right] \\
&\quad - (i\omega/\omega_0)^{1/\beta}\, \Gamma\left(1 - 1/\beta,\; ix^{-\beta}\, \omega/\omega_0\right), \tag{A.99}
\end{aligned}
$$

where we define $x \equiv \mu B$. As μ increases with B fixed, the quantity x increases commensurately. The argument of the exponential in Eq. (A.99) then decreases, permitting the following simplification to be applied:

$$
\begin{aligned}
\ln\left[\phi_X(\omega)\right] \quad &\to \quad -x\left[1 - \left(1 - ix^{-\beta}\, \omega/\omega_0\right)\right] \\
&\quad - (i\omega/\omega_0)^{1/\beta}\, \Gamma\left(1 - 1/\beta,\; ix^{-\beta}\, \omega/\omega_0\right) \\
&= \quad -ix^{1-\beta}\, \omega/\omega_0 - (i\omega/\omega_0)^{1/\beta}\, \Gamma\left(1 - 1/\beta,\; ix^{-\beta}\, \omega/\omega_0\right) \\
&\to \quad 0 - (i\omega/\omega_0)^{1/\beta}\, \Gamma(1 - 1/\beta). \tag{A.100}
\end{aligned}
$$

The result is in the precise form of a stable characteristic function, as defined in Eq. (9.14).

A.6.3 Autocorrelation

We proceed to provide expressions for the shot-noise autocorrelation for specific parameter ranges.

- For $\beta = \frac{1}{2}$ and $0 \le |t| < B - A$, we have

$$R_h(t) \quad = \quad \mathrm{E}\left[K^2\right] \int_A^{B-|t|} (s^2 + |t|\, s)^{-1/2}\, ds$$

$$
\begin{aligned}
&= \; 2\mathrm{E}\big[K^2\big] \ln \Big[s^{1/2} + (s + |t|)^{1/2} \Big]_A^{B-|t|} \\
&= \; 2\mathrm{E}\big[K^2\big] \ln \left[\frac{B^{1/2} + (B - |t|)^{1/2}}{A^{1/2} + (A + |t|)^{1/2}} \right], \quad\quad (\mathrm{A}.101)
\end{aligned}
$$

finite if $B < \infty$ and either $A > 0$ or $t \neq 0$.

- For $\beta = 1$ and $t = 0$,

$$
\begin{aligned}
R_h(t) &= \mathrm{E}\big[K^2\big] \int_A^B (s^2)^{-1}\, ds \\
&= \mathrm{E}\big[K^2\big] \big[A^{-1} - B^{-1} \big] \quad\quad (\mathrm{A}.102)
\end{aligned}
$$

is finite when $A > 0$. For $\beta = 1$ and $0 < |t| < B - A$, another logarithmic form emerges,

$$
\begin{aligned}
R_h(t) &= \mathrm{E}\big[K^2\big] \int_A^{B-|t|} (s^2 + |t|\, s)^{-1}\, ds \\
&= \mathrm{E}\big[K^2\big] \, |t|^{-1} \ln \left[\frac{|t|}{s + |t|} \right]_A^{B-|t|} \\
&= \mathrm{E}\big[K^2\big] \, |t|^{-1} \ln\Big[(1 - |t|/B)\,(1 + |t|/A) \Big], \quad (\mathrm{A}.103)
\end{aligned}
$$

which is finite if $A > 0$.

- For $\beta = 2$ and $t = 0$ we obtain

$$
\begin{aligned}
R_h(t) &= \mathrm{E}\big[K^2\big] \int_A^B (s^2)^{-2}\, ds \\
&= \tfrac{1}{3}\mathrm{E}\big[K^2\big] \big[A^{-3} - B^{-3} \big], \quad\quad (\mathrm{A}.104)
\end{aligned}
$$

finite for $A > 0$. For $\beta = 2$ and $0 < |t| < B - A$, we have

$$
\begin{aligned}
R_h(t) &= \mathrm{E}\big[K^2\big] \int_A^{B-|t|} (s^2 + |t|\, s)^{-2}\, ds \\
&= -\mathrm{E}\big[K^2\big] \left[\frac{2s + |t|}{|t|^2\, s(s + |t|)} + \frac{2}{|t|^3} \ln \frac{s}{s + |t|} \right]_A^{B-|t|} \\
&= \mathrm{E}\big[K^2\big] \left\{ \frac{2A + |t|}{|t|^2\, A(A + |t|)} - \frac{2B - |t|}{|t|^2\, B(B - |t|)} \right. \\
&\quad\quad \left. + 2|t|^{-3} \ln\Big[(1 - |t|/B)\,(1 + |t|/A) \Big] \right\}, \quad\quad (\mathrm{A}.105)
\end{aligned}
$$

finite when $A > 0$.

- For general $\beta > 1$, $A > 0$, and $B = \infty$, a simple form for $R_h(t)$ emerges in the limit $|t| \to \infty$. We first find an upper bound for the integral:

$$
\begin{aligned}
\int_A^\infty \left(s^2 + |t|\, s\right)^{-\beta} ds &= \int_A^\infty s^{-\beta} \left(s + |t|\right)^{-\beta} ds \\
&< \int_A^\infty s^{-\beta} \left(|t|\right)^{-\beta} ds \\
&= \frac{A^{1-\beta}}{\beta - 1} |t|^{-\beta}.
\end{aligned}
\tag{A.106}
$$

For the lower bound we truncate the integral at some value T,

$$
\begin{aligned}
\int_A^\infty \left(s^2 + |t|\, s\right)^{-\beta} ds &= \int_A^\infty s^{-\beta}(s + |t|)^{-\beta} ds \\
&> \int_A^T s^{-\beta} \left(T + |t|\right)^{-\beta} ds \\
&= \frac{A^{1-\beta} - T^{1-\beta}}{\beta - 1} \left(T + |t|\right)^{-\beta},
\end{aligned}
\tag{A.107}
$$

valid for any $T > A$. We choose $T = (A|t|)^{1/2}$, so that

$$
\begin{aligned}
\int_A^\infty \left(s^2 + |t|\, s\right)^{-\beta} ds &> \frac{A^{1-\beta} - (A|t|)^{(1-\beta)/2}}{\beta - 1} \left[(A|t|)^{1/2} + |t|\right]^{-\beta} \\
&= \frac{A^{1-\beta}}{\beta - 1} |t|^{-\beta} \left[1 + (A/|t|)^{1/2}\right]^{-\beta} \\
&\quad \times \left[1 - (A/|t|)^{(\beta-1)/2}\right].
\end{aligned}
\tag{A.108}
$$

Finally, combining limits yields

$$
\left[1 + (A/|t|)^{1/2}\right]^{-\beta} \left[1 - (A/|t|)^{(\beta-1)/2}\right]
$$

$$
< \frac{\displaystyle\int_A^\infty \left(s^2 + |t|\, s\right)^{-\beta} ds}{|t|^{-\beta}\, A^{1-\beta}/(\beta - 1)} < 1,
\tag{A.109}
$$

for all t such that $|t| > A$. In the limit $|t| \to \infty$, the lower bound approaches unity, so that

$$
\int_A^\infty \left(s^2 + |t|\, s\right)^{-\beta} ds \;\to\; |t|^{-\beta} A^{1-\beta}/(\beta - 1)
\tag{A.110}
$$

and

$$
R_h(t) \;\to\; \mathrm{E}\!\left[K^2\right] \frac{A^{1-\beta}}{\beta - 1} |t|^{-\beta}.
\tag{A.111}
$$

A.7 FRACTAL-SHOT-NOISE-DRIVEN POINT PROCESSES

A.7.1 Integrals for counting statistics

The counting distribution for the fractal-shot-noise-driven Poisson process derives from a recursion relation. We consider the case for deterministic K. Reiterating Eqs. (10.4) and (10.5), we have

$$p_Z(n+1;T) = \frac{1}{n+1} \sum_{k=0}^{n} c_k \, p_Z(n-k;T),$$ (A.112)

and

$$c_k \equiv \frac{\mu}{k!} \int_{-\infty}^{\infty} \left[h_T(K,t) \right]^{k+1} \exp[-h_T(K,t)] \, dt.$$ (A.113)

The recursion coefficients c_k assume four different forms, depending on the value of β and the relative magnitudes of A, B, and T.

- For $\beta \neq 1$ and $B > A + T$:

$$
\begin{aligned}
c_k \;=\; & \frac{\mu \, K^{k+1}}{k! \, (1-\beta)^{k+1}} \\
& \times \Bigg(\int_{A}^{A+T} \left[u^{1-\beta} - A^{1-\beta} \right]^{k+1} \exp\left\{ -\frac{K}{1-\beta} \left[u^{1-\beta} - A^{1-\beta} \right] \right\} du \\
& + \int_{A}^{B-T} \left[(u+T)^{1-\beta} - u^{1-\beta} \right]^{k+1} \\
& \qquad \times \exp\left\{ -\frac{K}{1-\beta} \left[(u+T)^{1-\beta} - u^{1-\beta} \right] \right\} du \\
& + \int_{B-T}^{B} \left[B^{1-\beta} - u^{1-\beta} \right]^{k+1} \\
& \qquad \times \exp\left\{ -\frac{K}{1-\beta} \left[B^{1-\beta} - u^{1-\beta} \right] \right\} du \Bigg).
\end{aligned}
$$ (A.114)

- For $\beta \neq 1$ and $B \leq A + T$:

$$
\begin{aligned}
c_k \;=\; & \frac{\mu \, K^{k+1}}{k! \, (1-\beta)^{k+1}} \\
& \times \Bigg(\int_{A}^{B} \left[u^{1-\beta} - A^{1-\beta} \right]^{k+1} \\
& \qquad \times \exp\left\{ -\frac{K}{1-\beta} \left[(u+T)^{1-\beta} - A^{1-\beta} \right] \right\} du \\
& + \int_{A}^{B} \left[B^{1-\beta} - u^{1-\beta} \right]^{k+1} \exp\left\{ -\frac{K}{1-\beta} \left[B^{1-\beta} - u^{1-\beta} \right] \right\} du
\end{aligned}
$$

$$+ (T + A - B) \left[B^{1-\beta} - A^{1-\beta}\right]^{k+1}$$

$$\times \exp\left\{-\frac{K}{1-\beta}\left[B^{1-\beta} - A^{1-\beta}\right]\right\}\right). \tag{A.115}$$

- For $\beta = 1$ and $B > A + T$:

$$\begin{aligned}
c_k &= \frac{\mu K^{k+1}}{k!}\left\{\int_A^{B-T}\left[\ln\left(\frac{u+T}{u}\right)\right]^{k+1}\left(\frac{u}{u+T}\right)^K du\right. \\
&\quad + \int_A^{A+T}\left[\ln\left(\frac{u}{A}\right)\right]^{k+1}\left(\frac{A}{u}\right)^K du \\
&\quad \left. + \int_{B-T}^{B}\left[\ln\left(\frac{B}{u}\right)\right]^{k+1}\left(\frac{u}{B}\right)^K du\right\}.
\end{aligned} \tag{A.116}$$

- Finally, for $\beta = 1$ and $B \leq A + T$:

$$\begin{aligned}
c_k &= \frac{\mu K^{k+1}}{k!}\left\{(T + A - B)\left[\ln\left(\frac{B}{A}\right)\right]^{k+1}\left(\frac{A}{B}\right)^K\right. \\
&\quad + \int_A^B\left[\ln\left(\frac{u}{A}\right)\right]^{k+1}\left(\frac{A}{u}\right)^K du \\
&\quad \left. + \int_A^B\left[\ln\left(\frac{B}{u}\right)\right]^{k+1}\left(\frac{u}{B}\right)^K du\right\}.
\end{aligned} \tag{A.117}$$

The count moments of the fractal-shot-noise-driven Poisson process also derive from a recursion relation, but in this case we can easily consider stochastic as well as deterministic K. Reiterating Eqs. (10.8) and (10.9), we have

$$E\left\{\frac{[Z(t)]!}{[Z(t) - (n+1)]!}\right\} = \sum_{k=0}^n b_k \binom{n}{k} E\left\{\frac{[Z(t)]!}{[Z(t) - (n-k)]!}\right\}, \tag{A.118}$$

with

$$E\left\{\frac{[Z(t)]!}{[Z(t)]!}\right\} \equiv 1 \quad \text{and} \quad b_k \equiv \mu E\left[\int_{-\infty}^{\infty}\left[h_T(K,t)\right]^{k+1} dt\right]. \tag{A.119}$$

The recursion coefficients b_k also have four different forms, depending on β and the relative magnitudes of A, B, and T:

- For $\beta \neq 1$ and $B > A + T$:

$$b_k = \frac{\mu E[K^{k+1}]}{k!\,(1-\beta)^{k+1}}\left\{\int_A^{B-T}\left[(u+T)^{1-\beta} - u^{1-\beta}\right]^{k+1} du\right.$$

$$+ \int_A^{A+T} \left[u^{1-\beta} - A^{1-\beta} \right]^{k+1} du$$

$$+ \int_{B-T}^B \left[B^{1-\beta} - u^{1-\beta} \right]^{k+1} du \Bigg\}. \qquad (A.120)$$

- For $\beta \neq 1$ and $B \leq A + T$:

$$b_k = \frac{\mu \, E[K^{k+1}]}{k! \, (1-\beta)^{k+1}} \Bigg\{ (T + A - B) \left[B^{1-\beta} - A^{1-\beta} \right]^{k+1}$$

$$+ \int_A^B \left[u^{1-\beta} - A^{1-\beta} \right]^{k+1} du$$

$$+ \int_A^B \left[B^{1-\beta} - u^{1-\beta} \right]^{k+1} du \Bigg\}. \qquad (A.121)$$

- For $\beta = 1$ and $B > A + T$:

$$b_k = \frac{\mu \, E[K^{k+1}]}{k!} \Bigg\{ \int_A^{B-T} \left[\ln\left(\frac{u+T}{u} \right) \right]^{k+1} du$$

$$+ \int_A^{A+T} \left[\ln\left(\frac{u}{A} \right) \right]^{k+1} du$$

$$+ \int_{B-T}^B \left[\ln\left(\frac{B}{u} \right) \right]^{k+1} du \Bigg\}. \qquad (A.122)$$

- Finally, for $\beta = 1$ and $B \leq A + T$:

$$b_k = \frac{\mu \, E[K^{k+1}]}{k!} \Bigg\{ (T + A - B) \left[\ln\left(\frac{B}{A} \right) \right]^{k+1}$$

$$+ \int_A^B \left[\ln\left(\frac{u}{A} \right) \right]^{k+1} du$$

$$+ \int_A^B \left[\ln\left(\frac{B}{u} \right) \right]^{k+1} du \Bigg\}. \qquad (A.123)$$

A.7.2 Expressions for normalized variance

General closed-form expressions for the fractal-shot-noise-driven Poisson-process normalized variance $F(T)$ do not exist. However, in some special cases and limits one can indeed find such forms, and we present their detailed derivations below.

- For $\beta = \frac{1}{2}$ the normalized variance becomes

$$F(T) = 1 + \frac{\mathrm{E}[K^2]}{T\,\mathrm{E}[K]\,(B^{1/2} - A^{1/2})} \int_0^\Phi (T - u) \int_A^{B-u} (t^2 + u\,t)^{-1/2}\,dt\,du, \tag{A.124}$$

where we define the upper limit of the outer integral as

$$\Phi \equiv \min(T, B - A), \tag{A.125}$$

namely the smaller of T and $B - A$. For the inner integral we have

$$\int_A^{B-u} (t^2 + u\,t)^{-1/2}\,dt = 2\left\{\ln\left[t^{1/2} + (t + u)^{1/2}\right]\right\}_A^{B-u}, \tag{A.126}$$

so that the outer integral simplifies to

$$2\int_0^\Phi (T - u)\,\ln\left[1 + (1 - u/B)^{1/2}\right]\,du$$
$$-\,2\int_0^\Phi (T - u)\,\ln\left[1 + (1 + u/A)^{1/2}\right]\,du$$
$$+\,\ln\left(\frac{B^{1/2}}{A^{1/2}}\right)(2T\Phi - \Phi^2). \tag{A.127}$$

The remaining integrals in Eq. (A.127) follow the form

$$2\int_0^\Phi (T - u)\,\ln\left[1 + (1 + u/c)^{1/2}\right]\,du$$
$$= 4\,c\,(T + c)\int_1^{(1+\Phi/c)^{1/2}} v\,\ln(1 + v)\,dv$$
$$-\,4\,c^2\int_1^{(1+\Phi/c)^{1/2}} v^3\,\ln(1 + v)\,dv \tag{A.128}$$
$$= \Phi(2T - \Phi)\,\ln\left[1 + (1 + \Phi/c)^{1/2}\right]$$
$$+\,c\left[\tfrac{2}{3}c + 2T - \tfrac{1}{3}\Phi\right](1 + \Phi/c)^{1/2}$$
$$-\,\tfrac{2}{3}c^2 - 2cT - T\Phi + \tfrac{1}{4}\Phi^2, \tag{A.129}$$

where Eq. (A.128) derives from the substitution $v \equiv (1 + u/c)^{1/2}$. This yields the following expression for the normalized variance itself when $\beta = \frac{1}{2}$:

$$F(T) = 1 + \frac{\mathrm{E}[K^2]}{T\,\mathrm{E}[K]\left(B^{1/2} - A^{1/2}\right)}$$

$$\times \left\{ \Phi(2T - \Phi) \ln \left[\frac{B^{1/2} + (B - \Phi)^{1/2}}{A^{1/2} + (A + \Phi)^{1/2}} \right] \right.$$

$$+ \tfrac{1}{3}(\Phi - 6T + 2B)(B^2 - B\Phi)^{1/2}$$

$$+ \tfrac{1}{3}(\Phi - 6T - 2A)(A^2 + A\Phi)^{1/2}$$

$$\left. - \tfrac{2}{3}(B^2 - A^2) + 2T(B + A) \right\}. \qquad (A.130)$$

- For $\beta = 2$ the normalized variance becomes

$$F(T) = 1 + \frac{2AB\,\mathrm{E}[K^2]}{T(B - A)\,\mathrm{E}[K]} \int_0^{\Phi} (T - u) \int_A^{B-u} (t^2 + u\,t)^{-2} \, dt \, du, \quad (A.131)$$

where Φ is defined as above. For the inner integral we have

$$\int_A^{B-u} (t^2 + u\,t)^{-2} \, dt$$

$$= \int_A^{B-u} \frac{1}{u^3} \left[\frac{u}{t^2} + \frac{u}{(t + u)^2} - \frac{2}{t} + \frac{2}{t + u} \right] dt$$

$$= u^{-3} \left[-\frac{u}{t} - \frac{u}{t + u} - 2\ln(t) + 2\ln(t + u) \right]_A^{B-u}, \quad (A.132)$$

so that the outer integral simplifies to

$$\left(\frac{T}{u^2} - \frac{2}{u} - \frac{1}{A} \right) \ln(1 + u/A) - \frac{T}{A\,u}$$

$$+ \left(\frac{T}{u^2} - \frac{2}{u} + \frac{1}{B} \right) \ln(1 - u/B) + \frac{T}{B\,u} \Bigg|_{u=0}^{\Phi}. \quad (A.133)$$

The quantity in Eq. (A.133) is not defined in the limit $u \to 0$, so we use l'Hôpital's rule to obtain

$$\lim_{u \to 0} \left[\left(\frac{T}{u^2} - \frac{2}{u} - \frac{1}{A} \right) \ln(1 + u/A) - \frac{T}{A\,u} \right]$$

$$= T \lim_{u \to 0} \frac{\ln(1 + u/A) - u/A}{u^2\,r}$$

$$- 2 \lim_{u \to 0} \frac{\ln(1 + u/A)}{u} - \frac{1}{A} \lim_{u \to 0} \ln(1 + u/A)$$

$$= -\frac{T}{2A^2} - \frac{2}{A} - 0. \quad (A.134)$$

Similarly, we obtain $-T/2B^2 + 2/B$ for the second pair of terms in Eq. (A.133). This yields the following expression for the normalized variance itself when

$\beta = 2$:

$$F(T) = 1 + \frac{2AB\,\mathrm{E}[K^2]}{T(B-A)\,\mathrm{E}[K]} \left[\frac{T}{2A^2} + \frac{2}{A} - \frac{T}{A\Phi} + \frac{T}{2B^2} - \frac{2}{B} + \frac{T}{B\Phi} \right.$$
$$+ \left(\frac{T}{\Phi^2} - \frac{2}{\Phi} - \frac{1}{A} \right) \ln(1 + \Phi/A)$$
$$\left. + \left(\frac{T}{\Phi^2} - \frac{2}{\Phi} + \frac{1}{B} \right) \ln(1 - \Phi/B) \right]. \tag{A.135}$$

In contrast to the exact expressions for the normalized variance that are available for the two specific values of β considered above, approximate expressions can be obtained for arbitrary β in the following limits: $T \ll A$, $A \ll T \ll B$, and $T \gg B$. Rather than considering limits of the entire normalized variance expression

$$F(T) = 1 + \frac{2\mathrm{E}[K^2]}{T\,\mathrm{E}[K] \displaystyle\int_A^B t^{-\beta}\, dt} \int_0^\Phi (T-u) \int_A^{B-u} (t^2 + u\,t)^{-\beta}\, dt\, du, \tag{A.136}$$

we obtain limits for the integrals within this expression.

- For $T \ll A$, we have $\Phi = T$. By using l'Hôpital's rule twice we obtain

$$\lim_{T \to 0} \int_0^T (T-u) \int_A^{B-u} (t^2 + u\,t)^{-\beta}\, dt\, du \Big/ T^2 = \frac{1}{2} \int_A^B (t^2)^{-\beta}\, dt, \tag{A.137}$$

so that for small T,

$$\begin{aligned} F(T) &\approx 1 + \frac{\mathrm{E}[K^2] \displaystyle\int_A^B t^{-2\beta}\, dt}{\mathrm{E}[K] \displaystyle\int_A^B t^{-\beta}\, dt}\, T \\ &= 1 + \frac{\mathrm{E}[K^2]}{a} \left[\int_A^B t^{-2\beta}\, dt \right] T, \end{aligned} \tag{A.138}$$

as provided in Eq. (10.15).

- For $A \ll T \ll B$, again we have $\Phi = T$, but now the limiting expression depends on β. Since in this case $A \ll B$, the integral in the denominator of the normalized-variance expression provided in Eq. (A.136) tends to a simple limit as $B/A \to \infty$:

$$\int_A^B t^{-\beta}\, dt \to \begin{cases} B^{1-\beta}/(1-\beta) & \beta < 1 \\ \ln(B/A) & \beta = 1 \\ A^{1-\beta}/(\beta-1) & \beta > 1. \end{cases} \tag{A.139}$$

The double integral in Eq. (A.136), henceforth denoted Ω, has a more complex form; we consider in turn five expressions for different ranges of β.

1. For $0 < \beta < \frac{1}{2}$, we define $x \equiv A/T$ and $y \equiv B/T$, so that

$$\Omega = T^{3-2\beta} \int_0^1 (1 - u) \int_x^{y-u} (t^2 + ut)^{-\beta} \, dt \, du. \qquad \text{(A.140)}$$

Setting $x = 0$ and using l'Hôpital's rule leads to

$$\lim_{y \to \infty} \int_0^1 (1 - u) \int_0^{y-u} (t^2 + ut)^{-\beta} \, dt \, du \bigg/ y^{1-2\beta} = [2(1 - 2\beta)]^{-1}, \qquad \text{(A.141)}$$

so that the normalized variance becomes

$$F(T) \approx 1 + \frac{E[K^2] (1 - \beta)}{E[K] (1 - 2\beta)} B^{-\beta} T, \qquad \text{(A.142)}$$

as provided in Eq. (10.16).

2. For $\beta = \frac{1}{2}$, we again set $x = 0$ and use l'Hôpital's rule to obtain

$$\lim_{y \to \infty} \int_0^1 (1 - u) \int_0^{y-u} (t^2 + ut)^{-1/2} \, dt \, du \bigg/ \ln(y) = \frac{1}{2}, \qquad \text{(A.143)}$$

so that

$$F(T) \approx 1 + \frac{E[K^2]}{E[K]} \frac{1}{2} B^{-1/2} \left[\ln(B/T)\right] T. \qquad \text{(A.144)}$$

3. For $\frac{1}{2} < \beta < 1$, we consider the limits in which both $x \to 0$ and $y \to \infty$. Here the integral in the numerator becomes

$$\begin{aligned}
\Omega &= T^{3-2\beta} \int_0^1 (1 - u) \int_0^\infty (t^2 + ut)^{-\beta} \, dt \, du \\
&= T^{3-2\beta} \int_0^1 (1 - u) \int_0^\infty (u^2 x^2 + u^2 x)^{-\beta} u \, dx \, du \quad \text{(A.145)} \\
&= T^{3-2\beta} \int_0^1 (1 - u) u^{1-2\beta} \, du \int_0^\infty (x^2 + x)^{-\beta} \, dx \\
&= \frac{T^{3-2\beta}}{2(1 - \beta)(3 - 2\beta)} \frac{\Gamma(1 - \beta) \Gamma(2\beta - 1)}{\Gamma(\beta)}, \qquad \text{(A.146)}
\end{aligned}$$

where Eq. (A.145) derives from the substitution $x \equiv t/u$ in the inner integral. The normalized variance then becomes

$$F(T) \approx 1 + \frac{E[K^2] \Gamma(1 - \beta) \Gamma(2\beta - 1)}{E[K] (3 - 2\beta) \Gamma(\beta)} B^{\beta-1} T^{2(1-\beta)}, \qquad \text{(A.147)}$$

which concurs with Eq. (10.16) when the definition $\alpha \equiv 2(1 - \beta)$ is used.

4. For $\beta = 1$ we define $x \equiv T/A$ and $y \equiv T/B$ to obtain

$$
\begin{aligned}
\Omega &= \int_0^T (T - u) \int_A^{B-u} (t^2 + u\,t)^{-1}\, dt\, du \\
&= T \int_0^1 \frac{1-u}{u}\, \ln(1 - y\,u)\, du \\
&\quad + T \int_0^1 \frac{1-u}{u}\, \ln(1 + x\,u)\, du. \tag{A.148}
\end{aligned}
$$

The first term in Eq. (A.148) approaches zero as $y \to 0$ since

$$
\begin{aligned}
0 > T \int_0^1 \frac{1-u}{u}\, \ln(1 - y\,u)\, du \\
> T \int_0^1 \frac{1-u}{u}\, u \ln(1 - y)\, du \\
= \tfrac{1}{2} T \ln(1 - y), \tag{A.149}
\end{aligned}
$$

and $\ln(1 - y) \to 0$ as $y \to 0$. For the second term in Eq. (A.148), two applications of l'Hôpital's rule and some simplification yield

$$
\lim_{x \to \infty} \int_0^1 \frac{1-u}{u}\, \ln(1 + x\,u)\, du \Big/ \ln^2(x) = 1. \tag{A.150}
$$

The normalized variance therefore becomes

$$
F(T) \approx 1 + \frac{2\mathrm{E}[K^2]}{\mathrm{E}[K]}\, \frac{\ln^2(T/A)}{\ln(B/A)}. \tag{A.151}
$$

5. For $\beta > 1$, we define $x \equiv T/A$ and $y \equiv B/T$ to obtain

$$
\Omega = A^{3-2\beta} \int_0^x (x - u) \int_1^{xy-u} (t^2 + u\,t)^{-\beta}\, dt\, du. \tag{A.152}
$$

Setting $y > 1$ and using l'Hôpital's rule yields

$$
\lim_{x \to \infty} \int_0^x (x - u) \int_1^{xy-u} (t^2 + u\,t)^{-\beta}\, dt\, du \Big/ x = \left[2(\beta - 1)^2\right]^{-1}, \tag{A.153}
$$

whereupon the normalized variance becomes

$$
F(T) \approx 1 + \frac{\mathrm{E}[K^2]\, A^{1-\beta}}{\mathrm{E}[K]\,(\beta - 1)} \approx 1 + \frac{\mathrm{E}[K^2]}{\mathrm{E}^2[K]}\, a. \tag{A.154}
$$

• Finally, in the third region, where $T \gg B$, we have $\Phi = B - A$. Using the substitution $v \equiv t + u$ and interchanging the order of integration in the numerator of Eq. (A.136) yields

$$
F(T) = 1 + \frac{2\mathrm{E}[K^2]}{\mathrm{E}[K]} \int_A^B t^{-\beta} \int_t^B \left(1 + \frac{t - v}{T}\right) v^{-\beta}\, dv\, dt \Big/ \int_A^B t^{-\beta}\, dt. \tag{A.155}
$$

In the limit $T \gg B$, the $(t - v)$ term in the numerator vanishes so that

$$
\begin{aligned}
F(T) &\approx 1 + \frac{2\mathrm{E}[K^2]}{\mathrm{E}[K]} \int_A^B t^{-\beta} \int_t^B v^{-\beta}\, dv\, dt \Big/ \int_A^B t^{-\beta}\, dt \\
&= 1 + \frac{2\mathrm{E}[K^2]}{\mathrm{E}[K]} \frac{1}{2} \left[\int_A^B t^{-\beta}\, dt \right]^2 \Big/ \int_A^B t^{-\beta}\, dt \\
&= 1 + \frac{\mathrm{E}[K^2]}{\mathrm{E}[K]} \int_A^B t^{-\beta}\, dt \\
&= 1 + \frac{\mathrm{E}[K^2]}{\mathrm{E}^2[K]}\, \mathrm{a},
\end{aligned}
\tag{A.156}
$$

in agreement with Eq. (10.17).

A.7.3 Expressions for normalized Haar-wavelet variance

Equation (3.41) provides a relation that permits us to obtain the normalized Haar-wavelet variance $A(T)$ directly from the normalized variance $F(T)$. This direct route is suitable for all forms of $F(T)$ except those in which its leading term is linear in T [see Eqs. (5.37)–(5.39)]. This latter condition arises for $T \ll A$, and for $A \ll T \ll B$ with $\beta < \frac{1}{2}$. In these two cases, we obtain $A(T)$ using other methods, as described below.

- For $T \ll A$, we form three derivatives of the double integral in Eq. (A.136), denoted Ω, which yields

$$
\begin{aligned}
\Omega &= \int_0^T (T - u) \int_A^{B-u} (t^2 + u\,t)^{-\beta}\, dt\, du \\
\frac{d\Omega}{dT} &= \int_0^T \int_A^{B-u} (t^2 + u\,t)^{-\beta}\, dt\, du \\
\frac{d^2\Omega}{dT^2} &= \int_A^{B-T} (t^2 + Tt)^{-\beta}\, dt \\
\frac{d^3\Omega}{dT^3} &= -\left[B(B - T) \right]^{-\beta} - \beta \int_A^{B-T} t^{-\beta}(t + T)^{-(1+\beta)}\, dt \\
\left. \frac{d^3\Omega}{dT^3} \right|_{T=0} &= -B^{-2\beta} - \beta \int_A^B t^{-(1+2\beta)}\, dt \\
&= -\tfrac{1}{2}\left[A^{-2\beta} + B^{-2\beta} \right] \\
F(T) &\approx 1 + \frac{\mathrm{E}[K^2]}{\mathrm{a}} \int_A^B t^{-2\beta}\, dt\, T - \frac{\mathrm{E}[K^2]}{6\mathrm{a}} \left[A^{-2\beta} + B^{-2\beta} \right] T^2 \\
A(T) &\approx 1 + \frac{\mathrm{E}[K^2]}{3\mathrm{a}} \left[A^{-2\beta} + B^{-2\beta} \right] T^2.
\end{aligned}
\tag{A.157}
$$

- For $A \ll T \ll B$ and $\beta < \frac{1}{2}$, we make use of results from Eq. (A.157) to obtain

$$
\frac{d^3\Omega}{dT^3} = -\left[B(B-T)\right]^{-\beta}
$$
$$
-\beta \int_A^{B-T} t^{-\beta} (t+T)^{-(1+\beta)} dt
$$

$$
T^{2\beta} \frac{d^3\Omega}{dT^3} = -\left[\frac{B}{T}\left(\frac{B}{T}-1\right)\right]^{-\beta}
$$
$$
-\beta \int_{A/T}^{B/T-1} x^{-\beta} (x+1)^{-(1+\beta)} dx
$$

$$
\lim_{\substack{A/T\to 0 \\ B/T\to\infty}} T^{2\beta} \frac{d^3\Omega}{dT^3} = -\beta \int_0^\infty x^{-\beta} (x+1)^{-(1+\beta)} dx
$$

$$
= -\beta \frac{\Gamma(1-\beta)\,\Gamma(2\beta)}{\Gamma(1+\beta)}. \tag{A.158}
$$

Finally, then, we obtain

$$
F(T) \approx 1 + \frac{\mathrm{E}\left[K^2\right]}{a} \int_A^B t^{-2\beta}\,dt\,T - \frac{2\mathrm{E}\left[K^2\right](1-\beta)}{\mathrm{E}[K]\,B^{1-\beta}\,T}
$$
$$
\times \frac{T^{3-2\beta}}{(1-2\beta)(2-2\beta)(3-2\beta)}\,\beta\,\frac{\Gamma(1-\beta)\,\Gamma(2\beta)}{\Gamma(1+\beta)}
$$
$$
= 1 + \frac{\mathrm{E}\left[K^2\right]}{a} \int_A^B t^{-2\beta}\,dt\,T
$$
$$
- \frac{\mathrm{E}\left[K^2\right]}{\mathrm{E}[K]}\,\frac{\beta\,\Gamma(1-\beta)\,\Gamma(2\beta)}{(1-2\beta)(3-2\beta)\,\Gamma(1+\beta)}\,\frac{T^{2-2\beta}}{B^{1-\beta}}
$$
$$
= 1 + \frac{\mathrm{E}\left[K^2\right]}{a} \int_A^B t^{-2\beta}\,dt\,T
$$
$$
- \frac{\mathrm{E}\left[K^2\right]}{\mathrm{E}[K]}\,\frac{\beta\,\Gamma(1-\beta)\,(2\beta-1)\,\Gamma(2\beta-1)}{(1-2\beta)(3-2\beta)\beta\,\Gamma(\beta)}\,\frac{T^{2-2\beta}}{B^{1-\beta}}
$$
$$
= 1 + \frac{\mathrm{E}\left[K^2\right]}{a} \int_A^B t^{\alpha-2}\,dt\,T
$$
$$
+ \frac{\mathrm{E}\left[K^2\right]}{\mathrm{E}[K]}\,\frac{\Gamma(\alpha/2)\,\Gamma(1-\alpha)}{(1+\alpha)\,\Gamma(1-\alpha/2)}\,\frac{T^\alpha}{B^{\alpha/2}}
$$
$$
A(T) \approx 1 + (2-2^\alpha)\,\frac{\mathrm{E}\left[K^2\right]}{\mathrm{E}[K]}\,\frac{\Gamma(\alpha/2)\,\Gamma(1-\alpha)}{(1+\alpha)\,\Gamma(1-\alpha/2)}\,\frac{T^\alpha}{B^{\alpha/2}}
$$
$$
= 1 + \frac{\mathrm{E}\left[K^2\right]}{\mathrm{E}[K]}\,\frac{(2^\alpha-2)\,\Gamma(\alpha/2)\,\Gamma(2-\alpha)}{(\alpha^2-1)\,\Gamma(1-\alpha/2)}\,\frac{T^\alpha}{B^{\alpha/2}}, \tag{A.159}
$$

in accordance with Eq. (10.19). This result applies for all $\beta < 1$.

A.7.4 Integrals for time statistics

Calculation of the forward-recurrence-time and interevent-interval statistics begins with $p_Z(0;T)$, the probability that there are zero events in an interval of duration T chosen independently of the process, and proceeds to its first two derivatives.

From Eq. (10.3), in the special case of deterministic K, we have

$$p_Z(0;T) = \exp\left(\mu \int_{-\infty}^{\infty} \{\exp[-h_T(u)] - 1\}\, du\right) = \exp[\mu f(T)], \qquad (A.160)$$

where the quantity $f(T)$, implicitly defined in Eq. (A.160), serves to simplify the notation. In accordance with the results provided in Sec. 3.3.1, the forward-recurrence-time probability density then becomes

$$p_\vartheta(t) = -\frac{d}{dT}\Big[p_Z(0;T)\Big]_{T=t} = -\mu p_Z(0;t)\frac{df(t)}{dt} \qquad (A.161)$$

while a second derivative yields the interevent-interval density:

$$\begin{aligned}
p_\tau(t) &= -\frac{1}{E[X]}\frac{d^2}{dT^2}\Big[p_Z(0;T)\Big]_{T=t} \\
&= \frac{1}{a}\, p_Z(0;t)\left\{\mu\left[\frac{df(t)}{dt}\right]^2 + \frac{d^2f(t)}{dt^2}\right\}. \qquad (A.162)
\end{aligned}$$

Thus, $p_Z(0;T)$, $p_\vartheta(t)$, and $p_\tau(t)$ depend, in turn, on $f(t)$ and its first two derivatives, which we calculate below.

The function $f(t)$ assumes four different forms, depending on the value of β and the relative magnitudes of A, B, and t.

- For $\beta \neq 1$ and $B > A + t$:

$$\begin{aligned}
f(t) &= \int_A^{A+t}\left(\exp\left\{-\frac{K}{1-\beta}\left[u^{1-\beta} - A^{1-\beta}\right]\right\} - 1\right)du \\
&\quad + \int_A^{B-t}\left(\exp\left\{-\frac{K}{1-\beta}\left[(u+t)^{1-\beta} - u^{1-\beta}\right]\right\} - 1\right)du \\
&\quad + \int_{B-t}^{B}\left(\exp\left\{-\frac{K}{1-\beta}\left[B^{1-\beta} - u^{1-\beta}\right]\right\} - 1\right)du \\
-\frac{df(t)}{dt} &= K\int_A^{B-t}(u+t)^{-\beta}\exp\left\{-\frac{K}{1-\beta}\left[(u+t)^{1-\beta} - u^{1-\beta}\right]\right\}du \\
&\quad + \left(1 - \exp\left\{-\frac{K}{1-\beta}\left[(A+t)^{1-\beta} - A^{1-\beta}\right]\right\}\right) \\
\frac{d^2f(t)}{dt^2} &= KB^{-\beta}\exp\left\{-\frac{K}{1-\beta}\left[B^{1-\beta} - (B-t)^{1-\beta}\right]\right\} \\
&\quad - K(A+t)^{-\beta}\exp\left\{-\frac{K}{1-\beta}\left[(A+t)^{1-\beta} - A^{1-\beta}\right]\right\}
\end{aligned}$$

$$+ K \int_{A+t}^{B} \left[\beta\, u^{-\beta-1} + K u^{-2\beta} \right]$$

$$\times \exp\left\{ -\frac{K}{1-\beta} \left[u^{1-\beta} - (u-t)^{1-\beta} \right] \right\} du. \qquad \text{(A.163)}$$

- For $\beta \neq 1$ and $B \leq A + t$:

$$f(t) = \int_{A}^{B} \left(\exp\left\{ -\frac{K}{1-\beta} \left[u^{1-\beta} - A^{1-\beta} \right] \right\} - 1 \right) du$$

$$+ \int_{A}^{B} \left(\exp\left\{ -\frac{K}{1-\beta} \left[B^{1-\beta} - u^{1-\beta} \right] \right\} - 1 \right) du$$

$$+ (t + A - B)\, (e^{-a} - 1)$$

$$-\frac{df(t)}{dt} = 1 - e^{-a}$$

$$\frac{d^2 f(t)}{dt^2} = 0. \qquad \text{(A.164)}$$

- For $\beta = 1$ and $B > A + t$:

$$f(t) = \int_{A}^{B-t} \left[\left(\frac{u}{u+t} \right)^K - 1 \right] du + \int_{A}^{A+t} \left[\left(\frac{A}{u} \right)^K - 1 \right] du$$

$$+ \int_{B-t}^{B} \left[\left(\frac{u}{B} \right)^K - 1 \right] du$$

$$-\frac{df(t)}{dt} = K \int_{A}^{B-t} \frac{u^K}{(u+t)^{K+1}}\, du - \left[\left(\frac{A}{A+t} \right)^K - 1 \right]$$

$$\frac{d^2 f(t)}{dt^2} = K \frac{(B-t)^K}{B^{K+1}} - K \frac{A^K}{(A+t)^{K+1}}$$

$$- K(K+1) \int_{A}^{B-t} \frac{u^K}{(u+t)^{K+2}}\, du. \qquad \text{(A.165)}$$

- Finally, for $\beta = 1$ and $B < A + t$:

$$f(t) = \int_{A}^{B} \left[\left(\frac{A}{u} \right)^K - 1 \right] du + \int_{A}^{B} \left[\left(\frac{u}{B} \right)^K - 1 \right] du$$

$$+ (t + A - B) \left[\left(\frac{A}{B} \right)^K - 1 \right]$$

$$-\frac{df(t)}{dt} = 1 - \left(\frac{A}{B} \right)^K = 1 - e^{-a}$$

$$\frac{d^2 f(t)}{dt^2} = 0. \qquad \text{(A.166)}$$

A.8 ANALYSIS AND ESTIMATION

A.8.1 Fourier-transform effects

For practical reasons, we estimate the spectrum via the Fourier transform of the sequence of counts $\{Z_k\}$. A simple factor of T^{-2} connects the count-based and rate-based spectral estimates. This gives rise to an estimated spectrum whose expected value differs from that of the point-process spectrum, as we now proceed to show (Thurner et al., 1997).

As previously, consider a set of M counts, each of duration T, with $0 \le k < M$. Define the Fourier transform of the counts via

$$X(n) \equiv \sum_{k=0}^{M-1} Z_k \, e^{-i2\pi kn/M}. \tag{A.167}$$

The estimate of the spectrum then becomes

$$
\begin{aligned}
\widehat{S}_Z(n) &= M^{-1} \, |X(n)|^2 \\
&= M^{-1} \sum_{k=0}^{M-1} \sum_{m=0}^{M-1} Z_k \, Z_m \, e^{i2\pi(k-m)n/M},
\end{aligned} \tag{A.168}
$$

with an expected value

$$\mathrm{E}\!\left[\widehat{S}_Z(n)\right] = M^{-1} \sum_{k=0}^{M-1} \sum_{m=0}^{M-1} e^{i2\pi(k-m)n/M} \, \mathrm{E}[Z_k Z_m]. \tag{A.169}$$

We can express the correlation between the counts in terms of the spectrum of the point process itself by means of

$$
\begin{aligned}
\mathrm{E}[Z_k Z_m] &= \mathrm{E}\!\left[\int_{s=0}^{T} \int_{t=0}^{T} dN(s+kT) \, dN(t+mT)\right] \\
&= \int_{s=0}^{T} \int_{t=0}^{T} G\!\left[s - t + (k-m)T\right] ds \, dt \\
&= \int_{u=-T}^{T} \int_{v=|u|}^{2T-|u|} G\!\left[u + (k-m)T\right] \frac{du \, dv}{2} \\
&= \int_{u=-T}^{T} (T - |u|) \, G\!\left[u + (k-m)T\right] du \\
&= \int_{u=-T}^{T} (T - |u|) \int_{f=-\infty}^{\infty} S_N(f) \, e^{i2\pi f[u+(k-m)T]} \, df \, du. \tag{A.170}
\end{aligned}
$$

Finally, combining Eqs. (A.169) and (A.170) yields

$$\mathrm{E}\!\left[\widehat{S}_Z(n)\right] = \frac{1}{M} \sum_{k=0}^{M-1} \sum_{m=0}^{M-1} e^{i2\pi(k-m)n/M} \int_{u=-T}^{T} (T - |u|)$$

$$\times \int_{f=-\infty}^{\infty} S_N(f)\, e^{i2\pi f[u+(k-m)T]}\, df\, du$$

$$= \frac{1}{M} \int_{f=-\infty}^{\infty} S_N(f) \left| \sum_{k=0}^{M-1} e^{ik2\pi(n/M+fT)} \right|^2$$

$$\times \int_{u=-T}^{T} (T - |u|)\, e^{i2\pi fu}\, du\, df$$

$$= \frac{1}{\pi^2 M} \int_{f=-\infty}^{\infty} S_N(f) \frac{\sin^2(\pi n + M\pi fT)}{\sin^2(\pi n/M + \pi fT)} \frac{\sin^2(\pi fT)}{f^2}\, df$$

$$= \frac{T}{\pi M} \int_{-\infty}^{\infty} S_N\!\left(\frac{x}{\pi T}\right) \frac{\sin^2(x)}{x^2} \frac{\sin^2(Mx)}{\sin^2(x + \pi n/M)}\, dx. \quad \text{(A.171)}$$

For a fractal-based point process in which $S_N(f)$ takes the form of Eq. (5.44a), we obtain

$$\mathrm{E}\!\left[\widehat{S}_Z(n)\right] = \frac{\mathrm{E}[\mu]\, T}{\pi M} \int_{-\infty}^{\infty} \left[1 + (\pi f_S T)^\alpha |x|^{-\alpha}\right] \frac{\sin^2(x)}{x^2} \frac{\sin^2(Mx)}{\sin^2(x + \pi n/M)}\, dx. \tag{A.172}$$

To proceed further, we consider the case $0 < \alpha < 2$, which encompasses the vast majority of fractal-based point processes observed in practice, as discussed in Sec. 5.2.2. The fraction inside the integral in Eqs. (A.171) and (A.172) then only becomes important within the range $-\pi(n+1)/M < x < -\pi(n-1)/M$. For large values of M, this range becomes quite narrow. Substituting $y \equiv x + \pi n/M$ we obtain

$$\lim_{x \to -\pi n/M} \frac{\sin^2(Mx)}{\sin^2(x + \pi n/M)} = \lim_{y \to 0} \frac{\sin^2[M(y - \pi n/M)]}{\sin^2(y)}$$

$$= \lim_{y \to 0} \frac{\sin^2(My)}{\sin^2(y)}$$

$$= M^2. \tag{A.173}$$

For large M we can therefore insert Eq. (A.173) into Eq. (A.172) to obtain

$$\mathrm{E}\!\left[\widehat{S}_Z(n)\right]$$

$$\approx \frac{\mathrm{E}[\mu]\, T}{\pi M} \int_{-\infty}^{\infty} \left[1 + (\pi f_S T)^\alpha |x|^{-\alpha}\right] \frac{\sin^2(x)}{x^2} M^2\, \delta(x + \pi n/M)\, dx$$

$$= \frac{\mathrm{E}[\mu]\, MT}{\pi} \left[\frac{\sin^2(\pi n/M)}{(\pi n/M)^2}\right] \left[1 + (f_S MT/n)^\alpha\right], \tag{A.174}$$

which essentially reproduces the dominant term of Eq. (3.67).

Finally, for low frequencies such that $n \ll M$, the factor in large brackets in Eq. (A.174) approaches unity, so that we recover the canonical form of Eq. (5.44a). For other values of n, this factor presents a confounding effect in estimating the fractal exponent.

Appendix B
Problem Solutions

B.1	Introduction	398
B.2	Scaling, Fractals, and Chaos	401
B.3	Point Processes: Definition and Measures	404
B.4	Point Processes: Examples	412
B.5	Fractal and Fractal-Rate Point Processes	427
B.6	Processes Based on Fractional Brownian Motion	441
B.7	Fractal Renewal Processes	447
B.8	Alternating Fractal Renewal Process	454
B.9	Fractal Shot Noise	459
B.10	Fractal-Shot-Noise-Driven Point Processes	463
B.11	Operations	473
B.12	Analysis and Estimation	486
B.13	Computer Network Traffic	494

B.1 INTRODUCTION

Prob. 1.1.1 Figure B.1 displays the measured coastline length d as a function of the measurement scale s used.

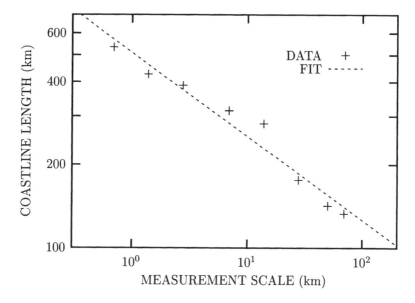

Fig. B.1 Coastline length d vs. measurement scale s (+ symbols). The data are well described by a straight line on this doubly logarithmic plot, revealing that $d \propto s^c$. A least-squares fit based on the logarithms, shown as the dashed line, exhibits a slope $c \approx -0.30$.

Prob. 1.1.2 Taking the logarithm of both sides of Eq. (1.1) yields $\ln(d) = c\ln(s) + \ln(b)$, for some constant b. This suggests fitting a straight line to a plot of $\ln(d)$ vs. $\ln(s)$ to obtain the slope, c. This also explains the use of logarithmic coordinates for both axes of Fig. B.1. Such a least-squares fit yields $c \approx -0.30$.

Prob. 1.1.3 Consulting an atlas reveals that the South African coastline is exceptionally smooth, the coastline of Britain is much rougher, and that of Australia lies somewhere between the two. Iceland's coastline appears the roughest of all. Evidently, the power-law exponent c provides an index of roughness, with larger negative magnitudes signifying more irregular coastlines.

Prob. 1.2.1 Place n points evenly spaced about the perimeter of a circle of unit circumference, and connect adjacent points to form a regular polygon of n sides, each of length s, as shown in Fig. B.2. Now draw line segments from each point on the perimeter to the center of the circle; all have a length equal to the radius $r = 1/2\pi$. Considering one of the isosceles triangles thus generated, let θ_1 denote the value of

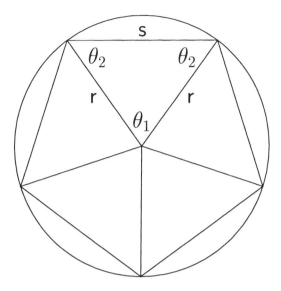

Fig. B.2 A regular polygon inscribed within a circle provides a means for approximating the circle's perimeter with a given resolution. Standard trigonometry yields a relation between the number of edges of the polygon and their size, and therefore yields the perimeter of the polygon.

the angle formed at the center of the circle. Since all n of these angles together subtend an angle of 2π, we have $\theta_1 = 2\pi/n$. For isosceles triangles, the other two angles θ_2 must each equal $\pi/2 - \pi/n$, since all three angles must sum to π. The sine theorem for triangles states that the ratio of two sides equals the ratio of the sines of the opposite angles, yielding

$$
\begin{aligned}
\frac{s}{r} &= \frac{\sin(\theta_1)}{\sin(\theta_2)} \\
\frac{s}{(2\pi)^{-1}} &= \frac{\sin(2\pi/n)}{\sin(\pi/2 - \pi/n)} \\
s &= \frac{\sin(2\pi/n)}{2\pi \sin(\pi/2 - \pi/n)} \\
&= \frac{2\sin(\pi/n)\,\cos(\pi/n)}{2\pi \cos(\pi/n)} \\
ns &= (n/\pi)\,\sin(\pi/n).
\end{aligned}
\tag{B.1}
$$

Graphs of this function indeed increase as n increases.

Prob. 1.2.2 A circle is *not* a fractal, however, since the perimeter does not increase significantly as the resolution increases. In the limit of a large number of sides n,

Eq. (B.1) becomes

$$
\begin{aligned}
\lim_{n \to \infty} ns &= \lim_{n \to \infty} (n/\pi)\sin(\pi/n) \\
&= \lim_{x \to 0} \sin(x)/x \\
&= 1.
\end{aligned}
\tag{B.2}
$$

As the number of sides n increases, s decreases in concert. Concurrently, the perimeter approaches unity, the value of the circle itself, as it must. Therefore, the estimated perimeter varies between $2/\pi$ for two segments (each of length $1/\pi$) and unity for infinitesimal segments: an increase of only 57%. For a fractal, in contrast, the perimeter changes dramatically over a broad range of measurement resolutions.

B.2 SCALING, FRACTALS, AND CHAOS

Prob. 2.1

1. Yes: circulatory systems comprise branches and sub-branches, with parts resembling the whole over a range of branchings.

2. Yes: again, each branching yields smaller random copies of the original.

3. No: with all hair strands essentially the same diameter and length, these two scales dominate and no scaling behavior emerges.

4. No: except for possible surface roughness or percolating pores, a brick essentially forms a simple rectangular prism.

5. No: steady winds of constant velocity sort sand grains by size and impart regular ripples to the sand; with only these two scales present, fractal characteristics do not occur.

6. Yes: clouds generally have borders with fractal characteristics and resemble coastlines in many respects.

7. Yes: mountain ranges have mountains and foothills, each with smaller features of similar shape, and so forth; horizontal slices (contours) of mountain ranges themselves take fractal form — similar slices at sea level determine coastlines, which take fractal forms as we have seen.

8. No: with or without air resistance included, the path of the ball forms a simple curve without fractal features; air turbulence results in slight variations in the path, but only negligible ones.

9. No: the added noise effectively destroys the self-similar structure of the set of line segments; examining the final result *a posteriori*, one cannot know of the exact self-similarity employed in the construction of the set C_3 *a priori*, before adding the noise.

Prob. 2.2.1 Proceeding directly from Eq. (2.31) yields

$$
\begin{aligned}
y &\equiv \pi^{-1}\arccos(1 - 2x) \\
\cos(\pi y) &= 1 - 2x \\
x &= \tfrac{1}{2}[1 - \cos(\pi y)].
\end{aligned}
\tag{B.3}
$$

Substituting Eq. (B.3) into Eq. (2.20) with $c = 4$ leads to

$$
\begin{aligned}
x_{n+1} &= 4x_n(1 - x_n) \\
\tfrac{1}{2}[1 - \cos(\pi y_{n+1})] &= 4 \times \tfrac{1}{2}[1 - \cos(\pi y_n)] \times \left(1 - \tfrac{1}{2}[1 - \cos(\pi y_n)]\right)
\end{aligned}
$$

$$\tfrac{1}{2}\big[1 - \cos(\pi y_{n+1})\big] \;=\; 1 - \cos^2(\pi y_n)$$
$$\cos(\pi y_{n+1}) \;=\; 2\cos^2(\pi y_n) - 1$$
$$=\; \cos^2(\pi y_n) - \sin^2(\pi y_n)$$
$$=\; \cos(2\pi y_n), \tag{B.4}$$

which yields results identical to Eq. (2.32).

Prob. 2.2.2 Equation (2.32) states that the values of y effectively experience a doubling at each iteration. This ignores the sign of the transformation in Eq. (2.32), which becomes irrelevant for estimates of absolute values in any case. This simply yields $|\epsilon_{n+1}/\epsilon_n| = 2$, which is exactly the same as for the original expression in x, Eq. (2.20). The two must assume the same values, since a monotonic transformation cannot change the ratio of perturbations (Ott, 2002).

Prob. 2.3.1 Essentially the same argument applies to C' as to C. Since we remove the middle half from each interval, the total width or Lebesgue measure of C'_n assumes the value 2^{-n}, which also approaches zero as n increases. In the limit $n \to \infty$, we have a width of zero for C' just as for C.

Prob. 2.3.2 Again, similar arguments apply. This time a *quaternary* expansion proves useful, with points having only 0 and 3 in their expansions (except for endpoints) belonging to C', and others not. The same one-to-one mapping to the original unit interval exists, showing that the number of points in the unit interval, in C', and in C for that matter, all coincide.

Prob. 2.3.3 Since each stage in the construction of C' yields twice as many intervals, each one quarter the size of the original, we find that decreasing ϵ by a factor of 4 yields a corresponding increase in $N(\epsilon)$ of a factor of 2. Employing the scaling equation $N(\epsilon) \sim \epsilon^{-D_0}$, we obtain $D_0 = \log(2)/\log(4) = \tfrac{1}{2}$, exactly.

Prob. 2.4.1 Solving for x we obtain

$$x \;\equiv\; 0.002002\ldots_3$$
$$27x \;=\; 2.002002\ldots_3$$
$$26x \;=\; 2 \tag{B.5}$$
$$x \;=\; 2/26$$
$$=\; 1/13, \tag{B.6}$$

where Eq. (B.5) results from the difference between the two lines above it.

Prob. 2.4.2 All endpoints of C have terminating ternary expansions. Since the ternary expansion for x does not terminate, but rather persists indefinitely, x does not belong to the endpoints of C.

Prob. 2.4.3 Since the ternary expansion of x has no 1's in it, x must belong to C. Since we have established that x does not belong to the endpoints of C, x must belong to the interior of C.

Prob. 2.4.4 An uncountably infinite number of irrational values belong to C, since C comprises an uncountably infinite number of values and only a countable number of them belong to the rational numbers. The value $x = 0.022020200020\ldots_3$, where $a_k = 2$ if k is prime and zero if not, forms one such example.

Prob. 2.5 Setting $x = 1$ in Eq. (2.4) yields

$$f(a) = g(a) f(1) \tag{B.7}$$

Substituting Eq. (B.7) back into Eq. (2.4), we obtain

$$\begin{aligned} g(ax) f(1) &= g(a) g(x) f(1) \\ g(ax) &= g(a) g(x). \end{aligned} \tag{B.8}$$

Defining

$$\begin{aligned} g_2(x) &\equiv \ln\{g[\exp(x)]\} \\ a' &\equiv \ln(a) \\ x' &\equiv \ln(x) \end{aligned} \tag{B.9}$$

and substituting Eq. (B.9) into Eq. (B.8) yields

$$\begin{aligned} \exp\{g_2[\ln(ax)]\} &= \exp\{g_2[\ln(a)]\} \exp\{g_2[\ln(x)]\} \\ g_2[\ln(ax)] &= g_2[\ln(a)] + g_2[\ln(x)] \\ g_2(a' + x') &= g_2(a') + g_2(x'). \end{aligned} \tag{B.10}$$

Equation (B.10) demonstrates that $g_2(\cdot)$ belongs to the class of linear functions, so that

$$\begin{aligned} g_2(x') &= \ln(c) x' \\ g(x) &= x^c \\ f(x) &= b x^c, \end{aligned} \tag{B.11}$$

in accord with Eqs. (2.5) and (2.6).

B.3   POINT PROCESSES: DEFINITION AND MEASURES

Prob. 3.1 An orderly, one-dimensional point process provides a useful model for examples 2, 3, 7, 8, and 9, despite the extremes of example 8, which has an average rate of well over 1 billion events/sec, and example 9, in which there is only one interval. The orderly, one-dimensional point-process model fails for the other examples for a variety of reasons. A *two*-dimensional point process describes example 1, and considering either the latitude or longitude (but not both) would yield an example amenable to the model of an orderly, one-dimensional point-process. For example 4, the failure lies in the lack of precise time localization, since thunderstorms arise and dissipate gradually. For the model to apply, one could instead consider the times of lightning strikes. A similar argument applies to example 5; considering the times at which cars pass a specified point, such as the toll booth, would render the model applicable. Example 6 contains an infinite number of events in a finite interval, and therefore has an infinite rate; in particular, any rate estimate diverges near the origin. No orderly point process can model this set. Restricting the integers n to lie below a certain maximum value would obviate this limitation. Finally, example 10 has no events at all, either well- or ill-defined. For the model to apply, one could consider the times at which the sign of the difference changes.

Prob. 3.2 In the limit as $T \to 0$, Eq. (3.33) indicates that the probability of two or more events occurring within a counting time becomes very small. This makes sense, since smaller durations tend to contain fewer events. Given the approximation $\Pr\{Z(T) > 1\} \approx 0$, the count random variable $Z(T)$ almost always takes one of two values: zero or unity. In both cases, $Z^2(T) = Z(T)$. Including the rare cases when $Z(T) > 1$ gives $Z^2(T) \approx Z(T)$. Taking expectations of both sides then yields $E[Z^2(T)] \approx E[Z(T)]$, as we set out to prove.

Prob. 3.3 Combine Eqs. (3.11) and (3.29), and proceed to take two derivatives:

$$
\begin{aligned}
\Pr\{Z(t) = 0\} &= 1 - P_\vartheta(t) \\
&= 1 - \frac{1}{E[\tau]} \int_0^t [1 - P_\tau(x)]\, dx \\
\frac{d}{dt} \Pr\{Z(t) = 0\} &= -\frac{1}{E[\tau]} [1 - P_\tau(t)] \\
\frac{d^2}{dt^2} \Pr\{Z(t) = 0\} &= -\frac{1}{E[\tau]} [-p_\tau(t)] \\
E[\tau] \frac{d^2}{dt^2} \Pr\{Z(t) = 0\} &= p_\tau(t). \tag{B.12}
\end{aligned}
$$

Prob. 3.4 For $Z(t) = 0$, we require that no events occur in an interval of duration t. Alternately, we can say that the time to the next event, starting at the beginning of the interval of duration t, exceeds t. In terms of the forward recurrence time ϑ, we have $\vartheta > t$. Taking probabilities, we have that $\Pr\{Z(t) = 0\} = \Pr\{\vartheta > t\}$. But by the definition of the probability distribution function, $P_\vartheta(t) \equiv \Pr\{\vartheta \le t\} = 1 - \Pr\{\vartheta > t\}$. Combining these two expressions yields Eq. (3.29).

Prob. 3.5 For arbitrary intervals, there is no restriction on the skewness, which can take any value from negative to positive infinity, inclusive. The kurtosis can also attain arbitrarily large positive values, but a lower limit exists. Since the kurtosis does not depend on the absolute scale of a distribution, we can choose zero mean and unit variance without loss of generality. We define a reduced variable x with these statistics,

$$x \equiv \frac{\tau - \mathrm{E}[\tau]}{\sigma_\tau}, \tag{B.13}$$

so that the kurtosis of τ simplifies to

$$\mathrm{E}[(\tau - \mathrm{E}[\tau])^4]/\sigma_\tau^4 - 3 \to \mathrm{E}[x^4] - 3. \tag{B.14}$$

Now define $y \equiv x^2$, and consider

$$\begin{aligned} \mathrm{E}[x^4] &= \mathrm{E}[y^2] \\ &= \mathrm{E}^2[y] + \mathrm{Var}[y] \\ &\le \mathrm{E}^2[y] \\ &= \mathrm{E}^2[x^2] \\ &= 1. \end{aligned} \tag{B.15}$$

Combining Eqs. (B.14) and (B.15) therefore provides

$$\mathrm{E}[(\tau - \mathrm{E}[\tau])^4]/\sigma_\tau^4 - 3 \le 1 - 3 = -2. \tag{B.16}$$

To achieve this lower limit, we require a random variable with zero mean, unit variance, and (constant) unit square. We then have $x = +1, -1$, with equal probability, as the only solution. In terms of the original variable τ, we may choose any two values as long as they each occur with probability $1/2$. In this case, the general probability density function takes the form

$$p_\tau(t) = \frac{\delta(t - a) + \delta(t - b)}{2} \tag{B.17}$$

for arbitrary (but distinct) values a and b.

Restricting τ to assume nonnegative values actually changes these ranges very little. The kurtosis retains its negative limit of -2 as well as its upper limit of positive infinity. Positive values of skewness derive from tails in positive values of τ, which does not change. To achieve negative skewness values, we can truncate a distribution of τ at some large negative number, and then increase the mean by that same number, resulting in a distribution with nearly the same negative skewness but limited to positive interevent intervals. In this manner one can generate any desired negative skewness except for negative infinity.

Prob. 3.6 Proceeding directly from the definition in Eq. (3.13), and using a change of variable $x = t_0/t$, we have

$$
\begin{aligned}
\mathrm{E}[\tau^c] &= \int_0^\infty t^c \sqrt{t_0/\pi}\; t^{-3/2} \exp(-t_0/t)\, dt \\
&= \int_0^\infty t_0^c\, x^{-c} \sqrt{t_0/\pi}\; t_0^{-3/2}\, x^{3/2}\, e^{-x}\, t_0\, x^{-2}\, dx \\
&= \frac{t_0^c}{\sqrt{\pi}} \int_0^\infty x^{-1/2-c}\, e^{-x}\, dx && (B.18) \\
&= \pi^{-1/2}\, \Gamma(\tfrac{1}{2} - c)\, t_0^c, && (B.19)
\end{aligned}
$$

for c sufficiently small, where $\Gamma(\cdot)$ again denotes the Eulerian gamma function [see Eq. (4.44)]. Moments of τ do not exist for positive integers c since the integral in Eq. (B.18) diverges near the origin for those exponents. For exponents less than or equal to -1, the integral in Eq. (B.18) has infinite area near the origin and therefore diverges. For convergence, we thus require $-\frac{1}{2} - c > -1$, or $c < \frac{1}{2}$. Hence, all moments of order c less than one half exist, including fractional moments between zero and one half as well as negative-integer moments.

Prob. 3.7 Employing the properties of regular Brownian motion, as discussed in Sec. 2.4.2, proves especially helpful. The rescaled range statistic and detrended fluctuation analysis both employ a summed version of the input series. For large k, and intervals with finite variance, the resulting sums will converge to a Gaussian distribution as a result of the central limit theorem. Thus, the sums approach regular Brownian motion. We know from Sec. 2.4.2 that scaling the time by a factor c is equivalent to scaling the amplitude by a factor $\sqrt{c}$. In terms of a sum of intervals, the case of interest here, the independent variable changes from time to the number of intervals. Both the rescaled range statistic and detrended fluctuation analysis yield results that derive from the amplitude of the process: a normalized difference for the former statistic, and an average root-mean-square deviation for the latter. So increasing the independent variable by a factor c will similarly increase the resulting statistics by a factor $\sqrt{c}$. Setting $c = k$, we conclude that both statistics scale as $\sqrt{k}$.

Prob. 3.8 The forward recurrence time is defined as the time remaining to the next event, starting at a time t_0 independent of the process. This time t_0 lies within some interevent interval τ, with the probability density of that interval proportional to $s p_\tau(s)$; the form $p_\tau(s)$ itself denotes the probability density of the times between events. The additional factor of s arises because this form of sampling (time-based rather than interval-based) preferentially selects longer intervals, in proportion to their duration. (To see this, consider a simple example where interevent intervals of durations 1 and 2 exist in equal numbers. A time selected at random will lie within an interval of duration 2 twice as often as one of duration 1.) Normalizing this interevent interval yields a probability density of $s p_\tau(s)/\mathrm{E}[\tau]$.

Since we selected the time t_0 independently of the point process, given the interval τ, we have no other information about where within this interval t_0 occurs. Thus, the time remaining between t_0 and the next event of the point process (the forward recurrence time) has a uniform distribution $1/s$, given that $s = \tau$. To evaluate the forward recurrence-time probability density at t, we integrate over all times that remain possible (greater than t). Taken together, we have

$$
\begin{aligned}
p_\vartheta(t) &= \int_t^\infty (1/s)\left\{sp_\tau(s)/\operatorname{E}[\tau]\right\} ds \\
&= \frac{1}{\operatorname{E}[\tau]} \int_t^\infty p_\tau(s)\, ds \\
&= \frac{1}{\operatorname{E}[\tau]}\left[1 - P_\tau(t)\right] \\
P_\vartheta(t) &= \frac{1}{\operatorname{E}[\tau]} \int_0^t \left[1 - P_\tau(s)\right] ds, \tag{B.20}
\end{aligned}
$$

which is precisely Eq. (3.11). See Cox & Isham (1980, pp. 7–8) for a related approach.

Prob. 3.9 Imagine an almost periodic series of events. We represent the interevent intervals τ_k as a perturbation about the mean value

$$
\tau_k = \operatorname{E}[\tau]\left(1 + \epsilon_k\right), \tag{B.21}
$$

where the sequence of dimensionless random variables $\{\epsilon_k\}$ represents the relative deviations of each interevent interval from the mean. Since we assume that the sequence $\{\epsilon_k\}$ remains small, we can adequately describe the point process in terms of a local rate, λ_k:

$$
\begin{aligned}
\lambda_k(\operatorname{E}[\tau]) &= 1/\tau_k \\
&= \left\{\operatorname{E}[\tau]\left(1 + \epsilon_k\right)\right\}^{-1} \\
&= \operatorname{E}[\mu]/(1 + \epsilon_k) \\
&\approx \operatorname{E}[\mu] - \operatorname{E}[\mu]\,\epsilon_k. \tag{B.22}
\end{aligned}
$$

Ignoring the constant term $\operatorname{E}[\mu]$ (which only affects the spectrum at zero frequency), and recalling that multiplying a sequence by a constant changes its spectrum by the square of that constant, we arrive at

$$
S_\lambda\left(f/\operatorname{E}[\mu]\right) \approx \operatorname{E}^2[\mu]\, S_\epsilon(f). \tag{B.23}
$$

For the interval-based spectrum, Eq. (B.21) leads to a similar result:

$$
S_\tau(f) = \operatorname{E}^2[\tau]\, S_\epsilon(f). \tag{B.24}
$$

Combining Eqs. (3.67), (B.23), and (B.24) leads to

$$
S_\tau(f) \approx \pi^{-2}\operatorname{E}^2[\tau]\, f^{-2}\left(\pi f \operatorname{E}[\tau]\right)^2 S_N(f/\operatorname{E}[\mu]), \tag{B.25}
$$

where we have made use of the relation $\sin(\pi f \, E[\tau]) \approx \pi f \, E[\tau]$ for low frequencies f (see Prob. 3.13). Finally, Eq. (B.25) simplifies to

$$S_\tau(f) \approx E^4[\tau] \, S_N(f/E[\mu]). \tag{B.26}$$

Prob. 3.10 Beginning with the definition of $F(T)$ provided in Eq. (3.32), we have

$$F(T) \equiv \frac{E[Z^2(T)] - E^2[Z(T)]}{E[Z(T)]}$$

$$E[Z(T)] \, F(T) = E[Z^2(T)] - E^2[Z(T)]$$

$$E[\mu] \, T \, F(T) = E\left[\int_{s=0}^{T} \int_{t=0}^{T} dN(s) \, dN(t)\right] - E^2[\mu] \, T^2$$

$$= \int_{s=0}^{T} \int_{t=0}^{T} G(t - s) \, ds \, dt - E^2[\mu] \, T^2. \tag{B.27}$$

Defining

$$u \equiv t - s \quad \text{and} \quad v \equiv t + s, \tag{B.28}$$

we obtain

$$ds \, dt = du \, dv / 2. \tag{B.29}$$

Substituting Eqs. (B.28) and (B.29) into Eq. (B.27) we obtain

$$E[\mu] \, T \, F(T) = \int_{u=-T}^{T} \int_{v=|u|-T}^{T-|u|} G(u) \, du \, dv / 2 - E^2[\mu] \, T^2$$

$$= \int_{u=-T}^{T} (T - |u|) \, G(u) \, du - E^2[\mu] \, T^2$$

$$= \int_{u=-T}^{T} (T - |u|) \left\{G(u) - E^2[\mu]\right\} \, du. \tag{B.30}$$

Prob. 3.11 We begin with the inverse Fourier transform relationship between $G(t)$ and $S_N(f)$ provided in Eq. (3.58), and keep in mind that the constant term $E^2[\mu]$ in the coincidence rate corresponds to a delta function $E^2[\mu] \, \delta(f)$ at zero frequency in the spectrum

$$G(u) = \int_{f=-\infty}^{\infty} S_N(f) \, \exp(i2\pi f u) \, df$$

$$= \int_{f=-\infty}^{\infty} S_N(f) \, \cos(2\pi f u) \, df$$

$$G(u) - E^2[\mu] = \int_{f=-\infty}^{\infty} \left\{S_N(f) - E^2[\mu] \, \delta(f)\right\} \cos(2\pi f u) \, df$$

$$= 2 \int_{f=0+}^{\infty} S_N(f) \, \cos(2\pi f u) \, df. \tag{B.31}$$

Reiterating Eq. (B.30) and making use of Eq. (B.31) yields

$$
\begin{aligned}
\mathrm{E}[\mu]\, T\, F(T) &= \int_{u=-T}^{T} (T - |u|) \left\{ G(u) - \mathrm{E}^2[\mu] \right\} du \\
&= 2 \int_{u=-T}^{T} (T - |u|) \int_{f=0+}^{\infty} S_N(f) \cos(2\pi f u)\, df\, du \\
&= 2 \int_{f=0+}^{\infty} S_N(f) \int_{u=-T}^{T} (T - |u|) \cos(2\pi f u)\, du\, df \\
&= \int_{f=0+}^{\infty} 4T^2\, S_N(f) \int_{x=0}^{1} (1 - x) \cos(2\pi f T x)\, dx\, df \\
&= \int_{f=0+}^{\infty} 4T^2\, S_N(f) \Big[(2\pi f T)^{-1} (1 - x) \sin(2\pi f T x) \\
&\qquad\qquad\qquad - (2\pi f T)^{-2} \cos(2\pi f T x) \Big]_{x=0}^{1} df \\
&= \int_{f=0+}^{\infty} \frac{4T^2}{(2\pi f T)^2} S_N(f) \big[1 - \cos(2\pi f T) \big]\, df \\
&= \int_{f=0+}^{\infty} \frac{1}{\pi^2 f^2} S_N(f)\, 2\sin^2(\pi f T)\, df \\
&= \frac{2}{\pi^2} \int_{0+}^{\infty} S_N(f)\, \sin^2(\pi f T)\, f^{-2}\, df. \tag{B.32}
\end{aligned}
$$

Prob. 3.12 Here we employ Eq. (3.41), and make use of Eq. (3.61), which we proved immediately above [see Eq. (B.32)]:

$$
\begin{aligned}
A(T) &= 2F(T) - F(2T) \\
&= \frac{4}{\pi^2\, \mathrm{E}[\mu]\, T} \int_{0+}^{\infty} S_N(f)\, \sin^2(\pi f T)\, f^{-2}\, df \\
&\quad - \frac{2}{\pi^2\, \mathrm{E}[\mu]\, 2T} \int_{0+}^{\infty} S_N(f)\, \sin^2(2\pi f T)\, f^{-2}\, df \\
&= \frac{1}{\pi^2\, \mathrm{E}[\mu]\, T} \int_{0+}^{\infty} S_N(f) \left[4\sin^2(\pi f T) - \sin^2(2\pi f T) \right] f^{-2}\, df \\
&= \frac{1}{\pi^2\, \mathrm{E}[\mu]\, T} \int_{0+}^{\infty} S_N(f) \Big[4\sin^2(\pi f T) \\
&\qquad\qquad\qquad - 4\sin^2(\pi f T)\cos^2(\pi f T) \Big] f^{-2}\, df \\
&= \frac{4}{\pi^2\, \mathrm{E}[\mu]\, T} \int_{0+}^{\infty} S_N(f)\, \sin^2(\pi f T) \left[1 - \cos^2(\pi f T) \right] f^{-2}\, df \\
&= \frac{4}{\pi^2\, \mathrm{E}[\mu]\, T} \int_{0+}^{\infty} S_N(f)\, \sin^4(\pi f T)\, f^{-2}\, df. \tag{B.33}
\end{aligned}
$$

Prob. 3.13 Conversion of the point process $dN(t)$ into a rate $\lambda_k(T)$ involves two operations taken in sequence: averaging and sampling. We consider each in turn, and examine how they affect the spectrum.

The first step involves averaging $dN(t)$ over a time T. We can cast this operation as filtering by a simple rectangular form of duration T and height $1/T$, which has an impulse response function

$$h(t) = \begin{cases} 1/T & 0 \leq t < T \\ 0 & \text{otherwise.} \end{cases} \tag{B.34}$$

Linear systems theory (Papoulis, 1991) tells us that filtering a point process changes the power spectral density by a factor equal to the square magnitude of the Fourier transform of the impulse response function [see Eq. (9.35)]. In this case we have

$$\begin{aligned}
\mathcal{F}\{h(t)\} &\equiv \int_{t=-\infty}^{\infty} h(t) \exp(-i2\pi ft)\, df \\
&= \int_0^T (1/T) \exp(-i2\pi ft)\, df \\
&= [1 - \exp(-i2\pi fT)]/(i2\pi fT) \\
&= \exp(-i\pi fT)\, [\exp(i\pi fT) - \exp(-i\pi fT)]/(i2\pi fT) \\
&= \exp(-i\pi fT)\, \sin(\pi fT)/(\pi fT) \\
|\mathcal{F}\{h(t)\}|^2 &= \sin^2(\pi fT)/(\pi fT)^2. \tag{B.35}
\end{aligned}$$

(The same result can be obtained, but with additional effort, by using the methods employed in Probs. 3.10 and 3.11.)

If we let $X(t)$ denote the averaged process thus defined, its spectrum becomes

$$S_X(f) = S_N(f) \sin^2(\pi fT)/(\pi fT)^2. \tag{B.36}$$

But $X(t)$ exists for all times t, while we have defined $\lambda_k(T)$ as a discrete-time process, defined only for integer values of k. Converting $X(t)$ into $\lambda_k(T)$ involves sampling at a time interval of T. Discrete-time processes have periodic spectra, spaced a frequency $1/T$ apart; this leads to aliasing when the component spectra overlap. In terms of the spectrum of the averaged process, we have

$$S_\lambda(f, T) = \sum_{k=-\infty}^{\infty} S_X(f + k/T). \tag{B.37}$$

Combining Eqs. (B.36) and (B.37) yields Eq. (3.67), where we have made use of the simplification

$$\sin(\theta + n\pi) = (-1)^n \sin(\theta). \tag{B.38}$$

To evaluate Eq. (3.67) for small values of fT, we employ the relation $\sin(\theta) \approx \theta$ for small θ, verified by l'Hôpital's rule. Reiterating Eq. (3.67) and employing this approximation yields

$$
\begin{aligned}
S_\lambda(f, T) &= \sum_{k=-\infty}^{\infty} S_N(f + k/T) \frac{\sin^2(\pi fT)}{(\pi fT + \pi k)^2} \\
&\approx \sum_{k=-\infty}^{\infty} S_N(f + k/T) \frac{(\pi fT)^2}{(\pi fT + \pi k)^2} \\
&= \sum_{k=-\infty}^{\infty} \frac{S_N(f + k/T)}{[1 + k/(fT)]^2}.
\end{aligned}
\tag{B.39}
$$

For small values of the product fT, the denominator in Eq. (B.39) assumes quite large values for $k \neq 0$, making these terms quite small. The term corresponding to $k = 0$ then dominates, and we have $S_\lambda(f, T) \approx S_N(f)$.

B.4 POINT PROCESSES: EXAMPLES

Prob. 4.1.1 A counting window of duration $T = n\tau$ contains exactly n events regardless of the starting time of the window. The randomness inherent in the point process makes interpolation between integer multiples of τ possible, so that for arbitrary T we have

$$Z(T) = \begin{cases} \text{int}(T/\tau) & \text{with probability} \quad 1 - p \\ \text{int}(T/\tau) + 1 & \text{with probability} \quad p, \end{cases} \tag{B.40}$$

where $\text{int}(x)$ denotes the largest integer not exceeding x and

$$p \equiv T/\tau - \text{int}(T/\tau) \tag{B.41}$$

represents the probability of the larger count. For the mean number of counts, we have the simple result

$$E[Z(T)] = T/\tau \tag{B.42}$$

common to all point processes (with $E[\tau] = \tau$ in this case). We employ Eqs. (B.40)–(B.42) to find the variance

$$\begin{aligned} \text{Var}[Z(T)] &= (1-p)\left[\text{int}(T/\tau) - T/\tau\right]^2 + p\left[\text{int}(T/\tau) + 1 - T/\tau\right]^2 \\ &= (1-p)p^2 + p(1-p)^2 \\ &= p(1-p), \end{aligned} \tag{B.43}$$

where p is defined in Eq. (B.41). The function $p(1-p)$ indeed achieves a maximum value of $\frac{1}{4}$ for $p = \frac{1}{2}$.

Prob. 4.1.2 Dividing Eq. (B.43) by Eq. (B.42) yields the normalized variance $F(T)$:

$$\begin{aligned} F(T) &= \text{Var}[Z(T)]/E[Z(T)] \\ &= p(1-p)\tau/T \\ &= [T/\tau - \text{int}(T/\tau)]\left[\text{int}(T/\tau) + 1 - T/\tau\right][\tau/T]. \end{aligned} \tag{B.44}$$

Prob. 4.2.1 To find the forward recurrence-time probability density, we make use of Eq. (3.12). Substituting Eq. (4.3) for $p_\tau(t)$ yields the result for the homogeneous Poisson process:

$$\begin{aligned} p_\vartheta(t) &= [1 - P_\tau(t)]/E[\tau] \\ &= \left[1 - \int_0^t \mu \exp(-\mu s)\,ds\right]\mu \\ &= \mu\left\{1 - [1 - \exp(-\mu t)]\right\} \\ &= \mu \exp(-\mu t) \\ &= p_\tau(t), \end{aligned} \tag{B.45}$$

so that the forward recurrence-time probability density is identical to the interevent-interval density for this process.

Prob. 4.2.2 Since v can lie anywhere within an interevent interval, the time to the events at either end must have the same statistics. Hence, the backward recurrence-time probability density is the same as the forward recurrence-time probability density, which was derived in Eq. (B.45) above.

Prob. 4.2.3 The forward recurrence time and the backward recurrence time together comprise the interevent interval surrounding v; their sum becomes that interval. As a result of the memoryless nature of the homogeneous Poisson process, these two times are independent. The convolution of the probability densities of two independent random variables provides the probability density of their sum. Recalling that τ_* denotes the interval within which v lies, we obtain

$$
\begin{aligned}
p_{\tau*}(t) &= p_\tau(t) \star p_\tau(t) \\
&= \int_0^t p_\tau(s)\, p_\tau(t-s)\, ds \\
&= \int_0^t \mu \exp(-\mu s)\, \mu \exp[-\mu(t-s)]\, ds \\
&= \mu^2 \int_0^t \exp(-\mu t)\, ds \\
&= \mu^2 t \exp(-\mu t).
\end{aligned} \tag{B.46}
$$

Prob. 4.2.4 The two probability densities, Eqs. (B.46) and (4.3), do indeed differ. For example, the mean interevent interval associated with Eq. (B.46) calculates to

$$
\begin{aligned}
\mathrm{E}[\tau_*] &= \int_0^\infty t\, p_{\tau*}(t)\, dt \\
&= \int_0^\infty t\, \mu^2 t \exp(-\mu t)\, dt \\
&= \mathrm{E}[\tau] \int_0^\infty x^2 \exp(-x)\, dx \\
&= 2\mathrm{E}[\tau],
\end{aligned} \tag{B.47}
$$

twice the value for the homogeneous Poisson process. The difference lies in how the intervals are selected. For the conventional interevent-interval probability density, $p_\tau(t)$, each interval is weighted equally. But to obtain the statistics of the interval spanning the time v, larger intervals are selected preferentially. In fact, the larger the interval, the more likely we select it, leading to the extra factor of t in Eq. (B.46); normalization to unity area yields the exact form of Eq. (B.46). Additional information pertinent to this issue can be found in the solution to Prob. 3.8.

Prob. 4.3 We begin with $q = 0$. For a particular value of T, the sum will have L/T terms. Each $Z_k(T)$ assumes a value of zero with probability $\exp(-\mu T)$, as provided by Eq. (4.7); this corresponds to an empty box. A full box then occurs with probability $1 - \exp(-\mu T)$. Multiplying this latter probability by the number of terms yields

$$\mathrm{E}\left[\sum_k Z_k^0(T)\right] = (L/T)\,[1 - \exp(-\mu T)]. \tag{B.48}$$

In the limit of small T, using Eq. (B.48) together with l'Hôpital's rule gives rise to

$$\begin{aligned}
\lim_{T\to 0} \mathrm{E}\left[\sum_k Z_k^0(T)\right] &= \lim_{T\to 0} \frac{1 - \exp(-\mu T)}{T/L} \\
&= \lim_{T\to 0} \frac{\mu \exp(-\mu T)}{1/L} \\
&= \mu L \\
&= \mathrm{E}[N(L)], \tag{B.49}
\end{aligned}$$

a constant value. In connection with Eq. (3.70) we obtain $D_0 = 0$, as we must for a finite collection of points. For large T the exponential function in Eq. (B.48) vanishes so that we have L/T. Using this latter form in conjunction with Eq. (3.72) yields $D_0 = 1$ for this scaling region, as expected for a nonfractal process.

For $q = 2$ we obtain

$$\begin{aligned}
\mathrm{E}\left[\sum_k Z_k^2(T)\right] &= (L/T)\,\mathrm{E}[Z_k^2(T)] \\
&= (L/T)\left\{\mathrm{E}^2[Z_k(T)] + \mathrm{Var}[Z_k(T)]\right\} \\
&= (L/T)\left\{(\mu T)^2 + \mu T\right\} \\
&= \mu L\,(1 + \mu T) \\
&= \mathrm{E}[N(L)]\,(1 + \mu T), \tag{B.50}
\end{aligned}$$

where we have used the particular cases listed immediately after Eq. (4.4). Here the limiting forms emerge more readily; for small T we again obtain $\mathrm{E}[N(L)]$, and for large T the unity term in Eq. (B.50) disappears. Equations (3.70) and (3.72) yield $D_2 = 0$ and $D_2 = 1$, respectively, in agreement with the values obtained for D_0 above.

Prob. 4.4.1 We begin with Eq. (3.6), the characteristic function of the interevent intervals:

$$\phi_\tau(\omega) \equiv \int_0^\infty p_\tau(t)\,\exp(-i\omega t)\,dt$$

$$
\begin{aligned}
&= \mu(\mu + i\omega)^{-1} \int_0^\infty \exp(-x)\, dx \\
&= \frac{\mu}{\mu + i\omega}. \tag{B.51}
\end{aligned}
$$

For the spectrum, we substitute Eq. (B.51) into Eq. (4.16), to obtain

$$
\begin{aligned}
S_N(f) &= \mathrm{E}^2[\mu]\,\delta(f) + \mathrm{E}[\mu]\,\mathrm{Re}\left\{\frac{1 + \phi_\tau(2\pi f)}{1 - \phi_\tau(2\pi f)}\right\} \\
&= \mu^2\,\delta(f) + \mu\,\mathrm{Re}\left\{\frac{1 + \mu(\mu + 2\pi i f)^{-1}}{1 - \mu(\mu + 2\pi i f)^{-1}}\right\} \\
&= \mu^2\,\delta(f) + \mu\,\mathrm{Re}\left\{\frac{(\mu + 2\pi i f) + \mu}{(\mu + 2\pi i f) - \mu}\right\} \\
&= \mu^2\,\delta(f) + \mu\,\mathrm{Re}\left\{\frac{2\mu + 2\pi i f}{2\pi i f}\right\} \\
&= \mu^2\,\delta(f) + \mu\,\mathrm{Re}\left\{-i\mu/(\pi f) + 1\right\} \\
&= \mu^2\,\delta(f) + \mu, \tag{B.52}
\end{aligned}
$$

in agreement with Eq. (4.9c). In accordance with Eq. (3.59), the mean rate of the process $\mathrm{E}[\mu] = \lim_{f\to\infty} S_N(f) = \mu$, using the results in Eq. (B.52).

Prob. 4.4.2 Since constants and delta functions interchange under Fourier transforms, Eqs. (4.9c) and (4.9d) form a Fourier-transform pair. Therefore, since Eq. (4.9c) is valid for this process, so must Eq. (4.9d) be valid. Now consider the expectation of two nonoverlapping intervals. For any t_1, t_2, t_3, t_4 satisfying $t_1 < t_2 \le t_3 < t_4$, we have

$$
\begin{aligned}
&\mathrm{E}\big\{[N(t_4) - N(t_3)]\,[N(t_2) - N(t_1)]\big\} \\
&= \mathrm{E}\left[\int_{s=t3}^{t4} dN(s) \int_{t=t1}^{t2} dN(t)\right] \\
&= \int_{s=t3}^{t4} \int_{t=t1}^{t2} \mathrm{E}[dN(s)\, dN(t)] \\
&= \int_{s=t3}^{t4} \int_{t=t1}^{t2} G(s - t)\, ds\, dt \\
&= \int_{s=t3}^{t4} \int_{t=t1}^{t2} \mu^2\, ds\, dt \tag{B.53} \\
&= \left[\int_{s=t3}^{t4} \mu\, ds\right] \left[\int_{t=t1}^{t2} \mu\, dt\right] \\
&= \mathrm{E}\big\{[N(t_4) - N(t_3)]\big\}\,\mathrm{E}\big\{[N(t_2) - N(t_1)]\big\}, \tag{B.54}
\end{aligned}
$$

where Eq. (B.53) follows from the lack of overlap in the two intervals, which prevents the delta function from appearing. Equation (B.54) thus establishes that the expectation of the product equals the product of the expectations, and therefore that the counts in these disjoint intervals are uncorrelated.

Prob. 4.4.3 Equation (4.2) asserts independence, which implies lack of correlation, but not vice versa. However, a similar argument to that used in the solution of Prob. 4.4.2 yields results for expectations of arbitrary orders:

$$E\{[N(t_4) - N(t_3)]^m \, [N(t_2) - N(t_1)]^n\}. \tag{B.55}$$

Taken together, these results establish independence.

Prob. 4.5 Examining Eq. (4.16), we see that low frequencies correspond to small arguments of the characteristic function $\phi_\tau(\omega)$, enabling us to employ a power-series expansion. We next examine the derivatives

$$\frac{d^n}{d\omega^n} \phi_\tau(\omega) = \int_0^\infty p_\tau(t) \, (-it)^n \, e^{-i\omega t} \, dt$$

$$\frac{d^n}{d\omega^n} \phi_\tau(\omega)_{\omega=0} = \int_0^\infty p_\tau(t) \, (-it)^n \, dt$$

$$= (-i)^n \, E[\tau^n], \tag{B.56}$$

which leads to a general property of characteristic functions with finite moments. Turning now to the power-series expansion, we arrive at

$$\phi_\tau(\omega) = \sum_{n=0}^\infty (-i\omega)^n \, E[\tau^n]/n!$$

$$\approx \sum_{n=0}^2 (-i\omega)^n \, E[\tau^n]/n!$$

$$= 1 - i\omega \, E[\tau] - \omega^2 \, E[\tau^2]/2. \tag{B.57}$$

Finally, substituting Eq. (B.57) into Eq. (4.16) yields

$$\lim_{f \to 0} S_N(f)$$

$$= E[\mu] \lim_{f \to 0} \text{Re}\left\{\frac{1 + \phi_\tau(2\pi f)}{1 - \phi_\tau(2\pi f)}\right\}$$

$$= E[\mu] \lim_{\omega \to 0} \text{Re}\left\{\frac{2 - i\omega \, E[\tau] - \omega^2 \, E[\tau^2]/2}{i\omega \, E[\tau] + \omega^2 \, E[\tau^2]/2}\right\}$$

$$= E[\mu] \lim_{\omega \to 0} \text{Re}\left\{\frac{2 - i\omega \, E[\tau] - \omega^2 \, E[\tau^2]/2}{i\omega \, E[\tau] + \omega^2 \, E[\tau^2]/2} \times \frac{E[\tau^2]/2 - i \, E[\tau]/\omega}{E[\tau^2]/2 - i \, E[\tau]/\omega}\right\}$$

$$= E[\mu] \lim_{\omega \to 0} \text{Re}\left\{\frac{E[\tau^2] - 2i \, E[\tau]/\omega - E^2[\tau] - \omega^2 \, E^2[\tau^2]/4}{E^2[\tau] + \omega^2 \, E^2[\tau^2]/4}\right\}$$

$$= E[\mu] \lim_{\omega \to 0} \frac{E[\tau^2] - E^2[\tau] - \omega^2 \, E^2[\tau^2]/4}{E^2[\tau] + \omega^2 \, E^2[\tau^2]/4}$$

$$
\begin{aligned}
&= \; E[\mu] \, \frac{E[\tau^2] - E^2[\tau]}{E^2[\tau]} \\
&= \; E^3[\mu] \, \mathrm{Var}[\tau],
\end{aligned}
\tag{B.58}
$$

which is identical to Eq. (4.17).

Prob. 4.6 We begin by substituting $k = 2$ in Eq. (4.19), yielding

$$
\begin{aligned}
E\{Z(T)\,[Z(T) + 1]\} \\
&= \; 2 \int_{0-}^{T} (T - t)\, G(t)\, dt \\
&= \; \int_{-T}^{T} (T - |t|)\, G(t)\, dt + \int_{0-}^{0+} (T - |t|)\, G(t)\, dt \\
&= \; \int_{-T}^{T} (T - |t|)\, G(t)\, dt + \int_{0-}^{0+} (T - |t|)\, E[\mu]\, \delta(t)\, dt \\
&= \; \int_{-T}^{T} (T - |t|)\, G(t)\, dt + T\, E[\mu].
\end{aligned}
\tag{B.59}
$$

For the integral over the infinitesimal region near the origin in Eq. (B.59), we have ignored all of $G(t)$ except for the delta function; see Eq. (3.50). Expressing the variance in terms of Eq. (B.59) leads to

$$
\begin{aligned}
\mathrm{Var}[Z(T)] \;&= \; E\{Z(T)\,[Z(T) + 1]\} - E^2[Z(T)] - E[Z(T)] \\
&= \; \int_{-T}^{T} (T - |t|)\, G(t)\, dt + T\, E[\mu] - \int_{-T}^{T} (T - |t|)\, E^2[\mu]\, dt - T\, E[\mu] \\
&= \; \int_{-T}^{T} (T - |t|) \left\{ G(t) - E^2[\mu] \right\} dt,
\end{aligned}
\tag{B.60}
$$

in agreement with Eq. (4.21).

Prob. 4.7.1 We again begin with Eq. (3.6), the characteristic function of the interevent intervals:

$$
\begin{aligned}
\phi_\tau(\omega) \;&\equiv\; \int_0^\infty p_\tau(t)\, \exp(-i\omega\, t)\, dt \\
&= \; \int_0^\infty [\Gamma(m)]^{-1} \tau_0^{-m}\, t^{m-1}\, \exp(-t/\tau_0)\, \exp(-i\omega t)\, dt \\
&= \; [\Gamma(m)]^{-1} \tau_0^{-m} \int_0^\infty t^{m-1}\, \exp[-(1/\tau_0 + i\omega)t]\, dt \\
&= \; [\Gamma(m)]^{-1} \tau_0^{-m} \,(1/\tau_0 + i\omega)^{-m} \int_0^\infty x^{m-1}\, \exp(-x)\, dx \\
&= \; [\Gamma(m)]^{-1} \,(1 + i\omega\, \tau_0)^{-m}\, \Gamma(m) \\
&= \; (1 + i\omega\, \tau_0)^{-m}.
\end{aligned}
\tag{B.61}
$$

Proceeding to Eq. (3.8) we obtain

$$
\begin{aligned}
C_n &= i^n \frac{d^n}{d\omega^n} \ln\left[\phi_\tau(\omega)\right]_{\omega=0} \\
&= i^n \frac{d^n}{d\omega^n} \ln\left[(1 + i\omega\,\tau_0)^{-m}\right]_{\omega=0} \\
&= (-1)^{n+1}\, m\,\tau_0^n \frac{d^n}{dx^n} \ln\left[(1 + x)\right]_{x=0} \\
&= \Gamma(n)\, m\,\tau_0^n,
\end{aligned}
\tag{B.62}
$$

and in particular, we have

$$
\begin{array}{rclcl}
\text{mean} &=& C_1 &=& m\,\tau_0 \\
\text{variance} &=& C_2 &=& m\,\tau_0^2 \\
&& C_3 &=& 2m\,\tau_0^3 \\
&& C_4 &=& 6m\,\tau_0^4 \\
\text{skewness} &=& C_3/C_2^{3/2} &=& 2/\sqrt{m} \\
\text{kurtosis} &=& C_4/C_2^2 &=& 6/m.
\end{array}
\tag{B.63}
$$

Prob. 4.7.2 For the spectrum, we again substitute Eq. (B.61) into Eq. (4.16) to obtain

$$
\begin{aligned}
S_N(f) \\
&= \mathrm{E}^2[\mu]\,\delta(f) + \mathrm{E}[\mu]\,\mathrm{Re}\left\{\frac{1 + \phi_\tau(2\pi f)}{1 - \phi_\tau(2\pi f)}\right\} \\
&= (m\,\tau_0)^{-2}\,\delta(f) + (m\,\tau_0)^{-1}\,\mathrm{Re}\left\{\frac{1 + (1 + 2\pi i f\tau_0)^{-m}}{1 - (1 + 2\pi i f\tau_0)^{-m}}\right\} \\
&= (m\,\tau_0)^{-2}\,\delta(f) + (m\,\tau_0)^{-1}\,\mathrm{Re}\left\{\frac{(1 + i\theta)^m + 1}{(1 + i\theta)^m - 1} \times \frac{(1 - i\theta)^m - 1}{(1 - i\theta)^m - 1}\right\} \\
&= (m\,\tau_0)^{-2}\,\delta(f) + (m\,\tau_0)^{-1}\,\mathrm{Re}\left\{\frac{(1 + \theta^2)^m + (1 - i\theta)^m - (1 + i\theta)^m - 1}{(1 + \theta^2)^m - (1 - i\theta)^m - (1 + i\theta)^m + 1}\right\} \\
&= (m\,\tau_0)^{-2}\,\delta(f) + \frac{(m\,\tau_0)^{-1}\left[(1 + \theta^2)^m - 1\right]}{(1 + \theta^2)^m - 2\mathrm{Re}\{(1 + i\theta)^m\} + 1},
\end{aligned}
\tag{B.64}
$$

where we have defined $\theta \equiv 2\pi f\tau_0$ to simplify the notation. The two middle terms in the numerator on the line before Eq. (B.64) form a complex-conjugate pair, so that their difference has only an imaginary component; taking the real part yields zero. If desired, further simplification can be achieved by making use of

$$
\begin{aligned}
1 + i\theta &= \sqrt{1 + \theta^2}\left\{\frac{1}{\sqrt{1 + \theta^2}} + \frac{i\theta}{\sqrt{1 + \theta^2}}\right\} \\
&= \sqrt{1 + \theta^2}\left\{\cos\left[\arctan(\theta)\right] + i\,\sin\left[\arctan(\theta)\right]\right\}
\end{aligned}
$$

$$= \sqrt{1+\theta^2}\, \exp\bigl[i\, \arctan(\theta)\bigr]$$

$$(1+i\theta)^m = \bigl(1+\theta^2\bigr)^{m/2}\, \exp\bigl[im\, \arctan(\theta)\bigr]$$

$$\mathrm{Re}\bigl\{(1+i\theta)^m\bigr\} = \bigl(1+\theta^2\bigr)^{m/2}\, \cos\bigl[m\, \arctan(\theta)\bigr], \tag{B.65}$$

and substituting Eq. (B.65) into Eq. (B.64).

Prob. 4.7.3 We begin by substituting $m = 2$ into Eq. (B.64)

$$
\begin{aligned}
S_N(f) &= (2\tau_0)^{-2}\,\delta(f) + \frac{(2\tau_0)^{-1}\bigl[(1+\theta^2)^2 - 1\bigr]}{(1+\theta^2)^2 - 2\mathrm{Re}\bigl\{(1+i\theta)^2\bigr\} + 1} \\
&= (2\tau_0)^{-2}\,\delta(f) + \frac{(2\tau_0)^{-1}\bigl[1 + 2\theta^2 + \theta^4 - 1\bigr]}{1 + 2\theta^2 + \theta^4 - 2\mathrm{Re}\bigl\{1 + 2i\theta - \theta^2\bigr\} + 1} \\
&= (2\tau_0)^{-2}\,\delta(f) + (2\tau_0)^{-1}\,\frac{2\theta^2 + \theta^4}{4\theta^2 + \theta^4} \\
&= (2\tau_0)^{-2}\,\delta(f) + (2\tau_0)^{-1}\,\frac{2 + (2\pi f\tau_0)^2}{4 + (2\pi f\tau_0)^2} \\
&= (2\tau_0)^{-2}\,\delta(f) + (2\tau_0)^{-1}\Bigl[1 - \frac{1/2}{1 + (\pi f\tau_0)^2}\Bigr]. \tag{B.66}
\end{aligned}
$$

We next substitute Eq. (B.66) into Eq. (3.58) to obtain

$$
\begin{aligned}
G(t) &= \int_{-\infty}^{\infty} S_N(f)\, \exp(i2\pi ft)\, df \\
&= \int_{-\infty}^{\infty} \Bigl\{ (2\tau_0)^{-2}\,\delta(f) + (2\tau_0)^{-1}\Bigl[1 - \frac{1/2}{1 + (\pi f\tau_0)^2}\Bigr]\Bigr\}\, \exp(i\,2\pi ft)\, df \\
&= (2\tau_0)^{-2} + (2\tau_0)^{-1}\,\delta(t) - (4\tau_0)^{-1}\int_{-\infty}^{\infty} \frac{\exp(i2\pi ft)}{1 + (\pi f\tau_0)^2}\, df \\
&= (2\tau_0)^{-1}\,\delta(t) + (2\tau_0)^{-2} - (4\tau_0)^{-1}\,(\pi\tau_0)^{-1}\,\pi\, \exp\bigl(-|2t/\tau_0|\bigr) \\
&= (2\tau_0)^{-1}\,\delta(t) + (2\tau_0)^{-2}\Bigl[1 - \exp\bigl(-|2t/\tau_0|\bigr)\Bigr]. \tag{B.67}
\end{aligned}
$$

As a next step, substitution of Eq. (B.67) into Eq. (3.52) yields

$$
\begin{aligned}
F(T) &= \frac{1}{\mathrm{E}[\mu]\,T}\int_{-T}^{T} \bigl\{ G(t) - \mathrm{E}^2[\mu]\bigr\}\,(T - |t|)\, dt \\
&= \frac{1}{(2\tau_0)^{-1}\,T}\int_{-T}^{T} \Bigl\{ (2\tau_0)^{-1}\,\delta(t) + (2\tau_0)^{-2}\Bigl[1 - \exp\bigl(-|2t/\tau_0|\bigr)\Bigr] \\
&\quad - (2\tau_0)^{-2}\Bigr\}\,(T - |t|)\, dt \\
&= 1 - \frac{1}{2\tau_0\,T}\int_{-T}^{T} \exp\bigl(-|2t/\tau_0|\bigr)\,(T - |t|)\, dt
\end{aligned}
$$

$$= 1 - \frac{T}{\tau_0} \int_0^1 \exp(-2xT/\tau_0)(1-x)\,dx$$

$$= 1 - \frac{T}{\tau_0}\left[\frac{\tau_0}{2T} + \frac{\tau_0^2}{4T^2}\exp(-2T/\tau_0) - \frac{\tau_0^2}{4T^2}\right]$$

$$= \frac{1}{2} + \frac{\tau_0}{4T}\left[1 - \exp(-2T/\tau_0)\right], \tag{B.68}$$

and, finally, Eqs. (B.68) and (3.41) lead to

$$A(T) = 2F(T) - F(2T)$$

$$= 1 + \frac{\tau_0}{2T}\left[1 - \exp(-2T/\tau_0)\right] - \frac{1}{2} - \frac{\tau_0}{8T}\left[1 - \exp(-4T/\tau_0)\right]$$

$$= \frac{1}{2} + \frac{\tau_0}{8T}\left[3 + \exp(-4T/\tau_0) - 4\exp(-2T/\tau_0)\right]. \tag{B.69}$$

Prob. 4.8.1 For $f = 0$, the fraction in Eq. (3.67) vanishes for all k except $k = 0$, whereupon it assumes a value of unity. Thus, the sum collapses to a single term, and $S_\lambda(0, T) = S_N(0)$: the delta function in the point process spectrum carries forward unchanged into the rate spectrum. For $0 < |f| < 1/T$, we have $S_N(0) = \mu$, as specified in Eq. (4.9c). Equation (3.67) then becomes

$$S_\lambda(f, T) = \sum_{k=-\infty}^{\infty} S_N(f + k/T)\frac{\sin^2(\pi fT)}{(\pi fT + \pi k)^2}$$

$$= \sum_{k=-\infty}^{\infty} \mu \frac{\sin^2(\pi fT)}{(\pi fT + \pi k)^2}$$

$$= \mu \tag{B.70}$$

(see Gradshteyn & Ryzhik, 1994, Eq. 1.422.4). The periodicity of $S_\lambda(f, T)$ means that delta functions appear at frequencies $f = k/T$ for any integer k; the rate and point process spectra do not agree at these frequencies (hence the frequency limit in the problem specification). However, they do coincide for all other frequencies.

Prob. 4.8.2 From the solution of Prob. 4.7.3, we have

$$S_N(f) = (2\tau_0)^{-2}\delta(f) + (2\tau_0)^{-1} - \frac{(4\tau_0)^{-1}}{1 + (\pi f\tau_0)^2}. \tag{B.71}$$

The first two terms are identical to those for the case of the homogeneous Poisson process, and we know from the solution of Prob. 4.8.1 that the results for the two spectra coincide. Since Eq. (3.67) describes a linear relation, we can use these results for the problem at hand. This leads to

$$S_\lambda(f, T)$$

$$= \sum_{k=-\infty}^{\infty} S_N(f + k/T)\frac{\sin^2(\pi fT)}{(\pi fT + \pi k)^2}$$

$$= (2\tau_0)^{-2}\delta(f) + (2\tau_0)^{-1} - \sum_{k=-\infty}^{\infty} \frac{(4\tau_0)^{-1}}{1 + (\pi f \tau_0 + \pi k \tau_0/T)^2} \frac{\sin^2(\pi f T)}{(\pi f T + \pi k)^2}$$

$$= (2\tau_0)^{-2}\delta(f) + (2\tau_0)^{-1}$$

$$- (4\tau_0)^{-1} + (4\tau_0)^{-1} \frac{(\tau_0/T)\coth(T/\tau_0)}{\coth^2(T/\tau_0) + \cot^2(\pi f T)}$$

$$= (2\tau_0)^{-2}\delta(f) + (4\tau_0)^{-1} + \frac{(4T)^{-1}\coth(T/\tau_0)}{\coth^2(T/\tau_0) + \cot^2(\pi f T)}, \tag{B.72}$$

where we used symbolic math software to obtain the line before Eq. (B.72). Alternatively, one could substitute Eq. (B.67) into Eq. (3.54), and that result, in turn, into Eq. (3.47), but this approach requires far more algebra.

Prob. 4.8.3 To establish agreement, we take the limit $T \to 0$ in Eq. (B.72):

$$\lim_{T \to 0} S_\lambda(f, T)$$

$$= (2\tau_0)^{-2}\delta(f) + (4\tau_0)^{-1} + \lim_{T \to 0} \frac{(4T)^{-1}\coth(T/\tau_0)}{\coth^2(T/\tau_0) + \cot^2(\pi f T)}$$

$$= (2\tau_0)^{-2}\delta(f) + (4\tau_0)^{-1} + \frac{(4T)^{-1}(T/\tau_0)^{-1}}{(T/\tau_0)^{-2} + (\pi f T)^{-2}}$$

$$= (2\tau_0)^{-2}\delta(f) + (4\tau_0)^{-1} + (4\tau_0)^{-1}\frac{(\pi f \tau_0)^2}{1 + (\pi f \tau_0)^2}$$

$$= (2\tau_0)^{-2}\delta(f) + (2\tau_0)^{-1}\left[1 - \frac{1/2}{1 + (\pi f \tau_0)^2}\right], \tag{B.73}$$

which accords with $S_N(f)$ in Eq. (B.66).

Prob. 4.9.1 We begin by reiterating Eq. (4.38), which is valid when the rate process $\mu(t)$ varies slowly over the time scale of a single event,

$$\mathrm{E}[\tau^n] = \mathrm{E}[\mu^{1-n}]/\mathrm{E}[\mu]$$

$$\mathrm{E}[\tau] = 1/\mathrm{E}[\mu]$$

$$\mathrm{E}[\tau^2] = \mathrm{E}[\mu^{-1}]/\mathrm{E}[\mu]$$

$$\mathrm{Var}[\tau] = \mathrm{E}[\tau^2] - \mathrm{E}^2[\tau]$$

$$= \mathrm{E}[\mu^{-1}]/\mathrm{E}[\mu] - 1/\mathrm{E}^2[\mu]$$

$$= \frac{\mathrm{E}[\mu]\,\mathrm{E}[\mu^{-1}] - 1}{\mathrm{E}^2[\mu]}$$

$$\frac{\mathrm{Var}[\tau]}{\mathrm{E}^2[\tau]} = \mathrm{E}[\mu]\,\mathrm{E}[\mu^{-1}] - 1$$

$$C_\tau = \sqrt{E[\mu] E[\mu^{-1}] - 1}.$$ (B.74)

Prob. 4.9.2 For a gamma-distributed rate, the associated probability density function takes the form of Eq. (4.45),

$$p_\mu(x) = \begin{cases} [\Gamma(m)]^{-1} \mu_0^{-m} x^{m-1} \exp(-x/\mu_0) & x > 0 \\ 0 & x \le 0, \end{cases}$$ (B.75)

with its associated moments

$$
\begin{aligned}
E[\mu^n] &\equiv \int_{-\infty}^{\infty} x^n p_\mu(x)\, dx \\
&= \int_{0}^{\infty} x^n [\Gamma(m)]^{-1} \mu_0^{-m} x^{m-1} \exp(-x/\mu_0)\, dx \\
&= [\Gamma(m)]^{-1} \mu_0^{-m} \int_{0}^{\infty} x^{n+m-1} \exp(-x/\mu_0)\, dx \\
&= [\Gamma(m)]^{-1} \mu_0^{-m} \mu_0^{n+m} \int_{0}^{\infty} y^{n+m-1} \exp(-y)\, dy \\
&= [\Gamma(m)]^{-1} \mu_0^{n} \Gamma(n+m)
\end{aligned}
$$

$$E[\mu] = m\mu_0$$ (B.76)

$$E[\mu^{-1}] = \mu_0^{-1}/(m-1).$$ (B.77)

Substituting Eqs. (B.76) and (B.77) into Eq. (B.74) yields

$$
\begin{aligned}
C_\tau &= \sqrt{E[\mu] E[\mu^{-1}] - 1} \\
&= \sqrt{m\, \mu_0\, \mu_0^{-1}/(m-1) - 1} \\
&= \sqrt{m/(m-1) - (m-1)/(m-1)} \\
&= 1/\sqrt{m-1}.
\end{aligned}
$$ (B.78)

Prob. 4.9.3 We require $m > 1$ so that the integral defining $E[\mu^{-1}]$ converges, thereby making Eqs. (B.77) and (B.78) meaningful.

Prob. 4.10.1 For the mean rate we have

$$
\begin{aligned}
E[\mu(t)] &= E\Big\{\mu_0 \big[1 + \cos(\omega_0 t + \theta)\big]\Big\} \\
&= \mu_0 + \mu_0\, E\big[\cos(\omega_0 t + \theta)\big] \\
&= \mu_0 + \mu_0 \int_{0}^{2\pi} (2\pi)^{-1} \cos(\omega_0 t + \theta)\, d\theta \\
&= \mu_0.
\end{aligned}
$$ (B.79)

For the autocorrelation of the rate we first set forth the trigonometric identity

$$\frac{\cos(x+y) + \cos(x-y)}{2}$$

$$= \tfrac{1}{2}\big[\cos(x)\,\cos(y) - \sin(x)\,\sin(y) + \cos(x)\,\cos(y) + \sin(x)\,\sin(y)\big]$$

$$= \cos(x)\,\cos(y). \tag{B.80}$$

We then proceed as follows:

$$\begin{aligned}
E[\mu(s)\,\mu(t)] &= E\big\{\mu_0[1 + \cos(\omega_0 s + \theta)]\,\mu_0[1 + \cos(\omega_0 t + \theta)]\big\}\\
&= \mu_0^2 + \mu_0^2\,E\big[\cos(\omega_0 s + \theta)\big] + \mu_0^2\,E\big[\cos(\omega_0 t + \theta)\big]\\
&\quad + \mu_0^2\,E\big[\cos(\omega_0 s + \theta)\,\cos(\omega_0 t + \theta)\big]\\
&= \mu_0^2 + 0 + 0\\
&\quad + \tfrac{1}{2}\mu_0^2\,E\big\{\cos\big[\omega_0(s+t) + 2\theta\big] + \cos\big[\omega_0(s-t)\big]\big\}\\
&= \mu_0^2 + \tfrac{1}{2}\mu_0^2\big\{0 + \cos\big[\omega_0(s-t)\big]\big\}\\
R_\mu(t) &= \mu_0^2 + \tfrac{1}{2}\mu_0^2\cos(\omega_0 t), \tag{B.81}
\end{aligned}$$

where we have made use of Eq. (B.80), and noted that the expectation of any sine or cosine function with θ or 2θ in its argument becomes zero, as in Eq. (B.79).

For the coincidence rate we use Eq. (4.24) with Eq. (B.81) to obtain

$$\begin{aligned}
G(t) &= R_\mu(t) + E[\mu]\,\delta(t)\\
&= \mu_0^2 + \tfrac{1}{2}\mu_0^2\cos(\omega_0 t) + \mu_0\,\delta(t). \tag{B.82}
\end{aligned}$$

Finally, for the normalized variance we employ Eqs. (3.52), (B.79), and (B.82), to obtain

$$\begin{aligned}
F(T) &= \frac{1}{E[\mu]\,T}\int_{-T}^{T}\big\{G(t) - E^2[\mu]\big\}\,(T - |t|)\,dt\\
&= \frac{1}{\mu_0\,T}\int_{-T}^{T}\left\{\frac{\mu_0^2}{2}\cos(\omega_0 t) + \mu_0\,\delta(t)\right\}(T - |t|)\,dt\\
&= 1 + \mu_0\,T\int_0^1\cos(\omega_0 T x)\,(1 - x)\,dx\\
&= 1 + \mu_0\,T\Big[(\omega_0 T)^{-1}\sin(\omega_0 T x)\,(1 - x) - (\omega_0 T)^{-2}\cos(\omega_0 T x)\Big]_0^1\\
&= 1 + \mu_0\,\omega_0^{-2}\,T^{-1}\big[1 - \cos(\omega_0 T)\big]\\
&= 1 + 2\mu_0\,\omega_0^{-2}\,T^{-1}\sin^2(\omega_0 T/2). \tag{B.83}
\end{aligned}$$

Prob. 4.10.2 With $2\pi\mu_0/\omega_0$ irrational, we sample all possible phases of the rate and can avoid troublesome cases such as $\mu_0 = n\omega_0/(2\pi)$ where interevent intervals cycle through n different values that sum to $2\pi/\omega_0$, but the n values depend on θ. The condition $\mu_0/\omega_0 \gg 1$ also permits the use of Eqs. (4.37) and (4.38), but first we require an expression for the probability density of the rate. Restricting ourselves to $0 \le x \le 2\mu_0$, the limits of possible values of the rate, we have

$$
\begin{aligned}
\Pr\{\mu < x\} &= \Pr\left\{\mu_0[1 + \cos(\omega_0 t + \theta)] < x\right\} \\
&= \Pr\{\cos(\omega_0 t + \theta) < x/\mu_0 - 1\} \\
&= \Pr\left\{\omega_0 t + \theta \bmod \pi > \arccos(x/\mu_0 - 1)\right\} \\
&= \Pr\left\{\frac{\omega_0 t + \theta}{\pi} \bmod 1 > \frac{1}{\pi}\arccos(x/\mu_0 - 1)\right\} \\
&= 1 - \frac{1}{\pi}\arccos(x/\mu_0 - 1) \\
p_\mu(x) &= \frac{d}{dx}\Pr\{\mu < x\} \\
&= \frac{d}{dx}\left[1 - \frac{1}{\pi}\arccos(x/\mu_0 - 1)\right] \\
&= \frac{1}{\pi}\left[1 - (x/\mu_0 - 1)^2\right]^{-1/2}\mu_0^{-1} \\
&= \pi^{-1}\left[x(2\mu_0 - x)\right]^{-1/2},
\end{aligned}
\tag{B.84}
$$

where $\bmod(\cdot)$ indicates the modulo operation. Substituting Eq. (B.84) into Eq. (4.37) yields

$$
\begin{aligned}
p_\tau(t) &= E[\mu]^{-1} t^{-3} p_\mu(1/t) \\
&= \mu_0^{-1} t^{-3} \pi^{-1}\left[(1/t)(2\mu_0 - 1/t)\right]^{-1/2} \\
&= (\pi\mu_0)^{-1} t^{-2} (2\mu_0 t - 1)^{-1/2}.
\end{aligned}
\tag{B.85}
$$

Prob. 4.10.3 An attempt to calculate the second moment of τ from Eq. (B.85) leads to

$$
\begin{aligned}
E[\tau^2] &= \int_{1/(2\mu_0)}^{\infty} t^2 (\pi\mu_0)^{-1} t^{-2} (2\mu_0 t - 1)^{-1/2}\, dt \\
&= (2/\pi)\int_0^{\infty} x^{-1/2}\, dx \\
&= \infty,
\end{aligned}
\tag{B.86}
$$

where we use the substitution $x \equiv t/(2\mu_0 - 1)$. Alternatively, we could employ Eqs. (4.38) and (B.84), but the numerator $E[\mu^{-1}]$ would suffer from the corresponding problem of a singularity ($\mu^{-3/2}$) near the origin.

The faulty assumption lies in the rate varying an infinitesimal amount during an interevent interval. This remains true except for t near the minima of the rate, at which times the effective rate approaches zero; here $\mu(t)/\omega_0 \gg 1$ does not hold (although $\mu_0/\omega_0 \gg 1$ still does), and the effective interevent time approaches infinity, which leads to a diverging second moment.

One way to eliminate this problem is to decrease the modulation depth a below unity ($a < 1$) by making use of the rate function

$$\mu(t) = \mu_0 \left[1 + a \cos(\omega_0 t + \theta) \right] \tag{B.87}$$

in place of Eq. (4.46). Equation (4.46) then becomes a special case of Eq. (B.87) with $a = 1$.

Prob. 4.11.1 We begin by defining

$$\begin{aligned} a_1 &= \mu_1 T \\ a_2 &= \mu_2 \tau_0 \end{aligned} \tag{B.88}$$

to simplify the notation. We note that the number of events in each secondary counting process $N_{2,k}(\tau_0)$ has a mean equal to the rate μ_2 times its duration τ_0, or a_2. Equation (4.41) then simply yields the mean number of events in $dN_3(t)$ as

$$\begin{aligned} E[N_3(T)] &= \mu_3 T \\ &= E[\mu_1] E[M_k] T \\ &= \mu_1 a_2 T \\ &= a_1 a_2. \end{aligned} \tag{B.89}$$

Since we have $T/\tau_0 \gg 1$ and $\mu_1 \tau_0 \ll 1$, we can assume that the events counted in an interval T will include full clusters of events, and effectively none of the secondary processes will span the edges of the counting window T. The secondary processes form identical segments of homogeneous Poisson point processes with identical rates for identical times. Therefore, the numbers of events in the secondary counting processes $N_{2,k}(\tau_0)$ follow a Poisson counting distribution with mean value a_2. Similarly, the numbers of events in the primary point process $dN_1(t)$ also follow a Poisson counting distribution, but with mean value a_1.

We now proceed to calculate the variance, beginning by conditioning on the number of events in the primary process

$$\begin{aligned} \mathrm{Var}&\left[N_3(T) \,\middle|\, N_1(T) = n \right] \\ &= n^2 \,\mathrm{Var}\left[N_{2,k}(\tau_0) \right] \\ &= n^2 a_2 \\ \mathrm{Var}&\left[N_3(T) \right] \\ &= \sum_{n=0}^{\infty} \mathrm{Var}\left[N_3(T) \,\middle|\, N_1(T) = n \right] \, \mathrm{Pr}\{ N_1(T) = n \} \\ &= \sum_{n=0}^{\infty} n^2 a_2 \exp(-a_1) a_1^n / n! \end{aligned}$$

$$
\begin{aligned}
&= a_2 \exp(-a_1) \sum_{n=1}^{\infty} n\, a_1^n / (n-1)! \\
&= a_2 \exp(-a_1) \sum_{m=0}^{\infty} (m+1)\, a_1^{m+1} / m! \\
&= a_2 \exp(-a_1)\, a_1 \left[\exp(a_1) + a_1 \exp(a_1)\right] \\
&= a_1 a_2 (1 + a_1) \\
&= \mu_1 \mu_2 \tau_0 T (1 + \mu_1 T). \qquad\qquad\qquad\text{(B.90)}
\end{aligned}
$$

Finally, we obtain the normalized variance as the ratio of Eqs. (B.90) and (B.89):

$$
F(T) = \frac{\mu_1 \mu_2 \tau_0 T (1 + \mu_1 T)}{\mu_1 \mu_2 \tau_0 T} = 1 + \mu_1 T. \qquad\qquad\text{(B.91)}
$$

This point process thus has a normalized variance that increases linearly with the counting time, unlike the Poisson process for which the normalized variance assumes a constant value of unity. The associated counting distribution, known as the Neyman Type-A distribution (Neyman, 1939), is discussed in Sec. 10.2.1.

Prob. 4.11.2 When the rate is zero, the primary process generates no events; with no primary events, no secondary events occur either. Therefore, transmitting a "zero" and receiving a "one" cannot occur.

To determine the probability of transmitting a "one" and receiving a "zero," we first condition on the number of events generated by the primary point process $dN_1(t)$. For each event in $dN_1(T)$, the probability of generating zero events in the associated counting process $N_{2,k}(\tau_0)$ simply becomes e^{-a_2} from the definition of the Poisson counting distribution [see Eq. (4.7)]. With $N_1(T) = n$, for $N_3(T) = 0$ we require that all n events fail to generate secondary events; employing the mutual independence of the secondary point process $dN_{2,k}(t)$ yields a probability $[e^{-a_2}]^n = e^{-na_2}$. The unconditional probability is obtained by summing over the possible conditions for $N_1(T)$:

$$
\begin{aligned}
&\Pr\{N_3(T) = 0 \,|\, \mu_1(t) = \mu_1\} \\
&= \sum_{n=0}^{\infty} \Pr\{N_3(T) = 0 \,|\, N_1(T) = n\}\, \Pr\{N_1(T) = n\} \\
&= \sum_{n=0}^{\infty} e^{-n a_2}\, e^{-a_1}\, a_1^n / n! \\
&= \exp(-a_1) \sum_{n=0}^{\infty} \left(e^{-a_2}\, a_1\right)^n / n! \\
&= \exp(-a_1)\, \exp\!\left(e^{-a_2}\, a_1\right) \\
&= \exp\!\left[-a_1 \left(1 - e^{-a_2}\right)\right]. \qquad\qquad\qquad\text{(B.92)}
\end{aligned}
$$

B.5   FRACTAL AND FRACTAL-RATE POINT PROCESSES

Prob. 5.1.1 Using a straight edge on the descending low-frequency portions of the COMPUTER and GENICULATE data presented in Fig. 5.1, we find that $\widehat{\alpha}_S \approx 0.8$ for the COMPUTER curve and $\widehat{\alpha}_S \approx 2.0$ for the GENICULATE curve. Using the same straight edge on the ascending large-counting-time portions of the COMPUTER and GENICULATE data presented in Fig. 5.2, we find that $\widehat{\alpha}_A \approx 0.7$ for the COMPUTER curve and $\widehat{\alpha}_A \approx 1.9$ for the GENICULATE curve. Thus, $\widehat{\alpha}_A + (-\widehat{\alpha}_S) \approx 0$ for both sets of data. The two exponents do not sum precisely to zero because both estimators exhibit variance, as discussed in Chapter 12.

Prob. 5.1.2 The normalized-variance scaling exponent $\widehat{\alpha}_F \approx 0.8$ for the COMPUTER data accords well with the normalized Haar-wavelet variance result $\widehat{\alpha}_A \approx 0.7$ as well as with the spectral result $\widehat{\alpha}_S \approx 0.8$. For the GENICULATE data, on the other hand, $\widehat{\alpha}_F \approx 1.0$; this lies significantly below the spectral and wavelet values $\widehat{\alpha}_S \approx 2.0$ and $\widehat{\alpha}_A \approx 1.9$, respectively, which are computed using the same data. The normalized-variance scaling exponent cannot exceed unity, in accordance with the restriction provided in Eq. (5.27). This example signals that one must use caution when making use of the normalized variance.

Another example in which the scaling exponent provided by the normalized variance can be misleading arises when nonstationarity is present. Careful analysis reveals that nonstationarity in a nonfractal point process can masquerade as fractal behavior with $\widehat{\alpha}_F = 1.0$, as shown in Eq. (5.33). A demonstration of this is provided by the spike train recorded at the lateral superior olivary complex in the mammalian auditory system (Teich et al., 1990).

Prob. 5.2.1 Using a straight edge to measure the slopes of the curves in Figs. 5.1 and 5.7, we find that $\widehat{\alpha}_S$ and $\widehat{\alpha}_{S_T}$ lie below unity for four sets of data: CORTEX, COMPUTER, COCHLEA, and RETINA; and lie above unity for the remainder.

Prob. 5.2.2 (i) Because of the decreasing power-law dependence of the rate spectrum, it appears that all seven data sets can be represented by fractal-based point processes. However, one measure alone is seldom sufficient to reach such a conclusion. (ii) In accordance with the criteria set forth in Sec. 5.5.1, data sets that exhibit $\widehat{\alpha}_S > 1$ cannot be fractal point processes. Thus, use of the point-process spectrum alone leaves four point processes as possible fractal point processes: CORTEX, COMPUTER, COCHLEA, and RETINA. (iii) Based only on the data presented in Fig. 5.1, these four processes could represent fractal renewal processes. (iv) According to the characteristics specified in Sec. 5.5.2, all seven point processes are possibly fractal-rate point processes.

Prob. 5.2.3 (i) Again, because of the decreasing power-law dependence of the rate spectrum, it appears that all seven data sets represent fractal-based point processes. However, this conclusion is strengthened substantially for all data sets because two different fractal measures exhibit the power-law dependence. (ii) The fractal exponents extracted from the interval spectrum are quite similar to those obtained from the point-process spectrum. This confirms that the same four point processes specified above could be fractal point processes: CORTEX, COMPUTER, COCHLEA, and RETINA. (iii) However, the power-law behavior in the interval spectrum tells us that these four fractal-based point processes *cannot* be fractal renewal processes. As discussed in Sec. 5.5.4, the interval spectrum of a fractal renewal process resembles that of a nonfractal renewal process, such as the homogeneous Poisson, which does not have power-law behavior. (iv) All seven point processes remain candidate fractal-rate point processes. In fact, since fractal point processes other than the fractal renewal process are encountered only infrequently, it is likely that all seven point processes are indeed fractal-rate point processes. This hypothesis can be tested in other ways, such as by examining the generalized dimension (see Prob. 5.5) and by using surrogate-data techniques (see Secs. 11.4 and 11.5).

Prob. 5.2.4 The close agreement between $\widehat{\alpha}_S$ and $\widehat{\alpha}_{S_\tau}$ suggests that $\widehat{S}_\tau(f)$ can indeed be used to provide a good estimate of the fractal exponent, at least for some fractal-based point processes (assuredly not for fractal renewal processes). The limitations of using interval-based measures as identification tools for fractal-based point processes are addressed in more detail in Sec. 12.3.1.

Prob. 5.2.5 Computation of the interval spectrum treats the data as a collection of discrete-time samples so that the maximum interval frequency $f = \frac{1}{2}$ by construction.

Prob. 5.3.1 Using a straight edge to measure the slopes of the curves in Figs. 5.2 and 5.8, we find that $\widehat{\alpha}_A$ and $\widehat{\alpha}_{A_\tau}$ both lie below unity for four sets of data, CORTEX, COMPUTER, COCHLEA, and RETINA, and above unity for the remainder. The result matches that obtained by examining Figs. 5.1 and 5.7 (see Prob. 5.2.1).

Prob. 5.3.2 (i) Because the normalized Haar-wavelet variance increases as a power-law function of the counting time, it seems that all seven data sets can be represented by fractal-based point processes. However, again, one measure alone is seldom sufficient to assure such a conclusion. (ii) Data sets that exhibit $\widehat{\alpha}_A > 1$ cannot be fractal point processes, according to Sec. 5.5.1. Using this measure alone again leaves only four point processes as possible fractal point processes: CORTEX, COMPUTER, COCHLEA, and RETINA. (iii) These same four processes could be fractal renewal processes. (iv) According to Sec. 5.5.2, all seven point processes could be fractal-rate point processes.

Prob. 5.3.3 (i) All seven data sets are likely representable by fractal-based point processes since two separate fractal measures indicate the presence of power-law behavior. (ii) The fractal exponents extracted from the normalized interval wavelet variance are similar to those from the normalized Haar-wavelet variance. This confirms that the same four point processes with $\widehat{\alpha} < 1$ could be fractal point processes. (iii) However, the presence of power-law behavior in the normalized interval wavelet variance indicates that the four possible fractal point processes indicated above cannot be fractal renewal processes. Like all interval-based measures, the normalized interval wavelet variance for a fractal renewal process resembles that for a nonfractal renewal process, which is devoid of power-law behavior. (iv) All seven point processes still remain candidates as fractal-rate point processes. Together with the information provided by the point-process and interval spectra (see Prob. 5.2), this is very likely the case.

Prob. 5.3.4 The reasonable agreement between the values of $\widehat{\alpha}_A$ and $\widehat{\alpha}_{A\tau}$ again suggests that $\widehat{A}_\tau(k)$ can be used to provide a good estimate of the fractal exponent, subject to the various limitations discussed in Sec. 12.3.1.

Prob. 5.3.5 As with the interval spectrum (see Prob. 5.2.5), the data are taken as discrete-time samples so that the minimum value of k is 2 by construction.

Prob. 5.4.1 The interevent-interval histogram, in isolation, reveals little about the nature of the underlying point process. In particular, it cannot be used to determine whether a point process represents a fractal-based point process, a fractal point process, a fractal renewal process, a fractal-rate point process, or indeed a nonfractal point process in many cases. However, it does allow elimination of the options of fractal point process and fractal renewal process if the histogram fails to exhibit a power-law region. However, the converse is not true: the presence of power-law behavior in the interevent-interval histogram does *not* definitively indicate that the point process is a fractal renewal process. This is because the ordering of the intervals can conspire to create a power-law interevent-interval histogram even though the intervals are not independent and identically distributed. Indeed, we will examine just such an example in Probs. 7.8 and 11.12.

Examining all of the histograms in Fig. 5.9 reveals that only three of the point processes exhibit what could be even remotely considered a region of power-law behavior: SYNAPSE, CORTEX, and COMPUTER. Each of these histograms displays approximately straight-line behavior over a range of interevent intervals whose ratio is about 30. While this is not sufficient to provide a convincing argument for the presence of power-law behavior, the other four point processes exhibit far less, if any at all. Of all the point processes, therefore, only these three can conceivably belong to the class of fractal renewal processes.

Prob. 5.4.2 The solution to Prob. 5.2 tells us that $\widehat{\alpha}_S$ and $\widehat{\alpha}_{S\tau}$ lie below unity for four sets of data: CORTEX, COMPUTER, COCHLEA, and RETINA. The use of the rate and interval spectra thus leave these four point processes as possible fractal point processes. However, as explained in the solution to Prob. 5.2, the presence of power-law behavior in the interval spectrum for all of these data sets indicates that these four processes *cannot* be fractal renewal point processes. These conclusions are confirmed by the results for the normalized Haar-wavelet variance and normalized interval wavelet variance, as reported in Prob. 5.3.

Combining this information, we conclude that all seven data sets very likely represent fractal-based point processes. None can be a fractal renewal process but four (CORTEX, COMPUTER, COCHLEA, and RETINA) can conceivably be fractal point processes other than fractal renewal processes. The generalized dimension D_q provides a more definitive test (see Prob. 5.5).

Prob. 5.5.1 Since we normalize by the estimated mean interval, we can use any theoretical mean values to simulate the data sets. We choose unity for convenience. Figure B.3 presents the results of these simulations. For the fractal renewal process,

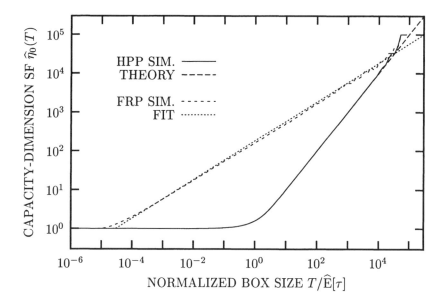

Fig. B.3 Capacity-dimension scaling functions (SF) $\widehat{\eta}_0(T)$, based on Eq. (3.74), for two simulated point processes: a homogeneous Poisson process (solid curve) and a fractal renewal process (short-dash curve). We also include the theoretical result for the former process (long-dash curve), discussed in Prob. 4.3, and a power-law fit given by $\sqrt{T/T_0}$, with $T_0 \equiv 3 \times 10^{-5}$ for this particular simulation (dotted curve).

the capacity-dimension scaling function provided in Eq. (3.74) increases as $T^\gamma = \sqrt{T}$ over a broad range of counting times T, yielding $\widehat{D}_0 = \frac{1}{2}$ and identifying the data as a fractal point process rather than a fractal-rate point process. The homogeneous Poisson process, in contrast, yields a scaling function that varies only as integer powers of the counting time: zero for $T \ll \widehat{E}[\tau]$, and unity for $T \gg \widehat{E}[\tau]$.

Prob. 5.5.2 All of the experimental curves displayed in Fig. 5.10 bear a strong similarity to those associated with the homogeneous Poisson process shown in Fig. B.3, exhibiting slopes (power-law exponents) of zero for short times and unity for longer times. None of these data sets has a fractional value of $\widehat{D}_0$. As expected for any data set, some variability exists for larger values of $T/\widehat{E}[\tau]$, where the sum extends over relatively few boxes. The principal distinction among the curves in Fig. 5.10 occurs near $T = \widehat{E}[\tau]$, where the sharpness of the transition region varies. The HEARTBEAT data set displays the sharpest transition (sharper even than the homogeneous Poisson process) whereas the SYNAPSE data set exhibits the most gradual transition.

The variation in sharpness of the curves near $T = \widehat{E}[\tau]$ stems from the local variability inherent in the corresponding point processes. For a perfectly periodic point process ($\text{Var}[\tau] = 0$) and $T \geq \tau$, the point-process events will be distributed as evenly as possible among the counting windows. This gives rise to the minimum number of windows that are devoid of events $[Z_k(T) = 0]$, which maximizes the sum in Eq. (3.74) and thereby minimizes $\eta_0(T)$. The HEARTBEAT data most closely follow such a periodic form, and therefore exhibit the lowest value of $\widehat{\eta}_0(\widehat{E}[\tau])$. The largest value of $\widehat{\eta}_0(\widehat{E}[\tau])$ obtains for the SYNAPSE data, which exhibits significant variability, departing most strongly from a periodic form. A number of measures can be used to index point-process variability as it affects $\widehat{\eta}_0(T)$. The count-based coefficient of variation evaluated at $T = \widehat{E}[\tau]$ turns out to accurately predict the behavior. Although not quite as reliable, the interval-based coefficient of variation C_τ [Eq. (3.5)] also provides a useful guide near $T = \widehat{E}[\tau]$. Visual inspection of Fig. 5.9 confirms that the SYNAPSE and HEARTBEAT data have large and small values of $\widehat{C}_\tau$, respectively. The interval-based serial correlation coefficient [Eq. (3.17)] has little relevance to $\widehat{\eta}_0(\widehat{E}[\tau])$ for the data sets examined. We confirmed this by randomly shuffling the intervals (see Sec. 11.5) of the SYNAPSE data and recomputing $\widehat{\eta}_0(T)$; the results nearly coincided with those of the original data.

Prob. 5.5.3 Except for transition regions of various widths, and variability associated with small numbers of boxes, the curves lie parallel to each other so that $\widehat{D}_q$ has essentially the same value, whatever the value of q. All curves exhibit power-law exponents of zero for short times, and unity for longer times, just as for the homogeneous Poisson process, a nonfractal object (see Prob. 4.3 and Fig. B.3). We conclude that neither a fractal point process nor a multifractal point process provides a good description for these data. A fractal-rate point process, rather, describes these data.

To determine why the curves for different values of q are parallel, rather than coincident, we examine the transition times $\widehat{T}_{Dq}$ at which the slopes change. Although $\widehat{T}_{Dq}$ lies within an order of magnitude of $\widehat{E}[\tau]$ for the SYNAPSE data displayed in Fig. 5.11, it clearly decreases as q increases: it diminishes by a factor of about 7 between $q = -1$ (solid curve) and $q = 2$ (dash-dot curve). On the other hand, for the INTERNEURON data displayed in Fig. 11.18, as well as for a simulated homogeneous Poisson process (not shown), $\widehat{T}_{Dq}$ essentially coincides with $\widehat{E}[\tau]$, regardless of q.

As part of a diagnostic process to determine the role played by interevent-interval ordering with respect to the dependence of $\widehat{T}_{Dq}$ on q, we shuffle the SYNAPSE data (see Sec. 11.5) before computing the generalized-dimension scaling functions $\widehat{\eta}_q(T)$. The values of $\widehat{T}_{Dq}$ then turn out *not* to depend on q, and the behavior of the shuffled SYNAPSE data is similar to that of the (unshuffled) INTERNEURON and simulated-Poisson-process data. This indicates that interval ordering plays an important role in determining the dependence of $\widehat{T}_{Dq}$ on q. In contrast, interval ordering plays essentially no role in determining the sharpness of the transition in $\widehat{\eta}_0(T)$ near $T = \widehat{E}[\tau]$, as discussed in Prob. 5.5.2.

Calculation of the interval serial correlation coefficient [Eq. (3.17)] proves useful in elucidating this role. As mentioned above, this measure does not prove useful for explaining the sharpness of the transition in $\widehat{\eta}_0(T)$ near $T = \widehat{E}[\tau]$. The outcome is $\widehat{\varrho}_\tau(1) \approx 0.3$ for the SYNAPSE data, while it is far smaller for the INTERNEURON data $[\widehat{\varrho}_\tau(1) \approx 0.08]$ and for the simulated homogeneous Poisson process with the same number of intervals (2644) as the SYNAPSE data $[\widehat{\varrho}_\tau(1) \approx 1/\sqrt{2644}\,]$. The decrease of $\widehat{T}_{Dq}$ with increasing q appears to be linked to the presence of correlated intervals; it also depends on the presence of heavy tails.

To confirm this connection, we generated a correlated random sequence from a homogeneous Poisson process, with unity mean rate and 2644 intervals, by adding $0.01k$ to the kth interval. If $\{\tau_k\}$ is the sequence of independent, identically distributed intervals comprising the initial process, the modified process then has the sequence $\{\tau_k + 0.01k\}$. The modified process has a large positive interval serial correlation coefficient (0.98), although it does not have nearly the range of values that the SYNAPSE data does (the modified-Poisson process has skewness and kurtosis near zero, whereas for the SYNAPSE data these values are 6 and 69, respectively). Nevertheless, the progression of $\widehat{T}_{Dq}$ with q for the correlated random sequence and for the SYNAPSE data are similar.

What is the underlying origin of this progression? As q increases, the quantity $\eta_q(T) = [\sum_k Z_k^q(T)/N(L)]^{1/(q-1)}$ depends more and more strongly on the largest value of $Z_k(T)$ over k, which $\eta_q(T)$ approaches for very large values of q. Similarly, for large-magnitude negative q, the smallest value of $Z_k(T)$ dominates $\eta_q(T)$. We now compare the expected results for a clustered point process on the one hand, with those for a nonclustered point process (such as the homogeneous Poisson process) on the other hand: the clustered process will have a greater number of larger and smaller values of $Z_k(T)$ than will the nonclustered process. For large q, this will result in a larger effective value of $\widehat{\eta}_q(T)$, whereas for large-magnitude negative q, it will result in a smaller effective value of $\eta_q(T)$. We conclude that for clustered point processes, the curves for large positive q will be shifted to the left, while those for large-magnitude negative q will be shifted to the right. This is precisely the behavior observed in the SYNAPSE data displayed in Fig. 5.11. The effect is more pronounced for intervals with heavy tails since the clustering effect is then that much stronger.

Prob. 5.6 The difference between these two expressions for $R_Z(k,T)$ lies in the evaluation of the integral. We have

$$\int_{k-1}^{k+1} x^{\alpha-1}\left(1 - |x - k|\right) dx$$

$$= \int_{k-1}^{k} x^{\alpha-1}\left[1 - (k - x)\right] dx + \int_{k}^{k+1} x^{\alpha-1}\left[1 - (x - k)\right] dx$$

$$= \int_{k-1}^{k} \left[(1 - k)x^{\alpha-1} + x^{\alpha}\right] dx + \int_{k}^{k+1} \left[(1 + k)x^{\alpha-1} - x^{\alpha}\right] dx$$

$$= (1 - k)\frac{k^{\alpha} - (k - 1)^{\alpha}}{\alpha} + \frac{k^{\alpha+1} - (k - 1)^{\alpha+1}}{\alpha + 1}$$

$$+ (1 + k) \frac{(k+1)^\alpha - k^\alpha}{\alpha} - \frac{(k+1)^{\alpha+1} - k^{\alpha+1}}{\alpha+1}$$

$$= \frac{(\alpha+1) k^\alpha - (\alpha+1) k^{\alpha+1} + (\alpha+1)(k-1)^{\alpha+1} + \alpha k^{\alpha+1}}{\alpha(\alpha+1)}$$

$$+ \frac{-\alpha(k-1)^{\alpha+1} + (\alpha+1)(k+1)^{\alpha+1} - (\alpha+1) k^\alpha}{\alpha(\alpha+1)}$$

$$+ \frac{-(\alpha+1) k^{\alpha+1} - \alpha(k+1)^{\alpha+1} + \alpha k^{\alpha+1}}{\alpha(\alpha+1)}$$

$$= \frac{(k-1)^{\alpha+1} - 2k^{\alpha+1} + (k+1)^{\alpha+1}}{\alpha(\alpha+1)}, \tag{B.93}$$

in accordance with Eq. (5.13).

Prob. 5.7 The expressions in the domains $\alpha < 1$ and $\alpha > 1$ differ only in the signs of two of the factors. These changes ensure that each factor is positive. Since the two sign changes cancel for the product, this leaves the result unchanged. For simplicity, we chose the form for $\alpha < 1$ for comparison with that for $\alpha = 1$:

$$\lim_{\alpha \to 1} \frac{\cos(\pi\alpha/2)\,\Gamma(\alpha+2)}{(2 - 2^\alpha)} = \Gamma(1+2) \lim_{\alpha \to 1} \frac{\cos(\pi\alpha/2)}{(2 - 2^\alpha)}$$

$$= 2 \lim_{\alpha \to 1} \frac{-(\pi/2)\sin(\pi\alpha/2)}{-2^\alpha \ln(2)}$$

$$= 2 \frac{(\pi/2)\sin(\pi/2)}{2\ln(2)}$$

$$= \frac{\pi}{2\ln(2)}, \tag{B.94}$$

as promised, where we have made use of l'Hôpital's rule.

Prob. 5.8 Equation (3.61) yields the normalized variance given the spectrum

$$F(T) = \frac{2}{\pi^2 \, \mathrm{E}[\mu]\, T} \int_{0+}^{\infty} S_N(f) \sin^2(\pi f T)\, f^{-2}\, df$$

$$= 1 + \frac{2}{\pi^2 \, \mathrm{E}[\mu]\, T} \int_{1/B}^{\infty} \mathrm{E}[\mu]\,(f_S/f) \sin^2(\pi f T)\, f^{-2}\, df$$

$$= 1 + \frac{2 f_S}{\pi^2 \, T} \int_{1/B}^{\infty} \sin^2(\pi f T)\, f^{-3}\, df$$

$$= 1 + \frac{2 f_S}{\pi^2 T} (\pi T)^2 \int_{\pi T/B}^{\infty} \sin^2(x)\, x^{-3}\, dx. \tag{B.95}$$

Now consider the limit

$$\lim_{\epsilon \to 0} \frac{\int_\epsilon^\infty \sin^2(x)\, x^{-3}\, dx}{\ln(1/\epsilon)} = \lim_{\epsilon \to 0} \frac{-\sin^2(\epsilon)\, \epsilon^{-3}}{-1/\epsilon}$$

$$= \lim_{\epsilon \to 0} \left[\frac{\sin(\epsilon)}{\epsilon} \right]^2$$

$$= 1. \tag{B.96}$$

Combining Eqs. (B.95) and (B.96), we obtain the approximation

$$
\begin{aligned}
F(T) &\approx 1 + \frac{2f_S}{\pi^2 T} (\pi T)^2 \ln(B/\pi T) \\
&= 1 + 2f_S T \ln(B/\pi T) \\
&\approx 1 + 2f_S T \ln(B/T),
\end{aligned}
\tag{B.97}
$$

where the last step in Eq. (B.97) obtains because in the limit $B/T \gg 1$, the term $\ln(B/T)$ dominates $\ln(\pi)$.

Substituting Eq. (B.97) into Eq. (3.41) yields the normalized Haar-wavelet variance

$$
\begin{aligned}
A(T) &= 2F(T) - F(2T) \\
&= 2 + 4f_S T \ln(B/T) - 1 - 4f_S T \ln(B/2T) \\
&= 1 + 4f_S T \left[\ln(B/T) - \ln(B/2T) \right] \\
&= 1 + 4f_S T \ln[(B/T)(2T/B)] \\
&= 1 + 4\ln(2) f_S T,
\end{aligned}
\tag{B.98}
$$

linear in T as we expect for $A(T)$ with $\alpha = 1$. Interestingly, retaining the factor of π within the logarithm in Eq. (B.97) leaves the result of Eq. (B.98) unchanged [see Eq. (A.37)].

Proceeding to the normalized coincidence rate, we use Eq. (3.55) with Eq. (B.97), yielding

$$
\begin{aligned}
G(t) &= E^2[\mu] + \frac{E[\mu]}{2} \frac{d^2}{dT^2} [TF(T)]_{T=t} \\
G(t) - E^2[\mu] &\approx \frac{E[\mu]}{2} \frac{d^2}{dt^2} \left[t + 2f_S t^2 \ln(B/t) \right] \\
&= E[\mu] f_S \frac{d^2}{dt^2} \left[t^2 \ln(B) - t^2 \ln(t) \right] \\
&= E[\mu] f_S \left[2\ln(B) - 3 - 2\ln(t) \right] \\
&= E[\mu] f_S 2\ln(B/e^{3/2} t) \\
&\approx 2E[\mu] f_S \ln(B/t),
\end{aligned}
\tag{B.99}
$$

where we ignore the factor of $e^{3/2} \doteq 4.481689$, for the same reason as in Eq. (B.97).

Finally, we substitute Eq. (B.99) into Eq. (3.54), and obtain

$$R_Z(k,T) = \int_{-T}^{T} G(kT + t)(T - |t|) \, dt$$

$$\approx \int_{-T}^{T} \left\{ \mathrm{E}^2[\mu] + 2\mathrm{E}[\mu]\, f_S \ln[B/(kT+t)] \right\} (T - |t|)\, dt$$

$$= \mathrm{E}^2[\mu]\, T^2 + 2\mathrm{E}[\mu]\, f_S \int_{-T}^{T} \ln[B/(kT+t)](T - |t|)\, dt. \quad \text{(B.100)}$$

We assume that $k \gg 1$ as before, and that the delta function in the coincidence rate does not lie within the limits of integration in Eq. (B.100). Furthermore, since $k \gg 1$, the argument of the logarithm in Eq. (B.100) changes little over the range of integration, so that we can ignore the t term in the denominator, whereupon

$$R_Z(k, T) \approx \mathrm{E}^2[\mu]\, T^2 + 2\mathrm{E}[\mu]\, f_S \int_{-T}^{T} \ln[B/(kT)]\, (T - |t|)\, dt$$

$$= \mathrm{E}^2[\mu]\, T^2 + 2\mathrm{E}[\mu]\, f_S \ln[B/(kT)]\, T^2. \quad \text{(B.101)}$$

Prob. 5.9 Calculation of the spectrum proceeds by substituting Eq. (5.50) into Eq. (3.57), which yields

$$S_N(f) = \int_{-\infty}^{\infty} G(t)\, e^{-i2\pi ft}\, df$$

$$= \mathrm{E}[\mu] + \mathrm{E}^2[\mu]\, \delta(f) + \mathrm{E}^2[\mu]\, \mathrm{sgn}(t_G)\, |t_G|^{1-\alpha}$$

$$\times \int_{-\infty}^{\infty} |t|^{\alpha-1}\, e^{-|t|/B}\, e^{-i2\pi ft}\, df. \quad \text{(B.102)}$$

Rearranging Eq. (B.102) leads to

$$S_N(f) - \mathrm{E}^2[\mu]\, \delta(f)$$

$$= \mathrm{E}[\mu] + \mathrm{sgn}(t_G)\, \mathrm{E}^2[\mu]\, |t_G|^{1-\alpha} \int_0^{\infty} t^{\alpha-1}\, e^{-t/B} \left[e^{i2\pi ft} + e^{-i2\pi ft} \right] df$$

$$= \mathrm{E}[\mu] + 2\, \mathrm{sgn}(t_G)\, \mathrm{E}^2[\mu]\, |t_G|^{1-\alpha} \, \mathrm{Re}\left\{ \int_0^{\infty} t^{\alpha-1}\, e^{-t/B}\, e^{i2\pi ft}\, df \right\}$$

$$= \mathrm{E}[\mu] + 2\, \mathrm{sgn}(t_G)\, \mathrm{E}^2[\mu]\, |t_G|^{1-\alpha} \, \mathrm{Re}\left\{ (1/B + i2\pi f)^{-\alpha} \int_0^{\infty} x^{\alpha-1} e^{-x}\, dx \right\}$$

$$= \mathrm{E}[\mu] + 2\, \mathrm{sgn}(t_G)\, \mathrm{E}^2[\mu]\, |t_G|^{1-\alpha}\, \Gamma(\alpha)\, \mathrm{Re}\left\{ (1/B + i2\pi f)^{-\alpha} \right\}. \quad \text{(B.103)}$$

Focusing now on the scaling region $f \gg 1/B$, the delta function at zero frequency can be ignored since we are concerned with positive frequencies only. Equation (B.103) then becomes

$$S_N(f) \approx \mathrm{E}[\mu] + 2\, \mathrm{sgn}(t_G)\Gamma(\alpha)\mathrm{E}^2[\mu]|t_G|^{1-\alpha}\mathrm{Re}\left\{(i2\pi f)^{-\alpha}\right\}$$

$$= \mathrm{E}[\mu] + 2\, \mathrm{sgn}(t_G)\Gamma(\alpha)\mathrm{E}^2[\mu]|t_G|^{1-\alpha}(2\pi f)^{-\alpha}\mathrm{Re}\left\{(e^{i\pi/2})^{-\alpha}\right\}$$

$$= \mathrm{E}[\mu] + 2\, \mathrm{sgn}(t_G)\Gamma(\alpha)\mathrm{E}^2[\mu]|t_G|^{1-\alpha}(2\pi f)^{-\alpha}\mathrm{Re}\left\{e^{i\alpha\pi/2}\right\}$$

$$= \mathrm{E}[\mu] + 2\, \mathrm{sgn}(t_G)\Gamma(\alpha)\mathrm{E}^2[\mu]|t_G|^{1-\alpha}(2\pi f)^{-\alpha}\cos(\alpha\pi/2)$$

$$S_N(f)/\mathrm{E}[\mu] = 1 + \underbrace{2\, \mathrm{sgn}(t_G)\Gamma(\alpha)\mathrm{E}[\mu]|t_G|}\times \underbrace{\cos(\alpha\pi/2)}\times\underbrace{(2\pi f|t_G|)^{-\alpha}}. \quad \text{(B.104)}$$

We display the last term on the right-hand-side of Eq. (B.104) as a product of three factors (delineated by braces). The first has the same sign as t_G. Its magnitude could easily exceed unity, and would lie near unity for a process with non-negligible fractal behavior. The second factor assumes negative values for $1 < \alpha < 3$, and lies within an order of magnitude of unity for $1.06377 < \alpha < 2.93623$. The final factor can be much larger than unity, given a sufficiently large value of B. The product of these three factors, then, can easily achieve large magnitudes, certainly exceeding unity. If t_G were to have a positive value, then the spectrum would lie below unity for low frequencies, which is not possible. We conclude that t_G cannot be positive.

We now compute the normalized variance $F(T)$. This proceeds from Eq. (3.52), which leads to

$$
\begin{aligned}
F(T) &= \frac{1}{\mathrm{E}[\mu]T} \int_{-T}^{T} \left\{ G(t) - \mathrm{E}^2[\mu] \right\} (T - |t|)\, dt \\
&= \frac{1}{\mathrm{E}[\mu]T} \int_{-T}^{T} \Big\{ \mathrm{E}[\mu]\, \delta(t) \\
&\quad + \mathrm{E}^2[\mu]\, \mathrm{sgn}(t_G)\, (|t/t_G|)^{\alpha-1}\, e^{-|t|/B} \Big\} (T - |t|)\, dt \\
&= 1 + 2\,\mathrm{sgn}(t_G)\, \mathrm{E}[\mu] |t_G|^{1-\alpha}\, T^{-1} \int_0^T t^{\alpha-1}\, e^{-|t|/B}\, (T - t)\, dt. \text{ (B.105)}
\end{aligned}
$$

Focusing on the range of times $0 < T \ll B$ permits us to ignore the exponential factor in Eq. (B.105), which leads to

$$
\begin{aligned}
F(T) &\approx 1 + 2\,\mathrm{sgn}(t_G)\, \mathrm{E}[\mu]\, |t_G|^{1-\alpha}\, T^{-1} \int_0^T t^{\alpha-1}(T - t)\, dt \\
&= 1 + 2\,\mathrm{sgn}(t_G)\, \mathrm{E}[\mu]\, |t_G|^{1-\alpha}\, T^{-1} \frac{T^{\alpha+1}}{\alpha(\alpha + 1)} \\
&= 1 + \mathrm{sgn}(t_G) \frac{2\mathrm{E}[\mu]\, |t_G|^{1-\alpha}}{\alpha(\alpha + 1)}\, T^{\alpha}. \qquad\qquad\text{(B.106)}
\end{aligned}
$$

By a similar argument as that provided for the spectrum, t_G cannot have a negative sign. Since t_G can be neither negative nor positive, it must be zero, indicating that the analytical form of the coincidence rate postulated in Eq. (5.50) is not suitable. The form of the normalized variance also confirms that something is awry; it exhibits an exponent that exceeds unity, which is impermissible.

Prob. 5.10 We begin with the normalized variance. The frequency range $f \ll f_S$ corresponds to the time scale $1/T \ll f_S$ or $f_S T \gg 1$; this leads to the limit $y_n \to \infty$. For large values of y_n, the y_n^2 term inside the parentheses in Eq. (5.17) dominates the other, constant terms, which we can then ignore. Substituting directly, we obtain

$$
F(T) - 1 \quad \to \quad \frac{\sqrt{8}}{\sqrt{\pi}\, y_n} \sqrt{\sqrt{y_n^2}}
$$

$$= \frac{\sqrt{8}}{\sqrt{\pi y_n}}$$

$$= \frac{2}{\pi \sqrt{fs}} T^{-1/2}, \tag{B.107}$$

in accordance with Eq. (5.20). Equation (3.41) provides $A(T)$ in terms of $F(T)$, leading to

$$A(T) = 2F(T) - F(2T)$$

$$A(T) - 1 = 2\frac{2}{\pi\sqrt{fs}} T^{-1/2} - \frac{2}{\pi\sqrt{fs}} (2T)^{-1/2}$$

$$= \frac{4}{\pi\sqrt{fs}} T^{-1/2} - \frac{2/\sqrt{2}}{\pi\sqrt{fs}} T^{-1/2}$$

$$= \frac{4 - \sqrt{2}}{\pi\sqrt{fs}} T^{-1/2}, \tag{B.108}$$

which is Eq. (5.21). Finally, we can determine the normalized coincidence rate from the normalized variance through Eq. (3.55), yielding Eq. (5.19):

$$G(t) = \mathrm{E}^2[\mu] + \frac{\mathrm{E}[\mu]}{2} \frac{d^2}{dT^2} [TF(T)]_{T=t}$$

$$G(t) - \mathrm{E}^2[\mu] \approx \frac{\mathrm{E}[\mu]}{2} \frac{2}{\pi\sqrt{fs}} \frac{d^2}{dt^2} \left(t \times t^{-1/2} \right)$$

$$= \frac{\mathrm{E}[\mu]}{\pi\sqrt{fs}} \frac{d^2}{dt^2} \left(t^{1/2} \right)$$

$$= \frac{\mathrm{E}[\mu]}{\pi\sqrt{fs}} \frac{-t^{-3/2}}{4}$$

$$= -\frac{\mathrm{E}[\mu]}{4\pi\sqrt{fs}} t^{-3/2}, \tag{B.109}$$

where we can safely ignore the delta function at zero delay because the pertinent range of times does not include this value.

Prob. 5.11 We begin by computing the normalized variance. Substituting Eq. (5.15) into Eq. (3.61) yields

$$F(T) = \frac{2}{\pi^2 \mathrm{E}[\mu] T} \int_{0+}^{\infty} S_N(f) \sin^2(\pi fT) f^{-2}\, df$$

$$= \frac{2}{\pi^2 \mathrm{E}[\mu] T} \int_{0}^{\infty} \mathrm{E}[\mu] \left[1 + \sqrt{f/fs}\, \exp(-f/fs) \right] \sin^2(\pi fT) f^{-2}\, df$$

$$= \frac{2}{\pi} \int_{0}^{\infty} \sin^2(u)\, u^{-2}\, du + \frac{4}{\pi y_n} \int_{0}^{\infty} v^{-3/2} \exp(-v) \sin^2(y_n v/2)\, dv$$

$$= 1 + \frac{4}{\pi y_n} \int_0^\infty v^{-3/2} \exp(-v) \frac{\exp(iy_n v) + \exp(-iy_n v) - 2}{-4} \, dv$$

$$= 1 - \frac{1}{\pi y_n} \int_0^\infty v^{-3/2} \left\{ \exp\left[-(1 - iy_n)v\right] \right.$$

$$\left. + \exp\left[-(1 + iy_n)v\right] - 2\exp(-v) \right\} dv$$

$$= 1 - \frac{1}{\pi y_n} \left(\sqrt{1 - iy_n} + \sqrt{1 + iy_n} - 2 \right)$$

$$\times \int_0^\infty w^{-3/2} \exp(-w) \, dw \tag{B.110}$$

$$= 1 - \frac{1}{\pi y_n} \left(2\sqrt{\frac{\sqrt{1 + y_n^2} + 1}{2}} - 2 \right) \Gamma\left(-\tfrac{1}{2}\right)$$

$$= 1 - \frac{1}{\pi y_n} \left(\sqrt{2}\sqrt{\sqrt{1 + y_n^2} + 1} - 2 \right) \left(-2\sqrt{\pi} \right)$$

$$= 1 + \frac{\sqrt{8}}{\sqrt{\pi} y_n} \left(\sqrt{\sqrt{1 + y_n^2} + 1} - \sqrt{2} \right), \tag{B.111}$$

in accordance with Eq. (5.17), where we have defined $y_n \equiv 2\pi f_S T$. The evaluation of the square roots in Eq. (B.110) follows from the relation provided in Eq. (A.38). The use of a gamma function with negative argument in Eq. (B.110) is permissible because the integrals from which it derives, in the lines above, are bounded near the origin. Since $\sin(\epsilon)$ varies as ϵ near the origin, the integral varies as $v^{-3/2+2} = \sqrt{v}$ for small arguments. Equations (5.16) and (5.18) for the coincidence rate and normalized Haar-wavelet variance follow from Eq. (5.17) in a straightforward manner, via the use of Eqs. (3.55) and (3.41), respectively.

Prob. 5.12 Data with nonstationary rates generally yield rate spectra that decay as f^{-2}. Consider a candidate data set of duration L that has a large nonstationary component, such that the rate changes significantly over the course of the data set. Estimating the spectrum involves calculating a Fourier transform of the data (for example, the familiar fast Fourier transform); this forces the data segment into a periodic form. The difference between the rate at the beginning and at the end of the signal becomes a jump discontinuity in the now-periodic signal. Such a discontinuity generally imparts a f^{-2} character to the estimated spectrum, so that $\hat{\alpha}_S = 2$, by virtue of the basic properties of the Fourier transform. To illustrate this, consider the signal $x(t) = t/L$, defined for $0 \le t < L$. Its Fourier series has the form

$$x(t) = \frac{1}{2} - \sum_{n=1}^\infty \frac{\sin(2\pi nt/L)}{n\pi}, \tag{B.112}$$

and thus decays with (discrete) frequency as $1/n$, or $1/f$ since f corresponds to n/L. The corresponding spectrum therefore decays as f^{-2}. A similar argument applies to any signal with a jump discontinuity[1]; the nonstationarity need not follow the simple linear form used for the illustration provided here.

Data sets with rate nonstationarities therefore generally yield spectra that decay as f^{-2}. Three caveats apply, however. First, the effect depends on the size of the nonstationarity. For a point process with robust fractal behavior but a very small nonstationary component, the former effect will dominate the f^{-2} term generated by the latter effect. Second, not all nonstationarities engender this effect. Consider a point process $dN(t)$ and a counting time T such that the sequence of counts $\{Z_k(T)\}$ is essentially independent, all counts have the same mean, and the variance increases with count number k. As a result of this independence, the count autocorrelation $R_Z(k, T)$ is constant for $k > 0$, so that the spectrum has no f^{-2} component despite the large nonstationarity. However, point processes that take the form of this somewhat artificial example seldom occur in practice. Finally, employing explicit window functions ameliorates the effects of nonstationarities. As discussed in Sec. 12.3.9, the use of a Hanning window reduces the deleterious effects of rate nonstationarities, causing a spectral artifact to decay as f^{-6} rather than as f^{-2}, at the cost of effectively losing half the data.

Further discussion of these issues appears in Secs. 12.3.9 and A.8.1.

Prob. 5.13.1 We begin by dividing the unit interval into 3^k equal segments, with $0 < k < m$, each of length $T = 3^{-k}$; of these, a proportion $p = 2^k/3^k$ contains $N = 2^m/2^k$ points each of the Cantor-set approximation.

By the binomial theorem, the resulting normalized variance becomes

$$
\begin{aligned}
F(T) &= N(1-p) \\
&= (2^m/2^k)(1 - 2^k/3^k) \\
&= 2^m(2^{-k} - 3^{-k}) \\
&= 2^m(T^D - T),
\end{aligned} \tag{B.113}
$$

where $D \equiv \log(2)/\log(3) \doteq 0.630930$ is the fractal dimension of the classic triadic Cantor set. Choosing different starting times for the counting durations (other than the origin) serves to reduce the normalized variance somewhat, but does not qualitatively change the argument.

Employing Eq. (3.41) leads to an equation for the normalized Haar-wavelet variance,

$$
\begin{aligned}
A(T) &= 2F(T) - F(2T) \\
&= 2^m(2T^D - 2T - 2^D T^D - 2T) \\
&= 2^m(2 - 2^D)T^D.
\end{aligned} \tag{B.114}
$$

---

[1]We could theoretically specify an infinite-extent signal, equal to the data set where it exists and to a constant value otherwise. We could then calculate the Fourier transform of this signal for all frequencies, not just for integer multiples of the inverse of the data set duration. However, this infinite-extent signal still has at least one discontinuity since the constant value cannot match both the beginning and the end of the data set. As a consequence, the f^{-2} form emerges for this case as well.

This expression does indeed vary in a power-law fashion with T, in accord with the fractal nature of the Cantor set itself.

Prob. 5.13.2 To consider the spectral properties of the modified Cantor set, we take Fourier transforms of Eq. (5.52). Convolutions become multiplications, yielding

$$\mathcal{F}\{dN_{m+1}(t)\} = 2e^{-i2\pi f/3^m} \cos(2\pi f/3^m)\,\mathcal{F}\{dN_m(t)\}\,. \qquad \text{(B.115)}$$

Since

$$\mathcal{F}\{dN_0(t)\} = \mathcal{F}\{\delta(t)\} = 1, \qquad \text{(B.116)}$$

an explicit result emerges:

$$\mathcal{F}\{dN_m(t)\} \;=\; \prod_{k=1}^{m} 2e^{-i2\pi f/3^k}\cos(2\pi f/3^k)$$

$$|\mathcal{F}\{dN_m(t)\}|^2 \;=\; 2^m \prod_{k=1}^{m} \cos^2(2\pi f/3^k). \qquad \text{(B.117)}$$

The same results have been obtained by others via different methods (Dettmann, Frankel & Taucher, 1994).

   This expression does not exhibit scaling so that the spectral density does not reveal fractal behavior in the Cantor set; in particular, Eq. (3.62) does not hold. Thus, for nonstationary collections of points, we can no longer rely on the validity of relationships that are central to the description of fractal and fractal-rate point processes.

   Nevertheless, the fractal character of the Cantor set does leave its imprint on this measure. Consider Eq. (B.117) evaluated over a range of frequencies extending from 0 to some maximum frequency $f_0 \gg 1$, where we require $f_0/3^m \ll 1$ to eliminate effects of finite m. Suppose we now change the maximum frequency to $f_0/3$. The result will resemble the unscaled version (Lowen & Teich, 1995, Fig. 6), as a result of the geometric progression of cutoff frequencies in the product; the form $|\mathcal{F}\{dN_m(t)\}|^2$ appears to contain copies of itself (Zhukovsky et al., 2001). Each factor $\cos^2(2\pi f/3^k)$ in Eq. (B.117) maps to $\cos^2(2\pi f/3^{k-1})$ in the scaled version. However, the mapping is not exact, since the factor $\cos^2(2\pi f)$ has no corresponding factor at a lower frequency; thus, plots of Eq. (B.117) are not true fractal objects.

B.6   PROCESSES BASED ON FRACTIONAL BROWNIAN MOTION

Prob. 6.1 Simply substitute the left-hand side of Eq. (6.6) into Eq. (6.1), to obtain

$$
\begin{aligned}
\mathrm{E}[B_H(as)\,B_H(at)] &= \tfrac{1}{2}\mathrm{E}[B_H^2(1)]\left(|at|^{2H} + |as|^{2H} - |at - as|^{2H}\right)\\
&= |a|^{2H}\,\tfrac{1}{2}\mathrm{E}[B_H^2(1)]\left(|t|^{2H} + |s|^{2H} - |t - s|^{2H}\right)\\
&= |a|^{2H}\,\mathrm{E}[B_H(s)\,B_H(t)], \qquad\qquad\text{(B.118)}
\end{aligned}
$$

in agreement with the right-hand side of Eq. (6.6).

Prob. 6.2 Since $B_H(t)$ belongs to the family of Gaussian processes, the difference between values of the process at different times forms a Gaussian random variable; it thus suffices to prove that the increments have identical means and variances. Since $\mathrm{E}[B_H(t)] = 0$ by construction, we need only prove that the variance of an increment depends only on the difference between the two times. We therefore have

$$
\begin{aligned}
\mathrm{E}&\Big\{[B_H(t+s) - B_H(t)]^2\Big\}\\
&= \mathrm{E}\big[B_H^2(t+s)\big] + \mathrm{E}\big[B_H^2(t)\big] - 2\mathrm{E}[B_H(t+s)\,B_H(t)]\\
&= \mathrm{E}\big[B_H^2(1)\big]\left[|t+s|^{2H} + |t|^{2H} - 2\cdot\tfrac{1}{2}\left(|t+s|^{2H} + |t|^{2H} - |s|^{2H}\right)\right]\\
&= \mathrm{E}\big[B_H^2(1)\big]\,|s|^{2H}, \qquad\qquad\text{(B.119)}
\end{aligned}
$$

which is independent of t. Thus, the increment statistics indeed depend only on the time span over which the increment exists (s) and not on the absolute time of the two values (t).

Prob. 6.3.1 Substituting Eq. (6.1) into Eq. (6.28) three times yields

$$
\begin{aligned}
\rho(s,t) &\equiv \frac{\mathrm{E}[B_H(s)\,B_H(t)]}{\left(\mathrm{E}[B_H^2(s)]\,\mathrm{E}[B_H^2(t)]\right)^{1/2}}\\
&= \frac{\tfrac{1}{2}\mathrm{E}[B_H^2(1)]\left(|t|^{2H} + |s|^{2H} - |t - s|^{2H}\right)}{\left(\mathrm{E}[B_H^2(1)]\,|s|^{2H}\,\mathrm{E}[B_H^2(1)]\,|t|^{2H}\right)^{1/2}}\\
&= \frac{|t|^{2H} + |s|^{2H} - |t - s|^{2H}}{2|s|^H\,|t|^H}\\
&= \tfrac{1}{2}\left(|t/s|^H + |s/t|^H - \left|\sqrt{|t/s|} - \mathrm{sgn}(s/t)\sqrt{|s/t|}\right|^{2H}\right), \text{(B.120)}
\end{aligned}
$$

where $\mathrm{sgn}(x)$ denotes the sign of the argument x. Equation (B.120) indeed depends only on the ratio of the two times, s/t.

Prob. 6.3.2 For the special case $s = -t$, we make this substitution into Eq. (B.120) to obtain

$$
\begin{aligned}
\rho(-t,t) &= \tfrac{1}{2}\left(|-1|^{H} + |-1|^{H} - \left|\sqrt{1} - \mathrm{sgn}(-1)\sqrt{1}\,\right|^{2H}\right) \\
&= \tfrac{1}{2}\left(1 + 1 - |1 - (-1)|^{2H}\right) \\
&= 1 - 2^{2H-1}.
\end{aligned}
\tag{B.121}
$$

To make $B_H(t)$ and $B_H(-t)$ independent, it suffices to make them uncorrelated since first- and second-order statistics determine a Gaussian process, and $B_H(t)$ has zero mean. So we require that the autocorrelation coefficient assume a value of zero

$$
\begin{aligned}
\rho(-t,t) &= 0 \\
1 - 2^{2H-1} &= 0 \\
0 &= 2H - 1 \\
H &= \tfrac{1}{2},
\end{aligned}
\tag{B.122}
$$

so that only for regular Brownian motion ($H = \tfrac{1}{2}$) do we have $B_H(t)$ and $B_H(-t)$ independent.

Prob. 6.4 The simplest method of proving this result invokes Eq. (2.15) directly, and leads to

$$
E\big[B^2(1)\big] = \min(1,1) = 1.
\tag{B.123}
$$

However, substituting $H = \tfrac{1}{2}$ into Eq. (6.2) yields the same result, as we now show. We first rewrite Eq. (6.2), making use of a property of the Gamma function presented in Eq. (5.8). Substituting $x = 2H$ in Eq. (5.8), and in turn substituting this into Eq. (6.2), leads to

$$
\begin{aligned}
E[B_H^2(1)] &= \Gamma(1 - 2H)\cos(\pi H)/(\pi H) \\
&= \frac{\pi}{\sin(2\pi H)\,\Gamma(2H)} \frac{\cos(\pi H)}{(\pi H)} \\
&= \frac{\pi}{2\sin(\pi H)\,\cos(\pi H)\,\Gamma(2H)} \frac{\cos(\pi H)}{(\pi H)} \\
&= \big[2H\sin(\pi H)\,\Gamma(2H)\big]^{-1} \\
E[B_{1/2}^2(1)] &= 1,
\end{aligned}
\tag{B.124}
$$

as it must.

Prob. 6.5.1 We begin by defining

$$
\begin{aligned}
x_0 &\equiv B(k\tau_0) \\
x_1 &\equiv B[(k+1)\tau_0] \\
x_h &\equiv B[(k+\tfrac{1}{2})\tau_0].
\end{aligned}
\tag{B.125}
$$

We then have

$$x_h = a\mathcal{N}(0, 1) + b + \tfrac{1}{2}(x_0 + x_1) \tag{B.126}$$

and it remains to find the constants a and b. Taking expectations of both sides of Eq. (B.126) immediately yields $b = 0$, so we may remove it from further consideration. Accounting for this result, we double, square, and take expectations of both sides of Eq. (B.126), to obtain

$$\begin{aligned} 4\mathrm{E}[x_h^2] &= 4a^2\,\mathrm{E}[\mathcal{N}^2(0,1)] + \mathrm{E}[x_0^2] + \mathrm{E}[x_1^2] \\ &\quad + 2\mathrm{E}[x_0\,x_1] + 2\mathrm{E}[x_0\,\mathcal{N}(0,1)] + 2\mathrm{E}[x_1\,\mathcal{N}(0,1)]. \end{aligned} \tag{B.127}$$

Since we assume that $\mathcal{N}(0, 1)$ is independent of both x_0 and x_1, we can replace the expectation of the product with the product of the expectations in each case. But since at least one of the quantities has zero mean (actually, both do), the product of the expectations assumes a value of zero. Thus, the last two terms in Eq. (B.127) vanish. The first term on the right-hand side of Eq. (B.127) involves the mean square of $\mathcal{N}(0, 1)$, which assumes a value of unity by construction. Eq. (B.127) therefore becomes

$$4\mathrm{E}[x_h^2] = 4a^2 + \mathrm{E}[x_0^2] + \mathrm{E}[x_1^2] + 2\mathrm{E}[x_0\,x_1]. \tag{B.128}$$

For the remaining terms, we employ Eq. (2.15). Assuming $k \geq 0$ for positive times, and substituting all of these simplifications into Eq. (B.128) yields

$$\begin{aligned} 4[(k + \tfrac{1}{2})\tau_0] &= 4a^2 + k\tau_0 + (k + 1)\tau_0 + 2k\tau_0 \\ (4k + 2)\tau_0 &= 4a^2 + (4k + 1)\tau_0 \\ \tau_0 &= 4a^2 \\ a &= \sqrt{\tau_0}/2, \end{aligned} \tag{B.129}$$

which, in turn, we substitute into Eq. (B.126) to provide the final answer.

Prob. 6.5.2 This method is suitable for regular Brownian motion since $B(t)$ has independent increments; in particular, $B[(k + \tfrac{1}{2})\tau_0] - B(k\tau_0)$ and $B[(n + \tfrac{1}{2})\tau_0] - B(n\tau_0)$ are independent for $k \neq n$.

Prob. 6.5.3 Similarly, the method fails for fractional Brownian motion with $H \neq \tfrac{1}{2}$ because $B_H(t)$ does not have independent increments; in particular, $B_H[(k + \tfrac{1}{2})\tau_0] - B_H(k\tau_0)$ and $B_H[(n + \tfrac{1}{2})\tau_0] - B_H(n\tau_0)$ are *not* independent for $H \neq \tfrac{1}{2}$, regardless of the value of k or n.

Prob. 6.6.1 If $dN_1(t)$ has a spectrum that decays as $\sim f^{-\alpha_X}$ where $\tfrac{1}{2} < \alpha_X < 1$ over some large range of frequencies, then Eq. (5.44) indicates that $dN_1(t)$ must have a coincidence rate that decays as $\sim t^{\alpha_X - 1}$. Equation (4.24) reveals that the autocorrelation of the rates $X_k(t)$ must, over a wide range of delay times t and for some cutoff time t_1, have a similar form, such as

$$R_X(t) \approx (t/t_1)^{\alpha_X - 1}. \tag{B.130}$$

Subtracting the mean does not change Eq. (B.130).

Direct substitution of Eq. (B.130) into Eq. (6.16) yields

$$
\begin{aligned}
R_\mu(t) &= 2M\, R_X^2(t) + \mathrm{E}^2[\mu] \\
&\approx 2M(t/t_1)^{2\alpha_X - 2} + M^2\, \mathrm{Var}^2[X] \\
&= c_1\left[1 + (t/t_2)^{2\alpha_X - 2}\right],
\end{aligned}
\tag{B.131}
$$

where c_1 and t_2 are constants implicitly defined by Eq. (B.131). Equation (B.131) shows that $\mu(t)$ has a different exponent than $X(t)$; since $2\alpha_X - 2 = \alpha_\mu - 1$, we have $\alpha_\mu = 2\alpha_X - 1$. Finally, using Eqs. (4.24) and (5.44) once again, we see that $dN_2(t)$ exhibits behavior of the form

$$
S_{N_2}(f) \sim f^{-\alpha_\mu} = f^{1 - 2\alpha_X},
\tag{B.132}
$$

which stands in contrast to the $S_{N_1}(f) \sim f^{-\alpha_X}$ behavior attendant to $dN_1(t)$. Since $\frac{1}{2} < \alpha_X < 1$, we have $0 < \alpha_\mu < 1$, and Eq. (5.44) indeed applies.

Prob. 6.6.2 A similar argument to that used for Prob. 6.6.1 shows that $R_X(t)$ must vary as $\sim t^{\alpha_X - 1}$ as before. But $R_X(t)$ approaches a nonzero value for large delay times t, so it must have a form similar to

$$
R_X(t) \approx \mathrm{E}^2[X]\left[1 + (t/t_1)^{\alpha_X - 1}\right],
\tag{B.133}
$$

for some cutoff time t_1. Direct substitution of Eq. (B.133) into Eq. (6.16) yields

$$
\begin{aligned}
R_\mu(t) &= 2M\, R_X^2(t) + \mathrm{E}^2[\mu] \\
&\approx 2M\left\{\mathrm{E}^2[X]\left[1 + (t/t_1)^{\alpha_X - 1}\right]\right\}^2 + M^2\, \mathrm{E}^2[X^2] \\
&= 2M\, \mathrm{E}^4[X]\left[1 + 2(t/t_1)^{\alpha_X - 1} + (t/t_1)^{2\alpha_X - 2}\right] \\
&\quad + M^2\, \mathrm{E}^2[X^2].
\end{aligned}
\tag{B.134}
$$

The important times here lie in the range $t/t_1 \gg 1$, since the tail of the autocorrelation determines the fractal behavior of the process. The autocorrelation behavior for $t/t_1 \ll 1$ cannot follow the form above, since that would lead to a large autocorrelation that exceeds the mean square, an impossibility. Furthermore, $t/t_1 \ll 1$ corresponds to high frequencies, where the constant term $\mathrm{E}[X]$ dominates the spectrum of $dN_1(t)$, making other, subtle contributions to the spectrum irrelevant.

Taking the limit $t/t_1 \gg 1$, we see that the first two terms within the square brackets in Eq. (B.134) dominate the last, which we therefore neglect. Continuing, we obtain

$$
\begin{aligned}
R_\mu(t) &\approx 2M\, \mathrm{E}^4[X]\left[1 + 2(t/t_G)^{\alpha_X - 1}\right] + M^2\, \mathrm{E}^2[X^2] \\
&= c_2\left[1 + c_3(t/t_1)^{\alpha_X - 1}\right] \\
&= c_2\left[1 + (t/t_3)^{\alpha_X - 1}\right],
\end{aligned}
\tag{B.135}
$$

where c_2, c_3, and t_3 are constants implicitly defined by Eq. (B.135).

Equation (B.135) shows that $\mu(t)$ has the same exponent as $X(t)$ for large mean values $E[X]$. Equations (4.24) and (5.44) thus reveal that the spectrum for $dN_2(t)$ takes the form

$$S_{N2}(f) \sim f^{-\alpha_\mu} = f^{-\alpha_X}, \tag{B.136}$$

which is the same as that for $dN_1(t)$.

Prob. 6.7 The exponential transform of a Gaussian process gives rise to a lognormal process with a mean given by Eq. (6.21) with $n = 1$:

$$E[\mu] = \exp\left(E[X] + \text{Var}[X]/2\right). \tag{B.137}$$

As shown in Eq. (3.59), this value $E[\mu]$ becomes the high-frequency asymptote of the point-process spectrum. Turning now to the autocorrelation, Eq. (6.22) provides

$$\begin{aligned}
R_\mu(t) &\approx E^2[\mu]\, \exp\left\{R_X(t) - E^2[X]\right\} \\
&= E^2[\mu]\, \exp\left\{E^2[X] + c\ln(t_0/|t|) - E^2[X]\right\} \\
&= E^2[\mu]\, \exp\left[c\ln(t_0/|t|)\right] \\
&= E^2[\mu]\, (|t|/t_0)^{-c}. \tag{B.138}
\end{aligned}$$

However, since the limit of the autocorrelation for large arguments must approach the square of the mean, we write

$$R_\mu(t) \approx E^2[\mu]\left[1 + (|t|/t_0)^{-c}\right], \tag{B.139}$$

where we acknowledge the approximate nature of Eq. (6.29).

Taking the Fourier transform of the autocorrelation yields

$$\begin{aligned}
S_\mu(f) &\approx \int_{-\infty}^{\infty} R_\mu(t)\, \exp(-i2\pi ft)\, dt \\
&= \int_{-\infty}^{\infty} E^2[\mu]\left[1 + (|t|/t_0)^{-c}\right]\exp(-i2\pi ft)\, dt \\
&= E^2[\mu]\,\delta(f) \\
&\quad + 2E^2[\mu]\, t_0^c\, \cos[\pi(c-1)/2]\,(2\pi f)^{c-1}\,\Gamma(1-c). \tag{B.140}
\end{aligned}$$

Combining Eqs. (B.140) and (4.25) yields

$$\begin{aligned}
S_N(f) &= S_\mu(f) + E[\mu] \\
&= E^2[\mu]\,\delta(f) \\
&\quad + E[\mu]\left\{1 + 2\cos[\pi(c-1)/2]\,\Gamma(1-c)\,(2\pi)^{c-1}\,E[\mu]\,t_0^c\, f^{c-1}\right\} \\
&= E^2[\mu]\,\delta(f) + E[\mu]\left[1 + (f/f_S)^{-\alpha}\right], \tag{B.141}
\end{aligned}$$

where we make the connection to Eq. (5.44a) by identifying

$$\begin{aligned}
\alpha &= 1 - c \\
(2\pi f_S)^\alpha &= 2\cos(\pi\alpha/2)\,\Gamma(\alpha)\,E[\mu]\,t_0^{1-\alpha}. \tag{B.142}
\end{aligned}$$

Prob. 6.8.1 For a fixed membrane voltage V, spontaneous vesicular exocytosis would be expected to lack memory; knowledge of previous event occurrence times would yield no additional information about the future, beyond that provided by the mean rate μ of the exocytic events. This lack of memory dictates that the homogeneous Poisson process characterizes the exocytic events.

Prob. 6.8.2 Different *fixed* membrane voltages V would lead to spontaneous exocytic patterns different only in their mean rates; the homogeneous Poisson process would serve as a suitable model for each of these sequences of events, whatever the voltage. The Poisson rate for each cell would be exponentially related to its cellular membrane voltage, as prescribed by Eq. (6.30).

Prob. 6.8.3 Since the resting membrane voltage V exhibits $1/f$-type noise with a Gaussian amplitude distribution, we accommodate the attendant voltage fluctuations by replacing V by the fractal Gaussian process $V(t)$ in Eq. (6.30):

$$\mu(t) = A \exp\{-[E_A - qV(t)]/RT\}. \tag{B.143}$$

Because the rate $\mu(t)$ is the exponential transform of the Gaussian process $V(t)$, Eq. (B.143) dictates that the rate behave as fractal lognormal noise. A model for the discrete exocytic sequence therefore belongs to the Poisson-process family, but is governed by a rate that varies in accordance with lognormal statistics with a fractal spectrum. This leads to the fractal-lognormal-noise-driven Poisson process described in Sec. 6.5.

We connect the results developed here with those provided in Sec. 6.5 by defining $X(t)$ as a normalized[2] version of the membrane voltage $V(t)$:

$$X(t) \equiv \ln(A) + [qV(t) - E_A]/RT, \tag{B.144}$$

so that the rate in Eq. (B.143) can become

$$\mu(t) = \exp[X(t)]. \tag{B.145}$$

Since it is a linear transform of the Gaussian membrane voltage $V(t)$, the auxiliary process $X(t)$ also behaves as a Gaussian process, with associated mean, variance, and autocorrelation given by:

$$
\begin{aligned}
E[X] &= \ln(A) + (q E[V] - E_A)/RT \\
\text{Var}[X] &= (q/RT)^2 \text{Var}[V] \\
R_X(t) &= E^2[X] + (q/RT)^2 \{R_V(t) - E^2[V]\}.
\end{aligned}
\tag{B.146}
$$

The quantities in Eq. (B.146), when substituted into Eqs. (6.20)–(6.25), yield results for the point process that describes the exocytic events, and its rate, in terms of the statistics of the membrane voltage (Lowen et al., 1997a,b).

---

[2] The process $X(t)$ has mixed units. We use this somewhat unusual approach to simplify notation; making use of a dimensionless process for $X(t)$ yields the same results, albeit with more cumbersome algebra.

B.7 FRACTAL RENEWAL PROCESSES

Prob. 7.1 We can achieve $\alpha = \frac{1}{2}$ by employing either $\gamma = \frac{1}{2}$ or $\gamma = 2 - \frac{1}{2} = \frac{3}{2}$. For the former, we make use of Eq. (5.3) together with the first line of Eq. (7.8), which yield

$$
\begin{aligned}
\mathrm{E}[\mu]\left[1 + (f/f_S)^{-\gamma}\right] &= \mathrm{E}[\mu]\left\{1 + 2\left[\Gamma(1-\gamma)\right]^{-1}\cos(\pi\gamma/2)\,(2\pi f A)^{-\gamma}\right\} \\
(f/f_S)^{-1/2} &= 2\left[\Gamma(\tfrac{1}{2})\right]^{-1}\cos(\pi/4)\,(2\pi f A)^{-1/2} \\
2\pi f_S A &= \left\{2\left[\sqrt{\pi}\,\right]^{-1}2^{-1/2}\right\}^2 \\
A &= \pi^{-2}/f_S \\
&\doteq 0.0101321 \text{ sec.}
\end{aligned}
\tag{B.147}
$$

Equation (7.2) then provides

$$
\begin{aligned}
1/\mathrm{E}[\mu] &= \frac{1/2}{1 - 1/2}\,(A/B)^{1/2}\,B\,\frac{1 - (A/B)^{1-1/2}}{1 - (A/B)^{1/2}} \\
&= \sqrt{AB} \\
&= \sqrt{(\pi^{-2}/f_S) \times 10^6(\pi^{-2}/f_S)} \\
\mathrm{E}[\mu] &= 10^{-3}\pi^2 f_S \\
&\doteq 0.0986960 \text{ events/sec.}
\end{aligned}
\tag{B.148}
$$

The calculation for $\gamma = \frac{3}{2}$ proceeds along the same lines. Using the third line of Eq. (7.8) yields

$$
\begin{aligned}
2\pi f_S A &= \left\{2(\tfrac{3}{2})^{-2}(\tfrac{3}{2} - 1)\,\Gamma(2 - \tfrac{3}{2})\left[-\cos(3\pi/4)\right]\right\}^2 \\
&= \left\{(\tfrac{4}{9})\,\Gamma(\tfrac{1}{2})\,2^{-1/2}\right\}^2 \\
A &= (4/81)/f_S \\
&\doteq 0.00493827 \text{ sec.}
\end{aligned}
\tag{B.149}
$$

For this value of γ, Eq. (7.2) returns

$$
\begin{aligned}
1/\mathrm{E}[\mu] &= \frac{3/2}{1 - 3/2}\,(A/B)^{3/2}\,B\,\frac{1 - (A/B)^{1-3/2}}{1 - (A/B)^{3/2}} \\
&= \frac{3A}{1 + \sqrt{A/B} + A/B} \\
&= \frac{3(4/81)/f_S}{1 + 10^{-3} + 10^{-6}} \\
\mathrm{E}[\mu] &\doteq 67.5676 \text{ events/sec.}
\end{aligned}
\tag{B.150}
$$

Prob. 7.2 For the range of times $A \ll t \ll B$, both exponentials in Eq. (7.5) have small arguments and therefore approach unity. Furthermore, we have $K_\gamma(z) \approx 2^{\gamma-1} \Gamma(\gamma) z^{-\gamma}$, valid for small values of the argument z (Gradshteyn & Ryzhik, 1994, Secs. 8.445 and 8.485). Employing these simplifications in Eq. (7.5) yields

$$
\begin{aligned}
p_\tau(t) &= \frac{(AB)^{\gamma/2}}{2K_\gamma\left(2\sqrt{A/B}\right)} e^{-A/t} e^{-t/B} t^{-(\gamma+1)} \\
&\approx \frac{(AB)^{\gamma/2}}{2\left(2^{\gamma-1}\right) \Gamma(\gamma) \left(2\sqrt{A/B}\right)^{-\gamma}} t^{-(\gamma+1)} \\
&= \frac{(AB)^{\gamma/2}}{\Gamma(\gamma) (A/B)^{-\gamma/2}} t^{-(\gamma+1)} \\
&= \frac{A^\gamma}{\Gamma(\gamma)} t^{-(\gamma+1)} \\
&= \gamma A^\gamma \left[\Gamma(1+\gamma)\right]^{-1} t^{-(\gamma+1)},
\end{aligned}
\tag{B.151}
$$

which, for $A \ll B$, differs from Eq. (7.1) only by the factor $\left[\Gamma(1+\gamma)\right]^{-1}$. By scaling arguments, the expression corresponding to Eq. (7.4) must also differ by this same factor, so that

$$
1 - \phi_\tau(\omega) \approx \frac{\Gamma(1-\gamma)}{\Gamma(1+\gamma)} (i\omega A)^\gamma.
\tag{B.152}
$$

Prob. 7.3 Substituting Eq. (7.4) into Eq. (4.16) yields

$$
\begin{aligned}
S_N(f)/E[\mu] &= \mathrm{Re}\left\{\frac{1+\phi_\tau(2\pi f)}{1-\phi_\tau(2\pi f)}\right\}, \quad B^{-1} \ll f \ll A^{-1} \\
&\approx \mathrm{Re}\left\{\frac{1+1-\Gamma(1-\gamma)(i2\pi f A)^\gamma}{1-\left[1-\Gamma(1-\gamma)(i2\pi f A)^\gamma\right]}\right\} \\
&\approx \mathrm{Re}\left\{\frac{2}{\Gamma(1-\gamma)(2\pi f A)^\gamma} \left[\exp(i\pi/2)\right]^{-\gamma}\right\} \\
&= 2\left[\Gamma(1-\gamma)\right]^{-1} (2\pi f A)^{-\gamma} \mathrm{Re}\{\exp(-i\gamma\pi/2)\} \\
&= 2\left[\Gamma(1-\gamma)\right]^{-1} \cos(\gamma\pi/2) (2\pi f A)^{-\gamma},
\end{aligned}
\tag{B.153}
$$

where we have made use of Eq. (A.35).

Prob. 7.4.1 For $1 < \gamma < 2$ and $A \ll B$, we can take the limit $B \to \infty$ in Eq. (7.1) before calculating the mean interevent interval. The generalized Pareto result emerges:

$$
\begin{aligned}
1/E[\mu] &\approx \gamma A^\gamma \int_A^\infty t^{-\gamma} dt \\
&\approx \gamma(\gamma-1)^{-1} A
\end{aligned}
$$

$$A \approx \left(1 - \gamma^{-1}\right) / \mathrm{E}[\mu]. \tag{B.154}$$

Recall that the expression to the right of the large left brace in Eq. (7.8) assumes a value of unity at the fractal onset frequency f_S. Combining that result with Eq. (B.154) yields

$$2\gamma^{-2}\left(\gamma - 1\right)\Gamma(2 - \gamma)\left[-\cos(\pi\gamma/2)\right]\left[2\pi f_S\left(1 - \gamma^{-1}\right)/\mathrm{E}[\mu]\right]^{\gamma-2} \approx 1, \tag{B.155}$$

which relates $\mathrm{E}[\mu]$ to f_S, as promised.

Prob. 7.4.2 For $0 < \alpha < 1$, we again begin with the mean interevent time; Eq. (7.2) yields

$$
\begin{aligned}
1/\mathrm{E}[\mu] &= \frac{\gamma}{1 - \gamma}\,(A/B)^\gamma\,B\,\frac{1 - (A/B)^{1-\gamma}}{1 - (A/B)^\gamma} \\
&\approx \frac{\gamma}{1 - \gamma}\,(A/B)^\gamma\,B \\
&= \gamma(1 - \gamma)^{-1}\,(B/A)^{1-\gamma}\,A \\
1/\mathrm{E}[\mu] &\gg A \\
\mathrm{E}[\mu] &\ll 1/A. \tag{B.156}
\end{aligned}
$$

Ignoring factors of the order of unity in Eq. (7.8) we obtain

$$
\begin{aligned}
1 &= 2\gamma^{-2}\left(\gamma - 1\right)\Gamma(2 - \gamma)\left[-\cos(\pi\gamma/2)\right](2\pi f_S A)^{\gamma-2} \\
1 &\approx (2\pi f_S A)^{\gamma-2} \\
1 &\approx 2\pi f_S A \\
f_S &\approx 1/A. \tag{B.157}
\end{aligned}
$$

Finally, combining Eqs. (B.156) and (B.157) leads to the inequality

$$\mathrm{E}[\mu] \ll f_S. \tag{B.158}$$

Prob. 7.5 We immediately take the limit $B \to \infty$ in Eq. (7.1), as in Prob. 7.4.1. Equation (3.6) then provides

$$1 - \phi_\tau(\omega) \approx \gamma A^\gamma \int_A^\infty \left(1 - e^{-i\omega t}\right) t^{-(\gamma+1)}\,dt. \tag{B.159}$$

We proceed to integrate by parts, with $U = 1 - e^{-i\omega t}$ and $dV = \gamma t^{-(\gamma+1)}\,dt$, which leads to

$$
\begin{aligned}
1 - \phi_\tau(\omega) &\approx A^\gamma\,i\omega \int_A^\infty e^{-i\omega t}\,t^{-\gamma}\,dt \\
&= A^\gamma\,i\omega \int_{iA\omega}^\infty e^{-x}\,x^{-\gamma}\,dx\,(i\omega)^{\gamma-1} \\
&\approx (i\omega A)^\gamma \int_0^\infty e^{-x}\,x^{-\gamma}\,dx \\
&= \Gamma(1 - \gamma)\,(i\omega A)^\gamma. \tag{B.160}
\end{aligned}
$$

Prob. 7.6 For $\gamma = \frac{1}{2}$ we have

$$E[\tau] = \sqrt{AB} = \sqrt{10^{-3} \times 10^3} = 1; \qquad (B.161)$$

with a simulation duration of $L = 10^8$ the expected number of events is $E[N(L)] = 10^8$. For $\gamma = \frac{3}{2}$, on the other hand, the mean interevent time becomes

$$E[\tau] = \frac{3A}{1 + \sqrt{A/B} + A/B} = \frac{0.003}{1.001001} \qquad (B.162)$$

so that

$$E[N(L)] = 10^8 \times 1.001001/0.003 = 3.33667 \times 10^{10}. \qquad (B.163)$$

Simulating this fractal renewal point process would thus take 333.667 times as long as for $\gamma = \frac{1}{2}$. Hence, the total computation time would lie in the neighborhood of 333.667 days, or about a year. Moreover, at four bytes per interval the expected file size would reach 133.4668 GB, which greatly exceeds the memory of the computer employed to carry out the calculations. This would necessitate reading in the entire file for each value of the counting time T, which would, in turn, greatly increase the calculation time. An educated guess would then put the computation time in excess of 20 years. Decreasing the duration L serves to reduce simulation times at the expense of accuracy in the resulting estimates. For purposes of illustrating fractal behavior, rather than producing smooth and accurate curves, a duration $E[N(L)] = 10^6$ suffices, whereupon the estimated simulation time decreases to about 15 minutes (see Prob. 12.8).

Prob. 7.7 Since the overall error process has independent interevent intervals, we consider renewal-point-process models. It therefore remains only to find the form of the inter-error probability density function. Over long time scales the homogeneous Poisson process dominates, so that the probability density has an exponential tail. The long-time cutoff is thus $B = \tau_{\text{HPP}}$. For time scales shorter than τ_{clk}, additional events do not register, thereby imposing on the process a practical short-time cutoff of $A = \tau_{\text{clk}}$. For other time scales, however, the inter-error probability density function assumes a power-law form. Thus, Eq. (7.5) provides a good representation for the fractal renewal point process that characterizes the error events.

Prob. 7.8.1 As shown in Fig. B.4, for the upper and lower cutoffs we have $\widehat{B} \approx 5$ sec and $\widehat{A} \approx 20$ msec, respectively. For intermediate values of the interevent interval t, the estimated density function decreases roughly as a straight line on this doubly logarithmic plot. This indicates a power-law dependence on the interevent interval t; we estimate the exponent to be $-\widehat{\gamma} \approx -0.8$. Based on Eq. (7.9), a fractal renewal point process with $0 < \gamma < 1$ should exhibit $\alpha = \gamma$.

Prob. 7.8.2 The dashed curve shown in Fig. B.5 provides a fit of the power-law portion of Eq. (5.44c) to the normalized-Haar-wavelet-variance data. Relative refractoriness in the neural-spike-generation mechanism gives rise to the dip below unity of the

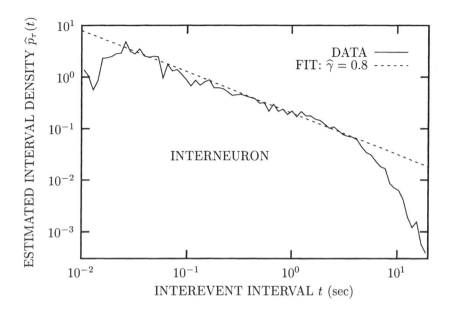

Fig. B.4 A decaying power-law function provides a good fit to the estimated interevent-interval density for the action-potential sequence recorded from the descending contralateral movement detector, a visual-system INTERNEURON in the locust (Turcott et al., 1995, Fig. 2, pp. 261–262, cell ADA062). The normalized Haar-wavelet variance for these same data is displayed in Fig. B.5.

normalized Haar-wavelet variance; it is also responsible for the soft lower cutoff A in the interevent-interval density (see Sec. 11.2.4). The curve increases with counting time T roughly as a straight line; estimating the slope and intercept of the curve gives $\widehat{\alpha}_A \approx 0.14$ and $\widehat{T}_A \approx 0.07$. Since the values of γ and α obtained from Figs. B.4 and B.5 differ so greatly, the data must not derive from a fractal renewal point process.

Prob. 7.8.3 The theoretical fractal-renewal-process power-law-decaying interevent-interval density fits the data well. In spite of this, the fractal renewal process is not a suitable model for characterizing this spike train, as will be definitively demonstrated in Prob. 11.12.

Prob. 7.9 There are four reasons why extensive data are not available for this process: First, a proper analysis requires comparative studies, which involve sequencing related proteins in a large number of different, related organisms, whereas most sequencing efforts (such as the Human Genome Project) focus on sequencing the entire genome of relatively few organisms. Second, other methods for estimating divergence dates yield imprecise results, making calibration of the model difficult. Third, evolution rates average about 10^{-7} substitutions per year for most proteins; in most cases, therefore, relatively few changes occur, making detailed calculation of the sequence-change

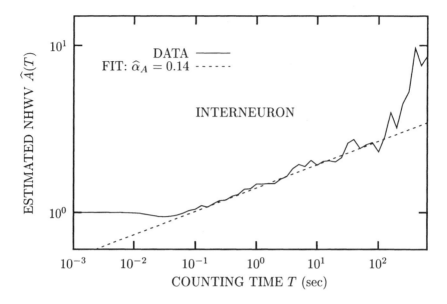

Fig. B.5 An increasing power-law function provides a good fit to the estimated normalized Haar-wavelet variance for the action-potential sequence recorded from the descending contralateral movement detector, a visual-system INTERNEURON in the locust (Turcott et al., 1995, Fig. 2, pp. 261–262, cell ADA062). However, the estimated exponent is in serious disagreement with that obtained from the estimated interevent-interval density shown in Fig. B.4.

statistics difficult. Finally, very few ancestral DNA samples exist, so substitution counts depend directly on the accuracy of cladistics, which is a particular method of phylogenetic analysis (Bickel, 2000). Nevertheless, there are sufficient data to rule out the homogeneous Poisson process, since it predicts fluctuations in the estimated substitution rate that are significantly smaller than those observed in real sequences (Gillespie, 1994). In fact, the data suggest a fractal model (West & Bickel, 1998). A variety of empirical fractal models exist for characterizing molecular evolution, including anomalous diffusion (West & Bickel, 1998), the fractal-shot-noise-driven Poisson process described in Chapter 10 (Bickel & West, 1998a), fractal Gaussian processes (Bickel, 2000), and fractal renewal point processes (Bickel & West, 1998b). The last proves simplest from a conceptual perspective so Occam's razor suggests its use, although the fractal Gaussian process is more tractable from an analytical point of view.

Prob. 7.10.1 Setting the integral of Eq. (7.26) to unity yields the normalization constant:

$$c \equiv E_0^{-1} \left[\exp(-E_L/E_0) - \exp(-E_H/E_0) \right]^{-1}. \qquad \text{(B.164)}$$

Prob. 7.10.2 If we define characteristic time cutoffs $A \equiv \tau_0 \exp(E_L/\kappa T)$ and $B \equiv \tau_0 \exp(E_H/\kappa T)$, and the power-law exponent $\gamma \equiv \kappa T/E_0$, the mean waiting time $q(E)$ is characterized by the power-law density

$$p_{q(E)}(s) = \frac{\gamma}{A^{-\gamma} - B^{-\gamma}} \times \begin{cases} s^{-(\gamma+1)} & A < s < B \\ 0 & \text{otherwise.} \end{cases} \tag{B.165}$$

Since the times spent in successive traps are independent, the fractal renewal process follows.

Prob. 7.10.3 Integrating over the conditioning yields

$$\begin{aligned} p_\tau(t) &= \int p_\tau[t|q(E) = s] \, p_{q(E)}(s) \, ds \\ &= \frac{\gamma}{A^{-\gamma} - B^{-\gamma}} \, t^{-(\gamma+1)} \int_{t/B}^{t/A} x^\gamma \, e^{-x} \, dx. \end{aligned} \tag{B.166}$$

For the case $A \ll t \ll B$, the limits in the integral of Eq. (B.166) can be approximated by zero and infinity, which provides

$$\begin{aligned} \lim_{\substack{A/t \to 0 \\ B/t \to \infty}} p_\tau(t) &= \frac{\gamma}{A^{-\gamma} - B^{-\gamma}} \, t^{-(\gamma+1)} \int_0^\infty x^\gamma \, e^{-x} \, dx \\ &= \frac{\gamma}{A^{-\gamma}} \, t^{-(\gamma+1)} \, \Gamma(\gamma+1) \\ &= \gamma \, \Gamma(\gamma+1) \, A^\gamma \, t^{-(\gamma+1)}. \end{aligned} \tag{B.167}$$

B.8   ALTERNATING FRACTAL RENEWAL PROCESS

Prob. 8.1 Since τ_a and τ_b have identical distributions, their means and characteristic functions must also coincide. We therefore have $E[\tau_a] = E[\tau_b] = E[\tau]$, and $\phi_{\tau a}(\omega) = \phi_{\tau b}(\omega) = \phi_\tau(\omega)$. Using Eq. (8.5), we have

$$
\begin{aligned}
S_X(f) &= E[X]\,\delta(f) + \frac{2(2\pi f)^{-2}}{E[\tau_a] + E[\tau_b]}\,\mathrm{Re}\left\{\frac{[1 - \phi_{\tau a}(2\pi f)]\,[1 - \phi_{\tau b}(2\pi f)]}{1 - \phi_{\tau a}(2\pi f)\,\phi_{\tau b}(2\pi f)}\right\} \\
&= \frac{E[\tau]}{E[\tau] + E[\tau]}\,\delta(f) + \frac{2(2\pi f)^{-2}}{E[\tau] + E[\tau]}\,\mathrm{Re}\left\{\frac{[1 - \phi_\tau(2\pi f)]^2}{1 - \phi_\tau^2(2\pi f)}\right\} \\
&= \frac{\delta(f)}{2} + \frac{(2\pi f)^{-2}}{E[\tau]}\,\mathrm{Re}\left\{\frac{1 - \phi_\tau(2\pi f)}{1 + \phi_\tau(2\pi f)}\right\}, \qquad\text{(B.168)}
\end{aligned}
$$

in accordance with Eq. (8.8).

Prob. 8.2 For the low frequency limit, Eq. (8.10) becomes

$$
\begin{aligned}
\lim_{f\to 0} S_X(f) &= \lim_{f\to 0} \frac{2\big(E[\tau_a] + E[\tau_b]\big)^{-1}}{(2\pi f)^2 + (2\pi f_S)^2} \\
&= \frac{2\big(E[\tau_a] + E[\tau_b]\big)^{-1}}{(2\pi f_S)^2} \\
&= \frac{2\big(E[\tau_a] + E[\tau_b]\big)^{-1}}{\big(1/E[\tau_a] + 1/E[\tau_b]\big)^2} \\
&= \frac{2E^2[\tau_a]\,E^2[\tau_b]}{\big(E[\tau_a] + E[\tau_b]\big)^3} \\
&= \frac{E^2[\tau_a]\,\mathrm{Var}[\tau_b] + E^2[\tau_b]\,\mathrm{Var}[\tau_a]}{\big(E[\tau_a] + E[\tau_b]\big)^3}, \qquad\text{(B.169)}
\end{aligned}
$$

in accordance with Eq. (8.6).

For the high-frequency limit, we demonstrate that as $f \to \infty$, the limit of the ratio of Eqs. (8.10) and (8.7) approaches unity:

$$
\begin{aligned}
\lim_{f\to\infty} \frac{2\big(E[\tau_a] + E[\tau_b]\big)^{-1}}{(2\pi f)^2 + (2\pi f_S)^2} &\Big/ 2(2\pi f)^{-2}\big(E[\tau_a] + E[\tau_b]\big)^{-1} \\
&= \lim_{f\to\infty} \frac{(2\pi f)^2}{(2\pi f)^2 + (2\pi f_S)^2} \\
&= \lim_{f\to\infty} \frac{1}{1 + (f_S/f)^2} \\
&= 1. \qquad\text{(B.170)}
\end{aligned}
$$

Prob. 8.3 We calculated the characteristic function of an exponential random variable in Eq. (B.51). To cast this in a form more suitable to our present needs we replace μ by $E^{-1}[\tau]$, which provides

$$\phi_\tau(\omega) = (1 + i\omega\, E[\tau])^{-1}. \tag{B.171}$$

Substituting Eq. (B.171) into Eq. (8.5), and using the shorthand notations $u \equiv E[\tau_u]$, $v \equiv E[\tau_v]$, and $\omega \equiv 2\pi f$ yields

$$
\begin{aligned}
S_X(f) &= E[X]\,\delta(f) + \frac{2\omega^{-2}}{u+v}\,\mathrm{Re}\left\{\frac{[1-\phi_{\tau a}(\omega)]\,[1-\phi_{\tau b}(\omega)]}{1-\phi_{\tau a}(\omega)\,\phi_{\tau b}(\omega)}\right\} \\
&= E[X]\,\delta(f) + \frac{2\omega^{-2}}{u+v}\,\mathrm{Re}\left\{\frac{[1-(1+i\omega u)^{-1}]\,[1-(1+i\omega v)^{-1}]}{1-(1+i\omega u)^{-1}\,(1+i\omega v)^{-1}}\right\} \\
&= E[X]\,\delta(f) + \frac{2\omega^{-2}}{u+v}\,\mathrm{Re}\left\{\frac{[(1+i\omega v)-1]\,[(1+i\omega u)-1]}{(1+i\omega u)(1+i\omega v)-1}\right\} \\
&= E[X]\,\delta(f) + \frac{2\omega^{-2}}{u+v}\,\mathrm{Re}\left\{\frac{-\omega^2 uv}{i\omega(u+v)-\omega^2 uv}\times\frac{-i(u+v)/\omega-uv}{-i(u+v)/\omega-uv}\right\} \\
&= E[X]\,\delta(f) + \frac{2\omega^{-2}}{u+v}\,\mathrm{Re}\left\{\frac{\omega^2 u^2 v^2 + i\omega uv(u+v)}{\omega^2 u^2 v^2 + (u+v)^2}\right\} \\
&= E[X]\,\delta(f) + \frac{2\omega^{-2}}{u+v}\times\frac{\omega^2}{\omega^2+(u+v)^2 u^{-2}v^{-2}} \\
&= E[X]\,\delta(f) + \frac{2(u+v)^{-1}}{\omega^2+(1/u+1/v)^2}, \tag{B.172}
\end{aligned}
$$

in accordance with Eq. (8.10).

Prob. 8.4 We make use of the Fourier relation between the spectrum of a real-valued process and its autocovariance. For algebraic convenience, we consider the case $t > 0$ for now. We begin by noting that in the range $1 < \gamma < 2$, and for $A/B \ll 1$, we have

$$
\begin{aligned}
E[\mu] &= \int_A^B t^{-(\gamma+1)}\,dt \Big/ \int_A^B t^{-\gamma}\,dt \\
&\approx \int_A^\infty t^{-(\gamma+1)}\,dt \Big/ \int_A^\infty t^{-\gamma}\,dt \\
&= \frac{A^{-\gamma}/\gamma}{A^{1-\gamma}/(\gamma-1)} \\
&= (\gamma-1)\,\gamma^{-1}\,A^{-1}. \tag{B.173}
\end{aligned}
$$

Using Eq. (8.11), and then Eq. (B.173), we obtain in the mid-frequency range

$$
\begin{aligned}
S_X(f) &\approx \tfrac{1}{2}(\gamma-1)^{-1}\,\Gamma(2-\gamma)\left[-\cos(\pi\gamma/2)\right]E[\mu]A^\gamma\,(2\pi f)^{\gamma-2} \\
&= (2\gamma)^{-1}\,\Gamma(2-\gamma)\left[-\cos(\pi\gamma/2)\right]A^{\gamma-1}\,(2\pi f)^{\gamma-2}
\end{aligned}
$$

$$
\begin{aligned}
R_X(t) - \mathrm{E}^2[X] &= 2 \int_{0+}^{\infty} S_X(f) \cos(2\pi f t)\, df \\
&= 2 \int_{0}^{\infty} (2\gamma)^{-1}\, \Gamma(2-\gamma) \left[-\cos(\pi\gamma/2) \right] A^{\gamma-1}\, (2\pi f)^{\gamma-2} \\
&\qquad \times \cos(2\pi f t)\, df \\
&= \gamma^{-1} \Gamma(2-\gamma) \left[-\cos(\pi\gamma/2) \right] A^{\gamma-1}\, t^{1-\gamma}\, (2\pi)^{-1} \\
&\qquad \times \int_{0}^{\infty} x^{\gamma-2}\, \cos(x)\, dx \\
&= \gamma^{-1} \Gamma(2-\gamma) \left[-\cos(\pi\gamma/2) \right] A^{\gamma-1}\, t^{1-\gamma}\, (2\pi)^{-1} \\
&\qquad \times \pi \Big/ \left\{ 2\Gamma(2-\gamma)\, \cos\big[(\pi/2)(2-\gamma)\big] \right\} \\
&= (4\gamma)^{-1}\, A^{\gamma-1}\, t^{1-\gamma}, \tag{B.174}
\end{aligned}
$$

in accordance with Eq. (8.13). Since the correlation must be an even function, the foregoing result applies for $t < 0$ as well as $t > 0$. The notation $0+$ signifies that the integral does not contain the delta function in $S_X(f)$ at $f = 0$.

Prob. 8.5.1 Proceeding directly from the definition, we have

$$
\begin{aligned}
\mathrm{E}[X_2^n] &= \mathrm{E}\{ [b + (a-b)X_1]^n \} \\
&= \sum_{m=0}^{n} \binom{n}{m} b^m\, (a-b)^{n-m}\, \mathrm{E}[X_1^{n-m}] \\
&= b^n + \sum_{m=0}^{n-1} \binom{n}{m} b^m\, (a-b)^{n-m}\, \mathrm{E}[X_1] \\
&= b^n (1 - \mathrm{E}[X_1]) + \mathrm{E}[X_1] \sum_{m=0}^{n} \binom{n}{m} b^m\, (a-b)^{n-m} \\
&= b^n (1 - \mathrm{E}[X_1]) + \mathrm{E}[X_1]\, [b + (a-b)]^n \\
&= b^n + (a^n - b^n)\, \mathrm{E}[X_1]. \tag{B.175}
\end{aligned}
$$

Prob. 8.5.2 The definition of the autocorrelation provides

$$
\begin{aligned}
R_{X2}(t) &\equiv \mathrm{E}[X_2(s)\, X_2(s+t)] \\
&= \mathrm{E}\{ [b + (a-b)\, X_1(s)]\, [b + (a-b)\, X_1(s+t)] \} \\
&= b^2 + b(a-b)\, \mathrm{E}[X_1(s) + X_1(s+t)] \\
&\qquad + (a-b)^2\, [X_1(s)\, X_1(s+t)] \\
&= b^2 + 2b(a-b)\, \mathrm{E}[X_1] + (a-b)^2 R_{X1}(t). \tag{B.176}
\end{aligned}
$$

Prob. 8.5.3 Suppose first that the times s and $s+t$ lie in the same "on" period. This will occur during a proportion p_1 of the times, and we then have $\mathrm{E}[X_2(s)\, X_2(s+t)] =$

$E[a^2]$. For another proportion p_2 of the times, s and $s+t$ lie in different "on" periods, in which case $E[X_2(s)\,X_2(s+t)] = E^2[a]$. Taking expectations over the conditions, we then have the overall result

$$E[X_2(s)\,X_2(s+t)] = p_1 E[a^2] + p_2 E^2[a]. \tag{B.177}$$

Equation (B.177) provides a meaningful result only if we impose the condition $E[a^2] < \infty$; we therefore require this of the random variable a.

Prob. 8.5.4 To find an expression for $R_{X2}(t)$, it suffices to determine the values of p_1 and p_2 in Eq. (B.177). We can express p_1 as a product of the probability that a random time, s, lies in an "on" period, multiplied by the probability that the remaining time in this "on" period exceeds t. For the first element of this product we have simply

$$E[X_1] = \frac{E[\tau_a]}{E[\tau_a] + E[\tau_b]}, \tag{B.178}$$

whereas for the second we have the recurrence time

$$p_\vartheta(t) = \frac{1 - P_{\tau a}(t)}{E[\tau_a]}. \tag{B.179}$$

The product of Eqs. (B.178) and (B.179) yields the probability p_1:

$$p_1 = \frac{E[\tau_a]}{E[\tau_a] + E[\tau_b]} \times \frac{1 - P_{\tau a}(t)}{E[\tau_a]} = \frac{1 - P_{\tau a}(t)}{E[\tau_a] + E[\tau_b]}. \tag{B.180}$$

For deterministic a fixed at $a = 1$, we see that both p_1 and p_2 contribute to $R_{X1}(t)$; this actually holds for any a, and we therefore have

$$p_1 + p_2 = R_{X1}(t). \tag{B.181}$$

Combining Eqs. (B.177), (B.180), and (B.181), we obtain the final result

$$\begin{aligned} R_{X2}(t) &= p_1 E[a^2] + p_2 E^2[a] \\ &= p_1\left(E[a^2] - E^2[a]\right) + R_{X1}(t)\,E^2[a] \\ &= \mathrm{Var}[a]\,\frac{1 - P_{\tau a}(t)}{E[\tau_a] + E[\tau_b]} + E^2[a]\,R_{X1}(t). \end{aligned} \tag{B.182}$$

We note that Eq. (B.176) with $b = 0$ agrees with Eq. (B.182) with deterministic a, as it must.

Prob. 8.6.1 Integrating the power-law density of Eq. (7.1) yields

$$P_\tau(t) = \begin{cases} 0 & t \le A \\ (A^{-\gamma} - t^{-\gamma})/(A^{-\gamma} - B^{-\gamma}) & A < t < B \\ 1 & t \ge B. \end{cases} \tag{B.183}$$

We can set Eq. (B.183) equal to X_U and solve for τ.

However, since this can involve the difference between two large numbers, it proves more useful to set Eq. (B.183) equal to $1 - X_U$ instead. Since $1 - X_U$ and X_U have identical distributions, this improves the computational accuracy without changing the underlying mathematics. Solving for τ then yields

$$\tau \equiv AB \left[(B^\gamma - A^\gamma) X_U + A^\gamma \right]^{-1/\gamma}. \tag{B.184}$$

Finally, we set $\gamma = 2 - \alpha$ in accordance with Eq. (8.12), and choose $A = 1/f_H$ and $B = 1/f_L$. An alternating fractal renewal process constructed from these three parameters (γ, A, B) will therefore satisfy the design requirements.

Prob. 8.6.2 Turning to Eq. (8.17) we have the design constraints

$$\frac{|1 - 2r|}{\sqrt{Mr(1 - r)}} < \epsilon \qquad \text{a)}$$

$$\frac{1}{M} \left| \frac{1}{r(1 - r)} - 6 \right| < \epsilon. \qquad \text{b)} \tag{B.185}$$

Setting $r = \frac{1}{2}$ satisfies Eq. (B.185a) for any value of M, while choosing $r = \frac{1}{2} \pm 1/\sqrt{12}$ provides the same for Eq. (B.185b). Since the two values of r differ, one of the constraints must be satisfied by adjusting the value of M. Setting $r = \frac{1}{2}$ and solving Eq. (B.185b) yields $M > 2/\epsilon$, while setting $r = \frac{1}{2} \pm 1/\sqrt{12}$ yields $M > 2/\epsilon^2$ via Eq. (B.185a). Since $\epsilon < 1$, the former constraint proves less stringent, and we therefore set $r = \frac{1}{2}$ and choose for M the smallest integer $> 2/\epsilon$. We therefore choose components that are symmetric alternating fractal renewal processes, since $r = \frac{1}{2}$, and sum at least $2/\epsilon$ of them together.

B.9 FRACTAL SHOT NOISE

Prob. 9.1.1 With identical impulse response functions for each $X_m(t)$, the sole difference between the resulting process $X_R(t)$ and the component processes $X_m(t)$ becomes the times at which the impulse response functions begin. Since these points derive from an independent homogeneous Poisson process for each process $X_m(t)$, there is no memory at all within or among these point processes. Their superposition must therefore also lack memory, and must also belong to the homogeneous Poisson point-process family. The total rate is given by $\mu_R = \sum_{m=1}^{M} \mu_m$, but otherwise $X_R(t)$ and $X_1(t)$ do not differ. This defines $X_R(t)$ as a shot-noise process.

Prob. 9.1.2 We can still make use of a shot noise framework for characterizing $X_R(t)$ if we employ random values of K to index the appropriate impulse response function. In other words, define $h_R(K, t)$ so that $h_R(m, t) = h_m(t)$, and define K so that $K = m$ for a proportion μ_m/μ_R of the homogeneous Poisson point-process events. The absence-of-memory argument of Prob. 9.1.1 remains applicable, so the result $X_R(t)$ remains a shot-noise process.

Prob. 9.1.3 If the fractal exponents differ for any two of the component shot noise processes $X_m(t)$, say m_1 and m_2, then the spectrum assumes the form

$$S_{XR}(f) \approx (f/f_{Sm1})^{-\alpha_{m1}} + (f/f_{Sm2})^{-\alpha_{m2}}. \tag{B.186}$$

Equation (B.186) must have a breakpoint at some value of f, with different effective slopes (values of α) on either side of the breakpoint. The result does not belong to the fractal shot-noise family of processes. Thus, for $X_R(t)$ to belong to this family, we require that all $X_m(t)$ have the same value of α.

Prob. 9.1.4 For this impulse response function we have

$$h_m(t) = \exp(-c_m t)$$

$$H_m(f) = \int_0^\infty \exp(-c_m t) \exp(-i2\pi f t)\, dt$$

$$= \frac{1}{c_m + i\,2\pi f}$$

$$|H_m(f)|^2 = \frac{1}{c_m^2 + (2\pi f)^2}$$

$$|S_{Xm}(f)|^2 = \mathrm{E}^2[X_m]\,\delta(f) + \frac{\mu_m}{c_m^2 + (2\pi f)^2}, \tag{B.187}$$

by virtue of Eq. (9.27).

We now attempt to construct an approximation to a fractal shot-noise process through the sum of a number of component processes $X_m(t)$, with parameters related in a power-law fashion. Clearly, the component spectra $S_{Xm}(f)$ must cross each other; otherwise, one would dominate over all frequencies and the trivial result $S_{XR}(f) \sim f^{-2}$ would ensue. Meaningful results thus obtain only if each component process dominates the whole over its own range of frequencies.

This occurs in the neighborhood of the crossover frequencies $f = c_m/(2\pi)$, whereupon

$$S_{XR}[c_m/(2\pi)] \approx \frac{\mu_m}{2c_m^2}. \tag{B.188}$$

Substituting Eq. (B.188) into the scaling equation $S_{XR}(f) \approx (f/f_S)^{-\alpha}$, and retaining the substitution $f = c_m/(2\pi)$, yields

$$\frac{\mu_m}{2c_m^2} \approx \left(\frac{c_m}{2\pi f_S}\right)^{-\alpha}$$

$$\mu_m\, c_m^{\alpha-2} \approx 2(2\pi f_S)^\alpha. \tag{B.189}$$

To minimize variation about the ideal $f^{-\alpha}$ behavior of $S_{XR}(f)$, we maintain a fixed ratio between adjacent values of c_m. We therefore set

$$c_m = c_1\, a^{m-1}, \quad a > 1, \tag{B.190}$$

so that Eqs. (B.189) and (B.190) yield

$$\begin{aligned} \mu_m &= \mu_1\, b^{m-1} \\ \mu_1 &= 2(2\pi f_S)^\alpha\, c_1^{2-\alpha} \\ b &= a^{2-\alpha}; \end{aligned} \tag{B.191}$$

we also have $b > 1$ for $0 < \alpha < 2$.

Finally, we choose a frequency ratio a; this value for a, in combination with Eqs. (B.190) and (B.191), defines the fractal shot noise process $S_{XR}(f)$. As an example, we plot the spectrum for $\alpha = 1$, $f_S = 1$, $c_1 = 2\pi$, $a = 10$, and $M = 4$. From this, and Eq. (B.191), we obtain $\mu_1 = 2$ and $b = 0.1$, and the spectrum simplifies to

$$S_{XR}(f) = \sum_{m=1}^{M} \frac{10^{m-1}}{10^{2m-2} + f^2}. \tag{B.192}$$

Figure B.6 provides a plot of $S_{XR}(f)$ vs. f; it demonstrates a close approximation to the desired $1/f$ form. The component spectra do indeed achieve a value of precisely $1/f$ at the breakpoints, as designed; however, their overlap causes the sum to exceed $1/f$ specification by up to a factor of 1.42, or 1.5 dB. Nevertheless, over the design frequency range $10^0 \le f \le 10^3$, $f \times S_{XR}(f)$ remains constant to within ± 0.34 dB.

Prob. 9.2.1 Using Eq. (9.36) we have

$$\int_{-\infty}^{\infty} h(K, s)\, h(K, s + |t|)\, ds = \begin{cases} K - |t| & |t| < K \\ 0 & \text{otherwise.} \end{cases} \tag{B.193}$$

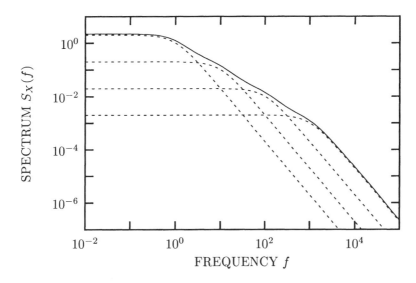

Fig. B.6 Spectrum as provided in Eq. (B.192) (solid curve). The graph also includes the four component spectra $S_{Xm}(f)$ (dashed curves).

Forming the expectation of Eq. (B.193) involves integrating the right-hand side against the probability density of K, for values of K in excess of $|t|$. We therefore have

$$R_h(t) \equiv E\left[\int_{-\infty}^{\infty} h(K,s)\,h(K,s+|t|)\,ds\right] = \int_{x=|t|}^{\infty} (x-|t|)\,p_K(x)\,dx. \quad \text{(B.194)}$$

Equations (B.194) and (9.19) together yield $R_X(t)$.

Prob. 9.2.2 Solving Eq. (B.194) for the special case of Eq. (9.37) leads to

$$
\begin{aligned}
R_h(t) &= \int_{s=|t|}^{\infty} (s-|t|)\,p_K(s)\,ds \\
&= \int_{s=t}^{\infty} (s-t)\,(\beta-1)\,A^{\beta-1}\,s^{-\beta}\,ds \\
&= (\beta-1)\,A^{\beta-1}\left[s^{2-\beta}/(2-\beta) - t\,s^{1-\beta}/(1-\beta)\right]_t^{\infty} \\
&= (\beta-1)\,A^{\beta-1}\,t^{2-\beta}\left[(\beta-2)^{-1} - (\beta-1)^{-1}\right] \\
&= (\beta-2)^{-1}\,A^{\beta-1}\,t^{2-\beta}. \quad \text{(B.195)}
\end{aligned}
$$

Since we specified that $2 < \beta < 3$, the corresponding exponent for t lies in the range $-1 < 2 - \beta < 0$, indicating a fractal form. Identifying this with the canonical form $t^{\alpha-1}$ for autocorrelation functions and related statistical measures provides

$$
\begin{aligned}
\alpha - 1 &= 2 - \beta \\
\alpha &= 3 - \beta, \quad \text{(B.196)}
\end{aligned}
$$

with α in the range $0 < \alpha < 1$.

Prob. 9.3 We begin with Eq. (9.38) and proceed by applying a monotonic transformation, and then substituting $x \equiv cm^{-D}$ and $\mathcal{M} \equiv Kt^{-\beta}$:

$$
\begin{aligned}
\Pr\{\mathcal{M} \geq m\} &= cm^{-D} \\
\Pr\{\mathcal{M}^{-D} \leq m^{-D}\} &= cm^{-D} \\
\Pr\left\{\left[Kt^{-\beta}\right]^{-D} \leq x/c\right\} &= x \\
\Pr\{c\,K^{-D}\,t^{\beta D} \leq x\} &= x \\
\Pr\{t \leq x\} &= x,
\end{aligned}
\tag{B.197}
$$

where we make the identifications $\beta = 1/D$ and $K = c^{1/D}$. Equation (B.197) has the form of a random variable uniformly distributed in the unit interval.

Hence, with t a random time chosen uniformly in the unit interval, the function $Kt^{-\beta} = (t/c)^{-1/D}$ has the same probability distribution as that given in Eq. (9.38). Returning to the original specification of the problem, to obtain the total mass in a given region of space, we sum a Poisson-distributed number of these random variables. Consequently, a fractal shot-noise process with impulse response function given by Eq. (9.2), $B = 1$, and appropriate mean Poisson rate μ will have an amplitude distribution equivalent to the total mass in a cluster. For $0 < D < 1$, we have $1 < \beta < \infty$, yielding a one-sided stable distribution.

B.10   FRACTAL-SHOT-NOISE-DRIVEN POINT PROCESSES

Prob. 10.1 If two events from a fractal-shot-noise-driven Poisson process $dN_2(t)$ span a large interval, such that $\tau > B - A$, then the events must derive from different impulse response functions. Therefore, any relation between the two times must derive from the primary homogeneous Poisson process $dN_1(t)$. This process has an exponential form for the times between events, and this form propagates to $dN_2(t)$ in the absence of any other connection between the events.

Prob. 10.2 Substituting directly into Eq. (10.14), we obtain

$$
\begin{aligned}
F(T) &= 1 + \frac{2\mathrm{E}[K^2]}{aT} \int_0^{\min(T, B-A)} (T - u) \int_A^{B-u} (t^2 + ut)^{-\beta}\, dt\, du \\
&= 1 + \frac{2\mathrm{E}[K^2]}{\mathrm{E}[K]\,(B-A)\,T} \int_0^{\min(T, B-A)} (T - u) \int_A^{B-u} dt\, du \\
&= 1 + \frac{2\mathrm{E}[K^2]}{\mathrm{E}[K]\,(B-A)\,T} \int_0^{\min(T, B-A)} (T - u)\big[B - (u + A)\big]\, du \\
&= 1 + \frac{2\mathrm{E}[K^2]}{\mathrm{E}[K]\,(B-A)\,T} \left[TBu - (B + T - A)\frac{u^2}{2} + \frac{u^3}{3}\right]_0^{\min(T, B-A)}. \quad \text{(B.198)}
\end{aligned}
$$

We now specialize to the two cases: $T \le B - A$ and $T > B - A$. In the former, Eq. (B.198) becomes

$$
\begin{aligned}
F(T) &= 1 + \frac{2\mathrm{E}[K^2]}{\mathrm{E}[K]\,(B-A)\,T} \left[T^2 B - (B + T - A)\frac{T^2}{2} + \frac{T^3}{3}\right] \\
&= 1 + \frac{2\mathrm{E}[K^2]}{\mathrm{E}[K]\,(B-A)\,T} \left[\frac{T^2\,(B + A)}{2} - \frac{T^3}{6}\right] \\
&= 1 + \frac{\mathrm{E}[K^2]\,T\,(B + A - T/3)}{\mathrm{E}[K]\,(B - A)}. \quad \text{(B.199)}
\end{aligned}
$$

Further specializing to the case $B/T \gg 1$ and $B/A \gg 1$, Eq. (B.199) simplifies to

$$
\begin{aligned}
F(T) &\approx 1 + \frac{\mathrm{E}[K^2]\,TB}{\mathrm{E}[K]\,B} \\
&= 1 + \frac{\mathrm{E}[K^2]}{\mathrm{E}[K]}\,T. \quad \text{(B.200)}
\end{aligned}
$$

For this value of β, Eq. (10.16) provides

$$
\begin{aligned}
F(T) &\approx 1 + \frac{\mathrm{E}[K^2]}{\mathrm{E}[K]}\frac{1 - \beta}{1 - 2\beta}\,B^{-\beta}\,T \\
&= 1 + \frac{\mathrm{E}[K^2]}{\mathrm{E}[K]}\frac{1}{1}\,B^0\,T
\end{aligned}
$$

$$= 1 + \frac{E[K^2]}{E[K]} T, \tag{B.201}$$

in accordance with Eq. (B.200). Finally, for $T > B - A$, Eq. (B.198) gives rise to

$$
\begin{aligned}
F(T) &= 1 + \frac{2E[K^2]}{E[K](B-A)T} \Big[TB(B-A) \\
&\quad - (B+T-A)\frac{(B-A)^2}{2} + \frac{(B-A)^3}{3} \Big] \\
&= 1 + \frac{2E[K^2]}{E[K]T} \Big[TB - T\frac{B-A}{2} - \frac{(B-A)^2}{2} + \frac{(B-A)^2}{3} \Big] \\
&= 1 + \frac{E[K^2]}{E[K]} \Big[B + A - \frac{(B-A)^2}{3T} \Big].
\end{aligned}
\tag{B.202}
$$

Prob. 10.3 We start with Fig. 10.6, and consider a single impulse response function. The parameters include $\beta = 2$, $A = 1$, $B = 10^5$, and a $= 100$, so we have

$$
\begin{aligned}
a &\equiv \int h(t)\,dt = \int_A^B K t^{-\beta}\,dt \\
100 &= K \int_1^{10^5} t^{-2}\,dt = K\left(1 - 10^{-5}\right) \\
K &\approx 100.
\end{aligned}
\tag{B.203}
$$

The rate then follows the form $100/t^2$ for $A < t < B$. At a general time t, a typical interevent interval generated by this single impulse response function will have a duration equal to the inverse of this rate, so $\tau(t) \approx t^2/100$. By this time t, we will have generated about $N_2(t)$ intervals, where we obtain the counting process $N_2(t)$ through

$$
\begin{aligned}
N_2(t) &\approx \int_A^t h(s)\,ds \\
&\approx \int_1^t 100\,s^{-2}\,ds \\
&= 100\,(1 - 1/t).
\end{aligned}
\tag{B.204}
$$

Combining this information, we can say that $N_2(t) \approx 100\,(1 - 1/t)$ intervals will lie below the limit $\tau \approx t^2/100$; eliminating t yields about

$$
N_2 = 100 \left[1 - \frac{1}{10\sqrt{\tau}} \right]
\tag{B.205}
$$

intervals less than τ. Normalizing Eq. (B.205) to unity as $\tau \to \infty$ leads to

$$
\Pr\{\tau < s\} = 1 - \frac{1}{10\sqrt{s}}.
\tag{B.206}
$$

Finally, taking the derivative of Eq. (B.206) yields the probability density

$$p_\tau(s) = (20)^{-1} s^{-3/2}. \tag{B.207}$$

The argument presented above remains valid down to interevent intervals $\tau \approx 1/(KA^{-2})$, the inverse of the maximum rate (which occurs at the onset of the impulse response function); at long interevent intervals, the times between the events of the primary point process $dN_1(t)$ dominate the interevent interval probability density, as discussed in Prob. 10.1. These limits, as well as the overall form of Eq. (B.207), are in good accord with the results displayed in Fig. 10.6. A similar argument, although with different power-law exponents, obtains for any value of $\beta > 1$; we require only that most of the area of the impulse response function reside near its onset.

Prob. 10.4.1 Equation (9.29) immediately provides a value for β. Given this value of β (or α) and the crossover frequency f_S, Eq. (10.27) links K and B. Equation (9.8), for $n = 1$, relates the average rate to the fractal-shot-noise-driven Poisson process parameters.

Prob. 10.4.2 For the specific case $\alpha = 1$ with $f_S = 1$ and fixed, deterministic K, we immediately have that $\beta = \frac{1}{2}$, whereupon Eq. (10.27) yields

$$
\begin{aligned}
\left(E[K]/E[K^2]\right)(2\pi f_S)^\alpha B^{\alpha/2} &= \alpha\Gamma^2(\alpha/2)/2 \\
(1/K)(2\pi)\sqrt{B} &= \left(\sqrt{\pi}\right)^2/2 \\
4\sqrt{B} &= K. \tag{B.208}
\end{aligned}
$$

Turning to the cumulants of the rate and employing Eq. (B.208), Eq. (9.8) provides

$$
\begin{aligned}
C_1 &= \mu E[K] \frac{B^{1-\beta} - A^{1-\beta}}{1 - \beta} \\
&\approx \mu 4\sqrt{B}\, \frac{B^{1/2}}{1/2} \\
&= 8\mu B \\
C_2 &= \mu E[K^2]\, \ln(B/A) \\
&= 16\mu B\, \ln(B/A) \tag{B.209} \\
C_3 &= \mu E[K^3] \frac{A^{1-3/2} - B^{1-3/2}}{3/2 - 1} \\
&\approx \mu\, 64 B^{3/2} \frac{A^{-1/2}}{1/2} \\
&= 128\mu B\sqrt{B/A}.
\end{aligned}
$$

Using Eq. (B.209) with Eq. (3.4) then leads to a square of the coefficient of variation of the rate given by

$$\frac{C_2}{C_1^2} = \frac{16\mu B\, \ln(B/A)}{(8\mu B)^2} = \frac{\ln(B/A)}{4\mu B}, \tag{B.210}$$

and a square of the skewness of the rate expressed as

$$\frac{C_3^2}{C_2^3} = \frac{\left(128\mu B\sqrt{B/A}\right)^2}{[16\mu B\ln(B/A)]^3} = \frac{4}{\mu A\left[\ln(B/A)\right]^3}. \tag{B.211}$$

For large values of the rate coefficient of variation we require small values of μB in Eq. (B.210), since the logarithm function varies so slowly. For small values of the rate skewness we require large values of μA in Eq. (B.211), again neglecting the logarithm. The conflict arises because the simultaneous specification of a small value of μB and a large value of μA leads to $A \gg B$, whereas we define $A < B$; in fact, obtaining a $1/f$ spectrum over an appreciable range of frequencies requires $B \gg A$. Thus, the rate either has a large coefficient of variation or a small skewness, but not both. Furthermore, we have not specified μ independently of A and B.

Prob. 10.4.3 Aside from $\beta = \frac{1}{2}$ which we established above, we have $A = 1/f_S = 1$, and from $B/A = 10^3$ we obtain $B = 10^3$. We also have a mean rate of unity, and Eq. (B.209) reveals that $C_1 = 8\mu B$; together this yields $\mu = (8B)^{-1} = 1/8000$. Also, Eq. (B.208) provides $K = 4\sqrt{B} = 4\sqrt{1\,000}$. In summary, we have

$$\begin{aligned}
\beta &= \tfrac{1}{2} \\
A &= 1 \\
B &= 1\,000 \\
K &= 40\sqrt{10} \\
\mu &= 1/8000.
\end{aligned} \tag{B.212}$$

For the coefficient of variation, we use the parameters provided in Eq. (B.212) in Eq. (B.210) to obtain

$$\frac{\sqrt{C_2}}{C_1} = \sqrt{\frac{\ln(B/A)}{4\mu B}} = \sqrt{\frac{\ln(10^3)}{8\mu B/2}} = \sqrt{6\ln(10)} \doteq 3.71692, \tag{B.213}$$

while Eq. (B.211) yields the skewness

$$\frac{C_3}{C_2^{3/2}} = \sqrt{\frac{4}{\mu A\left[\ln(B/A)\right]^3}} = \sqrt{\frac{32B/A}{8\mu B\left[\ln(10^3)\right]^3}} = \sqrt{\frac{32\,000}{[3\ln(10)]^3}} \doteq 9.85302. \tag{B.214}$$

Equation (B.212) defines a process with significant variation.

Prob. 10.5 We base our argument on the results provided in Table 9.1, which divides into two regions of interest. For $\beta \le 1$ and $B = \infty$, the impulse response function $h(t)$ has infinite area in its tail, and the results in Sec. A.6.1 indicate that the resulting shot noise process has an infinite value at all times with probability one. Using this as a rate for a Poisson point process yields a collection of points infinitely dense at every time. We have $N_2(t + \epsilon) - N_2(t) = \infty$, with probability one, for any t and $\epsilon > 0$, even if we use the process itself or its rate to select t or ϵ. This is far from an orderly process.

For $\beta \geq 1$ and $A = 0$, the infinite area now lies near the origin. This leads to shot noise with a stable amplitude density; the corresponding point process remains well-behaved except near the onset times of the impulse response functions. Here an infinite number of events exist after each impulse response function commences. If we let t_k denote any of the events in the primary Poisson point process $dN_1(t)$ from which the shot noise derives, then we have $N_2(t_k + \epsilon) - N_2(t_k) = \infty$, with probability one, for any $\epsilon > 0$. Again, the resulting point process $dN_2(t)$ is not orderly.

Prob. 10.6.1 This description applies for $t > 0$. For $t < 0$ the electromagnetic shock wave generated by the particle has not yet reached the observation point, so that all fields remain zero and no photons are yet present there.

Prob. 10.6.2 The electric field becomes

$$
\begin{aligned}
\mathbf{E} &= -\nabla\phi_L - c^{-1}\,\partial\mathbf{A}_L/\partial t \\
&= -2qJ^2n^{-2}\left[(x - vt)^2 - J^2(y^2 + z^2)\right]^{-3/2}\{x - vt,\, y,\, z\} \\
&= 2qJ^2n^{-2}\left[(vt)^2 + 2Jdvt\right]^{-3/2}\{vt + Jd,\, 0,\, -d\} \\
&= 2qJ^2n^{-2}v^{-2}[t^2 + 2t_1 t]^{-3/2}\{t + t_1,\, 0,\, -d/v\},
\end{aligned} \tag{B.215}
$$

where we define $t_1 \equiv Jd/v = d(n^2c^{-2} - v^{-2})^{1/2}$. The magnetic field becomes

$$
\begin{aligned}
\mathbf{H} &= \mathbf{B} \\
&= \nabla \otimes \mathbf{A}_L \\
&= 2qvc^{-1}\left[(x - vt)^2 - J^2(y^2 + z^2)\right]^{-3/2}\{0,\, J^2 z,\, -J^2 y\} \\
&= 2qdvJ^2c^{-1}\left[(vt)^2 + 2Jdvt\right]^{-3/2}\{0, 1, 0\} \\
&= 2qdJ^2c^{-1}v^{-2}[t^2 + 2t_1 t]^{-3/2}\{0, 1, 0\},
\end{aligned} \tag{B.216}
$$

where $\otimes$ again denotes the vector cross product.

This leads to expressions for the Poynting vector and the photon flux density given by

$$
\begin{aligned}
\mathbf{S} &\equiv (4\pi)^{-1}c\,\mathbf{E} \otimes \mathbf{H} \\
&= (4\pi)^{-1}c(2qJ^2n^{-2}v^{-2})(2qdJ^2c^{-1}v^{-2})[t^2 + 2t_1 t]^{-3} \\
&\quad \times \{t + t_1,\, 0,\, -d/v\} \otimes \{0, 1, 0\} \\
&= \pi^{-1}q^2dJ^4n^{-2}v^{-4}[t^2 + 2t_1 t]^{-3}\{d/v,\, 0,\, t + t_1\}
\end{aligned} \tag{B.217}
$$

$$
\begin{aligned}
h(t) &\approx |\mathbf{S}|/h\mathrm{E}[\nu] \\
&= (\pi h\,\mathrm{E}[\nu])^{-1}q^2dJ^4n^{-2}v^{-4}\left[t^2 + 2t_1 t\right]^{-3}\left[t^2 + 2t_1 t + t_1^2 + (d/v)^2\right]^{1/2} \\
&= (\pi h\mathrm{E}[\nu])^{-1}q^2dJ^4n^{-2}v^{-4}\left[t^2 + 2t_1 t\right]^{-3}\left[t^2 + 2t_1 t + (nd/c)^2\right]^{1/2} \\
&= (\pi h\mathrm{E}[\nu])^{-1}q^2dJ^4n^{-2}v^{-4}\left[t^2 + 2t_1 t\right]^{-3}\left[t^2 + 2t_1 t + t_2^2\right]^{1/2}, \tag{B.218}
\end{aligned}
$$

where we define $t_2 \equiv nd/c$.

Casting this photon-flux-density time function in the form of a simple power-law impulse response function, as in Eq. (9.2), yields the following results. The photon flux density exhibits a power-law decay with a power-law exponent that increases at the crossover time $t = t_1$, and decreases at time $t = t_2$. No real medium will pass frequency components of arbitrarily high frequency, and indeed all systems have practical limits to the frequency components that appear at the output (see Sec. 2.3.1 for a discussion pertinent to this issue). The difference between the upper ν_u and lower ν_l frequency limits forms the system bandwidth, $\nu_u - \nu_l$. Similarly, the onset time of the light pulse cannot exceed a value roughly equal to the inverse of the bandwidth; we therefore define $t_0 \equiv (\nu_u - \nu_l)^{-1}$. In addition, the nonzero size of the charged particle imposes a limit on the onset time (Zrelov, 1968), although this limit proves relatively unimportant since we can assume that the wavelength of the generated photons greatly exceeds the particle size.

The photon-flux-density time function $h(t)$ thus follows the form

$$
h(t) \sim \begin{cases} 0 & \text{for} \quad t < t_0 \\ t^{-3} & t_0 < t < t_1 \\ t^{-6} & t_1 < t < t_2 \\ t^{-5} & t > t_2. \end{cases} \tag{B.219}
$$

Even for relatively narrow bandwidths, the onset time t_0 often remains several orders of magnitude smaller than t_1, ensuring a large range of times for which t^{-3} behavior is observed. For example, using a 150-mCi radon source, Čerenkov (1938) studied radiation from a number of liquids in the wavelength range 536 to 556 nm, which gives the onset time

$$
t_0 \equiv (\nu_u - \nu_l)^{-1} = \left(\frac{3.00 \times 10^8 \text{ m / s}}{536 \times 10^{-9} \text{ m}} - \frac{3.00 \times 10^8 \text{ m / s}}{556 \times 10^{-9} \text{ m}} \right)^{-1} \approx 50 \text{ fsec.} \tag{B.220}
$$

Particles traveling close to the speed of light through materials with a refractive index as low as 1.2, with d as small as 1 cm to the observation point, yield a crossover time $t_1 \approx 22$ psec, almost three orders of magnitude slower. Most media have much larger bandwidths, and correspondingly larger differences between t_0 and t_1. For such particles we can make the approximation that $h(t) = 0$ for $t < t_0$, and similarly $h(t) = 0$ for $t > t_1$, since the power-law decay exponent increases at $t = t_1$. The energy-flow time-response functions associated with a single charged particle emitting Čerenkov radiation then closely follow

$$
h(t) \approx \begin{cases} Kt^{-3} & \text{for } A < t < B \\ 0 & \text{otherwise,} \end{cases} \tag{B.221}
$$

where we identify $A = t_0$ and $B = t_1$.

Prob. 10.6.3 In media whose index of refraction differs only slightly from unity, the power-law crossover time t_1 of the impulse response function $h(t)$ becomes very small, possibly smaller that the onset time t_0. In that case $h(t)$ lacks the t^{-3} portion, and instead exhibits a faster power-law decay,

$$h(t) \approx \begin{cases} Kt^{-5} & \text{for } t > t_0 \\ 0 & \text{otherwise.} \end{cases} \tag{B.222}$$

However, since the energy production is proportional to J^4, if the index of refraction differs only slightly from unity, then J assumes a small value, as does the total light energy. In that case, the charged particles generate few photons.

Prob. 10.6.4 The foregoing thus illustrates that a single particle gives rise to a photon flux density that follows a decaying power-law time function. If a number of particles travel along the x-axis, they stimulate noninterfering sets of photons, as long as these particles remain sufficiently separated so that their respective electric and magnetic fields do not overlap significantly. Since the form of the Poynting vector involves a vector multiplication, overlap means cross-products between the two sets of fields so that the resulting sequence of photons generated will not follow the simple linear superposition that results from two separately arriving charged particles. Radioactive sources, such as alpha- and beta-emitters, and particle accelerators operated at low current levels, generate Poisson time sequences of energetic charged particles with essentially identical positions and velocities. When these particles pass through a transparent medium under the conditions specified above, the point process resulting from the generated Čerenkov photon events will obey the fractal-shot-noise-driven Poisson process model.

Prob. 10.7.1 From the cluster start time, and the lack of any mechanism terminating the clusters, we have immediately $A = 2.3$ days and $B = \infty$. The average number of earthquakes per cluster is simply the area, so we also have a $= 6$. With a mean of $E[X] = 22$ earthquakes per year, we have a rate for the primary Poisson process given by

$$\mu = \frac{E[X]}{a} \approx \frac{22 \text{ yr}^{-1}}{6} \frac{1 \text{ yr}}{365 \text{ day}} \approx 0.01/\text{day}. \tag{B.223}$$

It remains to find K and β.

Postulating an impulse response function $h(t)$ of the form given in Eq. (9.2), the area remaining in the tail has the form

$$\begin{aligned} a - h_t(K, 0) &\equiv \int_t^\infty h(u)\, du \\ &= \int_t^\infty K u^{-\beta}\, du \\ &= \frac{K}{\beta - 1} t^{1-\beta}, \end{aligned} \tag{B.224}$$

where we consider $t > A$ and make use of the fact that $B = \infty$. We will justify the assumption that $\beta > 1$ shortly. Since we know that the number of earthquakes remaining in a cluster decays as $t^{-1/4}$, we have $t^{1-\beta} = t^{-1/4}$ so that $\beta = \frac{5}{4}$, which is indeed larger than unity, verifying our original assumption. Evaluating Eq. (B.224) at $t = A$ yields the area, which we know has a value of six earthquakes per cluster, so that

$$
\begin{aligned}
a &= \frac{K}{\beta - 1} A^{1-\beta} \\
6 &= \frac{K}{1/4} 2.3^{-1/4} \\
K &= 1.85.
\end{aligned}
\tag{B.225}
$$

Prob. 10.7.2 Equation (B.224) provides the average number of events remaining in a cluster. However, an exponential transform yields the probability of zero events remaining in the cluster, much as Eq. (10.3) does for the shot-noise-driven Poisson process. If we set the probability of zero events remaining at 0.8 or greater, we then have

$$
\begin{aligned}
\Pr\{\text{zero events after } t\} &= \exp\{-[a - h_t(K, 0)]\} \\
&= \exp\left(-\frac{K}{\beta - 1} t^{1-\beta}\right) \\
&= \exp\left[-a (t/A)^{1-\beta}\right] \\
0.8 &> \exp\left[-6 (t/2.3)^{-1/4}\right] \\
\ln(0.8)/(-6) &> (t/2.3)^{-1/4} \\
[\ln(0.8)/(-6)]^{-4} &< t/2.3 \\
t &> 2.3 [\ln(0.8)/(-6)]^{-4} \\
t &> 1\,200\,000 \text{ days} = 3\,300 \text{ years},
\end{aligned}
\tag{B.226}
$$

a surprisingly large number.

Prob. 10.8.1 We define $K \equiv u_0 (4\pi\Delta)^{-D_E/2}$ and $t_0 \equiv |\mathbf{x} - \mathbf{x}_0|^2/4\Delta$, and rewrite Eq. (10.38) as

$$
u(\mathbf{x}, t) = K \exp(-t_0/t) \, t^{-D_E/2}.
\tag{B.227}
$$

We further identify $A = t_0$ and $B = t_1$, yielding a response to a single deposit event that is essentially identical to that in Eq. (9.2), the only difference being in the nature of the lower cutoff. Fractal shot noise then describes the total concentration $u_\Sigma(\mathbf{x}, t)$ arising from the sequence of deposit events, which serves as the rate for the series of secondary events: the result is the fractal-shot-noise-driven Poisson process.

Prob. 10.8.2 The exponents $\beta = \frac{1}{2}$, 1, and $\frac{3}{2}$, corresponding to diffusion in one, two, and three dimensions, respectively, occur most often. In particular, the spectrum precisely follows a $1/f$ form for $\beta = \frac{1}{2}$; thus, diffusion in one dimension can give rise to a $1/f$-type spectrum.

Other values of β also obtain if the particles remain in a fractal set. In that case, the expression for the power-law exponent becomes $\beta = \frac{1}{2}D_s$, where D_s represents the spectral dimension of the fractal set, defined by

$$D_s \equiv 2D_{\mathrm{HB}}/(2 + D_d), \tag{B.228}$$

where D_{HB} is the Hausdorff–Besicovitch dimension and D_d is the exponent that describes the power-law variation of the diffusion constant with distance (Alexander & Orbach, 1982; Rammal & Toulouse, 1983). For percolation clusters at threshold, the spectral dimension D_s lies between 1 and 2, and approaches a limit of $\frac{4}{3}$ for an infinite-dimensional embedding space (Rammal & Toulouse, 1983).

Prob. 10.8.3 Employing stochastic values for K readily incorporates the effect of packets having various values of the initial concentration u_0, and using stochastic values of both K and A admits packets arriving at differing points x.

Prob. 10.8.4 Despite yielding mathematically plausible fractal exponents, the parameters associated with various diffusion processes sometimes make this process unrealistic as a physical model. For example, the diffusion of neurotransmitter across a synapse might appear to provide an explanation for the fractal behavior observed in a variety of neural firing patterns (Lowen & Teich, 1990). However, synapses typically span distances that are quite small (perhaps 5 nm), so that over millisecond time scales the Gaussian form of Eq. (10.38) becomes a linear concentration gradient. Moreover, neurotransmitter transport and metabolism impose a finite lifetime on the extracellular concentration at least as short as this time scale. Thus, no plausible mechanism exists whereby diffusion can impart power-law decay to neurotransmitter concentration over time scales of seconds or longer, as would be required if this process were to underlie fractal action-potential patterns.

Prob. 10.9.1 The solution to this semiconductor recombination problem is closely related to a similar problem: molecular reactions involving two species that combine in pairs (Burlatsky, Oshanin & Ovchinnikov, 1989; Oshanin, Burlatsky & Ovchinnikov, 1989). A cursory analysis for a diffusion process suggests that the concentration of electrons and holes would decay in time as $t^{-D_{\mathrm{E}}/2}$, where D_{E} represents the (integer) dimension of the space within which the electrons and holes move. The concentration would indeed follow this form if the distributions of the two types of carriers were highly correlated. However, often the two carrier distributions are independent of each other, at least over short distances. Consider a sub-volume of the depletion region which, as a result of the variance of the Poisson distribution, happens to have an excess of electrons at $t = 0$. The holes in this section readily recombine with local electrons, but the remaining excess electrons have to diffuse out of this region before encountering any additional holes. This requires more time, slowing the annihilation process. This effect appears on all time and length scales, and results in a concentration that decays as $t^{-D_{\mathrm{E}}/4}$ rather than $t^{-D_{\mathrm{E}}/2}$. If the particle concentrations exhibit correlation over distances longer than some dependence length l_d, then the concentration decays as $t^{-D_{\mathrm{E}}/2}$ for time $t > t_1 = l_d^2/\Delta$, where Δ is again a diffusion constant (Ovchinnikov & Zeldovich, 1978; Toussaint & Wilczek, 1983). At the creation of an electron-hole pair, there is a finite distance between the two particles, so the concentrations of electrons and holes will remain highly correlated over regions larger than an effective mean length.

Prob. 10.9.2 In the presence of drift, the distance traveled by a carrier along the direction of drift changes from $\sim t^{1/2}$ (diffusion alone) to $\sim t^1$ (with drift). Since we postulate that the electrons and holes diffuse through D_E dimensions, the total volume encountered increases as $t^{D_E/2}$ with diffusion alone. The inclusion of drift changes this to $D_E - 1$ dimensions, each varying as $t^{1/2}$, and one dimension varying as t^1, for a total volume that increases as $t^{(D_E+1)/2}$. Since the particle concentration decays as the inverse square-root of the volume encountered, it varies as $t^{-(D_E+1)/4}$ for independent electron and hole distributions, and as $t^{-(D_E+1)/2}$ for dependent distributions (Kang & Redner, 1984).

Prob. 10.9.3 We begin by considering the point process corresponding to the times of the electron-hole recombinations. These recombinations cause the decay in the number of electrons and holes, so the rate of recombination equals the rate of decrease in the number of particles. In the presence of drift and diffusion, the rate of recombination takes the form

$$h(t) \sim \begin{cases} 0, & t < A, \\ t^{-1-(D_E+1)/4} & A < t < B, \\ t^{-1-(D_E+1)/2} & t > B, \end{cases} \tag{B.229}$$

where we identify $A = x_0^2/\Delta$ and $B = x_c^2/\Delta$, where x_0 is a minimum separation for created electron-hole pairs, x_c is the maximum separation corresponding to a correlation length, and Δ is a combined effective diffusion constant. Equation (9.2) closely approximates this impulse response function, with $\beta = 1 + (D_E + 1)/4$.

Prob. 10.9.4 As indicated in the solution to Prob. 10.6.4, in many applications energetic particles impinge on the detector at discrete times corresponding to a one-dimensional Poisson point process. In these cases, the fractal-shot-noise-driven Poisson process is suitable for describing the resulting recombination process.

Prob. 10.10 Since the mean and variance of the numbers of conduction event onsets assume similar values, the homogeneous Poisson process provides the simplest explanation. In the absence of other evidence to the contrary, we rely on Occam's razor and choose this process as the point of departure. A similar argument holds for the numbers of conductance changes within a conduction event. Together, these suggest the Bartlett–Lewis point-process model. Since the spectrum varies as $1/f$, we further narrow our purview to the fractal version described in Sec. 10.6.4. In particular, we have $1/f = 1/f^\alpha$, so $\alpha = 1$. Equation (10.31) yields $\alpha = z + 3$ for $-3 < z < -1$, which gives rise to $z = -2$, which indeed falls within the range $-3 < z < -1$ and validates this result. With this choice of z, the process described in Sec. 10.6.4 successfully models the auxiliary process of conductance changes (Azhar & Gopala, 1992).

B.11 OPERATIONS

Prob. 11.1 Shuffling a point process yields a new process with completely independent intervals; this operation therefore essentially generates a renewal point process
with the same interevent-interval distribution as the original. On the other hand, the
transformation of the interevent intervals replaces the original interevent interval distribution with a new one, exponential in this case. Since shuffling does not affect the
interevent-interval distribution, and interevent-interval transformation does not affect
the relative ordering, the two operations do not influence each other. The same result
therefore obtains regardless of which operation is carried out first. Since the resultant
point process is renewal in nature, it is completely specified by its interevent-interval
distribution (see Sec. 4.2). A renewal process with exponentially distributed intervals defines the homogeneous Poisson process (see Sec. 4.1). Therefore, shuffling
and exponentialization, carried out in either order, leads to a homogeneous Poisson
process.

Prob. 11.2.1 A monofractal process can comprise only a single fractal exponent; we
thus require $\alpha_1 = \alpha_2$ for a true fractal-based point process.

Prob. 11.2.2 We immediately have $\alpha_R = \alpha_1 = \alpha_2$ and we also have $\mathrm{E}[\mu_R] = \mathrm{E}[\mu_1] + \mathrm{E}[\mu_2]$ from first principles. To determine f_{SR}, we transform Eq. (11.41) to
the frequency domain and set $M = 2$ to obtain

$$S_{NR}(f) = S_{N1}(f) + S_{N2}(f)$$

$$\mathrm{E}[\mu_R]\left[1 + (f/f_{SR})^{-\alpha}\right] = \mathrm{E}[\mu_1]\left[1 + (f/f_{S1})^{-\alpha}\right] + \mathrm{E}[\mu_2]\left[1 + (f/f_{S2})^{-\alpha}\right]$$

$$\mathrm{E}[\mu_R]\,(f/f_{SR})^{-\alpha} = \mathrm{E}[\mu_1]\,(f/f_{S1})^{-\alpha} + \mathrm{E}[\mu_2]\,(f/f_{S2})^{-\alpha}$$

$$f_{SR}^{\alpha} = \frac{\mathrm{E}[\mu_1]\,f_{S1}^{\alpha} + \mathrm{E}[\mu_2]\,f_{S2}^{\alpha}}{\mathrm{E}[\mu_1] + \mathrm{E}[\mu_2]}. \tag{B.230}$$

Equation (B.230) enables the calculation of the cutoff frequency f_{SR}.

Prob. 11.2.3 Although a true fractal-based point process can have only a single
fractal exponent, small contributions representing other exponents sometimes prove
undetectable in practice. Figure B.7 depicts the spectrum of two such fractal-based
point processes, with $\alpha_2 = 2\alpha_1$. For $f \gg f_{S2}$, we actually have $S_{N1}(f) > S_{N2}(f)$,
but since there are only small fractal fluctuations in $dN_2(t)$ at these frequencies, we
can model the contribution of $dN_1(t)$ to $dN_R(t)$ as a homogeneous Poisson point
process, which leads to what is effectively monofractal behavior in $dN_R(t)$.

Prob. 11.3 Table B.1 summarizes the nine possibilities. If shuffling completely
destroys any fractal behavior in a point process, while transforming the interevent
intervals does not alter the fractal qualities, then the ordering of the intervals must
completely account for this fractal behavior. Conversely, if transforming the intervals
eliminates the fractal characteristics of a process while shuffling does not, then the

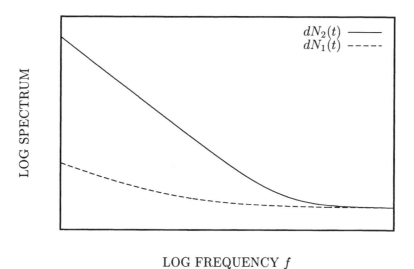

LOG FREQUENCY f

Fig. B.7 Spectrum for a superposed fractal-based point process. One of the component point processes imparts low-intensity fractal fluctuations (dashed curve), which are masked by the much larger contribution from the other process (solid curve).

| | | INTEREVENT-INTERVAL TRANSFORMATION | | |
		UNCHANGED	REDUCED	ELIMINATED
S H U F F L I N G	UNCHANGED	impossible	distribution dominant	distribution only
	REDUCED	ordering dominant	both important	impossible
	ELIMINATED	ordering only	impossible	impossible

Table B.1 Fractal behavior following shuffling and interval transformation.

distribution of the intervals must underlie the fractal nature of the process. If both operations decrease but do not eliminate the fractal behavior, then both effects prove important. For a monofractal process, both effects must have the same fractal exponent.

The situation is somewhat more complex when shuffling reduces the strength of the fractal fluctuations but interval transformation leaves them unchanged. The distribution of the intervals then contributes low-intensity fractal fluctuations, masked by the more robust fractal characteristics imparted by the ordering of the intervals. In this case, the two fractal contributions need not share the same fractal exponent (we considered a related scenario in Prob. 11.2.3, where the fractal behavior from one component dominated that of another). The converse holds when the results are unmodified by shuffling but reduced by interval transformation.

We conclude by considering the four possibilities that cannot occur. If shuffling completely eliminates any fractal behavior, then the interevent-interval distribution cannot play any role; similarly, if transforming the interevent intervals results in a nonfractal point process, then the ordering is irrelevant to the fractal characteristics of the process. Similarly, the two operations cannot both leave the same fractal-based point process unchanged.

Prob. 11.4 Beginning with Eq. (B.230) we have

$$f_{SR}^{\alpha} = \frac{E[\mu_1] f_{S1}^{\alpha} + E[\mu_2] f_{S2}^{\alpha}}{E[\mu_1] + E[\mu_2]}$$

$$f_{SR}^{\alpha}/E^{\alpha}[\mu_R] = \frac{E^{1+\alpha}[\mu_1] \left(f_{S1}^{\alpha}/E^{\alpha}[\mu_1]\right) + E^{1+\alpha}[\mu_2] \left(f_{S2}^{\alpha}/E^{\alpha}[\mu_2]\right)}{E^{1+\alpha}[\mu_R]}$$

$$c_{fR}^{\alpha} = c_{f1}^{\alpha} \left(\frac{E[\mu_1]}{E[\mu_R]}\right)^{1+\alpha} + c_{f2}^{\alpha} \left(\frac{E[\mu_2]}{E[\mu_R]}\right)^{1+\alpha}. \qquad (B.231)$$

For c_{fR} to exceed both c_{f1} and c_{f2}, it must exceed the larger of c_{f1} and c_{f2}. Assume that $c_{f1} \geq c_{f2}$ without loss of generality. Suppose we keep $E[\mu_1]$, $E[\mu_2]$, and c_{f1} fixed. Increasing c_{f2} increases c_{fR}, so to achieve the largest ratio c_{fR}/c_{f1} we increase c_{f2} until it equals c_{f1} (further increases would cause c_{f2} to assume the role of the larger fractal content). Setting $c_{f1} = c_{f2}$ then leads to a simplification of Eq. (B.231):

$$c_{fR}^{\alpha} = c_{f1}^{\alpha} \left[\left(\frac{E[\mu_1]}{E[\mu_R]}\right)^{1+\alpha} + \left(\frac{E[\mu_2]}{E[\mu_R]}\right)^{1+\alpha} \right]$$

$$c_{fR}/c_{f1} = \left[x^{1+\alpha} + (1-x)^{1+\alpha} \right]^{1/\alpha} \qquad (B.232)$$

$$x \equiv E[\mu_1]/E[\mu_R], \qquad (B.233)$$

where x can assume any value between zero and unity. For any value of α, the right-hand side of Eq. (B.232) achieves a maximum value of unity for $x = 0$ or $x = 1$, and a minimum value of $\frac{1}{2}$ for $x = \frac{1}{2}$. We conclude that c_{fR} can never exceed either c_{f1} or c_{f2}, although it can approach either one for small rates of the other process.

Prob. 11.5.1 Since we increase the rate of $dN_1(t)$ by a factor of two, and then keep only half of the resulting events in $dN_2(t)$, the mean rate remains unchanged: $E[\mu_R] = E[\mu_1]$. The simple scaling operations considered do not modify the overall fractal structure of the process, so we maintain $\alpha_R = \alpha_1$.

To determine the fractal onset frequency, we begin with Eq. (5.44a), and make use of Eqs. (11.5) and (11.7), in turn, to obtain

$$S_{N1}(f) = E[\mu_1]\left[1 + (f/f_{S,1})^{-\alpha}\right]$$

$$S_{N2}(f) = c^{-1}E[\mu_1]\left[1 + (cf/f_{S,1})^{-\alpha}\right]$$

$$S_{NR}(f) = c^{-1}E[\mu_1]\left[r + r^2(cf/f_{S,1})^{-\alpha}\right]$$

$$= E[\mu_1]\left[1 + \left(c^{1-1/\alpha}f/f_{S,1}\right)^{-\alpha}\right], \tag{B.234}$$

where the last step follows from the fact that $r = c$ since the mean rate remains unchanged. Comparing Eq. (B.234) with Eq. (5.44a) for $dN_R(t)$, we have

$$E[\mu_1]\left[1 + \left(c^{1-1/\alpha}f/f_{S,1}\right)^{-\alpha}\right] = E[\mu_R]\left[1 + (f/f_{S,R})^{-\alpha}\right]$$

$$\left(c^{1-1/\alpha}f/f_{S,1}\right)^{-\alpha} = (f/f_{S,R})^{-\alpha}$$

$$c^{1-1/\alpha}f/f_{S,1} = f/f_{S,R}$$

$$f_{S,R} = c^{1/\alpha-1}f_{S,1}. \tag{B.235}$$

In general, the fractal onset frequencies for $dN_1(t)$ and $dN_R(t)$ differ; however, for $\alpha = 1$ they coincide.

Prob. 11.5.2 Independent, random deletion of a homogeneous Poisson process, case (a1), results in another homogeneous Poisson process with reduced rate. Similarly, deleting every other event in an integrate-and-reset process, case (b2), does not alter the nature of the process but halves its rate. Retaining every other event in a Poisson process, case (a2), leads to the gamma renewal process (see Prob. 4.7). And finally, randomly and independently deleting points in an integrate-and-reset process, case (b1), gives rise to geometrically distributed interevent intervals and binomial counts. This latter point process finds use in auditory neurophysiology when using low-frequency tonal stimuli, provided that measurement times are sufficiently short so that fractal behavior does not affect the results (see, for example, Teich et al., 1993).

Prob. 11.6 Consider an event of the resulting point process $dN_R(t)$. Following this event we have a fixed interval of duration τ_f, followed by an exponentially distributed random interval of mean duration τ_r, followed by yet another exponentially distributed random interval of mean duration $1/\mu_1$. The next event of $dN_R(t)$ follows immediately after this last interval. The sum of three more intervals, one fixed and two exponentially distributed, comprises the next interevent interval, and all are independent of their counterparts in the previous interevent interval. Their sum must also be independent; continuing this argument we find no dependency among the interevent intervals of the output point process $dN_R(t)$, so that it belongs to the renewal family of point processes.

The mean interevent interval $E[\tau]$ is the simple sum of the components associated with the homogeneous Poisson point process and the two forms of dead time

$$E[\tau] = \mu_1^{-1} + \tau_f + \tau_r. \tag{B.236}$$

The inverse of this quantity yields the effective mean rate

$$
\begin{aligned}
E[\mu_R] &= 1/E[\tau] \\
&= \left(\mu_1^{-1} + \tau_f + \tau_r\right)^{-1} \\
&= \frac{\mu_1}{1 + \mu_1(\tau_f + \tau_r)}. \tag{B.237}
\end{aligned}
$$

The interevent-interval variance consists of the sum of the variances associated with the two stochastic components

$$\text{Var}[\tau] = \mu_1^{-2} + \tau_r^2. \tag{B.238}$$

Similarly, the convolution of the contributions from the three components yields the interevent-interval probability density $p_\tau(t)$. For $t \leq \tau_f$, this assumes a value of zero; for larger times we have

$$
p_\tau(t) = \begin{cases}
\mu_1(1 - \mu_1\tau_r)^{-1}\left[e^{-\mu_1(t-\tau_f)} - e^{-(t-\tau_f)/\tau_r}\right] & \mu_1\tau_r < 1 \\
\mu_1^2(t - \tau_f)\,e^{-\mu_1(t-\tau_f)} & \mu_1\tau_r = 1 \\
\mu_1(\mu_1\tau_r - 1)^{-1}\left[e^{-(t-\tau_f)/\tau_r} - e^{-\mu_1(t-\tau_f)}\right] & \mu_1\tau_r > 1.
\end{cases} \tag{B.239}
$$

The product of the characteristic functions of the individual quantities yields the characteristic function of the interevent interval τ itself

$$
\begin{aligned}
\phi_\tau(\omega) &= e^{-i\omega\tau_f}(1 + i\omega/\mu_1)^{-1}(1 + i\omega\tau_r)^{-1} \\
&= \frac{\mu_1 e^{-i\omega\tau_f}}{(\mu_1 + i\omega)(1 + i\omega\tau_r)}. \tag{B.240}
\end{aligned}
$$

Finally, substituting Eq. (B.240) into Eq. (4.16) yields the spectrum of the point process; for $\tau_r = 0$ this reduces to

$$
\begin{aligned}
S_{NR}(f) &= E^2[\mu_R]\,\delta(f) + E[\mu_R]\,\text{Re}\left[\frac{1 + \phi_\tau(2\pi f)}{1 - \phi_\tau(2\pi f)}\right] \\
&= E^2[\mu_R]\,\delta(f) + \frac{\pi^2 E[\mu_R] f^2}{\pi^2 f^2 + \pi\mu_1 f \sin(2\pi\tau_f f) + \mu_1^2 \sin^2(\pi\tau_f f)}. \tag{B.241}
\end{aligned}
$$

Prob. 11.7 We begin by substituting Eqs. (7.1) and (7.2) into Eq. (11.46). Making extensive use of the relations $A \ll t \ll B$ then yields

$$S_{\vartheta 1}(t) = E[\mu_1] \int_t^\infty (v - t)\,p_{\tau 1}(v)\,dv$$

$$\approx \frac{1-\gamma}{\gamma} A^{-\gamma} B^{\gamma-1} \gamma A^{\gamma} \int_t^B (v-t)\, v^{-(\gamma+1)}\, dv$$

$$= (1-\gamma)\, B^{\gamma-1} \left[\frac{B^{1-\gamma} - t^{1-\gamma}}{1-\gamma} - t\, \frac{t^{-\gamma} - B^{-\gamma}}{\gamma} \right]$$

$$= 1 - \gamma^{-1} (t/B)^{1-\gamma} + (\gamma^{-1} - 1)\, t/B$$

$$\approx 1 - \gamma^{-1} (t/B)^{1-\gamma}, \tag{B.242}$$

in accordance with Eq. (11.51).

We now make use of the relation $t \ll B$ and note that the second term in Eq. (11.51) lies much closer to zero than unity. We then employ the approximation $(1+x)^M \approx 1 + Mx$ for small x, which derives from the first two terms of the associated binomial series. This yields Eq. (11.52).

Taking the derivative of Eq. (3.12) provides

$$p_\vartheta(t) = S_\tau(t)/\mathrm{E}[\tau]$$

$$dp_\vartheta(t)/dt = -p_\tau(t)/\mathrm{E}[\tau]$$

$$p_\tau(t) = -\mathrm{E}[\tau]\, dp_\vartheta(t)/dt$$

$$= \mathrm{E}[\tau]\, d^2 S_\vartheta(t)/dt^2. \tag{B.243}$$

In the context of Eq. (11.52), Eqs. (B.243) and (7.2) lead to

$$p_{\tau R}(t) = \mathrm{E}[\tau]\, d^2 S_{\vartheta R}(t)/dt^2$$

$$p_{\tau R}(t) \approx M^{-1} \frac{\gamma}{1-\gamma} (A/B)^{\gamma} B \frac{d^2}{dt^2} \left[1 - M\gamma^{-1} (t/B)^{1-\gamma} \right]$$

$$= M^{-1} \frac{\gamma}{1-\gamma} (A/B)^{\gamma} B\gamma(1-\gamma)\, M\gamma^{-1} t^{-(1+\gamma)} B^{\gamma-1}$$

$$= \gamma A^{\gamma} t^{-(\gamma+1)} \tag{B.244}$$

as provided in Eq. (11.53); this indeed coincides with Eq. (7.1) over the range $A \ll t \ll B$.

Repeating this procedure for $\gamma > 1$ yields different results. Substituting Eqs. (7.1) and (7.2) for this range of γ into Eq. (11.46) yields

$$S_{\vartheta 1}(t) = \mathrm{E}[\mu_1] \int_t^\infty (v-t)\, p_{\tau 1}(v)\, dv$$

$$\approx (1 - 1/\gamma) A^{-1} \int_t^\infty (v-t)\gamma A^{\gamma} v^{-(\gamma+1)}\, dv$$

$$= (\gamma - 1) A^{\gamma-1} \int_1^\infty t(x-1) t^{-\gamma} x^{-(\gamma+1)}\, dx$$

$$= (\gamma - 1) A^{\gamma-1} t^{1-\gamma} \int_1^\infty \left[x^{-\gamma} - x^{-(\gamma+1)} \right] dx$$

$$
\begin{aligned}
&= (\gamma - 1)A^{\gamma-1}t^{1-\gamma}\left(\frac{1}{\gamma-1} - \frac{1}{\gamma}\right) \\
&= \gamma^{-1}(t/A)^{1-\gamma},
\end{aligned}
\tag{B.245}
$$

where we again make use of the inequalities $A \ll t \ll B$, and substitute $x \equiv v/t$. This gives Eq. (11.54). Proceeding, we obtain

$$
\begin{aligned}
\mathsf{S}_{\vartheta R}(t) &= \left[\mathsf{S}_{\vartheta 1}(t)\right]^M \\
&= \left[\gamma^{-1}(t/A)^{1-\gamma}\right]^M \\
&= \gamma^{-M}(t/A)^{M(1-\gamma)},
\end{aligned}
\tag{B.246}
$$

in accordance with Eq. (11.55). We note that in this regime of γ, the binomial expansion does not apply, but rather a power-law form with a changed exponent emerges. Finally, substituting Eq. (B.246) into Eq. (B.243), and employing the shorthand notation $c \equiv 1 + M(\gamma - 1)$, yields Eq. (11.56):

$$
\begin{aligned}
p_{TR}(t) &= \mathrm{E}[\tau_R]\, d^2\mathsf{S}_{\vartheta R}(t)/dt^2 \\
&= M^{-1}\gamma(\gamma-1)^{-1}A\frac{d^2}{dt^2}\left[\gamma^{-M}(t/A)^{1-c}\right] \\
&= M^{-1}\gamma^{1-M}(\gamma-1)^{-1}A^c\frac{d^2}{dt^2}\,t^{1-c} \\
&= M^{-1}\gamma^{1-M}(\gamma-1)^{-1}A^c(c-1)\,ct^{-(c+1)} \\
&= c\gamma^{1-M}A^c\,t^{-(c+1)}.
\end{aligned}
\tag{B.247}
$$

Prob. 11.8.1 Figure B.8 presents the resulting spectrum $S_{NR}(f)$. The block shuffling operation destroys correlations at times significantly smaller than the block size $T = 100/f_S$. This effectively reduces the spectrum to the high-frequency limit for frequencies much larger than the inverse of the block size; we then have $S_{NR}(f) \approx \mathrm{E}[\mu]$ for $f \gg 1/T = f_S/100$. However, correlations at times much larger than T remain essentially unchanged, so that $S_{NR}(f) \approx S_{N1}(f)$ for $f \ll 1/T$. In the transition region $f \approx 1/T$, the abrupt cutoff of the blocks leads to oscillations in the frequency domain. More precisely, the Fourier transform of a rectangular block of duration $1/T$ yields a sinc($\cdot$) function $\sin(\pi fT)/(\pi fT)$; its square modulates the spectrum. The overall form, verified by simulation, then becomes

$$
S_{NR}(f) \approx \mathrm{E}[\mu]\left\{1 + \left[\frac{\sin(\pi fT)}{\pi fT}\right]^2 (f/f_S)^{-\alpha}\right\}.
\tag{B.248}
$$

Prob. 11.8.2 Figure B.9 presents the resulting spectrum $S_{NR}(f)$. As with block shuffling, the displacement operation destroys correlations at times significantly smaller than the block size $T = 100/f_S$, so we again have $S_{NR}(f) \approx \mathrm{E}[\mu]$ for

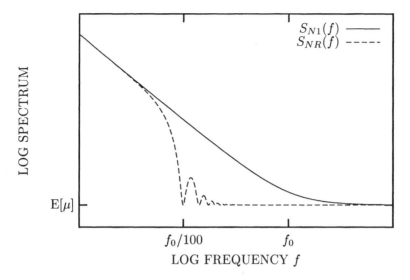

Fig. B.8 Spectrum for a fractal-rate point process $dN_1(t)$ that follows the form of Eq. (5.44a) (solid curve). Block shuffling with a block size $T = 100/f_S$ yields a point process $dN_R(t)$ with a spectrum reduced at frequencies above $1/T$, as provided in Eq. (B.248) (dashed curve).

$f \gg 1/T = f_S/100$. Also, as before, the slow components remain unchanged so that $S_{NR}(f) \approx S_{N1}(f)$ for $f \ll 1/T$. The transition region exhibits different behavior from that engendered by block shuffling. Focusing on the rate, the displacement closely resembles a noncausal filter of unit area, centered on $t = 0$, with a Gaussian shape and a standard deviation σ given by $100/f_S$ in this case.[3] In the spirit of Eq. (9.35), we then have

$$
h(t) = \frac{1}{\sqrt{2\pi}\,\sigma} \exp\left(-\frac{t^2}{2\sigma^2}\right)
$$

$$
H(2\pi f) = \int_{-\infty}^{\infty} \frac{1}{\sqrt{2\pi}\,\sigma} \exp\left(-\frac{t^2}{2\sigma^2}\right) \exp(-i2\pi ft)\,dt
$$

$$
= \frac{1}{\sqrt{\pi}} \int_{-\infty}^{\infty} \exp\left[-\left(\frac{t}{\sqrt{2}\,\sigma} + \frac{i2\pi\sigma f}{\sqrt{2}}\right)^2\right]
$$

$$
\times \exp\left[-\frac{(2\pi\sigma f)^2}{2}\right] \frac{dt}{\sqrt{2}\,\sigma}
$$

---

[3] We absorb the factor $E[\tau]$ into σ to simplify the notation instead of presenting it explicitly, as in Eq. (11.38).

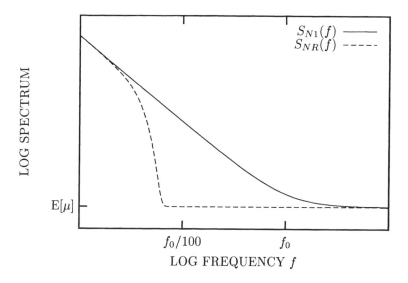

Fig. B.9   Spectrum for a fractal-rate point process $dN_1(t)$ that follows the form of Eq. (5.44a) (solid curve). Event-time displacement with a standard deviation of $T = 100/f_S$ yields a point process $dN_R(t)$ with a spectrum reduced at frequencies above $1/T$, as provided in Eq. (B.250) (dashed curve).

$$= \exp\left[-\frac{(2\pi\sigma f)^2}{2}\right] \frac{1}{\sqrt{\pi}} \int_{-\infty}^{\infty} \exp(-u^2)\, du$$

$$= \exp\left[-\frac{(2\pi\sigma f)^2}{2}\right]$$

$$\left|H(2\pi f)\right|^2 = \exp\left[-(2\pi\sigma f)^2\right]. \tag{B.249}$$

The overall form, verified by simulation, then becomes

$$S_{NR}(f) \approx \mathrm{E}[\mu]\left\{1 + \exp\left[-(2\pi\sigma f)^2\right](f/f_S)^{-\alpha}\right\}. \tag{B.250}$$

Prob. 11.9.1   The original process $dN_1(t)$ has independent intervals. Since we postulate an independent deletion operation, the interval following an event that survives the deletion remains independent of previous intervals. This defines a renewal point process.

Prob. 11.9.2   We condition on the number of deleted events between each adjacent surviving pair of events in $dN_R(t)$. Given no deletion, we have a probability density $p_{\tau 1}(t)$; this happens with probability r. For one deletion, we have $p_{\tau 1} \star p_{\tau 1}(t)$, where

$\star$ represents the convolution operation, and we expect this to occur with probability $r(1 - r)$. Summing over all possibilities, we have

$$
\begin{aligned}
p_{\tau R}(t) &= r p_{\tau 1}(t) + r(1 - r) p_{\tau 1}^{\star 2}(t) + r(1 - r)^2 p_{\tau 1}^{\star 3}(t) \cdots \\
&= r \sum_{n=1}^{\infty} (1 - r)^{n-1} p_{\tau 1}^{\star n}(t),
\end{aligned} \tag{B.251}
$$

where $p_{\tau 1}^{\star n}(t)$ denotes the n-fold convolution of $p_{\tau 1}(t)$ with itself.

Prob. 11.9.3 Taking the Fourier transform of Eq. (B.251) yields a related equation for the characteristic function

$$
\begin{aligned}
\phi_{\tau R}(\omega) &= r \sum_{n=1}^{\infty} (1 - r)^{n-1} \phi_{\tau 1}^n(\omega) \\
&= \frac{r \phi_{\tau 1}(\omega)}{1 - (1 - r) \phi_{\tau 1}(\omega)}.
\end{aligned} \tag{B.252}
$$

For the specific case of a fractal renewal point process with $0 < \gamma < 1$, Eq. (7.4) yields

$$
\begin{aligned}
1 - \phi_{\tau R}(\omega) &= 1 - \frac{r \phi_{\tau 1}(\omega)}{1 - (1 - r) \phi_{\tau 1}(\omega)} \\
&= \frac{1 - \phi_{\tau 1}(\omega)}{(1 - r) [1 - \phi_{\tau 1}(\omega)] + r} \\
&\approx \frac{\Gamma(1 - \gamma) (i\omega A)^{\gamma}}{(1 - r) \Gamma(1 - \gamma) (i\omega A)^{\gamma} + r} \\
&\approx r^{-1} \Gamma(1 - \gamma) (i\omega A)^{\gamma},
\end{aligned} \tag{B.253}
$$

where the last line results from the condition $\omega \ll A^{-1}$. Equation (B.253) differs from Eq. (7.4) only by a factor r^{-1}; thus, over the range $B^{-1} \ll \omega \ll A^{-1}$, corresponding to $A \ll t \ll B$, the randomly deleted process resembles the original process with A replaced by $r^{-1/\gamma} A$.

For $\gamma \geq 1$, the results provided in Eq. (B.253) are not valid, and in fact the probability densities $p_{\tau 1}^{\star n}(t)$ assume different shapes for different values of n. This leads to a process $dN_R(t)$ that does not have well-defined scaling regions: the process is, in general, nonfractal, except for $r \approx 0$ in which case $dN_R(t) \approx dN_1(t)$. The change in behavior at $\gamma = 1$ parallels the results presented at the end of Sec. 11.6.2 for the superposition of fractal renewal processes.

Prob. 11.10.1 Figure B.10a) is a simulation for various values of the dead-time parameter $\mu_1 \tau_e$. The mean of the initial Poisson counting distribution ($\mu_1 \tau_e = 0$) is $\mu_1 T = 15$. Increasing the dead time results in a decrease of both the mean and variance. The count mean decreases with dead time according to $\mathrm{E}[\mu_R] T = \mu_1 T / (1 + \mu_1 \tau_e)$, while the variance decreases more rapidly by virtue of the cube in the denominator of $\mathrm{Var}[n] \approx \mu_1 T / (1 + \mu_1 \tau_e)^3$.

COUNTING DISTRIBUTIONS $p_Z(n; T)$

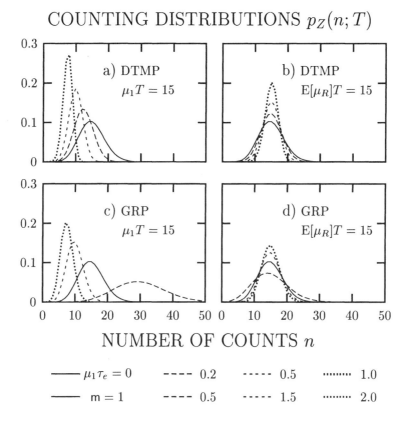

NUMBER OF COUNTS n

—— $\mu_1 \tau_e = 0$	- - - - 0.2	- - - - 0.5	········ 1.0
—— m = 1	- - - - 0.5	- - - - 1.5	········ 2.0

Fig. B.10 (a) Simulated dead-time-modified Poisson (DTMP) counting distributions with $\mu_1 T = 15$ and different dead-time parameters: $\mu_1 \tau_e = 0$ (solid curve, Poisson distribution), 0.2 (long-dash curve), 0.5 (short-dash curve), and 1.0 (dotted curve). The count mean and variance both decrease as the dead-time parameter increases, but at different rates. (b) Same as a) but with $\mu_1 T$ adjusted to compensate for the dead-time loss such that all dead-time modified counting distributions have the same final mean, $E[\mu_R] T = 15$. The counting distributions become narrower as the dead time increases. (c) Simulated gamma-renewal-process (GRP) counting distributions with m = 1 (solid curve, Poisson distribution), 0.5 (long-dash curve), 1.5 (short-dash curve), and 2.0 (dotted curve). The counting distribution is broader than the Poisson for m < 1 and narrower for m > 1. (d) Same as c) but with $\mu_1 T$ adjusted to compensate for the decimation so that all counting distributions have the same final mean, $E[\mu_R] T = 15$. All distributions are simulated using 10^8 intervals. Adapted from Fig. 3 of Teich & Vannucci (1978) and Fig. 14 of Teich et al. (1984).

Prob. 11.10.2 We illustrate dead-time-modified Poisson counting distributions with a final mean $E[\mu_R] T = 15$ in Fig. B.10b). The relative width clearly narrows as the dead-time parameter becomes larger, as a consequence of the increasing regularity of the event occurrences.

Prob. 11.10.3 Given that the two processes have the same initial mean rate, dead time is more effective in decreasing the mean and variance for the fractal-rate process. This occurs because the increased event clustering engendered by the rate modulation causes more events to be lost in each dead-time period.

Prob. 11.10.4 We present simulated counting distributions for the gamma renewal process in Fig. B.10c) for various values of the order parameter m. Again, the mean of the initial Poisson counting distribution (m $= 1$) is $\mu_1 T = 15$. Increasing the order parameter yields a decrease in both the count mean and variance. An order parameter that is less than unity results in an increase of both the mean and variance. The count mean varies with the order parameter in accordance with $\mathrm{E}[\mu_R]\, T = \mu_1 T/m$ while the variance changes more rapidly by virtue of the square in the denominator of $\mathrm{Var}[n] \approx \mu_1 T/m^2$.

Prob. 11.10.5 Simulated decimated-Poisson counting distributions with a final mean $\mathrm{E}[\mu_R]\, T = 15$ appear in Fig. B.10d). The relative width clearly narrows as m becomes larger, corresponding to the increasing regularity of the event occurrences. Values of m < 1 yield distributions broader than the Poisson by virtue of the increased clustering in the process.

Prob. 11.11 Although fractal fluctuations are present in the original point process, the small value of the coefficient of variation ($C_{\tau 1} \doteq 0.176$) dictates that even large variations in the rate result in rather modest changes in the numbers of counts. Since exponentialization imparts a coefficient of variation of unity to the process, it effectively amplifies the fractal character of the resulting process by the ratio of the interval variances, namely $(C_{\tau R}/C_{\tau 1})^2 = (1/C_{\tau 1})^2 \doteq 32.2$.

Prob. 11.12.1 Little information can be drawn from Fig. 11.16 since: (1) the shuffled surrogate, by construction, yields a curve identical to that for the original data; and (2) the exponentialized surrogate, by construction, yields an exponential interevent-interval histogram. Knowledge of the estimated interevent-interval histogram alone is thus of little help in identifying the underlying point process.

Prob. 11.12.2 All three curves in Fig. 11.17 differ. The power-law growth of the normalized Haar-wavelet variance for the original data, with increasing counting time, signals the likely presence of fractal behavior of some form in the point process. The exponentialized version of the data has similar fractal content although it *appears* to have an increased fractal onset time and a wavelet variance that is slightly reduced in magnitude. Taken together, these observations might suggest that we can ascribe at least a portion of the fractal behavior to the ordering of the intervals.

The normalized Haar-wavelet variance for the shuffled data increases a bit and then saturates at a value of $\widehat{A}(T) \approx 2$ [the results are similar to those for the cat striate cortex cell displayed in Fig. 11.14 and discussed by Teich et al. (1996, Fig. 2); this cell also has a quite low firing rate]. Behavior of this kind can arise from a nonfractal renewal process model such as the gamma (see Prob. 4.7 in the range $0 \le m \le 1$) or by a nonfractal cluster point process with slight clustering (see Sec. 4.5). So in view of the behavior of the shuffled data, it appears that very little, if any, of the fractal behavior in the original point process does derive from the form of the underlying interevent-interval density.

The apparent reduction of the fractal content of the exponentialized data is therefore simply a manifestation of removing the (essentially nonfractal) features of the interevent-interval density inherent in the original data. Since shuffling completely destroys the fractal behavior whereas exponentialization does not change the fractal qualities for this data set, we conclude that the ordering of the intervals is solely responsible for the fractal behavior.

These observations reinforce earlier caveats about attempting to infer the nature of a fractal-based point process from a single measure. Indeed, neural spike trains recorded from many loci, in many preparations, exhibit power-law interevent intervals similar to those seen in the solid curve of Fig. 11.16 (see, for example, Gerstein & Mandelbrot, 1964; Wise, 1981). Such data are often coopted to conclude that fractal renewal processes provide useful models for these action-potential sequences, but the interneuron example at hand makes it clear that drawing such conclusions requires caution.

Prob. 11.12.3 Adding the two curves results in something quite close to the curve for the original data. Indeed, for short counting times the correspondence is almost exact. This indicates that the surrogate data sets perform as designed; namely, they separate out two aspects of point-process behavior: fluctuations associated with the interval distribution and fluctuations associated with the interval ordering.

Prob. 11.12.4 It is not straightforward to make a definitive choice based on the limited number of measures studied. Considering the underlying neurophysiology of the preparation provides guidance in setting forth the options. A neurophysiologically plausible point process that generates sample functions that accord with the data is a nonfractal cluster process driven by fractal-rate fluctuations (Teich et al., 1997). A natural choice for the fractal-rate fluctuations is fractal binomial noise (Thurner et al., 1997), which, as discussed in Sec. 8.4, converges to the fractal Gaussian process. An alternative choice is a gamma-based fractal doubly stochastic Poisson process, such as the fractal-binomial-noise-driven gamma process mentioned in Sec. 8.4; however, this model is less appealing from a neurophysiological perspective (Teich et al., 1997). The behavior of the shuffled version of the data in Fig. 11.17 supports either interpretation. *A priori* information clearly plays an important role in point-process identification.

Prob. 11.12.5 The generalized-dimension scaling functions presented in Fig. 11.18 reveal that the INTERNEURON action-potential sequence is not a fractal point process. Given its fractal characteristics, it must belong to the family of fractal-rate point processes (see Sec. 3.5.4 and Prob. 5.5). This accords with the supposition set forth in Prob. 11.12.4.

B.12 ANALYSIS AND ESTIMATION

Prob. 12.1 We can understand the forms of the interevent-interval densities that characterize these processes by viewing them as modulated versions of the basic exponential form provided in Eq. (4.3), which, of course, applies exactly for a Poisson process driven by a fixed, deterministic rate (yielding the homogeneous Poisson process in that case). Modulating the rate in effect modulates the mean value of the exponential density. If the rate modulation is slow in comparison with the mean rate of events, and is sufficiently weak so that its coefficient of variation obeys $C_\mu \ll 1$, then the interval density remains essentially exponential, as provided in Eq. (4.33). However, for larger values of C_μ, perceptible broadening of the interval density can occur via Eq. (4.32). Since, by assumption, rate processes (2) and (3) do not differ substantially from rate process (1), the overall forms of the interevent-interval densities for all three processes should not differ greatly from each other.

In fact, distinguishing among these three processes proves nearly impossible without recourse to an extraordinarily long record. In Fig. 10.6 for the fractal-shot-noise-driven Poisson process, for example, the interevent-interval density for $\beta = \frac{1}{2}$, corresponding to $\alpha = 1$, deviates from the equivalent result for the fractal-Gaussian-process-driven Poisson process (not shown) only for about one interval in 10^8 [we do not consider the other two curves displayed in Fig. 10.6 since they do not yield spectra that follow the form of Eq. (5.44a)]. An exception to this general similarity occurs for shot-noise rates with large K and small A. The shot-noise rate then takes on relatively large values immediately following the onset of each impulse response function. This local augmentation of the rate, in turn, can lead to a significant increase in the proportion of small intervals, while leaving the remainder of the interval density essentially intact.[4] The results in Fig. 10.6 correspond to relatively small values of K, so that no such increase for small interevent intervals is apparent.

The nature of the rate process is far more accessible if the kernel has an integrate-and-reset form. In this case, a rate comprising fractal binomial noise generates a piecewise-periodic point process between transition times of the component alternating fractal renewal processes, if significant numbers of events occur between these transitions. A histogram of the interevent intervals would readily reveal this feature, thereby distinguishing a fractal-binomial-noise-driven integrate-and-reset process from its shot-noise and Gaussian-process brethren. Although histograms of the interevent intervals for these two latter processes may resemble each other, the point process deriving from fractal shot noise will exhibit steadily increasing intervals except at the onset times of the impulse response functions. The result for the fractal Gaussian process, in contrast, will exhibit as many increases as decreases and therefore lack this asymmetry.[5]

---

[4] Examples of this appear in Lowen et al. (2001, Figs. 3B and 4B on pp. 385 and 387, respectively) for fractal-rate visual-system action potentials. For nonfractal cathodoluminescence photon emissions, examples appear in Saleh & Teich (1982, Figs. 10 and 14 on pp. 236 and 239, respectively).

[5] Examples of the latter case, for the human-heartbeat point process, appear in Turcott & Teich (1993, Fig. 3 on p. 26) and in Turcott & Teich (1996, Fig. 3 on p. 278).

Prob. 12.2 Given additional *a priori* information about a process, we can fit increasingly more explicit and accurate functional forms to the various estimators we employ, such as the periodogram or the estimated normalized Haar-wavelet variance. If, for example, we knew that we had a fractal renewal process at hand, with an interevent-interval density given by Eq. (7.1) and $1 \leq \gamma \leq 2$, we could fit a suitable functional form to the periodogram or even use maximum-likelihood methods. We would know to expect oscillations in the periodogram resulting from the abrupt cutoff in the probability density, among other features (see Fig. 7.2). But this increase in accuracy comes at the expense of robustness. Such specificity would prove disastrous were we to encounter instead a point process that differs greatly from the prescribed form. A fractal-shot-noise-driven Poisson process, for example, has a periodogram with significantly different detailed structure (see Fig. 10.8), and the fractal renewal assumptions would lead to spurious estimates of α. Indeed, we already encountered just such a conundrum in examining the action-potential statistics generated by an insect visual-system interneuron (see Probs. 7.8 and 11.12).

Prob. 12.3 Computing the correlation coefficient between the absolute values in columns three and four of Table 12.1 yields $\rho = -0.620$, which is less than zero and therefore does indeed indicate a bias/variance tradeoff. However, carrying out this calculation using the entries in columns six and seven yields $\rho = +0.954$, in apparent contradiction to the bias/variance tradeoff hypothesis. Why might this arise?

Examining the entries in the latter two columns reveals that a positive bias appears only for the shortest time scales; on the other hand, the largest negative bias occurs at the largest time scales. The decrease in the magnitude of the bias as the time range increases results from a partial cancelation of the positive and negative bias values. The decreased standard deviation results from the larger numbers of normalized Haar-wavelet variance values over which we calculate the fractal-exponent estimate. Since these two sources of error vary together, we expect a positive correlation coefficient. In this case, we have substantially eliminated the bias by subtracting unity from the normalized Haar-wavelet variance so that we can fully utilize the entire range of time scales. The bias/variance tradeoff therefore does not apply.

Let us examine the variation in the bias a bit more carefully. Since the magnitude of the bias in column six lies well below the standard deviation in column seven, we might be tempted to attribute the entries in column six to random fluctuations about a true bias of zero; indeed, this would generate a positive correlation coefficient. However, this is not a likely explanation for the large positive value of ρ. The standard error (the standard deviation of the mean estimate) equals the standard deviation divided by the square root of the number of simulations; the bias exceeds this value in all cases, and in fact it exceeds it by a factor of 2.3 in most of the cases. This explanation might still have merit were the distribution of the estimated values of α to deviate significantly from Gaussian form, but, in fact, estimates of the skewness and kurtosis (not shown) confirm a Gaussian distribution. Hence, it is not likely that random Gaussian fluctuations about a true bias of zero provides an explanation for the values of the bias displayed in column six. Other, subtle effects offer a more cogent explanation, as discussed in Sec. 12.2.3.

Prob. 12.4 Given a finite data set, we find an estimate of the coincidence rate by assembling a list of all possible pairs of events (regardless of the presence or absence of intervening events), and note the delay times between each pair. We can say that a coincidence has occurred at each of these delay times, but we have no information about other delay times that might also represent coincidences; only those delay times that happened to occur in our data appear in the final result. Because of the sparseness of this statistics, the delay times almost surely do not include any particular time we choose *a priori*, and so we estimate $G(t) = 0$ for any delay time specified before collecting the data. To make use of this statistic we must average over time windows. Although other averaging methods exist, a rectangular filter of unit height and constant duration T seems most convenient. Applying this filter to a point process yields the sequence of counts $Z(k, T)$, and the filtered coincidence rate then becomes the count-based autocorrelation $R_Z(k, T)$. Section 12.3.1 addresses this issue.

This problem does not afflict the point-process spectrum. We can readily carry out the Fourier transform of the point process, take the absolute magnitude, square it, and divide by the duration of the data set to estimate it, as indicated in Sec. 3.5.2:

$$\widetilde{N}(f) \equiv \int_0^L e^{-i2\pi ft} \, dN(t)$$

$$= \int_0^L e^{-i2\pi ft} \sum_k \delta(t - t_k) \, dt$$

$$= \sum_k e^{-i2\pi ft_k}$$

$$\left|\widetilde{N}(f)\right|^2 = \left|\sum_k e^{-i2\pi ft_k}\right|^2$$

$$\widehat{S}_N(f) = \frac{1}{L}\left|\sum_k e^{-i2\pi ft_k}\right|^2. \tag{B.254}$$

Equation (B.254) does indeed provide an accurate estimate of the point-process spectrum, without the bias inherent in using the rate-based periodogram [see Eq. (3.67)]. However, as noted in Sec. 3.5.2, the estimate set forth in Eq. (B.254) suffers from a major drawback: the times $\{t_k\}$ span a continuous range of values, which precludes use of the fast Fourier transform algorithm. For a large data set this method can take several orders of magnitude longer to compute than the rate-based periodogram, with little advantage in accuracy. We therefore generally compute the rate spectrum (see Sec. 12.3.9).

Prob. 12.5.1 From Eq. (12.6) we have

$$F(T) \approx (T/T_F)^\alpha \left[1 - (T/L)^{1-\alpha}\right]$$

$$= T_F^{-\alpha} T^\alpha - T_F^{-\alpha} L^{\alpha-1} T$$

$$\frac{dF(T)}{dT} = \alpha T_F^{-\alpha} T^{\alpha-1} - T_F^{-\alpha} L^{\alpha-1} = 0$$

$$0 = \alpha T^{\alpha-1} - L^{\alpha-1}$$
$$T = \alpha^{1/(1-\alpha)} L, \qquad\qquad (\text{B.255})$$

which is independent of the fractal onset time T_F.

Prob. 12.5.2 For convenience, we employ the same simulation as that used through-out Chapter 12 (see Sec. 12.2.3), effectively extending Fig. 12.2 to larger count-ing times. Since estimating a variance requires at least two values, we must have $M \equiv \text{int}(L/T) \geq 2$, which, in turn, implies that $T \leq L/2$. To highlight the differ-ences between the predictions provided by Eqs. (5.44b) and (12.6), we present the results in the form of a doubly linear plot. As shown in Fig. B.11, the simulations lie far from Eq. (5.44b). Although they roughly follow the form of Eq. (12.6), they con-sistently lie above this prediction, suggesting that Eq. (12.6) slightly overestimates the bias.

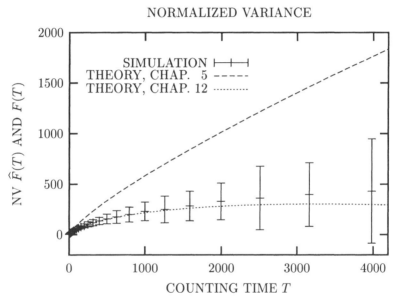

Fig. B.11   Two theoretical results for the normalized variance $F(T)$ vs. counting time T, along with an estimate of this quantity, $\widehat{F}(T)$, based on the same simulations as used to generate Figs. 12.1–12.7 and 12.9. We present the mean ± standard deviation of the simulated results (short horizontal lines ± error brackets), as well as the predicted theoretical curves from Chapter 5 [Eq. (5.44b), dashed curve] and from Chapter 12 [Eq. (12.6), dotted curve]. The simulations closely follow the latter result, but they consistently lie just above it.

Prob. 12.6 Our principal concern has heretofore been the accurate estimation of the fractal exponent α from the periodogram (spectrum estimate), rather than the estima-tion of the spectrum itself. Averaging adjacent computed values of the periodogram does indeed lead to spectral estimates with reduced variance. However, this comes at the cost of fewer spectral values so that the net variance does not change. But, since the averaged values derive from different frequencies, which have different expected values, the averaging process introduces bias. Although this bias is small at high frequencies, where adjacent values differ by small relative amounts, the averaging procedure strongly affects the results at low frequencies where the spectrum changes quickly with frequency. Since the averaging process introduces bias without reducing the variance, we decline to make use of it.

A similar argument applies for block averaging of the periodogram. Fourier transforming a block yields fewer values than does transforming the entire data record. Although the averaging then occurs across transforms at identical frequencies, which eliminates the bias problem, the division of the data into blocks precludes estimating components at frequencies that lie below the inverse of the block duration. Averaging again results in fewer values with proportionately reduced variance, but it destroys important information at the longest time scales. With the net variance unchanged, and a portion of the spectrum rendered unavailable, this method does not prove helpful in fractal-exponent estimation. Indeed, as a general rule, parametric-estimation practice eschews smoothing of any kind before parameter estimation.

Prob. 12.7 We focus on the spectrum, as estimated by the periodogram, although the following argument applies to any measure with an asymptote. A real data set has finite length so that the periodogram constructed from it exhibits random fluctuations. These fluctuations, an inherent part of constructing the estimator, depend on all features of the spectrum at any given frequency, not only on the fractal component. The fluctuations can thus remain significant even when the fractal component becomes small. In particular, near the high-frequency/short-time limit, the contribution from fluctuations can readily exceed that from the fractal component. Subtracting the asymptote from the estimated value might then give rise to a negative value for the adjusted periodogram at that frequency; computation of the logarithm then cannot proceed without a modification such as that used in Sec. 12.3.9. Even so, at high frequencies little usable information pertaining to the fractal content of the spectrum resides in its estimate.

Another potential problem with subtracting the high-frequency/short-time limit arises in connection with the intrinsic nature of certain types of point processes. Even without random effects in the periodogram, point processes often have intermediate-frequency effects (such as dead time) that cause the spectrum, over some range of frequencies, to lie below the high-frequency limit (see Fig. 5.1, particularly the curve labeled HEARTBEAT). Subtracting the high-frequency asymptote then leads to negative values in the periodogram, and thence to the absence of usable information over that range. The averaging methods employed in Sec. 12.3.9 do not render these spectral values useful, since they lie well below the high-frequency asymptote for an extended range of frequencies. The simulations used to generate Tables 12.1–12.10 had no dead time or other intermediate-frequency effects, by construction, so that this issue did not arise. We reiterate our caution in employing *a priori* information such as this when estimating an unknown process.

This issue also arises in connection with the normalized Haar-wavelet variance, as shown in Fig. 5.2 (see also Fig. 7.5 and Fig. B.13 in Prob. 12.8).

Prob. 12.8 Figures B.12 and B.13 present results for the periodogram (rate spectrum estimate) and normalized Haar-wavelet variance estimate, respectively, for the simulated fractal renewal process (FRP) and homogeneous Poisson process (HPP). For the fractal renewal point process, both of these count-based measures clearly reveal power-law variation over a large range of frequency and time, indicating fractal behavior. For the homogeneous Poisson process, both show no significant variation with frequency and time, indicating a nonfractal process. For both processes, therefore, the two measures reliably describe the presence or absence of fractal characteristics in the point process under study.

The dip in the normalized Haar-wavelet variance that occurs for the fractal renewal process in Fig. B.13 arises from the abrupt cutoff of small intervals in the interevent-interval density. This cutoff is equivalent to dead time in the underlying point process (Fig. 5.2 reveals how widespread behavior of this kind is for real data). The presence of this dip vividly illustrates one of the difficulties associated with subtracting asymptotic

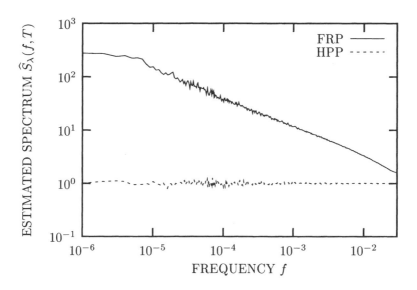

Fig. B.12 Estimated spectrum (periodogram) $\widehat{S}_\lambda(f, T)$ vs. frequency f for a simulated fractal renewal process (FRP, solid curve) and a homogeneous Poisson process (HPP, dashed curve) with the same mean rate, $E[\mu] = E[\tau] = 1$. The counting time $T = L/2^{16} \doteq 15.2588$. Parameters used in simulating the fractal renewal process are as follows: $\gamma = \frac{3}{2}$ $(\alpha = \frac{1}{2})$, $B/A = 10^6$, $A = 1.001001/3 = 0.333667$, $B = 0.333667 \times 10^6$, and $L = 10^6$. The spectrum reliably reports the presence of power-law behavior in the fractal process and its absence in the nonfractal process. Calculated point-process spectra for the fractal renewal process appear in Fig. 7.2.

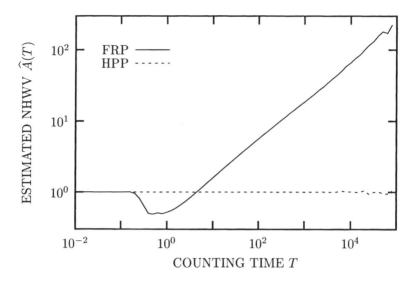

Fig. B.13 Estimated normalized Haar-wavelet variance, $\widehat{A}(T)$ vs. counting time T, for a fractal renewal point process (FRP, solid curve) and a homogeneous Poisson process (HPP, dashed curve) with the same mean rate of unity, $E[\mu] = E[\tau] = 1$. The caption of Fig. B.12 provides parameters for the fractal renewal process. Along with the results shown in Fig. B.12, this measure reliably reports power-law behavior in the fractal process and its absence in the nonfractal process. The dip in the curve for the fractal renewal process derives from the abrupt cutoff in the interevent-interval density for small intervals. A simulated version of $A(T)$ for $\gamma = \frac{1}{2}$, which also yields $\alpha = \frac{1}{2}$, appears in Fig. 7.5.

values in an attempt to improve fractal-exponent estimation, as discussed in Prob. 12.7. In the case at hand, the asymptote is unity; the decrease of $\widehat{A}(T)$ below unity therefore renders the adjusted normalized Haar-wavelet variance negative, an impossibility.

Figure B.14 shows the corresponding results for the estimated normalized rescaled range statistic, $\widehat{U}_2(k) = \widehat{U}^2(k)/k$. This measure should lie near unity for a nonfractal process, which indeed it does for the homogeneous Poisson process. However, it exhibits similar behavior for the fractal renewal process. Indeed, the two curves differ by less than a factor of 1.3 over the entire range examined, which spans five decades. This result confirms that this interval-based measure fails to reveal fractal behavior in the fractal renewal process.

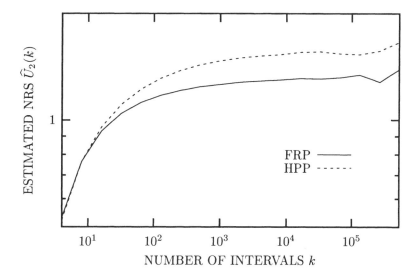

Fig. B.14 Normalized rescaled range estimate $\widehat{U}_2(k) = \widehat{U}^2(k)/k$ vs. number of intervals k, for a fractal renewal point process (FRP, solid curve) and a homogeneous Poisson process (HPP, dashed curve) with the same mean rate of unity, $E[\mu] = E[\tau] = 1$. The caption of Fig. B.12 provides parameters for the fractal renewal process. In contrast to the results demonstrated in Figs. B.12 and B.13, this measure fails to detect scaling behavior in the fractal process.

B.13   COMPUTER NETWORK TRAFFIC

Prob. 13.1 For both Eqs. (13.1) and (13.2) we simply set $Q_m \to \infty$. This leads to

$$\frac{dp\,(n,t)}{dt} = \begin{cases} -\mu_a\, p\,(n,t) & +\mu_s\, p\,(n+1,t) & n=0 \\ -(\mu_a+\mu_s)\, p\,(n,t) + \mu_a\, p\,(n-1,t) + \mu_s\, p\,(n+1,t) & & n>0 \end{cases}$$

(B.256)

and

$$p_\infty(n,t) \to (1-\rho_\mu)\,\rho_\mu^n, \tag{B.257}$$

as in Eq. (13.4).

Prob. 13.2 We wish to accommodate M servers, all handling requests from the same buffer. For large values of the queue length n, where all M servers are effective in decreasing it, we replace the service rate μ_s with $M\mu_s$. For smaller values of the buffer occupancy n, where not all servers can work at the same time, the service rate increases by a factor of only n, rather than M. The effective service rate therefore becomes $\min(n,M)\mu_s$, where the function $\min(\cdot,\cdot)$ returns the smaller of its two arguments. We must thus institute the following modifications in Eq. (13.1):

$$\begin{aligned} +\mu_s\, p\,(n+1,t) &\quad\to\quad +\min(n+1,M)\,\mu_s\, p\,(n+1,t) \\ -\mu_s\, p\,(n,t) &\quad\to\quad -\min(n,M)\,\mu_s\, p\,(n,t). \end{aligned}$$

(B.258)

Prob. 13.3 We obtain the queue-length distribution $p_\infty(n)$ directly from Eq. (13.4). From this we immediately write the four possibilities:

$$\begin{aligned} p_a &\equiv& p_\infty(Q_m) &=& (1-\rho_\mu)\,\rho_\mu^{Q_m} \\[4pt] p_b &\equiv& p_\infty(Q_m+1) &=& (1-\rho_\mu)\,\rho_\mu^{Q_m+1} \\[4pt] p_c &\equiv& \sum_{n=Q_m}^{\infty} p_\infty(n) &=& \sum_{n=Q_m}^{\infty}(1-\rho_\mu)\,\rho_\mu^n &=& \rho_\mu^{Q_m} \\[4pt] p_d &\equiv& \sum_{n=Q_m+1}^{\infty} p_\infty(n) &=& \rho_\mu^{Q_m+1}. \end{aligned}$$

(B.259)

To determine which of these lies closest to the true result $p_Q(Q_m)$, we divide each by $p_Q(Q_m)$, as provided in Eq. (13.8). We then compare results, considering the limit $Q_m \to \infty$:

$$\begin{aligned} p_a/p_Q(Q_m) &=& \left(1-\rho_\mu^{Q_m+1}\right) &\to& 1 \\ p_b/p_Q(Q_m) &=& \rho_\mu\left(1-\rho_\mu^{Q_m+1}\right) &\to& \rho_\mu \\ p_c/p_Q(Q_m) &=& (1-\rho_\mu)^{-1}\left(1-\rho_\mu^{Q_m+1}\right) &\to& (1-\rho_\mu)^{-1} \\ p_d/p_Q(Q_m) &=& \rho_\mu(1-\rho_\mu)^{-1}\left(1-\rho_\mu^{Q_m+1}\right) &\to& \rho_\mu(1-\rho_\mu)^{-1}. \end{aligned}$$

(B.260)

We conclude that $p_a \equiv p_\infty(Q_m)$ provides the closest approximation to $p_Q(Q_m)$.

Prob. 13.4 A mean interevent interval of 10 msec and a mean service time of 9 msec together yield a service ratio $\rho_\mu = 0.9$. Since both the arrival and service processes follow a homogeneous-Poisson form, and a single server handles the requests, we can make use of the results obtained for the M/M/1/Q_m queue. Inverting Eq. (13.9) yields

$$Q_m = -\log\left(\rho_\mu + \frac{1 - \rho_\mu}{P_B}\right) \Big/ \log(\rho_\mu). \tag{B.261}$$

Substituting the values $P_B = 10^{-3}, 10^{-6}$, and 10^{-9} into Eq. (B.261), and rounding up to the nearest integer, yields buffer sizes of 44, 110, and 175, respectively.

Prob. 13.5 We simulate this M/M/1 queue, effectively using two homogeneous Poisson processes. Since both the arrival and service processes lack memory, we can note the queue length after each event (arrival or departure from the queue), and base our results on that statistic. Alternatively, we can sample the queue at all times equal to integer multiples of a sampling time. Decreasing this sampling time provides better resolution. Indeed, we can effectively achieve an infinitesimal sampling time by recording the durations between events, which yields the proportion of the total time spent at each queue length. We use this approach in this simulation and in the ones following.

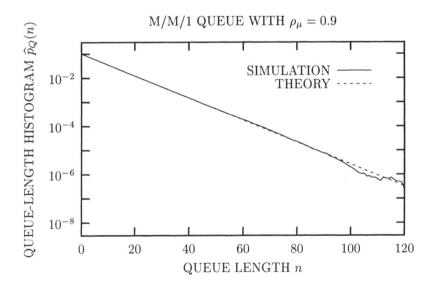

Fig. B.15 Simulated queue-length histogram for the M/M/1 queue with $\rho_\mu = 0.9$ (solid curve), and the geometric-distribution theoretical fit provided by Eq. (13.4) (dashed curve).

Figure B.15 presents the simulated queue-length histogram (solid curve), along with the theoretical result (dashed curve)

$$p_\infty(n) = (1 - \rho_\mu)\,\rho_\mu^n, \qquad (B.262)$$

which is the geometric distribution reported in Eq. (13.4). The simulation agrees well with the theory.

In Prob. 13.4 we used analytical formulas for the M/M/1/Q_m queue to determine that buffer sizes of 44 and 110 correspond to overflow probabilities of 10^{-3} and 10^{-6}, respectively. The queue-length distribution for the M/M/1/∞ queue plotted in Fig. B.15 reveals that $p_\infty(44) \approx 10^{-3}$ and $p_\infty(110) \approx 10^{-6}$, thereby confirming the validity of Eq. (13.12) for sufficiently large values of Q_m: $P_B \approx p_\infty(Q_m)$.

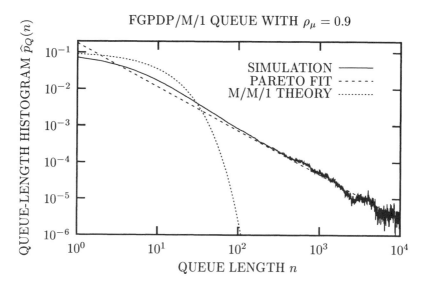

Fig. B.16   Simulated queue-length histogram $p_Q(n)$ for the FGPDP/M/1 queue (solid curve). Arrivals at the queue are described by a fractal-Gaussian-process-driven Poisson process (FGPDP) simulated using the following parameters: E[μ] = 100, duration $L = 10^6$, fractal exponent $\alpha = 0.8$, onset frequency $f_S = 0.2$, and fractal-Gaussian-process array size $M = 2^{24}$. With an expected service time of 0.009, this yields a service ratio $\rho_\mu = \mu_a/\mu_s = 100 \cdot 0.009 = 0.9$. We present an empirical power-law fit provided in Eq. (B.263) (dashed curve), and the M/M/1 theoretical geometric distribution from Eq. (13.4) (dotted curve). The simulated queue-length histogram closely follows the power-law decreasing form with the same fractal exponent, rather than the exponentially decreasing form of the M/M/1 distribution.

Prob. 13.6.1   Since we increase the length of the simulation by a factor of 100 over that used in Chapter 12, we increase the fractal Gaussian process array size M by a factor of $2^7 = 128$, the closest multiple of 2 to 100. Again, to ensure stationarity we prepend an additional 1% of the total simulation, the results of which we then discard before compiling the queue-length histogram.

Figure B.16 displays the estimated queue-length histogram obtained by using this fractal-Gaussian-process-driven Poisson-process simulation for the arrival process, coupled with exponential service times and a single server: the FGPDP/M/1 queue (solid curve). We also present a power-law fit (dashed curve)

$$p_Q(n) = c\, n^{\alpha-2},\tag{B.263}$$

with c chosen to yield a counting distribution that sums to unity (dashed curve), and α chosen to coincide with the design value $\alpha = 0.8$. The form of the exponent derives from asymptotic bounds on the queue (Likhanov, 2000). The distribution presented in Eq. (B.263) is known as the zeta distribution (see Sec. 2.7.1).

We also present the theoretical M/M/1 result set forth in Eq. (13.4) (dotted curve). The simulated queue-length histogram behaves as a decaying power-law function rather than as a decaying exponential function. The arrival process evidently imparts its fractal character to the queueing process, yielding the power-law behavior of the FGPDP/M/1 queue-length distribution (straight line on a doubly logarithmic plot).

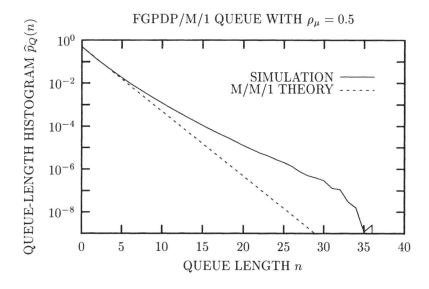

Fig. B.17 Simulated queue-length histogram $p_Q(n)$ for the FGPDP/M/1 queue (solid curve). Arrivals at the queue are described by a fractal-Gaussian-process-driven Poisson process (FG-PDP) simulated using the following parameters: $E[\mu] = 100$, duration $L = 10^6$, fractal exponent $\alpha = 0.8$, onset frequency $f_S = 0.2$, and fractal-Gaussian-process array size $M = 2^{24}$. Decreasing the expected service time to 0.005 reduces the service ratio to $\rho_\mu = \mu_a/\mu_s = 100 \cdot 0.005 = 0.5$. The M/M/1 theoretical geometric distribution from Eq. (13.4) appears as well (dashed curve). The simulated queue-length histogram follows the exponentially decreasing form of the M/M/1 distribution reasonably well.

Prob. 13.6.2 Reducing the mean service time from 0.009 to 0.005 serves to reduce the service ratio ρ_μ and to greatly diminish the queue length. With the average service rate nearly doubled, even large clusters of arrivals pass through the server without significantly burdening the queue. Figure B.17 presents the simulated FGPDP/M/1 queue-length histogram (solid curve), along with the theoretical M/M/1 queue-length distribution set forth in Eq. (13.4) (dashed curve). Though not perfect, the M/M/1 result provides a far superior fit than does a power-law form, which would exhibit significant curvature on this plot (not shown).

Prob. 13.7 Since the fractal-based point process used in Prob. 13.6 has a Poisson kernel and the rate process has a standard deviation smaller than the mean ($C_\mu < 1$), the probability density associated with the resulting interevent intervals does not depart greatly from an exponential form [see Eq. (4.33)]. Hence, shuffling the intervals results in a simulated renewal point process with an estimated interevent-interval density close to an exponential form. This corresponds to a homogeneous Poisson point process. We therefore expect that the SHUFFLED-FGPDP/M/1 traffic process will lead to results similar to those obtained for the M/M/1 queue discussed in Prob. 13.5.

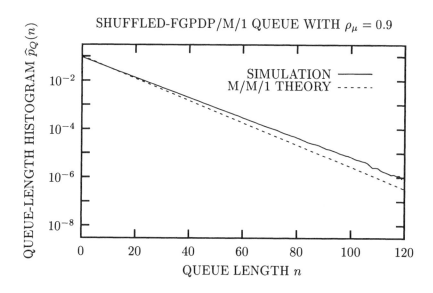

Fig. B.18 Simulated queue-length histogram $p_Q(n)$ for the FGPDP/M/1 queue, after shuffling (solid curve). Arrivals at the queue are described by a shuffled version of a fractal-Gaussian-process-driven Poisson process (FGPDP) simulated using the following parameters: $E[\mu] = 100$, duration $L = 10^6$, fractal exponent $\alpha = 0.8$, onset frequency $f_S = 0.2$, and fractal-Gaussian-process array size $M = 2^{24}$. The service ratio was $\rho_\mu = 0.9$. The M/M/1 theoretical geometric distribution (using the theoretical arrival rate) of Eq. (13.4) also appears (dashed curve). The simulated queue-length histogram closely follows the exponentially decreasing form of the M/M/1 distribution, rather than the power-law decreasing form followed by the unshuffled results (see Fig. B.16).

Figure B.18 displays the queue-length histogram resulting from the simulation (solid curve), along with a plot of the theoretical geometric M/M/1 result (dashed curve), as provided in Eq. (13.4). As before, we discard an added initial 1% before analyzing the queue-length statistics. The M/M/1 theoretical result fits the simulation reasonably well. The modulating effect of the fractal Gaussian process influences the ultimate mean value of the point process, so that we gain improved agreement by using the measured arrival rate (100.587) in Eq. (13.4), rather than the expected value $\mu_a = 100$ (not shown). Still better agreement would obtain from incorporating the deviation of the interevent-interval density from an exponential form.

Prob. 13.8 Figure B.19 displays the estimated queue-length histogram obtained by using a rectangular fractal-shot-noise-driven Poisson simulation for the arrival process, coupled with exponential service times and a single server (solid curve): this appears as the RFSNDP/M/1 queue. We also show the power-law form of Eq. (B.263) (dashed curve) using the design value $\alpha = 0.8$, and the theoretical M/M/1 queue re-

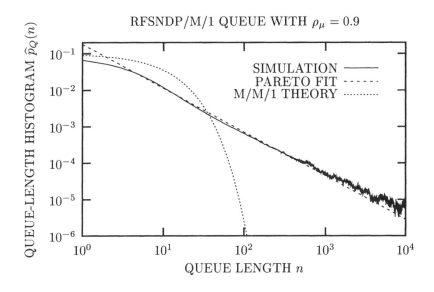

Fig. B.19 Simulated queue-length histogram $p_Q(n)$ for the RFSNDP/M/1 queue (solid curve). A rectangular fractal-shot-noise-driven Poisson process (RFSNDP) describes arrivals at the queue; the impulse-response-function duration B obeys a decaying power-law distribution $\sim B^{-1.2}$. The service ratio $\rho_\mu = \mu_a/\mu_s = 100 \cdot 0.009 = 0.9$. We also shown the simple power-law form provided in Eq. (B.263) (dashed curve), and the M/M/1 theoretical geometric distribution of Eq. (13.4) (dotted curve). The simulated queue-length histogram closely follows that of the FGPDP/M/1 queue displayed in Fig. B.16. A simple power-law decreasing fit with exponent $\alpha - 2 = -1.2$ thus describes both the FSNDP/M/1 and FGPDP/M/1 queues well for a broad range of queue lengths, while the exponentially decreasing M/M/1 queue-length distribution does not.

sult from Eq. (13.4) (dotted curve). The simulation follows a simple power law with exponent $\alpha - 2 = -1.2$ over a large range of queue lengths.

As expected, the outcome resembles that obtained for the FGPDP/M/1 queue (see Fig. B.16), where the arrival process at the queue is a fractal-Gaussian-process-driven Poisson process.

Prob. 13.9 Figure B.20 shows the estimated queue-length histogram obtained by using a modulated fractal-Gaussian-process-driven Poisson simulation for the arrival process, coupled with exponential service times and a single server (solid curve); we denote this the MODULATED-FGPDP/M/1 queue. The modulation is sinusoidal with unity modulation depth. We also show the result for the original (unmodulated) fractal-Gaussian-process-driven Poisson simulation (dashed curve).

In comparison with the original, the modulated version lies below it at small queue lengths, and above it at large queue lengths. The depression and bump in the modulated curve resemble the depression and bump in the sinusoidally modulated

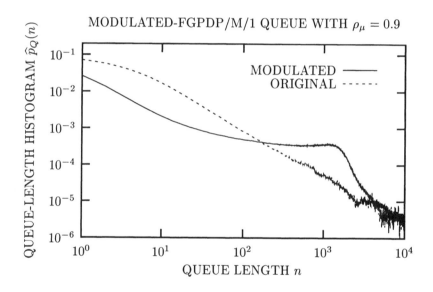

Fig. B.20 Simulated queue-length histogram $p_Q(n)$ for the MODULATED-FGPDP/M/1 queue (solid curve). We begin with an unmodulated simulation of the fractal-Gaussian-process-driven Poisson process (FGPDP). This process appears in connection with Prob. 13.6 and relies on the following parameters: $E[\mu] = 100$, duration $L = 10^6$, fractal exponent $\alpha = 0.8$, onset frequency $f_S = 0.2$, and fractal-Gaussian-process array size $M = 2^{24}$. The service ratio for both Prob. 13.6 and the problem at hand is $\rho_\mu = \mu_a/\mu_s = 100 \cdot 0.009 = 0.9$. To generate the modulated curve, we warped the time axis in a suitable manner (see text) to achieve sinusoidal modulation with $\mu_0/\omega_0 \gg 1$, $a = 1$, and $2\pi/\omega_0 = 1$ min [see Eq. (13.16)] and used this as the arrival process. We present a queue-length histogram for the FGPDP/M/1 queue as the dashed curve. Roughly speaking, the solid curve follows the power-law decaying trend of the dashed curve within an order of magnitude.

Poisson-process counting distribution; this result appears in both theory (Diament & Teich, 1970b, Fig. 7) and experiment (Teich & Vannucci, 1978, Fig. 1). These features carry through from the driving rate process, which heavily favors the peaks and troughs of the sinusoid at the expense of the mean. As in the absence of modulation, the arrival process imparts its fractal nature to the queue length.

The queue-length histogram roughly follows the power-law decaying trend of the unmodulated simulation and, in fact, the two curves lie within an order of magnitude of each other at all values of the queue length. The power-law form therefore provides a far closer approximation to modulated simulation than does the M/M/1 queue, although the modulation destroys the precise power-law behavior seen in the original.

Prob. 13.10.1 To summarize, we have

$$
\begin{aligned}
L &\geq 10\,T_{A4} \\
T_{A4} &\geq 10^3\,T_{A3} \\
T_{A3} &\geq T_{A2} \\
T_{A2} &\geq 10^3\,T_{A1} \\
T_{A1} &= 10\,\mathrm{E}[\tau],
\end{aligned}
\tag{B.264}
$$

which, taken together, provide $L \geq 10^8\,\mathrm{E}[\tau]$. Since the expected number of events $\mathrm{E}[N(L)] = L/\mathrm{E}[\tau]$, we have $\mathrm{E}[N(L)]/\mathrm{E}[\tau] \geq 10^8$, so that the average simulation must contain at least 10^8 events.

Prob. 13.10.2 First, suppose we assume equalities in Eq. (B.264). Then $T_{A3} = T_{A2}$. When T equals this transition time, the contribution of the α_1 term becomes $(T_{A2}/T_{A1})^{\alpha_1}$. For the α_2 term to dominate for $T > T_{A3} = T_{A2}$, it must achieve the same value at that same time; thus it becomes $(T_{A2}/T_{A1})^{\alpha_1} \times (T/T_{A2})^{\alpha_2}$. Recall again that we have set $T_{A3} = T_{A2}$. The resulting normalized Haar-wavelet variance then becomes

$$
A(T) = 1 + (T/T_{A1})^{\alpha_1} + (T_{A2}/T_{A1})^{\alpha_1}(T/T_{A2})^{\alpha_2}.
\tag{B.265}
$$

Least-squares fitting programs applied directly to doubly logarithmic curves yield results that proportionally follow large values of the ordinate far more closely than small ones. However, instead of employing Eq. (B.265) we can instead perform a least-squares fit to the modified equation

$$
f(x) = \log\left\{1 + \left[\exp(x)/T_{A1}\right]^{\alpha_1} + (T_{A2}/T_{A1})^{\alpha_1}\left[\exp(x)/T_{A2}\right]^{\alpha_2}\right\}.
\tag{B.266}
$$

To maintain equal weighting over all decades, we choose geometric spacing for T, which is equivalent to linear spacing for x.

Using this method, we obtain monofractal parameters $\alpha \doteq 0.615716$ and $T_A \doteq 24.7731$. Figure B.21 displays the bifractal curve and the monofractal fit, which follows it fairly well.

Were real teletraffic to follow the bifractal curve in Fig. B.21, a cursory analysis might lead us to conclude that the data follows a monofractal form. Indeed, Occam's

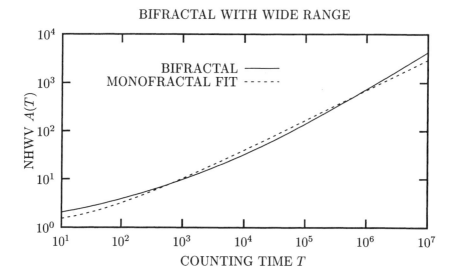

Fig. B.21 Bifractal form of the normalized Haar-wavelet variance provided in Eq. (B.265) (solid curve), and the monofractal logarithmic least-squares fit obtained via Eq. (B.266) (dashed curve), for the wide scaling range $T_{A2}/T_{A1} = T_{A4}/T_{A3} = 10^3$.

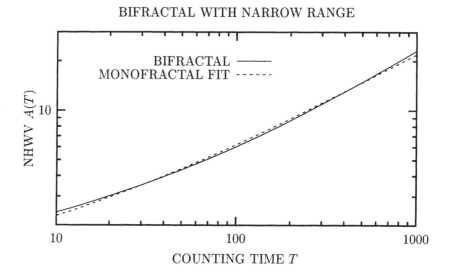

Fig. B.22 Bifractal form of the normalized Haar-wavelet variance provided in Eq. (B.265) (solid curve), and the monofractal logarithmic least-squares fit obtained via Eq. (B.266) (dashed curve), for the narrow scaling range $T_{A2}/T_{A1} = T_{A4}/T_{A3} = 10$.

razor would encourage us to draw this conclusion. However, Fig. B.21 illustrates that a significant difference exists between the two results, in the sense of Eq. (12.25), which favors the bifractal interpretation. Still, real data almost never follow precise forms, be they bifractal, monofractal, or otherwise; this makes a monofractal interpretation at least plausible, despite the relatively wide scaling ranges T_{A4}/T_{A3} and T_{A2}/T_{A1}.

For the narrow scaling range $T_{A2}/T_{A1} = T_{A4}/T_{A3} = 10$, the interpretation becomes quite murky. Here we obtain $\alpha \doteq 0.606821$ and $T_A \doteq 6.61332$; interestingly, the monofractal cutoff time T_A lies below both T_{A1} and T_{A3}. Figure B.22 displays the results. Deducing the presence of two scaling regions from data with these characteristics would prove nearly impossible.

Appendix C
List of Symbols

C.1	Roman Symbols	506
C.2	Greek Symbols	510
C.3	Mathematical Symbols	511

C.1   ROMAN SYMBOLS

Symbol	Description	Reference
a	Modulation depth	Eq. (13.16)
a–d	General continuous parameters	
a	Shot-noise impulse response function area	Eq. (10.12)
A	Short-time (high-frequency) cutoff	Eq. (7.1)
$A(T)$	Normalized Haar-wavelet variance	Sec. 3.4.3
$A_W(T)$	Normalized general-wavelet variance	Sec. 5.4.3
$A_\tau(k)$	Normalized interval Haar-wavelet variance	Sec. 3.3.4
$A^{(2)}(T)$	Normalized Haar-wavelet covariance	Eq. (3.75)
$\mathcal{A}$	Multiplicative rate constant	Prob. 6.8
$\mathbf{A}_L$	Lorentz vector potential	Prob. 10.6
b	Major axis of planetary elliptical orbit	Sec. 2.7.2
B	Long-time (low-frequency) cutoff	Eq. (7.1)
$B(t)$	Brownian motion	Sec. 2.4.2
$B^*(t), B^\dagger(t)$	Scaled versions of Brownian motion	Sec. 2.4.2
$B_H(t)$	Fractional Brownian motion	Sec. 6.1
$B'_H(t)$	Fractional Gaussian noise	Sec. 6.2.1
$B'_{H2}(t, v)$	Rectangularly filtered fractional Gaussian noise	Eq. (6.11)
$\mathsf{B}_H(t)$	Multifractal version of $B_H(t)$	Sec. 5.5.1
$\mathcal{B}$	Minimal covering width	Sec. 7.2.5
$\mathbf{B}$	Magnetic field vector	Prob. 10.6
c	Speed of light in free space	Prob. 10.6
C_{Euler}	Euler's constant ($\doteq 0.5772156649$)	Eq. (12.37)
C_n	Cumulant (semi-invariant)	Eq. (3.9)
C_μ	Rate coefficient of variation	Sec. 4.3
C_τ	Interevent-interval coefficient of variation	Eq. (3.5)
$C_{\psi,N}(a, b)$	Continuous-time wavelet transform	Sec. 3.4.3
$\mathcal{C}, \mathcal{C}_n$	Cantor set, nth-stage approximation	Sec. 2.4.1
$\mathcal{C}^F, \mathcal{C}_n^F$	Fat Cantor set, nth-stage approximation	Sec. 2.4.1
d	Distance	Sec. 2.7.2
$dN(t)$	Point process	Sec. 3.2
D	Fractal dimension	Sec. 3.5.4
D_d	Diffusion-constant exponent	Prob. 10.8
D_q	Generalized (Rényi) dimension	Sec. 3.5.4
D_s	Spectral-dimension exponent	Prob. 10.8
D_{E}	Euclidian dimension	Sec. 2.7.3
D_{HB}	Hausdorff–Besicovitch dimension	Sec. 3.5.4
D_0	Capacity (box-counting) dimension	Sec. 2.1.1
D_1	Information dimension	Sec. 3.5.4
D_2	Correlation dimension	Sec. 3.5.4
$\text{erfc}(\cdot)$	Complementary error function	Eq. (7.16)
E	Energy	Prob. 7.10

Symbol	Description	Reference
$E[\cdot]$	Expectation or mean	
$\mathbf{E}$	Electric field vector	Prob. 10.6
f	Frequency in cycles per unit time (Hz)	Sec. 3.3.3
$f(\cdot)$	General function	
f_S	Spectrum fractal cutoff frequency	Eq. (5.1)
f	Frequency in cycles per interval (dimensionless)	Sec. 3.3.3
$F(T)$	Normalized variance	Sec. 3.4.2
$\mathcal{F}\{\cdot\}$	Fourier transform	Sec. 9.4
F	Force or field	Sec. 2.7.2
$g(\cdot)$	General function	
$G(t)$	Coincidence rate	Sec. 3.5.1
h	Planck's constant	Prob. 10.6
$h(K,t)$	Shot-noise impulse response function	Sec. 9.1
$h_T(K,t)$	Integrated-shot-noise impulse response function	Eq. (10.2)
H	Hurst exponent	Chapter 6
$H(f)$	Fourier transform of impulse response function	Eq. (9.25)
$\mathbf{H}$	Magnetic induction vector	Prob. 10.6
i	$\sqrt{-1}$	
i	Space-charge-limited current	Sec. 2.7.2
$\mathrm{int}(\cdot)$	Integer function	
j	Quantum number of a simple system	Sec. 2.7.2
J	Čerenkov relativistic factor	Prob. 10.6
$k\!-\!n$	General counting variables	
k_2	Offset delay number for detrended fluctuations	Sec. 3.3.6
K	Shot-noise impulse response amplitude	Sec. 9.1
$\mathrm{K}_\gamma(x)$	Modified Bessel function of the second kind	Eq. (7.5)
l	Wavelet transform position index	Sec. 3.3.4
l_d	Correlation length	Prob. 10.9
ℓ	Decimation parameter	Sec. 11.2.2
$\ln(\cdot)$	Natural logarithm function (base e)	
$\log(\cdot)$	Logarithm function (arbitrary base)	
L	Duration of a data set	Sec. 3.5.2
$L(\cdot)$	Slowly varying function	Eq. (7.24)
$\mathcal{L}(\cdot)$	Lebesgue measure	Sec. 2.4.1
$\max(\cdot)$	Function returning largest argument	
$\min(\cdot)$	Function returning smallest argument	
$\mathrm{mod}(\cdot)$	Modulus function	
m	Order of the gamma renewal process	Prob. 4.7
M	General counting variable	
$M(\epsilon)$	Number of covering elements	Sec. 2.1.1
$\mathrm{M}(t)$	Multifractal process	Sec. 5.5.1
$\mathcal{M}$	Mass of an aggregated particle	Prob. 9.3

Symbol	Description	Reference
n_v	Number of contiguous vanishing moments of $\psi(t)$	Sec. 5.2.5
n	Refractive index	Prob. 10.6
$N(t)$	Counting process	Sec. 3.2
$N(L)$	Number of events from origin to time L	Sec. 3.5.4
$N_a(t)$	Arrival counting process at a queue	Sec. 13.1.1
$N_s(t)$	Service counting process at a queue	Sec. 13.1.1
$\widetilde{N}(f)$	Point-process Fourier transform	Eq. (B.254)
$\mathcal{N}(0,1)$	Normalized Gaussian random variable	Prob. 6.5
$p.(\cdot)$	Probability density function	Eq. (3.3)
$p_Q(n,t)$	Queue-length distribution	Sec. 13.1.1
$p_Z(n;T)$	Counting distribution	Eq. (3.28)
$p_\infty(n,t)$	Queue-length distribution (infinite buffer size)	Eq. (13.4)
$P.(\cdot)$	Probability distribution function	Eq. (3.3)
$\Pr\{\cdot\}$	Probability	
P_B	Buffer overflow (blocking) probability	Sec. 13.1.1
q	Generalized-dimension index	Sec. 3.5.4
$q(E)$	Trap waiting time	Prob. 7.10
q	Charge constant	Prob. 6.8
$Q(t)$	Queue length	Sec. 13.1.1
Q_m	Maximum queue length (buffer size)	Sec. 13.1.1
r	Bernoulli-trial success probability	Sec. 8.1.1
$r(t)$	Deletion recovery function	Chapter 11
r	Distance or deformation	Sec. 2.7.2
$\text{Re}\{\cdot\}$	Real part	
$R_Z(k,T)$	Count autocorrelation	Sec. 3.4.4
$R_\tau(k)$	Interval autocorrelation	Sec. 3.3.2
$R_2(k)$	Normalized count autocovariance	Sec. 12.3.3
$\mathcal{R}$	Thermodynamic gas constant	Prob. 6.8
s	Absolute time	Sec. 3.2
s	Measurement scale	Eq. (1.1)
$\text{sgn}(x)$	Sign of x	
$S_N(f)$	Point-process spectrum	Sec. 3.4.5
$S_{W,X}(t,f)$	Wigner–Ville spectrum	Eq. (6.9)
$S_Z(f,T)$	Count spectrum	Sec. 3.5.2
$S_\lambda(f,T)$	Rate spectrum	Sec. 3.4.5
$S_\tau(f)$	Interval spectrum	Sec. 3.3.3
$S_N^{(2)}(f)$	Point-process cross-spectrum	Eq. (3.76)
$S_\lambda^{(2)}(f,T)$	Rate cross-spectrum	Eq. (3.77)
$S_\tau(t)$	Interval survivor function	Sec. 3.3.1
$S_\vartheta(t)$	Recurrence-time survivor function	Prob. 11.7
$\mathcal{S}$	Linear-fit sum	Eq. (A.4)
S	Poynting vector	Prob. 10.6

Symbol	Description	Reference
t	Absolute time	Sec. 3.2
t_G	Coincidence-rate fractal cutoff time	Eq. (5.12)
T	Counting duration or counting window	Sec. 3.2
T_A	Normalized-Haar-wavelet-variance fractal cutoff time	Eq. (5.2)
T_{Dq}	Transition time for $\eta_q(T)$	Prob. 5.5.3
T_F	Normalized-variance fractal cutoff time	Eq. (5.11)
T_R	Autocorrelation fractal cutoff time	Eq. (5.14)
$\mathcal{T}$	Absolute temperature	Prob. 6.8
T	Planetary orbital period	Sec. 2.7.2
u–z	General continuous variables	
$u(\mathbf{x}, t)$	Particle concentration at position $\mathbf{x}$ and time t	Prob. 10.8
u_0	Initial particle concentration	Prob. 10.8
$U(k)$	Rescaled range analysis (R/S)	Sec. 3.3.5
$U_2(k)$	Normalized rescaled-range statistic	Sec. 12.3.4
v	Separation time	Sec. 6.2.1
v	Scalar velocity	Prob. 10.6
$\mathbf{v}$	Vector velocity	Prob. 10.6
V	Voltage	Prob. 6.8
$\mathrm{Var}[\,\cdot\,]$	Variance	
$W(\mathcal{B})$	Expected time between coverings	Eq. (7.20)
$W_{n,k}$	Multiplicative-process weighting factors	Sec. 5.5.1
$W_{\psi,\tau}(k, l)$	Discrete-time wavelet transform	Sec. 3.3.4
$\mathbf{x}$	Position vector	Prob. 10.8
$X(t)$	Continuous-time process	Sec. 6.1.2
$X_T(t)$	Integrated shot-noise process	Eq. (10.2)
$X_\Sigma(t)$	Binomial-noise process	Eq. (8.14)
$Y(k)$	Detrended fluctuation analysis	Sec. 3.3.6
$Y_2(k)$	Normalized detrended-fluctuation statistic	Sec. 12.3.5
z	Exponent in fractal Bartlett–Lewis process	Sec. 10.6.4
$Z_k(T)$	Event count	Sec. 3.2

C.2 GREEK SYMBOLS

Symbol	Description	Reference
α_x	Fractal exponent obtained from statistic x	Sec. 5.1.6
β	Fractal-shot-noise exponent	Sec. 9.1
γ	Fractal-renewal-process exponent	Chapter 7
$\Gamma(x)$	Complete Eulerian gamma function	Eq. (4.44)
$\Gamma(x, a)$	Incomplete Eulerian gamma function	Eq. (9.6)
$\delta(t)$	Dirac delta function	Sec. 3.5.1
Δ	Diffusion constant	Prob. 10.8
ϵ	Small number	Sec. 2.1.1
ε	Small number parameter	Sec. 12.3.9
ζ	Stable-distribution parameter	Sec. 9.2
$\eta_q(T)$	Generalized-dimension scaling function	Sec. 3.5.4
$\eta_0(T)$	Capacity-dimension scaling function	Sec. 3.5.4
θ	Phase angle	Prob. 4.10
$\vartheta(t)$	Forward recurrence time	Eq. (3.10)
κ	Boltzmann's constant	Prob. 7.10
$\lambda_k(t)$	Sample rate (measured value)	Eq. (3.27)
$\Lambda(t)$	Integrated rate (model property)	Eq. (4.28)
μ	Fixed rate of a point process (model property)	Sec. 4.1
$E[\mu]$	Expected rate of a point process (model property)	Sec. 3.5.1
$\mu(t)$	Varying rate of a point process (model property)	Sec. 3.5.1
μ_a	Arrival rate at a queue	Sec. 13.1.1
μ_s	Service rate at a queue	Sec. 13.1.1
ν	Frequency of electromagnetic radiation	Prob. 10.6
ξ	Weibull distribution parameter	Sec. 13.3.3
$\rho(s,t)$	Normalized autocorrelation	Eq. (6.28)
ρ_μ	Service ratio (server utilization)	Eq. (13.3)
$\varrho_\tau(k)$	Interval serial correlation coefficient	Eq. (3.17)
σ	Standard deviation	Eq. (3.4)
τ	Interevent interval	Sec. 3.2
τ_e	Effective dead time	Eq. (11.21)
τ_f	Fixed dead time	Eq. (11.16)
τ_w	Waiting time in a queue	Eq. (13.6)
ϕ_L	Lorentz scalar potential	Prob. 10.6
$\phi_\tau(\omega)$	Characteristic function	Eq. (3.6)
$\varphi(f)$	Fourier transform of wavelet $\psi(x)$	Eq. (5.41)
Φ	Smaller of counting and impulse-response times	Eq. (A.125)
χ^2	Linear-fit error	Eq. (A.3)
$\psi(x)$	Mother wavelet (time domain)	Sec. 3.3.4
Ψ	Integrate-and-reset threshold	Fig. 4.1
ω	Angular frequency (radians per unit time)	Eq. (4.46)
Ω	Double integral in normalized variance	Eq. (A.136)

C.3 MATHEMATICAL SYMBOLS

Symbol	Description	Reference
$*$	Complex conjugation	Sec. 3.5.5
$\star$	Convolution	Sec. 4.2
$\otimes$	Vector cross product	Prob. 10.6
$\widehat{x}$	Estimate of x	
$!$	Factorial of preceding expression	
$\equiv$	Definition	
$\doteq$	Very close to	
$\approx$	Approximately equal to	
$\sim$	Varies as	
$\ll$	Much less than	
$\gg$	Much greater than	
$0-$	Number infinitesimally smaller than zero	
$0+$	Number infinitesimally greater than zero	

Bibliography

Abbe, E. (1878). Über Blutkörper-Zählung. *Sitzungsberichte der Jenaischen Gesellschaft für Medizin und Naturwissenschaft, Jahrgang 1878*, 98–105. Reprinted: (1904). In S. Czapski (Ed.), *Gesammelte Abhandlungen von Ernst Abbe*, volume 1: *Abhandlungen über die Theorie des Mikroskops*. (pp. 173–180). Jena: Verlag von Gustav Fischer.

Abeles, M., de Ribaupierre, F., & de Ribaupierre, Y. (1983). Detection of single unit responses which are loosely time-locked to a stimulus. *IEEE Transactions on Systems, Man and Cybernetics, SMC-13*, 683–691.

Abry, P., Baraniuk, R., Flandrin, P., Riedi, R., & Veitch, D. (2002). Multiscale nature of network traffic. *IEEE Signal Processing Magazine, 19*(3), 28–46. Special issue on signal processing for networking.

Abry, P. & Flandrin, P. (1996). Point processes, long-range dependence and wavelets. In A. Aldroubi & M. Unser (Eds.), *Wavelets in Medicine and Biology* chapter 15, (pp. 413–437). Boca Raton, FL: CRC.

Abry, P., Flandrin, P., Taqqu, M. S., & Veitch, D. (2000). Wavelets for the analysis, estimation, and synthesis of scaling data. In K. Park & W. Willinger (Eds.), *Self-Similar Network Traffic and Performance Evaluation* chapter 2, (pp. 39–88). New York: Wiley–Interscience.

Abry, P., Flandrin, P., Taqqu, M. S., & Veitch, D. (2003). Self-similarity and long-range dependence through the wavelet lens. In P. Doukhan, G. Oppenheim, & M. S.

Taqqu (Eds.), *Theory and Applications of Long-Range Dependence* (pp. 527–556). Boston: Birkhäuser.

Abry, P. & Sellan, F. (1996). The wavelet-based synthesis of fractional Brownian motion proposed by F. Sellan and Y. Meyer: Remarks and fast implementation. *Applied and Computational Harmonic Analysis*, *3*, 377–383.

Aiello, W., Chung, F., & Lu, L. (2001). A random graph model for power law graphs. *Experimental Mathematics*, *10*, 53–66.

Aitchison, J. & Brown, J. A. C. (1957). *The Lognormal Distribution*. Cambridge, UK: Cambridge.

Aizawa, Y. (1984). On the f^{-1} spectral chaos. *Progress of Theoretical Physics (Kyoto)*, *72*, 659–661.

Aizawa, Y. & Kohyama, T. (1984). Asymptotically non-stationary chaos. *Progress of Theoretical Physics (Kyoto)*, *71*, 847–850.

Akay, M. (Ed.). (1997). *Time Frequency and Wavelets in Biomedical Signal Processing*. New York: IEEE Press.

Aks, D. J., Zelinsky, G. J., & Sprott, J. C. (2002). Memory across eye-movements: $1/f$ dynamic in visual search. *Nonlinear Dynamics, Psychology, and Life Sciences*, *6*, 1–25.

Albert, G. E. & Nelson, L. (1953). Contributions to the statistical theory of counter data. *Annals of Mathematical Statistics*, *24*, 9–22.

Albert, R. & Barabási, A.-L. (2002). Statistical mechanics of complex networks. *Reviews of Modern Physics*, *74*, 47–97.

Albert, R., Jeong, H., & Barabási, A.-L. (1999). Diameter of the world-wide web. *Nature (London)*, *401*, 130–131.

Aldroubi, A. & Unser, M. (Eds.). (1996). *Wavelets in Medicine and Biology*. Boca Raton, FL: CRC Press.

Alexander, S. & Orbach, R. (1982). Density of states on fractals: "fractons". *Journal de Physique: Lettres (Paris)*, *43*, L625–L631.

Allan, D. W. (1966). Statistics of atomic frequency standards. *Proceedings of the IEEE*, *54*, 221–230.

Alligood, K. T., Sauer, T. D., & Yorke, J. A. (1996). *Chaos: An Introduction to Dynamical Systems*. Berlin: Springer.

Anderson, C. M. (2001). From molecules to mindfulness: How vertically convergent fractal time fluctuations unify cognition and emotion. *Consciousness and Emotion*, *1*, 193–226.

Anderson, C. M., Lowen, S. B., Renshaw, P., Maas, L. C., & Teicher, M. H. (1999). State-of-mind-dependent fractal fluctuations in BOLD fMRI. In *Abstracts of Dynamical Neuroscience VII*, (pp. 027). Society for Neuroscience, Washington, DC.

Arecchi, F. T. & Califano, A. (1987). Noise-induced trapping at the boundary between two attractors: A source of $1/f$ spectra in nonlinear dynamics. *Europhysics Letters*, *3*, 5–10.

Arecchi, F. T. & Lisi, F. (1982). Hopping mechanism generating $1/f$ noise in nonlinear systems. *Physical Review Letters*, *49*, 94–98.

Argoul, F., Arneodo, A., Elezgaray, J., & Grasseau, G. (1989). Wavelet transform of fractal aggregates. *Physics Letters A*, *135*, 327–336.

Arlitt, M. & Jin, T. (1998). 1998 World-Cup Website Access Logs. Available at http://ita.ee.lbl.gov/html/contrib/WorldCup.html.

Arneodo, A., Grasseau, G., & Holschneider, M. (1988). Wavelet transform of multifractals. *Physical Review Letters*, *61*, 2281–2284.

Arrault, J. & Arneodo, A. (1997). Wavelet based multifractal analysis of rough surfaces: Application to cloud models and satellite data. *Journal of Statistical Physics*, *79*, 75–78.

Ashkenazy, Y., Lewkowicz, M., Levitan, J., Havlin, S., Saermark, K., Moelgaard, H., Bloch Thomsen, P. E., Moller, M., Hintze, U., & Huikuri, H. V. (2001). Scale-specific and scale-independent measures of heart rate variability as risk indicators. *Europhysics Letters*, *53*, 709–715.

Ashkenazy, Y., Lewkowicz, M., Levitan, J., Moelgaard, H., Bloch Thomsen, P. E., & Saermark, K. (1998). Discrimination of the healthy and sick cardiac autonomic nervous system by a new wavelet analysis of heartbeat intervals. *Fractals*, *6*, 197–203.

Asmussen, S. (2003). *Applied Probability and Queues* (Second ed.)., volume 51 of *Stochastic Modelling and Applied Probability*. Berlin: Springer.

Ayache, A. & Lévy Véhel, J. (1999). Generalized multifractional Brownian motion: Definition and preliminary results. In M. Dekking, J. Lévy Véhel, E. Lutton, & C. Tricot (Eds.), *Fractals: Theory and Applications in Engineering* (pp. 17–32). Berlin: Springer.

Azhar, M. A. & Gopala, K. (1992). Clustering Poisson process and burst noise. *Japanese Journal of Applied Physics*, *31*, 391–394.

Baccelli, F. & Brémaud, P. (2003). *Elements of Queueing Theory: Palm Martingale Calculus and Stochastic Recurrences* (Second ed.). Berlin: Springer.

Bachelier, L. (1900). *Théorie de la spéculation*. PhD thesis, La Sorbonne, Paris. Publication: (1900). *Annales Scientifiques de l'École Normale Supérieure*, *17*, 21–86.

Translation: (1964). *The Random Character of Stock Market Prices.* (P. H. Cootner, Ed.). Cambridge, MA: MIT Press.

Bachelier, L. (1912). *Calcul des probabilités.* Paris: Gauthier-Villars.

Bacry, E., Muzy, J. F., & Arneodo, A. (1993). Singular spectrum of fractal signals from wavelet analysis: Exact results. *Journal of Statistical Physics, 70,* 635–674.

Bak, P. (1996). *How Nature Works: The Science of Self-Organized Criticality.* New York: Copernicus/Springer.

Bak, P., Tang, C., & Wiesenfeld, K. (1987). Self-organized criticality: An explanation of the $1/f$ noise. *Physical Review Letters, 59,* 381–384.

Barakat, R. (1976). Sums of independent lognormally distributed random variables. *Journal of the Optical Society of America, 66,* 211–216.

Bardet, J.-M., Lang, G., Moulines, E., & Soulier, P. (2000). Wavelet estimator of long-range dependent processes. *Statistical Inference for Stochastic Processes, 3,* 85–99.

Bardet, J.-M., Lang, G., Oppenheim, G., Philippe, A., Stoev, S., & Taqqu, M. S. (2003). Semi-parametric estimation of the long-range dependence parameter: A survey. In P. Doukhan, G. Oppenheim, & M. S. Taqqu (Eds.), *Theory and Applications of Long-Range Dependence* (pp. 557–577). Boston: Birkhäuser.

Bardet, J.-M., Lang, G., Oppenheim, G., Philippe, A., & Taqqu, M. S. (2003). Generators of long-range dependent processes: A survey. In P. Doukhan, G. Oppenheim, & M. S. Taqqu (Eds.), *Theory and Applications of Long-Range Dependence* (pp. 579–623). Boston: Birkhäuser.

Barlow, H. B. (1957). Increment thresholds at low intensities considered as signal/noise discriminations. *Journal of Physiology (London), 136,* 469–488. Reprinted in Cohn (1993, pp. 204–223).

Barndorff-Nielsen, O. E., Blaesild, P., & Halgreen, C. (1978). First hitting time models for the generalized inverse Gaussian distribution. *Stochastic Processes and Their Applications, 7,* 49–54.

Barnes, J. A. & Allan, D. W. (1966). A statistical model of flicker noise. *Proceedings of the IEEE, 54,* 176–178.

Barnsley, M. F. (2000). *Fractals Everywhere* (Second ed.). San Francisco: Morgan Kaufmann/Elsevier.

Bartlett, M. S. (1955). *An Introduction to Stochastic Processes.* Cambridge, UK: Cambridge. Third edition: (1978).

Bartlett, M. S. (1963). The spectral analysis of point processes. *Journal of the Royal Statistical Society B, 25,* 264–296. Discussion of paper on pp. 281–296.

Bartlett, M. S. (1964). The spectral analysis of two-dimensional point processes. *Biometrika, 51*, 299–311.

Bartlett, M. S. (1972). Some applications of multivariate point processes. In P. A. W. Lewis (Ed.), *Stochastic Point Processes: Statistical Analysis, Theory, and Applications* (pp. 136–165). New York: Wiley–Interscience.

Barton, R. J. & Poor, H. V. (1988). Signal detection in fractional Gaussian noise. *IEEE Transactions on Information Theory, 34*, 943–959.

Bassingthwaighte, J. B., Liebovitch, L. S., & West, B. J. (1994). *Fractal Physiology*. New York: Oxford.

Bassingthwaighte, J. B. & Raymond, G. M. (1994). Evaluating rescaled range analysis for time series. *Annals of Biomedical Engineering, 22*, 432–444.

Bateman, H. (1910). Note on the probability distribution of α-particles. *Philosophical Magazine, 20*, 704–707.

Beggs, J. M. & Plenz, D. (2003). Neuronal avalanches in neocortical circuits. *Journal of Neuroscience, 23*, 11167–11177.

Beggs, J. M. & Plenz, D. (2004). Neuronal avalanches are diverse and precise activity patterns that are stable for many hours in cortical slice cultures. *Journal of Neuroscience, 24*, 5216–5229.

Bell, D. A. (1960). *Electrical Noise*, chapter 10. New York: Van Nostrand.

Bell, D. A. (1980). A survey of $1/f$ noise in electrical conductors. *Journal of Physics C, 13*, 4425–4437.

Benassi, A., Jaffard, S., & Roux, D. (1997). Elliptic Gaussian random processes. *Revista Matemática Iberoamericana, 13*, 19–90.

Beran, J. (1992). Statistical methods for data with long-range dependence. *Statistical Science, 7*, 404–427.

Beran, J. (1994). *Statistics for Long-Memory Processes*. New York: Chapman and Hall.

Beran, J., Sherman, R., Taqqu, M. S., & Willinger, W. (1995). Long-range dependence in variable-bit-rate video traffic. *IEEE Transactions on Communications, 43*, 1566–1579.

Berger, J. M. & Mandelbrot, B. B. (1963). A new model for the clustering of errors on telephone circuits. *IBM Journal of Research and Development, 7*, 224–236.

Berry, M. V. (1979). Diffractals. *Journal of Physics A: Mathematical and General, 12*, 781–797.

Berry, R. S., Rice, S. A., & Ross, J. (1980). *Physical Chemistry*. New York: Wiley.

Bertoin, J. (Ed.). (1998). *Lévy Processes*. Cambridge, UK: Cambridge.

Bharucha-Reid, A. T. (1997). *Elements of the Theory of Markov Processes and Their Applications*. Mineola, NY: Dover. Originally published by McGraw–Hill in 1960.

Bhattacharya, J., Edwards, J., Mamelak, A. N., & Schuman, E. M. (2005). Long-range temporal correlations in the spontaneous spiking of neurons in the hippocampal–amygdala complex of humans. *Neuroscience, 131*, 547–555.

Bickel, D. R. (1999). Estimating the intermittency of point processes with applications to human activity and viral DNA. *Physica A, 265*, 634–648.

Bickel, D. R. (2000). Implications of fluctuations in substitution rates: Impact on the uncertainty of branch lengths and on relative-rate tests. *Journal of Molecular Evolution, 50*, 381–390.

Bickel, D. R. & West, B. J. (1998a). Molecular evolution modeled as a fractal Poisson process in agreement with mammalian sequence comparisons. *Molecular Biology and Evolution, 15*, 967–977.

Bickel, D. R. & West, B. J. (1998b). Molecular evolution modeled as a fractal renewal point process in agreement with the dispersion of substitutions in mammalian genes. *Journal of Molecular Evolution, 47*, 551–556.

Biederman-Thorson, M. & Thorson, J. (1971). Dynamics of excitation and inhibition in the light-adapted *Limulus* eye *in situ*. *Journal of General Physiology, 58*, 1–19.

Bienaymé, I.-J. (1845). De la loi de multiplication et de la durée des familles. *L'Institut, Journal Universel des Sciences des Sociétés Savantes en France et à l'Étranger, 10 (Series 5)*, 37–39. These proceedings were periodically reissued as *Extraits des Procès-Verbaux des Séances — Société Philomathique de Paris*. This paper is reprinted at the conclusion of Kendall (1975).

Blair, E. A. & Erlanger, J. (1932). Responses of axons to brief shocks. *Proceedings of the Society for Experimental Biology and Medicine (New York), 29*, 926–927.

Blair, E. A. & Erlanger, J. (1933). A comparison of the characteristics of axons through their individual electrical responses. *American Journal of Physiology, 106*, 524–564.

Bouchaud, J.-P. & Georges, A. (1990). Anomalous diffusion in disordered media: Statistical mechanics, models and physical applications. *Physics Reports, 195*, 127–293.

Bovy, P. H. L. (Ed.). (1998). *Motorway Traffic Flow Analysis*. Delft, The Netherlands: Delft University Press.

Boxma, O. J. (1996). Fluid queues and regular variation. *Performance Evaluation, 27/28*, 699–712.

Bracewell, R. N. (1986). *The Fourier Transform and Its Applications* (Second ed.). New York: McGraw–Hill.

Brémaud, P. & Massoulié, L. (2001). Hawkes branching point processes without ancestors. *Journal of Applied Probability, 38*, 122–135.

Brichet, F., Roberts, J., Simonian, A., & Veitch, D. (1996). Heavy traffic analysis of a storage model with long range dependent on/off sources. *Queueing Systems, 23*, 197–215.

Brillinger, D. R. (1981). *Time Series: Data Analysis and Theory* (Second ed.). San Francisco: Holden–Day.

Brillinger, D. R. (1986). Some statistical methods for random process data from seismology and neurophysiology. Technical Report 84, Department of Statistics, University of California, Berkeley.

Brockmeyer, E., Halstrøm, H. L., & Jensen, A. (1948). *The Life and Works of A. K. Erlang*. Number 2 in Transactions of the Danish Academy of Technical Sciences. Copenhagen: The Copenhagen Telephone Company. This compilation contains a biography and bibliography, along with English translations of Erlang's principal papers.

Brown, R. (1828). A brief account of microscopical observations made in the months of June, July, and August, 1827, on the particles contained in the pollen of plants; and on the general existence of active molecules in organic and inorganic bodies. *London and Edinburgh Philosophical Magazine and Annals of Philosophy (London), 4*(21), 161–173.

Buckingham, M. J. (1983). *Noise in Electronic Devices and Systems*. Chichester: Ellis Horwood.

Buldyrev, S. V., Goldberger, A. L., Havlin, S., Mantegna, R. N., Matsa, M. E., Peng, C.-K., Simons, M., & Stanley, H. E. (1995). Long-range correlation properties of coding and noncoding DNA sequences: GenBank analysis. *Physical Review E, 51*, 5084–5091.

Burgess, R. E. (1959). Homophase and heterophase fluctuations in semiconducting crystals. *Discussions of the Faraday Society, 28*, 151–158.

Burlatsky, S. F., Oshanin, G. S., & Ovchinnikov, A. A. (1989). Fluctuation-dominated kinetics of incoherent excitation quenching. *Physics Letters A, 139*, 241–244.

Çinlar, E. (1972). Superposition of point processes. In P. A. W. Lewis (Ed.), *Stochastic Point Processes: Statistical Analysis, Theory, and Applications* (pp. 549–606). New York: Wiley–Interscience.

Caccia, D. C., Percival, D., Cannon, M. J., Raymond, G., & Bassingthwaighte, J. B. (1997). Analyzing exact fractal time series: Evaluating dispersional analysis and rescaled range methods. *Physica A, 246*, 609–632.

Campbell, N. (1909a). Discontinuities in light emission. *Proceedings of the Cambridge Philosophical Society, 15*, 310–328.

Campbell, N. (1909b). The study of discontinuous phenomena. *Proceedings of the Cambridge Philosophical Society, 15*, 117–136.

Campbell, N. (1939). The fluctuation theorem shot effect. *Proceedings of the Cambridge Philosophical Society, 35*, 127–129.

Cantor, B. I., Matin, L., & Teich, M. C. (1975). Photocounting distributions with variable dead time. *Applied Optics, 14*, 2819–2820.

Cantor, B. I. & Teich, M. C. (1975). Dead-time-corrected photocounting distributions for laser radiation. *Journal of the Optical Society of America, 65*, 786–791.

Cantor, G. (1883). Über unendliche, lineare Punktmannigfaltigkeiten. *Mathematische Annalen, 21*, 545–591. Part 5 of a 6-part series published between 1879 and 1884.

Carlson, J. M. & Doyle, J. (1999). Highly optimized tolerance: A mechanism for power laws in designed systems. *Physical Review E, 60*, 1412–1427.

Carlson, J. M. & Doyle, J. (2002). Complexity and robustness. *Proceedings of the National Academy of Sciences (USA), 99*, 2538–2545.

Carson, J. R. (1931). The statistical energy–frequency spectrum of random disturbances. *Bell System Technical Journal, 10*, 374–381.

Castaing, B. (1996). The temperature of turbulent flows. *Journal de Physique II (European Physical Journal), 6*, 105–114.

Čerenkov, P. A. (1934). Visible light from clear liquids under the action of gamma radiation. *Doklady Akademii Nauk SSSR, 2*, 451–454.

Čerenkov, P. A. (1937). Visible radiation produced by electrons moving in a medium with velocities exceeding that of light. *Physical Review, 52*, 378–379.

Čerenkov, P. A. (1938). Absolute output of radiation caused by electrons moving within a medium with super-light velocity. *Doklady Akademii Nauk SSSR, 21*(3), 116–121.

Chandler, R. E., Herman, R., & Montroll, E. W. (1958). Traffic dynamics: Studies in car following. *Operations Research, 6*, 165–184.

Chandrasekhar, S. (1943). Stochastic problems in physics and astronomy. *Reviews of Modern Physics, 15*, 1–89. Reprinted: (1954). In N. Wax (Ed.), *Selected Papers on Noise and Stochastic Processes*. (pp. 3–91). New York: Dover.

Chapman, K. M. & Smith, R. S. (1963). A linear transfer function underlying impulse frequency modulation in a cockroach mechanoreceptor. *Nature (London), 197*, 699–700.

Chistyakov, V. P. (1964). A theorem on sums of independent positive random variables and its applications to branching random processes. *Theory of Probability and Its Applications*, *9*, 640–648.

Christoph, G. & Wolf, W. (1992). *Convergence Theorems with a Stable Limit Law*. Berlin: Akademie.

Cohen, J. W. (1969). *The Single-Server Queue*. Amsterdam: North-Holland.

Cohen, J. W. (1973). Some results on regular variation for the distributions in queueing and fluctuation theory. *Journal of Applied Probability*, *10*, 343–353.

Cohn, T. E. (Ed.). (1993). *Visual Detection*, volume 3 of *Collected Works in Optics*. Washington, DC: Optical Society of America.

Cole, B. J. (1995). Fractal time in animal behaviour: The movement activity of Drosophila. *Animal Behaviour*, *50*, 1317–1324.

Collins, J. J., De Luca, C. J., Burrows, A., & Lipsitz, L. A. (1995). Age-related changes in open-loop and closed-loop postural control mechanisms. *Experimental Brain Research*, *104*, 480–492.

Conrad, M. (1986). What is the use of chaos? In A. V. Holden (Ed.), *Chaos: An Introduction* (pp. 3–14). Princeton, NJ: Princeton University Press.

Cooper, R. B. (1972). *Introduction to Queueing Theory*. New York: Macmillan.

Cox, D. R. (1948). The use of the correlogram in measuring yarn irregularity. *Proceedings of the International Wool Textile Organization*, *2*, 28–34.

Cox, D. R. (1955). Some statistical methods connected with series of events. *Journal of the Royal Statistical Society B*, *17*, 129–164.

Cox, D. R. (1962). *Renewal Theory*. London: Methuen.

Cox, D. R. (1963). Some models for series of events. *Bulletin of the International Statistical Institute*, *40*, 737–746.

Cox, D. R. (1984). Long-range dependence: A review. In David, H. A. & David, H. T. (Eds.), *Statistics: An Appraisal*, (pp. 54–74)., Ames, IA. Iowa State University Press. Proceedings of a conference marking the 50th anniversary of the Statistical Laboratory, Iowa State University, Ames, Iowa, June 13–15, 1983.

Cox, D. R. & Isham, V. (1980). *Point Processes*. London: Chapman and Hall.

Cox, D. R. & Lewis, P. A. W. (1966). *The Statistical Analysis of Series of Events*. London: Methuen.

Cox, D. R. & Smith, W. L. (1953). The superposition of several strictly periodic sequences of events. *Biometrika*, *40*, 1–11.

Cox, D. R. & Smith, W. L. (1954). On the superposition of renewal processes. *Biometrika, 41*, 91–99.

Crovella, M. E. & Bestavros, A. (1997). Self-similarity in World Wide Web traffic: Evidence and possible causes. *IEEE/ACM Transactions on Networking, 5*, 835–846.

Dal Negro, L., Oton, C. J., Gaburro, Z., Pavesi, L., Johnson, P., Lagendijk, A., Righini, R., Colocci, M., & Wiersma, D. S. (2003). Light transport through the band-edge states of Fibonacci quasicrystals. *Physical Review Letters, 90*, 055501.

Dal Negro, L., Stolfi, M., Yi, Y., Michel, J., Duan, X., Kimerling, L. C., LeBlanc, J., & Haavisto, J. (2004). Omnidirectional reflectance and optical gap properties of Si/SiO_2 Thue–Morse quasicrystals. *Materials Research Society Symposium Proceedings, 817*, L2.5.1–L2.5.7.

Dal Negro, L., Yi, J. H., Nguyen, V., Yi, Y., Michel, J., & Kimerling, L. (2005). Light emission in aperiodic Thue–Morse dielectrics. *Materials Research Society Symposium Proceedings, 832*, F1.3.1–F1.3.6.

Daley, D. J. (1974). Various concepts of orderliness for point processes. In E. F. Harding & D. G. Kendall (Eds.), *Stochastic Geometry* (pp. 148–161). Chichester: Wiley.

Daley, D. J. & Vere-Jones, D. (1988). *An Introduction to the Theory of Point Processes*. Berlin: Springer. Second edition in two volumes: (2002).

Dan, Y., Atick, J. J., & Reid, R. C. (1996). Efficient coding of natural scenes in the lateral geniculate nucleus: Experimental test of a computational theory. *Journal of Neuroscience, 16*, 3351–3362.

Daubechies, I. (1988). Orthonormal bases of compactly supported wavelets. *Communications on Pure and Applied Mathematics, 41*, 909–996.

Daubechies, I. (1992). *Ten Lectures on Wavelets*. Philadelphia: Society for Industrial and Applied Mathematics (SIAM).

Davenport, Jr, W. B. & Root, W. L. (1987). *An Introduction to the Theory of Random Signals and Noise*. New York: IEEE Press. Originally published by McGraw–Hill in 1958.

Davidsen, J. & Schuster, H. G. (2002). Simple model for $1/f^\alpha$ noise. *Physical Review E, 65*, 026120.

Davies, R. B. & Harte, D. S. (1987). Tests for Hurst effect. *Biometrika, 74*, 95–101.

Dayan, P. & Abbott, L. F. (2001). *Theoretical Neuroscience: Computational and Mathematical Modeling of Neural Systems*. Cambridge, MA: MIT Press.

DeBoer, R. W., Karemaker, J. M., & Strackee, J. (1984). Comparing spectra of a series of point events particularly for heart rate variability data. *IEEE Transactions on Biomedical Engineering, BME-31*, 384–387.

DeLotto, I., Manfredi, P. F., & Principi, P. (1964). Counting statistics and dead-time losses, Part 1. *Energia Nucleare, 11*, 557–564.

Dettmann, C. P., Frankel, N. E., & Taucher, T. (1994). Structure factor of Cantor sets. *Physical Review E, 49*, 3171–3178.

Devaney, R. L. (1986). *Introduction to Chaotic Dynamical Systems*. Menlo Park, CA: Benjamin–Cummings.

Diament, P. & Teich, M. C. (1970a). Photodetection of low-level radiation through the turbulent atmosphere. *Journal of the Optical Society of America, 60*, 1489–1494.

Diament, P. & Teich, M. C. (1970b). Photoelectron counting distributions for irradiance-modulated radiation. *Journal of the Optical Society of America, 60*, 682–689.

Ding, M. & Yang, W. (1995). Distribution of the first return time in fractional Brownian motion and its application to the study of on–off intermittency. *Physical Review E, 52*, 207–213.

Ditto, W. L., Spano, M. L., Savage, H. T., Rauseo, S. N., Heagy, J., & Ott, E. (1990). Experimental observation of a strange nonchaotic attractor. *Physical Review Letters, 65*, 533–536.

Doob, J. L. (1948). Renewal theory from the point of view of the theory of probability. *Transactions of the American Mathematical Society, 63*, 422–438.

Doob, J. L. (1953). *Stochastic Processes*. New York: Wiley.

Dorogovtsev, S. N. & Mendes, J. F. F. (2003). *Evolution of Networks: From Biological Nets to the Internet and WWW*. New York: Oxford.

Doukhan, P. (2003). Models, inequalities, and limit theorems for stationary sequences. In P. Doukhan, G. Oppenheim, & M. S. Taqqu (Eds.), *Theory and Applications of Long-Range Dependence* (pp. 43–100). Boston: Birkhäuser.

Doyle, J. & Carlson, J. M. (2000). Power laws, HOT, and generalized source coding. *Physical Review Letters, 84*, 5656–5659.

Duffy, D. E., McIntosh, A. A., Rosenstein, M., & Willinger, W. (1994). Statistical analysis of CCSN/SS7 traffic data from working CCS subnetworks. *IEEE Journal on Selected Areas in Communications, 12*, 544–552.

Dulea, M., Johannson, M., & Riklund, R. (1992). Localization of electrons and electromagnetic waves in a deterministic aperiodic system. *Physical Review B, 45*, 105–114.

Ebel, H., Mielsch, L.-I., & Bornholdt, S. (2002). Scale-free topology of e-mail networks. *Physical Review E*, *66*, 035103.

Eccles, J. C. (1957). *The Physiology of Nerve Cells*. Baltimore: Johns Hopkins University Press.

Efron, B. (1982). *The Jackknife, the Bootstrap, and Other Resampling Plans*. Philadelphia: Society for Industrial and Applied Mathematics (SIAM).

Efron, B. & Tibshirani, R. J. (1993). *An Introduction to the Bootstrap*. New York: Chapman and Hall.

Eguíluz, V. M., Chialvo, D. R., Cecchi, G. A., Baliki, M., & Apkarian, A. V. (2005). Scale-free brain functional networks. *Physical Review Letters*, *94*, 018102.

Einstein, A. (1905). Über die von der molekularkinetischen Theorie der Wärme geforderte Bewegung von in ruhenden Flüssigkeiten suspendierten Teilchen. *Annalen der Physik*, *17*, 549–560.

Ellis, R. L. (1844). On a question in the theory of probabilities. *Cambridge Mathematical Journal (Cambridge and Dublin Mathematical Journal)*, *4*(21), 127–133.

Embrechts, P., Klüppelberg, C., & Mikosch, T. (1997). *Modelling Extremal Events for Insurance and Finance*. Berlin: Springer.

Engset, T. (1915). Om beregningen av vælgere i et automatisk telefonsystem. Unpublished manuscript (in Norwegian), written in Kristiania (Oslo). This 130-page typed manuscript was discovered in the files of the Copenhagen Telephone Company (KTAS) in 1995 by Villy Bæk Iversen; the presumed original of the manuscript was subsequently located in the Norsk Telemuseum, Oslo. Translation: (1998). On the calculation of switches in an automatic telephone system. (A. Myskja, Trans.). *Telektronikk (Oslo)*, *94*, 99–142; translation reprinted in Myskja & Espvik (2002, pp. 40–148). A biography of Engset has been prepared by Myskja (1998b).

Engset, T. (1918). Die Wahrscheinlichkeitsrechnung zur Bestimmung der Wähleranzahl in automatischen Fernsprechämtern. *Elektrotechnische Zeitschrift*, *39*(31), 304–306. Translation: (1992). The probability calculation to determine the number of switches in automatic telephone exchanges. (E. Jensen, Trans.). *Telektronikk (Oslo)*, *88*, 90–93; translation reprinted in Myskja & Espvik (2002, pp. 175–183).

Erlang, A. K. (1909). Sandsynlighedsregning og Telefonsamtaler. *Nyt Tidsskrift for Matematik B (Copenhagen)*, *20*, 33–41. Translation: The theory of probabilities and telephone conversations. In Brockmeyer, Halstrøm & Jensen (1948, pp. 131–137).

Erlang, A. K. (1917). Løsning af nogle Problemer fra Sandsynlighedsregningen af Betydning for de automatiske Telefoncentraler. *Elektroteknikeren (Copenhagen)*, *13*, 5–13. Translation: Solution of some problems in the theory of probabilities of significance in automatic telephone exchanges. In Brockmeyer, Halstrøm & Jensen (1948, pp. 138–155).

Erlang, A. K. (1920). Telefon-Ventetider. Et Stykke Sandsynlighedsregning. *Matematisk Tidsskrift B (Copenhagen)*, *31*, 25–42. Translation: Telephone waiting times: An example of probability calculus. In Brockmeyer, Halstrøm & Jensen (1948, pp. 156–171).

Erramilli, A., Narayan, O., & Willinger, W. (1996). Experimental queueing analysis with long-range dependent packet traffic. *IEEE/ACM Transactions on Networking*, *4*, 209–223.

Evans, M., Smithies, O., & Capecchi, M. R. (2001). Mouse gene targeting. *Nature Medicine*, *7*, 1081–1090.

Evarts, E. V. (1964). Temporal patterns of discharge of pyramidal tract neurons during sleep and waking in the monkey. *Journal of Neurophysiology*, *27*, 152–171.

Fadel, P. J., Orer, H. S., Barman, S. M., Vongpatanasin, W., Victor, R. G., & Gebber, G. L. (2004). Fractal properties of human muscle sympathetic nerve activity. *American Journal of Physiology: Heart and Circulatory Physiology*, *286*, H1076–H1087.

Fairfield-Smith, H. (1938). An empirical law describing heterogeneity in the yields of agricultural crops. *Journal of Agricultural Science*, *28*, 1–23.

Fairhall, A. L., Lewen, G. D., Bialek, W., & de Ruyter van Steveninck, R. R. (2001a). Efficiency and ambiguity in an adaptive neural code. *Nature (London)*, *412*, 787–792.

Fairhall, A. L., Lewen, G. D., Bialek, W., & de Ruyter van Steveninck, R. R. (2001b). Multiple timescales of adaptation in a neural code. In T. K. Leen, T. G. Dietterich, & V. Tresp (Eds.), *Advances in Neural Information Processing Systems*, volume 13 (pp. 124 –130). MIT Press.

Falconer, K. (2003). *Fractal Geometry: Mathematical Foundations and Applications* (Second ed.). New York: Wiley.

Faloutsos, M., Faloutsos, P., & Faloutsos, C. (1999). On power-law relationships of the Internet topology. *ACM SIGCOMM Computer Communication Review*, *29*, 251–262.

Fano, U. (1947). Ionization yield of radiations. II. The fluctuations of the number of ions. *Physical Review*, *72*, 26–29.

Fatt, P. & Katz, B. (1952). Spontaneous subthreshold activity at motor nerve endings. *Journal of Physiology (London)*, *117*, 109–128.

Feder, J. (1988). *Fractals*. New York: Plenum.

Feldmann, A., Gilbert, A. C., & Willinger, W. (1998). Data networks as cascades: Investigating the multifractal nature of internet WAN traffic. *ACM SIGCOMM Computer Communication Review*, *28*, 42–55.

Feldmann, A., Gilbert, A. C., Willinger, W., & Kurtz, T. G. (1998). The changing nature of network traffic: Scaling phenomena. *ACM SIGCOMM Computer Communication Review*, *28*, 5–29.

Feller, W. (1941). On the integral equation of renewal theory. *Annals of Mathematical Statistics*, *12*, 243–267.

Feller, W. (1948). On probability problems in the theory of counters. In K. O. Friedrichs, O. E. Neugebauer, & J. J. Stoker (Eds.), *Studies and Essays Presented to R. Courant on his 60th Birthday, January 8, 1948* (pp. 105–115). New York: Interscience.

Feller, W. (1951). The asymptotic distribution of the range of sums of independent random variables. *Annals of Mathematical Statistics*, *22*, 427–432.

Feller, W. (1968). *An Introduction to Probability Theory and Its Applications* (Third ed.)., volume I. New York: Wiley.

Feller, W. (1971). *An Introduction to Probability Theory and Its Applications* (Second ed.)., volume II. New York: Wiley.

Feynman, R. P. (1965). *The Character of Physical Law*. Cambridge, MA: MIT Press. 1964 Messenger Lectures at Cornell University, subsequently broadcast by the BBC.

Feynman, R. P., Leighton, R. B., & Sands, M. (1963). *The Feynman Lectures on Physics*. Reading, MA: Addison–Wesley.

Fibonacci (1202). *Liber abaci*. Pisa. Leonardo Pisano was known as Fibonacci.

Field, A. J., Harder, U., & Harrison, P. G. (2004a). Measurement and modelling of self-similar traffic in computer networks. *IEE Proceedings–Communications*, *151*, 355–363.

Field, A. J., Harder, U., & Harrison, P. G. (2004b). Network traffic behaviour in switched Ethernet systems. *Performance Evaluation*, *58*, 243–260.

Fisher, L. (1972). A survey of the mathematical theory of multidimensional point processes. In P. A. W. Lewis (Ed.), *Stochastic Point Processes: Statistical Analysis, Theory, and Applications* (pp. 468–513). New York: Wiley–Interscience.

Flake, G. W. (2000). *The Computational Beauty of Nature: Computer Explorations of Fractals, Chaos, Complex Systems, and Adaptation*. Cambridge, MA: MIT Press.

Flandrin, P. (1989). On the spectrum of fractional Brownian motion. *IEEE Transactions on Information Theory*, *35*, 197–199.

Flandrin, P. (1992). Wavelet analysis and synthesis of fractional Brownian motion. *IEEE Transactions on Information Theory*, *38*, 910–917.

Flandrin, P. (1997). $1/f$ et ondelettes. In *GRETSI-97, Colloque sur le traitement du signal et des images*, (pp. 623–626)., Grenoble, France.

Flandrin, P. & Abry, P. (1999). Wavelets for scaling processes. In M. Dekking, J. Lévy Véhel, E. Lutton, & C. Tricot (Eds.), *Fractals: Theory and Applications in Engineering* (pp. 47–64). Berlin: Springer.

Fleckenstein, J. O. (Ed.). (1969). *Die Werke von Jakob Bernoulli*. Basel: Birkhäuser. Multiple volumes: (1969)–(1993).

Fourier, J. (1822). *Théorie analytique de la chaleur*. Paris: Chez Firmin Didot, Père et Fils. Reprinted: (1988). Paris: Editions Jacques Gabay.

Franken, P. (1963). Approximation durch Poissonsche Prozesse. *Mathematische Nachrichten*, *26*, 101–114.

Franken, P. (1964). Approximation der Verteilung von Summen unabhängiger nicht-negativer ganzzahliger Zufallsgrössen durch Poissonsche Verteilungen. *Mathematische Nachrichten*, *27*, 303–340.

Franken, P., König, D., Arndt, U., & Schmidt, V. (1981). *Queues and Point Processes*. Berlin: Akademie-Verlag.

Fréchet, M. (1940). *Les probabilités associées à un système d'événements compatibles et dépendents. Première partie: d'événements en nombre fini fixe*, (pp. 1–80). Number 859 in Actualités Scientifiques et Industrielles. Paris: Hermann.

Furry, W. H. (1937). On fluctuation phenomena in the passage of high energy electrons through lead. *Physical Review*, *52*, 569–581.

Gardner, M. (1978). White and brown music, fractal curves and $1/f$ fluctuations. *Scientific American*, *238*(4), 16–32.

Garrett, M. W. & Willinger, W. (1994). Analysis, modeling and generation of self-similar VBR video traffic. *ACM SIGCOMM Computer Communication Review*, *24*, 269–280.

Gauss, C. F. (1809). *Theoria motus corporum coelestium in sectionibus conicis solem ambientium*. Hamburg: Perthes und Besser. Translation: (1857). *Theory of the Motion of the Heavenly Bodies Moving about the Sun in Conic Sections*. (C. H. Z. Davis, Trans.). Boston: Little, Brown. Reprinted: (1963). New York: Dover.

Gellermann, W., Kohmoto, M., Sutherland, B., & Taylor, P. C. (1994). Localization of light waves in Fibonacci dielectric multilayers. *Physical Review Letters*, *72*, 633–636.

Gere, J. M. (2001). *Mechanics of Materials* (Fifth ed.). Pacific Grove, CA: Brooks/Cole.

Gerstein, G. L. & Mandelbrot, B. B. (1964). Random walk models for the spike activity of a single neuron. *Biophysical Journal, 4*, 41–68.

Ghulinyan, M., Oton, C. J., Dal Negro, L., Pavesi, L., Sapienza, R., Colocci, M., & Wiersma, D. S. (2005). Light-pulse propagation in Fibonacci quasicrystals. *Physical Review B, 71*, 094204.

Gilbert, E. N. (1961). Capacity of a burst-noise channel. *Bell System Technical Journal, 39*, 1253–1265.

Gilbert, E. N. & Pollak, H. O. (1960). Amplitude distribution of shot noise. *Bell System Technical Journal, 39*, 333–350.

Gilden, D. L. (2001). Cognitive emissions of $1/f$ noise. *Psychological Review, 108*, 33–56.

Gillespie, J. H. (1994). *The Causes of Molecular Evolution.* New York: Oxford.

Gisiger, T. (2001). Scale invariance in biology: Coincidence or footprint of a universal mechanism? *Biological Reviews, 76*, 161–209.

Glass, L. & Mackey, M. C. (1988). *From Clocks to Chaos: The Rhythms of Life.* Princeton, NJ: Princeton University Press.

Gnedenko, B. V. & Kolmogorov, A. N. (1968). *Limit Distributions for Sums of Independent Random Variables* (Second ed.). Reading, MA: Addison–Wesley. First edition: (1954); Translation by K. L. Chung.

Good, I. J. (1961). The real stable characteristic functions and chaotic acceleration. *Journal of the Royal Statistical Society B, 23*, 180–183.

Gottschalk, A., Bauer, M. S., & Whybrow, P. C. (1995). Evidence of chaotic mood variation in bipolar disorder. *Archives of General Psychiatry, 52*, 947–959. This article considers fractal activity as well as chaos.

Gradshteyn, I. S. & Ryzhik, I. M. (1994). *Table of Integrals, Series, and Products* (Fifth ed.). San Diego: Academic. (A. Jeffrey, Ed.).

Grandell, J. (1976). *Doubly Stochastic Poisson Processes.* Number 529 in Lecture Notes in Mathematics. Berlin: Springer. A. Dold and B. Eckmann, Series Editors.

Grassberger, P. (1985). On the spreading of two-dimensional percolation. *Journal of Physics A: Mathematical and General, 18*, L215–L219.

Grassberger, P. & Procaccia, I. (1983). Characterization of strange attractors. *Physical Review Letters, 50*, 346–349.

Grebogi, C., Ott, E., Pelikan, S., & Yorke, J. A. (1984). Strange attractors that are not chaotic. *Physica D*, *13*, 261–268.

Greenwood, M. & Yule, G. U. (1920). An inquiry into the nature of frequency distributions representative of multiple happenings with particular reference to the occurrence of multiple attacks of disease or of repeated accidents. *Journal of the Royal Statistical Society*, *83*, 255–279.

Greiner, M., Jobmann, M., & Klüppelberg, C. (1999). Telecommunication traffic, queueing models, and subexponential distributions. *Queueing Systems*, *33*, 125–153.

Greis, N. P. & Greenside, H. S. (1991). Implication of a power-law power-spectrum for self-affinity. *Physical Review A*, *44*, 2324–2334.

Grigelionis, B. (1963). On the convergence of sums of random step processes to a Poisson process. *Theory of Probability and Its Applications*, *8*, 177–182.

Gross, D. & Harris, C. M. (1998). *Fundamentals of Queueing Theory* (Third ed.). Hoboken, NJ: Wiley.

Grossglauser, M. & Bolot, J.-C. (1996). On the relevance of long-range dependence in network traffic. *ACM SIGCOMM Computer Communication Review*, *26*, 15–24.

Grüneis, F. (1984). A number fluctuation model generating $1/f$ pattern. *Physica A*, *123*, 149–160.

Grüneis, F. (1987). $1/f$ component in variance noise of a cluster process. In Van Vliet, C. M. (Ed.), *Proceedings of the Ninth International Conference on Noise in Physical Systems*, (pp. 347–350). World Scientific.

Grüneis, F. (2001). $1/f$ noise, intermittency and clustering Poisson process. In Bosman, G. (Ed.), *Proceedings of the Sixteenth International Conference on Noise in Physical Systems and $1/f$ Fluctuations*, (pp. 572–575). World Scientific.

Grüneis, F. & Baiter, H.-J. (1986). More detailed explication of a number fluctuation model generating $1/f$ pattern. *Physica A*, *136*, 432–452.

Grüneis, F. & Musha, T. (1986). Clustering Poisson process and $1/f$ noise. *Japanese Journal of Applied Physics*, *25*, 1504–1509.

Grüneis, F., Nakao, M., Mizutani, Y., Yamamoto, M., Meesmann, M., & Musha, T. (1993). Further study on $1/f$ fluctuations observed in central single neurons during REM sleep. *Biological Cybernetics*, *68*, 193–198.

Grüneis, F., Nakao, M., Yamamoto, M., Musha, T., & Nakahama, H. (1989). An interpretation of $1/f$ fluctuations in neuronal spike trains during dream sleep. *Biological Cybernetics*, *60*, 161–169.

Gumbel, E. J. (1958). *Statistics of Extremes*. New York: Columbia University Press.

Gurland, J. (1957). Some interrelations among compound and generalized distributions. *Biometrika, 44*, 265–268.

Gutenberg, B. & Richter, C. F. (1944). Frequency of earthquakes in California. *Bulletin of the Seismological Society of America, 34*, 185–188.

Haar, A. (1910). Zur Theorie der orthogonalen Funktionensysteme. *Mathematische Annalen, 69*, 331–371. Reprinted in Szőkefalvi-nagy (1959); this work formed the basis of Haar's doctoral dissertation, carried out under the supervision of David Hilbert and presented at Göttingen in 1909.

Haight, F. A. (1967). *Handbook of the Poisson Distribution*. New York: Wiley.

Halford, D. (1968). A general mechanical model for $|f|^\alpha$ spectral density random noise with special reference to flicker noise $1/|f|$. *Proceedings of the IEEE, 56*, 251–258.

Halley, J. M. & Inchausti, P. (2004). The increasing importance of $1/f$-noises as models of ecological variability. *Fluctuation and Noise Letters, 4*, R1–R26.

Halsey, T. C. (2000). Diffusion-limited aggregation: A model for pattern formation. *Physics Today, 53*(11), 36–41.

Harris, T. E. (1971). Random motions and point processes. *Zeitschrift für Wahrscheinlichkeitstheorie, 18*, 85–115.

Harris, T. E. (1989). *The Theory of Branching Processes*. New York: Dover. Originally published by Springer in 1963.

Hattori, H. T., Schneider, V. M., & Lisboa, O. (2000). Cantor set fiber Bragg grating. *Journal of the Optical Society of America A, 17*, 1583–1589.

Hattori, T., Tsurumachi, N., Kawato, S., & Nakatsuka, H. (1994). Photonic dispersion relation in one-dimensional quasicrystal. *Physical Review B, 50*, 4220–4223.

Hausdorff, J. M., Mitchell, S. L., Firtion, R., Peng, C.-K., Cudkowicz, M. E., Wei, J. Y., & Goldberger, A. L. (1997). Altered fractal dynamics of gait: Reduced stride-interval correlations with aging and Huntington's disease. *Journal of Applied Physiology, 82*, 262–269.

Hawkes, A. G. (1971). Spectra of some self-exciting and mutually exciting point processes. *Biometrika, 58*, 83–90.

Heath, D., Resnick, S., & Samorodnitsky, G. (1998). Heavy tails and long range dependence in on/off processes and associated fluid models. *Mathematics of Operations Research, 23*, 145–165.

Heneghan, C., Lowen, S. B., & Teich, M. C. (1996). Wavelet analysis for estimating the fractal properties of neural firing patterns. In J. M. Bower (Ed.), *Computational Neuroscience: Trends in Research 1995* (pp. 441–446). San Diego: Academic.

Heneghan, C., Lowen, S. B., & Teich, M. C. (1999). Analysis of spectral and wavelet-based measures used to assess cardiac pathology. In *Proceedings of the 1999 IEEE International Conference on Acoustics, Speech, and Signal Processing (ICASSP)*, Phoenix, AZ. Paper SPTM-8.2.

Heneghan, C. & McDarby, G. (2000). Establishing the relation between detrended fluctuation analysis and power spectral density analysis for stochastic processes. *Physical Review E, 62*, 6103–6110.

Hénon, M. (1976). A two-dimensional map with a strange attractor. *Communications in Mathematical Physics, 50*, 69–77.

Henry, M. & Zaffaroni, P. (2003). The long-range dependence paradigm for macroeconomics and finance. In P. Doukhan, G. Oppenheim, & M. S. Taqqu (Eds.), *Theory and Applications of Long-Range Dependence* (pp. 417–438). Boston: Birkhäuser.

Heyde, C. C. & Seneta, E. (Eds.). (2001). *Statisticians of the Centuries*. Berlin: Springer. Associate editors: P. Crépel, S. E. Fienberg, and J. Gani.

Hille, B. (2001). *Ionic Channels of Excitable Membranes* (Third ed.). Sunderland, MA: Sinauer.

Hohn, N., Veitch, D., & Abry, P. (2003). Cluster processes, a natural language for network traffic. *IEEE Transactions on Signal Processing, 51*, 2229–2244. Special Issue on Signal Processing in Networking.

Holden, A. V. (1976). *Models of the Stochastic Activity of Neurones*. Number 12 in Lecture Notes in Biomathematics. Berlin: Springer. S. Levin, Managing Editor.

Holtsmark, J. (1919). Über die Verbreiterung von Spektrallinien. *Annalen der Physik, 58*, 577–630.

Holtsmark, J. (1924). Über die Verbreiterung von Spektrallinien. II. *Physikalische Zeitschrift, 25*, 73–84.

Hon, E. H. & Lee, S. T. (1965). Electronic evaluations of the fetal heart rate patterns preceding fetal death, further observations. *American Journal of Obstetrics and Gynecology, 87*, 814–826.

Hooge, F. N. (1995). $1/f$ noise in semiconductor materials. In Bareikis, V. & Katilius, R. (Eds.), *Proceedings of the Thirteenth International Conference on Noise in Physical Systems and $1/f$ Fluctuations*, (pp. 8–13). World Scientific.

Hooge, F. N. (1997). 40 years of $1/f$ noise modelling. In Claeys, C. & Simoen, E. (Eds.), *Proceedings of the Fourteenth International Conference on Noise in Physical Systems and $1/f$ Fluctuations*, (pp. 3–10). World Scientific.

Hopcraft, K. I., Jakeman, E., & Matthews, J. O. (2002). Generation and monitoring of a discrete stable random process. *Journal of Physics A: Mathematical and General, 35*, L745–L751.

Hopcraft, K. I., Jakeman, E., & Matthews, J. O. (2004). Discrete scale-free distributions and associated limit theorems. *Journal of Physics A: Mathematical and General, 37*, L635–L642.

Hopcraft, K. I., Jakeman, E., & Tanner, R. M. J. (1999). Lévy random walks with fluctuating step number and multiscale behavior. *Physical Review E, 60*, 5327–5343.

Hopkinson, J. (1876). On the residual charge of the Leyden jar. *Philosophical Transactions of the Royal Society of London, 166*, 715–724.

Hsü, K. J. & Hsü, A. (1991). Self-similarity of the '$1/f$ noise' called music. *Proceedings of the National Academy of Sciences (USA), 88*, 3507–3509.

Hu, K., Ivanov, P. C., Chen, Z., Carpena, P., & Stanley, H. E. (2001). Effect of trends on detrended fluctuation analysis. *Physical Review E, 64*, 011114.

Huberman, B. A. & Adamic, L. A. (1999). Growth dynamics of the world-wide web. *Nature (London), 401*, 131.

Humbert, P. (1945). Nouvelles correspondances symboliques. *Bulletin de la Société Mathématique de France, 69*, 121–129.

Hurst, H. E. (1951). Long-term storage capacity of reservoirs. *Transactions of the American Society of Civil Engineers, 116*, 770–808.

Hurst, H. E. (1956). Methods of using long-term storage in reservoirs. *Proceedings of the Institution of Civil Engineers, General, Part I, 5*, 519–590. Discussion of paper on pp. 543–590.

Hurst, H. E., Black, R. P., & Simaika, Y. M. (1965). *Long Term Storage: An Experimental Study*. London: Constable.

Jaggard, D. L. (1997). Fractal electrodynamics: From super antennas to superlattices. In J. Lévy Véhel, E. Lutton, & C. Tricot (Eds.), *Fractals in Engineering: From Theory to Industrial Applications* (pp. 204–221). Berlin: Springer.

Jaggard, D. L. & Sun, X. (1990). Reflection from fractal multilayers. *Optics Letters, 15*, 1428–1430.

Jakeman, E. (1982). Fresnel scattering by a corrugated random surface with fractal slope. *Journal of the Optical Society of America, 72*, 1034–1041.

Jelenković, P. R. & Lazar, A. A. (1999). Asymptotic results for multiplexing subexponential on–off processes. *Advances in Applied Probability, 31*, 394–421.

Jelley, J. V. (1958). *Čerenkov Radiation and Its Applications*. London: Pergamon.

Jenkins, G. M. (1961). General considerations in the analysis of spectra. *Technometrics, 3*, 133–166.

Jensen, E. (1992). On the Engset loss formula. *Telektronikk (Oslo)*, *88*, 93–95.

Johnson, J. B. (1925). The Schottky effect in low frequency circuits. *Physical Review*, *26*, 71–85.

Jost, R. (1947). Bemerkungen zur mathematischen Theorie der Zähler. *Helvetica Physica Acta*, *20*, 173–182.

Kabanov, Y. M. (1978). The capacity of a channel of the Poisson type. *Theory of Probability and Its Applications*, *23*, 143–147.

Kagan, Y. Y. & Knopoff, L. (1987). Statistical short-term earthquake prediction. *Science*, *236*, 1563–1567.

Kallenberg, O. (1975). Limits of compound and thinned point processes. *Journal of Applied Probability*, *12*, 269–278.

Kang, K. & Redner, S. (1984). Scaling approach for the kinetics of recombination processes. *Physical Review Letters*, *52*, 955–958.

Kastner, M. A. (1985). The peculiar motion of electrons in amorphous semiconductors. In D. Alser, B. B. Schwartz, & M. C. Steele (Eds.), *Physical Properties of Amorphous Materials* (pp. 381–396). New York: Plenum.

Katz, B. (1966). *Nerve, Muscle, and Synapse*. New York: McGraw–Hill.

Kaulakys, B. (1999). On the intrinsic origin of $1/f$ noise. In Surya, C. (Ed.), *Proceedings of the Fifteenth International Conference on Noise in Physical Systems and $1/f$ Fluctuations*, (pp. 467–470). London: Bentham.

Kaye, B. H. (1989). *A Random Walk through Fractal Dimensions*. Weinheim, Germany: VCH.

Kelly, O. E., Johnson, D. H., Delgutte, B., & Cariani, P. (1996). Fractal noise strength in auditory-nerve fiber recordings. *Journal of the Acoustical Society of America*, *99*, 2210–2220.

Kendall, D. G. (1949). Stochastic processes and population growth. *Journal of the Royal Statistical Society B*, *11*, 230–264.

Kendall, D. G. (1953). Stochastic processes occurring in the theory of queues and their analysis by means of the imbedded Markov chain. *Annals of Mathematical Statistics*, *24*, 338–354.

Kendall, D. G. (1975). The genealogy of genealogy: Branching processes before (and after) 1873. *Bulletin of the London Mathematical Society*, *7*, 225–253.

Kendall, M. G. & Stuart, A. (1966). *The Advanced Theory of Statistics*, volume 3, (pp. 88–94). New York: Hafner.

Kenrick, G. W. (1929). The analysis of irregular motions with applications to the energy frequency spectrum of static and of telegraph signals. *Philosophical Magazine, 7 (Series 7)*, 176–196.

Kerner, B. S. (1998). Experimental features of self-organization in traffic flow. *Physical Review Letters, 81*, 3797–3800.

Kerner, B. S. (1999). The physics of traffic. *Physics World, 12*(10), 25–30.

Khinchin, A. Y. (1934). Korrelationstheorie der stationären stochastischen Prozesse. *Mathematische Annalen, 109*, 604–615.

Khinchin, A. Y. (1955). Mathematical Methods in the Theory of Queueing (in Russian). *Trudy Matematicheskogo Instituta imeni V. A. Steklova, 47*. Translation: (1960). (D. M. Andrews and M. H. Quenouille, Trans.) London: Griffin. Second edition: (1969).

Kiang, N. Y.-S., Watanabe, T., Thomas, E. C., & Clark, L. F. (1965). *Discharge Patterns of Single Fibers in the Cat's Auditory Nerve*. Research Monograph No. 35. Cambridge, MA: MIT Press.

Kingman, J. F. C. (1993). *Poisson Processes*. Oxford, UK: Oxford.

Klafter, J., Shlesinger, M. F., & Zumofen, G. (1996). Beyond Brownian motion. *Physics Today, 49*(2), 33–39.

Kleinrock, L. (1975). *Queueing Systems, Volume I: Theory*. New York: Wiley–Interscience.

Knoll, G. F. (1989). *Radiation Detection and Measurement* (Second ed.). New York: Wiley.

Kobayashi, M. & Musha, T. (1982). $1/f$ fluctuation of heartbeat period. *IEEE Transactions on Biomedical Engineering, BME-29*, 456–457.

Koch, C. (1999). *Biophysics of Computation: Information Processing in Single Neurons*. New York: Oxford.

Kodama, T., Mushiake, H., Shima, K., Nakahama, H., & Yamamoto, M. (1989). Slow fluctuations of single unit activities of hippocampal and thalamic neurons in cats. I. Relation to natural sleep and alert states. *Brain Research, 487*, 26–34.

Kogan, S. (1996). *Electronic Noise and Fluctuations in Solids*. Cambridge, UK: Cambridge.

Kohlrausch, R. (1854). Theorie des elektrischen Rückstandes in der Leidener Flasche. *Annalen der Physik und Chemie (Poggendorf), 91*, 179–214.

Kohmoto, M., Sutherland, B., & Tang, C. (1987). Critical wave functions and a Cantor-set spectrum of a one-dimensional quasicrystal model. *Physical Review B, 35*, 1020–1033.

Kolář, M., Ali, M. K., & Nori, F. (1991). Generalized Thue–Morse chains and their physical properties. *Physical Review B*, *43*, 1034–1047.

Kolmogorov, A. N. (1931). Über die analytischen Methoden in der Wahrscheinlichkeitsrechnung. *Mathematische Annalen*, *104*, 415–458.

Kolmogorov, A. N. (1940). Wienersche Spiralen und einige andere interessante Kurven im Hilbertschen Raum. *Doklady Akademii Nauk SSSR*, *26*(2), 115–118.

Kolmogorov, A. N. (1941). Über das logarithmisch normale Verteilungsgesetz der Dimensionen der Teilchen bei zerstückelung. *Doklady Akademii Nauk SSSR*, *31*(2), 99–101.

Kolmogorov, A. N. & Dmitriev, N. A. (1947). Branching stochastic processes (in Russian). *Doklady Akademii Nauk SSSR*, *56*(1), 7–10.

Komenani, E. & Sasaki, T. (1958). On the stability of traffic flow. *Journal of Operations Research (Japan)*, *2*, 11–26.

Kou, S. C. & Xie, X. S. (2004). Generalized Langevin equation with fractional Gaussian noise: Subdiffusion within a single protein molecule. *Physical Review Letters*, *93*, 180603.

Krapivsky, P. L., Redner, S., & Leyvraz, F. (2000). Connectivity of growing random networks. *Physical Review Letters*, *85*, 4629–4632.

Krapivsky, P. L., Rodgers, G. J., & Redner, S. (2001). Degree distributions of growing networks. *Physical Review Letters*, *86*, 5401–5404.

Krishnam, M. A., Venkatachalam, A., & Capone, J. M. (2000). A self-similar point process through fractal construction. In Pujolle, G., Perros, H. G., Fdida, S., Körner, U., & Stavrakakis, I. (Eds.), *Proceedings of International Conference NETWORKING 2000: Broadband Communications, High Performance Networking, and Performance of Communication Networks, Paris*, volume 1815 of *Lecture Notes in Computer Science*, (pp. 252–263). IFIP-TC6/European Union, Springer.

Kumar, A. R. & Johnson, D. H. (1993). Analyzing and modeling fractal intensity point processes. *Journal of the Acoustical Society of America*, *93*, 3365–3373.

Kurtz, T. G. (1996). Limit theorems for workload input models. In Kelly, F. P., Zachary, S., & Ziedins, I. (Eds.), *Stochastic Networks: Theory and Applications*, Royal Statistical Society Lecture Note Series 4, (pp. 119–140). Oxford.

Kuznetsov, P. I. & Stratonovich, R. L. (1956). On the mathematical theory of correlated random points. *Izvestiya Akademii Nauk Azerbaidzhanskoi SSR Seriya Fizika: Tekhnicheskikh I Matematicheskikh Nauk*, *20*, 167–178.

Kuznetsov, P. I., Stratonovich, R. L., & Tikhonov, V. I. (Eds.). (1965). *Non-Linear Transformations of Stochastic Processes*. Oxford, UK: Pergamon. Translation edited by J. Wise and D. C. Cooper.

Lapenna, V., Macchiato, M., & Telesca, L. (1998). $1/f^\beta$ fluctuations and self-similarity in earthquake dynamics: Observational evidences in southern Italy. *Physics of the Earth and Planetary Interiors, 106,* 115–127.

Lapicque, L. (1907). Recherches quantitatives sur l'excitation électrique des nerfs traitée comme une polarisation. *Journal de Physiologie et Pathologie Générale (Paris), 9,* 620–635.

Lapicque, L. (1926). *L'excitabilité en fonction du temps.* Paris: Presses Universitaires de France.

Latouche, G. & Remiche, M.-A. (2002). An MAP-based Poisson cluster model for web traffic. *Performance Evaluation, 49,* 359–370.

Läuger, P. (1988). Internal motions in proteins and gating kinetics of ionic channels. *Biophysical Journal, 53,* 877–884.

Lawrance, A. J. (1972). Some models for stationary series of univariate events. In P. A. W. Lewis (Ed.), *Stochastic Point Processes: Statistical Analysis, Theory, and Applications* (pp. 199–256). New York: Wiley–Interscience.

Lax, M. (1997). Stochastic processes. In G. L. Trigg & E. H. Immergut (Eds.), *Encyclopedia of Applied Physics* (pp. 19–60). New York: VCH–Wiley.

Leadbetter, M. R., Lindgren, G., & Rootzen, H. (1983). *Extremes and Related Properties of Random Sequences and Processes.* Berlin: Springer.

Leland, W. E., Taqqu, M. S., Willinger, W., & Wilson, D. V. (1993). On the self-similar nature of Ethernet traffic. In *Proceedings of the ACM SIGCOMM'93 Conference on Communications Architectures, Protocols and Applications,* (pp. 183–193)., San Francisco. New York: ACM Press.

Leland, W. E., Taqqu, M. S., Willinger, W., & Wilson, D. V. (1994). On the self-similar nature of Ethernet traffic (extended version). *IEEE/ACM Transactions on Networking, 2,* 1–15.

Leland, W. E. & Wilson, D. V. (1989). One million consecutive ethernet-packet arrivals at Bellcore Morristown Research and Engineering Facility. Data sets BC-pOct89 and BC-pAug89, as well as others, are available at the Internet traffic archive http://ita.ee.lbl.gov/html/contrib/BC.html.

Leland, W. E. & Wilson, D. V. (1991). High time-resolution measurement and analysis of LAN traffic: Implications for LAN interconnection. In *Proceedings of the Tenth Annual Joint Conference of the IEEE Computer and Communications Societies, Networking in the 90s (INFOCOM'91),* (pp. 1360–1366)., Bal Harbour, FL. Institute of Electrical and Electronic Engineers, New York.

Levy, J. B. & Taqqu, M. S. (2000). Renewal reward processes with heavy-tailed inter-renewal times and heavy-tailed rewards. *Bernoulli, 6,* 23–44.

Lévy, P. (1937). *Théorie de l'addition des variables aléatoires.* Paris: Gauthier-Villars. Second edition: (1954).

Lévy, P. (1940). Sur certains processus stochastiques homogènes. *Compositio Mathematica, 7,* 283–339.

Lévy, P. (1948). *Processus stochastiques et mouvement brownien.* Paris: Gauthier-Villars. Second edition: (1965).

Lévy Véhel, J., Lutton, E., & Tricot, C. (Eds.). (1997). *Fractals in Engineering: From Theory to Industrial Applications.* Berlin: Springer.

Lévy Véhel, J. & Riedi, R. (1997). Fractional Brownian motion and data traffic modeling: The other end of the spectrum. In J. Lévy Véhel, E. Lutton, & C. Tricot (Eds.), *Fractals in Engineering: From Theory to Industrial Applications* (pp. 185–202). Berlin: Springer.

Lewis, C. D., Gebber, G. L., Larsen, P. D., & Barman, S. M. (2001). Long-term correlations in the spike trains of medullary sympathetic neurons. *Journal of Neurophysiology, 85,* 1614–1622.

Lewis, P. A. W. (1964). A branching Poisson process model for the analysis of computer failure patterns. *Journal of the Royal Statistical Society B, 26,* 398–456.

Lewis, P. A. W. (1967). Non-homogeneous branching Poisson processes. *Journal of the Royal Statistical Society B, 29,* 343–354.

Lewis, P. A. W. (Ed.). (1972). *Stochastic Point Processes: Statistical Analysis, Theory, and Applications.* New York: Wiley–Interscience.

Li, T. & Teich, M. C. (1993). Photon point process for traveling-wave laser amplifiers. *IEEE Journal of Quantum Electronics, 29,* 2568–2578.

Li, W. (1991). Expansion-modification systems: A model for $1/f$ spectra. *Physical Review A, 43,* 5240–5260.

Libert, J. (1976). Comparaison des distributions statistiques de comptage des systèmes radioactifs. *Nuclear Instruments and Methods, 136,* 563–568.

Liebovitch, L. S. (1998). *Fractals and Chaos Simplified for the Life Sciences.* New York: Oxford.

Liebovitch, L. S., Fischbarg, J., & Koniarek, J. P. (1987). Ion channel kinetics: A model based on fractal scaling rather than multistate Markov processes. *Mathematical Biosciences, 84,* 37–68.

Liebovitch, L. S., Fischbarg, J., Koniarek, J. P., Todorova, I., & Wang, M. (1987). Fractal model of ion-channel kinetics. *Biochimica et Biophysica Acta, 896,* 173–180.

Liebovitch, L. S., Scheurle, D., Rusek, M., & Zochowski, M. (2001). Fractal methods to analyze ion channel kinetics. *Methods, 24* (Fractals in Neuroscience), 359–375.

Liebovitch, L. S. & Tóth, T. I. (1990). Using fractals to understand the opening and closing of ion channels. *Annals of Biomedical Engineering, 18,* 177–194.

Likhanov, N. (2000). Bounds on the buffer occupancy probability with self-similar input traffic. In K. Park & W. Willinger (Eds.), *Self-Similar Network Traffic and Performance Evaluation* chapter 8, (pp. 193–213). New York: Wiley–Interscience.

Likhanov, N., Tsybakov, B., & Georganas, N. D. (1995). Analysis of an ATM buffer with self-similar ("fractal") input traffic. In *Proceedings of the Fourteenth Annual Joint Conference of the IEEE Computer and Communication Societies,* volume 3, (pp. 985–992). IEEE Computer Society, Washington, DC.

Little, J. D. C. (1961). A proof for the queuing formula: $L = \lambda W$. *Operations Research, 9,* 383–387.

Liu, N. (1997). Propagation of light waves in Thue–Morse dielectric multilayers. *Physical Review B, 55,* 3543–3547.

Lotka, A. J. (1926). The frequency distribution of scientific productivity. *Journal of the Washington Academy of Sciences, 16,* 317–323.

Lotka, A. J. (1939). A contribution to the theory of self-renewing aggregates, with especial reference to industrial replacement. *Annals of Mathematical Statistics, 10,* 1–25.

Lowen, S. B. (1992). *Fractal Stochastic Processes.* PhD thesis, Columbia University, New York, NY.

Lowen, S. B. (1996). Refractoriness-modified doubly stochastic Poisson point process. Technical Report 449-96-15, Columbia University, Center for Telecommunications Research, New York.

Lowen, S. B. (2000). Efficient generation of fractional Brownian motion for simulation of infrared focal-plane array calibration drift. *Methodology and Computing in Applied Probability, 1,* 445–456. Erratum: $X(k)$ is improperly defined; the correct expression, for $0 < k < M$, is $X(k) = (G_{1,k} + iG_{2,k})\sqrt{S(k)/2}$, with all $G_{1,k}$ and $G_{2,k}$ independent, zero-mean, unit-variance Gaussian random variables.

Lowen, S. B., Cash, S. S., Poo, M.-m., & Teich, M. C. (1997a). Neuronal exocytosis exhibits fractal behavior. In J. M. Bower (Ed.), *Computational Neuroscience: Trends in Research 1997* (pp. 13–18). New York: Plenum.

Lowen, S. B., Cash, S. S., Poo, M.-m., & Teich, M. C. (1997b). Quantal neurotransmitter secretion rate exhibits fractal behavior. *Journal of Neuroscience, 17,* 5666–5677.

Lowen, S. B., Liebovitch, L. S., & White, J. A. (1999). Fractal ion-channel behavior generates fractal firing patterns in neuronal models. *Physical Review E*, *59*, 5970–5980.

Lowen, S. B., Ozaki, T., Kaplan, E., Saleh, B. E. A., & Teich, M. C. (2001). Fractal features of dark, maintained, and driven neural discharges in the cat visual system. *Methods*, *24* (Fractals in Neuroscience), 377–394.

Lowen, S. B., Ozaki, T., Kaplan, E., & Teich, M. C. (1998). Information exchange between pairs of spike trains in the mammalian visual system. In J. M. Bower (Ed.), *Computational Neuroscience: Trends in Research 1998* (pp. 447–452). New York: Plenum.

Lowen, S. B. & Teich, M. C. (1989a). Fractal shot noise. *Physical Review Letters*, *63*, 1755–1759.

Lowen, S. B. & Teich, M. C. (1989b). Generalised $1/f$ shot noise. *Electronics Letters*, *25*, 1072–1074.

Lowen, S. B. & Teich, M. C. (1990). Power-law shot noise. *IEEE Transactions on Information Theory*, *36*, 1302–1318.

Lowen, S. B. & Teich, M. C. (1991). Doubly stochastic Poisson point process driven by fractal shot noise. *Physical Review A*, *43*, 4192–4215.

Lowen, S. B. & Teich, M. C. (1992a). Auditory-nerve action potentials form a non-renewal point process over short as well as long time scales. *Journal of the Acoustical Society of America*, *92*, 803–806.

Lowen, S. B. & Teich, M. C. (1992b). Fractal renewal processes as a model of charge transport in amorphous semiconductors. *Physical Review B*, *46*, 1816–1819.

Lowen, S. B. & Teich, M. C. (1993a). Estimating the dimension of a fractal point process. In L. D. William (Ed.), *SPIE Proceedings: Chaos in Biology and Medicine*, volume 2036 (pp. 64–76). Bellingham, WA: SPIE.

Lowen, S. B. & Teich, M. C. (1993b). Fractal auditory-nerve firing patterns may derive from fractal switching in sensory hair-cell ion channels. In P. H. Handel & A. L. Chung (Eds.), *Proceedings of the Twelfth International Conference on Noise in Physical Systems and 1/f Fluctuations*, number 285 in AIP Conference Proceedings (pp. 745–748). New York: American Institute of Physics.

Lowen, S. B. & Teich, M. C. (1993c). Fractal renewal processes. *IEEE Transactions on Information Theory*, *39*, 1669–1671.

Lowen, S. B. & Teich, M. C. (1993d). Fractal renewal processes generate $1/f$ noise. *Physical Review E*, *47*, 992–1001. Errata: Several equations in this paper are incorrect; proper results are provided in Chapters 7 and 8.

Lowen, S. B. & Teich, M. C. (1995). Estimation and simulation of fractal stochastic point processes. *Fractals*, *3*, 183–210.

Lowen, S. B. & Teich, M. C. (1996a). The periodogram and Allan variance reveal fractal exponents greater than unity in auditory-nerve spike trains. *Journal of the Acoustical Society of America*, *99*, 3585–3591.

Lowen, S. B. & Teich, M. C. (1996b). Refractoriness-modified fractal stochastic point processes for modeling sensory-system spike trains. In J. M. Bower (Ed.), *Computational Neuroscience: Trends in Research 1995* (pp. 447–452). San Diego: Academic.

Lowen, S. B. & Teich, M. C. (1997). Estimating scaling exponents in auditory-nerve spike trains using fractal models incorporating refractoriness. In E. R. Lewis, G. R. Long, R. F. Lyon, P. M. Narins, C. R. Steele, & E. Hecht-Poinar (Eds.), *Diversity in Auditory Mechanics* (pp. 197–204). Singapore: World Scientific.

Lubberger, F. (1925). Die Theorie des Fernsprechverkehrs. *Elektrische Nachrichtentechnik*, *2*, 52–64.

Lubberger, F. (1927). Die Wahrscheinlichkeitsrechnung in der Fernsprechtechnik. *Zeitschrift für technische Physik*, *8*, 17–25.

Lukes, T. (1961). The statistical properties of sequences of stochastic pulses. *Proceedings of the Physical Society (London)*, *78*, 153–168.

Lundahl, T., Ohley, W. J., Kay, S. M., & Siffert, R. (1986). Fractional Brownian motion: A maximum likelihood estimator and its application to image texture. *IEEE Transactions on Medical Imaging*, *5*, 152–161.

Maccone, C. (1981). $1/f^x$ noises and Riemann–Liouville fractional integral/derivative of the Brownian motion. In Meijer, P. H. E., Mountain, R. D., & Soulen, Jr, R. J. (Eds.), *Proceedings of the Sixth International Conference on Noise in Physical Systems*, number 614 in National Bureau of Standards Special Publications, (pp. 192–195). US Government Printing Office, Washington, DC.

Machlup, S. (1954). Noise in semiconductors: Spectrum of a two-parameter signal. *Journal of Applied Physics*, *25*, 341–343.

Malamud, B. D. (2004). Tails of natural hazards. *Physics World*, *17*(8), 31–35.

Malik, M., Bigger, J. T., Camm, A. J., Kleiger, R. E., Malliani, A., Moss, A. J., Schwartz, P. J., & The Task Force of the European Society of Cardiology and the North American Society of Pacing and Electrophysiology (1996). Heart rate variability — Standards of measurement, physiological interpretation, and clinical use. *European Heart Journal*, *17*, 354–381. Also published in *Circulation*, *93*, 1043–1065.

Mandel, L. (1959). Fluctuations of photon beams: The distribution of the photoelectrons. *Proceedings of the Physical Society (London)*, *74*, 233–242.

Mandelbrot, B. B. (1960). The Pareto–Lévy law and the distribution of income. *International Economic Review*, *1*, 79–106.

Mandelbrot, B. B. (1964). The stable Paretian income distribution when the apparent exponent is near two. *International Economic Review*, *4*, 111–115.

Mandelbrot, B. B. (1965a). Self-similar error clusters in communication systems and the concept of conditional stationarity. *IEEE Transactions on Communication Technology*, *13*, 71–90.

Mandelbrot, B. B. (1965b). Une classe de processus stochastiques homothétiques à soi; application à la loi climatologique de H. E. Hurst. *Comptes Rendus de l'Académie des Sciences (Paris)*, *260 (Series 1)*, 3274–3277.

Mandelbrot, B. B. (1967a). How long is the coast of Britain? Statistical self-similarity and fractional dimension. *Science*, *156*, 636–638.

Mandelbrot, B. B. (1967b). Some noises with $1/f$ spectrum, a bridge between direct current and white noise. *IEEE Transactions on Information Theory*, *IT-13*, 289–298.

Mandelbrot, B. B. (1969). Long-run linearity, locally Gaussian processes, H-spectra and infinite variances. *International Economic Review*, *10*, 82–113.

Mandelbrot, B. B. (1972). Renewal sets and random cutouts. *Zeitschrift für Wahrscheinlichkeitstheorie*, *22*, 145–157.

Mandelbrot, B. B. (1974). Intermittent turbulence in self-similar cascades: Divergence of high moments and dimension of the carrier. *Journal of Fluid Mechanics*, *62*, 331–358.

Mandelbrot, B. B. (1975). Stochastic models for the earth's relief, the shape and the fractal dimension of the coastlines, and the number–area rule for islands. *Proceedings of the National Academy of Sciences (USA)*, *72*, 3825–3828.

Mandelbrot, B. B. (1982). *The Fractal Geometry of Nature* (Second ed.). New York: W. H. Freeman.

Mandelbrot, B. B. (1997). *Fractals and Scaling in Finance: Discontinuity, Concentration, Risk*. Berlin: Springer.

Mandelbrot, B. B. (1999). *Multifractals and $1/f$ Noise: Wild Self-Affinity in Physics (1963-1976)*. Berlin: Springer.

Mandelbrot, B. B. (2001). *Gaussian Self-Affinity and Fractals*. Berlin: Springer.

Mandelbrot, B. B. & Hudson, R. L. (2004). *The (Mis)Behavior of Markets: A Fractal View of Risk, Ruin, and Reward*. New York: Basic.

Mandelbrot, B. B. & Van Ness, J. W. (1968). Fractional Brownian motions, fractional noises and applications. *Society for Industrial and Applied Mathematics (SIAM) Review, 10*, 422–437.

Mandelbrot, B. B. & Wallis, J. R. (1969a). Computer experiments with fractional Gaussian noises. Part 1: Averages and variances. *Water Resources Research, 5*, 228–241.

Mandelbrot, B. B. & Wallis, J. R. (1969b). Computer experiments with fractional Gaussian noises. Part 2: Rescaled ranges and spectra. *Water Resources Research, 5*, 242–259.

Mandelbrot, B. B. & Wallis, J. R. (1969c). Robustness of the rescaled range R/S in the measurement of noncyclic long run statistical dependence. *Water Resources Research, 5*, 967–988.

Mannersalo, P. & Norros, I. (1997). Multifractal analysis of real ATM traffic: A first look. Technical Report COST257TD(97)19, VTT Information Technology, Espoo, Finland.

Marinari, E., Parisi, G., Ruelle, D., & Widney, P. (1983). On the interpretation of $1/f$ noise. *Communications in Mathematical Physics, 89*, 1–12.

Masoliver, J., Montero, M., & McKane, A. (2001). Integrated random processes exhibiting long tails, finite moments, and power-law spectra. *Physical Review E, 64*, 011110.

Matsuo, K., Saleh, B. E. A., & Teich, M. C. (1982). Cascaded Poisson processes. *Journal of Mathematical Physics, 23*, 2353–2364.

Matsuo, K., Teich, M. C., & Saleh, B. E. A. (1983). Thomas point process in pulse, particle, and photon detection. *Applied Optics, 22*, 1898–1909.

Matsuo, K., Teich, M. C., & Saleh, B. E. A. (1984). Poisson branching point processes. *Journal of Mathematical Physics, 24*, 2174–2185.

Matthes, K. (1963). Stationäre zufällige Punktfolgen. I. *Jahresbericht der Deutschen Mathematiker-Vereinigung, 66*, 66–79.

Matthews, J. O., Hopcraft, K. I., & Jakeman, E. (2003). Generation and monitoring of discrete stable random processes using multiple immigration population models. *Journal of Physics A: Mathematical and General, 36*, 11585–11603.

McGill, W. J. (1967). Neural counting mechanisms and energy detection in audition. *Journal of Mathematical Psychology, 4*, 351–376.

McGill, W. J. & Goldberg, J. P. (1968). A study of the near-miss involving Weber's law and pure-tone intensity discrimination. *Perception & Psychophysics, 4*, 105–109.

McGill, W. J. & Teich, M. C. (1995). Alerting signals and detection in a sensory network. *Journal of Mathematical Psychology*, *39*, 146–163. An earlier version of this paper was circulated as Technical Report 132, Center for Human Information Processing (CHIP), University of California San Diego, La Jolla, CA, January 1991.

McWhorter, A. L. (1957). $1/f$ noise and germanium surface properties. In R. H. Kingston (Ed.), *Semiconductor Surface Physics* (pp. 207–228). Philadelphia: University of Pennsylvania.

Merlin, R., Bajema, K., Clarke, R., Juang, F.-Y., & Bhattacharya, P. K. (1985). Quasiperiodic GaAs–AlAs heterostructures. *Physical Review Letters*, *55*, 1768–1770.

Mikosch, T., Resnick, S., Rootzén, H., & Stegeman, A. (2002). Is network traffic approximated by stable Lévy motion or fractional Brownian motion? *Annals of Applied Probability*, *12*, 23–68.

Millhauser, G. L., Salpeter, E. E., & Oswald, R. E. (1988). Diffusion models of ion-channel gating and the origin of power-law distributions from single-channel recording. *Proceedings of the National Academy of Sciences (USA)*, *85*, 1503–1507.

Mitchell, R. L. (1968). Permanence of the log-normal distribution. *Journal of the Optical Society of America*, *58*, 1267–1272.

Mitchell, S. L., Collins, J. J., De Luca, C. J., Burrows, A., & Lipsitz, L. A. (1995). Open-loop and closed-loop postural control mechanisms in Parkinson's disease: Increased mediolateral activity during quiet standing. *Neuroscience Letters*, *197*, 133–136.

Molchan, G. M. (2003). Historical comments related to fractional Brownian motion. In P. Doukhan, G. Oppenheim, & M. S. Taqqu (Eds.), *Theory and Applications of Long-Range Dependence* (pp. 39–42). Boston: Birkhäuser.

Montanari, A. (2003). Long-range dependence in hydrology. In P. Doukhan, G. Oppenheim, & M. S. Taqqu (Eds.), *Theory and Applications of Long-Range Dependence* (pp. 461–472). Boston: Birkhäuser.

Montgomery, H. C. (1952). Transistor noise in circuit applications. *Proceedings of the IRE*, *40*, 1461–1471.

Montroll, E. W. & Shlesinger, M. F. (1982). On $1/f$ noise and other distributions with long tails. *Proceedings of the National Academy of Sciences (USA)*, *79*, 3380–3383.

Moon, F. C. (1992). *Chaotic and Fractal Dynamics: An Introduction for Applied Scientists and Engineers*. New York: Wiley–Interscience.

Moran, P. A. P. (1967). A non-Markovian quasi-Poisson process. *Studia Scientiarum Mathematicarum Hungarica*, *2*, 425–429.

Morant, G. (1921). On random occurrences in space and time when followed by a closed interval. *Biometrika*, *13*, 309–337.

Moriarty, B. J. (1963). Blocks in error and Pareto's law. Technical Report 25G-15, MIT Lincoln Laboratory, Lexington, MA. Armed Services Technical Information Agency (ASTIA) Report AD-296490, unclassified.

Morse, M. (1921a). A one-to-one representation of geodesics on a surface of negative curvature. *American Journal of Mathematics*, *43*, 33–51.

Morse, M. (1921b). Recurrent geodesics on a surface of negative curvature. *Transactions of the American Mathematical Society*, *22*, 84–100.

Moskowitz, H. R., Scharf, B., & Stevens, J. C. (Eds.). (1974). *Sensation and Measurement: Papers in Honor of S. S. Stevens*. Boston: Reidel.

Moyal, J. E. (1962). The general theory of stochastic population processes. *Acta Mathematica*, *108*, 1–31.

Müller, J. W. (1973). Dead-time problems. *Nuclear Instruments and Methods*, *112*, 47–57.

Müller, J. W. (1974). Some formulae for a dead-time-distorted Poisson process. *Nuclear Instruments and Methods*, *117*, 401–404.

Müller, J. W. (1981). Bibliography on dead time effects. Technical Report BIPM-81/11, Bureau International des Poids et Mesures, Sèvres, France.

Musha, T. (1981). $1/f$ fluctuations in biological systems. In Meijer, P. H. E., Mountain, R. D., & Soulen, Jr, R. J. (Eds.), *Proceedings of the Sixth International Conference on Noise in Physical Systems*, number 614 in National Bureau of Standards Special Publications, (pp. 143–146). US Government Printing Office, Washington, DC.

Musha, T. & Higuchi, H. (1976). The $1/f$ fluctuation of a traffic current on an expressway. *Japanese Journal of Applied Physics*, *15*, 1271–1275.

Musha, T., Katsurai, K., & Teramachi, Y. (1985). Fluctuations of human tapping intervals. *IEEE Transactions on Biomedical Engineering*, *BME-32*, 578–582.

Musha, T., Kosugi, Y., Matsumoto, G., & Suzuki, M. (1981). Modulation of the time relation of action potential impulses propagating along an axon. *IEEE Transactions on Biomedical Engineering*, *BME-28*, 616–623.

Musha, T., Takeuchi, H., & Inoue, T. (1983). $1/f$ fluctuations in the spontaneous spike discharge intervals of a giant snail neuron. *IEEE Transactions on Biomedical Engineering*, *BME-30*, 194–197.

Myskja, A. (1998a). The Engset report of 1915: Summary and comments. *Telektronikk (Oslo)*, *94*, 143–153. Reprinted in Myskja & Espvik (2002, pp. 149–174).

Myskja, A. (1998b). The man behind the formula: Biographical notes on Tore Olaus Engset. *Telektronikk (Oslo)*, *94*, 154–164. Reprinted in Myskja & Espvik (2002).

Myskja, A. & Espvik, O. (Eds.). (2002). *Tore Olaus Engset (1865–1943): The Man Behind the Formula.* Trondheim, Norway: Tapir Akademisk Forlag.

Newell, G. F. & Sparks, G. A. (1972). Statistical properties of traffic counts. In P. A. W. Lewis (Ed.), *Stochastic Point Processes: Statistical Analysis, Theory, and Applications* (pp. 166–174). New York: Wiley–Interscience.

Newton, I. (1687). *Philosophiæ Naturalis Principia Mathematica.* London. Translation of Third (1726) Edition: (1999). *Mathematical Principles of Natural Philosophy.* (I. B. Cohen and A. Whitman, Trans.). Berkeley: University of California Press.

Neyman, J. (1939). On a new class of 'contagious' distributions, applicable in entomology and bacteriology. *Annals of Mathematical Statistics*, *10*, 35–57.

Neyman, J. & Scott, E. L. (1958). A statistical approach to problems of cosmology. *Journal of the Royal Statistical Society B*, *20*, 1–43.

Neyman, J. & Scott, E. L. (1972). Processes of clustering and applications. In P. A. W. Lewis (Ed.), *Stochastic Point Processes: Statistical Analysis, Theory, and Applications* (pp. 646–681). New York: Wiley–Interscience.

Norros, I. (1994). A storage model with self-similar input. *Queueing Systems*, *16*, 387–396.

Norros, I. (1995). On the use of fractional Brownian motion in the theory of connectionless networks. *IEEE Journal on Selected Areas in Communications*, *13*, 953–962.

Norsworthy, S. R., Schreier, R., & Temes, G. C. (Eds.). (1996). *Delta-Sigma Data Converters: Theory, Design, and Simulation.* New York: IEEE Press.

Olson, S. (2004). The genius of the unpredictable. *Yale Alumni Magazine*, *68*(2), 36–43.

Omori, F. (1895). On the aftershocks of earthquakes. *Journal of the College of Science, Imperial University of Tokyo*, *7*, 111–200.

Oppenheim, A. V. & Schafer, R. W. (1975). *Digital Signal Processing.* Englewood Cliffs, NJ: Prentice–Hall.

Orenstein, J., Kastner, M. A., & Vaninov, V. (1982). Transient photoconductivity and photo-induced optical absorption in amorphous semiconductors. *Philosophical Magazine B*, *46*, 23–62.

Orer, H. S., Das, M., Barman, S. M., & Gebber, G. L. (2003). Fractal activity generated independently by medullary sympathetic premotor and preganglionic sympathetic neurons. *Journal of Neurophysiology*, *90*, 47–54.

Oshanin, G. S., Burlatsky, S. F., & Ovchinnikov, A. A. (1989). Fluctuation-dominated kinetics of irreversible bimolecular reactions with external random sources on fractals. *Physics Letters A, 139*, 245–248.

Ott, E. (2002). *Chaos in Dynamical Systems* (Second ed.). Cambridge, UK: Cambridge.

Ott, E., Sauer, T., & Yorke, J. A. (Eds.). (1994). *Coping with Chaos: Analysis of Chaotic Data and the Exploitation of Chaotic Systems*. New York: Wiley.

Ovchinnikov, A. A. & Zeldovich, Y. B. (1978). Role of density fluctuations in bimolecular reaction kinetics. *Chemical Physics, 28*, 215–218.

Palm, C. (1937). Några undersökningar över väntetider vid telefonanläggningar (Some studies of waiting times at telephone exchanges). *Tekniska Meddelanden från Kungliga Telegrafstyrelsen*, (7–9).

Palm, C. (1943). Intensitätsschwankungen im Fernsprechverkehr. *Ericsson Technics, 44*, 1–189. Translation: (1988). *Intensity Variations in Telephone Traffic*. (North-Holland Studies in Telecommunication, *Vol. 10*, pp. 1–209, C. Jacobæus and G. Neovius, Eds.). Amsterdam: Elsevier.

Papangelou, F. (1972). Integrability of expected increments of point processes and a related random change of scale. *Transactions of the American Mathematical Society, 165*, 483–506.

Papoulis, A. (1991). *Probability, Random Variables, and Stochastic Processes* (Third ed.). New York: McGraw–Hill.

Pareto, V. (1896). *Cours d'économie politique professé à l'Université de Lausanne*, volume I and II. Lausanne: Rouge.

Park, J. & Gray, R. M. (1992). Sigma delta modulation with leaky integration and constant input. *IEEE Transactions on Information Theory, 38*, 1512–1533.

Park, K. (2000). Future directions and open problems in performance evaluation and control of self-similar network traffic. In K. Park & W. Willinger (Eds.), *Self-Similar Network Traffic and Performance Evaluation* chapter 21, (pp. 531–553). New York: Wiley–Interscience.

Park, K., Kim, G., & Crovella, M. (1996). On the relationship between file sizes, transport protocols, and self-similar network traffic. In *Proceedings of the 1996 International Conference on Network Protocols (ICNP '96)*, (pp. 171–180). IEEE Computer Society, Washington, DC.

Park, K., Kim, G., & Crovella, M. E. (2000). The protocol stack and its modulating effect on self-similar traffic. In K. Park & W. Willinger (Eds.), *Self-Similar Network Traffic and Performance Evaluation* chapter 14, (pp. 349–366). New York: Wiley–Interscience.

Park, K. & Willinger, W. (Eds.). (2000). *Self-Similar Network Traffic and Performance Evaluation.* New York: Wiley–Interscience.

Parzen, E. (1962). *Stochastic Processes.* San Francisco: Holden–Day. Reissued as Number 24 in Classics in Applied Mathematics. (1999). Philadelphia: Society for Industrial and Applied Mathematics (SIAM).

Pastor-Satorras, R. & Vespignani, A. (2004). *Evolution and Structure of the Internet: A Statistical Physics Approach.* Cambridge, UK: Cambridge.

Paulus, M. P. & Geyer, M. A. (1992). The effects of MDMA and other methylenedioxy-substituted phenylalkylamines on the structure of rat locomotor activity. *Neuropsychopharmacology, 7,* 15–31.

Paxson, V. & Floyd, S. (1995). Wide area traffic: The failure of Poisson modeling. *IEEE/ACM Transactions on Networking, 3,* 226–244.

Pecher, C. (1939). La fluctuation d'excitabilité de la fibre nerveuse. *Archives Internationales de Physiologie (Liège), 49,* 129–152.

Peitgen, H. & Saupe, D. (1988). *The Science of Fractal Images.* Berlin: Springer.

Peitgen, H.-O., Jürgens, H., & Saupe, D. (1997). *Chaos and Fractals: New Frontiers of Science* (Second ed.). Berlin: Springer.

Peltier, R.-F. & Lévy Véhel, J. (1995). Multifractional Brownian motion: Definition and preliminary results. Technical Report RR-2645, Projet Fractales, INRIA, Rocquencourt, France.

Penck, A. (1894). *Morphologie der Erdoberfläche*, volume 1. Stuttgart: Verlag von J. Engelhorn. Series: *Bibliothek geographischer Handbücher.* (F. Ratzel, Series Ed.).

Peng, C.-K., Havlin, S., Stanley, H. E., & Goldberger, A. L. (1995). Quantification of scaling exponents and crossover phenomena in nonstationary heartbeat time series. *Chaos, 5,* 82–87.

Peng, C.-K., Mietus, J., Hausdorff, J. M., Havlin, S., Stanley, H. E., & Goldberger, A. L. (1993). Long-range anticorrelations and non-Gaussian behavior of the heartbeat. *Physical Review Letters, 70,* 1343–1346.

Peřina, J. (1967). Superposition of coherent and incoherent fields. *Physics Letters, 24A,* 333–334.

Perkal, J. (1958a). O długości krzywych empirycznych (On the length of empirical curves). *Zastosowania Matematyki, 3,* 257–286.

Perkal, J. (1958b). Próba obiektywnej generalizacji. *Geodezja i Kartografia, 7,* 130–142. Translation: (1966). An attempt at objective generalization. *Michigan Inter-University Community of Mathematical Geographers, Discussion Paper No. 10,* 1–34. (J. D. Nystuen, Ed.). (R. Jackowski and W. Tobler, Trans.).

Perrin, J.-B. (1909). Mouvement brownien et réalité moléculaire. *Annales de Chimie et de Physique, 18*, 5–114. Translation: (1910). Brownian Movement and Molecular Reality. (F. Soddy, Trans.). London: Taylor and Francis.

Petropulu, A. P., Pesquet, J.-C., Yang, X., & Yin, J. (2000). Power-law shot noise and its relationship to long-memory α-stable processes. *IEEE Transactions on Signal Processing, 48*, 1883–1892.

Picinbono, B. (1960). Tendence vers le caractère gaussien par filtrage sélectif. *Comptes Rendus de l'Académie des Sciences (Paris), 250*, 1174–1176.

Pinsky, M. A. (1984). *Introduction to Partial Differential Equations with Applications.* New York: McGraw–Hill.

Pipiras, V. & Taqqu, M. S. (2003). Fractional calculus and its connections to fractional Brownian motion. In P. Doukhan, G. Oppenheim, & M. S. Taqqu (Eds.), *Theory and Applications of Long-Range Dependence* (pp. 165–201). Boston: Birkhäuser.

Poincaré, H. (1908). *Science et méthode.* Paris: Ernest Flammarion. Translation of the 1914 edition with a preface by Bertrand Russell: (2001). *Science and Method.* South Bend, IN: St. Augustine Press–Key Text Editions.

Poisson, S. D. (1837). *Recherches sur la probabilité des jugements en matière criminelle et en matière civile, précédées des règles générales du calcul des probabilités.* Paris: Bachelier.

Pollard, H. (1946). The representation of $e^{-x^{\lambda}}$ as a Laplace integral. *Bulletin of the American Mathematical Society, 52*, 908–910.

Pontrjagin, L. & Schnirelmann, L. (1932). Sur une propriété métrique de la dimension. *Annals of Mathematics, 33 (Second Series)*, 156–162.

Powers, N. L. & Salvi, R. J. (1992). Comparison of discharge rate fluctuations in the auditory nerve of chickens and chinchillas. In Lim, D. J. (Ed.), *Abstracts of the XV Midwinter Research Meeting, Association for Research in Otolaryngology*, (pp. 101)., St. Petersburg Beach, FL. Association for Research in Otolaryngology, Des Moines, IA. Abstract 292.

Press, W. H., Teukolsky, S. A., Vetterling, W. T., & Flannery, B. P. (1992). *Numerical Recipes in C* (Second ed.). Cambridge, UK: Cambridge.

Prucnal, P. R. & Saleh, B. E. A. (1981). Transformations of image-signal-dependent noise into image-signal-independent noise. *Optics Letters, 6*, 316–318.

Prucnal, P. R. & Teich, M. C. (1979). Statistical properties of counting distributions for intensity-modulated sources. *Journal of the Optical Society of America, 69*, 539–544.

Prucnal, P. R. & Teich, M. C. (1980). An increment threshold law for stimuli of arbitrary statistics. *Journal of Mathematical Psychology, 21*, 168–177.

Prucnal, P. R. & Teich, M. C. (1982). Multiplication noise in the human visual system at threshold: 2. Probit estimation of parameters. *Kybernetik (Biological Cybernetics)*, *43*, 87–96.

Prucnal, P. R. & Teich, M. C. (1983). Refractory effects in neural counting processes with exponentially decaying rates. *IEEE Transactions on Systems, Man and Cybernetics*, *SMC-13*, 1028–1033. Special issue on neural and sensory information processing.

Quenouille, M. H. (1949). A relation between the logarithmic, Poisson, and negative binomial series. *Biometrics*, *5*, 162–164.

Quine, M. P. & Seneta, E. (1987). Bortkiewicz's data and the law of small numbers. *International Statistical Review*, *55*, 173–181.

Rammal, R. & Toulouse, G. (1983). Random walks on fractal structures and percolation clusters. *Journal de Physique: Lettres (Paris)*, *44*, L13–L22.

Rana, I. K. (1997). *An Introduction to Measure and Integration*. London: Narosa.

Rangarajan, G. & Ding, M. (2000). Integrated approach to the assessment of long range correlation in time series data. *Physical Review E*, *61*, 4991–5001.

Raymond, G. M. & Bassingthwaighte, J. B. (1999). Deriving dispersional and scaled windowed variance analyses using the correlation function of discrete fractional Gaussian noise. *Physica A*, *265*, 85–96.

Reid, C. (1982). *Neyman — From Life*. New York: Springer.

Reiss, R.-D. (1993). *A Course on Point Processes*. Berlin: Springer.

Rényi, A. (1955). On a new axiomatic theory of probability. *Acta Mathematica Hungarica*, *6*, 285–335.

Rényi, A. (1956). A characterization of Poisson processes. *A Magyar Tudományos Akadémia Matematikai Kutatóintézetének Közleményei*, *1*, 519–527. Translation: (1976). In P. Turán (Ed.), *Selected Papers of Alfréd Rényi*, Vol. 1 (pp. 622–628). Budapest: Akadémiai Kiadó.

Rényi, A. (1970). *Probability Theory*, volume 10 of *Applied Mathematics and Mechanics*. Amsterdam: North-Holland.

Ricciardi, L. M. & Esposito, F. (1966). On some distribution functions for non-linear switching elements with finite dead time. *Kybernetik (Biological Cybernetics)*, *3*, 148–152.

Rice, S. O. (1944). Mathematical analysis of random noise. *Bell System Technical Journal*, *23*, 282–332. Reprinted: (1954). In N. Wax (Ed.), *Selected Papers on Noise and Stochastic Processes*. (pp. 133–294). New York: Dover.

Rice, S. O. (1945). Mathematical analysis of random noise. *Bell System Technical Journal, 24*, 4–156. Reprinted: (1954). In N. Wax (Ed.), *Selected Papers on Noise and Stochastic Processes.* (pp. 133–294). New York: Dover.

Rice, S. O. (1983). Private communication: Letter to M. C. Teich.

Richardson, L. F. (1960). *Statistics of Deadly Quarrels.* Pittsburgh, PA: Boxwood. (Q. Wright and C. C. Lienau, Eds.). Although Richardson completed this work in 1950 it was not published until 1960, seven years after his death.

Richardson, L. F. (1961). The problem of contiguity: An appendix to *Statistics of Deadly Quarrels.* In A. Rapoport, L. von Bertalanfly, & R. L. Meier (Eds.), *General Systems: Yearbook of the Society for General Systems Research*, volume VI (pp. 139–187). New York: Society for General Systems Research. This work was published eight years after Richardson's death in 1953.

Riedi, R. H. (2003). Multifractal processes. In P. Doukhan, G. Oppenheim, & M. S. Taqqu (Eds.), *Theory and Applications of Long-Range Dependence* (pp. 625–716). Boston: Birkhäuser.

Riedi, R. H. & Lévy Véhel, J. (1997). Multifractal properties of TCP traffic: A numerical study. Rapport de recherche 3129, Institut National de Recherche en Informatique et en Automatique (INRIA), Le Chesnay, France. Thème 4 — Simulation et optimisation des systèmes complexes: Project Fractales.

Riedi, R. H. & Willinger, W. (2000). Toward an improved understanding of network traffic dynamics. In K. Park & W. Willinger (Eds.), *Self-Similar Network Traffic and Performance Evaluation* chapter 20, (pp. 507–530). New York: Wiley–Interscience.

Rieke, F., Warland, D., de Ruyter van Steveninck, R., & Bialek, W. (1997). *Spikes: Exploring the Neural Code.* Cambridge, MA: MIT Press.

Roberts, A. J. & Cronin, A. (1996). Unbiased estimation of multi-fractal dimensions of finite data sets. *Physica A, 233*, 867–878.

Roughan, M., Veitch, D., & Rumsewicz, M. (1998). Computing queue-length distributions for power-law queues. In *Proceedings of the Seventeenth Annual Joint Conference of the IEEE Computer and Communications Societies, Gateway to the 21st Century (INFOCOM'98)*, (pp. 356–363)., San Francisco. Institute of Electrical and Electronics Engineers, New York.

Rudin, W. (1959). Some theorems on Fourier coefficients. *Proceedings of the American Mathematical Society, 10*, 855–859.

Rudin, W. (1976). *Principles of Mathematical Analysis* (Third ed.). New York: McGraw–Hill.

Ruszczynski, P. S., Kish, L. B., & Bezrukov, S. M. (2001). Noise-assisted traffic of spikes through neuronal junctions. *Chaos, 11*, 581–586.

Rutherford, E. & Geiger, H. (1910). The probability variations in the distribution of alpha particles. *Philosophical Magazine, 20 (Series 6)*, 698–707.

Ryu, B. K. & Elwalid, A. (1996). The importance of long-range dependence of VBR video traffic in ATM traffic engineering: Myths and realities. *ACM SIGCOMM Computer Communication Review, 26*, 3–14.

Ryu, B. K. & Lowen, S. B. (1995). Modeling, analysis, and simulation of self-similar traffic using the fractal-shot-noise-driven Poisson process. In Hamza, M. H. (Ed.), *Proceedings of the IASTED International Conference on Modeling and Simulation,* (pp. 45–48). IASTED-ACTA Press.

Ryu, B. K. & Lowen, S. B. (1996). Point process approaches to the modeling and analysis of self-similar traffic — Part I: Model construction. In *Proceedings of the Fifteenth Annual Joint Conference of the IEEE Computer and Communications Societies, Networking the Next Generation (INFOCOM'96),* (pp. 1468–1475)., San Francisco. Institute of Electrical and Electronics Engineers, New York.

Ryu, B. K. & Lowen, S. B. (1997). Point process approaches to the modeling and analysis of self-similar traffic — Part II: Queueing applications. In *Proceedings of the Fifth International Conference on Telecommunication Systems — Modeling and Analysis,* (pp. 62–70). American Telecommunications Systems Management Association, Nashville, TN.

Ryu, B. K. & Lowen, S. B. (1998). Point process models for self-similar network traffic, with applications. *Stochastic Models, 14*, 735–761.

Ryu, B. K. & Lowen, S. B. (2000). Fractal traffic models for internet simulation. In *Proceedings of the Fifth IEEE Symposium on Computers and Communications (ISCC'2000)*, Antibes–Juan-les-Pins, France. Institute of Electrical and Electronics Engineers, New York.

Ryu, B. K. & Lowen, S. B. (2002). Fractal traffic models for Internet traffic engineering. In A. N. Ince (Ed.), *Modeling and Simulation Environment for Satellite and Terrestrial Communication Networks* (pp. 65–103). Norwell, MA: Kluwer. Proceedings of the European COST Telecommunications Symposium.

Sakmann, B. & Neher, E. (Eds.). (1995). *Single-Channel Recording* (Second ed.). New York: Plenum.

Saleh, B. E. A. (1978). *Photoelectron Statistics*. Berlin: Springer.

Saleh, B. E. A., Stoler, D., & Teich, M. C. (1983). Coherence and photon statistics for optical fields generated by Poisson random emissions. *Physical Review A, 27*, 360–374.

Saleh, B. E. A., Tavolacci, J. T., & Teich, M. C. (1981). Discrimination of shot-noise-driven Poisson processes by external dead time: Application to radioluminescence from glass. *IEEE Journal of Quantum Electronics, QE-17*, 2341–2350.

Saleh, B. E. A. & Teich, M. C. (1982). Multiplied-Poisson noise in pulse, particle, and photon detection. *Proceedings of the IEEE*, *70*, 229–245.

Saleh, B. E. A. & Teich, M. C. (1983). Statistical properties of a nonstationary Neyman–Scott cluster process. *IEEE Transactions on Information Theory*, *IT-29*, 939–941.

Saleh, B. E. A. & Teich, M. C. (1985a). Multiplication and refractoriness in the cat's retinal-ganglion-cell discharge at low light levels. *Kybernetik (Biological Cybernetics)*, *52*, 101–107.

Saleh, B. E. A. & Teich, M. C. (1985b). Sub-Poisson light generation by selective deletion from cascaded atomic emissions. *Optics Communications*, *52*, 429–432.

Saleh, B. E. A. & Teich, M. C. (1991). *Fundamentals of Photonics*. New York: Wiley–Interscience.

Samorodnitsky, G. & Taqqu, M. S. (1994). *Stable Non-Gaussian Random Processes: Stochastic Models with Infinite Variance*. London: Chapman and Hall.

Sapoval, B., Baldassarri, A., & Gabrielli, A. (2004). Self-stabilized fractality of seacoasts through damped erosion. *Physical Review Letters*, *93*, 098501.

Sato, K.-i. (Ed.). (1999). *Lévy Processes and Infinitely Divisible Distributions*. Cambridge, UK: Cambridge.

Scharf, R., Meesmann, M., Boese, J., Chialvo, D. R., & Kniffki, K. (1995). General relation between variance–time curve and power spectral density for point processes exhibiting $1/f^\beta$-fluctuations, with special reference to heart rate variability. *Biological Cybernetics*, *73*, 255–263.

Schepers, H. E., van Beek, J. H. G. M., & Bassingthwaighte, J. B. (1992). Four methods to estimate the fractal dimension from self-affine signals. *IEEE Engineering in Medicine and Biology Magazine*, *11*(2), 57–64 & 71. Erratum: (1992). *11*(3), 79.

Scher, H. & Montroll, E. W. (1975). Anomalous transit-time dispersion in amorphous solids. *Physical Review B*, *12*, 2455–2477.

Schick, K. L. (1974). Power spectra of pulse sequences and implications for membrane fluctuations. *Acta Biotheoretica*, *23*, 1–17.

Schiff, S. J. & Chang, T. (1992). Differentiation of linearly correlated noise from chaos in a biologic system using surrogate data. *Biological Cybernetics*, *67*, 387–393.

Schmitt, F., Vannitsem, S., & Barbosa, A. (1998). Modeling of rainfall time series using two-state renewal processes and multifractals. *Journal of Geophysical Research*, *103*, 23181–23193.

Schönfeld, H. (1955). Beitrag zum $1/f$-Gesetz beim Rauschen von Halbleitern. *Zeitschrift für Naturforschung A, 10,* 291–300.

Schottky, W. (1918). Über spontane Stromschwankungen in verschiedenen Elektrizitätsleitern. *Annalen der Physik, 57,* 541–567.

Schreiber, T. & Schmitz, A. (1996). Improved surrogate data for nonlinearity tests. *Physical Review Letters, 77,* 635–638.

Schroeder, M. R. (1990). *Fractals, Chaos, Power Laws: Minutes from an Infinite Paradise.* New York: W. H. Freeman.

Schuster, H. G. (1995). *Deterministic Chaos: An Introduction* (Third ed.). Weinheim, Germany: VCH.

Seidel, H. (1876). Über die Probabilitäten solcher Ereignisse welche nur seiten vorkommen, obgleich sie unbeschränkt oft möglich sind. *Bayerische Akademie der Wissenschaften (München): Sitzungsberichte der Mathematisch-Physischen Klasse, 6,* 44–50.

Sellan, F. (1995). Synthèse de mouvements browniens fractionnaires à l'aide de la transformation par ondelettes. *Comptes Rendus de l'Académie des Sciences (Paris), 321 (Series 1),* 351–358.

Shapiro, H. S. (1951). Extremal problems for polynomials and power series. Master's thesis, Massachusetts Institute of Technology, Cambridge, MA.

Shimizu, Y., Thurner, S., & Ehrenberger, K. (2002). Multifractal spectra as a measure of complexity in human posture. *Fractals, 10,* 103–116.

Shlesinger, M. F. (1987). Fractal time and $1/f$ noise in complex systems. *Annals of the New York Academy of Sciences, 504,* 214–228. Discussion of paper on pp. 226–228.

Shlesinger, M. F. & West, B. J. (1991). Complex fractal dimension of the bronchial tree. *Physical Review Letters, 67,* 2106–2108.

Sigman, K. (1995). *Stationary Marked Poisson Processes: An Intuitive Approach.* New York: Chapman and Hall.

Sigman, K. (1999). Appendix: A primer on heavy-tailed distributions. *Queueing Systems, 33,* 261–275.

Sikula, J. (1995). Models for burst and RTS noise. In Bareikis, V. & Katilius, R. (Eds.), *Proceedings of the Thirteenth International Conference on Noise in Physical Systems and $1/f$ Fluctuations,* (pp. 343–348). World Scientific.

Simoncelli, E. P. & Olshausen, B. A. (2001). Natural image statistics and neural representation. *Annual Review of Neuroscience, 24,* 1193–1216.

Smith, W. L. (1958). Renewal theory and its ramifications. *Journal of the Royal Statistical Society B*, *20*, 243–302. Discussion of paper on pp. 284–302.

Snyder, D. L. & Miller, M. I. (1991). *Random Point Processes in Time and Space* (Second ed.). Berlin: Springer.

Solomon, S. & Richmond, P. (2002). Stable power laws in variable economies; Lotka–Volterra implies Pareto–Zipf. *European Physical Journal B*, *27*, 257–261.

Soma, R., Nozaki, D., Kwak, S., & Yamamoto, Y. (2003). $1/f$ noise outperforms white noise in sensitizing baroreflex function in the human brain. *Physical Review Letters*, *91*, 078101.

Song, C., Havlin, S., & Makse, H. A. (2005). Self-similarity of complex networks. *Nature (London)*, *433*, 392–395.

Sornette, D. (2004). *Critical Phenomena in Natural Sciences* (Second ed.). Berlin: Springer.

Srinivasan, S. K. (1974). *Stochastic Point Processes*. London: Griffin.

Steinhaus, H. (1954). Length, shape and area. *Colloquium Mathematicum (Wrocław)*, *3*, 1–13.

Stepanescu, A. (1974). $1/f$ noise as a two-parameter stochastic process. *Il Nuovo Cimento*, *23B*, 356–364.

Stern, E. A., Kincaid, A. E., & Wilson, C. J. (1997). Spontaneous subthreshold membrane potential fluctuations and action potential variability of rat corticostriatal and striatal neurons in vivo. *Journal of Neurophysiology*, *77*, 1697–1715.

Stevens, S. S. (1957). On the psychophysical law. *Psychological Review*, *64*, 153–181.

Stevens, S. S. (1971). Sensory power functions and neural events. In W. R. Loewenstein (Ed.), *Principles of Receptor Physiology*, volume 1 of *Handbook of Sensory Physiology* (pp. 226–242). Berlin: Springer.

Stoksik, M. A., Lane, R. G., & Nguyen, D. T. (1994). Accurate synthesis of fractional Brownian motion using wavelets. *Electronics Letters*, *30*, 383–384.

Stoyan, D. & Stoyan, H. (1994). *Fractals, Random Shapes and Point Fields: Methods of Geometrical Statistics*. Chichester: Wiley.

Strogatz, S. H. (1994). *Nonlinear Dynamics and Chaos: With Applications to Physics, Biology, Chemistry, and Engineering*. Reading, MA: Perseus.

Szökefalvi-nagy, B. (Ed.). (1959). *Haar Alfréd: Összegyűjtött munkái*. Budapest: Akadémiai Kiadó. Collected papers of Alfréd Haar in their original languages: German, Hungarian, and French.

Takács, L. (1960). *Stochastic Processes.* London: Methuen.

Takayasu, H., Nishikawa, I., & Tasaki, H. (1988). Power-law mass distribution for aggregation systems with injection. *Physical Review A, 37,* 3110–3117.

Taqqu, M. S. (2003). Fractional Brownian motion and long-range dependence. In P. Doukhan, G. Oppenheim, & M. S. Taqqu (Eds.), *Theory and Applications of Long-Range Dependence* (pp. 5–38). Boston: Birkhäuser.

Taqqu, M. S. & Levy, J. B. (1986). Using renewal processes to generate long-range dependence and high variability. In E. Eberlein & M. S. Taqqu (Eds.), *Dependence in Probability and Statistics: A Survey of Recent Results,* volume 11 of *Progress in Probability and Statistics* (pp. 73–89). Boston: Birkhäuser.

Taqqu, M. S. & Teverovsky, V. (1998). On estimating the intensity of long-range dependence in finite and infinite variance series. In R. J. Adler, R. E. Feldman, & M. S. Taqqu (Eds.), *A Practical Guide to Heavy Tails: Statistical Techniques and Applications* (pp. 177–217). Boston: Birkhäuser.

Taqqu, M. S., Teverovsky, V., & Willinger, W. (1995). Estimators for long-range dependence: An empirical study. *Fractals, 3,* 785–798.

Taqqu, M. S., Teverovsky, V., & Willinger, W. (1997). Is network traffic self-similar or multifractal? *Fractals, 5,* 63–73.

Taubes, G. (1998). Fractals reemerge in the new math of the Internet. *Science, 281,* 1947–1948.

Taylor, R. P. (2002). Order in Pollock's chaos. *Scientific American, 287*(6), 116–121.

Teich, M. C. (1981). Role of the doubly stochastic Neyman Type-A and Thomas counting distributions in photon detection. *Applied Optics, 20,* 2457–2467.

Teich, M. C. (1985). Normalizing transformations for dead-time-modified Poisson counting distributions. *Kybernetik (Biological Cybernetics), 53,* 121–124.

Teich, M. C. (1989). Fractal character of the auditory neural spike train. *IEEE Transactions on Biomedical Engineering, 36,* 150–160. Special issue on neurosystems and neuroengineering.

Teich, M. C. (1992). Fractal neuronal firing patterns. In T. McKenna, J. Davis, & S. F. Zornetzer (Eds.), *Single Neuron Computation* chapter 22, (pp. 589–625). Boston: Academic.

Teich, M. C. & Cantor, B. I. (1978). Information, error, and imaging in deadtime-perturbed doubly stochastic Poisson counting systems. *IEEE Journal of Quantum Electronics, QE-14,* 993–1003.

Teich, M. C. & Diament, P. (1980). Relative refractoriness in visual information processing. *Kybernetik (Biological Cybernetics), 38,* 187–191.

Teich, M. C. & Diament, P. (1989). Multiply stochastic representations for K distributions and their Poisson transforms. *Journal of the Optical Society of America A, 6,* 80–91.

Teich, M. C., Heneghan, C., Lowen, S. B., Ozaki, T., & Kaplan, E. (1997). Fractal character of the neural spike train in the visual system of the cat. *Journal of the Optical Society of America A, 14,* 529–546.

Teich, M. C., Heneghan, C., Lowen, S. B., & Turcott, R. G. (1996). Estimating the fractal exponent of point processes in biological systems using wavelet- and Fourier-transform methods. In A. Aldroubi & M. Unser (Eds.), *Wavelets in Medicine and Biology* chapter 14, (pp. 383–412). Boca Raton, FL: CRC.

Teich, M. C., Johnson, D. H., Kumar, A. R., & Turcott, R. G. (1990). Rate fluctuations and fractional power-law noise recorded from cells in the lower auditory pathway of the cat. *Hearing Research, 46,* 41–52.

Teich, M. C. & Khanna, S. M. (1985). Pulse-number distribution for the neural spike train in the cat's auditory nerve. *Journal of the Acoustical Society of America, 77,* 1110–1128.

Teich, M. C., Khanna, S. M., & Guiney, P. C. (1993). Spectral characteristics and synchrony in primary auditory-nerve fibers in response to pure-tone acoustic stimuli. *Journal of Statistical Physics, 70,* 257–279.

Teich, M. C. & Lowen, S. B. (1994). Fractal patterns in auditory nerve-spike trains. *IEEE Engineering in Medicine and Biology Magazine, 13*(2), 197–202.

Teich, M. C. & Lowen, S. B. (2003). Fractal integrate-and-reset models characterize retinal-ganglion-cell and lateral-geniculate-nucleus action-potential sequences in the cat visual system. Unpublished manuscript.

Teich, M. C., Lowen, S. B., Jost, B. M., Vibe-Rheymer, K., & Heneghan, C. (2001). Heart rate variability: Measures and models. In M. Akay (Ed.), *Dynamic Analysis and Modeling,* volume II of *Nonlinear Biomedical Signal Processing* chapter 6, (pp. 159–213). New York: IEEE.

Teich, M. C., Lowen, S. B., & Turcott, R. G. (1991). On possible peripheral origins of the fractal auditory neural spike train. In Lim, D. J. (Ed.), *Abstracts of the XIV Midwinter Research Meeting, Association for Research in Otolaryngology,* (pp.50)., St. Petersburg Beach, FL. Association for Research in Otolaryngology, Des Moines, IA. Abstract 154.

Teich, M. C., Matin, L., & Cantor, B. I. (1978). Refractoriness in the maintained discharge of the cat's retinal ganglion cell. *Journal of the Optical Society of America, 68,* 386–402.

Teich, M. C. & McGill, W. J. (1976). Neural counting and photon counting in the presence of dead time. *Physical Review Letters, 36,* 754–758. Erratum: (1976). *36,* 1473.

Teich, M. C., Prucnal, P. R., Vannucci, G., Breton, M. E., & McGill, W. J. (1982a). Multiplication noise in the human visual system at threshold: 1. Quantum fluctuations and minimum detectable energy. *Journal of the Optical Society of America*, *72*, 419–431.

Teich, M. C., Prucnal, P. R., Vannucci, G., Breton, M. E., & McGill, W. J. (1982b). Multiplication noise in the human visual system at threshold: 3. The role of non-Poisson quantum fluctuations. *Kybernetik (Biological Cybernetics)*, *44*, 157–165.

Teich, M. C. & Rosenberg, S. (1971). N-fold joint photocounting distribution for modulated laser radiation: Transmission through the turbulent atmosphere. *International Journal of Opto-Electronics*, *3*, 63–76. Errata: The correct expression for Eq. (28) on p. 69 is $\mathbf{B} = \mathbf{Q} - \mathbf{\Lambda}^{-1}$ where the elements of $\mathbf{Q}$ are simply $Q_{ij}^{(2)}$, with $i, j = 1, 2, \ldots, N$. The lettering on Figure 1, p. 71, should read $R = 0, 0$, 0.499, 0.998. Reference 9 should read *Applied Optics* **10** (1971) 1664. Reference 28 should read *J. Appl. Phys.* **43** (1972) 1256.

Teich, M. C. & Saleh, B. E. A. (1981a). Fluctuation properties of multiplied-Poisson light: Measurement of the photon-counting distribution for radioluminescence radiation from glass. *Physical Review A*, *24*, 1651–1654.

Teich, M. C. & Saleh, B. E. A. (1981b). Interevent-time statistics for shot-noise-driven self-exciting point processes in photon detection. *Journal of the Optical Society of America*, *71*, 771–776.

Teich, M. C. & Saleh, B. E. A. (1982). Effects of random deletion and additive noise on bunched and antibunched photon-counting statistics. *Optics Letters*, *7*, 365–367.

Teich, M. C. & Saleh, B. E. A. (1987). Approximate photocounting statistics of shot-noise light with arbitrary spectrum. *Journal of Modern Optics*, *34*, 1169–1178.

Teich, M. C. & Saleh, B. E. A. (1988). Photon bunching and antibunching. In E. Wolf (Ed.), *Progress in Optics*, volume 26 chapter 1, (pp. 1–104). Amsterdam: North-Holland/Elsevier.

Teich, M. C. & Saleh, B. E. A. (1998). Cascaded stochastic processes in optics. *Traitement du Signal (Grenoble)*, *15*, 457–465.

Teich, M. C. & Saleh, B. E. A. (2000). Branching processes in quantum electronics. *IEEE Journal of Selected Topics in Quantum Electronics*, *6*, 1450–1457.

Teich, M. C., Saleh, B. E. A., & Peřina, J. (1984). Role of primary excitation statistics in the generation of antibunched and sub-Poisson light. *Journal of the Optical Society of America B*, *1*, 366–389.

Teich, M. C., Tanabe, T., Marshall, T. C., & Galayda, J. (1990). Statistical properties of wiggler and bending-magnet radiation from the Brookhaven vacuum-ultraviolet electron storage ring. *Physical Review Letters*, *65*, 3393–3396.

Teich, M. C. & Turcott, R. G. (1988). Multinomial pulse-number distributions for neural spikes in primary auditory fibers: Theory. *Kybernetik (Biological Cybernetics)*, *59*, 91–102.

Teich, M. C., Turcott, R. G., & Lowen, S. B. (1990). The fractal doubly stochastic Poisson point process as a model for the cochlear neural spike train. In P. Dallos, C. D. Geisler, J. W. Matthews, M. A. Ruggero, & C. R. Steele (Eds.), *The Mechanics and Biophysics of Hearing*, number 87 in Lecture Notes in Biomathematics (pp. 354–361). Berlin: Springer.

Teich, M. C., Turcott, R. G., & Siegel, R. M. (1996). Temporal correlation in cat striate-cortex neural spike trains. *IEEE Engineering in Medicine and Biology Magazine*, *15*(5), 79–87.

Teich, M. C. & Vannucci, G. (1978). Observation of dead-time-modified photocounting distributions for modulated laser radiation. *Journal of the Optical Society of America*, *68*, 1338–1342.

Teicher, M. H., Ito, Y., Glod, C. A., & Barber, N. I. (1996). Objective measurement of hyperactivity and attentional problems in ADHD. *Journal of the American Academy of Child and Adolescent Psychiatry*, *35*, 334–342.

Telesca, L., Balasco, M., Colangelo, G., Lapenna, V., & Macchiato, M. (2004). Analyzing cross-correlations between earthquakes and geoelectrical extreme events, measured in a seismic area of Southern Italy. *Physics and Chemistry of the Earth*, *29*, 289–293.

Telesca, L., Cuomo, V., Lanfredi, M., Lapenna, V., & Macchiato, M. (1999). Investigating clustering structures in time-occurrence sequences of seismic events observed in the Irpinia-Basilicata region (southern Italy). *Fractals*, *7*, 221–234.

Telesca, L., Cuomo, V., Lapenna, V., & Macchiato, M. (2002a). Fractal characterization of the temporal distribution of aftershocks associated with the 1994 M_W 6.7 Northridge earthquake. *Fractals*, *10*, 67–76.

Telesca, L., Cuomo, V., Lapenna, V., & Macchiato, M. (2002b). On the methods to identify clustering properties in sequences of seismic time-occurrences. *Journal of Seismology*, *6*, 125–134.

Terman, F. E. (1947). *Radio Engineering* (Third ed.). New York: McGraw–Hill.

Tewfik, A. H. & Kim, M. (1992). Correlation structure of the discrete wavelet coefficients of fractional Brownian motion. *IEEE Transactions on Information Theory*, *38*, 904–909.

Theiler, J. (1990). Estimating fractal dimension. *Journal of the Optical Society of America A*, *7*, 1055–1073.

Theiler, J., Eubank, S., Longtin, A., Galdrikian, B., & Farmer, J. D. (1992). Testing for nonlinearity in time series: The method of surrogate data. *Physica D, 58,* 77–94.

Theiler, J. & Prichard, D. (1996). Constrained-realization Monte-Carlo method for hypothesis testing. *Physica D, 94,* 221–235.

Thiébaut, D. (1988). From the fractal dimension of the intermiss gaps to the cache-miss ratio. *IBM Journal of Research and Development, 32,* 796–803.

Thomas, M. (1949). A generalization of Poisson's binomial limit for use in ecology. *Biometrika, 36,* 18–25.

Thompson, J. M. T. & Stewart, H. B. (2002). *Nonlinear Dynamics and Chaos* (Second ed.). New York: Wiley.

Thorson, J. & Biederman-Thorson, M. (1974). Distributed relaxation processes in sensory adaptation. *Science, 183,* 161– 172.

Thue, A. (1906). Über unendliche Zeichenreihen. *Kongelige Norske Videnskabers Selskab Skrifter. I. Matematisk-Naturvitenskapelig Klasse 1906, 7,* 1–22. Reprinted: (1977). In T. Nagell, A. Selberg, S. Selberg, & K. Thalberg (Eds.), *Selected Mathematical Papers of Axel Thue.* (pp. 139–158). Oslo: Universitetsforlaget.

Thue, A. (1912). Über die gegenseitige Lage gleicher Teile gewisser Zeichenreihen. *Kongelige Norske Videnskabers Selskab Skrifter. I. Matematisk-Naturvitenskapelig Klasse 1912, 1,* 1–67. Reprinted: (1977). In T. Nagell, A. Selberg, S. Selberg, & K. Thalberg (Eds.), *Selected Mathematical Papers of Axel Thue.* (pp. 413–477). Oslo: Universitetsforlaget.

Thurner, S., Feurstein, M. C., Lowen, S. B., & Teich, M. C. (1998). Receiver-operating-characteristic analysis reveals superiority of scale-dependent wavelet and spectral measures for assessing cardiac dysfunction. *Physical Review Letters, 81,* 5688–5691.

Thurner, S., Feurstein, M. C., & Teich, M. C. (1998). Multiresolution wavelet analysis of heartbeat intervals discriminates healthy patients from those with cardiac pathology. *Physical Review Letters, 80,* 1544–1547.

Thurner, S., Lowen, S. B., Feurstein, M., Heneghan, C., Feichtinger, H. G., & Teich, M. C. (1997). Analysis, synthesis, and estimation of fractal-rate stochastic point processes. *Fractals, 5,* 565–595.

Tiedje, T. & Rose, A. (1980). A physical interpretation of dispersive transport in disordered semiconductors. *Solid State Communications, 37,* 49–52.

Timmer, J. & König, M. (1995). On generating power law noise. *Astronomy and Astrophysics, 300,* 707–710.

Toib, A., Lyakhov, V., & Marom, S. (1998). Interaction between duration of activity and time course of recovery from slow inactivation in mammalian brain Na$^+$ channels. *Journal of Neuroscience, 18*, 1893–1903.

Toussaint, D. & Wilczek, F. (1983). Particle–antiparticle annihilation in diffusive motion. *Journal of Chemical Physics, 78*, 2642–2647.

Tuan, T. & Park, K. (2000). Congestion control for self-similar network traffic. In K. Park & W. Willinger (Eds.), *Self-Similar Network Traffic and Performance Evaluation* chapter 18, (pp. 447–480). New York: Wiley–Interscience.

Tuckwell, H. C. (1988). *Introduction to Theoretical Neurobiology. Volume 1: Linear Cable Theory and Dendritic Structure.* Cambridge, UK: Cambridge.

Tukey, J. W. (1957). On the comparative anatomy of transformations. *Annals of Mathematical Statistics, 28*, 602–632.

Turcott, R. G., Barker, P. D. R., & Teich, M. C. (1995). Long-duration correlation in the sequence of action potentials in an insect visual interneuron. *Journal of Statistical Computation and Simulation, 52*, 253–271.

Turcott, R. G., Lowen, S. B., Li, E., Johnson, D. H., Tsuchitani, C., & Teich, M. C. (1994). A nonstationary Poisson point process describes the sequence of action potentials over long time scales in lateral-superior-olive auditory neurons. *Kybernetik (Biological Cybernetics), 70*, 209–217.

Turcott, R. G. & Teich, M. C. (1993). Long-duration correlation and attractor topology of the heartbeat rate differ for healthy patients and those with heart failure. In L. D. William (Ed.), *SPIE Proceedings: Chaos in Biology and Medicine*, volume 2036 (pp. 22–39). Bellingham, WA: SPIE.

Turcott, R. G. & Teich, M. C. (1996). Fractal character of the electrocardiogram: Distinguishing heart-failure and normal patients. *Annals of Biomedical Engineering, 24*, 269–293.

Turcotte, D. L. (1997). *Fractals and Chaos in Geology and Geophysics* (Second ed.). Cambridge, UK: Cambridge.

Turner, M. J., Blackledge, J. M., & Andrews, P. R. (1998). *Fractal Geometry in Digital Imaging*. San Diego: Academic.

Usher, M., Stemmler, M., & Olami, Z. (1995). Dynamic pattern formation leads to $1/f$ noise in neural populations. *Physical Review Letters, 74*, 326–329.

van der Waerden, B. L. (Ed.). (1975). *Die Werke von Jakob Bernoulli*, volume 3: *Wahrscheinlichkeitsrechnung*. Basel: Birkhäuser.

van der Ziel, A. (1950). On the noise spectra of semi-conductor noise and of flicker effect. *Physica, 16*, 359–372.

van der Ziel, A. (1979). Flicker noise in electronic devices. In L. Marton (Ed.), *Advances in Electronics and Electron Physics*, volume 49 (pp. 225–297). New York: Academic.

van der Ziel, A. (1986). *Noise in Solid State Devices and Circuits*. New York: Wiley–Interscience.

van der Ziel, A. (1988). Unified presentation of $1/f$ noise in electronic devices: Fundamental $1/f$ noise sources. *Proceedings of the IEEE, 76*, 233–258.

Vannucci, G. & Teich, M. C. (1978). Effects of rate variation on the counting statistics of dead-time-modified Poisson processes. *Optics Communications, 25*, 267–272.

Vannucci, G. & Teich, M. C. (1981). Dead-time-modified photocount mean and variance for chaotic radiation. *Journal of the Optical Society of America, 71*, 164–170.

Veitch, D. & Abry, P. (1999). A wavelet-based joint estimator of the parameters of long-range dependence. *IEEE Transactions on Information Theory, 45*, 878–897. Special issue on multiscale statistical signal analysis and its applications.

Vere-Jones, D. (1970). Stochastic models for earthquake occurrence. *Journal of the Royal Statistical Society B, 32*, 1–62.

Verhulst, P. (1845). Recherches mathématiques sur la loi d'accroissement de la population. *Nouveaux Mémoires de l'Académie Royale des Sciences et Belles-Lettres de Bruxelles, 18*, 1–41.

Verhulst, P. (1847). Deuxième mémoire sur la loi d'accroissement de la population. *Mémoires de l'Académie Royale des Sciences, des Lettres et des Beaux-Arts de Belgique, 20*, 1–32.

Verveen, A. A. (1960). On the fluctuation of threshold of the nerve fibre. In Tower, D. B. & Schadé, J. P. (Eds.), *Structure and Function of the Cerebral Cortex*, (pp. 282–288). Elsevier.

Verveen, A. A. & Derksen, H. E. (1968). Fluctuation phenomena in nerve membrane. *Proceedings of the IEEE, 56*, 906–916.

Vicsek, T. (1992). *Fractal Growth Phenomena* (Second ed.). Singapore: World Scientific.

Vicsek, T. (Ed.). (2001). *Fluctuations and Scaling in Biology*. Oxford, UK: Oxford.

Ville, J. (1948). Théorie et applications de la notion de signal analytique. *Cables et Transmission, 2A*, 61–74.

Viswanathan, G. M., Afanasyev, V., Buldyrev, S. V., Murphy, E. J., Prince, P. A., & Stanley, H. E. (1996). Lévy flight search patterns of wandering albatrosses. *Nature (London), 381*, 413–415.

Voldman, J., Mandelbrot, B. B., Hoevel, L. W., Knight, J., & Rosenfeld, P. (1983). Fractal nature of software-cache interaction. *IBM Journal of Research and Development*, 27, 164–170.

von Bortkiewicz, L. (1898). *Das Gesetz der kleinen Zahlen*. Leipzig: Teubner.

von Schweidler, E. R. (1907). Studien über die Anomalien im Verhalten der Dielektrika. *Annalen der Physik*, 24, 711–770.

Voss, R. F. (1989). Random fractals: Self-affinity in noise, music, mountains and clouds. *Physica D*, 38, 362–371.

Voss, R. F. & Clarke, J. (1978). '$1/f$ noise' in music: Music from $1/f$ noise. *Journal of the Acoustical Society of America*, 63, 258–263.

Watson, H. W. & Galton, F. (1875). On the probability of the extinction of families. *The Journal of the Anthropological Institute of Great Britain and Ireland*, 4, 138–144.

Weber, W. (1835). Über die Elastizität der Seidenfaden. *Annalen der Physik und Chemie (Poggendorf)*, 34, 247–257.

Weiss, G. (1973). *Filtered Poisson Processes as Models for Daily Streamflow Data*. PhD thesis, University of London.

Weiss, G. H., Dishon, M., Long, A. M., Bendler, J. T., Jones, A. A., Inglefield, P. T., & Bandis, A. (1994). Improved computational methods for the calculation of Kohlrausch–Williams–Watts (KWW) decay functions. *Polymer*, 35, 1880–1883.

Weissman, M. B. (1988). $1/f$ noise and other slow, nonexponential kinetics in condensed matter. *Reviews of Modern Physics*, 60, 537–571.

Wescott, M. (1976). Simple proof of a result on thinned point processes. *The Annals of Probability*, 4, 89–90.

West, B. J. (1990). Physiology in fractal dimension: Error tolerance. *Annals of Biomedical Engineering*, 18, 135–149.

West, B. J. & Bickel, D. R. (1998). Molecular evolution modeled as a fractal stochastic process. *Physica A*, 249, 544–552.

West, B. J., Bologna, M., Grigolini, P., & MacLachlan, C. C. (2003). *Physics of Fractal Operators*. Berlin: Springer.

West, B. J. & Deering, W. (1994). Fractal physiology for physicists: Lévy statistics. *Physics Reports*, 246, 1–100.

West, B. J. & Deering, W. (1995). *The Lure of Modern Science: Fractal Thinking*, volume 3 of *Studies of Nonlinear Phenomena in Life Sciences*. Singapore: World Scientific.

West, B. J. & Shlesinger, M. F. (1989). On the ubiquity of $1/f$ noise. *International Journal of Modern Physics B*, *3*, 795–819.

West, B. J. & Shlesinger, M. F. (1990). The noise in natural phenomena. *American Scientist*, *78*(1), 40–45.

West, B. J., Zhang, R., Sanders, A. W., Miniyar, S., Zuckerman, J. H., & Levine, B. D. (1999). Fractal fluctuations in transcranial Doppler signals. *Physical Review E*, *59*, 3492–3498.

Whittle, P. (1962). Topographic correlation, power-law covariance functions, and diffusion. *Biometrika*, *49*, 305–314.

Wickelgren, W. A. (Ed.). (1977). *Learning and Memory*. Englewood Cliffs, NJ: Prentice–Hall.

Wiener, N. (1923). Differential space. *Journal of Mathematics and Physics (Studies in Applied Mathematics)*, *2*, 131–174.

Wiener, N. (1930). Generalized harmonic analysis. *Acta Mathematica*, *55*, 117–258.

Willinger, W., Alderson, D., & Li, L. (2004). A pragmatic approach to dealing with high-variability in network measurements. In *Proceedings of the Fourth ACM SIGCOMM Conference on Internet Measurement*, (pp. 88–100)., Taormina, Italy. New York: ACM Press.

Willinger, W., Paxton, V., Riedi, R. H., & Taqqu, M. S. (2003). Long-range dependence and data network traffic. In P. Doukhan, G. Oppenheim, & M. S. Taqqu (Eds.), *Theory and Applications of Long-Range Dependence* (pp. 373–407). Boston: Birkhäuser.

Willis, J. C. (1922). *Age and Area: A Study in Geographical Distribution and Origin of Species*. Cambridge, UK: Cambridge.

Wise, M. E. (1981). Spike interval distributions for neurons and random walks with drift to a fluctuating threshold. In C. E. A. Taillie, G. P. Patil, & B. A. Baldessari (Eds.), *Statistical Distributions in Scientific Work*, volume 6 (pp. 211–231). Boston: Reidel.

Witten, Jr, T. A. & Sander, L. M. (1981). Diffusion-limited aggregation, a kinetic phenomenon. *Physical Review Letters*, *47*, 1400–1403.

Wixted, J. T. (2004). On common ground: Jost's (1897) law of forgetting and Ribot's (1881) law of retrograde amnesia. *Psychological Review*, *111*, 864–879.

Wixted, J. T. & Ebbesen, E. B. (1991). On the form of forgetting. *Psychological Science*, *2*, 409–415.

Wixted, J. T. & Ebbesen, E. B. (1997). Genuine power curves in forgetting: A quantitative analysis of individual subject forgetting functions. *Memory and Cognition*, *25*, 731–739.

Wold, H. (1948). On stationary point processes and Markov chains. *Skandinavisk Aktuarietidskrift, 31*, 229–240.

Wold, H. (1949). Sur les processus stationnaires ponctuels. In *Le calcul des probabilités et ses applications*, volume 13 of *Colloques Internationaux CNRS* (pp. 75–86). Paris: Centre National de la Recherche Scientifique.

Wold, H. (Ed.). (1965). *Bibliography on Time Series and Stochastic Processes*. Cambridge, MA: MIT Press.

Wolff, R. W. (1982). Poisson arrivals see time averages. *Operations Research, 30*, 223–231.

Yamamoto, M. & Nakahama, H. (1983). Stochastic properties of spontaneous unit discharges in somatosensory cortex and mesencephalic reticular formation during sleep–waking states. *Journal of Neurophysiology, 49*, 1182–1198.

Yamamoto, M., Nakahama, H., Shima, K., Kodama, T., & Mushiake, H. (1986). Markov-dependency and spectral analyses on spike-counts in mesencephalic reticular neurons during sleep and attentive states. *Brain Research, 366*, 279–289.

Yang, X., Du, S., & Ma, J. (2004). Do earthquakes exhibit self-organized criticality? *Physical Review Letters, 92*, 228501.

Yang, X. & Petropulu, A. P. (2001). The extended alternating fractal renewal process for modeling traffic in high-speed communication networks. *IEEE Transactions on Signal Processing, 49*, 1349–1363.

Yu, J., Petropulu, A. P., & Sethu, H. (2005). Rate-limited EAFRP — A new improved model for high-speed network traffic. *IEEE Transactions on Signal Processing, 53*, 505–522.

Yu, Y., Romero, R., & Lee, T. S. (2005). Preference of sensory neural coding for $1/f$ signals. *Physical Review Letters, 94*, 108103.

Yule, G. U. (1924). A mathematical theory of evolution, based on the conclusions of Dr J. C. Willis. *Philosophical Transactions of the Royal Society of London B, 213*, 21–87.

Zhukovsky, S. V., Gaponenko, S. V., & Lavrinenko, A. V. (2001). Spectral properties of fractal and quasi-periodic multilayered media. *Nonlinear Phenomena in Complex Systems, 4*, 383–389.

Zipf, G. K. (1949). *Human Behavior and the Principle of Least Effort: An Introduction to Human Ecology*. Cambridge, MA: Addison–Wesley. Second edition: (1965). New York: Hafner.

Zrelov, V. P. (1968). *Cherenkov Radiation in High-Energy Physics, Part I*. Moscow: Atomizdat. English translation: Israel Program for Scientific Translations, Jerusalem, 1970.

Zucker, R. S. (1993). Calcium and transmitter release. *Journal of Physiology (Paris)*, *87*, 25–36.

Zuckerkandl, E. & Pauling, L. (1962). Molecular disease, evolution, and genic heterogeneity. In M. Kasha & B. Pullman (Eds.), *Horizons in Biochemistry: Albert Szent-Györgyi Dedicatory Volume* (pp. 189–225). New York: Academic.

Zuckerkandl, E. & Pauling, L. (1965). Molecules as documents of evolutionary history. *Journal of Theoretical Biology*, *8*, 357–366.

Zumofen, G., Hohlbein, J., & Hübner, C. G. (2004). Recurrence and photon statistics in fluorescence fluctuation spectroscopy. *Physical Review Letters*, *93*, 260601.

Author Index

Abbe (1878), 64, 513
Abeles et al. (1983), 227, 513
Abry et al. (2002), 314, 513
Abry & Flandrin (1996), 58, 68, 513
Abry et al. (2000), 58, 274, 331, 332, 513
Abry et al. (2003), 58, 114, 274, 513
Abry & Sellan (1996), 139, 514
Aiello et al. (2001), 321, 514
Aitchison & Brown (1957), 36, 147, 514
Aizawa (1984), 173, 176, 514
Aizawa & Kohyama (1984), 173, 514
Akay (1997), 58, 514
Aks et al. (2002), 45, 514
Albert & Nelson (1953), 236, 514
Albert & Barabási (2002), 37, 38, 321, 514
Albert et al. (1999), 16, 321, 514
Aldroubi & Unser (1996), 58, 514
Alexander & Orbach (1982), 471, 514
Allan (1966), 68, 276, 514
Alligood et al. (1996), 25, 514
Anderson (2001), 16, 44, 45, 514
Anderson et al. (1999), 16, 44, 515
Arecchi & Califano (1987), 173, 515
Arecchi & Lisi (1982), 173, 515
Argoul et al. (1989), 75, 515

Arlitt & Jin (1998), 330, 515
Arneodo et al. (1988), 58, 515
Arrault & Arneodo (1997), 75, 515
Ashkenazy et al. (2001), 275, 515
Ashkenazy et al. (1998), 44, 275, 515
Asmussen (2003), 316, 328, 515
Ayache & Lévy Véhel (1999), 123, 515
Azhar & Gopala (1992), 224, 472, 515
Baccelli & Brémaud (2003), 51, 515
Bachelier (1900), 19, 515
Bachelier (1912), 19, 516
Bacry et al. (1993), 75, 516
Bak (1996), 37, 516
Bak et al. (1987), 37, 516
Barakat (1976), 36, 516
Bardet et al. (2000), 280, 516
Bardet et al. (2003), 114, 139, 280, 516
Barlow (1957), 43, 516
Barndorff-Nielsen et al. (1978), 156, 516
Barnes & Allan (1966), 68, 144, 516
Barnsley (2000), 16, 39, 516
Bartlett (1955), 50, 82, 87, 516
Bartlett (1963), 4, 72, 88, 94, 516
Bartlett (1964), 72, 94, 95, 202, 517
Bartlett (1972), 4, 517
Barton & Poor (1988), 137, 143, 517

Bassingthwaighte et al. (1994), 16, 41, 517
Bassingthwaighte & Raymond (1994), 60, 517
Bateman (1910), 64, 517
Beggs & Plenz (2003), 42, 517
Beggs & Plenz (2004), 42, 517
Bell (1960), 116, 517
Bell (1980), 116, 517
Benassi et al. (1997), 123, 517
Beran (1992), 309, 517
Beran (1994), 60, 309, 517
Beran et al. (1995), 325, 517
Berger & Mandelbrot (1963), 154, 167, 517
Berry (1979), 16, 40, 517
Berry et al. (1980), 151, 517
Bertoin (1998), 35, 174, 518
Bharucha-Reid (1997), 236, 518
Bhattacharya et al. (2005), 42, 518
Bickel (1999), 122, 123, 518
Bickel (2000), 168, 452, 518
Bickel & West (1998a), 168, 452, 518
Bickel & West (1998b), 168, 452, 518
Biederman-Thorson & Thorson (1971), 42, 518
Bienaymé (1845), 95, 518
Blair & Erlanger (1932), 92, 518
Blair & Erlanger (1933), 92, 518
Bouchaud & Georges (1990), 35, 518
Bovy (1998), 4, 16, 518
Boxma (1996), 334, 518
Bracewell (1986), 51, 519
Brémaud & Massoulié (2001), 217, 519
Brichet et al. (1996), 328, 519
Brillinger (1981), 51, 519
Brillinger (1986), 78, 519
Brockmeyer et al. (1948), 316, 519, 524, 525
Brown (1828), 19, 519
Buckingham (1983), 16, 39, 107, 116, 172, 173, 196, 519
Buldyrev et al. (1995), 290, 519
Burgess (1959), 231, 519
Burlatsky et al. (1989), 471, 519
Çinlar (1972), 84, 256, 258, 519
Caccia et al. (1997), 60, 519
Campbell (1909a), 186, 520
Campbell (1909b), 186, 520

Campbell (1939), 50, 520
Cantor et al. (1975), 237, 520
Cantor & Teich (1975), 237, 263, 520
Cantor (1883), 17, 40, 520
Carlson & Doyle (1999), 37, 520
Carlson & Doyle (2002), 37, 520
Carson (1931), 195, 520
Castaing (1996), 123, 520
Čerenkov (1934), 220, 520
Čerenkov (1937), 220, 520
Čerenkov (1938), 220, 468, 520
Chandler et al. (1958), 4, 520
Chandrasekhar (1943), 193, 520
Chapman & Smith (1963), 42, 520
Chistyakov (1964), 57, 521
Christoph & Wolf (1992), 35, 521
Cohen (1969), 316, 328, 521
Cohen (1973), 328, 521
Cohn (1993), 516, 521
Cole (1995), 45, 521
Collins et al. (1995), 16, 43, 521
Conrad (1986), 25, 521
Cooper (1972), 316, 521
Cox (1948), 33, 521
Cox (1955), 88, 521
Cox (1962), 23, 51, 82, 85, 231, 264, 521
Cox (1963), 241, 521
Cox (1984), 15, 336, 521
Cox & Isham (1980), vi, 23, 51, 54, 66, 71, 77, 82, 83, 85, 88, 95, 97, 231, 258, 259, 264, 407, 521
Cox & Lewis (1966), 51, 70, 82, 88, 273, 521
Cox & Smith (1953), 84, 521
Cox & Smith (1954), 84, 522
Crovella & Bestavros (1997), 325, 329, 335, 336, 522
Dal Negro et al. (2003), 40, 522
Dal Negro et al. (2004), 40, 522
Dal Negro et al. (2005), 40, 522
Daley (1974), 54, 522
Daley & Vere-Jones (1988), 51, 85, 94, 522
Dan et al. (1996), 45, 522
Daubechies (1988), 120, 522
Daubechies (1992), 58, 522
Davenport & Root (1987), 186, 189, 522
Davidsen & Schuster (2002), 149, 522

Davies & Harte (1987), 139, 142, 522
Dayan & Abbott (2001), 77, 522
DeBoer et al. (1984), 57, 64, 282, 523
DeLotto et al. (1964), 237, 263, 523
Dettmann et al. (1994), 440, 523
Devaney (1986), 25, 523
Diament & Teich (1970a), 36, 523
Diament & Teich (1970b), 501, 523
Ding & Yang (1995), 138, 523
Ditto et al. (1990), 30, 523
Doob (1948), 85, 523
Doob (1953), 189, 523
Dorogovtsev & Mendes (2003), 37, 321, 523
Doukhan (2003), 16, 523
Doyle & Carlson (2000), 37, 523
Duffy et al. (1994), 317, 523
Dulea et al. (1992), 40, 523
Ebel et al. (2002), 321, 524
Eccles (1957), 91, 524
Efron (1982), 255, 524
Efron & Tibshirani (1993), 255, 524
Eguíluz et al. (2005), 38, 524
Einstein (1905), 19, 524
Ellis (1844), 97, 524
Embrechts et al. (1997), 57, 524
Engset (1915), 315, 316, 524
Engset (1918), 316, 524
Erlang (1909), 315, 524
Erlang (1917), 316, 318, 319, 524
Erlang (1920), 316, 525
Erramilli et al. (1996), 328, 525
Evans et al. (2001), 227, 525
Evarts (1964), 16, 42, 525
Fadel et al. (2004), 42, 525
Fairfield-Smith (1938), 33, 525
Fairhall et al. (2001a), 42, 525
Fairhall et al. (2001b), 42, 525
Falconer (2003), 16, 39, 525
Faloutsos et al. (1999), 321, 525
Fano (1947), 66, 525
Fatt & Katz (1952), 41, 44, 45, 151, 525
Feder (1988), 16, 39, 525
Feldmann et al. (1998), 323, 331, 335, 526
Feller (1941), 85, 526
Feller (1948), 50, 237, 526
Feller (1951), 60, 526
Feller (1968), vi, 87, 93, 174, 179, 526

Feller (1971), vi, 21, 35, 51, 56, 82, 86, 138, 146, 157, 164, 165, 192, 237, 526
Feynman (1965), 34, 526
Feynman et al. (1963), 24, 526
Fibonacci (1202), 40, 526
Field et al. (2004a), 334, 526
Field et al. (2004b), 334, 526
Fisher (1972), 82, 526
Flake (2000), 16, 39, 526
Flandrin (1989), 138, 141, 526
Flandrin (1992), 139, 143, 527
Flandrin (1997), 15, 527
Flandrin & Abry (1999), 15, 16, 39, 527
Fleckenstein (1969), vii, 527
Fourier (1822), 102, 527
Franken (1963), 84, 527
Franken (1964), 84, 527
Franken et al. (1981), 84, 527
Fréchet (1940), 50, 527
Furry (1937), 95, 527
Gardner (1978), 116, 527
Garrett & Willinger (1994), 330, 527
Gauss (1809), 36, 173, 527
Gellermann et al. (1994), 40, 527
Gere (2001), 34, 528
Gerstein & Mandelbrot (1964), 485, 528
Ghulinyan et al. (2005), 40, 528
Gilbert (1961), 154, 167, 528
Gilbert & Pollak (1960), 186, 187, 189, 190, 378, 528
Gilden (2001), 116, 528
Gillespie (1994), 168, 452, 528
Gisiger (2001), 37, 528
Glass & Mackey (1988), 25, 249, 528
Gnedenko & Kolmogorov (1968), 35, 528
Good (1961), 192, 528
Gottschalk et al. (1995), 16, 43, 528
Gradshteyn & Ryzhik (1994), 104, 166, 420, 448, 528
Grandell (1976), 88, 90, 197, 528
Grassberger (1985), 199, 528
Grassberger & Procaccia (1983), 75, 528
Grebogi et al. (1984), 30, 529
Greenwood & Yule (1920), 64, 147, 529
Greiner et al. (1999), 57, 529
Greis & Greenside (1991), 126, 529
Grigelionis (1963), 84, 529

Gross & Harris (1998), 317, 529
Grossglauser & Bolot (1996), 327, 529
Grüneis (1984), 218, 335, 529
Grüneis (1987), 218, 529
Grüneis (2001), 218, 335, 529
Grüneis & Baiter (1986), 218, 335, 529
Grüneis & Musha (1986), 218, 219, 529
Grüneis et al. (1993), 42, 529
Grüneis et al. (1989), 42, 529
Gumbel (1958), 36, 57, 147, 529
Gurland (1957), 95, 530
Gutenberg & Richter (1944), 33, 530
Haar (1910), 67, 74, 102, 104, 530
Haight (1967), 23, 82, 530
Halford (1968), 39, 530
Halley & Inchausti (2004), 41, 530
Halsey (2000), 35, 530
Harris (1971), 242, 530
Harris (1989), 95, 530
Hattori et al. (2000), 40, 530
Hattori et al. (1994), 40, 530
Hausdorff et al. (1997), 43, 530
Hawkes (1971), 217, 530
Heath et al. (1998), 334, 335, 530
Heneghan et al. (1996), 113, 296, 530
Heneghan et al. (1999), 59, 531
Heneghan & McDarby (2000), 62, 531
Hénon (1976), 28, 531
Henry & Zaffaroni (2003), 39, 531
Heyde & Seneta (2001), vii, 531
Hille (2001), 151, 531
Hohn et al. (2003), 335, 336, 346, 531
Holden (1976), 91–93, 149, 151, 531
Holtsmark (1919), 193, 531
Holtsmark (1924), 193, 531
Hon & Lee (1965), 275, 531
Hooge (1995), 169, 531
Hooge (1997), 169, 531
Hopcraft et al. (2002), 36, 531
Hopcraft et al. (2004), 36, 38, 532
Hopcraft et al. (1999), 36, 532
Hopkinson (1876), 38, 532
Hsü & Hsü (1991), 116, 532
Hu et al. (2001), 62, 532
Huberman & Adamic (1999), 321, 532
Humbert (1945), 192, 532
Hurst (1951), 59, 60, 116, 137, 287, 532
Hurst (1956), 59, 60, 137, 532
Hurst et al. (1965), 59, 60, 137, 532

Jaggard (1997), 40, 532
Jaggard & Sun (1990), 40, 532
Jakeman (1982), 40, 532
Jelenković & Lazar (1999), 334, 532
Jelley (1958), 221, 222, 532
Jenkins (1961), 78, 532
Jensen (1992), 316, 533
Johnson (1925), 115, 533
Jost (1947), 237, 533
Kabanov (1978), 77, 533
Kagan & Knopoff (1987), 222, 533
Kallenberg (1975), 232, 533
Kang & Redner (1984), 472, 533
Kastner (1985), 169, 533
Katz (1966), 41, 151, 533
Kaulakys (1999), 149, 533
Kaye (1989), 16, 533
Kelly et al. (1996), 42, 533
Kendall (1949), 95, 533
Kendall (1953), 317, 533
Kendall (1975), 95, 518, 533
Kendall & Stuart (1966), 249, 533
Kenrick (1929), 172, 534
Kerner (1998), 4, 534
Kerner (1999), 4, 534
Khinchin (1934), 172, 534
Khinchin (1955), 84, 534
Kiang et al. (1965), 249, 534
Kingman (1993), 51, 534
Klafter et al. (1996), 19, 534
Kleinrock (1975), 316, 317, 534
Knoll (1989), 223, 534
Kobayashi & Musha (1982), 43, 116, 275, 534
Koch (1999), 91, 534
Kodama et al. (1989), 42, 534
Kogan (1996), 16, 116, 534
Kohlrausch (1854), 33, 534
Kohmoto et al. (1987), 16, 40, 534
Kolář et al. (1991), 40, 535
Kolmogorov (1931), 19, 535
Kolmogorov (1940), 136, 137, 535
Kolmogorov (1941), 36, 535
Kolmogorov & Dmitriev (1947), 95, 535
Komenani & Sasaki (1958), 4, 535
Kou & Xie (2004), 35, 535
Krapivsky et al. (2000), 38, 535
Krapivsky et al. (2001), 38, 535
Krishnam et al. (2000), 179, 535

Kumar & Johnson (1993), 123, 147, 535
Kurtz (1996), 335, 535
Kuznetsov & Stratonovich (1956), 70, 535
Kuznetsov et al. (1965), 70, 535
Lapenna et al. (1998), 155, 222, 536
Lapicque (1907), 91, 93, 536
Lapicque (1926), 91, 536
Latouche & Remiche (2002), 336, 536
Läuger (1988), 16, 41, 536
Lawrance (1972), 95, 202, 249, 536
Lax (1997), 186, 536
Leadbetter et al. (1983), 51, 536
Leland et al. (1993), 325, 536
Leland et al. (1994), 16, 325, 536
Leland & Wilson (1989), 117, 325, 340, 341, 345, 350, 536
Leland & Wilson (1991), 117, 325, 340, 341, 345, 350, 536
Levy & Taqqu (2000), 334, 335, 536
Lévy (1937), 35, 192, 537
Lévy (1940), 35, 192, 537
Lévy (1948), 19, 537
Lévy Véhel et al. (1997), 16, 39, 537
Lévy Véhel & Riedi (1997), 331, 335, 537
Lewis et al. (2001), 42, 537
Lewis (1964), 94, 537
Lewis (1967), 94, 537
Lewis (1972), 51, 82, 88, 537
Li & Teich (1993), 147, 537
Li (1991), 37, 537
Libert (1976), 236, 263, 264, 537
Liebovitch (1998), 16, 41, 537
Liebovitch et al. (1987), 41, 537
Liebovitch et al. (2001), 41, 173, 538
Liebovitch & Tóth (1990), 41, 538
Likhanov (2000), 497, 538
Likhanov et al. (1995), 337, 538
Little (1961), 318, 538
Liu (1997), 40, 538
Lotka (1926), 33, 538
Lotka (1939), 50, 85, 538
Lowen (1992), xxi, 86, 87, 155, 157, 169, 175–177, 538
Lowen (1996), 73, 239, 240, 538
Lowen (2000), 139, 142, 538
Lowen et al. (1997a), 41, 42, 45, 148, 149, 152, 446, 538

Lowen et al. (1997b), 16, 41, 42, 45, 117, 132, 147–149, 152, 360, 446, 538
Lowen et al. (1999), 178, 539
Lowen et al. (2001), 42, 77, 78, 117, 121, 486, 539
Lowen et al. (1998), 77, 539
Lowen & Teich (1989a), 186, 187, 191, 192, 539
Lowen & Teich (1989b), 186, 187, 196, 539
Lowen & Teich (1990), 186, 187, 189– 191, 193–196, 216, 376–378, 471, 539
Lowen & Teich (1991), 90, 95, 186, 193, 204–209, 212–214, 336, 377, 539
Lowen & Teich (1992a), 42, 117, 249, 539
Lowen & Teich (1992b), 169, 173, 539
Lowen & Teich (1993a), 73, 115, 308, 539
Lowen & Teich (1993b), 42, 145, 174, 539
Lowen & Teich (1993c), 41, 173, 539
Lowen & Teich (1993d), 41, 86, 155– 157, 164, 173, 178, 539
Lowen & Teich (1995), 13, 41, 42, 66, 115, 133, 173, 174, 273, 306, 440, 540
Lowen & Teich (1996a), 16, 42, 68, 540
Lowen & Teich (1996b), 145, 540
Lowen & Teich (1997), 145, 249, 540
Lubberger (1925), 50, 540
Lubberger (1927), 50, 540
Lukes (1961), 86, 197, 540
Lundahl et al. (1986), 139, 540
Maccone (1981), 144, 540
Machlup (1954), 172, 540
Malamud (2004), 33, 540
Malik et al. (1996), 274, 540
Mandel (1959), 146, 540
Mandelbrot (1960), 33, 154, 541
Mandelbrot (1964), 154, 541
Mandelbrot (1965a), 16, 154, 167, 541
Mandelbrot (1965b), 137, 541
Mandelbrot (1967a), 4, 541
Mandelbrot (1967b), 144, 541
Mandelbrot (1969), 335, 541

Mandelbrot (1972), 154, 541

Mandelbrot (1974), 331, 541

Mandelbrot (1975), 4, 541

Mandelbrot (1982), 3, 11, 13, 16, 33, 35, 36, 39, 40, 59, 60, 75, 95, 116, 137, 140, 143, 150, 154, 541

Mandelbrot (1997), 39, 123, 154, 330, 541

Mandelbrot (1999), 15, 123, 541

Mandelbrot (2001), 16, 39, 60, 541

Mandelbrot & Hudson (2004), 39, 154, 541

Mandelbrot & Van Ness (1968), 136–138, 141, 143, 144, 542

Mandelbrot & Wallis (1969a), 142, 542

Mandelbrot & Wallis (1969b), 60, 542

Mandelbrot & Wallis (1969c), 60, 542

Mannersalo & Norros (1997), 331, 542

Marinari et al. (1983), 34, 542

Masoliver et al. (2001), 189, 542

Matsuo et al. (1982), 95, 542

Matsuo et al. (1983), 95, 206, 542

Matsuo et al. (1984), 95, 542

Matthes (1963), 54, 542

Matthews et al. (2003), 36, 542

McGill (1967), 146, 147, 202, 542

McGill & Goldberg (1968), 43, 542

McGill & Teich (1995), 43, 543

McWhorter (1957), 39, 169, 173, 543

Merlin et al. (1985), 16, 40, 543

Mikosch et al. (2002), 336, 543

Millhauser et al. (1988), 16, 41, 543

Mitchell (1968), 36, 543

Mitchell et al. (1995), 43, 543

Molchan (2003), 136, 543

Montanari (2003), 39, 543

Montgomery (1952), 172, 543

Montroll & Shlesinger (1982), 36, 116, 543

Moon (1992), 25, 543

Moran (1967), 249, 543

Morant (1921), 237, 544

Moriarty (1963), 154, 544

Morse (1921a), 40, 544

Morse (1921b), 40, 544

Moskowitz et al. (1974), 43, 544

Moyal (1962), 50, 544

Müller (1973), 236, 237, 263, 264, 544

Müller (1974), 236, 263, 264, 544

Müller (1981), 236, 237, 263, 544

Musha (1981), 16, 43, 45, 116, 544

Musha & Higuchi (1976), 16, 116, 544

Musha et al. (1985), 116, 544

Musha et al. (1981), 42, 544

Musha et al. (1983), 16, 42, 116, 544

Myskja (1998a), 316, 544

Myskja (1998b), 524, 545

Myskja & Espvik (2002), 524, 544, 545

Newell & Sparks (1972), 4, 545

Newton (1687), 24, 34, 545

Neyman (1939), 202, 426, 545

Neyman & Scott (1958), 93, 94, 202, 545

Neyman & Scott (1972), 93–95, 202, 545

Norros (1994), 328, 545

Norros (1995), 325, 331, 335, 545

Norsworthy et al. (1996), 91, 545

Olson (2004), 4, 545

Omori (1895), 33, 545

Oppenheim & Schafer (1975), 304, 306, 545

Orenstein et al. (1982), 169, 545

Orer et al. (2003), 42, 545

Oshanin et al. (1989), 471, 546

Ott (2002), 25, 402, 546

Ott et al. (1994), 25, 227, 546

Ovchinnikov & Zeldovich (1978), 472, 546

Palm (1937), 315, 546

Palm (1943), 50, 82, 84, 227, 231, 237, 256, 258, 316, 318, 546

Papangelou (1972), 228, 546

Papoulis (1991), 64, 186, 197, 375, 410, 546

Pareto (1896), 33, 154, 155, 546

Park & Gray (1992), 93, 546

Park (2000), 328, 546

Park et al. (1996), 33, 329, 335, 336, 546

Park et al. (2000), 329, 333, 546

Park & Willinger (2000), 16, 39, 314, 547

Parzen (1962), 23, 51, 82, 85, 97, 192, 197, 231, 236, 237, 264, 547

Pastor-Satorras & Vespignani (2004), 37, 321, 547

Paulus & Geyer (1992), 16, 44, 547

Paxson & Floyd (1995), 325, 335, 336, 547

Pecher (1939), 92, 149, 547

Peitgen & Saupe (1988), 16, 39, 139, 275, 547

Peitgen et al. (1997), 16, 25, 28, 29, 39, 547

Peltier & Lévy Véhel (1995), 123, 547

Penck (1894), 2, 547

Peng et al. (1995), 43, 61, 547

Peng et al. (1993), 43, 547

Peřina (1967), 147, 547

Perkal (1958a), 2, 547

Perkal (1958b), 2, 547

Perrin (1909), 19, 548

Petropulu et al. (2000), 192, 197, 548

Picinbono (1960), 186, 187, 189, 548

Pinsky (1984), 35, 223, 548

Pipiras & Taqqu (2003), 144, 548

Poincaré (1908), 25, 174, 548

Poisson (1837), 63, 83, 548

Pollard (1946), 192, 548

Pontrjagin & Schnirelmann (1932), 12, 75, 548

Powers & Salvi (1992), 42, 548

Press et al. (1992), 356, 548

Prucnal & Saleh (1981), 249, 548

Prucnal & Teich (1979), 110, 548

Prucnal & Teich (1980), 249, 548

Prucnal & Teich (1982), 202, 549

Prucnal & Teich (1983), 237, 263, 549

Quenouille (1949), 95, 549

Quine & Seneta (1987), 83, 549

Rammal & Toulouse (1983), 471, 549

Rana (1997), 18, 549

Rangarajan & Ding (2000), 126, 549

Raymond & Bassingthwaighte (1999), 144, 549

Reid (1982), vii, 549

Reiss (1993), 51, 549

Rényi (1955), 74, 549

Rényi (1956), 232, 549

Rényi (1970), 74, 549

Ricciardi & Esposito (1966), 263, 549

Rice (1944), 147, 172, 176, 186, 190–192, 194, 195, 549

Rice (1945), 147, 172, 176, 186, 190–192, 194, 195, 550

Rice (1983), 175, 550

Richardson (1960), 3, 550

Richardson (1961), 2, 6, 550

Riedi (2003), 123, 550

Riedi & Lévy Véhel (1997), 331, 550

Riedi & Willinger (2000), 331, 550

Rieke et al. (1997), 77, 550

Roberts & Cronin (1996), 15, 550

Roughan et al. (1998), 328, 550

Rudin (1959), 40, 550

Rudin (1976), 13, 550

Ruszczynski et al. (2001), 45, 550

Rutherford & Geiger (1910), 24, 64, 551

Ryu & Elwalid (1996), 327, 551

Ryu & Lowen (1995), 204, 336, 551

Ryu & Lowen (1996), 259, 260, 334, 551

Ryu & Lowen (1997), 204, 334, 336, 551

Ryu & Lowen (1998), 115, 204, 334, 336, 551

Ryu & Lowen (2000), 189, 551

Ryu & Lowen (2002), 336, 551

Sakmann & Neher (1995), 41, 551

Saleh (1978), 51, 82, 88, 89, 146, 147, 551

Saleh et al. (1983), 88, 202, 551

Saleh et al. (1981), 202, 237, 551

Saleh & Teich (1982), 88, 90, 94, 95, 186, 189, 202, 205, 207–209, 486, 552

Saleh & Teich (1983), 93, 94, 202, 552

Saleh & Teich (1985a), 202, 552

Saleh & Teich (1985b), 227, 552

Saleh & Teich (1991), 34, 222, 552

Samorodnitsky & Taqqu (1994), 35, 174, 552

Sapoval et al. (2004), 173, 552

Sato (1999), 35, 174, 552

Scharf et al. (1995), 68, 552

Schepers et al. (1992), 60, 552

Scher & Montroll (1975), 169, 552

Schick (1974), 173, 552

Schiff & Chang (1992), 227, 253, 552

Schmitt et al. (1998), 123, 173, 331, 552

Schönfeld (1955), 195, 553

Schottky (1918), 186, 553

Schreiber & Schmitz (1996), 253, 553

Schroeder (1990), 16, 28, 34, 39, 46, 116, 553

Schuster (1995), 25, 553
Seidel (1876), 63, 553
Sellan (1995), 139, 553
Shapiro (1951), 40, 553
Shimizu et al. (2002), 16, 43, 553
Shlesinger (1987), 36, 116, 553
Shlesinger & West (1991), 13, 553
Sigman (1995), 54, 553
Sigman (1999), 57, 553
Sikula (1995), 173, 553
Simoncelli & Olshausen (2001), 45, 553
Smith (1958), 85, 554
Snyder & Miller (1991), 51, 82, 88, 197, 554
Solomon & Richmond (2002), 37, 554
Soma et al. (2003), 45, 554
Song et al. (2005), 321, 554
Sornette (2004), 15, 16, 35, 554
Srinivasan (1974), 51, 82, 554
Steinhaus (1954), 2, 554
Stepanescu (1974), 169, 173, 554
Stern et al. (1997), 41, 151, 554
Stevens (1957), 43, 554
Stevens (1971), 43, 554
Stoksik et al. (1994), 139, 554
Stoyan & Stoyan (1994), 16, 554
Strogatz (1994), 25, 554
Szőkefalvi-nagy (1959), viii, 530, 554
Takács (1960), 85, 555
Takayasu et al. (1988), 199, 555
Taqqu (2003), 136, 555
Taqqu & Levy (1986), 177, 335, 555
Taqqu & Teverovsky (1998), 62, 289, 555
Taqqu et al. (1995), 274, 309, 555
Taqqu et al. (1997), 331, 555
Taubes (1998), 325, 555
Taylor (2002), 45, 555
Teich (1981), 202, 206, 555
Teich (1985), 249, 555
Teich (1989), 16, 42, 45, 555
Teich (1992), 42, 145, 249, 555
Teich & Cantor (1978), 227, 263, 555
Teich & Diament (1980), 237, 555
Teich & Diament (1989), 95, 556
Teich et al. (1997), 16, 42, 183, 204, 485, 556

Teich et al. (1996), 16, 42, 51, 58, 74, 113, 117, 296, 351, 484, 556, 558
Teich et al. (1990), 42, 145, 202, 204, 249, 427, 556–558
Teich & Khanna (1985), 227, 249, 556
Teich et al. (1993), 227, 232, 476, 556
Teich & Lowen (1994), 42, 249, 556
Teich & Lowen (2003), 145, 217, 556
Teich et al. (2001), 16, 44, 145, 227, 275, 556
Teich et al. (1991), 42, 556
Teich et al. (1978), 227, 237, 556
Teich & McGill (1976), 147, 237, 556
Teich et al. (1982a), 202, 557
Teich et al. (1982b), 202, 557
Teich & Rosenberg (1971), 36, 557
Teich & Saleh (1981a), 202, 557
Teich & Saleh (1981b), 202, 557
Teich & Saleh (1982), 227, 232, 233, 236, 557
Teich & Saleh (1987), 202, 206, 557
Teich & Saleh (1988), 88, 202, 557
Teich & Saleh (1998), 202, 557
Teich & Saleh (2000), 88, 202, 557
Teich et al. (1984), 202, 264, 483, 557
Teich & Turcott (1988), 42, 558
Teich & Vannucci (1978), 227, 237, 483, 501, 558
Teicher et al. (1996), 16, 44, 558
Telesca et al. (2004), 77, 558
Telesca et al. (1999), 155, 222, 558
Telesca et al. (2002a), 155, 558
Telesca et al. (2002b), 222, 558
Terman (1947), 34, 558
Tewfik & Kim (1992), 114, 139, 296, 297, 300, 558
Theiler (1990), 74, 75, 558
Theiler et al. (1992), 227, 253, 559
Theiler & Prichard (1996), 253, 559
Thiébaut (1988), 155, 559
Thomas (1949), 206, 559
Thompson & Stewart (2002), 25, 559
Thorson & Biederman-Thorson (1974), 42, 559
Thue (1906), 40, 559
Thue (1912), 40, 559
Thurner et al. (1998), 43, 275, 559

Thurner et al. (1997), 16, 41, 58, 66, 68, 71, 115, 146, 174, 217, 242, 273, 310, 394, 485, 559
Tiedje & Rose (1980), 169, 559
Timmer & König (1995), 139, 559
Toib et al. (1998), 41, 560
Toussaint & Wilczek (1983), 472, 560
Tuan & Park (2000), 329, 560
Tuckwell (1988), 91, 93, 560
Tukey (1957), 249, 560
Turcott et al. (1995), 42, 167, 168, 265–267, 451, 452, 560
Turcott et al. (1994), 112, 560
Turcott & Teich (1993), 16, 43, 275, 487, 560
Turcott & Teich (1996), 16, 43, 117, 227, 265, 275, 282, 487, 560
Turcotte (1997), 16, 39, 560
Turner et al. (1998), 16, 39, 560
Usher et al. (1995), 37, 560
van der Waerden (1975), 231, 560
van der Ziel (1950), 39, 560
van der Ziel (1979), 195, 561
van der Ziel (1986), 16, 561
van der Ziel (1988), 16, 116, 561
Vannucci & Teich (1978), 237, 238, 561
Vannucci & Teich (1981), 237, 238, 561
Veitch & Abry (1999), 274, 280, 561
Vere-Jones (1970), 94, 202, 204, 222, 223, 561
Verhulst (1845), 26, 561
Verhulst (1847), 26, 561
Verveen (1960), 16, 41, 92, 116, 173, 561
Verveen & Derksen (1968), 41, 149, 151, 173, 561
Vicsek (1992), 16, 35, 561
Vicsek (2001), 16, 561
Ville (1948), 138, 561
Viswanathan et al. (1996), 45, 561
Voldman et al. (1983), 155, 562
von Bortkiewicz (1898), 83, 562
von Schweidler (1907), 39, 562
Voss (1989), 116, 562
Voss & Clarke (1978), 116, 562
Watson & Galton (1875), 95, 562
Weber (1835), 33, 562
Weiss (1973), 197, 562
Weiss et al. (1994), 192, 562

Weissman (1988), 16, 116, 562
Wescott (1976), 232, 562
West (1990), 45, 562
West & Bickel (1998), 168, 452, 562
West et al. (2003), 16, 39, 562
West & Deering (1994), 16, 41, 562
West & Deering (1995), 16, 41, 45, 116, 562
West & Shlesinger (1989), 36, 563
West & Shlesinger (1990), 36, 563
West et al. (1999), 116, 563
Whittle (1962), 34, 563
Wickelgren (1977), 43, 563
Wiener (1923), 19, 563
Wiener (1930), 172, 563
Willinger et al. (2004), 330, 563
Willinger et al. (2003), 314, 325, 329, 563
Willis (1922), 33, 563
Wise (1981), 485, 563
Witten & Sander (1981), 35, 199, 563
Wixted (2004), 43, 563
Wixted & Ebbesen (1991), 43, 563
Wixted & Ebbesen (1997), 43, 563
Wold (1948), 50, 564
Wold (1949), 50, 564
Wold (1965), 51, 564
Wolff (1982), 319, 564
Yamamoto & Nakahama (1983), 42, 564
Yamamoto et al. (1986), 42, 564
Yang et al. (2004), 222, 564
Yang & Petropulu (2001), 181, 184, 335, 564
Yu et al. (2005), 45, 335, 564
Yule (1924), 95, 564
Zhukovsky et al. (2001), 40, 440, 564
Zipf (1949), 33, 564
Zrelov (1968), 221, 222, 468, 564
Zucker (1993), 151, 565
Zuckerkandl & Pauling (1962), 168, 565
Zuckerkandl & Pauling (1965), 168, 565
Zumofen et al. (2004), 173, 565

Subject Index

absolute refractoriness, *See* event deletion

action potentials, 50, 204
  amygdala, 42
  auditory nerve fiber, 42, 68, 117–
    119, 128–131, 145, 147, 233–
    235, 244–246, 249–251, 253,
    254, 344, 345
  central nervous system, 42
  hippocampus, 42
  integrate-and-reset model, *See* integrate-and-reset process(es)
  lateral geniculate nucleus, 42, 77,
    117–121, 126–131, 183, 233–
    235, 244–246, 250, 251, 253,
    254
  medulla, 42
  reticular formation, 42
  retinal ganglion cell, 42, 77, 117–
    119, 128–131, 183, 233–235,
    244–246, 250, 251, 253, 254,
    344
  somatosensory cortex, 42
  striate cortex, 42, 117–119, 128–
    131, 233–235, 244–246, 250,
    251, 253, 254, 344, 345, 351

  surrogate data analysis, 227
  thalamus, 42
  visual-system interneuron, 42, 167–
    168, 265–267, 344, 345
Allan, David W., viii, 68, 269, 276
Allan factor, 68
Allan variance, 68, 269
alternating fractal renewal process(es),
    172–173, 177–182
  autocorrelation, 183–184
  autocovariance, 178, 183
  chain of Markov processes, 178–
    179
  computer network traffic, 173, 334–
    335
  dwell times, 174
  fractal binomial noise, 173, 181
  fractal Gaussian process, 173, 181–
    182
  fractal test signals, 173, 184
  ion channels, 41, 173
  nanoparticle fluorescence fluctuations, 173
  nerve-membrane voltage fluctuations,
    173
  rainfall, 173

semiconductor noise, 173
 spectrum, 177, 183
 sums of, 173, 181
 systems with fractal boundaries, 173
alternating renewal process(es), 172–182
 alternating fractal renewal process,
  See alternating fractal renewal
  process(es)
 autocorrelation, 175
 Bernoulli random variables, 174
 binomial noise, 173, 179–181
 burst noise, 172
 characteristic function, 175, 183
 dwell times, 174
 exponential dwell times, 176–177
 extreme asymmetry, 176
 Gaussian process, 181–182
 moments, 174–175
 on–off process, 172
 random telegraph signal, 172
 relation to renewal point process,
  176
 spectrum, 175–177, 183
 sums of, 173, 179–181
amygdala, *See* action potentials
attention-deficit hyperactivity disorder,
 44
auditory nerve fiber, *See* action potentials

Barnes, James, 269, 276
Bartlett, Maurice, 94, 201, 202
Bartlett–Lewis process, *See* cascaded pro-
 cess(es)
Berger, Jay, 153, 154
Bernoulli random deletion, *See* event deletion
Bernoulli, Jakob, 225, 226
binomial noise
 as a sum of alternating renewal processes, 173, 179–181
 autocorrelation, 179–181
 binomial distribution, 179
 convergence to a Gaussian process,
  181–182
 fractal, *See* fractal binomial noise
 moments, 179
bivariate point process, *See* point pro-
 cess(es)

block shuffling, *See* operations on point
 processes
blocked counter, *See* event deletion
bootstrap method, *See* operations on point
 processes
box-counting dimension, *See* dimension
branching process, *See* cascaded process(es)
Brownian motion, 19–21
 as a neuronal threshold, 149–150
 Bachelier process, 19
 definition, 19–20
 diffusion process, *See* power-law
  behavior
 fractal-based point process from,
  149–150
 generation of, 150
 history, 19
 relation to fractional Brownian motion, 137
 Wiener–Lévy process, 19
 zero crossings, 15
Burgess variance theorem, 231
burst noise, *See* alternating renewal pro-
 cess(es)

Cantor, Georg, 9
Cantor set, 17–19
 fat, 18
 Hausdorff–Besicovitch dimension,
  75
 membership, 47
 photonic multilayer-structure version, 40
 randomized version, 95
 semiconductor multilayer-structure
  version, 40
 triadic, 17
 variant, 46, 133–134
capacity dimension, *See* dimension
capacity-dimension scaling function, *See*
 dimension
cascaded process(es), 93–95
 applications of, 202, 323
 Bartlett–Lewis, 94, 98–99, 324
 branching, 95
 cluster, 93
 compound, 93
 doubly stochastic Poisson process
  version, 95, 333, 336, 346

fractal Bartlett–Lewis, 218–219, 335–336, 345–351
fractal Neyman–Scott, 336–337, 345–351
Neyman–Scott, 93, 202, 204, 324
Poisson branching, 95
Thomas, 95, 206
Yule–Furry, 95
central limit theorem, 36, 173, 174, 181, 188, 191, 216, 306
central nervous system, *See* action potentials
Čerenkov radiation, *See* photon statistics
chaos, 25–32
    fractal attractors, 25
    fractals, connection to, 24–32
    functional roles, 25
    phase-randomization surrogate, 227
    strange attractors, 25
characteristic function, *See* interval statistics
cluster process, *See* cascaded process(es)
coastline(s)
    Australian, 6
    British, 6
    fractal, 2–4
    Höfn, 2, 6
    Icelandic, 2, 6, 13–15
    length of, 2–4, 6, 14
    Seyðisfjörður, 2, 6
    South African, 6
compound process, *See* cascaded process(es)
computer cache misses, 155
computer communication networks, *See* computer network traffic
computer network traffic, 313–354
    alternating fractal renewal process, 173, 334–335
    analysis and synthesis, 332
    applications layer, 323
    arrival process, 317
    as a point process, 50
    bit transmission, 323
    blocking probability, 319
    buffer occupancy, 316
    buffer overflow probability, 319–321, 325, 334, 351–352
    buffer size, 316

CAIDA, 322
capacity-dimension scaling function, 343
cascaded-process models, 335–337, 345–351
    characteristic features of, 342–343
    computer communication networks, 320–323, 328, 332
    data sources, 117, 325
    detrended fluctuations, 338, 340, 341, 348, 349
    drop probability, 319
    Ethernet traffic, 117–119, 315, 325, 337, 340–342, 345, 350, 351
    event clustering, 345
    exponentialized data, 250, 251, 326, 327, 337, 343, 345
    extended alternating fractal renewal process, 335
    feedback, 332
    file transfers, 323
    flow control, 333
    fluid-flow models, 331
    forward Kolmogorov equation, 318, 351
    fractal Bartlett–Lewis process, 218, 335–336, 345–351
    fractal exponents, 44, 126, 343–344
    fractal features, 16, 40, 324–332
    fractal-Gaussian-process-driven Poisson process, 335, 352–353
    fractal Neyman–Scott process, 204, 328, 333, 335–337, 345–353
    fractal-rate point process, 342
    fractal renewal process, 334
    fractal-shot-noise-driven Poisson process, 204, 335–337
    fractional Brownian motion, 325, 331
    FTP, 323, 329, 335, 336
    general arrival and service processes, 317
    generalized dimension, 343
    geometric queue-length distribution, 318, 319, 328, 351–352
    heavy-tailed service times, 328, 333, 337
    HTTP, 323, 335

internetwork layer, 323
interval histogram, 130, 131, 233, 244, 338, 340–344, 348, 349, 351
interval sequence, 337, 340, 341, 348, 349
interval spectrum, 128, 339–341, 344, 348, 349
interval statistics, 350–351
interval wavelet variance, 129, 338, 340, 341, 344, 348, 349
IP, 322–323
ISP, 322
link layer, 323
Little's law, 318
local-area network, 325
Markov process, 317, 325, 327, 328
message-loss probability, 328, 332
model complexity, 332–333
modeling, 332–337, 345–351
modulated fractal-Gaussian-process-driven Poisson process, 353
monofractal approximation, 353–354
multifractal features, 16, 329, 331–332, 353–354
multiple data sets, 341–342
multiple servers, 317, 319, 351
multiple statistical measures, 337–340
normalized Haar-wavelet variance, 119, 235, 325–327, 330, 339–341, 344, 348, 349, 351, 353
normalized variance, 126, 127
packets, 50, 315, 323, 325
PASTA, 319
periodicities, 342
persistence, 329
physical layer, 323
point-process description, 330–331
point-process identification, 271, 337–351
power-law file sizes, 33, 323, 329–330, 335, 336
power-law queue-length distribution, 328, 352
predictability, 329
queue length, 316–318
queue waiting-time jitter, 328

queue-length distribution, 317, 318, 325, 328, 334, 351, 353
queueing theory, 316–320, 327–329, 336–337
randomly deleted data, 234, 235, 344
randomly displaced data, 245, 246
rate-process description, 330–331, 334, 338, 340, 341, 348, 349
rate spectrum, 118, 234, 325–326, 339–341, 344, 348, 349, 351
rescaled range, 338, 340, 341, 348, 349
resemblance to striate-cortex action potentials, 345, 351
scale-free networks, 37–38, 40, 321–322
scaling cutoffs, 330
second-order statistics, 325–328, 340, 341, 348, 349
server utilization, 318
service process, 317
service ratio, 318, 328
shuffled data, 253, 254, 326, 327, 337, 342–344, 351, 352
simulations, 332–333, 345–353
SSH, 323, 329
static representation, 322
TCP, 323, 329, 334
teletraffic theory, 315
TELNET, 323, 335
transport layer, 323
UDP, 329
vertical layers, 323
video traffic, 325, 330
waiting number mean, 318
waiting time mean, 318
wide-area network, 325
World Cup access log, 330
WWW, 325
correlation dimension, *See* dimension
counting statistics, 63–70
    α-particle counting, 64
    accidents, 64, 147
    Allan factor, 68
    Allan variance, 68, 269
    autocorrelation, 69, 71, 72, 106–107, 121, 124, 132, 222, 232, 282, 296, 311

autocovariance, normalized form, 69, 285–287
count sequence, 51, 63
counting distribution, 64, 65, 146–147, 163, 205–206, 218, 263–264
counting-time increments, 297–298
counting-time oversampling, 302–304
counting-time weighting, 298–302
cross-spectrum, 78
dead-time-modified point process, 145, 227, 238, 240, 263–264
decimated point process, 263–264
dispersion ratio, 66
doubly stochastic Poisson process, *See* doubly stochastic Poisson process(es)
factorial moments, 65, 83, 87, 161
Fano factor, 66
fractal renewal process, *See* fractal renewal process(es)
fractal-shot-noise-driven point process, *See* fractal-shot-noise-driven point process(es)
generalized rates, 64
generalized version of normalized Haar-wavelet variance, 69
homogeneous Poisson process, *See* homogeneous Poisson process
index of dispersion, 66
integrate-and-reset process, *See* integrate-and-reset process(es)
kurtosis, 65
moments, 65–66
negative binomial distribution, 146
Neyman Type-A distribution, 202, 206, 209, 212
noncentral negative binomial distribution, 147
normalized Daubechies-wavelet variance, 120
normalized general-wavelet variance, 74, 113–114, 120, 296–297
normalized Haar-wavelet covariance, 77–78
normalized Haar-wavelet variance, 62, 66–69, 71, 73, 80, 97, 103–105, 107, 111–112, 114, 115, 117–119, 121, 122, 124–126, 133, 167, 168, 232, 235, 236, 240, 246, 247, 249, 251, 254, 265, 266, 270, 275–282, 284–289, 291, 293, 296–304, 306, 307, 309–312, 325–327, 330, 339–344, 348, 349, 351, 353
normalized Haar-wavelet variance, relation to normalized variance, 62, 68–69, 112, 209–211, 284, 296, 344
normalized variance, 66, 68–69, 71, 73, 79, 80, 96–99, 105, 107, 109–110, 112, 121, 124, 126, 127, 133, 134, 212, 219, 222, 231, 234, 236, 282–284, 286–288, 296, 311, 312, 344
normalized wavelet cross-correlation function, 77
periodic processes, 64
periodogram, 70
rate-based measures, 64
relation to interval statistics, 65, 344
relationship among measures, 114–115
renewal process, *See* renewal process(es)
sample rate, 64
shot-noise-driven Poisson process, *See* doubly stochastic Poisson process(es)
skewness, 65
spectrum, 39, 43, 64, 70, 73, 80, 87, 88, 97, 103, 111, 113–117, 121, 122, 124–126, 133, 134, 234, 236, 245, 247, 249, 250, 253, 254, 271, 275, 282, 304–312, 325–326, 339–341, 344, 348, 349, 351
Thomas distribution, 206
variance-to-mean ratio, 66
Cox, David R., vi, viii, 81, 88
Cox process, *See* doubly stochastic Poisson process(es)
cross-spectrum, *See* counting statistics

data-transmission errors, *See* fractal renewal process(es)

dead-time deletion, *See* event deletion
decimation, *See* event deletion
deletion, *See* event deletion
detrended fluctuation analysis, *See* interval statistics
developmental disorders, 44
developmental insults, 44
diffusion processes, *See* power-law behavior
dilation, *See* operations on point processes
dimension
    box-counting, 12, 14, 75, 164, 237
    Cantor set, 18
    Cantor-set variant, 46
    capacity, 12, 14, 75, 96, 131, 164, 237
    capacity-dimension scaling function, 131, 132, 343
    correlation, 75, 96
    Euclidian, 11, 23, 35, 75, 223
    generalized, 74–76, 96, 130–132, 256, 332, 343
    generalized-dimension scaling function, 76, 131, 132, 266, 267
    Hausdorff–Besicovitch, 75
    information, 75
    Kolmogorov entropy, 75
    monofractal, 18, 75, 121
    multifractal, 75
    of a space, 11
    of an object, 11
    of diffusion processes, 34–35
    of point processes, 75–76, 96, 111, 121, 126, 130–132, 256, 272, 332, 343
    Rényi entropy, 74
    topological, 11, 75
    wavelet estimate of, 75
Dirac delta function, special property of, 92, 228
dispersion ratio, *See* counting statistics
displacement, *See* operations on point processes
doubly stochastic Poisson process(es), 87–90
    autocorrelation, 88
    cascaded-process isomorph, 95, 333, 336, 346

coincidence rate, 88
counting statistics, 88–89
    dead-time-modified, 237–241
    exponential interval density, 89–90, 125, 249, 272, 281
    factorial moments, 88
    fractal-binomial-noise-driven, 174, 182, 183, 272
    fractal-Gaussian-process-driven, 124–125, 145, 183, 217, 229, 249, 270, 275–281, 310, 328, 335, 352–353
    fractal-lognormal-noise-driven, 250
    fractal-rate-driven, 124, 262
    fractal-shot-noise-driven, *See* fractal-shot-noise-driven point process(es)
    integrated rate, 88
    interval density, 89
    interval statistics, 89–90
    multistage shot-noise-driven, 95
    random deletion of, 236
    rate coefficient of variation, 89–90, 281
    renewal version of, 90
    shot-noise-driven, 90, 202–204
    simulation of, 270, 310
    spectrum, 88
    superposition of, *See* superposition
drug abuse, 44

earthquakes, 33, 40, 77, 150, 155, 204, 222–223
emotional state, 44
equilibrium counter, *See* event deletion
Erlang, Agner Krarup, 97, 313, 315, 316, 319
Euclidian dimension, *See* dimension
event deletion, 226, 229–241
    Bernoulli random deletion, 226, 227, 230–236, 262–263, 344
    blocked counter, 236, 264
    Burgess variance theorem, 231
    dead-time deletion, 226, 227, 230, 236–241, 262–264
    decimation, 97, 226, 227, 230–232, 262–264
    decimation parameter, 231

doubly stochastic Poisson process,
See doubly stochastic Poisson process(es)
effects on fractal features, 229–231, 236
equilibrium counter, 236, 264
experimental interval histograms, 232–233
experimental normalized Haar-wavelet-variance curves, 235
experimental rate spectra, 234
fractal onset frequency, 232, 241
fractal onset time, 232, 241
fractal renewal process, See fractal renewal process(es)
general results, 229–231
homogeneous Poisson process, See homogeneous Poisson process
limit of a homogeneous Poisson process, 232
periodic process, 232–236
renewal process, See renewal process(es)
type-p dead time, 236
unblocked counter, 236, 263, 264
excitable-tissue recordings, 41
expansion-modification systems, 37
exponentialization, See operations on point processes
extended dead time, See event deletion

Fano factor, 66
Fatt & Katz, 45
Feller, William, vi, 225, 237
fern, 22
Fibonacci sequences, See photonic materials, See semiconductors
fixed dead time, See event deletion
fluorescence fluctuations of nanoparticles, See alternating fractal renewal process(es)
Fourier, Jean-Baptiste, 101–102
fractal analysis, See fractal parameter estimation
fractal-based point processes, See point process(es)
fractal binomial noise
as a rate function, 174, 182–183, 272, 334

as a sum of alternating fractal renewal processes, 173, 181
convergence to a Gaussian process, 181–182
fractal-binomial-noise-driven gamma process, 183
fractal chi-squared noise, 145–147
as a rate function, 150–151
fractal exponential noise, 146
fractal noncentral chi-squared noise, 147
fractal noncentral Rician-squared noise, 147
negative binomial counting distribution, 146
noncentral negative binomial counting distribution, 147
fractal exponent(s)
auditory nerve fiber, See action potentials
computer network traffic, See computer network traffic
cutoffs, 14–15, 36, 103, 105, 107–108, 133, 274, 330
diffusion, See power-law behavior
estimation of, See fractal parameter estimation
for fractal Bartlett–Lewis process, 219
for fractal point process, 121
for fractal-rate process, 124
for fractal shot noise, See fractal shot noise
for multifractals, 15, 75, 331
for nonstationary nonfractal processes, 110, 112, 133
for normalized general-wavelet variance, 113–114
for normalized Haar-wavelet variance, 111–114
from autocorrelation, 110–111
from count-based autocovariance, 287
from interval spectrum, 126–128, 295
from normalized Daubechies-wavelet variance, 120
from normalized detrended fluctuations, 291

from normalized Haar-wavelet variance, 117–119, 235, 246, 251, 254, 278, 299, 303

from normalized interval wavelet variance, 127–129, 293

from normalized rate spectrum, 116–118, 234, 245, 250, 253

from normalized rescaled range, 289

from normalized variance, 109–110, 126, 127, 285

from rate spectrum, 307

human heartbeat, *See* heartbeat

Hurst exponent, 137, 143–144, 287, 289

lateral geniculate nucleus, *See* action potentials

limited range of, 109–111

negative values of, 107–109, 133

observed values of, 109

range of values, 107–114

relations among, 105, 107, 114–115, 133

relative strength of fluctuations, 103, 273

retinal ganglion cell, *See* action potentials

same exponent from different fractal renewal processes, 166

spectrum, 133

striate cortex, *See* action potentials

superposition, *See* superposition

time varying, 331

under exponentialization, 250, 251, 264

under general deletion, 229–231

under random deletion, 234, 235

under random displacement, 245, 246

under shuffling, 253, 254

values in biological systems, 34

vesicular exocytosis, *See* vesicular exocytosis

visual-system interneuron, *See* action potentials

fractal exponential noise, 146

fractal Gaussian process(es), 144–145

as a rate function, 145, 216–217

as a sum of alternating fractal renewal processes, 173, 181–182

nomenclature for fractional processes, 143–145

fractal lognormal noise, 147–149

as a rate function, 148–149, 151–152

rate statistics, 147–148

fractal networks, *See* scale-free networks

fractal noncentral chi-squared noise, 147

fractal noncentral Rician-squared noise, 147

fractal parameter estimation, 269–312

asymptote subtraction, 312

autocovariance, 285–287

bias from cutoffs, 274

bias/variance tradeoff, 311

choice of scaling range, 274

coincidence-rate limitations, 311

comparison of measures, 309–310

count-based measures, 282–287

counting-time increments, 297–299, 303

counting-time oversampling, 302–304

counting-time weighting, 298–302

detrended fluctuations, 289–291

discrete-time processes, 274

estimator bias, 274, 278–280, 284, 285, 287, 289, 291, 293, 295, 307

estimator root-mean-square error, 278, 285, 287, 289, 291, 293, 295, 299, 303, 307

estimator standard deviation, 278, 285, 287, 289, 291, 293, 295, 307

estimator variance, 273

fractal exponents, 107, 126, 270, 273–281, 285, 287, 289, 291, 293, 295, 299, 303, 307

heart rate variability, 274–275

interval-based measures, 287–296

interval spectrum, 294–296

interval wavelet variance, 291–293

limitations of, 310

maximum-likelihood approach, 274

nonparametric approach, 273–274

normalized general-wavelet variance, 296–297
normalized Haar-wavelet variance, 276–281, 296–304, 344
normalized variance, 127, 282–285, 296, 311, 344
optimal measures, 271, 309
rate spectrum, 133, 304–309, 311
rescaled range, 287–289
robustness/error tradeoff, 311
simulations, 270, 275–278, 284–295, 297, 299, 303, 305, 307, 310, 312
speed/accuracy tradeoff, 274
fractal point processes, *See* point process(es)
fractal-rate point processes, *See* point process(es)
fractal renewal process(es), 87, 124, 131, 132, 154–166, 281
capacity dimension, 164
characteristic function, 155, 156, 166
coincidence rate, 159–160
comparison with homogeneous Poisson process, 122
computer cache misses, 155
computer network traffic, 334
counting distribution, 163
data-transmission errors, 40, 154, 166–167
earthquake occurrences, 155
effect of interval-density exponent, 157
factorial moments, 160–161
features of, 122
forward recurrence time, 262
fractal exponents, 158
fractal onset frequency, 166
generalized inverse Gaussian density, 156
generalized Pareto density, 165
interneuron counterexample, 167–168, 265–267
interval density, 155–157
interval density with abrupt cutoffs, 155
interval density with smooth transitions, 156–157
interval moments, 155, 156
molecular evolution, 168–169
nondegenerate realization, 164–166
normalized Haar-wavelet variance, 162
normalized variance, 160–162
Pareto density, 154–155
point-process spectrum, 157–159, 166
random deletion of, 236, 263
same fractal exponent from different interval densities, 166
simulation time, 166, 312
stable distribution, 157
superposition of, *See* superposition
trapping in semiconductors, 169, 224
Wald's Lemma, 164
fractal shot noise, 186–197
amplitude statistics, 189–193
as a rate function, 90, 202–205
autocorrelation, 194–195
characteristic function, 189–190
cumulants, 190
degenerate, 188, 193
fractal exponents, 195–197
Gaussian limit, 145, 188
impulse response function, 187–188, 202, 205
integrated, 204–205
mass distributions, 198–199
multifractal impulse response function, 331
parameter ranges, 188, 189
point processes from, *See* doubly stochastic Poisson process(es)
power-law-duration variant, 188–189, 198, 336, 352
spectrum, 188, 195–197
stable distribution, 188, 192, 193, 197
sums of, 198
fractal-shot-noise-driven integrate-and-reset process, *See* fractal-shot-noise-driven point process(es)
fractal-shot-noise-driven point process(es), 202–217
applications of, 204

applications of the Neyman Type-
A distribution, 202
applications of the shot-noise-driven
Poisson process, 202
Čerenkov radiation, 220–222
coincidence rate, 214
computer network traffic, 328, 335–
337, 352–353
counting distribution, 205–206
counting statistics, 205–212
design of, 220
diffusion, 223
earthquakes, 222–223
factorial moments, 207–208
forward recurrence time, 212–213
fractal exponents, 209, 211, 214,
215
fractal-Gaussian-process-driven limit,
216–217
fractal-shot-noise-driven integrate-
and-reset process, 217
fractal-shot-noise-driven Poisson process, 90, 202–217
Hawkes point process, 217
impulse response function without
cutoffs, 220
interval density, 212–213, 219, 272
multifractal version, 331
Neyman–Scott process, 202, 204
Neyman Type-A distribution, 202,
206
normalized Haar-wavelet variance,
210–212
normalized variance, 208–209, 219
self-exciting point process, 217
semiconductor particle detectors, 223–
224
spectrum, 215–216
fractal-shot-noise-driven Poisson process,
See fractal-shot-noise-driven
point process(es)
fractals
and Kant, 33
and Kohlrausch, 33
and Laplace, 33
and Leibniz, 33
and Weber, 33
and Weierstraß, 33
artificial, 16–21

chaos, connection to, 24–32
coastlines, 2–4, 6
convergence to stable distributions,
35–36
deterministic, 13, 16–19, 21–22
diffusion processes, 34–35
dynamical processes, 13
examples of fractals, 16–23, 28–
30, 33, 115–120
examples of nonfractals, 23–24, 26–
28
expansion-modification systems, 37
highly optimized tolerance, 37
historical antecedents, 32–33
in art, 45
in ecology, 26, 41
in human behavior, 43–44
in mathematics, 39–40
in medicine, 43–44
in music, 116
in the biological sciences, 41–44
in the neurosciences, 41–43
in the physical sciences, 39–40
in the psychological sciences, 42,
43, 45, 116
in vehicular-traffic flow, 4, 44, 45,
50, 116
laws of physics, 33–34
lognormal distribution, 36, 147
long-range dependence, 14–15
natural, 16, 21–23
noninteger dimension, 14
objects, 4
onset frequencies, 114–115
onset times, 114–115
origins of fractal behavior, 32–39,
329–330
Pareto's Law, 33
pink noise, 115–116
power-law behavior, connection to,
14, 32–39
putative exponential cutoff, 39
random, 13, 16, 19–23
range of time constants, 38–39, 332
recognizing the presence of fractal
behavior, 44–45
salutary features of fractal behavior, 41, 45
scale-free networks, 37–38, 45

scaling, connection to, 13–15
self-organized criticality, 37
static, 13
ubiquity of fractal behavior, 39–44
fractals in human behavior
attention-deficit hyperactivity disorder, 44
developmental disorders, 44
developmental insults, 44
drug abuse, 44
mood fluctuations, 43
fractals in mathematics
convergence to stable distributions, 35–36
fractal geometry, 40
lognormal distribution, 36
fractals in medicine
blood flow, 116
congestive heart failure, 275
fluctuations in human standing, 43
heart rate variability, 43–44, 270, 274–275
pain relief, 45
sensitization of baroreflex function, 45
fractals in the neurosciences
action potentials in auditory nerve fibers, 42, 131, 145, 147, 249
action potentials in central-nervous-system neurons, 42, 131
action potentials in isolated preparations, 41–42
action potentials in visual-system neurons, 42, 77, 131, 183, 217, 267
cognitive processes, 43, 116
electroencephalogram fluctuations, 116
excitable-tissue fluctuations, 41, 92, 116, 149, 151, 173
ion-channel transitions, 41, 151, 173
neuronal avalanches in slice preparations, 42
sensory detection and estimation, 42–43, 45
vesicular exocytosis, 41, 131, 132, 149, 151–152
fractals in the physical sciences

Čerenkov radiation, 34, 40, 204, 220–222
computer network traffic, 40, 313–354
data-transmission errors, 40, 154, 166–167
diffusion processes, 34–35, 204, 223
earthquake occurrences, 33, 40, 77, 150, 155, 204, 222–223
highly optimized tolerance, 37
laws of physics, 33–34
light scattering, 36, 40, 173
photonics, 40
self-organized criticality, 37, 222
semiconductors, 34, 39, 40, 116, 169, 172, 173, 223–224
fractional Brownian motion, 136–141
as a model for computer network traffic, 325, 331
as a rate function, 140–141
autocorrelation, 137, 150
autocorrelation coefficient, 150
definition, 21, 137
generalized dimensions, 139–140
generation by fractional integration, 144
history, 136
Hurst exponent, 137
level crossings, 138
nomenclature for fractional processes, 143–145
ordinary Brownian motion, *See* Brownian motion
properties, 138–139
realizations, 139–140
relation of Hurst and scaling exponents, 143–144
relation to fractional Gaussian noise, 141
relation to ordinary Brownian motion, 137
self-similarity, 138
stationary increments, 137, 150
synthesis, 139
Wigner–Ville spectrum, 138–139
zero crossings, 15
fractional Gaussian noise, 141–142
as a rate function, 142
definition, 141

generalized dimensions, 142
generation by fractional integration, 144
in a Langevin equation, 35
nomenclature for fractional processes, 143–145
properties, 141–142
realizations, 142–143
relation of Hurst and scaling exponents, 143–144
relation to fractional Brownian motion, 141
synthesis, 142
Wigner–Ville spectrum, 141–142

gamma renewal process, *See* renewal process(es)
Gauss, Carl Friedrich, 36, 171, 173
generalized dimension, *See* dimension
generalized-dimension scaling function, *See* dimension
generalized inverse Gaussian density, 156
Grand Canyon river network, 22
Greenwood, Major, 49, 64, 147
Gutenberg–Richter Law, 33

Haar, Alfréd, 68, 101, 102
Hausdorff–Besicovitch dimension, *See* dimension
heart rate variability, 44, 270, 274–275
heartbeat, 43, 50, 64, 79, 116–119, 128–131, 145, 232–235, 244–246, 250, 251, 253, 254, 264, 274–275, 293, 345
heavy-tailed distributions, *See* interval statistics
highly optimized tolerance, 37
hippocampus, *See* action potentials
Holtsmark distribution, 193
homogeneous Poisson process, 3, 23–24, 71, 82–85, 96–97, 108, 122, 124, 125, 131, 132, 167, 198, 231, 236, 246, 249, 255, 258, 260, 276, 278, 281, 316–320, 323, 325, 328, 335, 352
dead-time-modified, 237–238, 249, 262–264
decimated, 97, 263–264
factorial moments, 84

moments, 83
human standing, 43
Hurst, Harold Edwin, 59, 269, 287
Hurst exponent, *See* fractal exponent(s)
hypothesis testing, *See* operations on point processes

Icelandic coastline, 14
index of dispersion, *See* counting statistics
information dimension, *See* dimension
integrate-and-reset process(es), 91–93
dead-time-modified, 237, 241
decimated, 231
fractal-binomial-noise-driven, 174, 183
fractal-Gaussian-process-driven, 145, 243
fractal-shot-noise-driven, 217
gamma-distributed rate, 98
identification of, 273
interval density, 92
interval moments, 92
interval statistics, 91–92
kernel for heartbeat model, 293, 310
leaky, 93
model for action potentials, 91
modulated rate, 98, 110, 112
normalized variance, 96
oversampled sigma-delta modulator, 91
packet generation, 334
point-process spectrum, 91
randomly deleted, 236
time-varying threshold, 92–93, 149
interevent-interval transformation, *See* operations on point processes
interneuron, *See* action potentials
interval statistics, 54–62
autocorrelation, 20, 57, 83, 282
characteristic function, 55–56
coefficient of variation, 55, 231, 233, 236, 345
cumulants, 55, 56
density, 55, 89–90, 121, 129–130, 227, 281
detrended fluctuation pseudocode, 62

detrended fluctuation statistic, 61–
62, 79, 282
detrended fluctuation statistic, normalized form, 62, 289–291
discriminating among fractal-rate
processes, 310–311
distribution, 281
doubly stochastic Poisson process,
See doubly stochastic Poisson process(es)
exponential density, 83, 89–90, 92,
97, 125
fractal renewal process, *See* fractal
renewal process(es)
fractal-shot-noise-driven point process, *See* fractal-shot-noise-
driven point process(es)
heavy-tailed distributions, 13, 56,
57, 328, 333, 337
homogeneous Poisson process, *See*
homogeneous Poisson process
infinite moments, 56, 79, 165
integrate-and-reset process, *See* integrate-
and-reset process(es)
interval ordering, 90, 227, 247–254,
256, 281, 345
kurtosis, 55, 79, 175, 179
limitations of, 122–124, 281–282,
344
moments, 55, 60
normalized wavelet variance, 59
Pareto distribution, 57, 138, 154,
155, 165
periodic processes, 64, 96
periodogram, 70
power-law distribution, *See* fractal
renewal process(es)
recurrence time, 56, 65, 79, 80, 96
relation to counting statistics, 65,
344
rescaled range pseudocode, 60
rescaled range statistic, 59–60, 79,
282
rescaled range statistic, normalized
form, 60, 287–289
semi-invariants, 55
serial correlation coefficient, 57

shot-noise-driven Poisson process,
See doubly stochastic Poisson process(es)
skewness, 55, 79, 175, 179
spectrum, 42, 43, 58–59, 62, 64,
80, 83, 116, 126–128, 271,
275, 282, 294–296, 339–341,
344, 348, 349, 351
subexponential distributions, 57
survivor function, 55–57, 165, 238,
259–262
wavelet transform, 58
wavelet variance, 58–59, 62, 127–
129, 275, 282, 291–293, 344
Weibull distribution, 57, 328
interval transformation, *See* operations
on point processes
ion channels, *See* alternating fractal renewal process(es)
Isham, Valerie, vi

Kenrick, Gleason W., 172
knockout mice, 227
Kolmogorov, Andrei, 135–136

Lapicque, Louis, 81, 91
lateral geniculate nucleus, *See* action potentials
laws of physics, *See* power-law behavior
Leyden-jar discharge, 33, 39
light scattering, *See* fractals
logistic equation, 26, 37
logistic map, 26, 28, 30, 46
lognormal distribution, 36, 57, 147
long-range dependence, 15
Lévy, Paul, 35, 36, 171, 174
Lévy dust, *See* point process(es)
Lévy-stable distributions, *See* stable distributions

Mandelbrot, Benoit, viii, 4, 135, 136,
153, 154
marked point process, *See* point process(es)
medulla, *See* action potentials
mixed Poisson process, *See* doubly stochastic Poisson process(es)
molecular evolution, *See* fractal renewal
process(es)
monofractals, 15–16, 274, 331

mood fluctuations, 43
multidimensional point process, *See* point process(es)
multifractals, 15–16, 75, 188, 331–332, 353–354
multivariate point process, *See* point process(es)

Newton's Law, 34
Neyman, Jerzy, 94, 201, 202
Neyman Type-A distribution, *See* counting statistics
Neyman–Scott process, *See* cascaded process(es)
Nile river flow patterns, 59, 116, 269
noncentral limit theorem, 174
nonextended dead time, *See* event deletion
nonfractal(s)
    Euclidian shapes, 6–7, 23
    examples of, 14, 23–24, 26–28, 46
    generalized dimensions, 75, 96
    heart rate variability measures, 43, 275
    homogeneous Poisson process, *See* homogeneous Poisson process
    influences, 279
    orbits in a two-body system, 24
    point processes, *See* point process(es)
    radioactive decay, 24, 50
nonparalyzable dead time, *See* event deletion
nonstationary point process, *See* point process(es)
normalization, *See* operations on point processes
normalized Haar-wavelet covariance, *See* counting statistics
normalized Haar-wavelet variance, *See* counting statistics
normalized variance, *See* counting statistics
normalized wavelet cross-correlation function, 77
normalizing transformation, *See* operations on point processes

Omori's Law, 33

on–off process, *See* alternating renewal process(es)
operations on point processes
    block shuffling, 255, 262
    bootstrap method, 255
    event deletion, *See* event deletion
    event-time displacement, 145, 226, 242–247, 251, 254, 255, 262
    hypothesis testing, 126, 247, 253, 254, 271
    imposed by experimenter, 227
    imposed by measurement system, 227
    interval displacement, 242
    interval exponentialization, 226, 227, 249–251, 255, 261, 264–267, 326, 327, 337, 343, 345
    interval normalization, 249
    interval shuffling, 226–227, 252–256, 261–262, 265–267, 271, 326, 327, 337, 342–344, 351, 352
    interval transformation, 226, 247–251, 261–262
    intrinsic to underlying process, 227
    phase randomization, 227
    point-process identification, 255–256
    superposition, *See* superposition
    surrogate data, 15, 126, 227, 247, 253, 265–267, 271
    time dilation, 226, 228–229, 262

Palm, Conny, 50, 82, 257, 313, 315, 316
paralyzable dead time, *See* event deletion
Pareto, Vilfredo, 33, 153–154
Pareto distribution, 33, 138, 154–155, 165, 198, 346
Pareto's Law, 138, 154
Penck, Albrecht, 1–2
periodogram, *See* counting statistics, *See* interval statistics
phase randomization, *See* operations on point processes
photon statistics
    betaluminescence, 202
    cathodoluminescence, 202

Čerenkov radiation, 34, 40, 204, 220–222
  in presence of atmospheric turbulence, 36
  in presence of dead time, 263
  radioluminescence, 202
  scattered light, 40
  superposed coherent and thermal light, 147
  thermal light, 146
photonic materials
  diffractals, 40
  fractal reflectance, 40
  fractal transmittance, 40
  group-velocity reduction, 40
  light scattering, 40
  multilayer structures, 40
  phase screen, 40
  pseudo-bandgaps, 40
Poincaré, Henri, 9, 25, 174
point process(es), 4–5, 50–80, 82–99
  Bartlett–Lewis, *See* cascaded point process(es)
  bivariate, 77
  branching, *See* cascaded process(es)
  capacity dimension, 164
  cascaded, *See* cascaded process(es)
  coincidence rate, 70–72, 74, 80, 105–106, 110–111, 133, 159–160, 214
  computer network traffic, *See* computer network traffic
  correlation in a bivariate process, 77–78
  count-based measures, 63–70
  deleted, *See* event deletion
  doubly stochastic Poisson, *See* doubly stochastic Poisson process(es)
  early work, 50
  estimation of, *See* fractal parameter estimation
  examples of, 4, 79, 82–99
  filtered general, 197–198
  fractal, 76, 121–123, 131, 255
  fractal-based, 4–5, 120–124, 130
  fractal behavior in, 115–120
  fractal parameter estimation, *See* fractal parameter estimation
  fractal-rate, 76, 123–124, 251, 255, 345
  fractal renewal process, *See* fractal renewal process(es)
  fractal-shot-noise-driven, *See* fractal-shot-noise-driven point process(es)
  from Brownian motion, 149–150
  from fractal binomial noise, 182–183
  general measures of, 70–76
  Hawkes, 217
  homogeneous Poisson, *See* homogeneous Poisson process
  identification of, 125–131, 255–256, 270–273, 337–351
  infinitely divisible cascade, 123, 331
  integrate-and-reset, *See* integrate-and-reset process(es)
  intermittency, 122
  interval-based measures, 54–62
  limitations of measures, 282
  Lévy dust, 15, 95–96, 138
  marked, 54, 77, 176, 222, 334
  measures of fractal behavior, 103–107
  modulated integrate-and-reset, *See* integrate-and-reset point process(es)
  monofractal, 131
  multidimensional, 82
  multifractal, 123
  multivariate, 77
  Neyman–Scott, *See* cascaded point process(es)
  nonfractal, 96, 124–125, 131
  nonstationary, 71, 110, 112, 133–134
  operations on, *See* operations on point processes
  orderly, 52–54, 66, 75, 79, 95, 96, 110, 174, 197, 206, 220
  periodic, 91, 232–236
  renewal, *See* renewal process(es)
  right-continuous, 51
  self-exciting, 217
  sinusoidally modulated, 98
  spectrum, 72–74, 80, 88, 91, 97, 103, 111, 115, 121, 122, 124,

125, 133, 149, 151, 157–159, 166, 183, 215–216, 218–220, 230–232, 261, 262, 311, 312
spectrum, normalized form, 72
superposed, *See* superposition
Poisson, Siméon Denis, 49, 63
Poisson process, *See* homogeneous Poisson process
power-law behavior
anharmonic-oscillator energy, 34
Čerenkov radiation, 34
computer file sizes, 33, 323, 329–330, 335, 336
Coulomb's Law, 34
diffusion processes, 34–35, 204, 223
dipole field, 34
expansion-modification systems, 37
fractal exponent, *See* fractal exponent(s)
fractals, connection to, 14, 32–39
Gutenberg–Richter Law, 33
harmonic-oscillator energy, 34
highly optimized tolerance, 37
Hooke's Law, 34
hydrogen-atom energy, 34
infinite-quantum-well energy, 34
interval distribution, *See* fractal renewal process(es)
Kepler's Third Law, 34
Langmuir–Childs Law, 34
laws of physics, 33–34
line of charge, 34
logistic equation, 37
lognormal distribution, 36
mass distributions, 198–199
Newton's Law, 34
Omori's Law, 33
Pareto's Law, 33, 138, 154–155, 165, 198, 346
preservation of, 103
quadrupole field, 34
quantum number, 34
relationships among measures, 114–115
Richardson's Law, 3
rigid-rotor energy, 34
scale-free networks, 37–38
scaling functions, 3, 12–13
self-organized criticality, 37

stable distributions, 35–36
superposed relaxation processes, 38–39
time functions, 34
van der Waals force, 34

queueing theory, *See* computer network traffic

random deletion, *See* event deletion
random telegraph signal, *See* alternating renewal process(es)
rate spectrum, *See* counting statistics
recovery function, *See* event deletion
refractoriness, *See* event deletion
relative dead time, *See* event deletion
relative refractoriness, *See* event deletion
renewal process(es), 85–87
alternating, *See* alternating renewal process(es)
coincidence rate, 85–86
decimated Poisson process, 97, 263–264
doubly stochastic Poisson version of, 90
event deletion, *See* event deletion
exponential density, 97
factorial moments, 87
fractal, *See* fractal renewal process(es)
gamma density, 97
gamma density for computer network traffic, 336
history of, 85
invariance to shuffling, 271
operations on, *See* operations on point processes
random deletion of, 236, 263
relation between interval and counting statistics, 87
spectrum, 86–87, 97
superposition of, *See* superposition
Rényi dimension, *See* dimension
rescaled range analysis, *See* interval statistics
reticular formation, *See* action potentials
retinal ganglion cell, *See* action potentials
Rice, Steven O., 147, 185, 186

Richardson, Lewis Fry, 1–3
Rudin–Shapiro sequences, *See* semiconductors

scale-free networks, 37–38, 40, 321–322
scaling, *See* fractals
scaling cutoffs, *See* fractal exponent(s)
scaling exponents, *See* fractal exponent(s)
Schottky, Walter, 185, 186
Scott, Elizabeth, 201, 202
self-organized criticality, 37
semiconductors
    fractional scaling exponents, 34
    multilayer structures, 40
    noise in, 39–40, 116, 169, 172, 173
    particle detectors, 223–224
    range of time constants, 39
    trapping in, 169, 224
semi-experiments, *See* operations on point processes
shot noise, 186–187
    amplitude, 186–187
    as a rate function, 90
    filtered general point process, 197–198
    fractal, *See* fractal shot noise
    Gaussian limit, 186, 191
    generalized, 187
    impulse response function, 202
shuffling, *See* operations on point processes
sick time, *See* event deletion
somatosensory cortex, *See* action potentials
spectral smoothing, 117, 128, 326, 339
spectrum, *See* counting statistics, *See* interval statistics, *See* point process(es)
spike trains, *See* action potentials
stable distributions, 35–36, 79, 157, 174, 192, 193
stochastic dead time, *See* event deletion
striate cortex, *See* action potentials
superposition
    alternating renewal processes, 41, 173, 179–182
    doubly stochastic Poisson processes, 258–259

fractal-based and homogeneous Poisson processes, 273
fractal-based point processes, 258, 261, 310
fractal content, 262
fractal Gaussian process and modulating stimulus, 145
fractal ion-channel transitions, 42
fractal renewal processes, 260–261
harmonic functions, 101
packet arrival times, 323
periodic series of events, 81
point processes, 84–85, 227, 256–261
Poisson-process limit, 85
relaxation processes, 38–39
renewal processes, 259–260, 334
secondary events comprising, 218
surrogate data, *See* operations on point processes
survivor function, *See* interval statistics
synapse, *See* vesicular exocytosis

telephone network traffic, 40, 84, 97, 154, 166–167, 313, 315–320, 324
tent map, 46
thalamus, *See* action potentials
thinning, *See* event deletion
Thomas distribution, *See* counting statistics
Thomas process, *See* cascaded process(es)
Thue–Morse sequences, *See* photonic materials, *See* semiconductors
time dilation, *See* operations on point processes
time series, 4
topological dimension, *See* dimension
translation, *See* operations on point processes
triadic Cantor set, *See* Cantor set
type-*p* dead time, *See* event deletion

unblocked counter, *See* event deletion

Van Ness, John W., viii, 135, 136
variance-to-mean ratio, *See* counting statistics

vesicular exocytosis, 41–43, 117–119, 128–132, 149, 151–152, 233–235, 244–246, 250, 251, 253, 254, 344, 345
visual-system interneuron, *See* action potentials

Wald's Lemma, 164, 237
wavelet(s)
    computer-network-traffic analysis, 336
    Daubechies, 120
    estimating the generalized dimension, 75
    generating fractional Brownian motion, 139
    Haar, 101, 269
    higher-order moments, 332
    interval wavelet variance, *See* interval statistics
    normalized Daubechies-wavelet variance, *See* counting statistics
    normalized general-wavelet variance, *See* counting statistics
    normalized Haar-wavelet covariance, *See* counting statistics
    normalized Haar-wavelet variance, *See* counting statistics
    removing trends, 62, 113
    transform, 58, 67, 74
Weibull distribution, 57, 328
Wiener–Khintchine theorem, 73

Yule, G. Udny, 49, 64, 95, 147
Yule–Furry branching process, *See* cascaded process(es)

zeta distribution, 33, 38

WILEY SERIES IN PROBABILITY AND STATISTICS
ESTABLISHED BY WALTER A. SHEWHART AND SAMUEL S. WILKS

Editors: *David J. Balding, Noel A. C. Cressie, Nicholas I. Fisher,*
Iain M. Johnstone, J. B. Kadane, Geert Molenberghs. Louise M. Ryan,
David W. Scott, Adrian F. M. Smith, Jozef L. Teugels
Editors Emeriti: *Vic Barnett, J. Stuart Hunter, David G. Kendall*

The *Wiley Series in Probability and Statistics* is well established and authoritative. It covers many topics of current research interest in both pure and applied statistics and probability theory. Written by leading statisticians and institutions, the titles span both state-of-the-art developments in the field and classical methods.

Reflecting the wide range of current research in statistics, the series encompasses applied, methodological and theoretical statistics, ranging from applications and new techniques made possible by advances in computerized practice to rigorous treatment of theoretical approaches.

This series provides essential and invaluable reading for all statisticians, whether in academia, industry, government, or research.

ABRAHAM and LEDOLTER · Statistical Methods for Forecasting
AGRESTI · Analysis of Ordinal Categorical Data
AGRESTI · An Introduction to Categorical Data Analysis
AGRESTI · Categorical Data Analysis, *Second Edition*
ALTMAN, GILL, and McDONALD · Numerical Issues in Statistical Computing for the
    Social Scientist
AMARATUNGA and CABRERA · Exploration and Analysis of DNA Microarray and
    Protein Array Data
ANDĚL · Mathematics of Chance
ANDERSON · An Introduction to Multivariate Statistical Analysis, *Third Edition*
* ANDERSON · The Statistical Analysis of Time Series
ANDERSON, AUQUIER, HAUCK, OAKES, VANDAELE, and WEISBERG ·
    Statistical Methods for Comparative Studies
ANDERSON and LOYNES · The Teaching of Practical Statistics
ARMITAGE and DAVID (editors) · Advances in Biometry
ARNOLD, BALAKRISHNAN, and NAGARAJA · Records
* ARTHANARI and DODGE · Mathematical Programming in Statistics
* BAILEY · The Elements of Stochastic Processes with Applications to the Natural
    Sciences
BALAKRISHNAN and KOUTRAS · Runs and Scans with Applications
BARNETT · Comparative Statistical Inference, *Third Edition*
BARNETT and LEWIS · Outliers in Statistical Data, *Third Edition*
BARTOSZYNSKI and NIEWIADOMSKA-BUGAJ · Probability and Statistical Inference
BASILEVSKY · Statistical Factor Analysis and Related Methods: Theory and
    Applications
BASU and RIGDON · Statistical Methods for the Reliability of Repairable Systems
BATES and WATTS · Nonlinear Regression Analysis and Its Applications
BECHHOFER, SANTNER, and GOLDSMAN · Design and Analysis of Experiments for
    Statistical Selection, Screening, and Multiple Comparisons
BELSLEY · Conditioning Diagnostics: Collinearity and Weak Data in Regression

*Now available in a lower priced paperback edition in the Wiley Classics Library.
†Now available in a lower priced paperback edition in the Wiley–Interscience Paperback Series.

† BELSLEY, KUH, and WELSCH · Regression Diagnostics: Identifying Influential Data and Sources of Collinearity

BENDAT and PIERSOL · Random Data: Analysis and Measurement Procedures, *Third Edition*

BERRY, CHALONER, and GEWEKE · Bayesian Analysis in Statistics and Econometrics: Essays in Honor of Arnold Zellner

BERNARDO and SMITH · Bayesian Theory

BHAT and MILLER · Elements of Applied Stochastic Processes, *Third Edition*

BHATTACHARYA and WAYMIRE · Stochastic Processes with Applications

† BIEMER, GROVES, LYBERG, MATHIOWETZ, and SUDMAN · Measurement Errors in Surveys

BILLINGSLEY · Convergence of Probability Measures, *Second Edition*

BILLINGSLEY · Probability and Measure, *Third Edition*

BIRKES and DODGE · Alternative Methods of Regression

BLISCHKE AND MURTHY (editors) · Case Studies in Reliability and Maintenance

BLISCHKE AND MURTHY · Reliability: Modeling, Prediction, and Optimization

BLOOMFIELD · Fourier Analysis of Time Series: An Introduction, *Second Edition*

BOLLEN · Structural Equations with Latent Variables

BOROVKOV · Ergodicity and Stability of Stochastic Processes

BOULEAU · Numerical Methods for Stochastic Processes

BOX · Bayesian Inference in Statistical Analysis

BOX · R. A. Fisher, the Life of a Scientist

BOX and DRAPER · Empirical Model-Building and Response Surfaces

* BOX and DRAPER · Evolutionary Operation: A Statistical Method for Process Improvement

BOX, HUNTER, and HUNTER · Statistics for Experimenters: Design, Innovation, and Discovery, *Second Editon*

BOX and LUCEÑO · Statistical Control by Monitoring and Feedback Adjustment

BRANDIMARTE · Numerical Methods in Finance: A MATLAB-Based Introduction

BROWN and HOLLANDER · Statistics: A Biomedical Introduction

BRUNNER, DOMHOF, and LANGER · Nonparametric Analysis of Longitudinal Data in Factorial Experiments

BUCKLEW · Large Deviation Techniques in Decision, Simulation, and Estimation

CAIROLI and DALANG · Sequential Stochastic Optimization

CASTILLO, HADI, BALAKRISHNAN, and SARABIA · Extreme Value and Related Models with Applications in Engineering and Science

CHAN · Time Series: Applications to Finance

CHARALAMBIDES · Combinatorial Methods in Discrete Distributions

CHATTERJEE and HADI · Sensitivity Analysis in Linear Regression

CHATTERJEE and PRICE · Regression Analysis by Example, *Third Edition*

CHERNICK · Bootstrap Methods: A Practitioner's Guide

CHERNICK and FRIIS · Introductory Biostatistics for the Health Sciences

CHILÈS and DELFINER · Geostatistics: Modeling Spatial Uncertainty

CHOW and LIU · Design and Analysis of Clinical Trials: Concepts and Methodologies, *Second Edition*

CLARKE and DISNEY · Probability and Random Processes: A First Course with Applications, *Second Edition*

* COCHRAN and COX · Experimental Designs, *Second Edition*

CONGDON · Applied Bayesian Modelling

CONGDON · Bayesian Statistical Modelling

CONOVER · Practical Nonparametric Statistics, *Third Edition*

COOK · Regression Graphics

COOK and WEISBERG · Applied Regression Including Computing and Graphics

*Now available in a lower priced paperback edition in the Wiley Classics Library.

†Now available in a lower priced paperback edition in the Wiley–Interscience Paperback Series.

COOK and WEISBERG · An Introduction to Regression Graphics

CORNELL · Experiments with Mixtures, Designs, Models, and the Analysis of Mixture Data, *Third Edition*

COVER and THOMAS · Elements of Information Theory

COX · A Handbook of Introductory Statistical Methods

* COX · Planning of Experiments

CRESSIE · Statistics for Spatial Data, *Revised Edition*

CSÖRGŐ and HORVÁTH · Limit Theorems in Change Point Analysis

DANIEL · Applications of Statistics to Industrial Experimentation

DANIEL · Biostatistics: A Foundation for Analysis in the Health Sciences, *Eighth Edition*

* DANIEL · Fitting Equations to Data: Computer Analysis of Multifactor Data, *Second Edition*

DASU and JOHNSON · Exploratory Data Mining and Data Cleaning

DAVID and NAGARAJA · Order Statistics, *Third Edition*

* DEGROOT, FIENBERG, and KADANE · Statistics and the Law

DEL CASTILLO · Statistical Process Adjustment for Quality Control

DeMARIS · Regression with Social Data: Modeling Continuous and Limited Response Variables

DEMIDENKO · Mixed Models: Theory and Applications

DENISON, HOLMES, MALLICK and SMITH · Bayesian Methods for Nonlinear Classification and Regression

DETTE and STUDDEN · The Theory of Canonical Moments with Applications in Statistics, Probability, and Analysis

DEY and MUKERJEE · Fractional Factorial Plans

DILLON and GOLDSTEIN · Multivariate Analysis: Methods and Applications

DODGE · Alternative Methods of Regression

* DODGE and ROMIG · Sampling Inspection Tables, *Second Edition*

* DOOB · Stochastic Processes

DOWDY, WEARDEN, and CHILKO · Statistics for Research, *Third Edition*

DRAPER and SMITH · Applied Regression Analysis, *Third Edition*

DRYDEN and MARDIA · Statistical Shape Analysis

DUDEWICZ and MISHRA · Modern Mathematical Statistics

DUNN and CLARK · Basic Statistics: A Primer for the Biomedical Sciences, *Third Edition*

DUPUIS and ELLIS · A Weak Convergence Approach to the Theory of Large Deviations

* ELANDT-JOHNSON and JOHNSON · Survival Models and Data Analysis

ENDERS · Applied Econometric Time Series

ETHIER and KURTZ · Markov Processes: Characterization and Convergence

EVANS, HASTINGS, and PEACOCK · Statistical Distributions, *Third Edition*

FELLER · An Introduction to Probability Theory and Its Applications, Volume I, *Third Edition,* Revised; Volume II, *Second Edition*

FISHER and VAN BELLE · Biostatistics: A Methodology for the Health Sciences

FITZMAURICE, LAIRD, and WARE · Applied Longitudinal Analysis

* FLEISS · The Design and Analysis of Clinical Experiments

FLEISS · Statistical Methods for Rates and Proportions, *Third Edition*

FLEMING and HARRINGTON · Counting Processes and Survival Analysis

FULLER · Introduction to Statistical Time Series, *Second Edition*

FULLER · Measurement Error Models

GALLANT · Nonlinear Statistical Models

GEWEKE · Contemporary Bayesian Econometrics and Statistics

GHOSH, MUKHOPADHYAY, and SEN · Sequential Estimation

GIESBRECHT and GUMPERTZ · Planning, Construction, and Statistical Analysis of Comparative Experiments

GIFI · Nonlinear Multivariate Analysis

GIVENS and HOETING · Computational Statistics

GLASSERMAN and YAO · Monotone Structure in Discrete-Event Systems

GNANADESIKAN · Methods for Statistical Data Analysis of Multivariate Observations, *Second Edition*

GOLDSTEIN and LEWIS · Assessment: Problems, Development, and Statistical Issues

GREENWOOD and NIKULIN · A Guide to Chi-Squared Testing

GROSS and HARRIS · Fundamentals of Queueing Theory, *Third Edition*

† GROVES · Survey Errors and Survey Costs

* HAHN and SHAPIRO · Statistical Models in Engineering

HAHN and MEEKER · Statistical Intervals: A Guide for Practitioners

HALD · A History of Probability and Statistics and their Applications Before 1750

HALD · A History of Mathematical Statistics from 1750 to 1930

† HAMPEL · Robust Statistics: The Approach Based on Influence Functions

HANNAN and DEISTLER · The Statistical Theory of Linear Systems

HEIBERGER · Computation for the Analysis of Designed Experiments

HEDAYAT and SINHA · Design and Inference in Finite Population Sampling

HELLER · MACSYMA for Statisticians

HINKELMANN and KEMPTHORNE · Design and Analysis of Experiments, Volume 1: Introduction to Experimental Design

HINKELMANN and KEMPTHORNE · Design and Analysis of Experiments, Volume 2: Advanced Experimental Design

HOAGLIN, MOSTELLER, and TUKEY · Exploratory Approach to Analysis of Variance

HOAGLIN, MOSTELLER, and TUKEY · Exploring Data Tables, Trends and Shapes

* HOAGLIN, MOSTELLER, and TUKEY · Understanding Robust and Exploratory Data Analysis

HOCHBERG and TAMHANE · Multiple Comparison Procedures

HOCKING · Methods and Applications of Linear Models: Regression and the Analysis of Variance, *Second Edition*

HOEL · Introduction to Mathematical Statistics, *Fifth Edition*

HOGG and KLUGMAN · Loss Distributions

HOLLANDER and WOLFE · Nonparametric Statistical Methods, *Second Edition*

HOSMER and LEMESHOW · Applied Logistic Regression, *Second Edition*

HOSMER and LEMESHOW · Applied Survival Analysis: Regression Modeling of Time to Event Data

† HUBER · Robust Statistics

HUBERTY · Applied Discriminant Analysis

HUNT and KENNEDY · Financial Derivatives in Theory and Practice

HUSKOVA, BERAN, and DUPAC · Collected Works of Jaroslav Hajek— with Commentary

HUZURBAZAR · Flowgraph Models for Multistate Time-to-Event Data

IMAN and CONOVER · A Modern Approach to Statistics

† JACKSON · A User's Guide to Principle Components

JOHN · Statistical Methods in Engineering and Quality Assurance

JOHNSON · Multivariate Statistical Simulation

JOHNSON and BALAKRISHNAN · Advances in the Theory and Practice of Statistics: A Volume in Honor of Samuel Kotz

JOHNSON and BHATTACHARYYA · Statistics: Principles and Methods, *Fifth Edition*

JOHNSON and KOTZ · Distributions in Statistics

JOHNSON and KOTZ (editors) · Leading Personalities in Statistical Sciences: From the Seventeenth Century to the Present

---

*Now available in a lower priced paperback edition in the Wiley Classics Library.

†Now available in a lower priced paperback edition in the Wiley–Interscience Paperback Series.

JOHNSON, KOTZ, and BALAKRISHNAN · Continuous Univariate Distributions, Volume 1, *Second Edition*

JOHNSON, KOTZ, and BALAKRISHNAN · Continuous Univariate Distributions, Volume 2, *Second Edition*

JOHNSON, KOTZ, and BALAKRISHNAN · Discrete Multivariate Distributions

JOHNSON, KOTZ, and KEMP · Univariate Discrete Distributions, *Second Edition*

JUDGE, GRIFFITHS, HILL, LÜTKEPOHL, and LEE · The Theory and Practice of Econometrics, *Second Edition*

JUREČKOVÁ and SEN · Robust Statistical Procedures: Aymptotics and Interrelations

JUREK and MASON · Operator-Limit Distributions in Probability Theory

KADANE · Bayesian Methods and Ethics in a Clinical Trial Design

KADANE AND SCHUM · A Probabilistic Analysis of the Sacco and Vanzetti Evidence

KALBFLEISCH and PRENTICE · The Statistical Analysis of Failure Time Data, *Second Edition*

KASS and VOS · Geometrical Foundations of Asymptotic Inference

† KAUFMAN and ROUSSEEUW · Finding Groups in Data: An Introduction to Cluster Analysis

KEDEM and FOKIANOS · Regression Models for Time Series Analysis

KENDALL, BARDEN, CARNE, and LE · Shape and Shape Theory

KHURI · Advanced Calculus with Applications in Statistics, *Second Edition*

KHURI, MATHEW, and SINHA · Statistical Tests for Mixed Linear Models

* KISH · Statistical Design for Research

KLEIBER and KOTZ · Statistical Size Distributions in Economics and Actuarial Sciences

KLUGMAN, PANJER, and WILLMOT · Loss Models: From Data to Decisions, *Second Edition*

KLUGMAN, PANJER, and WILLMOT · Solutions Manual to Accompany Loss Models: From Data to Decisions, *Second Edition*

KOTZ, BALAKRISHNAN, and JOHNSON · Continuous Multivariate Distributions, Volume 1, *Second Edition*

KOTZ and JOHNSON (editors) · Encyclopedia of Statistical Sciences: Volumes 1 to 9 with Index

KOTZ and JOHNSON (editors) · Encyclopedia of Statistical Sciences: Supplement Volume

KOTZ, READ, and BANKS (editors) · Encyclopedia of Statistical Sciences: Update Volume 1

KOTZ, READ, and BANKS (editors) · Encyclopedia of Statistical Sciences: Update Volume 2

KOVALENKO, KUZNETZOV, and PEGG · Mathematical Theory of Reliability of Time-Dependent Systems with Practical Applications

LACHIN · Biostatistical Methods: The Assessment of Relative Risks

LAD · Operational Subjective Statistical Methods: A Mathematical, Philosophical, and Historical Introduction

LAMPERTI · Probability: A Survey of the Mathematical Theory, *Second Edition*

LANGE, RYAN, BILLARD, BRILLINGER, CONQUEST, and GREENHOUSE · Case Studies in Biometry

LARSON · Introduction to Probability Theory and Statistical Inference, *Third Edition*

LAWLESS · Statistical Models and Methods for Lifetime Data, *Second Edition*

LAWSON · Statistical Methods in Spatial Epidemiology

LE · Applied Categorical Data Analysis

LE · Applied Survival Analysis

LEE and WANG · Statistical Methods for Survival Data Analysis, *Third Edition*

LePAGE and BILLARD · Exploring the Limits of Bootstrap

LEYLAND and GOLDSTEIN (editors) · Multilevel Modelling of Health Statistics

---

*Now available in a lower priced paperback edition in the Wiley Classics Library.
†Now available in a lower priced paperback edition in the Wiley–Interscience Paperback Series.

LIAO · Statistical Group Comparison

LINDVALL · Lectures on the Coupling Method

LINHART and ZUCCHINI · Model Selection

LITTLE and RUBIN · Statistical Analysis with Missing Data, *Second Edition*

LLOYD · The Statistical Analysis of Categorical Data

LOWEN and TEICH · Fractal-Based Point Processes

MAGNUS and NEUDECKER · Matrix Differential Calculus with Applications in Statistics and Econometrics, *Revised Edition*

MALLER and ZHOU · Survival Analysis with Long Term Survivors

MALLOWS · Design, Data, and Analysis by Some Friends of Cuthbert Daniel

MANN, SCHAFER, and SINGPURWALLA · Methods for Statistical Analysis of Reliability and Life Data

MANTON, WOODBURY, and TOLLEY · Statistical Applications Using Fuzzy Sets

MARCHETTE · Random Graphs for Statistical Pattern Recognition

MARDIA and JUPP · Directional Statistics

MASON, GUNST, and HESS · Statistical Design and Analysis of Experiments with Applications to Engineering and Science, *Second Edition*

McCULLOCH and SEARLE · Generalized, Linear, and Mixed Models

McFADDEN · Management of Data in Clinical Trials

* McLACHLAN · Discriminant Analysis and Statistical Pattern Recognition

McLACHLAN, DO, and AMBROISE · Analyzing Microarray Gene Expression Data

McLACHLAN and KRISHNAN · The EM Algorithm and Extensions

McLACHLAN and PEEL · Finite Mixture Models

McNEIL · Epidemiological Research Methods

MEEKER and ESCOBAR · Statistical Methods for Reliability Data

MEERSCHAERT and SCHEFFLER · Limit Distributions for Sums of Independent Random Vectors: Heavy Tails in Theory and Practice

MICKEY, DUNN, and CLARK · Applied Statistics: Analysis of Variance and Regression, *Third Edition*

* MILLER · Survival Analysis, *Second Edition*

MONTGOMERY, PECK, and VINING · Introduction to Linear Regression Analysis, *Third Edition*

MORGENTHALER and TUKEY · Configural Polysampling: A Route to Practical Robustness

MUIRHEAD · Aspects of Multivariate Statistical Theory

MULLER and STOYAN · Comparison Methods for Stochastic Models and Risks

MURRAY · X-STAT 2.0 Statistical Experimentation, Design Data Analysis, and Nonlinear Optimization

MURTHY, XIE, and JIANG · Weibull Models

MYERS and MONTGOMERY · Response Surface Methodology: Process and Product Optimization Using Designed Experiments, *Second Edition*

MYERS, MONTGOMERY, and VINING · Generalized Linear Models. With Applications in Engineering and the Sciences

† NELSON · Accelerated Testing, Statistical Models, Test Plans, and Data Analyses

† NELSON · Applied Life Data Analysis

NEWMAN · Biostatistical Methods in Epidemiology

OCHI · Applied Probability and Stochastic Processes in Engineering and Physical Sciences

OKABE, BOOTS, SUGIHARA, and CHIU · Spatial Tesselations: Concepts and Applications of Voronoi Diagrams, *Second Edition*

OLIVER and SMITH · Influence Diagrams, Belief Nets and Decision Analysis

PALTA · Quantitative Methods in Population Health: Extensions of Ordinary Regressions

PANKRATZ · Forecasting with Dynamic Regression Models

PANKRATZ · Forecasting with Univariate Box-Jenkins Models: Concepts and Cases

*Now available in a lower priced paperback edition in the Wiley Classics Library.

†Now available in a lower priced paperback edition in the Wiley–Interscience Paperback Series.

* PARZEN · Modern Probability Theory and Its Applications

PEÑA, TIAO, and TSAY · A Course in Time Series Analysis

PIANTADOSI · Clinical Trials: A Methodologic Perspective

PORT · Theoretical Probability for Applications

POURAHMADI · Foundations of Time Series Analysis and Prediction Theory

PRESS · Bayesian Statistics: Principles, Models, and Applications

PRESS · Subjective and Objective Bayesian Statistics, *Second Edition*

PRESS and TANUR · The Subjectivity of Scientists and the Bayesian Approach

PUKELSHEIM · Optimal Experimental Design

PURI, VILAPLANA, and WERTZ · New Perspectives in Theoretical and Applied Statistics

† PUTERMAN · Markov Decision Processes: Discrete Stochastic Dynamic Programming

QIU · Image Processing and Jump Regression Analysis

* RAO · Linear Statistical Inference and Its Applications, *Second Edition*

RAUSAND and HØYLAND · System Reliability Theory: Models, Statistical Methods, and Applications, *Second Edition*

RENCHER · Linear Models in Statistics

RENCHER · Methods of Multivariate Analysis, *Second Edition*

RENCHER · Multivariate Statistical Inference with Applications

* RIPLEY · Spatial Statistics

RIPLEY · Stochastic Simulation

ROBINSON · Practical Strategies for Experimenting

ROHATGI and SALEH · An Introduction to Probability and Statistics, *Second Edition*

ROLSKI, SCHMIDLI, SCHMIDT, and TEUGELS · Stochastic Processes for Insurance and Finance

ROSENBERGER and LACHIN · Randomization in Clinical Trials: Theory and Practice

ROSS · Introduction to Probability and Statistics for Engineers and Scientists

† ROUSSEEUW and LEROY · Robust Regression and Outlier Detection

* RUBIN · Multiple Imputation for Nonresponse in Surveys

RUBINSTEIN · Simulation and the Monte Carlo Method

RUBINSTEIN and MELAMED · Modern Simulation and Modeling

RYAN · Modern Regression Methods

RYAN · Statistical Methods for Quality Improvement, *Second Edition*

SALTELLI, CHAN, and SCOTT (editors) · Sensitivity Analysis

* SCHEFFE · The Analysis of Variance

SCHIMEK · Smoothing and Regression: Approaches, Computation, and Application

SCHOTT · Matrix Analysis for Statistics, *Second Edition*

SCHOUTENS · Levy Processes in Finance: Pricing Financial Derivatives

SCHUSS · Theory and Applications of Stochastic Differential Equations

SCOTT · Multivariate Density Estimation: Theory, Practice, and Visualization

* SEARLE · Linear Models

SEARLE · Linear Models for Unbalanced Data

SEARLE · Matrix Algebra Useful for Statistics

SEARLE, CASELLA, and McCULLOCH · Variance Components

SEARLE and WILLETT · Matrix Algebra for Applied Economics

SEBER and LEE · Linear Regression Analysis, *Second Edition*

† SEBER · Multivariate Observations

† SEBER and WILD · Nonlinear Regression

SENNOTT · Stochastic Dynamic Programming and the Control of Queueing Systems

* SERFLING · Approximation Theorems of Mathematical Statistics

SHAFER and VOVK · Probability and Finance: It's Only a Game!

SILVAPULLE and SEN · Constrained Statistical Inference: Inequality, Order, and Shape Restrictions

SMALL and McLEISH · Hilbert Space Methods in Probability and Statistical Inference

---

*Now available in a lower priced paperback edition in the Wiley Classics Library.

†Now available in a lower priced paperback edition in the Wiley–Interscience Paperback Series.

SRIVASTAVA · Methods of Multivariate Statistics

STAPLETON · Linear Statistical Models

STAUDTE and SHEATHER · Robust Estimation and Testing

STOYAN, KENDALL, and MECKE · Stochastic Geometry and Its Applications, *Second Edition*

STOYAN and STOYAN · Fractals, Random Shapes and Point Fields: Methods of Geometrical Statistics

STYAN · The Collected Papers of T. W. Anderson: 1943–1985

SUTTON, ABRAMS, JONES, SHELDON, and SONG · Methods for Meta-Analysis in Medical Research

TANAKA · Time Series Analysis: Nonstationary and Noninvertible Distribution Theory

THOMPSON · Empirical Model Building

THOMPSON · Sampling, *Second Edition*

THOMPSON · Simulation: A Modeler's Approach

THOMPSON and SEBER · Adaptive Sampling

THOMPSON, WILLIAMS, and FINDLAY · Models for Investors in Real World Markets

TIAO, BISGAARD, HILL, PEÑA, and STIGLER (editors) · Box on Quality and Discovery: with Design, Control, and Robustness

TIERNEY · LISP-STAT: An Object-Oriented Environment for Statistical Computing and Dynamic Graphics

TSAY · Analysis of Financial Time Series

UPTON and FINGLETON · Spatial Data Analysis by Example, Volume II: Categorical and Directional Data

VAN BELLE · Statistical Rules of Thumb

VAN BELLE, FISHER, HEAGERTY, and LUMLEY · Biostatistics: A Methodology for the Health Sciences, *Second Edition*

VESTRUP · The Theory of Measures and Integration

VIDAKOVIC · Statistical Modeling by Wavelets

VINOD and REAGLE · Preparing for the Worst: Incorporating Downside Risk in Stock Market Investments

WALLER and GOTWAY · Applied Spatial Statistics for Public Health Data

WEERAHANDI · Generalized Inference in Repeated Measures: Exact Methods in MANOVA and Mixed Models

WEISBERG · Applied Linear Regression, *Third Edition*

WELSH · Aspects of Statistical Inference

WESTFALL and YOUNG · Resampling-Based Multiple Testing: Examples and Methods for p-Value Adjustment

WHITTAKER · Graphical Models in Applied Multivariate Statistics

WINKER · Optimization Heuristics in Economics: Applications of Threshold Accepting

WONNACOTT and WONNACOTT · Econometrics, *Second Edition*

WOODING · Planning Pharmaceutical Clinical Trials: Basic Statistical Principles

WOODWORTH · Biostatistics: A Bayesian Introduction

WOOLSON and CLARKE · Statistical Methods for the Analysis of Biomedical Data, *Second Edition*

WU and HAMADA · Experiments: Planning, Analysis, and Parameter Design Optimization

YANG · The Construction Theory of Denumerable Markov Processes

* ZELLNER · An Introduction to Bayesian Inference in Econometrics

ZHOU, OBUCHOWSKI, and McCLISH · Statistical Methods in Diagnostic Medicine

*Now available in a lower priced paperback edition in the Wiley Classics Library.

†Now available in a lower priced paperback edition in the Wiley–Interscience Paperback Series.